Selective Toxicity

D0218311

CHART OF SIZES

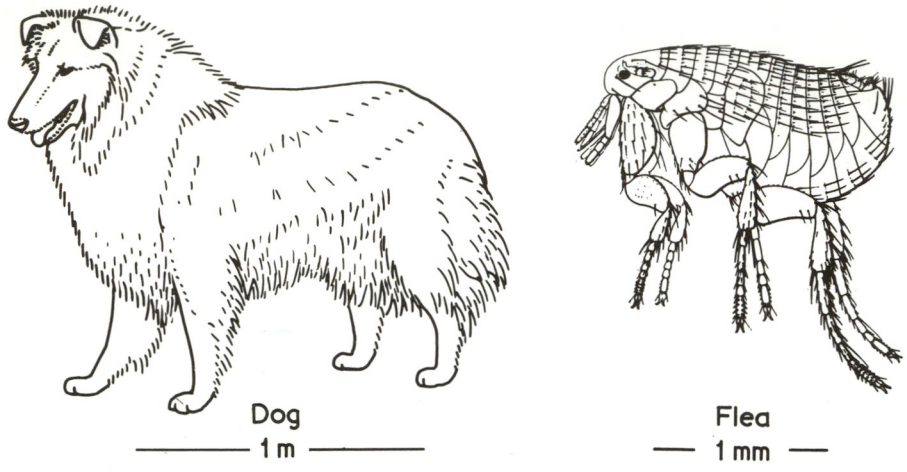

Dog
—— 1 m ——

Flea
— 1 mm —

Streptococci
– 1 µm –

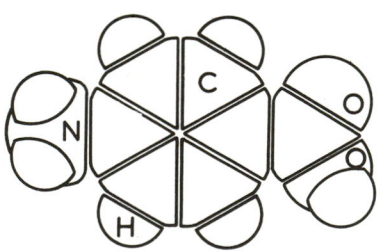

Molecule
(*p*–aminobenzoic acid)
———————— 1 nm ————————

This chart is to help one remember the relative sizes of Mammals, Insects, Microbes, and Molecules. In each case an example of medium size has been chosen (e.g. Dog and not Whale). Each object is drawn with a magnification 1000 times greater than the object preceding it.

Selective Toxicity

The physico-chemical basis of therapy

ADRIEN ALBERT

D.Sc. (Lond.), Ph.D. Medicine (Lond.),
Fellow of the Australian Academy of Science

Professor Emeritus, Department of Chemistry,
Australian National University, Canberra;
Research Professor, Department of
Pharmacological Sciences, School of Medicine,
State University of New York, Stony Brook.

SEVENTH EDITION

London   New York

Chapman and Hall

First published 1951
Second edition 1960
Third edition 1965
Fourth edition 1968
Reprinted once
Fifth edition 1973
Sixth edition 1979
Seventh edition 1985
First paperback edition 1981

Published by Chapman and Hall Ltd
11 New Fetter Lane, London EC4P 4EE
Published in the USA by
Chapman and Hall
733 Third Avenue, New York NY 10017

© 1985 Adrien Albert

Printed in Great Britain at the
University Press, Cambridge

ISBN 0 412 26010 7 (cased edition)
ISBN 0 412 26020 4 (paperback edition)

This title is available in both hardbound and paperback editions. The paperback edition is sold subject to the condition that it shall not, by way of trade or otherwise, be lent, re-sold, hired out, or otherwise circulated without the publisher's prior consent in any form of binding or cover other than that in which it is published and without a similar condition including this condition being imposed on the subsequent purchaser.

All rights reserved. No part of this book may be reprinted, or reproduced or utilized in any form or by any electronic, mechanical or other means, now known or hereafter invented, including photocopying and recording, or in any information storage and retrieval system, without permission in writing from the publisher.

British Library Cataloguing in Publication Data

Albert, Adrien
  Selective toxicity.——
  1. Chemotherapy   2. Chemistry, Pharmaceutical
  3. Drugs——Physiological effect
  I. Title
  615'.7   RM262

  ISBN 0-412-26010-7
  ISBN 0-412-26020-4 Pbk

Library of Congress Cataloging in Publication Data

Albert, Adrien.
  Selective toxicity.

  Bibliography: p.
  Includes indexes.
  1. Pharmacology.   2. Biological chemistry.
3. Chemotherapy.   I. Title.   [DNLM: 1. Biochemistry.
2. Drug Therapy.   3. Pharmacology.   QV 38 A333s]
  RM301.A36 1984   615'.7   84–7781
  ISBN 0–412–26010–7
  ISBN 0–412–26020–4 (pbk.)

Contents

Preface to the seventh edition

This book is about selectively toxic agents. That is to say, it is about those substances that affect certain cells without harming others, even when they are close neighbours. Toxicity need not be fatal. It can be made easily reversible, as is the case with general anaesthetics. Selective toxicity covers an immense field: most of the drugs used for treating illness in man and his economic animals, as well as all of the fungicides, insecticides, and weed killers that are used in agriculture. Essentially, this book is a discussion of the physical and chemical means which contribute to selectivity, and this is the basis of molecular pharmacology.

Selective Toxicity began as a course of lectures that Professor F.G. Young encouraged me to give in University College London, in 1948 and again in 1949. The first edition appeared in 1951, as a very small book because little was then known about the factors that provide selectivity. Since those early days, the subject has undergone tremendous development. At first, industry was unreceptive to the word 'toxicity', however qualified! Yet the market was being supplied with biologically powerful substances of which several had the potential to cause harm. This aspect was brought to light by two events of the early 1960s. The first of these was the discovery that a sedative, thalidomide, administered to expectant mothers, after what was then considered to be adequate testing, had caused permanent deformities in about 10 000 children. The other event was the demonstration that our environment was becoming contaminated through unrestrained use of the chlorinated insecticides. The resulting outcry led to introduction of regulatory laws of considerable severity.

Since that turning point, have we become overcautious? Is there now a tendency to condemn familiar chemicals on insufficient grounds? It may be hard for us, but it is essential, to hold an objective frame of mind while solid evidence is collected and discussed. Obviously the safe use of drugs and other agents depends on a detailed knowledge of their mode of action. Given this, an acceptable level of hazard can be defined, as is done for other dangers (e.g.

traffic accidents), that living in a community imposes. In other words, between benefits and risks a sensible balance must be struck. This calls for more education in these matters, and it can profitably be started at an earlier age, and to a wider spectrum of students, than has been traditional.

From 1975, I have been lecturing on 'The Selectivity of Drugs' to final-year undergraduates in Britain, Holland, and the United States. Often the classes were joined by graduate students in the first year of their Ph.D. course. The undergraduates were directing their careers variously to medicine, pharmacy, dentistry, veterinary science, agriculture, and the drug industry. For enrolment, the prerequisites were two years of chemistry and one of cell biology, at University level. These classes, somewhat to my surprise initially, took well to *Selective Toxicity*, finding that its *Part One* ran parallel to the 12 lecture courses, whereas the 30 lecture courses, given in the State University of New York at Stony Brook, were found to be covered by this and selected topics from *Part Two*. In both cases, *Part Two* provided material and references to the literature for essays and term papers. The ground plan of all these courses was to begin with an outline of the World's illnesses and what selective agents are doing to help overcome them, followed by a lecture on the correlation of structure with biological action (as set out in Chapters 1 and 2 respectively). The remainder of the shorter courses dealt with the *three principles* which enable such biological action to be made selective. Apart from these courses, *Part Two* has independent value as background for the research worker.

Previous editions of *Selective Toxicity* have been well received, and translated into German, Italian, Japanese, and Russian. This new edition, which reviews the literature up to September 1984, has been thoroughly revised to reflect current awareness of the subject. Many additions have been made, some of them quite substantial, but all blended into the original framework. There are new sections on drugs that influence the immune process, on inhibitors based on the transition state of the molecule, and on IMBIs, which are the irreversible mechanism-based inhibitors. There are new taxonomic tables of bacteria and protozoa, and an index of the 650 structural formulae used in the text. A new section (17.4) provides help in searching for the physical and biological properties of drugs, whether in compilations (books) or in computerized databanks.

Other topics whose treatment has been restructured and expanded are: the nature of drug receptors including a photomicrograph of a receptor, the evolution of ideas on structure–action relationships, conformational and other steric factors that influence drug action, the GABA receptor and the benzo-diazepines, narcotics and the opioid receptors, the catecholamines; chemotherapy, anti-protozoal drugs, the deviant purine and carbohydrate metabolism of protozoa and helminths, fungicides and anthelmintics for human use, anti-viral drugs, anti-cancer agents, chelating agents in tuberculosis, cancer, and virus diseases, drugs that act by inhibiting synthesis (or use) of nucleic acids, agrochemicals (insecticides, fungicides, weed killers), drug resistance; the classification of pharmacodynamic drugs, pro-drugs, drug metabolism, the new self-cancelling drugs, pharmacokinetics, diuretics, anti-metabolites (including stereoscopic diagrams of inhibitors inside enzymes), inhibitors of the histamine receptors, psychotherapeutic agents, diuretics, the calcium channel blockers, regression equations, and the use of nuclear magnetic resonance to pinpoint the sites of recognition.

Important new information is also included on antibiotics (especially bleomycin, doxocycline, the aminoglycosides, and the β-lactam family), mode of action of hypnotics and anaesthetics, therapy of hypoxic malignant cells, sub-cellular architecture, the biological consequences of the various bond-types, radiopharmaceuticals, tetrahydrocannabinol, anti-inflammatory drugs, benzodioxan, prostaglandins, insect hormones and pheromones, pyrethroids, nitro-compounds as drugs, cardiac glucosides, and use of the Evans & Suther-land picture system to visualize how drugs interact with their receptors.

Some changes in nomenclature follow from WHO's *International Nonproprietary Names for Pharmaceutical Substances* (1982). Adrenaline, oestrone, and sulphanila-mide have become epinephrine, estrone, and sulfanilamide, respectively. American readers will note the following changes in that list (the internationally agreed name is in parentheses): acetaminophen (paracetamol), albuterol (salbutamol), chlorpheniramine (chlorphenamine), dibucaine (cinchocaine), isoproterenol (isoprenaline), mechlorethamine (chlormethine), melphalan (sarcolysin), meperidine (pethidine), metaproterenol (orciprenaline), pyril-amine (mepyramine), methimazole (thiamazole) and rifampin (rifampicin).

There are 3000 references to the literature, many of them new. It is hoped that this book will be enjoyed by all researchers who combine chemical and bio-logical interests. Two useful 'brush-up' books are Jepson and Smith (1974) for the essentials of organic chemical nomenclature, and Howland (1973) for an introduction to cell physiology. For portraits of the earlier drug scientists and accounts of their work, see Holmstedt and Liljestrand (1963).

In conclusion, I thank Professor Arthur P. Grollman M.D., Chairman of the Department of Pharmacological Sciences, State University of New York at Stony Brook, for constant encouragement; Drs D.F. Waterhouse F.R.S. and J.N. Phillips both of CSIRO (Canberra) for current information on insecticides, fungicides, and herbicides; and to the following who, severally, read one or more chapters of the manuscript: Professor R.N. Warrener, Drs W.L.F. Armarego, D.J. Brown, W.D.L. Crow, J.A. Elix, N.S. Gill, A.J. Jones, R.J. Pace, Y.T. Pang, D.D. Perrin, M. Rasmussen, and B.K. Selinger. I also thank Dr Jean Cartaud of the Centre Nationale de la Recherche Scientifique, Paris, for the photomicrograph of an acetylcholine receptor; also the Wellcome Institute Library (London) for photographs of Ehrlich and Hata and of Ehrlich's Institute in Frankfurt.

A.A.
Canberra

PART ONE

Topics of general interest

1
Selectivity in the service of man*

Throughout the countless millenia of evolution and under the strong pressure of natural selection, nature has evolved many small, highly selective molecules to make the cell work. These *natural agonists*† govern its nutrition, growth, and reproduction. Examples are the vitamins, coenzymes, hormones, neuro-transmitters, inorganic ions, pigments (respiratory and photosynthetic), and metabolic fragments, such as acetyl. The most remarkable, because the most highly conserved, of all agonists are the polyaza-heterocycles: (a) adenosine triphosphate (ATP) which stores the energy provided by the breakdown of nutrients and releases it on demand, (b) the purine and pyrimidine bases of deoxyribonucleic acid (DNA) which encode all information for regulating the cell's moment-to-moment metabolism as well as defining its individuality and heredity, and (c) pteridines of the folic acid type which regulate the biosynthesis of purines and pyrimidines.

These agonists interact with their complementary biopolymers to generate every response needed for the cell's continuity both as an individual and as a species.

1.0 What is 'selectivity'?

A remedy is said to have selectivity if it can influence one kind of living cell without affecting others, even when these cells are close neighbours. Man has found many selective agents for treating his diseases and those of his farm animals and field crops. Most of the chemical substances that man uses in therapy differ from those evolved by nature, and yet are often related to them because both kinds may act on the same receptors. The substances employed by

---

*As in Charles Darwin's immortal writings 'Man', as used here, embraces women (and children).
†*Agonist* is from the Ancient Greek word for participant, or competitor (in games). This term was introduced by Reuse (1948).

Man are called *drugs* when used for the treatment of human beings or farm animals, and *agricultural agents* when designed to suppress weeds, insects, or fungi in crops. Collectively they are known as biologically active agents, or simply *agents,* and the principles governing their actions are identical. Most of what is written about drugs in this book is applicable to agents generally.

Man's therapeutic agents are of three kinds, namely replacements, agonists, and antagonists. *Replacements,* which consist mainly of vitamins, hormones, and minerals, are used simply to restore what has been lost by depletion. Examples: insulin in diabetes, vitamin B-12 in pernicious anaemia, and calcium during pregnancy. Man's *agonists* are formed by modifying natural agonists, usually hormones or neurotransmitters, in order to get a more prolonged action. This is done by making small molecular changes to provide slower release, or to slow down destruction or elimination. Examples can be found in Sections 3.6, 9.4, 12.4, and 12.6.

However, the majority of Man's drugs are of a third kind called *antagonists*. Essentially, these are inhibitors of natural agonists. Antagonists are designed either to eliminate invading organisms, or to suppress pain or control those biochemically unbalanced states so typical of the sick body. All antagonists are toxic (poisonous) and yet this can be their most valuable property because antagonists can enlist toxicity in the *service* of Man. It must be understood that toxicity in a drug is valuable only when it is selective. *Selective toxicity* enables one kind of cell to be influenced strongly without any effect on other cells, even when the two kinds are growing side by side. Toxic effects can be either reversible or permanent, just as required. The cells that are to be affected by the drug (even if only temporarily, as in anaesthesia) are conveniently called *uneconomic cells* in contrast to the others, the *economic* cells, which have to remain completely unaffected.

In *chemotherapy*, where the aim is to kill all the uneconomic cells, because they are invaders, it is evident that uneconomic and economic cells are related to one another as parasite and host. On the other hand, *pharmacodynamics** has as its goal the eventual normalization of the uneconomic cells because the economic and uneconomic cells with which it has to deal form parts of the same organism.

Many antagonists produce such striking manifestations in Man that we can easily understand how our forefathers classed them, mistakenly, as agonists. For example, ethanol (alcohol) was spoken of as a 'stimulant' whereas it is actually an inhibitor. What it does is to damp down some inhibitory nerves, in the central nervous system, so that any excitement reigns unchecked (see Sections 7.5.1 and 7.6.1). Strychnine, strongest of the convulsants, acts similarly, but lower down in the central nervous system.

General anaesthetics admirably illustrate the selective use of toxicity. The more toxic the anaesthetic, the more valuable it is, but only if the toxicity is

---

*Chemotherapy and pharmacodynamics, together, constitute the branch of science known as *pharmacology*.

selective for the central nervous system and completely reversible with time. Morton's success with ether as a general anaesthetic in 1846 was an early and convincing demonstration of selective toxicity. The accepted general anaesthetics combine a high toxicity for the central nervous system with negligible toxicity to other tissues; *all* toxicity rapidly and completely disappears when administration is halted. So, too, with local anaesthetics, muscle relaxants, and (but less rapidly) the antagonists of histamine and of neurotransmitters. Antiparasitic agents, on the other hand, although they must be selective against the parasite and sparing to the host, should be irreversible.

Man should be proud to have discovered so many drugs that are at least moderately selective for most of his illnesses (also for those of his farm animals) and similar agents for his crops and forests. To improve on this degree of selectivity and to extend the benefits more widely is still likely to require many decades of intense and devoted research.

Man's agonists and antagonists are usually synthetic, and of low molecular weight (usually less than 500). Sometimes, though, they are not synthetic but of natural origin. When this occurs, they are being used outside their natural physiological context, as is the case with alkaloids and antibiotics. Immunochemicals, as in vaccines and sera, are also used, but these are distanced from our subject by their enormous molecular weights.

In giving replacement therapy, care is taken with the dose because natural agonists are often poisonous in excess. For example, calciferol (vitamin D) in excess causes calcification of arteries and the kidneys, whereas vitamin A causes vomiting, generalized peeling of the skin, and pathological change in the liver. Many young children die each year from quite small doses of iron, usually taken as ferrous sulphate. Thyroid hormones, in small excess, cause muscular tremor, and adrenaline (epinephrine) can precipitate tachycardia when injected before a dental extraction. Not even food is free from toxicity. Through millenia of enforced experimentation, Man has gradually learnt to avoid eating acutely toxic species. However, choice or necessity can restrict the diet to foods whose feeble toxicity may not otherwise be apparent. Chronic toxicity can then arise, as from the natural goitrogens of cabbage and cauliflower, the liver-injuring pigment lycopene in tomatoes, the convulsant alkaloid in yams, the biotin-depleting white of eggs, and calcium deprivation from the phytic acid of oatmeal. (For many other hazardous factors in common foods, see National Academy of Sciences, 1973.)

Biological control provides a frequently discussed alternative to selective toxicity. In one of its manifestations, economic species are bred, or trained, to become more disease-resistant. Also, specific parasites can be sought for the uneconomic species. For example, the cactus known as prickly pear, which deprived Australian farmers of great areas of valuable pasture, was eliminated in the 1930s by the release of a beetle (*Cactoblastis*) which attacked no other form of life. Again the Japanese beetle, which became a serious pest to crops on the Atlantic seaboard of the USA in about 1916, has been kept in check by the introduction of a parasitic wasp (*Tiphia vernalis*) and a bacterium (*Bacillus popilliae*), both of which are harmless to other forms of life. Spores of *Bacillus thuringiensis* are used, in the United States, to limit the gipsy moth's attack on trees. The same bacterium is also undergoing large-scale trials by the World Health Organization to kill the larvae of mosquito and blackfly in the control of malaria and onchocerciasis, respectively. An alternative and well-established biological weapon against mosquitoes is to release fish (e.g. *Gambusia*) to feed on their larvae in stagnant waters (WHO, 1971).

What can be done with virus infections is exemplified by the strong killing action of the myxomatosis virus in rabbits following its careful liberation in Australia (1950) and France (1952). However, the use of viruses for biological control is out of favour because of the risk of adding malign genes to the chromosomes of economic species, Man included.

Genetic control differs from these forms of biological control in that no new species is introduced. It usually takes the form of releasing huge numbers of sterilized male insects, to outnumber the fertile ones and thus diminish fertilization. Sterilization can be effected with X-rays or, more often, by chemosterilants that are selectively toxic for spermatogenesis. A typical agent is tepa, which is tris(1-aziridinyl)phosphine oxide. If genetic control could be perfected, it would endanger no other type of living organism, whereas ordinary biological control is risky. Sterilized males have been used to exterminate screw-worm, which is a serious insect pest of livestock. Despite a few successes with genetic control, WHO has this opinion: 'The possibilities are for the distant future' (WHO, 1971). For more on chemosterilants, see Borkevec (1976).

At the present time, the most successful examples of biological control are those effected with selectivity toxic agents. Thus, trypanosomiasis, a protozoal disease of Man and cattle, is controlled by chemical defoliation of those areas of the African jungle where tsetse flies breed, and then spraying organophosphorus insecticides on the exposed breeding sites. In this way, by attacking the insect vectors, which transfer trypanosomes to their mammalian hosts with every bite, the biological life cycle of these parasites is broken. For the same reason, houses in malarial areas and swampy grounds which harbour anopheline mosquitoes are regularly sprayed with insecticides to kill these insect vectors of plasmodia (the protozoa which, transmitted to humans by the mosquito's bite, produce the disease malaria). Needless to say, drainage of the swamps where practicable has helped to control this disease. Another example of the use of selectively toxic agents to break a life cycle is the spraying of streams with molluscicides to kill snails that are the intermediate host to the worm that causes bilharziasis in Man.

Although these three examples are of tropical diseases, the principle of exterminating the vectors of disease by selectively toxic agents is also fundamental to maintaining good health in temperate climates. Two universally dreaded diseases are kept in check only by constant vigilance over rats and insects: typhus (rat → louse → Man) and bubonic plague (rat → flea → Man).

Another way of combining biological and toxic methods is to use pheromones (the natural insect sex-attractants) as lures to bring insects to poisoned baits (for pheromones, see Section 4.7). This method looks promising. Another contemporary search is for substances which could make crops unattractive to insects, or impair their appetites.

1.1 Beneficial results from the use of selectively toxic agents

Ability to resist change is inherent in even the simplest physical system, such as a cup of water, as Le Chatelier showed in 1880. Any external effect, such as heat or pressure, always displaces the equilibrium in the direction that tends to restore the original state. Small wonder, then, that living organisms resist change, particularly as they have stores of energy to apply to the task. This homeostasis of living cells enables them to fight Man's best efforts to control them, and though he has won some notable victories, some of these have been only temporary. This is not surprising because even the humblest species* has been in existence much longer than Man and, in the course of that time, has built into its genome much information on how to survive almost every imaginable type of catastrophe. The real surprise, then, is that Man has, in many cases, discovered how to influence, injure, or even eliminate a pathological form of life without endangering his own. These remarks are particularly applicable to selective toxicity.

Let us enquire what selectively toxic agents have accomplished, where they have lost ground, and what they have yet to do. A useful summary can be found in the *Biennial Report of the Director-General to the World Health Assembly and to the United Nations*. Issued in the May of even-numbered years, this Report is available as a separate publication, or it can be consulted as part of the annual, the *Work of WHO*. A summary will be found in the monthly *WHO Chronicle*. All of these sources are published in Geneva, Switzerland. The following account has drawn on these reports, e.g. (WHO, 1977, 1982) and on WHO's *World Health Statistics Annual*.

1.1.1 Infectious diseases

The duration of human life in the industrialized countries used to be limited by infection. However, since 1935, such diseases have come under good control, largely through remarkable discoveries in chemotherapy. Hence the span of life has increased in such lands where most people now live long enough to die of *non*-infectious diseases. In contrast, the non-industrial (or 'developing') countries are overrun by infectious diseases of which they have an enormous variety. Their people enjoy only a relatively short span of life.

---

*For example, bacteria have been in existence for at least 3000 million years, from the fossil record.

In what follows, the infectious diseases of the World will be broken down into five groups caused by protozoa, bacteria, viruses, fungi, and worms.

(a) *Diseases caused by protozoa.* These organisms (see Table 1.1) are one-celled animals, many of them with a complex life cycle. Malaria, caused by various species of the genus *Plasmodium*, is the worst of several highly dangerous protozoal diseases. In fact, malaria causes more debility, illness, and death than any other known disease. The World Health Organization gives its top priority to advising the various nations on how to work for eventual elimination of this disease, by draining swamps and spraying houses to eliminate the mosquito which is the insect vector (carrier), and by medication both prophylactic and (when that aspect has been neglected) curative. WHO also labours constantly to discover improvements in all these approaches.

Malaria is a chronic illness characterized by severe anaemia and attacks of high fever (often every second day) which result in extreme debilitation. The complex life cycle of the parasite begins when a biting female mosquito ingests human blood containing the sexual form (gametocytes) of the parasite. These mate inside the mosquito, and the progeny (sporozoites) reside in the salivary glands and so enter Man when he is bitten. The sporozoites multiply in the human liver and their progeny (merozoites) enter red blood corpuscles where they mature to schizonts, which later burst the corpuscles and escape into the host's bloodstream (this erythrocytic cycle usually takes about 48 hours). Most of the escaped schizonts migrate to other erythrocytes and repeat the cycle, but a few become gametocytes. Only partial immunity is acquired after repeated attacks.

Malaria has been eliminated from Europe and the USA within living memory. For example Italy had 8407 deaths from this disease in 1919, but none since 1948. WHO's worldwide programme, begun in the late 1940s, had, by 1976, eradicated malaria from about 20 countries with the result that about 436 million people are now freed from the risk of infection. Moreover, the incidence of the disease has been greatly reduced in areas inhabited by another 1260 million. Unfortunately, about 200 million people are infected with malaria at present, some of them in countries which at first had the disease under good control but have since lost ground through diminished vigilance (Noguer *et al.*, 1978). In 1966, WHO estimated the world's annual death rate from malaria as about one million, and it has not greatly fallen, for the following reason.

Deaths from malaria were becoming quite rare wherever drainage and spraying were carried out, and prophylaxis and treatment were available. [For the effective agents, see Section 6.3.3, 10.3.5 (drugs) and 6.4.1 (insecticides).] However, in 1961, chloroquine-resistant strains of *Plasmodium falciparum* (the most pathogenic of the malarial parasites) began to develop, and have now grown to alarming proportions. (For the phenomenon of drug-resistance, see Section 6.5.) Because chloroquine was the standard drug for treating malaria, this was a serious matter. The situation became more serious when resistance

developed to the principal prophylactics, pyrimethamine and cycloguanil. The
drug-resistant areas are Southeast Asia, East Africa, and South America. The
problem is complicated by resistance of the mosquitoes to the commonly used
insecticides. Naturally, these problems have led to renewed research, and
acceptable substitutes seem to be forthcoming. Moreover, many malarial areas
can still be managed with the old agents. However, the near-disaster provides a
warning of the possibly temporary nature of any triumph in fighting disease.

After malaria, the most serious of the diseases caused by protozoa are
trypanosomiasis, leishmaniasis (including kala-azar), and amoebiasis
(amoebic dysentery).

Whereas malaria is endemic in the majority of tropical countries, trypanosomiasis is confined to Africa (in a wide belt between latitudes 10°N and 25°S)
and to Latin America. The acute and the chronic forms of African trypanosomiasis are caused, respectively, by *Trypanosoma brucei rhodesiense* and *T. brucei
gambiense*. A third sub-species, *T. brucei brucei,* causes great economic loss among
the cattle, but is non-infectious to humans. In all types, the life cycle begins
when a tsetse fly (*Glossina*) takes a blood meal from an infected mammal. In the
insect gut, the trypanosomes change into what is called a procyclic form with a
very different kind of metabolism. After about three weeks, the parasites
migrate to the insect's salivary glands where they slowly change to forms similar
to those found in the mammalian bloodstream to which they are transferred by
the next bite of the fly. The mammalian form divides rapidly by binary fission
(sexual mating seems relatively unimportant). Both mammalian and insect
forms can, of recent years be readily cultured in the laboratory.

The *rhodesiense* sub-species of this parasite often causes death within one year,
whereas the *gambiense* type produces a chronic infection that may last for many
years. In both types the central nervous system becomes infected giving rise to
extreme lassitude, hence the popular name 'sleeping sickness'. About 45 million
victims suffer in this way and also lose their horses, cattle, and camels from
related trypanosomal infections. Several reasonably selective remedies are
known (see Section 6.3.3), and WHO is encouraging further research into drugs
for prophylaxis and treatment, and into insecticides and defoliants. Some
drug-resistance exists, but the principal factor preventing elimination of this
disease is economic, and the incidence of infection does not seem to be decreasing.

In America, from Mexico down to the Argentine, a different species (*T. cruzi*)
is transmitted by a face-biting nocturnal bug. The result, Chagas' disease, is
caused by trypanosomes lodging in the heart muscle, leading often to sudden
heart failure, especially in children. About 12 million cases a year occur. For
some effective drugs, see Section 6.3.3. Chagas' disease is confined to under-
developed areas, and is intensified by under-nutrition.

Leishmaniasis is caused by a trypanosome-like organism, *Leishmania*, which is
conveyed to Man by female sandflies. The disease may take the form of a
generalized visceral infection (kala-azar) or of a purely cutaneous infection

(oriental sore). The visceral form occurs from the south and east shores of the Mediterranean down to East Africa and southern Arabia, also parts of India, China, and South America. Multiplication of the parasite takes place in the liver and spleen (which enlarge) and in the bone-marrow, causing anaemia, wasting, and death. Both forms of leishmaniasis, which together affect 12 million victims, respond to antimonials (Section 4.4), with diamidines as a standby for resistant cases (Section 10.3.5).

Amoebiasis, due to infection with *Entamoeba histolytica,* is conveyed between humans by its cysts which are 10 microns in diameter. They survive well outside the body and are ingested in water and uncooked food. In the colon, the larger vegetative forms (trophozoites) emerge and cause chronic diarrhoea and, often, ulceration of the bowel wall. Unlike bacterial dysentery, this disease is seldom self-limiting without proper medication. Abscesses in the liver form a common complication. A ready cure can be effected with metronidazole (Section 6.3.3).

In general, the human body makes only a feeble immune response to invasion by protozoa. This fact indicates the great importance of selectively toxic agents for preventing and treating diseases caused by these parasites.

(*b*) *Diseases caused by bacteria.* Concerning *bacterial* diseases, it is interesting to contrast conditions today in the developed countries with those encountered by medical students in the 1930s. The medical wards always had several patients severely ill (and some dying) with pneumonia; there were special wards for patients with tuberculosis, and at the outskirts of the city there were special TB hospitals. In the surgical wards, severe and disabling bacterial infections of the hands and limbs were common and difficult to treat, bacterial infection of the bladder was almost inevitable in elderly men with prostatic enlargement, and peritonitis was a dreaded complication of abdominal surgery for which little could be done. Mothers, in childbirth, often acquired septicaemia from which very many died. In the children's wards osteomyelitis was an intractable disease, and there were always cases of severe middle ear infection. After the discovery and application of sulfonamides, penicillin, the tetracyclines, and isoniazid, these severe bacterial infections almost completely disappeared because chemotherapy either prevented or cured them. Bacterial epidemics are now less dreaded; children are no longer immunized against scarlet fever (a streptococcal infection) because penicillin so rapidly cures it. Similarly, although travellers can be immunized against typhoid and paratyphoid fevers, treatment of the non-immune with chloramphenicol is simple and rapid.

The least-controlled bacterial diseases in the world today are cholera, trachoma, leprosy, tuberculosis, brucellosis, urethritis, and the trepo-nematoses. Indonesia, North Africa, and (especially) the Indian subcontinent are the most severely afflicted with cholera, but thanks to the aeroplane, no part of the world is safe. Cholera is easily cured with tetracycline and intravenous saline. Cholera epidemics begin through poor hygiene, and spread explosively when the number of sufferers exceeds the nursing resources. Trachoma is an infection of the eyes by a minute, intracellular bacterium of the *Chlamydia* division (see Table 1.3). There are estimated to be 400 million sufferers in tropical and sub-tropical regions and, according to WHO, it is the greatest cause of blindness in the World. Although this disease is easily cured by tetracycline drops, reinfection is common in lands where water is scarce. Leprosy affects about 12 million people in tropical Asia, the Far East, tropical Africa, Central and South America, and some of the Pacific Islands. It is one of the most seriously disabling diseases known, but many sufferers have no access to the standard treatment, which requires several years' medication with dapsone (9.17), supplemented by rifampicin (4.37).

Tuberculosis, although easily prevented and rapidly cured (e.g. with isoniazid), is rampant in Latin America, Africa, Asia, and the Western Pacific. In these countries it is a public health problem of the first order. WHO reports: 7 million infectious cases in the world, and half a million deaths annually.

Brucellosis is a very worrying disease in all countries, Cattle, sheep, pigs, and goats are commonly infected and in turn infect the men who handle them; also whole families can be stricken through drinking the unsterilized milk of an infected animal. Treatment with tetracycline is effective in Man, but prophylaxis still presents a problem. This disease is a typical zoonosis, i.e. a disease transmitted to Man by animals (there are many zoonoses).

The incidence of urethritis, whether gonococcal (decreasing) or chlamydial (increasing) is largely due, in the developed countries, to failure to seek treatment, rapid and painless though this is. Fifty per cent of all cases occur in the age-group 15 to 24.

Of the treponematoses, syphilis (about 40 times less prevalent than urethritis) is slightly increasing, and yaws (mainly affecting children in the less hygienic of the tropical countries) is diminishing; penicillin remains the best treatment for both diseases. Louse-born typhus, caused by a very small bacterium called a rickettsia, has declined greatly through the use of DDT and rodenticides: two African countries (Ethiopia and Burundi) have 95% of the World's cases, according to WHO.

(c) *Diseases caused by viruses.* Smallpox, once the most feared and fatal of all virus-caused diseases, has been wiped out, thanks to WHO's strategy of world-wide prophylactic vaccination. Although such immuno-prevention remains the cornerstone of dealing with several other virus diseases, it is inoperative in many others, and in any case useless in *treating* an infection. Fortunately, several very promising clues have appeared in recent years for designing selective drugs against viral diseases (Section 6.3.2), a few of which, notably herpes, are already being successfully treated in this way. Many investigators think that the immediate future will provide the same success for anti-viral therapy as occurred for anti-bacterial therapy when sulfonamide drugs were discovered in 1935. Meanwhile, selectively toxic agents are being sought against hepatitis, yellow fever, rabies, dengue fever, mumps, influenza, and the common cold. For a review of the chemotherapy of virus infections, see Came and Caliguiri (1982).

(*d*) *Diseases caused by fungi*. Systemic fungal diseases, although sometimes life-threatening, are comparatively rare in Man. A few moderately successful drugs have been found for treating them. On the other hand, localized fungal infections occur in all countries, and are treated by several highly selective drugs (Section 6.3.4).

(*e*) *Diseases caused by worms*. By far the worst of many highly damaging diseases caused by worms is schistosomiasis. In fact it is second only to malaria for causing prolonged debilitating illness and economic loss. The species that lodges in the intestinal veins, where it causes prodigious pain and swelling, is called *Schistosoma mansoni*. *S. japonicum* causes a similar but more severe infection. A genito-urinary form of the disease, called bilharziasis, is caused by *S. haematobium*. All species follow a similar life cycle by passing through a water-snail. Of the 600 million people at risk in Egypt, Arabia, West Africa, China, and neighbouring lands, about 180 million are severely infected. Other sufferers have been located in Brazil, Venezuela, and the Caribbean.

The life cycle begins with a larval stage (cercaria) in freshwater snails, which penetrates the skin of anyone working or bathing in the same stream. The larvae mate, then lay eggs in the victim's intestinal veins. The eggs, due to an immunochemical effect, are intensely irritating and cause large, painful swellings. Eggs, passing out in the urine or faeces, hatch in the streams, releasing embryos called miracidia. Magnesium ions, emitted by the snail, attract these embryos which enter the snail's liver where they give rise to larvae. Schistosomiasis tends to occur in those hot, dry countries where sanitation is poor because of lack of water. New irrigation schemes, population growth (particularly migrants), and innate poverty all tend to increase the prevalence of infection.

Many selectively toxic drugs have become available in recent years for treating schistosomiasis (Section 6.3.5). This is most fortunate because, as in worm-induced diseases generally, the host's immune response is, at best, only feeble where not actually counterproductive. Further help comes from the use of molluscicides against the snails.

Filariasis is caused by a worm, *Wuchereria*, which is carried by a mosquito and injected into the human victim. It might be supposed that nature would require an uncommonly large insect for this task, yet it has, with characteristic ingenuity, solved the problem by miniaturizing the worms which grow to full length (10 cm) only in the lymphatics of the human. There it causes allergic inflammation which leads to obstruction, and it eventually brings about enlargement of a leg (elephantiasis). There are thought to be about 200 million sufferers in tropical Africa, Asia, Indonesia, the Americas, and the West Indies. A selectively toxic drug, diethylcarbamazine (*6.35*) has proved to be an efficient prophylactic and very useful in treatment. Extra help is available by spraying insecticides against mosquito larvae.

Onchocerciasis is a related disease, caused by the worm *Onchocerca*, which is transmitted by biting flies in tropical Africa, Southern Arabia, South Mexico, and Guatemala. It is estimated that 20 million people are infected. In parts of West and Central Africa, it affects the whole adult population of which from 10 to 35% are blinded by it (African River Blindness). Because of this disease, huge tracts of fertile land stay untilled. No adequate drug is known for prophylaxis or treatment (diethylcarbamazine is only moderately effective) but the cycle can be broken by spraying insecticides.

Ancylostomiasis is the parasitization of the small intestine by *Necator* or *Ankylostoma* hookworms which are nematodes (Table 1.1), about 1 cm long, with a fascinating life cycle. The larvae develop in warm moist soil, penetrate the skin of field labourers, and are carried to the lungs. From this vantage point, they ascend the bronchi, are swallowed (when not actually coughed out), and mature in the wall of the small intestine where they prosper by sucking the host's blood, and are finally voided in the faeces. Hookworm is widespread, where conditions are insanitary, throughout the tropics and subtropics including Africa, the Americas, and the Far East. About 600 million sufferers endure a gross debility which leads, if untreated, to severe anaemia and death. Children are most often the victims. A rapid cure can be effected by tetrachloroethylene; other selectively toxic remedies are known (see Section 6.3.5).

Of the pathogenic intestinal worms, the human race is estimated to be harbouring, at any moment of time, about 1000 million cases *each* of *Ascaris* (roundworm), *Enterobius* (pin or thread worms), and *Trichuris* (whipworm). The roundworms, common in the tropics, have a life cycle shared between lungs and intestines. They, and the other two cited species which are common in every country of the World, are easily abolished by selectively toxic anthelmintics (see Section 6.3.5), as are the universally occurring *Taenia saginata* (beef tapeworm) several metres in length but causing few symptoms in the human intestine. Two other tapeworm infestations, found in the temperate climates but difficult to treat, are trichinellosis, which begins in undercooked pork and finishes in the diner's muscles, and hydatid disease, which often follows the course sheep → dog → Man; promising drugs are available.

Many farm animals suffer from severe worm diseases which sap their vitality and decrease their market value. In most cases, effective anthelmintics are known (Section 6.3.5), but many good ones are too uneconomic to use. This state of affairs illustrates the well-known fact that veterinary remedies have to be inexpensive or they cannot be afforded.

1.1.2 *Economic factors*

Over and over, in the above account, it can be seen that even more important than good selective agents to prevent infectious diseases is a good water supply and waste-disposal system. At present, about three-quarters of the World's population lacks an adequate and safe water supply, and they depend on the most primitive methods for sewage disposal. In 1980, the United Nations

launched a programme to provide clean water and waste disposal for the entire
World by 1990.

Malnutrition, too, plays a large part in perpetuating the diseases of poorer
countries. The Director-General of the UN Food and Agricultural Organization (FAO), in his report to the 1979 conference in Rome, said that about 400
million people in the World suffer from serious malnutrition. He added that,
even with the most energetic efforts to increase food and agricultural production, about 250 million will be starving by the year 2000. Fortunately, much
is being done, both by FAO and by private enterprise, to prevent crop spoilage
and to increase yields by the use of pesticides (see Section 6.4).

Meanwhile the fight against infections must be waged with selectively toxic
agents. Since 1976, WHO has urged search for new chemotherapeutic agents
against malaria, trypanosomiasis, leprosy, and schistosomiasis to cope with
likely emergence of resistant strains of the causative organisms. Leishmaniasis
and filariasis come next on their list of priorities.

1.1.3 Non-infectious diseases

Whereas in undeveloped countries, most cases of illness stem from infectious
diseases that require chemotherapeutic agents, most of the illness in the more
prosperous countries is metabolic in origin and hence requires pharmacodynamic agents. In such countries, the principal cause of death is cardiovascular (heart disease and stroke), followed by cancer. Together these account
for 70% of all deaths, with accidents coming next on the list. Mental ill-health
and rheumatoid diseases account for a high percentage of incapacitating illness.
Common diseases in industrialized countries, but almost unknown in communities untouched by Western urbanization and Western dietary habits are:
coronary heart disease, cancer of the large bowel, diabetes, gallstones, and
obesity. For each of these diseases some pharmacodynamic agents are available,
but still more effective ones are sought.

Yet the versatility of pharmacodynamic medication is remarkable. Patients
can be relieved of pain of all types and degrees of severity, put to sleep or made
more alert, prevented from having convulsions or caused to have them for their
therapeutic value. All of these things can be done with simple selectively toxic
agents. Similarly, the patient's temperature can be raised or lowered, his
sympathetic or his parasympathetic nervous system can be selectively stimulated or depressed, his basal metabolic rate raised or lowered, and the clotting
power of his blood can be made greater or less. Moreover, deficiency or
hyperactivity in the action of muscles (including the heart) has come under
control, and so have the activities of several of the endocrine glands. Excessive
secretion of histamine, the cause of so many distressing symptoms, can be
counteracted.

For the last several decades, mental illness has been treated far more successfully by medication than by the psychological approach. Countless otherwise

hopeless cases have been able to return to their families, and to employment too, on maintenance doses of drugs which counter the underlying biochemical lesion. The prospect for yet better psychotherapeutic drugs is very bright. For details of effective drugs, see Section 12.9.

Cancer is a collective name for at least 200 different diseases characterized by unrestrained growth. From 1942, it began to be controlled by medication. Cancers are of two major kinds: (a) solid tumours and (b) the leukaemias and lymphomas of the blood and lymphatic systems respectively. In the Western nations, the largest causes of cancer-related deaths are lung cancer, colonic (including rectal) cancer in men, and breast cancer in women.

Drug treatment of the leukaemia of young people has attracted much attention because lessons learned in this area have resulted in major advances in the chemotherapy of other types of malignancy. In contrast to leukaemia in adults, that in children is almost always acute (97%) and lymphocytic (five out of six cases), so that, before selectively toxic drugs were found, death usually occurred within 3 months. Since 1948, developments in the chemotherapy of this disease changed the outlook dramatically. The 5-year survival rate has improved from 5% in 1960 to 50% today. In most of this group, an actual cure seems to have been attained.

Each year in the USA, with its population of 220 million, about 800 000* new cases of cancer are diagnosed each year. (This figure excludes about 400 000 cases of easily cured skin cancer, e.g. basal cell carcinoma.) These, and the following US figures, are taken from the American Cancer Society (1983). It is expected, using current treatments, that 40% of these patients will be cured (about 5% by chemotherapy alone, the rest mainly by surgery, supported by radiation and chemotherapy). In 1982, 431 000 Americans died of cancer but many could have been saved by earlier diagnosis and prompt treatment. Few links have been found between cancer and viral infections.

Most *solid* cancers begin in epithelial layers of the skin, stomach, gut, breast, bladder, lungs, or womb. If neglected, some 50% of malignant tumours in Man can initiate colonies in remote sites. Hence chemotherapy, which used to be reserved for terminal cases, is now often introduced at the beginning of treatment, as soon as the mass of the tumour has been removed by surgery or radiation. This is done because drugs can reach out, far beyond the surgeon's knife and the radiotherapist's rays, to destroy unidentifiable cancer cells anywhere in the body.

Chemotherapy, *on its own,* has proved effective in the following conditions. In choriocarcinoma, a womb tumour of young, pregnant women who used to suffer 90% mortality within a year but now have a 90% chance of complete cure, thanks to selectively toxic drugs; in Burkitt's lymphoma of children in Africa, in testicular and ovarian cancer, in bone cancers (including rhabdomyosarcoma, Ewing's sarcoma, osteogenic sarcoma), muscle cancers, histiocytic lymphoma,

*This figure excludes about 400 000 cases of easily cured skin cancers, e.g. basal cell carcinoma.

nodular mixed lymphoma, adult myelogenous leukaemia (see above for leukaemia of childhood).

Chemotherapy (after radiation) cures Hodgkin's disease, a cancer of the lymph glands: 90% of early and 70% of late cases. Chemotherapy after surgery improves the outlook in melanoma. Surgery, radiotherapy, and chemotherapy, together, cure 80% of Wilms's disease, a kidney cancer in children. In early, localized breast cancer, there is an 87% chance of survival, but this figure is reduced to 47% for established cases. Some 20% of breast cancers respond to hormone therapy and another 48% to anti-metabolites, the figures referring to 5-year survival times (Brulé *et al.*, 1973). Advanced cases usually receive surgery or radiation, or both, as well.

In lung cancer, treated only by surgery, about 10% of patients survive for 5 years. In colorectal cancer, for which surgery is the principal treatment, the cure rate is about 50%.

For the whole World (about 4000 million people), WHO estimates 5 million new cases of cancer occur each year.

Some 50 anti-cancer drugs have now been established as clinically useful (see Sections 5.1, 6.3.6, and 13.4 for the principal examples). Several very promising new anti-cancer drugs are now undergoing clinical trial. The solid malignancies, once thought so hard to treat, are beginning to yield to bleomycin, amsacrine, doxorubicin, cyclophosphamide, and the alkyl nitrosoureas; also to methotrexate in heroic doses followed by 'rescue' with citrovorin. In general, anti-cancer drugs have not shown the high degree of selectivity that has become the norm for other chemotherapeutic remedies, but steady improvements are being effected.

Some of the best cancer treatments use a *mixture of drugs*, each of which has a *different* kind of toxicity for the patient. Because only small doses of each drug are used, the patient can tolerate the diffuse, minor assaults. However, the drugs are chosen to give the *same* kind of toxicity for the malignancy, which therefore suffers a major assault.

For further reading on the chemotherapy of cancer, see Carter *et al.* (1981).

1.1.4 *Clinical trials*

Every stage in the search for a new drug is subject to Cost–Benefit Analysis (Cavalla, 1981). Every unnecessary test or delay adds to the cost of the eventual treatment. Hence, as soon as a substance has shown it is both promising and harmless in two laboratory species, nothing short of its administration to Man can give useful new information. Many a seemingly specific and potentially useful substance, chosen on the basis of animal trials, has had to be rejected in the clinic for such reasons as: too brief an action, not absorbed from the gut, or serious side-effects not shown earlier. (Parenthetically, the member of a series of new compounds that turned out best in Man has not always been the member that excelled in laboratory experiments.)

When different test-species are compared, little connection can be found between dosage and activity, but activity is usually well correlated with blood level. Hence the first task of a clinical unit is cautiously to find what dose in healthy human beings will produce the blood level found effective in laboratory animals. From kinetic data, obtained from the analysis of blood and urine samples as described in Section 3.7, a safe and effective probable dose for patients can be calculated.

The next step, provided that the drug is unquestionably more promising than any existing remedy, is to introduce it to a selected group of volunteer patients, using the necessary precautions of placebos and cross-over tests. Where any element of risk exists, much can be done on (a) human post-mortem material, (b) excised samples (biopsy, or necessary surgery), and (c) tissue cultures. However, human trials (where possible) are preferable, and should be conducted within the strict ethical framework laid down by the Declaration of Helsinki made by the World Medical Association in 1964.

It is interesting to speculate how many selectively toxic agents are in use at the present time. Figures are available for the United States where the Food and Drugs Administration (FDA) has estimated that about 4000 drugs were used there in 1981. Likewise, the Environmental Protection Agency (Washington) considers that 1500 pesticides were used in the USA in that year.

For details of reference books and databanks for gaining access to information on biologically active chemicals, see Section 17.4.

1.2 The physical basis of selectivity: the three principles

The present author is of the opinion that there are three main principles by which a biologically active agent can exert selectivity. Either it can be accumulated principally by the *uneconomic* species, *or*, utilizing comparative biochemistry, it may injure a chemical system important for the *uneconomic* (but not for the economic) species, *or* it may react exclusively with a cytological feature that exists only in the *uneconomic* species. Often two or all of these principles can be seen functioning together, but with one preponderating.

(*a*) *Selectivity through comparative distribution (accumulation)* is sometimes only a matter of gross morphology. Thus the comparative hairiness of weeds in a crop of grain, or the comparatively large surface area (per unit weight) of an insect resting on a mammal, brings about a greater retention of sprayed material by the uneconomic species. In other cases, selective accumulation is achieved in a more positive way. This topic is developed in Chapter 3.

(*b*) *Selectivity through comparative biochemistry*. It used to be thought that all living matter, whether animal, plant, or microbial, had a common biochemistry which, if universal, would offer no basis for selective toxicity. Some of the more important items in this common groundplan are as follows. Life, in all its

Table 1.1 Epitome of animal classification

Sub-kingdom 1: Protozoa
Unicellular animals: Subdivided into ciliates (for example, *Paramecium*), flagellates (for example, *Trypanosoma*), sporozoa (for example, *Plasmodium*), and amoebae.

Sub-kingdom 2: Porifera
Multicellular animals without a nervous system (for example, sponges)

Sub-kingdom 3: Metazoa
A   Coelenterates (for example, jellyfish)
B   Platyhelminthes (flat worms, occasionally segmented)
    (a) Cestodes (tapeworms, for example, *Taenia, Echinococcus*)
    (b) Trematodes (for example, *Schistosoma, Fasciola*)
C   Nematoda (round worms, unsegmented; for example, *Ascaris, Nippostrongylus, Haemonchus, Litomosoides, Wuchereria, Trichuris*)
D   Annelida (the typical segmented worms)
    (a) Polychaeta (for example, lugworms)
    (b) Oligochaeta (for example, earthworms)
    (c) Hirudinea (for example, leeches)
E   Mollusca
    (a) Gastropoda (for example, snails, slugs)
    (b) Lamellibranchia (for example, clams)
    (c) Cephalopoda (for example, squids)
F   Arthropoda
    (a) Crustacea (for example, crabs, barnacles)
    (b) Insecta (for example, flies, lice, fleas, beetles, roaches)
    (c) Arachnida (for example, spiders, mites, ticks)
G   Echinodermata (five-rayed animals, for example, starfish, sea urchins, *Arbacia, Echinus*)
H   Chordata
    (a) Urochordata (for example, tunicate sea-squirts, ascidians)
    (b) Craniata (Vertebrates)
       (i) Pisces (fish)
      (ii) Amphibia (for example, frogs)
     (iii) Reptilia (for example, turtles, lizards)
      (iv) Aves (birds)
      (v) Mammalia (including Man)

(Compiled from Clark and Panchen, 1971.)

Table 1.2 Epitome of plant classification

A   Phycophyta (green, brown, and red algae)
B   Mycophyta (fungi)
C   Bryophyta (mosses, liverworts)
D   Pteridophyta (ferns, lycopodia)
E   Spermatophyta (seed plants)
    (a) Gymnospermae (conifers and allies)
    (b) Angiospermae (flowering plants)
       (i) Monocotyledons
      (ii) Dicotyledons

aspects, depends on the cell as a unit (even viruses require cells to parasitize, to effect their reproduction). All forms of life have nucleic acid on which is encoded the genetic information required for the functioning of the particular organism. Agents, such as colchicine, which can interfere with mitosis do so at one particular stage in *all* the species examined, and this indicates a universal biochemical pathway of cell-division.

Moreover, a great resemblance in catabolic processes is shown by all cells. There is no essential difference between glycolysis in such a lowly form of life as yeast, and in some of the most highly organized tissues such as human muscle and liver. This has been shown conclusively by the use of inhibitors and by the actual isolation of enzymes and intermediates. Adenosine triphosphate, too, is an almost universal 'currency' by which cells exchange large increments of energy between the various parts of the metabolic cycle, balancing anabolism against catabolism. Certain vitamins, notably thiamine, riboflavine, and nicotinamide, form essential parts of coenzymes in all living cells.

Yet, remarkable as these similarities are, the very fact that one species functions differently from another indicates that there must actually be marked biochemical differences, and many of these are now known. Similarly, marked differences in the biochemistry of various tissues within a single organism have been found. A discussion of selectivity through biochemical differences constitutes Chapter 4.

Table 1.3 Epitome of prokaryote classification

1.	Cyanobacteria (formerly called 'blue-green algae') (aerobic photosynthesizers)
2.	Photrophic bacteria (anaerobic photosynthesizers)
3.	Inorganic-sheathed bacteria
4.	Budding bacteria
5.	Spirochetes (fast moving) (for example, *Treponema*)
6.	Rigidly spiral bacteria (for example, *Vibrio*)
7.	Gram-negative aerobic rods (for example, *Brucella*)
8.	Gram-negative facultatively anaerobic rods (for example, *Escherichia*, *Salmonella*)
9.	Gram-negative anaerobic rods
10.	Gram-negative aerobic cocci (for example, *Neisseria*)
11.	Gram-negative anaerobic cocci
12.	Gram-negative chemolithotrophes (oxidizers of inorganic matter)
13.	Methane producers
14.	Gram-positive cocci (for example, *Staphylococcus*)
15.	Gram-positive endospore-forming rods (for example, *Clostridium*, an anaerobe)
16.	Gram-positive rods (for example, *Lactobacillus*)
17.	Actinomycetales (Gram-positive rods with tendency to branch) (for example, *Mycobacterium*, *Corynebacterium*, and *Streptomyces*, the source of many antibiotics)
18.	Rickettsias (small intracellular bacteria) (for example, *Rickettsia*, *Chlamydia*)
19.	Mycoplasma (lack a cell wall; the plasma membrane contains sterols otherwise unknown in bacteria)

(Compiled from Bergey, 1974.)

(c) *Selectivity through comparative cytology.* It has long been known that plants and animals have outstanding cytological differences. Thus cell walls and photosynthetic apparatus are found in plants but not in animals; likewise nerve and muscle cells are found in animals but not in plants. With the help of the electron microscope, it has been found that the cell itself is full of component parts (called organelles) and that each kind of these components displays strong species differences; also there are differences between cells from different tissues in the same species. How these differences can assist selectivity in toxic agents is discussed in Chapter 5.

A classification (very condensed) of animals and plants, from the simplest to the most highly evolved, is given in Tables 1.1 and 1.2 respectively. Alternative taxonomic systems exist. The phrase 'higher animals' usually refers to the vertebrates, and 'higher plants' to the spermatophytes, but many of the physiological characteristics of higher forms are already well developed lower down in the scale of evolution. All the foregoing are *eukaryotes*, namely organisms with a nucleus in each cell. Less complicated are the *prokaryotes* which have a single chromosome but no nucleus. The classification adopted in Table 1.3 is that of the widely accepted Manual of Determinative Bacteriology (Bergey, 1974), in which Gram* staining plays a prominent part. (Bacterial classification is sure to change as more biochemical and genetic information accumulates.) Viruses present an even simpler form of life in so far as they lack a metabolism of their own, and do not exist in the form of cells (see Section 5.5).

---

*Discovered by Christian Gram, Danish bacteriologist, in 1884. Gram-positive bacteria are those which, exposed in turn to crystal violet, iodine, and ethanol, are stained purple.

2

Steps in the correlation of structure with biological action

The three best known sources of selectivity that are available for controlling uneconomic cells were outlined at the end of Chapter 1. Before going on to an expanded treatment of these principles, the narrative must pause to review, in this Chapter, the means by which a foreign substance can influence living matter, whether selectively or not. Experience has shown that to neglect this step and proceed straight into discussion of the principles of selectivity is an unreasonably large jump, as a result of which too many independent variables compete for attention at the same time. It seems better to begin with a simple examination of the sources of foreign molecules' biological activity, which can assume many forms. This activity is, in fact, the primary force, one that can be tamed in the service of Man by application of the principles of selectivity. Ill advised is the investigator who, esteeming his new candidate drug 'too toxic' (meaning, really, 'insufficiently selective') changes the molecule in a way that extinguishes the toxicity, and thereby loses the force that could have been made selective by thoughtful molecular modification.

Fundamental to any study of correlations, between structure and biological activity, is knowledge that the messengers and coenzymes of each organism depend strongly on small details of their chemical structure without which their characteristic biological effect is lost. If these details are varied, even slightly, the degree of action is usually radically changed. For example, the vitamin activity of thiamine (*2.1*) (tested on pigeons) drops to 5% if the methyl-group is removed from the pyrimidine ring, and to <1% if the methyl-group is removed from the thiazole ring (Schultz, 1940). Finally, if an extra methyl-group is inserted into the thiazole ring (between nitrogen and sulphur), the vitamin activity completely disappears (Bergel and Todd, 1937). This rule of the essential nature of every part of a molecule need not apply to a side-chain. For example, the long aliphatic side-chain in the 3-position of vitamin K_1 (*2.2*) can be pruned without affecting the principal action of this vitamin. In such a case, it is evident that the side-chain lacks every atom responsible for adsorption of this vitamin.

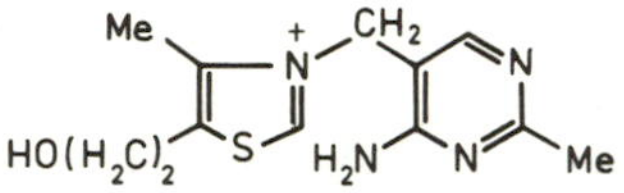

Thiamine
(2.1)

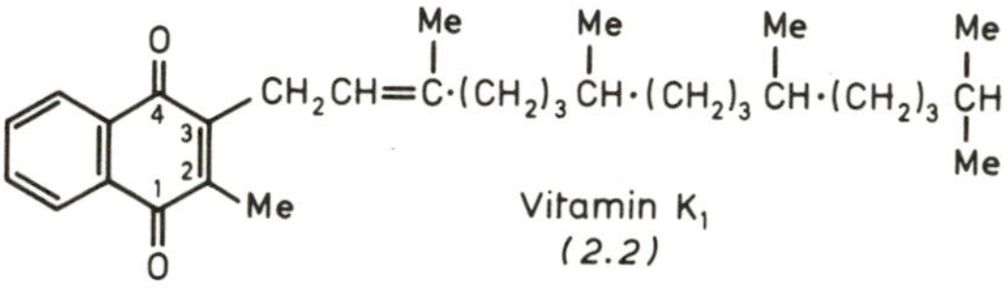

Vitamin K₁
(2.2)

Synthetic drugs show a similar dependence on minute detail. In benzene-sulfonamide (*2.3*), an amino-group can be inserted in three different positions: in two of these it gives rise to an inactive substance, in the other it becomes the highly anti-bacterial substance sulfanilamide. In acridine (*2.4*), an amino-group can be inserted in five different positions: in three of these it gives aminoacridines that are almost inactive, but in the other two it gives powerful anti-bacterials. In quinoline (*2.5*), a hydroxy-group can be inserted in seven different positions: in six of these, completely inert substances arise, but in the remaining position a strong anti-bacterial and anti-fungal substance is produced. What is more, the reasons why the active isomerides are active, and the inactive ones inactive, are well understood and will be described in what follows. The marked biological differences shown by optical isomers (Section 2.1) further illustrate this point.

Before plunging into these details, it will help to stand back a little and take a broader view of how the ideas of structure–activity relationships arose.

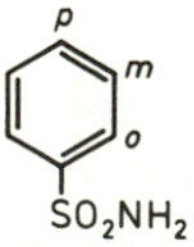

Benzenesulfonamide
(2.3)

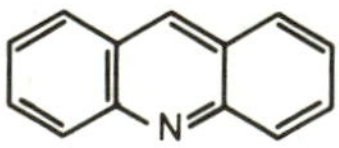

Acridine
(2.4)

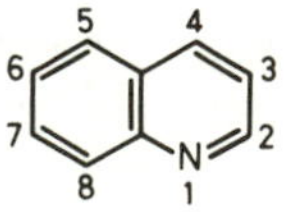

Quinoline
(2.5)

2.0 The earliest correlations

In Renaissance times, Paracelsus (1493–1541) taught: 'All substances are poisons; there is none which is not a poison. Only the right dose differentiates between a poison and a remedy'. This somewhat nihilistic view began to undergo modification, very slowly it is true, through scientific observation and reasoning in the 19th century. Thus in 1848, Blake, in the United States, published his opinion that the biological activity of a salt was due to its basic *or* its acidic component, and not to the whole salt. Thus the poisonous entity in lead acetate and lead nitrate was the lead moiety and not the acetate or nitrate part. Similarly, the toxicity of sodium, potassium, and calcium arsenites resided only in the arsenite portion of these salts. This was, for its time, a daring thought, because it was not until 1884 that Arrhenius introduced his theory of electrolytic dissociation (namely: salts dissolved in water are dissociated into oppositely charged ions).

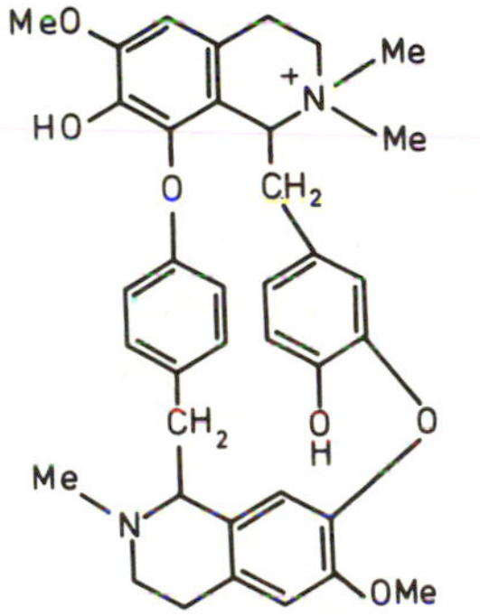

Tubocurarine
(revised formula, 1970)
(2.6)

The next correlation was found in Scotland where Crum Brown and Fraser (1869) made a major discovery. They showed that a series of alkaloids, including convulsive ones, lost their typical pharmacological actions and were all converted to muscle relaxants when their tertiary nitrogen atom was quaternized by methylation. In fact, this simple chemical change had converted strychnine, bruceine, codeine, morphine, thebaine, nicotine, atropine, and coniine into substances with the biological property of the alkaloid tubocurarine (2.6), itself a quaternary amine. The site of action of this alkaloid had already been shown to lie at the junction between nerve and voluntary muscle, by Claude Bernard (1856). The Scottish authors wrote: 'There can be no reasonable doubt that a relation exists between the physiological action of a substance and its chemical composition and constitution, understanding by the latter term the mutual relations of the atoms in the substance.' These discoverers of the first structure–action relationship among organic substances hopefully entitled their paper 'On the Connection between Chemical Constitution and Physiological Action'.

We must not underestimate the stimulating effect of this discovery on pharmacologists and medicinal chemists. Like rain in the desert, a simple connexion had at last been found between a constitution and a biological property. This correlation started the search for other chemical groups or nuclei (ring systems) to which a unique pharmacological action might be assigned. However, even as late as 1910, all that could be added was that organic arsenicals (but only some of them) could cure syphilis. No other example came to light of a single chemical group being able to confer a single pharmacological action on a variety of complicated nuclei. An explanation of Crum Brown's puzzling correlation came only in the present century, for it had to await development of the idea of drug receptors and the discovery of some analogous phenomena in enzyme chemistry.

Today, we would say that current insight into structure–action relationships has come principally from a study of physical properties and the receptor theory, and particularly from the reconciliation of these two approaches originally seen as rivals. These two concepts will now be expanded, in turn.

Attention was first drawn to the overriding importance of a physical property when, at the turn of the present century, Ernest Overton and Hans Meyer independently put forward a 'Lipoid Theory of Cellular Depression' (Meyer, 1899; Overton, 1901). This stated that chemically inert substances, of widely different molecular structures, exert depressant properties on those cells (particularly those of the central nervous system) that are rich in lipids; and that the higher the partition coefficient (between any lipid solvent and water) the greater the depressant action. This statement requires only insertion of the words, 'up to the point where hydrophilic properties are almost extinguished' after 'partition coefficient' to outline the present day viewpoint. Overton and

Table 2.1 Correlation of lipid/water partition coefficients with biological depression (suppression of motility of tadpoles)

Substance	Partition coefficient olive oil/water	Minimal immobilizing concentration. mol/l (water)
Trional	4.46	0.0018
Butylchloral hydrate	1.59	0.0020
Sulfonal	1.11	0.0060
Triacetin	0.30	0.010
Diacetin	0.23	0.015
Chloral hydrate	0.22	0.020
Ethyl urethane	0.14	0.040
Monacetin	0.06	0.050

(Meyer, 1899; Baum, 1899)

Meyer appreciated that the cells of the central nervous system are particularly rich in lipids (see Section 15.0). Table 2.1 offers an example of the original results (see Table 15.2 for more recent data).

Depressants may be hydrocarbons, halogenated hydrocarbons, alcohols, ethers, ketones, weak acids (like the barbiturates), weak bases, or sulphones. They are the selectively toxic agents used in medicine as hypnotics and general anaesthetics. This is the only kind of biological activity in which structure simply does not matter (there is much more about this in Chapter 15). See Section 3.3 for the general function of partition effects in securing selective distribution of drugs.

2.1 The concept of 'receptors'. The receptor as part of an enzyme, permease, or other protein

Three striking characteristics of the action of drugs indicate very strongly that they are concentrated by cells on small, specific areas known as receptors. The *first* of these three characteristics is the *high dilution* (even 10^{-9} M) at which solutions of many drugs retain their potency. This implies that a complementary structure exists in the cell to rescue the drug from such a drowning. The *second* characteristic is seen in those substances that form pairs of optically active isomers, as do atropine, morphine, and epinephrine (adrenaline). The two forms of each of these bases, namely the dextro- and the laevo-rotatory isomers, differ strikingly in biological activity. Because the two members of such pairs have otherwise identical physical and chemical properties and differ only in that their molecules are built as mirror images of one another, it became evident that the *shape* of a drug molecule can be crucial for its action, and that a part of the molecule was obliged to fit a structure complementary to itself. For example, Arthur Cushny showed in 1909 that $(-)$-epinephrine acted about 15 times as powerfully as $(+)$-epinephrine in raising blood-pressure, in causing glycosuria, and in sheer lethality. He collected and discussed all known examples of such stereochemical differences in his monograph *Biological Relations of Optically Isomeric Substances* (Cushny, 1926) (see Section 12.1). The *third* characteristic is the high biological specificity of drugs, e.g. epinephrine has a powerful effect on cardiac muscle, but very little on striated muscle.

The idea that drugs act upon receptors began with John Langley in Cambridge who, after studying the opposing actions of atropine and pilocarpine on the flow of saliva in the cat, wrote: 'We may, I think, without much rashness, assume that there is some substance or substances in the nerve endings or gland cells with which both atropine and pilocarpine are capable of forming compounds. On this assumption, then, the atropine or pilocarpine compounds are formed according to some law of which their relative mass and chemical affinity for the substance are factors' (Langley, 1878).

We now know that pilocarpine (12.81) is an agonist [imitating acetylcholine (ACh)] and atropine (*7.16*) is its antagonist. What Langley had done was to start the flow of saliva with a small dose of pilocarpine and to stop it with a little atropine. Larger doses of the agonist restarted the flow which still more atropine was able to stop, and so on. Langley interpreted these results according to the Law of Mass Action (Guldberg and Waage, 1864), examples of whose equilibria had been confined to non-living situations, previously.

Later, Langley (1905) coined the term 'receptive substances', visualizing one for each of: pilocarpine, atropine, nicotine, tubocurarine, strychnine, epinephrine, and thyroxine. Paul Ehrlich, in Germany, was already using the word 'receptor' but only for his immunochemical work (Ehrlich and Morgenroth, 1900): 'For the sake of brevity in what follows we shall in general always designate as *Receptor* that binding group of the protoplasmic molecule to which a foreign, newly-introduced group binds'.* In an attempt to adapt this idea to the action of drugs, Ehrlich began to speak of the 'receptor apparatus' of trypanosomes (Ehrlich and Shiga, 1904). Later Ehrlich stated that some drugs are attached to the protoplasm of a cell by atom groups clinging to simplified analogues of antigenic receptors, and he called them 'chemoreceptors' (Ehrlich, 1907). In his Nobel prize address, Ehrlich outlined the receptor as a small, chemically defined area (on a large protoplasmic molecule) which was normally occupied with the cell's nutrition and metabolism, but which could take up specific antigens or drugs instead (Ehrlich, 1908).

Ehrlich's clearest exposition of a drug receptor occurs in his address to the German Chemical Society in which he stated that the high specificity of chemotherapeutic drugs arose in a purely chemical way: by the drug satisfying groups of specific affinity in the cell, groups which he called 'chemoreceptors'. He added that the chemoreceptors for arsenical drugs in trypanosomes were, apparently, mercapto ($-SH$) groups, and that the formation of $As-S$ bonds led to the death of the parasite (Ehrlich, 1909). Thus, we see the origins of our current concept of *receptor* as a chemical group, normally active in the cell's metabolism, which, by combining with a drug, triggers the observed physiological response. For more about Ehrlich's work, see Section 6.1.

At first the receptor concept was received with scepticism because of repeated failure to isolate any such substance. However, the idea of receptors became more firmly established by the work of Alfred Clark (1926, 1927) who showed that the combination of a drug with a receptor quantitatively followed the Law of Mass Action. He also demonstrated that a great deal of the most precise quantitative data on drug action could be interpreted as the result of the formation of a non-covalent bond between a drug and a receptor specific for that drug. This relationship is discussed further in Section 7.5. Because no receptor-bearing molecule had ever been isolated, Clark was obliged to work with, at best, single cells; but the quantitative and repeatable nature of his work, which used a variety of drugs and many different tissues, created a much wider

*'Wir wollen im Folgenden stets, um eine grössere Kürze des Ausdrucks zu ermöglichen, diejenige bindende Gruppe im Protoplasmamolekül, an welche eine fremde, neu eingeführte Gruppe angreift, allgemein als "Receptor" bezeichnen.'

acceptance of receptor theory. He summed up his work in a monograph (Clark, 1937) a few years before his death in 1941.

In the late 1920s, pharmacologists began to visualize an agonist as relating to a receptor in much the way that a coenzyme is related to an enzyme. No analogy, of course, can exist between an agonist and a substrate, for the latter is chemically altered by the enzyme whereas a drug is not. Similarly, the same pharmacologists compared an antagonistic drug to an enzyme inhibitor. By 1910, it was well known that many enzymes could be blocked by substances that structurally resembled their substrates. For example, amylase, which normally hydrolyses starch to maltose, could easily be blocked by glucose (Wohl and Glimm, 1910). As we shall see, a connexion between a drug-receptor and the active patch on an enzyme was first demonstrated in 1926.

That acetylcholine, isolated from vegetable material, had a strong physiological action in animals, was known early in the present century. That it might be the mammalian body's principal neurotransmitter was suspected by Otto Loewi (1921) working in Vienna. Yet the world of pharmacology was unprepared, and shaken, when, 5 years later, Loewi and Navratil (1926) showed that this was indeed the case. Most workers were surprised that transmission of nervous impulses across a synapse (gap) should be effected by a chemical, because so many believed it would be electrical.

Greater than this surprise was a shock which arose from the following paradox. Although a quaternary amine, acetylcholine was no muscle relaxant (like tubocurarine and all of Crum Brown's artifacts) but was actually Nature's number one muscle activator! This discovery should have put an end to any more advocacy of the dogma 'one chemical group gives one biological action', but this school of thought proved to be conservative!

That a given chemical group could produce either an agonist or an antagonist, depending on its chemical setting was explained by H.R. Ing (London) as follows: acetylcholine and tubocurarine act on the same receptor, but the smaller molecule exactly fits the site and activates it, whereas the larger molecule covers the receptor and blocks it (Ing, 1936). We now know of many series where the lower members are agonists, but the homologues of higher molecular weight are antagonists (see Section 7.5.2).

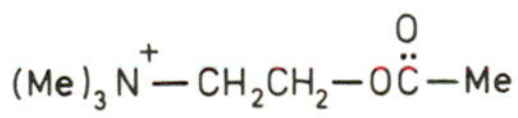

$$(Me)_3 \overset{+}{N} - CH_2CH_2 - O\overset{O}{\overset{\|}{C}} - Me$$

Acetylcholine (cation)

(2.7)

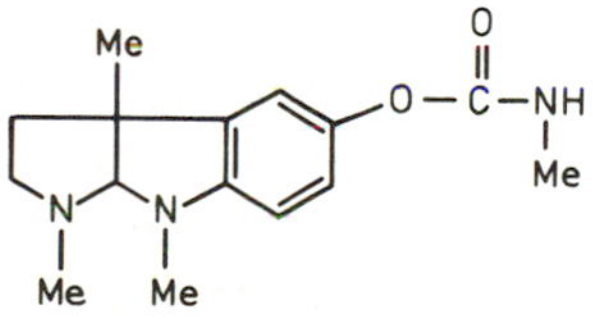

Physostigmine (eserine)

(2.8)

Let us resume the account of enzymes as bearers of drug receptors. The alkaloid physostigmine (2.8), although without effect on leech muscle, was found to potentiate one million-fold the action of acetylcholine (ACh) on this organ (Fühner, 1918). This was not interpreted at the time, but the slowing of the mammalian heart by physostigmine was traced to its protecting the destruction of ACh by a bodily esterase which was not, at that time, purified or even named (Loewi and Navratil, 1926). It was named cholinesterase by Stedman, Stedman and Easson (1932), and today we call it acetyl-cholinesterase, or AChase for short. Physostigmine, although it mimics the action of ACh, is not a true agonist because it does not act on the ACh receptor (which lies on the muscle). Hence it is classed as a *pseudo-agonist*. In clinical use, physostigmine enables the patient to have a much longer-acting dose of his own acetylcholine.

2.1.1 *Receptors on permeases*

Since then, many receptors have been found to be the active sites of enzymes. However, the receptor for ACh is not on an enzyme but on a different type of protein, one that regulates the passage of sodium, potassium, and calcium ions in and out of the muscle cell, without effecting any chemical change (Karlin, 1974). This *permease*, as such a protein is called, was isolated in 1969 by Jean-Pierre Changeux (Paris) from the electric organ of a fish (Changeux, 1969). The permease is situated in the plasma membrane of the muscle cell, at the synapse. Apparently, permeases provide the graded response needed at synapses. The homogeneity and richness in cholinergic synapses of the electric tissue (electroplax) of two fish species, *Torpedo* and *Electrophorus*, made it the ideal source of ACh receptor. Purification began with homogenization, differential centrifugation, and then sedimentation in sucrose density gradient solutions. The membrane fragments obtained in this way contained the receptor, free from acetylcholinesterase. These fragments quickly changed into vesicles whose permeability to $^{22}Na^+$, $^{42}K^+$, and $^{45}Ca^{2+}$ was greatly increase by the ACh analogue, carbachol (*2.11*) (Kasai and Changeux, 1971). It was reassuring to find characteristic properties of the ACh receptor surviving so much manipulation.

This preparation could be freed from membrane fragments, leading to a fairly pure sample of the receptor–protein complex, by the action of aqueous detergents, both anionic and neutral, followed by ultracentrifugation; and so it became available in decigram quantities (Changeux, 1980). Kinetic measurements of agonist and antagonist uptakes were almost identical with those obtained on the intact organ (electroplax).

The purified receptor-carrier was then found to be a glycoprotein of molecular weight about 50 000. Under the electron microscope, it was seen as rows of rosettes, at a density of about a dozen rosettes to the linear 0.1 μm. This was first illustrated by Changeux (1980) and a more recent photomicrograph will be seen in Fig. 2.1. The sequence of all the amino acids is known. Several residues of carbohydrate are bound to the protein which has a somewhat lipophilic character. More about this ACh receptor is presented in Section 12.6.

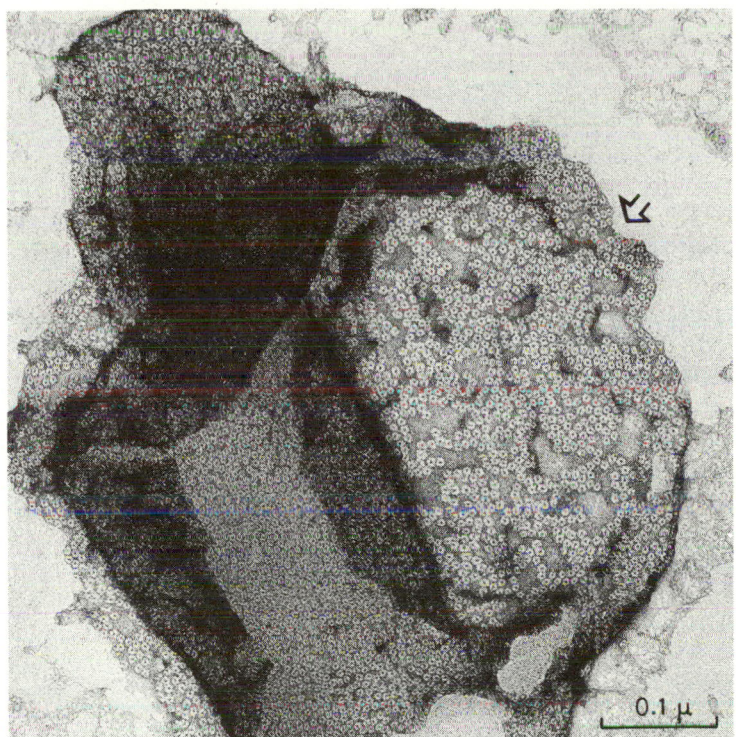

Fig. 2.1 Massed receptors for acetylcholine in the postsynaptic membrane of the electric organ in the fish, *Torpedo*. The centre of each rosette is a Na^+/K^+ channel (the arrow points to an area where the membrane is single). (Courtesy of Dr J. Cartaud, Paris.)

The foregoing account is of the 'nicotinic' variety of ACh receptor from the voluntary neuromuscular junction. This work has been extended by isolation of the 'muscarinic' ACh receptor from ox brain by extraction with sodium cholate. 'Nicotinic' and 'muscarinic' are defined at the beginning of Section 12.6. The brain receptor was characterized by its intense power to bind the specific muscarinic antagonist, [³H]atropine, with the high inhibitory constant of 2×10^{-9}. This label can be displaced by unlabelled atropine, and also by ACh and its agonists (Carson *et al.*, 1977; Carson, 1982).

The permeases of other neurotransmitters have been isolated, but not in quantity. All appear to be glycoproteins, highly phosphorylated. The isolation of catecholamine receptors is described in Section 12.4. The receptors for the natural polypeptide analgesics, which also bind morphine, are discussed in Section 12.8.

2.1.2 *Receptors on enzymes*

Let us return to the blocking of AChase by physostigmine, the first location of a drug receptor on an enzyme (p. 28). Synthesis of analogues showed that the two portions of the physostigmine molecule needed to inhibit the esterase were the basic group and the methylcarbamoyloxy group (*2.9*) (Stedman, 1926; Stedman and Stedman, 1931). Neostigmine (*2.10*), synthesized as a simplified analogue of physostigmine (Aeschlimann and Reinert, 1931), was found to be a more useful drug than the latter for treating myasthenia gravis, a muscle-wasting disease with symptoms similar to those of tubocurarine poisoning. Neostigmine, like physostigmine, carries a carbamoyloxy group, but the basic group of the latter is that of acetylcholine. These two groups together make neostigmine an excellent substitute for ACh at the active site of acetylcholinesterase, but unlike physostigmine it also has a direct true agonistic effect on the ACh receptor of muscle.

These indications that the carbamoyloxy (or 'urethane') group could mimic the acetyl group of ACh, led to the introduction of carbachol (carbamoylcholine) (*2.11*) whose actions resemble those of acetylcholine but are much longer lasting (Molitor, 1936). Although its action is, in part, truly agonistic (replacing ACh on the muscle receptor) carbachol acts principally as a pseudo-agonist, yet in a way totally different from that of physostigmine: namely, carbachol drives ACh from its stores in the nerve terminal. Carbachol neither blocks nor is hydrolysed by acetylcholinesterase. It is valuable for restoring tone to the bladder and intestines of patients recovering from surgery.

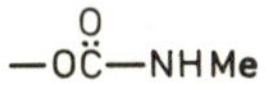

$$-O\overset{O}{\overset{\|}{C}}-NHMe$$

Methylcarbamoyloxy-group

(*2.9*)

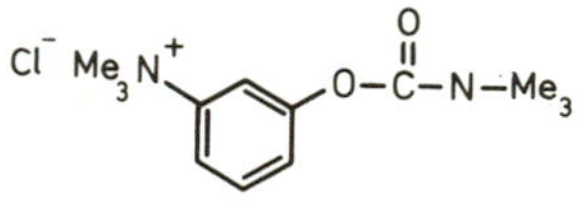

Neostigmine

(*2.10*)

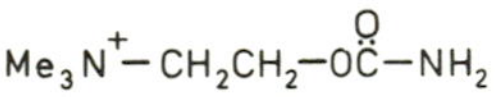

$$Me_3\overset{+}{N}-CH_2CH_2-O\overset{O}{\overset{\|}{C}}-NH_2$$

Carbachol (cation)

(*2.11*)

Among other pharmacodynamic drugs that are enzyme inhibitors are the cardiac glycosides which act on the heart by inhibiting sodium–potassium adenosine triphosphatase (Bonting, 1970) (see Section 14.1).

It seems curious that no extension, from pharmacodynamics to chemotherapy, took place until 1940, when Donald Woods (London) demonstrated the reversal, by *p*-aminobenzoic acid (PAB) (*2.12*), of the anti-bacterial action of sulfanilamide (*2.13*) (Woods, 1940). He attributed this effect to a general steric and electronic similarity between the two substances, two kinds of relationship later quantified by Paul Bell and Richard Roblin (1942). Very small changes in these physical properties were then shown to alter the action of the drug for worse, if the properties became too divergent, but for better if they converged. This correlation was confirmed in 1962 when Gene Brown (Massachusetts) isolated the enzyme dihydrofolate synthetase which makes the important co-enzyme, dihydrofolic acid (2.14), of which PAB forms the central part. This enzyme has a receptor site for PAB which the sulfonamide anti-bacterials fit well. The enzyme comes to equilibrium, quite reversibly, with either the substrate, the inhibitor, or both, depending on the concentrations present. When the sulfonamide is in excess, the enzyme is blocked and no dihydrofolic acid is made (Brown, 1962).

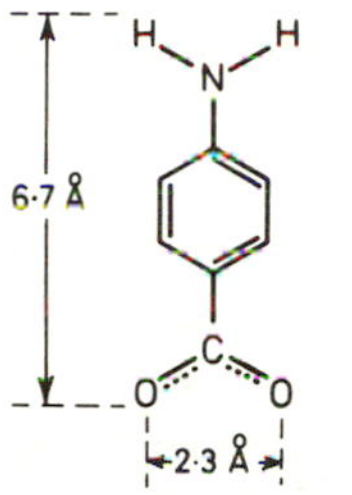

p-Aminobenzoic acid (anion)
(2.12)

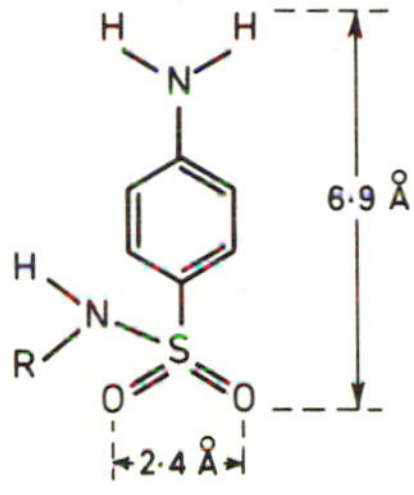

Sulfanilamide (molecule) (R=H)
(2.13)

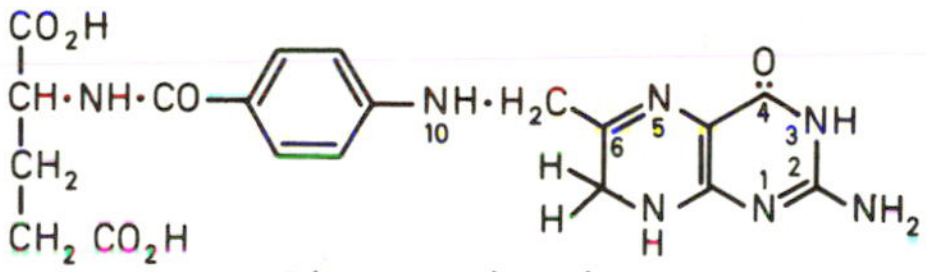

Dihydrofolic acid
(2.14)

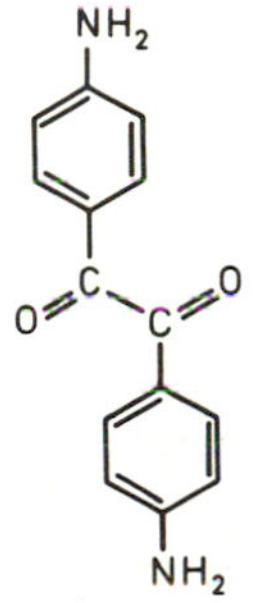

4,4'-Diaminobenzil
(2.15)

    The sulfonamide anti-bacterials, which, soon after their discovery in the late 1930s, seemed to reinforce the 'one-group one-action' hypothesis, ended by dealing it a mortal blow. It soon became evident that the presence of a sulfonamide group would not introduce anti-bacterial properties unless the other conditions (such as *para* substitution) for fitting the enzyme-receptor were observed (see Section 9.3.1). If these conditions were met, the sulfonamide group could be replaced by a different group of similar polarity. For example,

4,4'-diaminobenzil (*2.15*), which is much more anti-bacterial than sulfanilamide (Kuhn, Weygand and Möller, 1943), acts similarly by inhibiting the
enzymic uptake of PAB. Moreover, many useful anti-diabetic and diuretic
drugs, such as tolbutamide (*12.61*) and chlorothiazide (*14.1*) respectively, have
been launched, each with a sulfonamide group attached to a benzene ring.
However, these drugs do not have the dimensions and the charge-pattern
suitable for becoming attached to dihydrofolate synthetase, and hence they
have no anti-bacterial action. The anti-bacterial blockers of dihydrofolate
synthetase, on the other hand, can meet the enzyme specifications without even
containing sulphur!

Many other drugs have receptors on enzymes, for example the antibiotic
rifampicin (*4.37*) which acts by inhibiting RNA polymerase (Section 4.0).
Another example, allopurinol (*9.51*) was carefully designed to fit, and block, two
consecutive enzymes: hypoxanthine oxidase and xanthine oxidase (Elion *et al.*,
1966). It has provided the best of all treatments for chronic gout, by preventing
the formation of uric acid (Section 9.4.4). The anti-malarial action of pyrimethamine depends on its blocking the enzyme dihydrofolate reductase which is also
the target for the anti-bacterial drug, trimethoprim (Section 9.3.3). Penicillin
acts against bacteria by blocking the polymerase that they need to form new cell
wall (Section 13.1). Many other examples will be encountered in Chapter 9.

Extension to insecticides of the idea that receptors could occur on enzymes was made soon after the end of the Second World War, when organic
phosphates began to be used for their inhibition of AChase in insects. The
hydroxy-group of serine in this enzyme was pinpointed as the receptor site for
organophosphate insecticides (Section 13.3).

How enzymes work is discussed, at the molecular level, in Section 9.0.

2.2 The receptor as part of a nucleic acid

The period of the Second World War (1939–1944) was a turning point in the
study of structure–action relationships. Mere 'paper resemblances' between
two formulae began to lose favour as a guide to biological action. Less often were
groups (or nuclei) assumed to be the direct source of some pharmacological
effect, and more attention was given to the physical properties which these
groups (and nuclei) introduced and maintained. This slow, evolutionary
movement gathered pace from what was being learnt about the anti-bacterial
sulfonamides (Section 2.1).

The chief physical properties studied were (a) electron distribution, particularly as manifest in *ionization*, the degree of which could facilitate or forbid the
combination of a drug with its receptor, and (b) *steric properties*, which governed
access to the correct receptor and a good fit upon arrival there. Both of these
factors came to the fore in studies of the aminoacridines described briefly here,
and more fully in Section 10.3.1.

This work was undertaken by my colleagues and me for the Australian Army, during the Second World War. At that time, the aminoacridines, because of their high selectivity, were much used to irrigate deep, infected wounds (Poate, 1944). In short, they killed bacteria without harm to either leucocytes or abraded tissues, and the wounds rapidly healed. This work proved to be interesting in two ways. It established, quantitatively for the first time, the importance of ionization in the action of a chemotherapeutic drug. Also, it led the way to recognition that some drug receptors were not on proteins but, as here, on nucleic acids.

First, it was shown that the bacteriostatic action of these topical anti-bacterials was proportional to the fraction ionized as cation (Albert, Rubbo and Goldacre, 1941). At first, it seemed puzzling that, of the five possible amino-acridines, two were highly anti-bacterial, whereas three had little activity. At that time, almost nothing was known about the ionization of heterocyclic bases, so we began to determine the pK_a values of a great many examples, and later we published the main rules that connect structures with basic strength (Albert, Goldacre and Phillips, 1948). Fortunately, the correlation that we needed for our acridine work was found early.

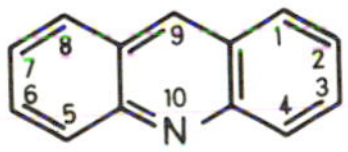

Acridine
(International Union of Chemistry numbering)
(2.16)

(2.17)   (2.18)

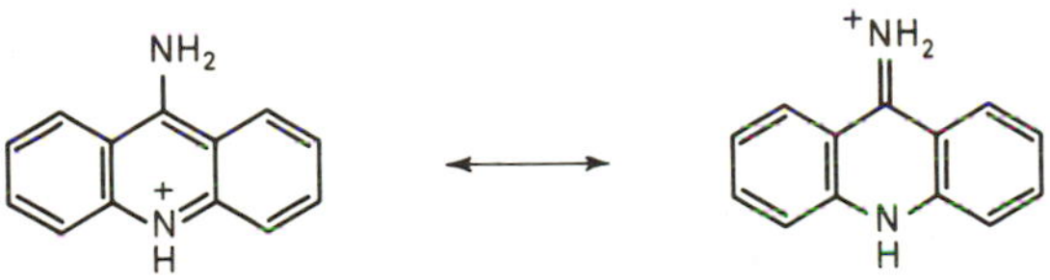

(2.19)   (2.20)

Cationic resonances which make 3- and
9-aminoacridines strong bases

Acridine (2.16) is a weak base. With its pK_a of 5.3, it is only 1% ionized at pH 7.3. It turned out that the 3-amino- and 9-amino-acridines had a resonance in their cation that was lacking in the neutral species [e.g. (2.17) ⟷ (2.18) and (2.19) ⟷ (2.20), respectively] and this made them very strong bases. The other three aminoacridines could not, for reasons of valence, acquire this resonance stabilization and, hence, were very little stronger than acridine itself (see Table 2.2). We called this the 4-aminopyridinium type of base-strengthening resonance, because we had demonstrated it first in 4-aminopyridine.

Using a wide range of bacterial species (22 kinds, embracing aerobes and anaerobes, Gram-positive and -negative) and 102 different acridines, we showed that the anti-bacterial action of aminoacridines increased with the proportion that was ionized (at the pH and temperature of the bacteriostatic test). The nature of any non-amino substituent, whether electron-attracting or electron-releasing, made absolutely no difference so long as it permitted at least 50% cationic ionization under these conditions (Albert *et al.*, 1945). Typical examples are offered in Tables 10.6 and 10.7.

Table 2.2 Dependence of anti-bacterial action on ionization

-acridine	*Minimal bacteriostatic concentration* (Streptococcus pyogenes) *1 part in*	*Percentage ionized as cation* pH 7.3; 37°C
1-Amino-	10 000	2
2-Amino-	10 000	2
3-Amino-	80 000	73
4-Amino-	5 000	< 1
9-Amino-	160 000	100
2,7-Diamino-	20 000	3
3,6-Diamino-	160 000	99
4,5-Diamino-	< 5 000	< 1

As a result of this work, we were able to replace the Army's favoured deep-yellow wound irrigant, proflavine (3,6-diaminoacridine), by the more selective and non-staining analogue, aminacrine (9-aminoacridine).

The anti-bacterial action of aminoacridines has these unusual characteristics: it takes place even at high dilution, is unchanged by the presence of serum proteins, and is without harm to mammalian tissues. Because no other cations known at that time had this combination of properties, it was tempting to ascribe them to the presence of the acridine nucleus. Nevertheless, we made several stepwise alterations to this molecule, to discover the parameters of its efficacy. What we found was that any nucleus would do as well as acridine, so long as it was (a) basic enough to be ionized at least 50% at the pH of our test and (b) had no less than 38 Å^2 of flat area (Albert, 1944). 4-Aminopyridine and 4-aminoquinoline (which may be seen as 9-aminoacridine without respectively one and two benzene rings), although ionized enough, had too little flat area. However, when they were supplied with more of this desirable property by inserting a *co-planar* substituent, as in 4-amino-2-styrylquinoline (*2.21*), the anti-bacterial activity returned. Moreover, the order of the rings comprising the acridine molecule was found relatively unimportant, because many highly anti-bacterial amino benzoquinolines and phenanthridines were soon brought to light. Not surprisingly, our standard (9-aminoacridine) was deprived of its activity by hydrogenating one of the outer rings, an operation that necessarily deprived one-third of the molecule of its flatness.

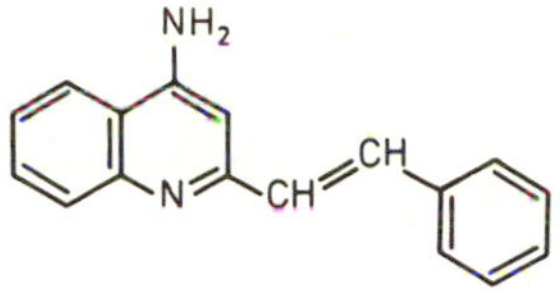

4-Amino-2-styryl-quinoline
(1:80 000)
(2.21)

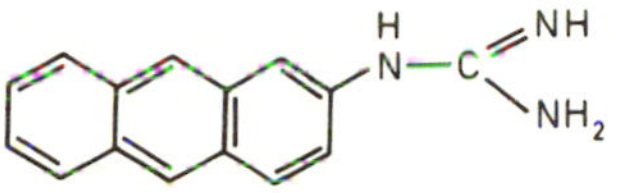

2-Anthrylguanidine
(2.22)

Finally, we boldly left the heterocycles behind and began to make basic anthracenes, such as 2-guanidinoanthracene (2.22). This has enough of both requirements: ionization and flatness. Moreover, it had the typical amino-acridine-like bacteriostatic properties, namely, activity against that same wide range of bacteria at high dilution, even in the presence of serum, and without harm to phagocytes (Albert, Rubbo and Burvill, 1949). Here, just as with the anti-bacterial sulfonamides, the required biological action depends on the correct steric and electronic properties and not on the presence of a particular nucleus or substituent. Needless to say, such a conclusion was far less acceptable at that time than it is today.

That aminoacridines are accumulated only by the *nucleic acids* of the living cell became known through their use in vital staining (cf. Strugger, 1940). The reason for the requirement of molecular flatness was explained in 1961 when Leonard Lerman (Colorado) showed that aminoacridine molecules became *intercalated* into DNA by stacking between the layers of base-pairs, to which they clung by van der Waals forces supplemented by stronger ionic bonds to the phosphate ions of the DNA backbone. The resultant increase of 20°C in the T_M ('melting temperature') showed that intercalation had interfered with separation of the strands and, hence, with normal functioning of the DNA (Lerman, 1961).

In the next year, Hurwitz and his colleagues in New York demonstrated that aminoacridines injure bacteria by blocking the DNA template required by the polymerases that synthesize bacterial DNA and RNA (Hurwitz *et al.*, 1962).

These studies of structure–action relationships in the aminoacridines established that nucleic acids can be the site of receptors. In fact, the drug–receptor interaction was observed here in unusual detail, much of it at the level of molecular biology (for further details, see Section 10.3).

It is now recognized that DNA is also the target for the many other kinds of drugs that act by intercalation. These include the anti-cancer drugs daunorubicin, doxorubicin (adriamycin), actinomycin D, anthramycin, amsacrine, and ellipticine; the anti-malarials mepacrine (quinacrine, atebrin) and chloroquine; and the anti-trypanosomal drugs ethidium and quinapyramine (more about these drugs in Section 10.3.5).

The DNA-cleaving anti-cancer drugs such as bleomycin and camptothecin are bound by what they later destroy (Section 4.0).

Furanocoumarins are given orally (a) to restore skin pigment lost in vitiligo, and (b) to sensitize cells to u.v. light in treating psoriasis. Methoxsalen [8-methoxy-psoralen (2.23)] is used for both purposes. Re-pigmentation (Africk and Fulton, 1971) follows interstrand linking of two pyrimidine residues by the drug (Musajo and Rodighiero, 1970). Pigmentation, undesired in treating psoriasis, can be avoided by using sterically-hindered derivatives which bind to one strand only (Rodighiero and Dall' Acqua, 1984).

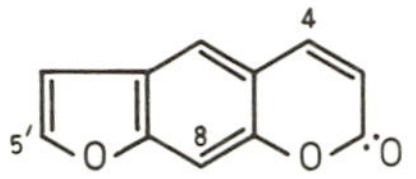

Psoralen
(2.23)

DNA also provides the eventual receptor for steroid hormones, as is explained under 'Sequential hormones' in Section 2.4. Cholecalciferol (vitamin D_3), a seco-steroid, indirectly affects calcium absorption through its inductive action on DNA in the cells lining the intestine (Haussler and Norman, 1969). Ecdysone (4.77), the steroid insect hormone, induces DNA to make (through RNA) the enzyme synthesizing N-acetyl-3,4-dihydroxyphenylethylamine, which then interacts with cutaneous protein to bring about moulting (Karlson and Sekeris, 1962). The nitrogen mustard class of anti-cancer drugs functions by locking together the two strands of DNA through their guanine residues (Section 13.4). All the purine and pyrimidine analogues that become built into DNA (as idoxuridine does) or RNA (e.g. 8-azaguanine) belong here (Section 4.0).

Taking all the above examples together, it is evident that a very large number of drugs find their receptors on a nucleic acid.

2.3 The receptor as a coenzyme or other small molecule

Another kind of drug–receptor relationship was brought to light by the next project that my colleagues and I carried out for the Australian Army. We had been asked to look at the mode of action of oxine (8-hydroxyquinoline) (2.24) which they were using for its strong fungicidal and bactericidal properties. Typical of earlier ways of thinking, Hata (1932) had supposed that oxine owed its antibacterial properties to a combination of those of quinoline and phenol in the one molecule. Yet neither quinoline nor phenol is at all anti-bacterial at a dilution of 1:5000, whereas oxine is active at 2 parts/million. That the biological properties of two substances could be combined by introducing their individual groups into a single molecule strikes us today as absurd, because the favourable distribution of electrons in each of the constituent molecules (here, phenol and quinoline) must, more often than not, be incompatible in the hybrid.

In the new study (Albert *et al.*, 1947), it was found that the six isomers of oxine, obtained by moving the hydroxy group in (*2.24*) to each other possible position, were non-anti-bacterial. This suggested to us that the biological properties of oxine were linked to its ability to chelate, i.e. to bind metal cations tightly by two or more atoms so that a 5- or 6-membered ring (no other size is stable) is formed. When tested, the six inert isomers failed to chelate, whereas oxine chelated strongly. Oxine had long been used in the analysis of metals because of this property, but chelation was not accepted as the mode of action for any chemotherapeutic agent at that time. A typical chelated product, with ferrous iron as an example, is shown as (*2.25*).

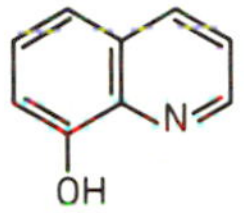

8-Hydroxyquinoline (oxine)   The 1:1-ferrous complex of oxine
(2.24)   (2.25)

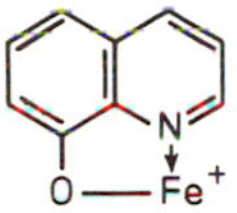

The question was then posed: does oxine act on bacteria by removing metals essential to bacterial welfare, or does it cause traces of metal ions to become more toxic to the bacteria? The latter proved to be the case. When we incubated *Staphylococcus aureus* in distilled water, with oxine, with iron, and then with both chemicals together, subsequent plating out on nutritive medium showed that only bacteria that had been exposed to *both* oxine and iron were killed (see Table 2.3) (Albert, Gibson and Rubbo, 1953). In further work, it was found that iron is a necessary co-toxicant for all kinds of bacteria, whereas copper takes over this function for fungi.

Table 2.3 Necessity for a co-toxicant metal for the bactericidal action of oxine, as shown by incubating in distilled water at 20°C, and plating out after 1 hour

Oxine (1/M)	Ferrous sulphate (1/M)	Growth (Staphylococcus aureus)
Nil	Nil	Prolific
100 000	Nil	Prolific
Nil	100 000	Prolific
100 000	100 000	Undetectable

(Albert, Gibson and Rubbo, 1953.)

Before long, several substances with chemical constitutions very different from that of oxine, and using quite different atoms to form the chelate ring, were found to have the typical oxine mode of action on bacteria. This typical action is definable as the rapid killing of bacteria and fungi at high dilutions, requiring the presence of iron or copper (respectively), and preventable by a trace of cobalt (but no other metal) (Rubbo, Albert and Gibson, 1950). Examples are: pyrithione (*2.26*) (Albert, Rees and Tomlinson, 1956) which is much used in dermatology of the scalp; and dimethyldithiocarbamic acid (*2.27*) (Sijpesteijn and Janssen, 1959) whose salts are common agricultural fungicides.

Cobalt is well known for its ability to break an oxidatively destructive chain reaction catalysed by another metal (cf. Baur and Preis, 1936). This suggested to the Dutch workers that the iron and copper complexes of oxine, pyrithione, and dimethyldithiocarbamic acid were oxidatively destroying thioctic acid (dihydrolipoic acid) (*2.28*) which is the essential coenzyme for the oxidative decarboxylation of pyruvic acid. This was confirmed when they found pyruvic acid accumulating in the medium (Sijpesteijn and Janssen, 1959; also personal communications from these authors). The receptor in all three examples is the small molecule (*2.28*) although, at the time, it caused surprise to find one of such low molecular weight.

Pyrithione
(*2.26*)

$$S$$
$$Me_2N \cdot \overset{S}{\underset{}{C}} \cdot SH$$

Dimethyldithiocarbamic acid
(*2.27*)

$$\overset{SH}{\underset{|}{} } \quad \overset{SH}{\underset{|}{} }$$
$$CH_2 \cdot CH_2 \cdot CH(CH_2)_4 \cdot CO_2H$$

Thioctic acid
(Dihydrolipoic acid)
(*2.28*)

An earlier example of a coenzyme acting as a receptor is the porphyrin molecule of cytochrome oxidase. The lethal action of hydrogen cyanide, counter-selective for mammals, follows directly from the binding of this poison to the free valence of the chelated iron in the porphyrin. Many bacteria, lacking this enzyme, are not affected.

More recent examples of small molecules as receptors are furnished by the polyene antibiotics such as nystatin (*14.19*) and amphotericin B (*5.14*). These anti-fungal drugs have sufficient physical resemblance to steroids to be accepted by the fungal plasma membrane alongside the ergosterol molecules (Hamilton-Miller, 1973). This makes the cell porous and leaky, a therapeutic advantage. The action of phenols, polypeptide antibiotics, and long-chain quaternary amines in disintegrating the plasma membrane of bacteria (Section 14.3) follows from a similar binding to the lipoid components.

A series of glycolipid (ganglioside) receptors on human cell walls cause conformational changes in (a) glycoprotein hormones such as the luteinizing hormone (LH) and the thyroid-stimulating hormone (TSH), (b) bacterial toxins, such as those of cholera and tetanus, and (c) interferon. The change enables each biopolymer to transmit its message through the cell membrane whereas contact with a non-matching oligosaccharide (the coded portion of the ganglioside) effects no conformational change. Chemically related gangliosides, usually present, compete with their activating analogues for the biopolymer and (reversibly) inactivate it, a regulatory process with therapeutic potential (Kohn, 1977).

2.4 Reversibility, and other aspects of receptors

(*a*) *Reversibility of combination with receptors.* The majority of drugs and other selectively toxic agents combine very loosely with their receptors. They can usually be easily washed off the receptors, which then cease to register the adverse effects produced by the agent. It is comparatively rare for agents to form covalent bonds with their receptors (covalent bonds are defined in Section 8.0), but those that can do so inflict a change that is difficult to reverse. Examples, of which the most important are penicillin and the organic phosphates, will be found in Chapter 13.

Ehrlich was the first to note that other, more easily reversible, bonds were concerned in the action of most drugs. He wrote: 'If alkaloids, aromatic amines, antipyretics or aniline dyes be introduced into the animal body, it is a very easy matter, by means of water, alcohol or acetone, to remove all these substances quickly and easily from the tissues' (Ehrlich, 1900).

The chemotherapeutic worker, with the example of penicillin in mind, may tend to think that agents generally act by forming covalent bonds. But the worker in pharmacodynamics, used to washing organ-preparations and using them over and over again, is better informed in this matter. Fig. 2.2 gives a typical example of the ease with which drugs can be removed from their receptors by washing. There is no doubt that most of the known drugs react with receptors by readily reversible bonds, namely ionic, hydrogen, and van der Waals bonds (see Section 8.0), and that combination with receptors by covalent bonds is a rarity.

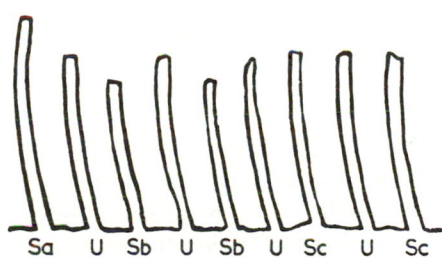

Sa   U   Sb   U   Sb   U   Sc   U   Sc

Fig. 2.2 Easy displacement of drug from receptors by washing. This example is the assay of histamine by immersing guinea-pig gut alternately in unknown and standard solutions, made up to 2 ml with isotonic saline.
    S. Standard solution (1 in 5 000 000)
        (a) 0.1 ml, (b) 0.07 ml, (c) 0.09 ml.
    U. Unknown solution (0.1 ml).
Between each reading, the tissue was washed with saline, which restored it to its original condition. (Result of assay: 0.1 ml of unknown has the potency of 0.09 ml of standard.) (Gaddum, 1936.)

(*b*) *The vulnerability of pacemaker enzymes.* Any drug-combining enzyme (in a series of enzymatic reactions), is more susceptible if it is a relatively unabundant enzyme and hence the pacemaker of the series (Krebs, 1957). The triose phosphate dehydrogenase system in glycolysis is an example. It follows that one stands a better chance of success when designing drugs to inhibit enzymes if these are *pacemaker* enzymes.

(*c*) *Conformational changes in drugs and their receptors.* Because substrates often change the *conformation* of enzymes, and enzymes that of substrates (Koshland, 1964), it is likely that receptors and drugs can change one another's conformations. Of course some drugs have quite rigid structures, but others could be deformed by the receptor, and a protein receptor should be deformable by the drug, so that in some cases the drug may be acting in a cavity that its presence has created, a case of 'induced fit'. This concept is developed further in Sections 7.5.2 and 9.0. Originally stereochemical specificity (see earlier in this chapter) made people think, wrongly, that receptors must be quite rigid. For more data on conformation, see Section 12.3.

In the past, there was much puzzlement as to why drugs of quite different chemical and physical properties could bring about the same physiological result. Surely they could not all act on the same receptor? Actually, there is often a *succession of receptors*, each of which can be controlled by a drug selective for it. For example, in treating high blood-pressure, it is possible to medicate peripherally by relaxing the muscle of blood-vessels with hydralazine (*11.47*) or nitrites. However, one can proceed a little higher up the chain of command by blocking the sympathetic nerve-endings at their junction with these muscles, using a β-receptor blocker such as propranolol (*12.56*). It is possible to work at a still higher stage of organization by blocking the sympathetic ganglia with hexamethonium (*7.30*). Finally one can act on the highest level of all, the central nervous system, with clonidine (*12.53*). Naturally, these four kinds of drug do not share any physical or chemical properties, for they act on four chemically distinct receptors.

(*d*) *Analogous receptors.* Just as there are analogous enzymes (Section 4.2), so there are *analogous receptors*. For instance, the acetylcholinesterase in sheep's gut combines more reversibly with organic phosphates than the analogous enzyme (the receptor) does in worms parasitizing the intestine. Highly selective deworming can be accomplished thanks to this difference (Lee and Hodsden, 1963). Again, the AChase of the domestic fowl is, unlike its mammalian counterpart, incapable of being inhibited by bisquaternary amines such as ambenon, or of hydrolysing bisquaternary esters such as suxamethonium (*7.29*). The human body has at least four main types of receptor for catecholamines (Section 12.4), two for histamines (Section 9.4.5), and two (called nicotinic and muscarinic respectively) for acetylcholine (Section 12.6). Selective agonists and antagonists have been found for all of them.

(*e*) *Sequential receptors.* Some examples, few but important, are known where the first receptor for an agonist is not the final one. For example, the thyroid hormone, L-thyroxine (*11.14*), finds its first receptor in a protein, prealbumin, which transports it through the bloodstream to the thyroid gland where it is transferred to its final receptor. X-ray studies on the molecule of prealbumin (molecular weight 55 000) show two identical channels each of which accommodates one molecule of thyroxine bound with an affinity of 10^7 M^{-1}. The hormone is placed so that the phenolic group is near the centre of the protein molecule while the amino acid part remains near the port of entry, with its CO_2^- group pressed hard against the NH_3^+ group of lysine. The central (diphenyl ether) moiety of the hormone is held rigidly in the channel by the close fit of its three main portions into well-defined complementary pockets in the protein. In D-thyroxine, on the contrary, the NH_3^+ group of the hormone lies towards the NH_3^+ group of the receptor, and this diminishes the affinity because of coulombic repulsion (Andrea *et al.*, 1980).

Although DNA is the eventual receptor for steroid hormones, as is shown by the antagonistic action of actinomycin D (a specific inhibitor of the DNA-primed synthesis of RNA), their first receptor is a protein. The pioneering studies of Jensen and Jacobson (1962) revealed that a receptor protein, specific for estrogens, existed, *inter alia*, in the uterus, vagina, and anterior pituitary gland. They had applied 17β-[^{3}H]estradiol to various tissues and found it was selectively bound to these few. The receptor was isolated by ultracentrifugation. Similarly the oviduct was shown to have specificity for progesterone.

These steroid–receptor complexes are formed in, or diffuse into, the nucleus where they bind to the chromatin (no steroid is bound in the absence of its protein). In this way, a length of DNA (codon) is de-repressed and produces the required mRNA which, in turn, uses cytoplasmic ribosomes to make the characteristic protein. So rapid is this process that the physiological response characteristic of the hormone can sometimes be seen within an hour. These are very typical examples of the *induction of* a characteristic protein (for a review of induction, see Cohen, 1966).

Many other examples were then discovered. The principal androgenic steroid, dihydrotestosterone, was found to be quite specifically bound by an acidic protein in androgen-dependent tissues such as the prostate gland (Anderson and Liao, 1968). Corticosteroids, too, are bound by specific protein in the cytoplasm of liver and some other cells, and the complex enters the nucleus where it is bound by DNA. This leads to the appearance of a specific mRNA (its formation suppressable by actinomycin D) which, in turn, produces enzymes characteristic of the corticosteroid (Sekeris, 1971). A similar sequence governs the diuretic effect of aldosterone (Edelman, Bogoroch and Porter, 1963).

2,3,7,8-Tetrachlorodibenzo-*p*-dioxin (*2.28a*), the much-discussed contaminant of trichlorophenol, is taken up by a receptor protein in the cytosol of liver, then transported to the nucleus where it induces cytochrome *P*-450 and aryl hydrocarbon hydroxylase. These seem to be beneficial properties, but in some strains of mice it induces expression of other genes which effects cellular involution, division, and differentiation, presumably using a normal cell-

regulating system of which we had not been aware. The pathology of this action
of 'dioxin' is that it is *sustained*. Most molecules that are reasonably flat, and
occupy a rectangle 3×10 Å are taken up by this receptor (Poland and Knutson,
1982). It can also induce δ-laevulinic acid synthetase which is rate-determining
in the biosynthesis of porphyrins. The extent of its toxicity to man has not been
exactly defined, but may have been greatly exaggerated. However, as far as
susceptible strains of mice go, dioxin must be rated as the most toxic known
substance of low molecular weight (Whitlock and Israel, 1984).

'Dioxin'
(2.28a)

(*f*) *Special receptors for antibiotics?* The introduction of penicillin into medical
practice in 1941, and of streptomycin, chloramphenicol, and the tetracyclines in
the following 8 years, opened up a new vista in chemotherapy. Antibiotics were
at first regarded with much awe, and a completely new mode of action was
predicted for them. In the course of time, though, they have been found to use
the established types of receptor. A property that sets many antibiotics apart
from synthetic drugs is their complex stereochemistry which, being laid down in
contact with living matter, often makes a remarkably effective fit in a
therapeutic situation, but not always a selective one.

(*g*) *Targets that are non-receptors.* Rarely, targets of drugs can be *non-receptors*.
Several chelating agents are routinely used in hospitals as antidotes for poisoning by inorganic substances; examples are dimercaprol (dithioglycerol) (*11.23*)
used to treat poisoning caused by compounds of arsenic, antimony, mercury,
and gold; and the calcium salt of EDTA (*11.27*), an effective remedy for lead
poisoning (see Section 11.6). Because these metals are not normal body constituents, they are not receptors.

(*h*) *Purification of receptor-carriers.* When a receptor is on a macromolecule, it is no
more conceivable to isolate it than to isolate the active site of an enzyme. In fact,
as we have seen in Section 2.1, many receptors *are* the active sites of enzymes.
Except when the receptor is a small molecule (as in Section 2.3), the goal is to
isolate and purify the biopolymer which carries the receptor. This is usually
done by incubation of a tissue homogenate with a radioactively labelled agonist
or antagonist of the normal ligand (say, a neurotransmitter). If the marker is
chosen for the strength of its binding, it can guide the concentration process. It
can later be exchanged for weaker agonists or antagonists, or removed by
dialysis or affinity chromatography.

Molecular probes are also useful. These are small molecules with a measurable physical property such as fluorescence, visible absorption, n.m.r., or

electron spin resonance from, usually, a nitroxide radical. All that is required is that the physical property be altered by absorption on the receptor. Fluorescent probes are sensitive to the polarity of the binding site; energy can be transferred from the binding site to the probe, thus changing the fluorescence. A fluorescent probe for the ACh receptor is 1-dimethylaminonaphthalene-5-sulfonamido-ethyl trimethylammonium perchlorate. This can be competitively displaced by ACh, dimethyltubocurarine, and decamethonium, thus providing a basis for identification and analysis (Weber *et al.*, 1971). For further reading on probes, their use, advantages, and disadvantages, see Lee (1982). A multivolume work on receptors is now being issued at intervals (O'Brien, 1979).

2.5 How a small change in a molecule can lead to a large change in biological properties

It is quite common to find a pair of closely related molecules, of which the first has much biological action but the other has none. How can two such substances, which may differ in composition by only a single methyl-group, perform so differently in a biological test?

In this Section, a study of methyl- and methylene-groups will be made as examples of what are commonly classed as 'chemically inert' groups. If suitably placed, these groups can profoundly change the chemical behaviour of molecules, by well-understood steric and electronic effects. It can be shown that their altered biological properties are related to these changes.

2.5.1 *Steric influences*

The steric effects introduced by small, inert groups are of two kinds. Some are evident even in aqueous solution; others require a suitable, partly complementary surface for manifestation, as in enzyme reactions.

(*a*) *Steric influences on solubility*. It might be thought that the insertion into a given molecule of a methyl-group would always lower solubility in water. It is true that, because the methyl-group is water-repelling (see Section 17.1), it usually does lower solubility, but there are interesting exceptions. Table 2.4 lists the aqueous solubilities of some aliphatic alcohols. It can be seen that the solubility of the isomeric pentanols increases as the side-chain is broken up into smaller lengths. This is the expected behaviour, because of the strong hydrogen-bonding that exists between water molecules. In order that a substance may dissolve, these water molecules must be forced apart by breaking their hydrogen bonds. The lower alcohols, methanol and ethanol, readily do this because the hydroxyl-group forms such a large part of each molecule, and this group forms hydrogen bonds to water. But in higher alcohols, the paraffinic side-chain becomes a more dominant feature: it cannot be accommodated in interstices, it cannot force the water molecules apart, and hence it tends to be squeezed out of the water, dragging the whole molecule with it. This effect is considerably lessened by shifting the hydroxyl-group to the centre of the molecule (as in tertiary amyl alcohol) which is consequently more soluble than its lower homologue *n*-butanol. No less surprising, the 2-aminobutyric acids are more soluble than 2-aminopropionic acid (alanine) because of chain folding (Cohn *et al.*, 1934).

Table 2.4 Solubilities of alcohols (g per 100 g water at 20°C)

Pentanols	
CH_3—CH_2—CH_2—CH_2—CH_2—OH	2.4
CH_3—CH_2—CH_2—CH—CH_3 $\quad\quad\quad\quad\quad$ | $\quad\quad\quad\quad\quad\quad$ OH	4.9
$\quad\quad CH_3$ $\quad\quad$ | CH_3—C—CH_2—CH_3 $\quad\quad$ | $\quad\quad$ OH	12.2
n-Butanol (for comparison)	8.2

(Ginnings and Baum, 1937.)

The sulfonamidopyrimidine drugs are another series where the addition of methyl-groups strikingly increases solubility (see Table 2.5). At first sight this seems surprising, because these substances are acids and the addition of each methyl-group decreases the ionization of the acidic (SO_2NH—) group as would be predicted from the inductive effect of methyl-groups (see Section 17.2). Thus it would not be expected that sulfadiazine would be less soluble than its methyl-derivatives, because sulfadiazine is the most ionized member of the series and it is normal for an ion to be more soluble than the corresponding neutral species. This abnormal effect of methyl-groups, to be found also in other molecules having a similar degree of complexity and rigidity, is due to the protruding methyl-groups preventing the ready adsorption of dissolved solute on to the crystal-lattice of the solid phase, thus displacing the final equilibrium in the direction of increased solubility. The solubility measurements in Table 2.5 were made at pH 5.2 because this has practical significance in urinary disinfection and also in assessing the ease of clearance of the drug by the kidneys.* The same solubility sequences were found also at pH 6 and 7.

A further, and biologically very important, instance of methyl-groups increasing solubility is found in the triazine herbicide series. Non-symmetrical examples such as atrazine (*2.29*) are more soluble in water than symmetrical lower homologues such as simazine (*4.62*). These two examples have aqueous

*All three drugs are less soluble than their acetyl-derivatives, otherwise the solubility of the latter would be of more relevance.

Table 2.5 Increased solubility in water caused by insertion of methyl-groups

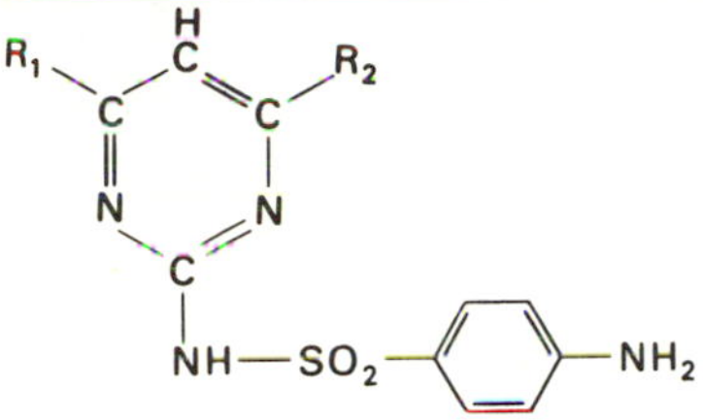

R_1	R_2	Drug	pK (acidic)	Per cent Ionized at pH 5.2	Solubility at pH 5.2 (g mol/litre) (37°C)
H	H	Sulfadiazine	6.5	3.9	0.0005
CH_3	H	Sulfamerazine	7.1	1.4	0.0013
CH_3	CH_3	Sulfadimidine	7.4	0.7	0.0024

(Gilligan and Plummer, 1943.)

solubilities of 70 and 5 p.p.m. respectively, at 25°C, and as a consequence atrazine is much the more toxic to leaves.

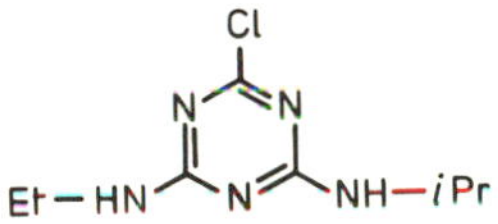

Atrazine
(2.29)

One of the most striking cases where an inert substituent exerts a profound change in biological action is *the homologous series*. In such a series, each member is usually found to be more biologically active than the previous one until, suddenly, the addition of just one more —CH_2-group severely diminishes, or even abolishes, the biological effect. This 'cut-off' occurs at a different place in the series for different test-objects. Now, it is not surprising that toxicity should increase as a series is ascended, because the addition of each methylene-group makes possible the creation of another van der Waals bond, thus adding to the adsorptive factors which bind each substance to the organism's receptors. At the same time, there is no increase in the desorptive forces, because the kinetic energy of every molecule is identical, regardless of its size. The 'cut-off', however, was unexplained until Ferguson (1939) showed that, in a homologous series, the equitoxic concentrations for various members usually fall on a straight line, if plotted logarithmically: solubilities also fall on a straight line, but these two lines are not quite parallel and hence they intersect. It is at those points where the lines intersect that the sharp cut-off occurs (Fig. 2.3). A discussion of this effect, from the viewpoint of thermodynamic activities versus concentrations, is given in Section 15.0.

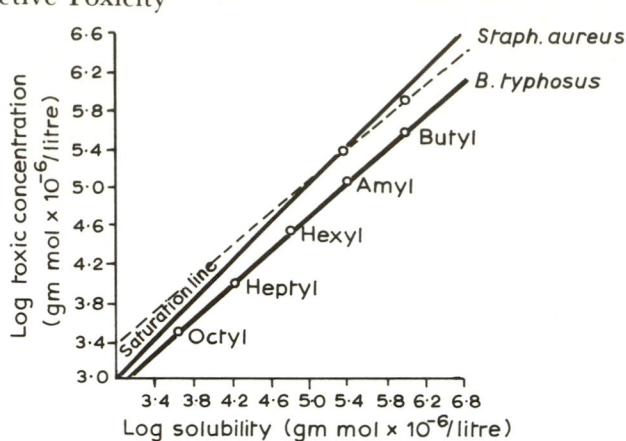

Fig. 2.3 Plot of bactericidal concentration versus solubility for normal primary alcohols.

In Fig. 2.3, where the toxic concentrations of primary alcohols have been plotted against their aqueous solubilities, it can be seen that the Gram-negative organism *B. typhosus* is very sensitive to these alcohols. Hence, even as high up in the series as octanol, the required lethal concentration is not in excess of the solubility. On the other hand, the Gram-positive organism *Staphylococcus aureus* is less sensitive to alcohols, so that a higher concentration is required for the killing. Consequently a sharp cut-off occurs at amyl alcohol because the predictable lethal concentration of hexanol is in excess of the solubility. In Fig. 2.3, the 'saturation line' is a diagonal, i.e. the plot of log solubility versus log solubility. It is the line on which all the points would fall for an imaginary series of substances that were optimally active in saturated solutions.

Sometimes the loss of biological activity in the upper part of a homologous series has a slightly different cause from that just discussed. The primary aliphatic amines, for example, have solubility and toxicity curves which are adequately spaced. Nevertheless, for most bacteria, a maximum in toxicity is reached somewhere about the C_{12} member (dodecylamine) and a rapid falling-off occurs when one or two more carbon atoms are added to the long paraffinic side-chain (see Fuller, 1942). It should be noted that this falling-off occurs in that part of the series where micelle-formation (Section 14.0) is beginning to increase rapidly with increase in chain-length. Thus, as the critical micelle concentration falls from 0.01 to 0.003 M in passing from the C_{12} to the C_{14} amine (Klevens, 1948), each higher homologue contributes fewer and fewer molecules of monomer, even in moderately dilute solutions. If we assume (and this is very likely) that bacterial receptors and micelles are in competition with one another for the unassociated molecules, it is easy to understand how this decreasing concentration of monomer can cause the decline in biological activity.

(*b*) *Steric influences on covalent hydration.* A methyl-group can prevent the addition of water to an adjacent double-bond, thus greatly increasing the lipophilicity of a substance. To place this remarkable effect in perspective, the phenomenon of *covalent hydration* will first be described. Discovery that the potentiometric titration of pteridin-6-one (6-hydroxypteridine) gave a loop instead of a line (Albert, Brown and Cheeseman, 1952a) led to the recognition that many nitrogenous heterocyclic substances add water, covalently across a double-bond, in at least one ionic species (for reviews, see Albert, 1967, 1976; Perrin, 1965a). Thus the neutral species of quinazoline in aqueous solution consists largely of the anhydrous form (*2.30*), but its cation is the hydrate (*2.31*) (Albert, Armarego and Spinner, 1961).

Quinazoline
(2.30)

Hydrated quinazoline (cation)
(2.31)

Pteridine
(2.32)

There are three reliable methods for detecting covalent hydration, and each reveals deletion of a double bond by the addition of a water molecule. Thus ultraviolet spectra are displaced to lower wavelengths; determination of ionization constants reveal acids to be weaker, and bases stronger, than normal; and proton magnetic resonance spectra show the signals displaced upfield. The first and third of these methods can indicate, in addition, which ionic species is the hydrated one. For all these methods, normal values can be predicted from isomers, and from molecules with one fewer ring-nitrogen atoms or one benzene ring less. Similarly, values for the hydrates are approximated by those for the corresponding dihydro-compounds.

An example: the spectrum of quinazoline is unexpectedly displaced to much shorter wavelengths when the solution is acidified (Fig. 2.4); also the pK_a is 3.51 (an equilibrium value) instead of 1.9, the pK_a for totally anhydrous species,

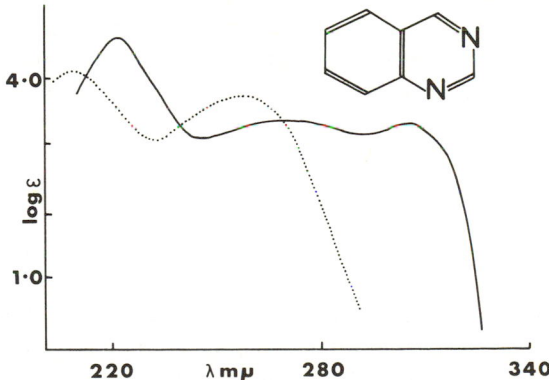

Fig. 2.4 Ultraviolet spectra of quinazoline in water. The solid line shows the neutral species, the dotted line the cation.

determined in rapid-reaction apparatus (Bunting and Perrin, 1967). Covalent hydration is largely suppressed if a methyl-group is present on the carbon atom which is attacked by the OH-group of water. This effect has been shown to be due to a steric effect reinforced by induction (Albert *et al.*, 1961). Thanks to this, 4-methylquinazoline shows a normal spectral shift on acidification (Fig. 2.5), and has a pK_a of 2.52 in agreement with calculations.

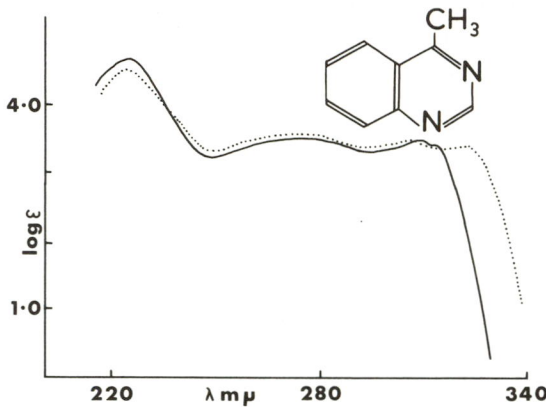

Fig. 2.5 Ultraviolet spectra of 4-methylquinazoline in water. The solid line shows the neutral species, the dotted line the cation.

Pteridine (*2.32*) has a higher tendency to hydrate than quinazoline, hence even the neutral molecule is 22% hydrated (in aqueous solution, at 20°). Here again a methyl-group in the 4-position substantially prevents hydration of this substance; it also prevents hydration of 2-aminopteridine (Albert, Howell and Spinner, 1962), and of pteridin-2-one (Albert and Howell, 1962). The hydration of pteridin-6-one, which occurs in the 7, 8-position (Brown and Mason, 1956), is similarly hindered by a methyl-group in the 7-position (Albert and Reich, 1961).

The extra hydroxyl-group furnished by covalent hydration, as in (*2.31*), reduces lipid/water partition coefficients and hence membrane-permeability. It is thus easy to visualize how the presence of a favourably situated methyl-group could increase permeability by diminishing hydration (see Section 17.1) and thus completely change physiological properties.

Several naturally occurring pteridines, such as xanthopterin, are hydrated, whereas 7-methylxanthopterin (also naturally occurring) is not (Albert and Reich, 1961; Inoue and Perrin, 1962). Several other natural products are in equilibrium with covalently hydrated forms, e.g. the aflatoxins with which the fungus *Aspergillus flavus* contaminates food (Patterson and Roberts, 1972); anthramycin (*4.43*), a pyrrolobenzodiazepine antibiotic (Goldberg and Friedman, 1971); the blood-pressure-lowering alkaloid hortiamine (Pachter *et al.*, 1960); tetrodotoxin, the sodium-channel blocker from Japanese puffer fish (Goto *et al.*, 1965); and the ergot alkaloids if exposed to sunlight (Hellberg,

1959). The oxidation of purines by xanthine oxidase is thought to require prior covalent hydration of these substances, although only a small percentage may be hydrated at equilibrium (Bergmann *et al.*, 1960).

(*c*) *Steric influences on chelation.* The anti-bacterial action of 8-hydroxyquinoline (discussed in Section 11.7.1) is seriously decreased by a methyl-group in the 2-position (Albert *et al.*, 1947). This deactivating effect is most likely exerted through a steric effect at a biological surface. Even in solution, this substance (2-methyl-8-hydroxyquinoline) has lost affinity for Al^{3+} (while retaining it for Fe^{3+}) because of the steric effect of the methyl-group. Other examples of steric hindrance by a methyl-group are described in Section 11.4.

(*d*) *Steric influences on receptors and enzymes.* Most molecules that fit the receptor of acetylcholine (and thus imitate the muscarinic action of this neurotransmitter) have a quaternary nitrogen atom of which one substituent is a straight chain of five atoms in length (see Section 12.6). The addition of just one more methylene-group to this chain causes a dramatic loss of biological effect. Two of the other substituents on the nitrogen atom must be methyl-groups to achieve maximal action; if one of these is substituted by either hydrogen or ethyl, a sharp drop in action takes place.

Enzymes, also, are often so specific that the addition or elimination of a single methyl-group in a substrate or coenzyme can cause a large, or even complete, loss of reactivity. How the biological action of thiamine is affected by the addition or loss of a single methyl-group was described at the beginning of this Chapter.

Indoleacetic acid (*4.82*), one of the natural growth-regulators of plants, is an example of a substance whose powerful action is abolished by the insertion of a methyl-group in the 2-position. On the other hand, in the K family of vitamins, e.g. menaphthone (*2.33*), the presence of a methyl-group in the 2-position is absolutely essential for biological activity (see also below).

Sometimes a methyl-group increases the biological effect of a drug by making it a poorer fit for a *destructive enzyme*. Thus amphetamine (*9.44*) (1-methyl-2-phenylethylamine) has a much more prolonged hypertensive effect than 2-phenylethylamine. This has been traced to the resistance of amphetamine to monoamine oxidase which rapidly destroys the lower homologue (Blaschko, 1952). Similarly the action of corticosteroids and the steroid sex hormones can be intensified by inserting a methyl- or fluorine-substituent, a device which has

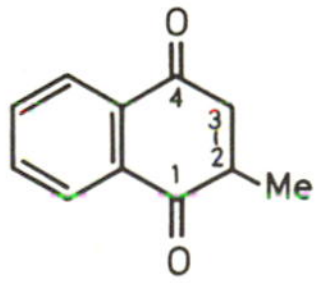

Menaphthone
(2.33)

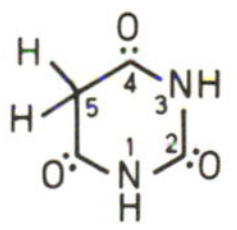

Barbituric acid
(2.34)

produced many clinically valuable drugs. Such seemingly inert substituents probably turn the steroids into poorly fitting substrates for their natural destructive enzymes (Ringold, 1961).

2.5.2 *Electronic influences*

The methyl-group is the commonest example of those few substituents which release electrons in any environment, no matter whether inductive or mesomeric mechanisms are operating (see Section 17.2).

(*a*) *Electronic influences on ionization.* Because of its electron-releasing nature, a methyl-group, if attached to a carbon atom, strengthens a base and weakens an acid. Also, a methyl-group attached to a nitrogen, so as to make a secondary amine, is base-strengthening, but most tertiary amines are weaker than secondary amines. Such changes in strength are usually less than one pK unit, but can influence biological results if the pK falls within one unit of the pH at which the biological test is made (see Section 10.0). Under these circumstances, a difference of one pK unit between two drugs can cause a tenfold increase in the ionization of one of them. When, as so often happens, one ionic species (say, the cation) is far more biologically active than another (say, the neutral species), this change in ionization can decide whether a substance is biologically active or not (see Section 10.3).

The triphenylmethane dyestuffs, whose somewhat complicated chemistry is

Table 2.6 Connexion between ionization and antibacterial activity in a set of triphenylmethane compounds

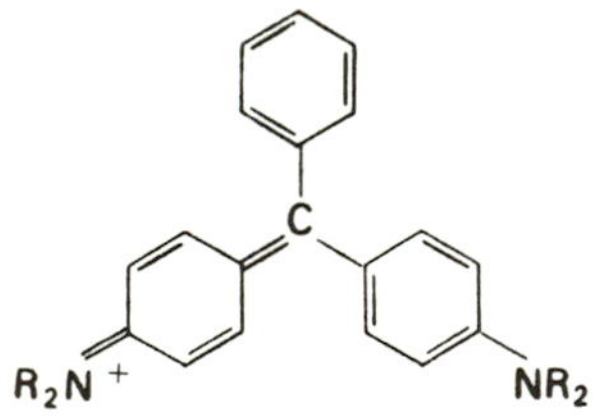

Substance	R (all four)	pK (equil.)	Per cent Ionized at pH 7.3	Min. bacteriostatic conc'n for Staph. aureus (24 h at 37°C and pH 7.3)
Doebner's Violet	H	5.38	2	1 in 20 000
Malachite Green	CH_3	6.90	28	1 in 80 000
Brilliant Green	C_2H_5	7.90	80	1 in 1 280 000

(Goldacre and Phillips, 1949.)

discussed in Section 10.2, show a very remarkable increase in basic strength on *N*-alkylation. It can be seen from Table 2.6 that anti-bacterial activity is strongly correlated with ionization for the substances examined, and that the anti-bacterial activity thus depends on the presence of 'chemically inert groups'.

It is obvious that methylation of an acidic —OH or —NH group must abolish its ability to ionize. The consequences of this in the barbituric acid series are particularly interesting. In aqueous solution, barbituric acid exists in the trioxo-form (*2.34*), and forms the mono-anion by loss of a proton from *C*-5. It is a fairly strong acid (pK_a 3.9). The insertion of one alkyl-group into the 5-position decreases the acidic strength only slightly, but two such groups remove any possibility of an anion being formed in the 5-position. In the latter case the anion is formed from *N*-3, but is much weaker. Thus barbitone (5,5-diethylbarbituric acid) has a pK_a of 7.9 and hence is 10^4 times weaker than barbituric acid. The consequences of the insertion of these inert ethyl groups, for structure–activity relationships, is momentous. A substance with a pK_a of 3.9 is completely ionized at pH 7.3 (see Section 17.0) and hence quite unlikely to pass the blood–brain barrier. But when the pK_a is 7.9, as in barbitone, the substance is 80% non-ionized at pH 7.3, and hence should pass without any difficulty. So essential is lipophilicity for hypnotic action that no less than four carbon atoms (altogether) must be present in the substituents of the 5-position (see further Chapter 15).

(*b*) *Electronic influences on reduction–oxidation* (redox) *potentials.* The electrons released to the rest of a molecule by a C-methyl substituent, lower the redox potential (E_o). As a result, the affected substance becomes a more active reducing agent and a less active oxidizing agent (i.e. it is more resistant to becoming reduced) than is the unmethylated homologue. Redox potentials are defined in Section 11.4 and refer to the equilibrium between oxidized and reduced forms (all values in this book are versus the normal hydrogen electrode).

To give an example of this lowering of E_o, a methyl-group inserted into the 2-position of 1,4-naphthoquinone, to give (*2.33*), depresses the potential (by 76 mV) to +408 mV (Fieser and Fieser, 1935). This 408 mV is a very workable potential for an oxidizing agent in the living cell. On the other hand, the reduction potential of NAD (nicotinamide adenine dinucleotide) is −280 mV, a value so low that a substituted NAD of slightly lower potential can probably not become reduced to its NADH. If it could not be reduced in the living cell, it could not act as an important hydrogen carrier. It is probably for this reason that 2-methylnicotinamide has no biological activity, although the effect of the methyl-group may also be partly steric.

(*c*) *Electronic influences on reactions where a covalent bond is broken.* The electron-releasing effects of a methyl-group described above were of a virtually instantaneous character. Some time-dependent (i.e. kinetically controlled) effects will now be mentioned. Methyl-groups, because of their electron-releasing properties, promote electrophilic substitution, e.g. they make neighbouring amino-groups readier to be acylated, or to form an azomethine (Schiff-base) with an aldehyde. A methyl-group also constitutes a side-chain that is often bio-degraded. Thus the metabolic oxidation of a methyl-group to a carboxylic acid in an aromatic hydrocarbon completely changes the distribution of the substance and leads to its rapid excretion (cf. Schultzen and Naunyn, 1867).

Methyl-groups can stabilize a molecule by replacing a hydrogen atom that would otherwise be split out in conjunction with a neighbouring atom. Thus dithiocarbamates split out hydrogen sulphide if at least one alkyl-group is replaced by hydrogen. In this way they produce *iso*thiocyanates. For example, the agricultural fungicide nabam breaks down in this way, and the *iso*thiocyanate that is produced on the plant seems to be the true active agent. However, fully alkylated dithiocarbamates, e.g. (*2.27*), are stable, and act as such (mode of action was described in Section 2.3).

(*d*) *Solubility*. In an aromatic nitrogen heterocycle, replacement by methyl of the hydrogen in an —OH, —NH$_2$ or —C(:O)-NH group can increase solubility in water dramatically, even in the pyridine series. Substituents with a bondable hydrogen atom (e.g. —NH$_2$) decrease the solubility in water through hydrogen-bonding of the substituent to the highly polar ring-nitrogen atom; in short, the attraction of one molecule for another is greater than its attraction for water molecules. As a result, the crystal lattice energy is increased, and the substance falls out of solution. Such an effect is rarely observed among non-heterocyclic compounds. The reversal of this insolubilizing trend is not strictly an electronic effect of a methyl-group, but rather the consequence of the ability of this group to prevent the strong, electron-mediated self-polymerization by hydrogen bonds.

Some examples will be given from the pteridine (*2.32*) series (all in water at 20°). 4-Aminopteridine is soluble only 1 in 1400, but 4-dimethylaminopteridine 1 in 2 (Albert, Brown and Cheeseman, 1952b). Similarly, 7-hydroxypteridine is soluble only 1 in 900, but both its *O*- and *N*-methyl-derivatives are soluble 1 in 50. Again, 6-aminopurine (*4.3*) (adenine) is only soluble 1 in 1100 but 6-dimethylaminopurine 1 in 120 (Albert and Brown, 1954).

2.6 Correlations: Today's perspective

This chapter has traced the slow, hesitant steps by which drug workers have advanced beyond the belief that structure–action relationships depended on particular substituents or nuclei, and how they gradually learnt to appreciate the importance of physical properties. At the present time it is considered that three properties are most relevant: (i) partition, (ii) ionization (or other measurement of electron distribution), and (iii) shape (steric fit). Both the aminoacridine and the oxine projects (Sections 2.2 and 2.3 respectively) were carried out in the 1940s when interest was awakening in the possibilities of physical chemistry as a basis for drug action. They not only provided some very striking examples of this correlation, but extended the current concept of a receptor in (at that time) unimagined directions, namely to nucleic acids and coenzymes.

Understanding the role of physical properties in designing new drugs introduced a new freedom to exchange nuclei so long as physical properties were conserved. On occasion this freedom can be breathtaking. We saw (p. 35) how the acridine ring of the aminoacridine anti-bacterials could be replaced by anthracene. This principle of replacing a heterocyclic ring by its physical equivalent was later applied to the anti-malarial alkaloid, quinine. First, the quinuclidine ring was replaced by the following aliphatic chain: $—CH(OH)—$ CH_2NR_2. Next, the second heterocyclic ring (quinoline) was replaced by naphthalene (Section 10.3.5). The anti-malarial properties were not only conserved by these changes but actually enhanced. In fact the goal of these changes is to convert a molecule into a non-substrate for its usual metabolizing enzyme. Needless to say, existing useful physicochemical properties must be conserved.

With these principles in mind, it has proved profitable sometimes to replace carbon by nitrogen as in the highly successful insecticide that has been made from octopamine (*2.35*), an important neurotransmitter in insects. Here — $N:CH—$ was substituted for the $—CH(OH)—CH_2$-group, creating a new series of insecticides of which chlordimeform (*2.36*) was the best. It gives the insect an overdose of agonistic activity (Evans and Gee, 1980).

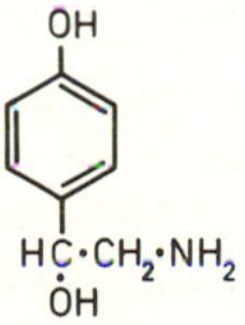

Octopamine
(2.35)

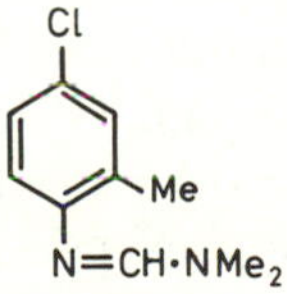

Chlordimeform
(2.36)

What liberties may be taken with the structure of a drug, so long as its stereochemistry is preserved, is exemplified by the many active analogues of the hypertension-lowering drug, captopril (*9.50*). Activity requires that a carbonyl-group must fit a particular hydrogen-bonding site in the receptor, but it is no matter whether the $—CO$ is derived from an amide, an ester, or a ketone fragment, and the $—SO_2$ of a sulfonamide-group will also serve (Condon *et al.*, 1982).

No less stimulating to the drug designer was the replacement of all five chlorine atoms in DDT (*2.37*) by methyl- or methoxy-groups [giving 1,1-

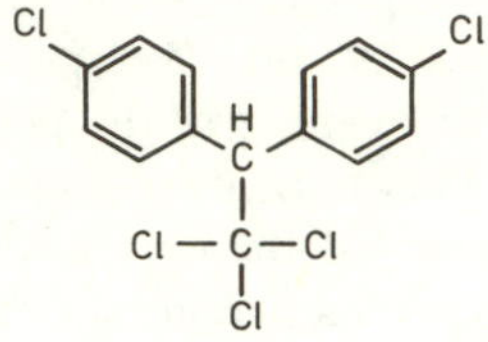

DDT
(2.37)

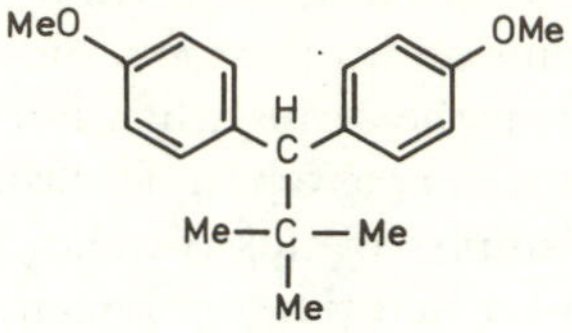

Dianisylneopentane
(2.38)

dianisyl*neo*pentane (2.38)] without loss of insecticidal properties (Brown and Rogers, 1950).

In a final analysis, we must ask ourselves what we mean when we speak of 'structure–activity relations'? When we say 'structure', in this connexion, we surely mean 'constitution', namely all the information on physical and chemical properties that is stored in the chemical formula, or that can be discovered by measurement and experimentation. Also when we say 'activity' we mean the action on the drug-receptor, but this effect, we know, is often connected to the desired physiological result only through a long chain of other reactions. It is only with reference to the action *at the receptor* that the constitution of the drug has any relevance.

As a consequence, all who work with drugs, today, must cultivate the ability to read the constitution (properties) from a glance at a structure. A rich harvest of chemical and physical properties can be gleaned from a chemical formula. Students of medicinal chemistry are taught to 'read out' properties when confronted by structures. For example, the formula of phenol shows it to be a weak acid of pK_a about 10; moreover, if it were a substituted phenol, the rise or fall in acidic properties could be read out at once by subtracting the Hammett σ-constant for that substituent (see Section 17.2) from the pK_a of phenol. It is clear that the molecule can be readily attacked at the electron-rich 2, 4, and 6-positions by electrophilic reagents. In addition, the hydroxy-group is available for esterification and ether-formation. The ease with which all these reactions can take place in a substituted phenol is governed by the sign and magnitude of Hammett σ-constants for the substituents; electron-attracting groups disfavour and electron-releasing groups aid the reaction. Nucleophilic attack and reactions of addition are clearly unlikely, reduction difficult, and oxidation likely to produce quinones and hence self-condensation. Powerful, electron-attracting substituents ($-CN$, $-NO_2$, $-CONH_2$) will favour nucleophilic attack, addition reactions, and reduction; but will disfavour oxidation. The distribution of a phenol between an aqueous and a lipoid solvent can be approximately calculated by use of the π-constants for each substituent, as derived by Hansch and Rekker (see Section 17.1).

Similarly the formula of pyridine reveals that it is a weak base, with a pK_a about 5, and the basic strength of substituted pyridines can quickly be calculated by the procedures of Perrin, Dempsey and Serjeant (1981). Further, the formula shows that there is a large deficiency of electrons in the 2, 4, and 6-positions, which are therefore liable to nucleophilic attack; only the nitrogen atom is open to addition reactions; electrophilic reactions are extremely unlikely; reduction, although not easy, is possible and oxidation unlikely. The rules, allowing these categories of information to be read out of the formulae of various kinds of heterocycles, are quite easy to memorize (Albert, 1968).

For help in using chemical names to search for physical and biological properties, whether in compilations (books) or on-line databanks, see Section 17.4. For information on the chemistry of drugs, see the standard works on medicinal chemistry edited by Doerge (1982) and Wolff (1981).

3

Comparative distribution:
the first principle of selectivity

Of the various principles that govern selectivity, that concerned with distribution is (even when its contribution is relatively small) never completely
absent. At the other extreme, it can be the principal source of the selective gain.
In what follows, we shall begin with examples of selectivity achieved by
exploiting natural differences in distribution. Next, we examine the various
mechanisms by which organisms manage to differentiate in this way. Finally,
we shall take a look at the quantitative basis for this selective process.

3.0 Examples of selectivity through distribution

It is remarkable that selectivity through distribution can take place even when
the agent is toxic to both economic and uneconomic cells, provided that it is
accumulated only by the latter. An example where the economic and
uneconomic cells are in different species is illustrated in Fig. 3.1. Some years ago
a Frenchman discovered that an aqueous solution of sulfuric acid could safely be
sprayed on cereal crops to destroy the weeds (Rabaté, 1927). The unsprayed
strip at the right shows wild radish in full bloom among the wheat, but the crop
is seen uncontaminated on the sprayed side of the field (Ball and French, 1935;
Robbins, Crafts and Raynor, 1952). The strength used was 10% (w/w) H_2SO_4 at
about 120 gallons per acre. Sulfuric acid is, of course, injurious to the cytoplasm

Fig. 3.1 Elimination, by early spraying with 10% sulfuric acid, of wild radish from a wheat crop. The strip on the left was sprayed at emergence. (Ball and French, 1935.)

of both wheat and weed, but it never reaches that of the former for two reasons. The exterior of cereal grasses is smooth and waxy, whereas that of dicotyledonous weeds is rough and wax-free; the acid runs off the former but is accumulated by the latter. Moreover the tender new shoots of cereals arise from the base of the plant and are protected by leaf-sheaths, whereas the growing point of the dicotyledons is exposed and vulnerable because it is the apex of the shoot (Fig. 3.2). Hence the weeds die, and the economic crop persists, because of a selective action that depends entirely on distribution (Blackman, 1946).

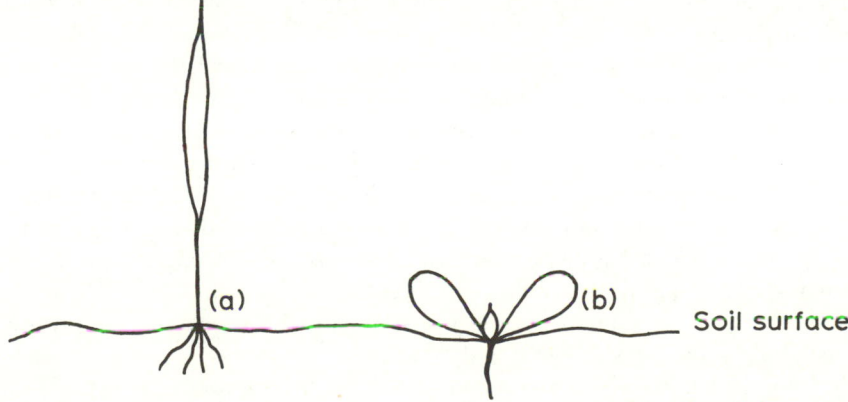

Fig. 3.2 Sketch of emergent seedlings of (a) a monocotyledon; (b) a dicotyledon.

The above example clearly demonstrates that even the smallest conceivable agent, namely the hydrogen ion, can exert strong and useful selectivity.

The selectivity of the tetracyclines, so much used in treating bacterial infections in mammals, depends on a similarly favourable distribution: the bacteria concentrate these antibiotics whereas mammalian cells do not. Concentration of tetracyclines by bacteria (both Gram-positive and -negative types) was found to be a function of the cytoplasmic membrane (Franklin, 1971). Because of this difference, tetracyclines inhibit ribosomal protein synthesis in bacteria at doses which do not affect it in higher organisms (Franklin, 1963b, 1966). This selectivity depends on the intactness of the mammalian cell membrane, because isolated ribosomes (rat liver was used) were found to be as subject to inhibition of protein synthesis as bacterial ones were (Franklin, 1963a; Section 11.8).

Other cases are known where an invading organism concentrates a poison which selectively injures it. A remarkable feature of trypanosomes is the speed and extent of their concentration of organic arsenicals. Within ten minutes of first contact, these drugs may reach a level several hundred times greater than in the surrounding fluid. This is thought to be due to the relatively great number of mercapto-groups on the organisms' cell surface: in one calculation, it was estimated that every fifth side-chain of the protein forming the surface bears a —SH group (Eagle, 1945).

Similarly phenothiazine, a highly effective oral anthelmintic in sheep, is accumulated by intestinal worms but not by the cells lining the sheep's gut: yet, when given intravenously, it is toxic for both species (Lazarus and Rogers, 1951).

By similar localization, difluoromethylornithine (*9.88*) cures coccidiosis in poultry. The protozoon (*Eimeria*), which causes this economically disastrous disease, multiplies in and damages the endothelium of the chicken gut. The drug, which works by inhibiting the parasite's ornithine decarboxylase, spares the host's enzyme by being unable to penetrate beyond the intestinal wall (see Section 9.7.2).

Examples will now be given of selective partitioning between the tissues of a single organism.

After oral administration, griseofulvin is specifically localized in a patient's keratinized cells, namely the epidermis, hair, and nails, and is used to injure fungi parasitizing these tissues. Griseofulvin blocks fungal mitosis, causing multinucleate cells to be formed (Gull and Trinci, 1973): mammalian (and plant) cells suffer similarly, so the selectivity depends on the initial distribution.

Cyanocobalamin (vitamin B_{12}), injected into a muscle, travels to the bone-marrow and is accumulated there after a dilution of 10^{10}-fold in the body fluids. Even a microgram, injected in this way, is enough to cause new reticulocytes to form in the marrow of a patient suffering from pernicious anaemia. The process of distribution has been followed with [57]Co.

Iodine, which is selectively accumulated by the thyroid gland, provides another good example of selective action through specific distribution. This process is easily followed with radioactive iodine (^{131}I, half-life 8 days), which is used for the treatment of thyrotoxicosis. Depending on the dosage, the radioactive chemical can merely inhibit the excessive metabolism of the gland or actually attack a tumour in it. The usual oral dose is only 10^{-12} g, yet 80% of this can be demonstrated in the gland soon after administration.

Inorganic phosphates are accumulated specifically by the trabecular bone-tissue which lies close to the haemopoietic red bone-marrow. The over-production of red cells in polycythaemia is treated clinically with ^{32}P (half-life 14 days).

The safe and accurate diagnosis of diseases by using radiopharmaceuticals represents an important contribution of atomic energy to human health. Thus, gallium (^{67}Ga), with its half-life of 78 h, has a specificity that makes it ideal for visualizing tumours in the lymphatics. Chromium (^{51}Cr, half-life 28 days) is specific for labelling red blood cells. Rubidium (^{82}Rb, half-life of only 75 seconds) shows promise for diagnosing heart defects, being extracted by the heart from the bloodstream (as though it were potassium) but by no other organ. It is available from a portable generator containing ^{82}Sr (strontium). Indium (^{111}In, half-life 67 h), as its bleomycin salt, is proving very useful in the detection of tumours. A different isotope (^{113}In, half-life 99 min) is used to scan the lungs by first adsorbing it on to particles of 30–60 μm, a size that the lung capillaries selectively retain. Technetium (as ^{99}Tc, half-life 6 h), a powerful but safe emitter of soft gamma rays, is also used in this way, and also (in simple solution) to localize brain tumours which selectively accumulate it. Similarly thallium (^{201}Tl, half-life 74 h) is used to locate damaged cells in the myocardium of a patient's heart. For more details of the high specificity of radiopharmaceuticals, see Reynolds (1982).

The daring therapy of brain tumours with boron rests on the isotope ^{10}B which is not radioactive, but has an extraordinary ability to capture neutrons. These are converted to alpha particles which have 100 million times the energy of the neutrons. Normal brain tissue is little affected by neutrons, and the alpha particle has a range of only 10 μm in tissues. Normally boron compounds penetrate the blood–brain barrier quite poorly, but the tumours have large gaps in this membrane: hence the boron-containing agent accumulates in the malignant tissue, as was first shown by Krüger (1955) using mouse brain.

The boron is injected into the patient usually as a carborane (e.g. the ring-compound $C_2H_{12}B_{10}$) which is furnished with two mercapto-groups to give affinity for protein. Although there is 19% of ^{10}B in natural boric acid (the rest is mainly ^{11}B), the therapeutic material is further enriched. One hour after the injection, a neutron beam from a thermonuclear reactor is lined up with an appropriately localized hole made in the patient's skull. The pioneer clinical work was done in Tokyo (Hatanaka and Sano, 1973; cf. Wong, Tolpin and Lipscomb, 1974).

Many organic compounds containing normal iodine (^{127}I) are used as radiopaques (X-ray contrast agents) in the radiography of the area where each is selectively accumulated. For delineating the gall-bladder and biliary ducts, iopanoic acid (*3.1*) or the related sodium ipodate, which is sodium 3-(3-dimethylaminomethyleneamino-2,4,6-tri-iodophenyl)propionate,   are given orally. For outlining the urinary tract, intravenous injection is made of diatrizoic acid (*3.2*) or its isomer, iothalamic acid, which is 5-acetamido-2,4,6-tri-iodo-*N*-methylisophthalamic acid.

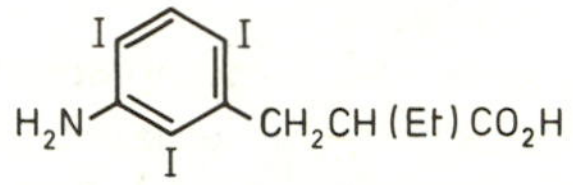

Iopanoic acid
(3.1)

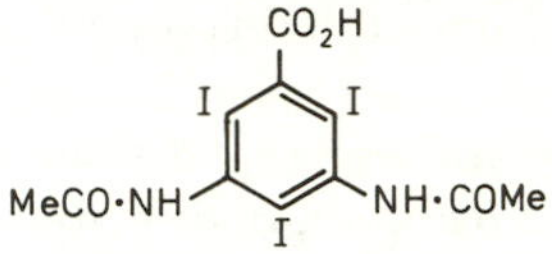

Diatrizoic acid
(3.2)

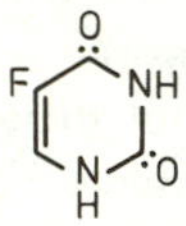

5-Fluorouracil
(3.3)

N-Substituted phenothiazines with basic side-chains (e.g. the anti-histamine promethazine, and the tranquillizer chlorpromazine) are accumulated in the eyes of all mammals. A concentration 50 times as great as in other tissues is often reached (Potts, 1962). Each naturally occurring steroid has a specific protein to transport it to the nuclear DNA in the cells which it selectively influences (see Section 2.4).

The next example of selectivity through favourable distribution is culled from anti-cancer therapy. Heidelberger, knowing that many malignant cells take up uracil more readily than normal cells, synthesized 5-fluorouracil (*3.3*) and introduced it into the clinic (Heidelberger, *et al.*, 1958). Today it is used by dermatologists to cure skin cancers (both squamous and basal cell carcinomas). So selective is this substance that the 5% cream is usually rubbed in with the bare hand without harmful effect, and there is no objection to covering the whole face or even trunk with it when the growths are widely distributed. The malignant tissue, only, becomes inflamed and finally disintegrates being replaced by healthy granulation tissue followed by new skin (Williams and Klein, 1970; Klein *et al.*, 1972). Surgery and diathermy act more quickly but produce scarring. The biochemistry of 5-fluorouracil action is discussed in Section 4.0, p. 125.

Unlike bacterial and mammalian cells, yeasts (including *Candida albicans* which is infective to man) readily take up di- and tri-peptides including some that are fungicidal, provided that the terminal amino-group is acetylated. However, if the terminal acid group is esterified, they will not absorb peptides, although bacteria can do so easily (Lichliter, Naider and Becker, 1976).

Selective distribution can even be effected by size, alone. When particles are inhaled, those above 5 μm diameter remain in the nasal passages, those of about 2 μm lodge in the larger bronchial areas, whilst particles must be narrower than 1 μm to reach the smallest bronchi and alveolar sacs, as is necessary for effective medication by aerosols (nasal sprays).

*   *   *   *   *

This Section has provided typical examples of selectivity at work. Many more instances could be cited, but it seems better to press forward and explore, at the cellular and molecular levels, the mechanisms by which selectivity is effected. The next Sections (3.1 to 3.6) deal with these phenomena qualitatively whereas Section 3.7 will discuss the quantitative aspects.

3.1 How drugs are absorbed, distributed, and eliminated

Whether an agent is given by mouth or injected, it must usually traverse one or several semi-permeable membranes before the required receptor is encountered. For example, an anti-malarial, such as chloroquine (*10.31*), given by mouth, must penetrate the barrier that lies between the gastrointestinal tract and the bloodstream, then the erythrocyte membrane, and finally that of the malarial parasite. On both sides of each membrane the concentration of a drug continually decreases through storage, excretion, and inactivation. Examples of storage are; (i) in lipoids for liposoluble substances such as thiopentone, (ii) on nucleic acids or chondroitin for cationic substances such as chloroquine, and (iii) on serum albumin for anionic substances such as suramin and sulfonamides (Brodie and Hogben, 1957). Storage is usually freely reversible and can be beneficial, e.g. when it tends to keep the blood-level of a drug constant in septicaemia; but it can be disadvantageous, e.g. when, after a bedtime dose of a hypnotic, the patient cannot keep awake at his work next day.

Excretion may be through the kidneys, the bile-duct (and hence into the gut), or through the lungs (as with general anaesthetics). Ether and strychnine are examples of drugs which are rapidly excreted (in the urine) without storage or chemical change. For many other drugs, excretion is preceded by inactivation, a process which involves the making or breaking of covalent bonds and hence is not freely reversible. When the drug has penetrated into the tissues through membranes that line the blood-vessels, it is subject again to loss by storage and inactivation, but in many situations it cannot be excreted until it penetrates back into the bloodstream. Finally, when the drug penetrates the last membrane surrounding the relevant receptor, combination occurs with the receptor, a signal is generated, and the physiological effect characteristic of the drug begins. This effect usually continues as long as the concentration is high enough to keep a significant number of receptors activated (Section 7.5). However, as the concentration is falling all the time, the receptors will not stay activated unless more of the drug is administered.

These competing effects are shown diagrammatically in Fig. 3.3, which shows how the frequency and size of the dose depend on all these factors, each arrow representing an equilibrium or a steady state (see Section 3.7, below). Sometimes there is an additional complication: the drug which acts on the receptor is not administered as such, but as a pro-drug which is turned into the drug by metabolism (Section 3.6). In this picture of distribution (Fig. 3.3), the reversibility of most of the steps should be noted.

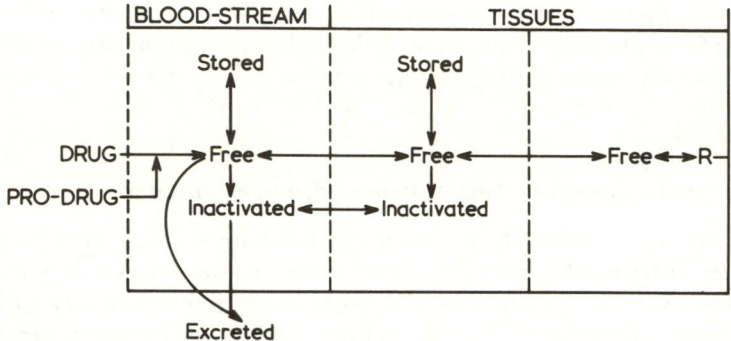

Fig. 3.3 Distribution of a drug or other biologically active substance. The broken vertical lines indicate selective membranes and R stands for a receptor.

The useful concept of *apparent volume of distribution* (V_D) of a drug is defined:

$$V_D = \frac{\text{Weight of drug in the body}}{\text{Plasma concentration of the free drug at equilibrium}}$$

Thus V_D is the apparent fluid volume in which the drug is dissolved. Values of V_D compatible with the known volume of a body compartment may suggest that the drug is confined to that compartment. Values of V_D greater than the total body volume indicate that the drug is deposited in a tissue (Goldstein, Aronow and Kalman, 1974). Some relevant volumes of body compartments are (in litres): circulating plasma of blood (3), erythrocytes (3), extracellular water other than blood (11), intracellular water (24); the total is 41 litres or 58% of body weight (average values) (Goldstein *et al.*, 1974).

A survey of 123 different drugs disclosed that the volume of distribution is proportional to the partition coefficient (P) as follows (Ritschel and Hammer, 1980):

$$V_D = 0.156P + 0.86$$

The picture of distribution would not be complete without mentioning some recycling mechanisms. Drugs in circulation enter the liver by the hepatic artery and portal vein. From the liver's two lobes, the drugs emerge (as such, or in a degraded state) with the bile which passes down the 'common duct' into the gall-bladder, which exists for bile storage. At intervals, bile leaves the gall-bladder by the bile-duct which discharges into the second part of the duodenum; this is a slender tube about 30 cm long which connects the stomach to the rest of the small intestine. Some drugs are absorbed from the small intestine by the portal vein and pass by the bloodstream to the liver, which secretes them into the small intestine again, through the bile. Phenolphthalein and bialamicol ('Camoform') are examples, and the last-named is also recycled by the following pathway: intestine–lungs–bronchi–trachea–pharynx–intestine. Both circuits involve a gradual decrement via the faeces, and the latter circuit also involves loss through expectoration (Dill *et al.*, 1957).

Redistribution in plants is a related phenomenon. When a field is sprayed, much of the plant surface is shadowed by other parts of the plant and hence does not receive the spray. However, redistribution of the sprayed agent is effected by wind and by moisture. Because the surface of plants is negatively charged, redistribution is efficient mainly for positively charged substances (such as Bordeaux Mixture, or streptomycin) (Dimond and Horsfall, 1959).

From the above it will be apparent that the physico-chemical properties of a potential agent must be very finely adjusted if it is to avoid the many casualties of storage, excretion, and degradation. In many cases, by accident or by design, substances have been obtained with properties favourable for their *accumulation* near an appropriate receptor; then, if their chemical structure is complementary to that of the receptor, the desired biological action takes place.

Apart from concentration at effector sites, high concentrations of drugs are commonly found in the liver and kidneys, arising from the functioning of these organs as centres of detoxication and excretion.

Compartmentalization, a word coined by A. Zaffaroni, is the restriction of agents to the target tissue by mechanical means. In its commoner form, a small piece of plastic, impregnated with the drug, is inserted at the desired site of action. The uses of this type of sustained release (see Section 3.7) range from controlling glaucoma to birth control. Alternatively, a tablet can be encased in a membrane through which a hole has been accurately drilled. As a result, the drug is steadily leached, as in the so-called 'osmotic tablets' of indometacin.

In another field of use, rubber sheets impregnated with organic tin compounds are fixed on ships' keels to prevent fouling by barnacles. Copper compounds are similarly being compartmentalized and anchored in tropical streams to kill schistosome-infected snails. In a different approach, artificial sweeteners have been covalently bonded to polymers which impart a sweet taste to food, but are not absorbed from the gastrointestinal tract.

For further general reading on the distribution of drugs, see La Du, Mandel and Way (1971) and Saunders (1974).

3.1.1 *The structure of water*

The role of water in all distribution phenomena is dominant. Because water is so familiar to us, we are not inclined to admit that it is one of the most complex of all liquids. Its irreplaceable role in all living processes calls for deeper understanding: water is not just an inert medium, accidentally present, of little more relevance than a reaction vessel. In fact, water has unique physicochemical properties: it has a broad domain of thermodynamic stability and can participate in acid–base equilibria over a wide range (actually over 16 pH units); and it can sustain redox equilibria over a potential range of more than 2 V (Section 11.4). The curious fact of a maximal density at 4°C, and the ability of water to absorb or release calories without much change in temperature, have profoundly influenced the distribution of life on earth. All these properties point to a structure immensely more complex than the common symbol H_2O would indicate, because water is, from the melting temperature of ice to the condensation temperature of steam, a large and complex polymer.

Because of its strong self-bonding properties, an isolated molecule of water can be found only in steam (Fig. 3.4a). Such a molecule is roughly triangular with an O—H distance of 0.96 Å and a H—O—H angle of 105°. The two protons and two lone-pairs of electrons occupy tetrahedrally disposed, hybridized orbitals.

In ice, hydrogen-bonding between molecules leads to the formation of a five-molecule structural unit, which is approximately tetrahedral (Fig. 3.4b). Each oxygen atom is surrounded by four others at a distance of 2.75 Å; three of them are in the same layer (forming, with other water molecules, puckered six-membered rings) and the fourth is in an adjacent layer. Large polyhedral cavities lie between these layers.

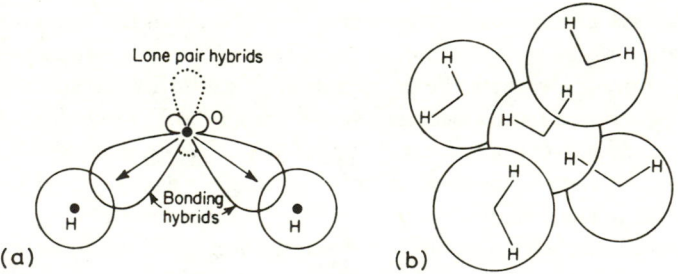

Fig. 3.4 (a) The water molecule, as in steam; (b) primary unit of water association, as in ice. (Ives and Lemon, 1968.)

Liquid water is unique in its ability to promote three-dimensional order, for it is a polymer of the only molecule which, from a single atomic centre, can give rise to *four* hydrogen bonds (two as a donor, and two as an acceptor). The hydrides most closely related to it (NH_3 and HF) can form only two hydrogen bonds (by acting once as a donor and once as an acceptor). As a rule, only *equal* numbers of donor and acceptor bonds are energetically stable in liquids. Molecular association in liquid water raises the dipole moment (1.84 D in steam) to 2.4 D.

Of the many hypotheses of the structure of liquid water, that of Pople (1951), as modified by Sceats, Stavola and Rice (1979), agrees very well with all experimentally determined properties. In this model, water is formulated as a *continuous* polymer in which H_2O units are united by a network of hydrogen bonds that extend throughout the whole liquid which becomes, in this sense, one large molecule. This formulation is compatible with all recorded physical properties. No support remains for older ideas of 'flickering clusters', 'icebergs', 'monomeric inclusions', or other types of discontinuity.

Chaotropic agents are those used to open up the structure of water. The best-known are urea, guanidine, and inorganic ions. They are much used to change the conformation of enzymes and other macromolecules in contact with water by removing the structure which is extending them. Some enzymes are inactivated by this treatment, others activated.

For further reading on the structure and properties of liquid water, see Stillinger (1980) for a short review, and Franks (1972–82) for a 7-volume treatise.

3.2 The permeability of natural membranes

It is evident that the distribution of an agent is highly dependent on its ability to penetrate semi-permeable membranes. The plasma membrane is discussed in Section 5.4. In the classical conception of this membrane, its behaviour was static, something like a dialysis sac. More recent work discovered its latent dynamic properties, such as phase-reversal and processes akin to enzymic activity (e.g. permease activity, and transport in response to the metabolism of glucose). In the following pages, membranes will be discussed according to their observed functions and classified as four main types.

3.2.1 'Type 1' membrane

This type of membrane ensures *simple diffusion*, namely: the transport velocity is directly proportional to the concentration difference across the membrane. Thus, when an equilibrium is reached, the internal and external concentrations of the drug are the same. The transport velocity is affected by molecular weight, lipid solubility, and charge, but not greatly by temperature.

This seems to be by far the commonest type of membrane. It hinders the passage of ions, and permits that of neutral molecules. Through this type of membrane, those molecules that have high oil/water partition coefficients (and hence are quite lipophilic) diffuse fastest. The half-time for equilibrium can vary from about 1 min to 30 days. The action of this membrane is exemplified in Table 3.1.

Table 3.1 The permeability of natural membranes

Non-electrolytes	Partition coefficient (olive oil/water) ×10^5	Permeability of living cells (Mol/s/μm²/molar conc. difference) ×10^20				
		A	B	C	D	E
1,2-Dihydroxypropane	570	—	13 200	13 000	4000	24 000
Propionamide	360	2200	—	23 000	—	36 000
Acetamide	83	800	—	10 000	—	15 000
Glycol	50	1100	6700	7300	2100	12 000
N-Methylurea	44	90	—	—	—	1900
Urea	15	15	2500	—	78 000	1000
Glycerol	7	18	180	50	17	210
Erythritol	3	3.1	—	—	—	13
Sucrose	3	0.8	—	—	—	8

A, *Curcuma* (flowering plant) (Collander, 1937); B, *Gregarina* (protozoon) (Adcock, 1940); C, *Arabacia* eggs (marine animal) (Stewart and Jacobs, 1936); D, Ox erythrocytes (Jacobs *et al.*, 1935); E, *Chara* (green alga) (Collander, 1937).
—signifies 'not determined'.

The partition coefficient (P) of a substance (S) is defined by the following expression for the equilibrium state:

$$P = [S]_o/[S]_w$$

where w (for 'water') is the aqueous phase and o (originally 'oil') is the non-aqueous phase. The greater the lipophilicity of the substance, the higher the P.

The influence of chemical constitution on partition coefficients is discussed in Section 3.3, and also the effects of varying the organic phase.

Table 3.1 illustrates a general trend: the higher the lipid/water coefficient of a substance, the greater the proportion of it found inside the cell after equilibration has taken place. For example, the insertion of another hydroxyl-group into 1,2-dihydroxypropane, giving glycerol, causes a large fall in partition coefficient, and it can be seen that there is a corresponding decrease in penetration. In fact, few molecules penetrate readily that have more than three water-attracting groups and a molecular weight over 150 (Davson and Danielli, 1952). The size of the molecules and the nature of the groups also exert an effect (usually only secondary). The outstanding exception in Table 3.1 is the easy penetration of urea into ox erythrocytes. This facilitated uptake of urea occurs in all mammalian red cells, but not in those of birds.

The Type 1 membrane appears to be about 5 nm thick and to consist mainly of lipoidal material mixed with some protein. The presence of this type is diagnosed if substances of similar molecular weight and molecular diameter are found to penetrate at a rate proportional to their partition coefficients. It is interesting to note that if the partition coefficient is too high, the substance enters the membrane freely but cannot leave it.

For a discussion on membrane permeability and equilibria see Wilbrandt (1959), for the kinetics of diffusion see Laidler and Shuler (1949), and Zwolinski, Eyring and Reese (1949).

The understandable desire to quantify the passage of drugs into cells has led to a search for permeability coefficients, keeping in mind that each one can apply only to a particular substance and a particular membrane. The starting point is Fick's diffusion law which allows us to find the rate of transport, namely dS/dt, where dS is the microscopic amount of a substance transferred in an exceedingly short time (dt). When applied to the living cell, Fick's law usually takes this form:

$$dS/dt = -DA(C_o - C_i)$$

where D is the diffusion coefficient*, A is the standard area (usually $1\ \mu m^2$), C_o is the outside (and C_i the inside) concentration.

Many other factors intervene. For example, hydrogen-bonding substances like glycerol cannot penetrate into the membrane until they have acquired enough kinetic energy to shed their bound water and also to push aside several lipid molecules. This is necessarily a slow process; yet all molecules of glycerol can lose energy by hydration as they quit the membrane for the cell interior, and this part of the process is rapid. By contrast, a lipophilic molecule like phenobarbital can enter the membrane rapidly but can leave it only slowly.

A useful *permeabiliy constant (K)* can be calculated, for a particular substance and membrane, by the following equation (Lueck et al., 1957). This equation describes a diffusion process under quasi steady-state conditions in which two agitated liquids are separated by a membrane permeable to a solute:

$$\log (C_o - 2C_i) = -\frac{2K}{2.303}t + \log C_o$$

where C_o is the initial concentration of the solute, C_i is the concentration beyond the membrane, and t is the time of sampling. A plot of $\log (C_o - 2C_i)$ against time gives a straight line of slope equal to $-2K/2.3$ from which K can readily be found. K (the permeability constant) also equals ADD_c/VL, where A is the cross-sectional area of the membrane, L is its thickness, V is the volume of each of the two chambers (fore and aft of the membrane), D is the diffusion coefficient, and D_c is the distribution coefficient between solution and membrane. This relationship enables other interesting data to be calculated.

*The diffusion coefficient represents the number of moles of solute that cross unit area of the membrane in unit time, under unit difference of concentration. It is not a constant outside of the system being studied.

Use can also be made of Graham's Law, in which $D\sqrt{M}$ is a constant in many cases where the rate is proportional to the square root of the molecular weight up to about 180 daltons, as with glucose (Thovert, 1910). See Section 3.7 for some other quantitative aspects of distribution.

The cardiac glycosides afford clinically important examples of the effect of partition coefficients on permeability. Digitoxin, the most readily accumulated of these glycosides, is the most lipophilic. It is excreted only slowly into the bile and is largely reabsorbed from this fluid. Related glycosides which are more hydrophilic because of the presence of extra sugar groups, or extra hydroxyl- or carboxyl-groups in the steroid nucleus, are more rapidly excreted into the bile. Digoxin and lanatoside C are examples of glycosides which have less clinical activity for this reason (Wright, 1960). For the mode of action of cardiac glycosides, see Section 14.1.

A simple apparatus has been designed for measuring the passive diffusion of drugs through an artificial lecithin membrane. It is claimed that this model system gives results similar to those obtained with natural Type 1 membranes (Misra, Hunger and Keberle, 1966). Other work with artificial membranes is outlined in Chapter 14.

3.2.2 'Type 2' membranes

These membranes differ by the presence of a specific carrier, to effect *facilitated diffusion* (also called mediated transport). As a result, transport is faster than can be calculated for simple diffusion, but at equilibrium no difference exists between the concentrations inside and outside the cell (contrast with Type 3 membranes). Type 2 membranes exist not only to hasten the absorption of a limited number of vital nutrients, but also to effect absorption of those metabolites which could not be expected to penetrate a Type 1 membrane because of ionization or an excessively hydrophilic structure.

The combination of the transported molecule with the carrier is reversible. Because of the thinness of the membrane, a small conformational change should suffice for the carrier to pick up the molecule on the outside and shed it on the inside of the cell. A mere shift of charge could suffice. The Type 2 membrane has the following properties: (a) the carrier soon becomes saturated after which the substance cannot be transported, even though the gradient is favourable, (b) each carrier molecule shows high chemical specificity, even towards optically active stereoisomers, (c) carriers can be inhibited by molecules that resemble the substrates, but not too closely, and (d) metabolic energy is not consumed by the act of transportation. This last property can further distinguish Type 2 from Type 3 membranes, using the test: can extra respiration be detected?

The most intensely studied example of facilitated diffusion is that of glucose in human erythrocytes (Wilbrandt and Rosenberg, 1961). A study of specificity showed that, apart from D-glucose, the carrier also transported D-mannose, D-xylose, D-ribose, and (rather poorly) D-arabinose. It also transports some unnatural, unmetabolizable sugars such as 3-O-methyl-D-glucose, 2-deoxy-D-glucose, 5-thio-D-glucose, and 3-chloro-3-deoxy-D-glucose. It also transports galactose in which the $-OH$ of the 4-position is reversed. Comparable L-sugars are not transported. Myoinositol and fructose are not transported, but have their own carriers. All kinds of glucosides are well transported, so that there must be a great deal of space around that part of the carrier where the 1-glucoside-forming position of the sugar comes to rest; space elsewhere seems quite limited (Barnett, Ralph and Munday, 1970).

Choline is an interesting example of a molecule dependent, for its cellular uptake, on facilitated diffusion. Unable, because of its permanently ionized and poorly lipophilic nature, to penetrate by simple diffusion through a Type 1 membrane, choline is rapidly transported into erythrocytes and several other kinds of cell by a specific carrier. Tetramethylammonium cations can also enter cells on this carrier, a facility denied to higher homologues which can, however, block the physiological uptake of choline (Martin, 1969).

Most amino acids require facilitated diffusion for their uptake, as do the purine and pyrimidine bases and their nucleosides. Several inhibitors of the uptake of nucleosides are known, such as 6-anilinopurine and the coronary vasodilator, dipyridamole. Folic acid and its derivatives, too, need facilitation to enter vertebrate cells, a form of aid that is completely lacking in one-celled organisms (Section 4.2).

At least seven carriers control entry through the mitochondrial membrane. One carrier facilitates entry of succinate, D- and L-malate, malonate, and *meso*-tartrate anions, but not tartrate, maleate, or fumarate. Another mediates the entry of citrate, *cis*-aconitate, *iso*citrate, and D- or L-tartrate, but not fumarate or maleate. A third carrier transports adenosine nucleotides. Also phosphate anions can enter mitochondria whereas other inorganic anions cannot (Chappell, 1966).

3.2.3 *'Type 3' membranes*

These membranes, the most complex of all, can concentrate substances against a gradient, if necessary. In this *active carrier transport*, as it is called, an energy-consuming process drives the operation, which is highly temperature-sensitive, and easy to saturate.

Examples of permeability through a Type 3 membrane are: (a) the transport of Na^+ and K^+ in mammalian cells (see Section 14.2), (b) the absorption and secretion of a wide range of ionized and non-ionized substances by the kidney tubules and, to a less extent, by the gastrointestinal membrane, (c) the uptake of inorganic ions, amino acids, and sugars by bacteria, (d) accumulation of the iodide anion by the thyroid gland, and (e) accumulation of K^+, Na^+, Ca^{2+}, and Mg^{2+} against a concentration gradient in mitochondria. The transport of glucose, which uses 'Type 2' in the human erythrocyte, needs 'Type 3' in the kidney and intestinal membranes. In the latter, only sugars with a 2-hydroxy group are transported, and it is thought that this group becomes phos-phorylated in the process. The glucose carrier in kidney tubules, cannot take up mannose or arabinose. It is easily inhibited by phloridzin, a poisonous (for this reason) glucoside from the bark of pear tree.

Activated processes also exist for some anions. More on the passage of inorganic ions through membranes will be found in Section 14.2.

For each 18 sodium ions transported through the frog's skin, one extra molecule of oxygen is consumed. Toad bladder and guinea-pig intestine behave similarly. Many cathartics (e.g. cascara, phenolphthalein, podophyllin) inhibit absorption of sodium by the membrane of the intestinal lumen (from experi-ments in living rabbits), thus causing accumulation of sodium salts in the colon, and hence retention of water (Phillips *et al.*, 1965).

Often 'Type 2' and 'Type 3' activities are found as scattered spots on a membrane that is otherwise a normal Type 1. It is not yet clear how the carriers in Types 2 and 3 membranes operate. Many examples isolated from eukaryotic membranes were found to be dimeric glycoproteins of mol. wt. about 100 000. Most think that they are transmembrane proteins (see Section 5.4) which undergo small conformational changes (Kyte, 1981). Examples are (a), the Na^+ and K^+ ATPase which transports sodium and potassium ions across membranes of all eukaryotic cells; (b), the Ca^{2+} and Mg^{2+} ATPase which transports calcium ions into muscle cells; (c), the anion-exchange protein that regulates the passage of CO_2 through the red blood-cell membrane; (d), rhodopsin, the retinal-binding visual pigment which governs permeability within the retinal rods; and (e), the acetylcholine receptors which change the ionic permeability of nerve and muscle cells after combination with acetyl-choline (Section 2.1). A protein (mol. wt. 50 000) actively transports both glucose and histidine from the jejunum across the brush border into the blood-stream (Faust and Shearin, 1974).

But a carrier can be fairly simple, as in the transport of amino acids from urine into proximal renal tubule cells. These convert the amino acid to its γ-glutamyl derivative by the action of glutamyltransferase and glutathione. Once inside the cell, the glutamyl group is excised and reconverted to glutathione (Meister and Tate, 1976).

Type 3 membranes are also active in the higher plants. For example, carrot roots can take up sodium chloride only when oxidizing glucose. Many carriers have been isolated from bacteria, including those for carbohydrates, amino acids, and several anions such as sulfate. They seem to be sugar-free proteins of molecular weight about 30 000.

When new drugs are being designed, the requisite permeability has usually been obtained by stepwise increases in liposolubility, a strategy aimed at Type 1 membranes. A different and more selective approach makes use of Types 2 and 3 membranes by designing a part of the drug to resemble a natural substance for which a specific transport mechanism exists in particular cells. That this approach can be highly successful is attested by the much-used anti-cancer drugs 6-mercaptopurine, 5-fluorouracil, and cytarabine (arabinosylcytosine) which enter the cells through carrier mechanisms provided for the normal metabolites which are, respectively, hypoxanthine, uracil, and deoxycytidine. To increase the penetration of nitrogen-mustard (anti-cancer drug) into cells, Bergel and Stock (1954) attached various amino acids, of which phenylalanine was most successful. The product, melphalan (*13.45*) has given excellent results in the clinic. As expected, the D-phenylalanine isomer was not taken up. This idea is capable of immense expansion, and the therapeutic rewards should be large. For a review of permeability through Types 2 and 3 membranes, see Wilbrandt and Rosenberg (1961).

3.2.4 'Type 4' membranes

These membranes are distinguished from Type 1 by the presence of pores, the size of which can be gauged from the size of the largest molecule which can penetrate them. As a homologous series is ascended, substances are obtained which penetrate Type 4 membranes less (and Type 1 membranes more) readily. One of the best-known examples of a Type 4 membrane is the glomerular tuft in Bowman's capsule of the kidney (see later). This tuft is permeable to all molecules smaller than albumin (mol. wt. 70000). The pore size is estimated as 3 nm, and inulin (mol. wt. 5000) passes through easily. Mammalian capillaries and liver parenchyma, too, have Type 4 membranes, which are not often met apart from these important examples.

3.2.5 Pinocytosis and phagocytosis

These are two energy-consuming processes by which a cell membrane can pass even larger particles than could go through a pore in a Type 4 membrane.

(*a*) *Pinocytosis* is a process in which a membrane (usually Type 1) develops invaginations that are then pinched off to form vesicles. It is a device for passing through a membrane those molecules which are too large to diffuse in the normal way, especially proteins. In this way material formerly outside the cell appears inside it, and also vice versa.

(*b*) *Phagocytosis*, a somewhat similar process, allows the passage of still larger particles. Thus the electron microscope clearly shows solid particles passing across the cell membranes of mammalian capillaries, the entire surface of which seems to be available for this purpose. Enzymes and hormones are often extruded from cells as vesicles enclosed in a lipid membrane. The five hydrolytic proenzymes of the pancreas are thus extruded together as 'zymogen granules'. The vesicles in which acetylcholine is secreted by nerve-endings (Whittaker, 1963), and the granules in which noradrenaline leaves the medulla of the adrenal gland (Blaschko, 1959), are excreted similarly.

3.2.6 *Permeability of the different mammalian tissues*

The absorption and distribution of foreign organic substances follow much simpler rules than those that obtain for natural metabolites and cell constituents. For example, a simple Type 1 lipid barrier regulates the passage of foreign molecules through the following structures: the gastrointestinal epithelia, the renal tubular epithelium, the blood–brain barrier, and the barrier between the bloodstream and cerebrospinal fluid (Schanker, 1961).

The following data on the permeability of individual tissues and organs will enable the reader to answer all questions in Box 3.1, with the help of ionization data in Sections 10.1 and 17.0. The answers are in Box 3.2, at the end of this Chapter.

Box 3.1 A quiz on the permeability of some mammalian membranes
Given:
    (a) An acid with pK_a −1.3 (toluene-4-sulfonic acid),
    (b) An acid with pK_a 4.2 (benzoic acid),
    (c) An acid with pK_a 7.6 (thiopental),
    (d) A base with pK_a 8.2 (codeine), and
    (e) A base with pK_a 10.0 (atropine).

Predict:
    (1) Which will be best absorbed from the stomach when the pH there is 2.0.
    (2) Which will be best absorbed from the small intestine, when the pH there is 6.0.
    (3) Which will show the lowest rate of renal elimination when the urine has pH 4.2.
    (4) Which will pass most readily from plasma into the brain (pH of plasma, 7.3).

Answers at end of this chapter.

(*a*) *Stomach.* In the rat's stomach, it was found that drugs were well absorbed only if non-ionized. Thus, when the pH of the stomach contents was raised, basic drugs were better absorbed, because more of their molecules became non-ionized; but this pH change decreased the absorption of acidic drugs, because less of the non-ionized form remained (see Section 10.0 for an outline of the chemistry of ionization). The value of lipophilic properties in assisting absorption was demonstrated with three barbiturates of similar pK_a but different lipid/water partition coefficients see Table 3.2): absorption rose proportionally as the coefficient rose (Schanker *et al.*, 1957; Brodie, Kurz and Schanker, 1960). The pattern of absorption from human stomach contents (pH 1) was very similar. Weakly acidic drugs, such as salicylic acid, aspirin, thiopental, and many other lipophilic barbiturates, were readily absorbed because they were not ionized at pH 1, whereas basic substances such as quinine, ephedrine, and aminophenazone (amidopyrine) were not absorbed because they were totally ionized at this pH (Hogben *et al.*, 1957).

(*b*) *Small intestine.* More drugs are absorbed from the small intestine than elsewhere. The epithelial lining of the small intestine (rat) was found to permit

Table 3.2 Correlation of gastric absorption
with liposolubility

	pK	A	P_c
Barbital	7.8	4	<0.001
Quinalbarbital	7.9	30	0.10
(secobarbital)			
Thiopental	7.6	46	3.30

Where A is the per cent absorption from rat stomach when
fed a solution (pH 1) orally, and P_c is the partition coefficient between heptane and water (pH 1): the higher values
of P_c are the more lipophilic. (Schanker *et al.*, 1957.)

the penetration of non-ionized drugs and impede the passage of the corresponding ions (Hogben *et al.*, 1959). Three permeability barriers have to be
passed, in this order: the lumen-facing membrane of intestinal cells, the
capillary-facing membrane of the same cells, and the basement membrane of a
capillary. Two experimental methods are available: a length of intestine can be
removed, the lower end tied, and the whole placed in an organ bath. Alternatively a sac of the intestine can be everted from the living rat and surrounded
by Ringer's solution in a glass jacket. Both methods gave similar results (Misra
et al., 1966). The kinetic and physiochemical factors underlying these absorption processes have been described by Nogami and Matsuzawa (1963). The
initial pH (8.0) and the average pH (6.6) of the small intestine membrane are
higher than that of the stomach, and hence it permits the passage of aromatic
amines but not of the stronger aliphatic amines. Absorption of drugs from the
small intestine must be through the lipid-rich areas rather than through the
aqueous channels, because liposoluble drugs of high molecular weight were
absorbed more rapidly than lipid-insoluble substances such as urea or deuterium oxide. Raising the pH increased the absorption of bases and decreased
that of acids, just as in the stomach. Moreover, substances whose degree of
ionization was not changed by such alterations in pH showed no change in the
rate of absorption.

The absorption of foreign ions from the small intestine (rat) was found to be
very slow, and declined with time. The absorption, from the small intestine, of
natural substrates, e.g. L-amino acids, glucose, and uracil, uses specific
activated transport systems which can work against a concentration gradient,
and can become saturated.

(*c*) *Colon*. The absorption of drugs from the rat's large intestine was found
similar to that from the small intestine (Schanker, 1959). All absorption from
the gastrointestinal tract can be slowed by atropine which decreases mobility of
the muscular lining.

After absorption from the bowel, drugs enter a phase known as *first pass*: they
are conveyed directly to the liver by the portal circulation and there many of
them are deactivated, wholly or partly. Drugs that are prone to inactivation in
the liver, or the gut wall, are usually given parenterally.

(*d*) *Between blood-plasma and the tissues*, the distribution of drugs follows essentially the same pattern as between gastrointestinal tract and the blood-plasma (see above) (Waddell and Butler, 1959). Only the fraction *not* strongly bound by plasma albumin is free to diffuse in this way. A weakly bound drug can be displaced from blood-proteins by another drug for which they have a higher affinity.

(*e*) *The skin* offers a major barrier to absorption because of the outer layer known as the stratum corneum which is composed of dead, tightly packed cells. A layer below this, the epidermis, is the selective barrier of the skin. It is mainly Type 1 and many neutral organic substances diffuse across it in proportion to their lipid/water partition coefficients (Treherne, 1956). For a treatise on skin permeability, see Schaefer, Zesch and Stüttgen (1982).

(*f*) *Erythrocytes*. The permeability of vertebrate erythrocytes has been extensively studied and has already been referred to in Table 3.1 and associated text. In general the membrane is Type 1, but the entry of particular substances is facilitated in certain species, e.g. glucose in humans and primates (Le Fèvre, 1961). A special peculiarity of red blood cells is the ease with which they admit inorganic anions, by exchanging them for bicarbonate ion, which is then dehydrated to (non-ionic) carbon dioxide.

(*g*) *Capillaries* of the circulating blood system, like those of the kidney's glomerulus (see below) and the liver, have porous Type 4, membranes whereas those of the brain are more tightly structured (see below).

(*h*) *Lungs* provide one of the body's most effective absorptive areas for liposoluble material, including general anaesthetics. This facility derives from the 200 sq.m surface area of the pulmonary alveoli, richly provided with capillaries.

(*i*) *The kidney*. The fundamental unit of the kidney is the nephron, of which about 1.2 million are present in each human being. If uncoiled and placed end to end, one individual's nephrons would stretch about 50 miles. Structurally each nephron consists of a porous tube within a non-porous tube and is U-shaped (see Fig. 3.5). At the beginning of each nephron is a little tuft of blood capillaries called the glomerulus. Blood flows into this tuft which has a Type 4 (porous) membrane that retains all the particles and most of the protein: it feeds the rest into the nephron. The tubule directs useful constituents back into the circulation, and sends the waste (urine) down the ureters into the bladder. The human kidney produces about 185 litres of glomerular filtrate each day, but the tubules resorb all but 1.5 litres of water and many dissolved substances of value in the body's economy. The renal tubules, which effect this resorption, have a normal Type 1 membrane, which permits the passage of liposoluble substances in either direction (i.e. to or from the bloodstream), depending upon the

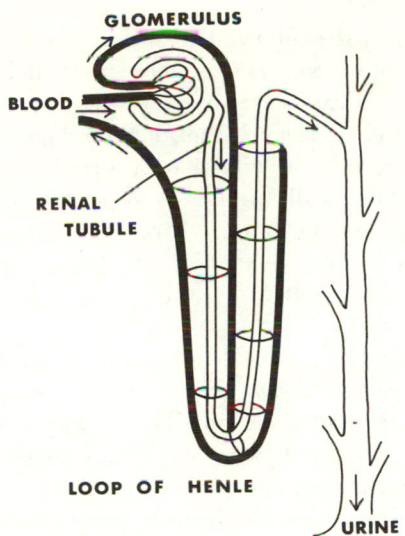

Fig. 3.5 Unit structure of the human kidney.

concentration gradient. The tubules and accompanying cells also have special-ized patches of activated transport for secreting many kinds of organic ions, even against a gradient. For cations, see Peters (1960), and for anions Sperber (1959). There is no corresponding process for the resorption of these organic ions. The most important physiological role of the tubules is the resorption of water, bicarbonate, chloride, and other inorganic ions.

Of two similar drugs, the one that is more strongly bound to blood-albumin is excreted by the glomerulus with more difficulty. This means that porosity is of little avail against the forces (electrostatic, van der Waals, or hydrogen-bond-ing) which tie the drug to a blood-protein. On the other hand, protein-binding is of little avail when the constitution of the drug makes it eligible for one of the activated transport mechanisms in the kidney tubule. These principles can be borne in mind when redesigning a drug so as to reduce the dose.

(*j*) *The liver*. This organ has large pores (without parallel in other mammalian cells) in the membranes of hepatic parenchymal cells (Schanker, 1961). The mechanism of *biliary excretion* of drugs is still insufficiently known. The pathway is blood → liver interstitial fluid → hepatic parenchyma cells → bile → small intestine. Active transport is provided for anions and cations whereas lipo-soluble neutral molecules diffuse through a Type 1 membrane. Molecules heavier than 250 daltons, even inulin (mol. wt. 5000), pass readily from blood into the bile. This facility seems designed to allow the passage of bilirubin glucuronide (Schanker, 1961). For substances in the molecular weight range of 350–500, strong species differences were found, and substances of lower molecular weight were poorly excreted in bile, seemingly because they were more quickly excreted by the kidney. The overall picture of biliary excretion of drugs is one in which foreign substances whose molecular weight is too high for excretion by the kidneys can be returned to the bowel. Unfortunately such substances are likely to be resorbed from the bowel and recycled as above, so that little loss from the body can take place (phenolphthalein is an example). Drugs that are enterohepatically circulated, especially if they are metabolized with difficulty, can build up dangerously large blood levels.

For further reading on biliary excretion see the book by Smith (1973), also Hirom, Millburn and Smith (1976).

(*k*) *Between blood and cerebrospinal fluid.* Many drugs were found to penetrate from the blood into the *cerebrospinal fluid* (dog, rabbit) by simple diffusion at rates roughly parallel to their lipid/water partition coefficients at pH 7.4 (Brodie *et al.*, 1960; Mayer, Maickel and Brodie, 1959; Mark *et al.*, 1958). When the pH gradient between blood and cerebrospinal fluid (CSF) was altered by changing the pH of plasma, those drugs which underwent no ionization change (e.g. sulfanilamide) retained their normal distribution ratio, whereas those drugs which became more highly ionized achieved a poorer CSF/plasma concentration ratio than before (Rall and Zubrod, 1960). Cations and anions penetrate the CSF and the brain exceedingly slowly; among the few exceptions are the phenylarsonic acids used in trypanosomiasis (Section 6.2) which may be taken up by the phosphate transport mechanism. Regardless of their rate of entry into the CSF, drugs often disappear quite rapidly from this fluid, a poorly explained phenomenon (Pappenheimer, Heissey and Jordan, 1961).

(*l*) *The blood–brain barrier.* Towards the end of the last century, Ehrlich injected the dye trypan blue intravenously and found that it entered all tissue except the central nervous system. Conversely, when he injected this dye into the cerebrospinal fluid, the brain was stained rapidly, but none of the dye entered the bloodstream. Diffusion of drugs from the blood to the brain is more restrictive than elsewhere. This barrier (known as the blood–brain barrier) is capillary endothelium in which intercellular clefts are more tightly sealed and the cell overlap is tighter than in other parts of the body (Rapoport, 1976). This barrier is unusually permeable to lipids, but particularly impermeable to ions. When the barrier membranes are inflamed, however, a wider range of substances can pass through. The permeability of this barrier to a drug is proportional to the latter's octanol/water partition coefficient and also inversely proportional to the molecular weight right up to a cut-off about 400. For substances of higher molecular weight, penetration is slow and depends on rare leaks occurring where capillary cells join one another (Levin, 1980). Section 15.0 (hypnotics and general anaesthetics) may be of interest in this connexion.

(*m*) *The brain.* After passing the barrier, a drug still has to penetrate the membranes of the brain. Transfer of drugs occurs between various regions of the brain. Highly lipophilic substances (such as thiopental, chlorpromazine, and DDT) were found in the grey regions (of high vascularity) soon after oral administration; but a few hours later they were seen to be localized in the white regions which are richer in lipids. This is a complex topic, for which few data are available.

3.2.7 *Permeability between organelles*

Barriers to *permeability within cells* largely consist of the membranes of mito-chondria and of the endoplasmic reticulum, both of which appear to be Type 1 with patches of Types 2 and 3. The nuclear membrane appears to be the porous Type 4.

3.2.8 *Penetrability of other types of cell*

Bacteria whether intact or freed from cell walls have, on the exterior of the cyto-plasm, a membrane which appears to be mainly Type 1 and is their sole perme-ability barrier. This has been studied in, for example, *Staph. aureus, Micrococcus lysodeikticus*, and *Sarcina lutea* (Mitchell and Moyle, 1959). Organic substances with more than four hydrogen-bonding groups (e.g. —OH groups), or ions that transport more than four water molecules, cannot penetrate the membrane at a significant rate. Whereas lysine can penetrate this membrane by diffusion, the more hydrophilic amino acids, such as glutamic acid and histidine, penetrate only by activated transport (Gale, 1947) as glucose does.

The membranes of *diatoms*, and the chitin-impregnated membranes of *arthropods*, are Type 4.

The common belief that the integument of *insects* is penetrated faster by chlorinated insecticides than is mammalian skin is incorrect (O'Brien, 1967).

The membranes of *plant cells* are remarkably like those of animals. The plasma membrane of plants seems not to offer so great a barrier as that of the mito-chondria, or of the cytoplasmic reticulum, or of the vacuole (a feature absent in animals). Most, if not all, movement into plant roots is physical diffusion unaided by biochemical processes. The plasma membrane of roots seems to have a special structure with unusual selectivity. Maleic hydrazide causes chromosome breakages in plant cells, but not in mammalian cells. As mammalian and plant chromosomes have the same chemical composition, a difference in permeability is indicated (Barnes *et al.*, 1957).

For more on the nature of membranes, see Section 5.4. For further reading on the movement of molecules across membranes, see Bittar (1980–81) and Stein (1967).

3.3 The significance of partition coefficients

Chemists and pharmacologists began to take an interest in partition coefficients at the turn of the present century, as outlined in Section 2.0. Their stimulus was the positive correlation that Overton and H. Meyer had demonstrated between the bio-depressant (e.g. hypnotic) action of many chemically unrelated substances, and their preference for the lipoid layer when partitioned between olive oil and water. This work was continued by K. Meyer and Hemmi (1935) who obtained more precise correlations by refinement of the technique.

Interest in partition coefficients then waned, but was renewed in the early 1960s when a group in Pomona College, California, under the leadership of Corwin Hansch, showed that this relationship is often parabolic and hence, after an optimal degree of lipophilicity had been built into the molecule, further increases diminished the biological action (Hansch and Fujita, 1964; expressed more clearly in Hansch, 1971). The Pomona school widened drug scientists' interest in partition coefficients by pointing out that, no matter what electronic or stereochemical factor was chiefly responsible for the action of a class of drugs, the partition coefficient usually played *some* part in the delivery of the drug to its site of action. For this reason, a favourable balance between lipophilic and hydrophilic properties should help any type of agent to reach its receptor. Even when partitioning is only a secondary property, they concluded, it cannot be neglected.

Recalculating the published data on bio-depressants, Hansch and his colleagues found that the following regression equation made a good fit (the squared term ensured a parabolic relationship):

$$\log 1/C = k(\log P) - k'(\log P)^2 + k''$$

where C is the concentration that produces a standard biological response, P is the partition coefficient, and k, k', and k'' are constants to be selected by computer (using the method of least squares) to make a 'best fit' of the experimental results (Hansch *et al.*, 1968; Hansch and Anderson, 1967). Because this equation describes an equilibrium, a *linear free-energy relationship* is said to exist between the logarithms of the two sets of data (dosage and liposolubility). It will be noted that the provision of no less than three kinds of k, allows for much adjustment in fitting the data to the equation.

This, and other statistical work, is reported with the following information:

n, the number of data points used,
s, the standard deviation for the regression, and
r, the correlation coefficient, which should be, ideally, 1.00.

For drugs which are not simple bio-depressants, it is necessary to insert one or more further descriptors (such as an electronic and a steric constant) into the above equation to convert it to *multiple* regression analysis. This modification, due to Hansch (1968, 1971), is discussed in Chapter 16.

(*a*) *Methods and solvents.* Partition coefficients are usually determined by a rapid shaking technique described in Section 17.1. Some workers use the retention indices obtained by high-performance liquid chromatography, which correlate well with log P (Könemann *et al.*, 1979; Mirlees *et al.*, 1976).

In seeking the ideal non-aqueous solvent for determinations, Meyer and Hemmi (1935) replaced the traditional olive oil by oleyl alcohol, thinking it nearer in properties to actual membrane components. In current terms, it had a better ratio of hydrophilic to lipophilic properties than oil, and the hydroxyl group also provided hydrogen-bonding properties, Collander experimented with a large number of non-aqueous phases (Collander, 1933, 1937, 1947, 1954). He found that the logarithm of the partition coefficient in a pair of solvents is *linearly* related to the logarithm of the partition coefficient in a second pair of solvents, one solvent being water in each case, thus:

$$\log P' = a \log P + b$$

where P and P' are partition coefficients of one substance in two different solvent pairs, and a and b are constants. The less the lipoidal solvent was capable of dissolving water, the more 'discriminatory' it was, spreading the values further apart but usually without altering their ratios (Collander, 1954). This work was extended by Leo and Hansch (1971), who compiled a list of a and b values.

Setting a at 1 for octanol, both chloroform and ether become 1.13, and butanol 0.70, and most other a values lie between these extremes. However, b varies more widely. Setting it at 0 for octanol, it ranges from -2.85 for heptane, through -1.34 for chloroform, -0.58 for oleyl alcohol, -0.17 for ether, right up to $+0.87$ for cyclohexanol. A substrate with P of 2.0 in octanol has $P-0.14$ in cyclohexane.

Some exceptions have been encountered where partition coefficients, between water and two non-aqueous solvents, do not run parallel. A common example is the solute that forms a hydrogen bond with one non-aqueous solvent but not with the other, e.g. phenol with oleyl alcohol but not with dodecane (Burton, Clarke and Gray, 1964). Similarly, carboxylic acids form dimers in hydrocarbons, but not in moist alcohols (Biagi, *et al.*, 1974).

In selecting octan-1-ol as their standard non-aqueous solvent, the Pomona school were influenced by its ease of handling, comparative freedom from anomalous results, and closeness of discriminatory power to at least some natural membranes. Typical $P_{octanol}$ values are: benzene 2.13 (± 0.01), nitrobenzene 1.85, aniline, 0.90, phenol 1.46, and benzyl alcohol 1.10. A table of 5806 experimentally determined partition coefficients, by many authors and from various non-aqueous phases, has been compiled (Leo, Hansch and Elkins, 1971). The use of octanol is now universal.

Changes in temperature of, say, 5°C, have little effect on a partition coefficient. When log P is above 4, the shaking method has little reliability and recourse must be made to chromatography, or to calculation from Rekker's fragmental constants (see below).

For hypnosis or general anaesthesia in mammals, the maximum effect is reached at log P 2.0 in many series of diverse chemical composition (see Table 15.1).

(*b*) *Log P and the regression equation.* Solution of the regression equation gives three term-coefficients (k, k', and k''). These are obtained by feeding only one-half of the examples through the equation. The solution is then tested by feeding the other half through the equation, with this difference; one must use the term-coefficients derived from the first half instead of allowing the computer to furnish new coefficients. The r (correlation coefficient) calculated on the second set of results should not differ appreciably from 1.0.

Each term in an equation needs five compounds to give significance to its coefficients. Thus the equation

$$\log 1/C = k(\log P) - k'(\log P)^2 + k''$$

although it uses only *one* descriptor (partition), has *three* terms, and hence three term-coefficients have to be found. This will require 15 examples for the experiment and another 15 for the confirmatory test.

Some experiments have yielded results that did not require a $(\log P)^2$ term in the regression equation. This could mean that only low values of log P were tested, because the early part of a parabola approximates to a straight line. More often, though, it has been a result of continuous (as opposed to single, or highly spaced) dosage. In model experiments on fish, which were designed to mirror the industrial pollution of lakes, the toxicity (narcosis) correlated *linearly* with log P. In one such experiment on fish, 50 toxicants were used, comprising aliphatic and aromatic hydrocarbons some of which were unsubstituted and others substituted by chlorine, hydroxyl or alkoxyl. Narcosis correlated linearly with log P over the range -1.3 to 6.4 (octanol/water), and the r coefficient had the unusually high value of 0.988 (Könemann, 1981).

(*c*) *The influence of ionization.* For substances that are at least partly ionized at the pH of the determination, the partition coefficient (P) has two more terms, as follows:

$$P = \frac{[B]_o + [BH^+]_o}{[B]_w + [BH^+]_w}$$

This equation is obviously constructed for substances (B) which ionize as cations (BH^+), and can easily be converted to one for those that ionize as anions. In the stated equation, $[BH^+]_o$ is usually so small that it can be neglected, but achieves significance for drugs that have been designed, by the addition of a short lipophilic side-chain, to have a lipophilic cation.

Lipophilicity falls by a factor of about 10 000 when ionization takes place (Hansch, Leo *et al.*, 1973). Provided that ion and molecule absorb in the ultraviolet with two distinct maxima, the above equation is easily solved.

Alternatively one may use the following equation of Fujita, Iwasa and Hansch (1964):

$$P = C_{octanol}/C_{water} \, (1-\alpha)$$

where α is the fraction ionized as worked out from the pK_a and the table in Section 17.0.

Scherrer and Howard (1977) list several interesting biological correlations discovered only by separating the contributions of ionized and non-ionized species to the partition coefficient.

(d) *How is solubility in water related to P?* A liquid becomes partitioned between itself and water just as it would between a foreign solvent (e.g. octanol) and water. Hence the partition coefficients of neutral liquids run parallel to their solubilities in water. Even so, this general observation has been found not to apply to saturated hydrocarbons (Hansch, Quinlan and Lawrence, 1968).

The solubility of a liquid in water has often been proposed as a substitute for log *P*, for purposes of ranking. However, it has the following disadvantages: it could not be used for liquids like ethanol that are completely miscible with water, it is more difficult to determine accurately for poorly soluble substances (ultracentrifugation is needed to remove traces of suspended particles), it does not apply to substances like carboxylic acids which are dimeric in water, but monomeric in lipoidal solvents, and it is particularly misleading for solids where a large factor in determining solubility is breaking the crystal-lattice. To overcome this difficulty with solids, Bannerjee, Yalkowsky and Valvoni (1980) have proposed the following equation:

$$\log P = 6.5 - 0.89(\log S) - 0.015 \, mpt.$$

where *S* is the solubility in water in micromoles per litre, and the melting point is used as a measure of crystal-lattice energy to be overcome.

(e) *The π values.* Hansch and colleagues formulated a substituent constant (π) that was intended to be proportional to the free energy of transfer of the substituent from one phase to another. This constant was defined as $\pi = \log P_X - \log P_H$, where P_X is the partition coefficient of a derivative and P_H of the parent (Fujita, Iwasa and Hansch, 1964). A table of 128 aromatic π values was published by Hansch, Leo *et al.* (1973), and their additivity was stressed. It was pointed out that when a partition coefficient could not be determined experimentally, it could be calculated by adding th π values of the substituents to the *P* of the nucleus (Hansch, 1971). Although enjoying great popularity among the more theoretical of contemporary drug designers, the concept of π values was adversely criticized by Davis (1973) and Nys and Rekker (1973).

Hansch and his colleagues obtained these π values by a process analogous to the derivation of Hammett's sigma constants, which the Pomona school was already applying to their investigation of plant growth-factors. Unfortunately, the analogy was not close enough, and this derivation robbed π values of thermodynamic significance. The fault in the derivation is evident in the following example:

$$\pi_{\text{methyl}} = \log P_{\text{toluene}} - \log P_{\text{benzene}}$$

because this calculation really furnishes only $\pi_{\text{methylene}}$. Arising, apparently unconsciously, from their earliest publications on π, the Pomona school have always taken π_{methyl} and $\pi_{\text{methylene}}$ to be identical (about 0.5). Similarly they assumed that $\log P_{\text{benzene}}$ is identical with π_{phenyl}, and (hence) that π_{H} is 0.

(*f*) *Rekker's f values.* Rekker, wishing to overcome this error, introduced *f*, the hydrophobic fragmental *constant*, to replace π which is the hydrophobic fragmental *substituent*. Thus:

$$f_{\text{methyl}} = \log P_{\text{toluene}} - f_{\text{phenyl}}$$

The calculations were performed on a computer using a multiple regression analysis programme, into which 128 published values of log *P*, directly determined in octanol, were processed statistically (Nys and Rekker, 1973; Rekker, 1977). Inspection of these *f* values shows that they tend to be about 0.2–0.3 higher than π values. The *f* value for methyl is 0.17 higher than that for methylene, and a hydrogen atom is now seen to contribute about 0.2 to lipophilicity. The Pomona school, conceding the merits of the *f* system, published their own experimentally determined *f* values (Leo *et al.*, 1975). Given *P* (experimentally determined) for hydrogen gas (0.45), they assigned half of this (0.225) to *f*H, and this value, when subtracted from P_{methane} (1.09) gave 0.865 for *f*CH$_3$, a value later refined to 0.89; similarly 0.66 for *f*CH$_2$. See Section 17.1 for a selection of these *f* values.

One immediate gain from adopting *f* values was the disappearence of former abnormalities in the values for omega-substituted phenylpropanes, which have erroneously been supposed to have *folded* molecules because of the seeming irregularities. Unfortunately other abnormalities remain, and some of these will now be mentioned.

Aromatic π or *f* values derived from *meta* and *para* substituents tend to be identical, but *ortho* substituents often give outlying values, e.g., when they permit internal hydrogen-bonding, lipophilicity is increased. Nevertheless, by much more complicated calculations, Fujita and Nishioka (1975) have made a special table for *ortho* substituents that integrate well with the *meta* and *para* values. Apart from this, these π and *f* values are very sensitive to polar environments. For example, π for chlorine substituted in benzene is 0.71, but this becomes (insertion is in all cases, *meta*) 0.61 in nitrobenzene, 0.68 in phenylacetic acid, 0.83 in benzoic acid (all ionizable substances are corrected for ionization), 0.98 in aniline, and 1.04 in phenol. This difference of 0.43 between extremes is increased to 0.90 when nitro-group replaces chlorine in the same nuclei (Hansch, Leo *et al.*, 1973). Two highly polar substituents, particularly if both are nucleophilic, show enhanced lipophilicity (often 0.8) if separated by only one carbon atom, and about half this enhancement if separated by two carbons (Leo, Hansch and Elkins, 1971; Rekker, 1977, pp. 49, 98, 293).*

Continuing with anomalies of the values: small molecules like ethane are highly irregular (Leo *et al.*, 1975). Then there is the strangely enhanced hydrophilicity of pyridinium salts after alkylation on the nitrogen atom (Leo, Hansch and Elkins, 1971; Rekker, 1977, p. 150). Covalent hydration (described in Section 2.5) introduces an alcoholic group, not allowed for in the usual calculations. The formation of micelles in the aqueous phase, by compounds bearing a long hydrocarbon chain, presents a further difficulty.

The Pomona school recommends drug designers to calculate, from a table of f values, the P value of a molecule before preparing it, to make sure that it lies within the range where the desired biological activity has already been found (Leo *et al.*, 1975).

(*g*) *Biological implications of partition coefficients.* Thanks to the initial stimulus of the Pomona school, experimental modification of the P values of common drugs was soon under way. The sulfonamide anti-bacterials were recognized as slightly hydrophilic substances whose performance improved slightly when they were made a little more lipophilic, but declined with further increases. The penicillins and cephalosporins, on the other hand, improved slightly in performance when made a little more hydrophilic (Biagi *et al.*, 1974), but declined when this quality was accentuated. These results seem to be typical of the small amount of fine-tuning that can be accomplished when lipophilicity is *not* the principal factor in the action of a class of drugs. As can be seen from Section 15.0, it *is* the principal factor mainly for hypnotics, general anaesthetics, and the weaker non-selective insecticides.

Partition coefficients can exert a local effect, particularly in the more rigid molecules. For example, benzamides that carry a large hydrocarbon group in the 4-position are strong inhibitors of alcohol dehydrogenase, but the same group in the 3-position does not make the molecules inhibitory (Hansch, Kim and Sarma, 1973). This example reminds us that it is futile to quote P values without further qualifying information.

The principal biological interest of P values is as models for the penetration of drugs through membranes, but to obtain this type of information presents great experimental difficulties. At least we know that the partition coefficients of a series of alcohols between red cell ghosts and water shows a free energy of about -690 cal/mol for the transfer of each methylene group, approximately the same result as between octanol and water (Seeman, Roth and Schneider, 1971).

*Rekker (personal communication) has observed that the difference between a predicted and an experimental fragmental constant, when not 0, is often 0.29 (or a multiple of this), the molecule being less hydrophilic than predicted, as though *one* of the expected hydrogen bonds to water failed to be made.

The buccal membrane is, so far, the best investigated of all living membranes. Healthy volunteers are asked to retain a drug in the mouth while its appearance in the bloodstream is measured at intervals (Beckett, Boyes and Triggs, 1968). These authors thought that their results were typical of a highly discriminating system, like heptane/water. However, rigorous statistical analysis by Rekker (1977) showed that it was less discriminating than octanol/water and more like the system butanol/water. Rekker thought that the gastrointestinal membrane behaved more like octanol/water, whereas the blood–brain barrier, most discriminating of all, had a resemblance to the heptane/water system (Rekker, 1977). It is to be hoped that the next decade will reveal what differences in the lipoid and protein content of various kinds of membranes are responsible for these differences in partitioning.

3.4 Mechanisms that produce loss. Storage and elimination

Three principal mechanisms whereby an active substance can be lost before it reaches the effective receptor are: storage, elimination, and chemical in-activation (Fig. 3.3, p. 62). Veldstra referred to the loci where these processes occur as 'sites of loss' and suggested that the well-known synergistic action of biologically inert substances is actually a blocking of such sites, which allows a higher concentration of the drug to reach the receptor (Veldstra, 1956a).

3.4.1 *Storage*

Three important storage sites of loss are lipids (for neutral substances), ribo-nucleic acid and α_1 acid glycoprotein* (for cations), and albumin (for anions). Body-fat, a lipid, stores drugs of high liposolubility, e.g. thiobarbiturates [such as thiopental ('Pentothal')], and dibenamine. This appears to be a simple lipid/water partition effect (Brodie and Hogben, 1957). Plant-growth acceler-ators, such as α-naphthylacetic acid, are taken up by the fatty reserves of pea shoots; these sites of loss can be blocked with a biologically inactive, but more liposoluble, analogue such as decahydronaphthylacetic acid (Veldstra, 1956b).

Ribonucleic acid combines with flat, well-ionized cationic drug molecules. Thus the principal storage of mepacrine (*6.10*) after intravenous injection is in the nuclei of capillaries where this anti-malarial does no harm and is available for replenishing the blood level (Hecht, 1936). Chondroitin, acid glycoprotein, and other anionic biopolymers also store cations.

Several proteins in the mammalian bloodstream are capable of binding drugs, but albumin is by far the most effective of these. Neither fibrinogen nor the γ-globulins combine with drugs; and α- and β-globulins are usually enzymes whose affinity is almost confined to their substrates (although β_1-globulin combines with iron, zinc, and copper). Lipoproteins, related to the

*This glycoprotein, in blood-plasma, binds 70% of all propranolol in the blood (Sager, Nilsen and Jacobsen, 1979).

Table 3.3 Binding of drugs by serum albumin

| Animal species | Percentage unbound | | | |
	Benzylpenicillin	Cloxacillin	Sulfadiazine	Sulfisoxazole
Man	49	7	67	16
Horse	59	30	—	—
Rabbit	65	22	45	18
Rat	—	—	55	16
Mouse	—	—	93	69

(Penicillins: Rolinson and Sutherland, 1965; sulfonamides: Anton, 1960.)

globulins, combine with steroid hormones (lipid–lipid attraction). However, the substances enumerated are natural metabolites, and suramin seems to be the only drug that is bound by a globulin.

Serum albumin, on the other hand, is a storage site for many drugs, most of which are weak acids. Table 3.3 shows how this affinity varies not only from species to species, but also among two series of chemically related substances. Man, who binds drugs by serum albumin more strongly than other mammals do, usually metabolizes drugs less readily than other mammals.

Human serum albumin, of which blood contains about 4%, has a mol. wt. of 69 000, and possesses 109 cationic and 120 anionic groups. At pH 7.3, although it contains a net negative charge, one of the cationic groups must be particularly accessible because this albumin binds mainly anions, and in a 1:1 molar ratio. Examples of anionic drugs that have been shown to bind at a single site of high affinity on serum albumin are: sulfonamides, pyrazolone analgesics, thiazide diuretics, and penicillins (Phillips et $al.$, 1970). Arginine seems to be the binding site (Jonas and Weber, 1971). The binding of anions to serum albumin increases with increasing aromatic nature in the drug bound. The presumed mode of binding is by van der Waals forces between the benzenoid rings of the aromatic molecules and the flat ring of the solitary tryptophan residue in the albumin, because the fluorescence arising from tryptophan is quenched by binding to, for example, warfarin (*9.40*) or dicoumarol (*9.35*) (Chignell, 1970). This binding increases as the partition coefficient is increased by the drug designer. Thus the relationship for penicillin, as measured by the affinity constant (K) is $\log K = 1.32 \log P + 0.37$, and that for the sulfonamide anti-bacterials is $\log K = 1.15 \log P + 1.23$ (Scholtan, 1968, 1978). For a series of penicillins that show this striking increase, see Bird and Marshall (1967). Binding follows the law of mass action.

The ratio (r) of the number of molecules of drug bound to each molecule of protein is given by the equation:

$$r = \frac{nKC}{1 + KC}$$

where n is the number of binding sites on each molecule of protein, C is the concentration of the drug, and K is the affinity constant. This equation can be rearranged to:

$$r/C = nK - rK$$

which enables construction of a graph of r/C against r, to give a straight line with a slope of $-K$. From this plot, r equals n when r/C equals zero (Barlow, 1980).

Often it is more convenient to work with the dissociation constant (K_D) which is $1/K$. This varies from 900 for sulfadiazine, which is poorly bound, to 11 for sulfadimethoxine ('Madribon', see Section 9.3), which is almost too well bound to be a useful drug. Unless more than 90% of a drug is bound by albumin, the renal clearance is not slowed, and the serum protein acts as a depot and not a site of loss.

In practice, the concentration of the free drug is obtained by ultrafiltration, equilibrium dialysis, or by measurement of the ultraviolet absorption spectrum which is often displaced by the binding. The use of radioactively labelled drugs increases the precision of some of these techniques.

The clinical effect of a drug is increased, sometimes dangerously, if a second drug displaces the first one from serum albumin (Oliver *et al.*, 1963). Thus aspirin chases phenylindanedione (a frequently prescribed anti-coagulant) out of its store in serum albumin and often precipitates a crisis of bleeding.

The percentage (of a drug) bound to serum albumin decreases as the concentration in the plasma is increased, i.e. as the albumin store becomes progressively more saturated, there is difficulty in loading it further. For this reason it is desirable, when quoting the percentage bound, to give the total drug concentration (bound and unbound).

The drugs most readily bound by serum albumin are aliphatic acids, aromatic acids (including salicylic acid, and related anti-inflammatory agents), sulfonamides, and barbiturates. The iodinated anions used as X-ray contrast agents (e.g. iopanoic acid) are also well bound. Neutral substances with a strong, localized, negative polar charge combine with serum albumin: naphthaquinones, coumarins [e.g. warfarin (9.40)], indanediones, lactones including the cardiac glycosides, and porphyrins. Among the many simple drugs and metabolites *not* bound by albumin are ether, glucose, and urea.

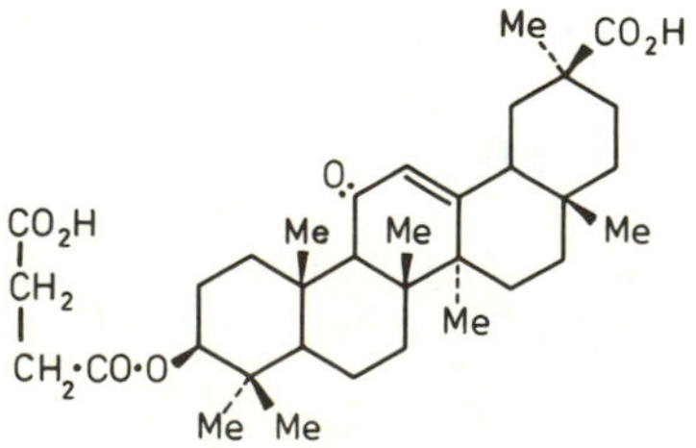

Carbenoxolone
(3.4)

The rather lipophilic triterpenoid drug carbenoxolone (*3.4*), which heals peptic ulcers, has a very high affinity for serum albumin (K 10^{-7} mol/l).

For further reading on the binding of agents to the proteins of blood, see Anton and Solomon (1973).

3.4.2 *Transfer and elimination*

In a great many cases the level of free drug in the tissue fluids eventually becomes the same as that in the plasma. This relationship is convenient because the plasma is more accessible for analysis. Among examples investigated are the anti-bacterial sulfonamide, sulfadoxine ('Fanasil') in the intraperitoneal cavity of rats after oral injection (McQueen, 1968), also various penicillins in lymph of dogs. The literature of this correlation has been reviewed by Robinson (1966).

Many hydrophilic drugs are excreted unchanged by the human body (see under kidney and bile in Section 3.2), but much excretion occurs only after metabolic alteration, as discussed in the next section.

Radioisotope labelling is the preferred method for locating and measuring drugs after administration, and for following their fate in the test animal or human subject. The products are usually isolated by high-performance liquid chromatography followed by mass spectroscopy. The electrophoresis of urine can be useful. After isolation, they can be measured by scintillation-counting, or by scanning, or autoradiography. The most used label is carbon-14, next comes the less expensive tritium, and sometimes 'double labelling' with both isotopes is practised. In planning a radiochemical synthesis, the first consideration is to choose the most useful labelling position in the molecule. Then the specific activity must be decided, according to the experimental dose levels planned. Carbon-13 (natural abundance spectroscopy) offers an easier, though less specific, approach.

For further reading on the storage and elimination of drugs, see Brodie, Gillette and Ackerman (1971).

3.5 Metabolic change as an early step in excretion. Synergism and antagonism

3.5.1 *Metabolism*

Metabolic alteration of drugs involves the making and breaking of covalent bonds (defined in Section 8.0) and hence is seldom reversible. Although many hydrophilic drugs are excreted unchanged by the mammalian body, others are 'conjugated', i.e. they are joined to small metabolites to assist excretion. These changes almost always increase the hydrophilic properties of xenobiotics (foreign substances), and this aids their excretion. Organic acids, for instance, which are too weak to be ionized at the pH of urine, and which therefore would be difficult to eliminate, are conjugated with glycine in the mitochondria of the liver. Thus, benzoic acid gives *N*-benzoylglycine, a type of substance that the kidney can readily excrete. Some amines become *N*-acetylated by acetyl-coenzyme A in the liver cytosol (cell sap). It is unfortunate when the acetylated substance is a drug, because some human beings are fast acetylators, the others slow; what is acetylated has no more therapeutic effect (see sulfonamides in Section 3.7 and, particularly, isoniazid in Section 9.9). Fortunately, most amines are conjugated with glucuronic or sulfuric acid, a more regular process. Thus *p*-acetamidophenol (paracetamol, acetaminophen) (*3.5*), the much-used headache remedy, is changed partly to the glucuronide (*3.6*), and partly to the sulfate ester (*3.7*). Glucuronides are formed in the endoplasmic reticulum (*e.r.*) of liver, kidney, and gut. Provided that they are liposoluble enough to get into the *e.r.*, amines, alcohols, phenols, amides, and carboxylic acids can be converted to glucuronides. Sulfation takes place in the liver cytosol (for sterols), and also in that of the kidney and gut cells (for hydrophilic alcohols and phenols).

NH·COMe	NH·COMe	NH·COMe
OH		O·SO$_2$·OH
p-Acetamidophenol (*3.5*)	Glucuronide of (*3.5*) (*3.6*)	Sulfate ester of (*3.5*) (*3.7*)

The glucuronides of amines and phenols are easily hydrolysed by the β-glucuronidase of the gut and bladder. Hence to a certain extent, the formation of glucuronides can become a step in a recycling mechanism. Thus *o*-amino-phenols, which are carcinogenic, are converted by the liver into innocuous glucuronides. However, this can work to our disadvantage: if β-glucuronidase activity is high in the bladder, the aminophenol will be re-formed and the risk of cancer is increased (Boyland, Wallace and Williams, 1955).

Drugs which are more lipophilic than the above examples are resorbed by the renal tubules. If they were not submitted to metabolic degradation, a single dose could remain in the body for many weeks. However, they are usually concentrated in a membranous organelle in the liver. This organelle, the *endoplasmic reticulum*, contains many kinds of scavenging enzymes, which alter them chemically so that they become more hydrophilic. Each altered drug is then excreted, either as such or (if it acquired a hydroxy-, carboxy-, or primary amino-group) in one of the conjugated forms favoured by hydrophilic drugs. For example toluene is oxidized by the *e.r.* to benzyl alcohol, the oxidation of which is continued in the cytoplasm to give benzoic acid which, as mentioned earlier, is then conjugated with glycine in the mitochondria, and the resultant benzoylglycine (hippuric acid) is rapidly eliminated in the urine.

Although this example shows that *e.r.* is not the sole site of metabolic degradation, this organelle is by far the most versatile. It can be separated from liver by differential ultracentrifugation and is similarly freed from the neigh-

bouring ribosomes. During the course of much purification, the *e.r.* is broken up
into spherules ('microsomes') without loss of enzyme activity. These enzymes,
which are numerous, are mainly oxidative, but a few of them perform reduc-
tions, hydrolyses, and at least one synthesis (Fouts, 1962; Gillette and Mitchell,
1975.)

That the *e.r.* of the liver contains a set of enzymes which can transform many
chemical substances into more hydrophilic (and therefore more excretable)
derivatives, stems from the observation of Axelrod (1955) that rabbits convert
amphetamine (*9.44*) to phenylacetone and ammonia. He found that the site of
this reaction lay in the *e.r.*, and it needed NADPH (hydrogenated nicotinamide
adenine dinucleotide) and molecular oxygen. Many other examples of microsomal metabolism were then discovered by Brodie and his associates (Brodie,
Gillette and LaDu, 1958). It was soon found that these microsomal enzymes
regulated how long a drug remained in a patient (Quinn, Axelrod and Brodie,
1958). Later it was found that most of these enzymes worked through a new kind
of cytochrome which was named '*P*-450' from the wavelength of its principal
optical absorption line (Omura *et al.*, 1965). Today we recognize that *P*-450 is a
set of isoenzymes, separable by electrophoresis. Some authors claim as many as
eight of these, and some of them are inducible by chemicals of widely differing
types, such as phenobarbitone, 2-naphthoflavone, cyanopregnenolone, and
isosafrole (Guengerich *et al.*, 1982; Ryan *et al.*, 1982).

Cytochrome *P*-448, whose normal function may lie in the oxidative biosynthesis of steroids, is induced further by some potential carcinogens such as
2-acetamidofluorene, 3-methylcholanthrene, and cigarette smoke. This cytochrome has a selective binding site that attacks hindered areas inaccessible to
P-450. It converts some potential carcinogens (such as benzo[*a*]pyrene and
1,2,5,6-dibenzanthracene) to ultimate carcinogens (Sect. 13.5). *P*-448 predominates over *P*-450 in malignant tissues (Ioannides, Lum and Parke, 1983).
For more on *P*-450, see Schenkman and Kupfer (1982).

The following typical oxidative processes are performed by the microsomal
enzymes; there is at least one enzyme for each process:

(i) Aliphatic *C*-hydroxylation ($R \cdot CH_3 \rightarrow R \cdot CH_2OH$), for which the side-chains
of barbiturates are common substrates.

(ii) Aromatic *C*-hydroxylation, e.g. the conversion of acetanilide to *p*-
hydroxyacetanilide. This aryl-hydrocarbon hydroxylase detoxifies some
hydrocarbons by oxidation to phenols, but converts others to carcinogenic
epoxides.

(iii) *N*-Oxidation ($R_3N \rightarrow R_3NO$), for which both aliphatic and aromatic tertiary
amines are good substrates.

(iv) *S*-Oxidation ($R_2S \rightarrow R_2SO$), as in the oxidation of chlorpromazine.

(v) *O*- and *S*-Dealkylations (e.g. $ROC_2H_5 \rightarrow ROH + CH_3 \cdot CHO$), for which phenacetin is a well-known substrate.

(vi) *N*-Dealkylation ($RNH \cdot CH_3 \rightarrow R \cdot NH_2 + H \cdot CHO$), as in the conversion of
methylaniline to aniline.

(vii) Deamination ($R \cdot CH(NH_2) \cdot CH_3 \rightarrow R \cdot CO \cdot CH_3 + NH_3$), as in the metabolism of the side-chain of amphetamine.

The *e.r.* can perform yet other reactions such as dechlorination of chlorinated aliphatic hydrocarbons. This can be either oxidative, in which case the products are ketones, or reductive, as with carbon tetrachloride which produces the highly toxic free radical: $CH_2^{\cdot \cdot}$ (Salmon, Jones and Mackrodt, 1981). Moreover, the *e.r.* carry at least two reducing enzymes: a nitro-reductase and an azo-reductase, both of which produce primary amines.

In the liver, many halogenated hydrocarbons, both aliphatic and aromatic, condense with the —SH group of glutathione, with liberation of hydrogen halide. The conjugate produced from reaction with bromobenzene (*3.8*) for example, is degraded to the arylcysteine (*3.9*) which is excreted either as such or, more often, acetylated to the 'phenylmercapturic acid' (*3.10*). This reaction occurs not only in the *e.r.*, but also in the cytosol and mitochondria (Hirom and Millburn, 1981).

Bromobenzene   S-Phenylcysteine   Phenylmercapturic acid
   (*3.8*)   (*3.9*)   (*3.10*)

Brodie (1956) has convincingly argued that the enzymes of the *e.r.* exist for degradation of toxic substances normally occurring in food or produced by bacterial decomposition in the gut. In addition, they have a normal role to play in steroid metabolism, particularly in the hydroxylative destruction of such hormones as estradiol, testosterone, progesterone, and the corticoids (Conney *et al.*, 1968). The *e.r.* enzymes are not very demanding about the structure of their substrates and hence are capable of attacking drugs not previously encountered. These *e.r.* enzymes, by their requirement for NADP, stand apart from the many NAD-requiring enzymes that the body's intermediary metabolism uses in its stepwise conversions of nutriment into energy. Conversely, the *e.r.* enzymes attack neither the raw materials nor the products of intermediary metabolism, partly because such substances are too hydrophilic to penetrate into the *e.r.*

The *e.r.* enzymes can be selectively inhibited, e.g. 6-aminochrysene inhibits the *N*-demethylation enzyme while increasing the activity of the hydroxylating and *O*-demethylating enzymes (Russo *et al.*, 1976). The *e.r.* drug-metabolizing enzymes are largely, but not entirely, confined to the liver. At least the *e.r.* of lung tissue similarly degrades inhaled foreign substances (Matsubara, Nakamura and Tochino, 1975). The destructive effects of *e.r.* enzymes can be avoided by inserting a group that makes the drug a poor substrate. Thus, insertion of ethinyl (—C⋮CH) into the 17-position of estradiol prolongs the contraceptive action from a minute to a day.

Metabolic alteration of foreign substances has often been called 'detoxification', but examples are known where the product of an *e.r.* enzyme is more

toxic than the substrate. These unfortunate instances will be discussed in
Section 3.6 (p. 106).

For a review of microsomal oxidation and reduction, see Gillette *et al.* (1969)
and Gillette and Mitchell (1975).

For most experiments on the enzymes of the *e.r.*, the source has been rat liver.
However, it has been shown that human *e.r.* enzymes are qualitatively similar to
those of the rat, but act at different rates, some faster and some slower
(Kuntzman *et al,.* 1966). It has been suggested that the large differences in
effective dosage that exist between man and laboratory animals depend more on
such species differences in the rate of destruction than on any species differences
in the sensitivity of target organs. Hence a given pharmacological effect should
appear at a similar bloodlevel in all mammals, even though the doses required
to produce this level are known to vary greatly from one species to another
(Brodie, 1964). Table 3.4 supports this argument; it shows some quantitative
data on carisoprodol (*3.11*), a muscle relaxant. However, too little information
is available to say how widely the above hypothesis is valid.

Table 3.4 Species differences and similarities
in the action of a drug, carisoprodol (*3.11*)
given intraperitoneally (0.2 g/kg)

Species	Duration of action (loss of righting reflex) (h)	Plasma level on recovery (μg/ml)
Cat	10	125
Rabbit	5	100
Rat	1.5	125
Mouse	0.2	130

(Brodie, 1964.)

Exceptions to the hypothesis obviously exist in those cases, possibly rare,
when the major pathway of metabolism is not the same in two species. Thus the
mouse simply hydrolyses 6-propylthiopurine (*3.13*) to 6-mercaptopurine (*3.14*),
and hence it has excellent carcinostatic properties in this animal. Man, on the
contrary, oxidizes the drug in two places, without hydrolysis, and the product
(*3.12*) is not carcinostatic (Elion *et al.*, 1963).

The possibility of species variations is recognized also in the narrower field of
conjugation. Phenylacetic acid, the classic example of divergent paths, is conjugated with glutamine in Man and chimpanzee (only), with glycine and with
glucuronic acids in most other mammals, and with ornithine in the hen
(Williams, 1959). Again, amphetamine is metabolized in the rat by *para*-
hydroxylation, but by deamination in Man, monkey, and guinea pig (Dring,
Smith and Williams, 1970).

Other drug changes that take place in the liver cytosol (not the *e.r.*) include

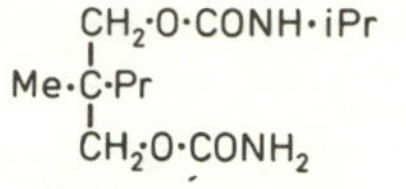

$$CH_2 \cdot O \cdot CONH \cdot iPr$$
$$Me \cdot C \cdot Pr$$
$$CH_2 \cdot O \cdot CONH_2$$

Carisoprodol
(3.11)

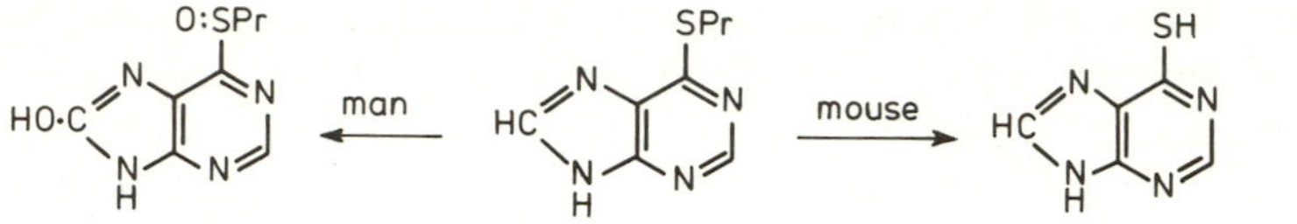

8-Hydroxy-6-propyl-sulfinylpurine
(3.12)

⟵ man ⟵

6-Propylthiopurine
(3.13)

⟶ mouse ⟶

6-Mercaptopurine
(3.14)

$$Pr \cdot C \cdot CO \cdot OCH_2 CH_2 \cdot NEt_2$$
with Ph above and Ph below

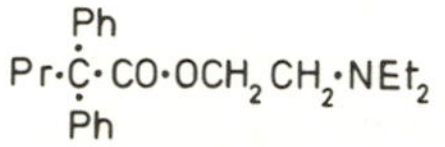

SKF 525-A
(3.15)

the oxidation of cyclohexanes to benzenes, of alcohols to aldehydes, and both the oxidation and reduction of aldehydes, also conjugation of bile acids (and aryl- and aryloxy-acetic acids) with taurine (2-aminoethanesulfonic acid), a reaction which is particularly prominent in canines (Caldwell, 1982).

Although the liver is the body's principal centre for metabolic degradation, this process also goes on elsewhere. For example the bloodstream carries non-specific esterases which rapidly hydrolyse almost every kind of ester, natural or fabricated, and they are accompanied by a rather less active non-specific amidase. Many degradative enzymes operate in the bowel-wall and the kidney. An inactivator of catecholamines, catechol O-methyltransferase, is found in many tissues, and Section 12.5 describes the enzyme that hydrolyses acetylcholine in the neuromuscular junction.

Whereas most metabolic change in the human body increases the hydrophilicity of drugs, there are a few cases, admittedly rare, where this property is decreased. Thus, drugs given to lower the lipid content of the bloodstream, if of the type of 4-benzyloxybenzoic acid, unite with Coenzyme A in the liver cytosol. The products, all thioesters, are converted by the enzyme, diglyceride acetyltransferase, into triglycerides which are simply stored in the body's fat (Fears and Richards, 1981).

In industry, it is becoming customary to follow the metabolic degradation of all candidate drugs and pesticides (see end of Section 3.4). The pathway and products of metabolic breakdown are followed and quantified. The toxicity of each isolated product is independently assessed. See Section 4.8 for more data on species differences.

For further reading: The metabolic alteration of foreign substances is a highly specialized topic with an enormous literature, of which the following give a modest cross-section.

Books: Williams (1959), Gillette and Mitchell (1975), Goodwin (1976), Jenner and Testa (1981), Parke and Smith (1976), Aito (1978); *Bound Periodicals* (with year of commencement): *Foreign Compound Metabolism in Mammals* (London: The Chemical Society, 1970 <), *Drug Metabolism Reviews* (1972 <), *Progress in Drug Metabolism* (1975 <), *Enzymatic Basis of Detoxification* (1980 <); *Journals: Xenobiotica* (1971 <), *Chemico-Biological Interactions* (1969 <), *Drug Metabolism and Disposition* (1973 <), and *European Journal of Drug Metabolism and Pharmacokinetics* (1976 <).

3.5.2 *Synergism*

Synergism is the phenomenon where two agents, applied simultaneously, have more effect than is shown by the sum of each agent acting alone. The frequent dependence of synergism on one agent's preserving the other from destruction, was brought to light in the 1950s by the candidate drug proadifen, which is the diphenylpropylacetic ester of diethylaminoethanol (*3.15*) and is often referred to as SKF 525-A. This substance can synergize the action of a wide variety of drugs by preventing their metabolism in the *e.r.* It seems to exert this effect, not by making the membrane of the *e.r.* impermeable to lipophilic drugs, but by non-competitive inhibition of all hydroxylation reactions and by competitive inhibition of hydrolytic reactions (Gillette, 1966). These actions of SKF 525-A are examples of a very common type of synergism, namely, blocking sites of loss (Veldstra, 1956). It is not used medicinally.

Metabolic inactivation, whether taking place in the *e.r.* or at other sites, is often accidentally inhibited by other drugs. Thus many patients have died as a result of the simultaneous administration of an inhibitor of monoamine oxidase (an enzyme present in mitochondria) and an amine drug which is not toxic on its own. These monoamine oxidase inhibitors, such as tranylcypromine (*9.47*), are prescribed as mood-elevators in depressive illnesses. Until their synergistic properties were realized, they caused many deaths after usually safe doses of amphetamine, pethidine, and amitriptyline, or after the patient had consumed food rich in 'pressor amines' such as tyramine; such foods are red wine, meat-extract, yeast-extract, broad beans, and particularly cheese. These are examples of *unfortunate* synergism, but many *favourable* cases are known, examples of which will now be given.

Loss by elimination can sometimes be blocked by an analogue of similar charge type. Thus the penicillins belong to the class of moderately liposoluble acids which get facilitated transport through the proximal tubules of the kidney. This elimination can be largely blocked by physiologically inert substances of similar physical properties, such as probenecid (*3.16*), 4-dipropylamino-sulfonylbenzoic acid. This substance is used clinically to increase the action of penicillins.

Loss by enzymic destruction can be overcome through use of a synergist. Thus pyrethrins in fly sprays are commonly formulated with a methylene-

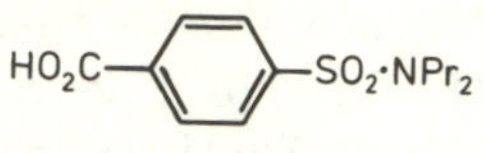

$$HO_2C\!-\!\langle\ \rangle\!-\!SO_2 \cdot NPr_2$$

Probenecid
3.16

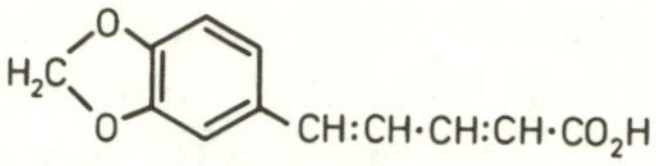

$$H_2C\big\langle^O_O\big\rangle \text{(benzene ring)} CH{:}CH \cdot CH{:}CH \cdot CO_2H$$

Piperic acid
(3.17)

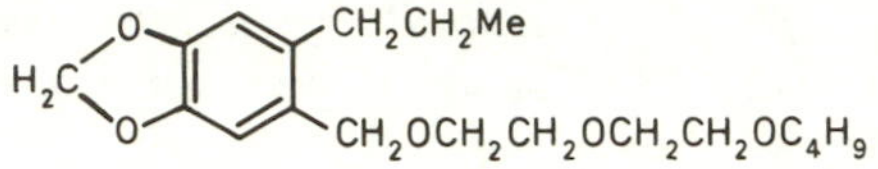

$$H_2C\big\langle^O_O\big\rangle \text{(benzene ring)} ^{CH_2CH_2Me}_{CH_2OCH_2CH_2OCH_2CH_2OC_4H_9}$$

Piperonyl butoxide
(3.18)

dioxybenzene synergist, often derived from piperic acid (*3.17*). One of the most used synergists of this class is piperonyl butoxide (*3.18*) but some quite simple methylenedioxy-derivatives of benzene (e.g. safrole) show this effect. Metabolites of these compounds are bound to cytochrome *P*-450 (the terminal oxidase of the microsomal electron-transport system) which is thus hindered in its customary destruction of pyrethrins, carbamates, and organic phosphates (Franklin, 1972). To avoid the use of a synergist, a blocking group can be inserted into the agent to hinder enzymic degradation: this has been done by designers of longer-acting steroid drugs in which a trifluoromethyl, or even a strategically placed methyl group, has been inserted (Ringold, 1961; Briggs and Christie, 1977).

Apart from the synergism that arises from blocking sites of loss, some other types are known. The first of these is sequential blocking (the inhibition of two or more consecutive metabolic processes) which is dealt with in Section 9.6. Another type of synergism is the use of two or more drugs to retard the growth of bacterial mutants, based on observations that a mutant resistant to one drug does not easily undergo further mutation to give a resistant strain. It is for this reason that a second anti-tubercular substance is always included in the isoniazid therapy of tuberculosis (see Section 6.5 for discussion of drug-resistance). Synergism of a different type is shown by penicillin and streptomycin. The inhibition of cell-wall synthesis, brought about by penicillin, mechanically weakens the plasma membrane, allowing easier access of the aminoglycoside.

Resembling synergism in its effects is the genetically determined lack (in an individual or a race) of a detoxifying enzyme, as a result of which a patient reacts to a small dose of a drug as though it were a large one (see Section 9.9). This phenomenon differs from gradual sensitization to a particular drug, which is a immune response.

3.5.3 Antagonism

Examples of unintentional overdosage were described above; in these a drug blocks an enzyme that normally detoxifies a second drug taken at the same time (e.g. phenelzine taken with the opioid pethidine). In contrast to this type of mishap a patient may be subjected to underdosage through a drug-induced induction of *e.r.* enzymes (Conney and Burns, 1962). The anti-rheumatic drug phenylbutazone (*10.37*) is one of several drugs known to induce the excessive production of these enzymes, so that a fixed daily dose eventually produces an ever-decreasing effect; this is a consequence of the faster rate of destruction. Thus, 25 hours after a dose of 0.1 g/kg of phenylbutazone, the plasma of a dog showed a concentration of 100 μg/ml; but after five consecutive daily doses, the level of the drug had fallen to 15 μg/ml. If medication is suspended for a week or more, the patient can regain good use of the drug.

Similarly, successive doses of various barbiturates in mice and rats produced shorter and shorter periods of sleep. A similar induction of the barbiturate-destroying enzyme occurs in Man, usually within a week of the patient's beginning to take a small, nightly dose. If the patient increases the dose, the accustomed sedation will return, but only for a few nights, because the induction of the barbiturate-destroying enzyme in the *e.r.* will have escalated. If he continues to increase the dose, habituation and withdrawal symptoms are to be expected. On the other hand, the patient may discontinue the drug as soon as he feels the first decrease in action. After one or two weeks, the excess of enzyme will have disappeared and the original potency of the drug will be available. During the period of abstinence, the patient is at no disadvantage because withdrawal symptoms do not occur at low doses of barbiturates.

That an inducing drug actually increases the amount of destructive enzyme in the *e.r.* has often been shown, e.g. by administering an azo-dye to laboratory animals for several days; then, when excretion of the dye was sharply diminished, hepatic *e.r.* was isolated, and the relevant enzyme found, by assay, to have increased greatly (Porter and Bruni, 1959). In one experiment in dogs, the amount of enzyme did not return to normal until 10 weeks had elapsed.

Examples of other substances which stimulate their own metabolic destruction are chlorcyclizine, probenecid, tolbutamide, aminopyrine, meprobamate, glutethimide, chlorpromazine, chlordiazepoxide, methoxyflurane, 3,4-benzpyrene, and DDT.

Moreover, heavy dosage with a drug can induce increased production of an enzyme capable of destroying a different drug introduced into the dosage scheme simultaneously, or many days later (Remmer, 1962). For example phenylbutazone, also barbiturates, speed up the metabolism of the coumarin anti-coagulants in Man. Hence a patient on anti-coagulant therapy may be worse off if (as often happens) a barbiturate is also prescribed. For example a patient, who was taking 75 mg daily of dicoumarol (alone) and was later given 60 mg daily of phenobarbitone as well, showed a large decrease in plasma level of the coumarin drug, and the anti-coagulant power fell. Yet soon after the phenobarbitone was discontinued, the coumarin drug regained its former level and so did the prothrombin time (Cucinell *et al.*, 1965; cf. Robinson and MacDonald, 1966). Examples of pairs of drugs, the first of which can accelerate the metabolic destruction of the other *in Man*, are: phenobarbitone and diphenylhydantoin, phenylbutazone and aminopyrine, and phenobarbitone and digitoxin. Similarly, administration of the following drugs has been found to accelerate destruction of steroid hormones by the *e.r.*: phenobarbitone, chlorcyclizine, and phenylbutazone (Conney, 1967).

Some of the most powerful inducers of these drug-destroying enzymes are found among the chlorinated insecticides. A small dose of DDT or benzene hexachloride can make laboratory animals highly resistant to the effects of other agents. Hence such insecticides must not be used on animals when drug tests are to be carried out on them. Moreover chlorinated insecticides increase the metabolism of progesterone, estradiol, and testosterone.

The primary site of action of inductive antagonists such as phenobarbitone and 3-methylcholanthrene, is on the DNA core of RNA polymerase. By increasing the activity of this enzyme, more RNA is synthesized, and finally more of the metabolizing enzymes (Gelboin, Wortham and Wilson, 1967).

A well-known effect of DDT and other chlorinated hydrocarbons is weakening the bony structure of fish and causing birds to lay thin-shelled eggs. It is thought to proceed as follows. The chlorinated hydrocarbon induces excess production of P-450 cytochrome at the expense of ascorbic acid, deficiency of which leads to a decreased collagen content in bone. It is the high ratio of calcium phosphate to collagen that renders scorbutic bones so fragile. For a discussion of this ratio in fish exposed to (a) polychlorinated biphenyls, see Mehrle *et al.* (1982), and (b) toxaphene (chlorinated camphene), see Mayer, Mehrle and Crutcher (1978).

For more on the induction of *e.r.* enzymes, see Jenner and Testa (1981).

3.5.4 *Self-cancelling drugs*

A new development is the self-cancelling drug, one that is automatically degraded after use and leaves behind no product that is biologically active at the concentration attained. The novelty of this concept is that the drug does not depend on any enzyme for its destruction nor on any organ for its elimination. This is a special advantage when, through adverse pharmacogenetics (Section 9.9), the destructive enzyme is lacking, or when the kidneys or liver are not properly functional. Given the presence of water, a self-cancelling drug simply disintegrates at a rate that had been designed into its structure by consideration of Hammett sigma constants (Section 17.2).

An outstanding example is atracurium (*3.19*), a muscle-relaxant used in general anaesthesia. It is 2,2'-(3,11-dioxo-4,10-dioxatridecylene)bis-[6,7-dimethoxy-(3,4-dimethoxybenzyl)-2-methyl-1,2,3,4-tetrahydroisoquinolinium] cation, administered as its benzenesulfonate (Hughes and Chapple, 1981). It was devised by J. Stenlake at Strathclyde University (Scotland). An average dose, given intravenously in water, is adequate for 30 minutes of surgery, and can be renewed if necessary. Its action is non-depolarizing, as with tubocurarine

(*2.6*), and it is interesting to compare the two structural formulae. In water, it slowly undergoes a Hofmann degradation and produces two inert substances by fission at the dotted line in (*3.19*). In surgical practice, it has proved effective and non-toxic to the patient; recovery is fast (Payne and Hughes, 1981).

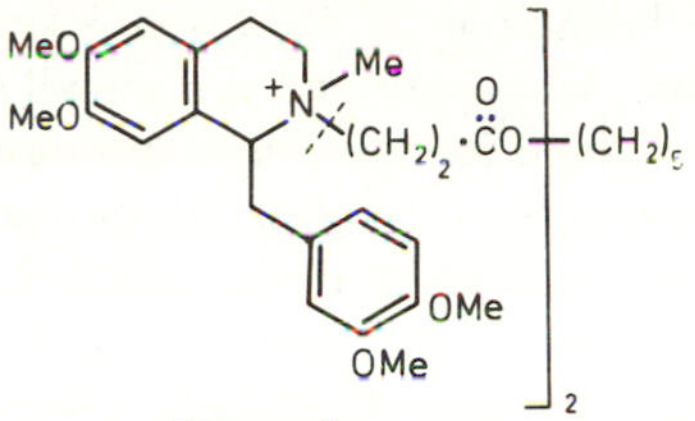

Atracurium
(*3.19*)

For patients with hyperhidrosis (uncontrollable sweating), N. Bodor and colleagues (1980) have devised the self-limiting drug (*3.20*). It has a muscarinic type of cholinergic action (Section 7.5) and, being a quaternary amine, is not absorbed through the skin. Chemically, it is 1-(α-cyclopentyl-α-phenyl-acetoxy)methyl-1-methylpyrrolidinium chloride. The quaternizing agent was an ester of chloromethanol, and the choice of cyclopentylphenylacetic acid as the esterifying acid was made after testing several other lipophilic acids. On the skin, it had a half-life of 20 hours. See, further, Bodor's review of self-limiting drugs (which he calls 'Soft Drugs') (Bodor, 1982).

An example of the new self-limiting anti-inflammatory drugs, is the methyl ester of prednisol-20-one-21-carboxylic acid (*3.21*). This is applied to the skin, e.g. in atopic dermatitis, but becomes hydrolysed when it diffuses out of the inflamed area into the bloodstream. Hence it does not show the undesirable systemic effects of earlier types of corticosteroids on prolonged use (Lee and Soliman, 1982).

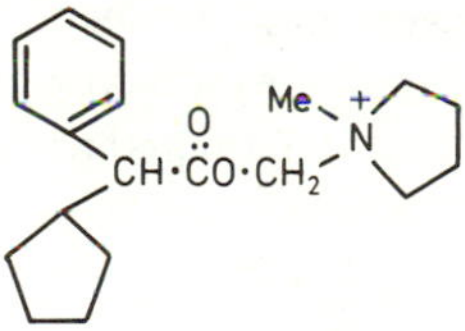

Topical anti-cholinergic drug
(*3.20*)

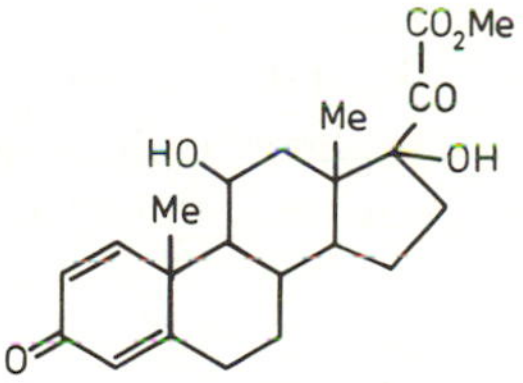

Methyl prednisol-20-one-21-oate
(*3.21*)

3.6 Metabolic changes as an early step in activation. Pro-drugs

As indicated in Fig. 3.3, it has sometimes happened that a substance thought to be a drug is really a pro-drug* which is converted to the actual agent after

---

*I apologize for having invented this term, now too widely used to alter, for literary purists tell me they would have preferred 'pre-drug'.

administration. In other cases, the decision to administer the drug in the form of a pro-drug is deliberate.

3.6.1   *A well-planned start*

The first pro-drug, methenamine (*3.22*), was introduced as far back as 1899 by the Berlin firm, Schering. It was devised as a source of formaldehyde liberated from it by the acidity of the urine, and hence was a potent urinary antiseptic. This substance, hexamethylenetetramine (hexamine) is a cage-shaped molecule, very easily made by mixing ammonia and formaldehyde. If taken before meals, it encounters no acid until it enters the urine, and so it passes blandly through the body up to that stage. Since that time, many other urinary antiseptics have been discovered, yet methenamine continues to hold an effective place in treatment and is included in 20 of the World's pharmacopoeias.

   In spite of this promising start, the introduction of pro-drugs took on a rather haphazard character until the 1960s when they came, once again, to be thoughtfully designed.

3.6.2 *The intermediate period*

Several pro-drugs were discovered accidentally. The anthracene glucoside purgatives had been used for centuries (in crude forms, as cascara, senna, rhubarb, etc.) before recognition that their aglycones (e.g. emodin) were the true active forms (Straub and Triendl, 1937). Castor oil, which acts on the bowel as ricinoleic acid, and sodium citrate, which, after partial combustion in the body to sodium bicarbonate, basifies the urine, are other early examples of pro-drugs.

   When acetanilide (*3.23*) was introduced as a mild analgesic and febrifuge in 1886, it was found to have a disadvantage, namely it turned the patients blue through promoting formation of methaemoglobinaemia. As soon as it was realized that acetanilide was only a pro-drug for *p*-acetamidophenol (*3.5*), other pro-drugs for this phenol were sought, and phenacetin (*3.24*) was introduced in 1887. Long experience with this analgesic showed that it damaged the kidneys,

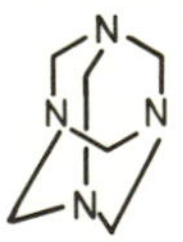

Methenamine
(hexamine)
(*3.22*)

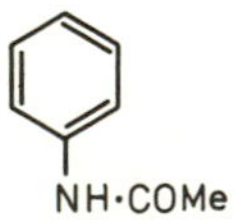

NH·COMe
Acetanilide
(*3.23*)

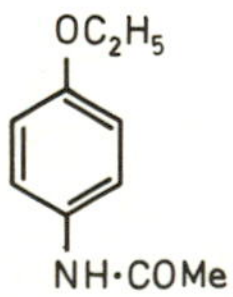

OC_2H_5

NH·COMe
Phenacetin
(*3.24*)

$CCl_3 \cdot CH(OH)_2$
Chloral hydrate
(*3.25*)

$CCl_3 \cdot CH_2OH$
Trichloroethanol
(*3.26*)

and it has been replaced, thanks to the work of Brodie and Axelrod (1949), by the true drug, *p*-acetamidophenol.

Chloral hydrate (*3.25*), another nineteenth century drug, has been shown to be reduced in the human body to trichloroethanol (*3.26*) (Butler, 1948): the well-known hypnotic action is shared by both substances.

Aspirin, one of the most heavily prescribed drugs of our day (4–8 g daily for arthritis), has a pro-drug background. The anti-febrile properties of willow tree bark (*Salix alba*), well-known to the ancients, were due to a glucoside called salicin. The patient's gastrointestinal tract hydrolysed this to glucose and salicyl alcohol; the latter underwent cytoplasmic oxidation to salicylic acid, the true drug. Following this train of thought, Buss (1875) introduced oral sodium salicylate, but this was found to be an irritant and became slowly superseded by aspirin (acetylsalicylic acid) put forward by Dreser in 1899.

Paul Ehrlich worked with organic arsenicals at three levels of oxidation, the phenylarsonic acids (pentavalent) as in (*3.27*), the phenylarsenoxides (*3.28*), which occupy the middle level, and the arsenobenzenes (*3.29*) which are at the lowest level of the three. He showed that the phenylarsonic acids were inactive until converted by the living cell to the corresponding arsenoxides (Ehrlich, 1909). That his famous arsenobenzene drug, arsphenamine ('Salvarsan') (*6.3*), was inactive until oxidized to the arsenoxide level was shown later, in the USA, by Carl Voegtlin (1925). This knowledge led to the arsenoxide oxidation product of arsphenamine replacing the latter in the treatment of syphilis, because the patient could be cured with much smaller doses, thus increasing the margin of safety (Tatum and Cooper, 1934). This arsenoxide, called oxophenarsine (*6.4*) ('Mapharsen'), was used increasingly from 1935 until arsenicals were totally replaced by penicillin for this disease, in the 1940s. In this case the true drug (oxophenarsine) was found preferable to the pro-drug (arsphenamine). However, in current treatment of trypanosomiasis in the central nervous system, an arsenical pro-drug, melarsoprol (*13.3*) has to be used because arsenoxides do not penetrate the blood–brain barrier. See Section 13.0 for more on arsenicals.

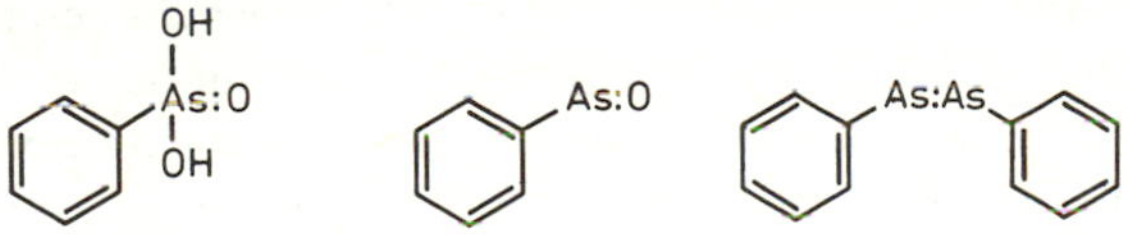

OH
|
As:O
|
OH

(*3.27*)   (*3.28*)   (*3.29*)

Three oxidation states of arsenic

Sulfachrysoidine ('Prontosil') (*3.30*), the first of the anti-bacterial sulfonamides, was thought to be the true drug when it was introduced into medicine in 1935. But workers in the Institut Pasteur were able to show, in the same year, that this substance was inactive, and established that the true drug was *p*-aminobenzenesulfonamide (sulfanilamide) (*3.31*) which was formed by reductive fission in the body (Tréfoüel *et al.*, 1935). Accordingly, sulfanilamide replaced 'Prontosil' in the clinic because it acted more promptly and directly. The reductive fission is effected by intestinal flora and by cells of the intestinal wall (Gingell and Bridges, 1973).

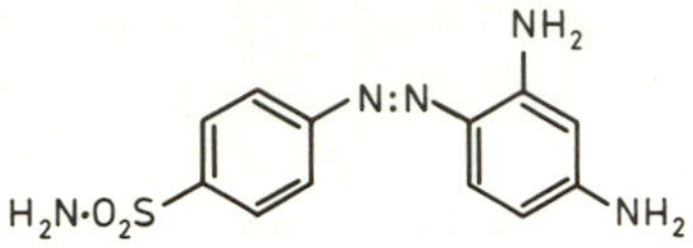

Sulfachrysoidine
'Prontosil'
(3.30)

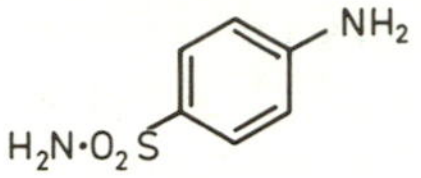

Sulfanilamide
(3.31)

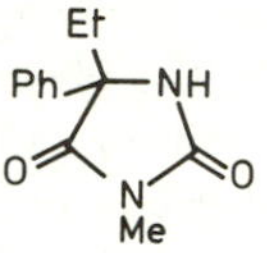

Mephenytoin
(3.32)

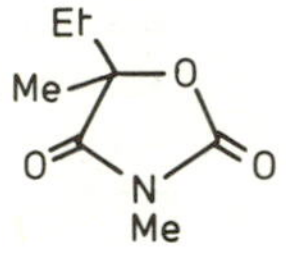

Paramethadione
(3.33)

Because *N*-demethylation takes place so readily in the *e.r.* of the liver, drug designers have sometimes incorporated an *N*-methyl-group to increase lipo-solubility, and hence absorption, the true drug then becomes available after a first pass of the blood through the liver (Butler, 1955). The pro-drugs of this type that have remained in the World's pharmacopoeias are mainly anti-epileptics. Thus mephobarbital (1-methylphenobarbitone) loses the 1-methyl-group and become phenobarbitone (*15.2*); methoin (mephenytoin) (*3.32*), which is 5-ethyl-3-methyl-5-phenylhydantoin, loses the 3-methyl-group to give nirvanol, a drug considered too toxic to give as such (Butler, 1953); 3,5,5-trimethyl-2,4-oxazolidinedione (troxidone, trimethadione) and 5-ethyl-3,5-dimethyl-2,4-oxa-zolidinedione (paramethadione) (*3.33*) both lose a 3-methyl-group during 'first pass' to give the true drug (Butler, Waddell and Poole, 1965).

Whenever the concentration of a drug in the plasma does not correlate with its therapeutic effect, it is very likely that the supposed drug was only a pro-drug. From such a clue it was established that the anti-malarial proguanil (*3.34*) acted only after cyclization in the body to the dihydrotriazine (cyclo-guanil) (*3.35*) (Crowther and Levi, 1953). The pro-drug is almost inactive against cultures of malarial parasites *in vitro*, whereas (*3.35*) is highly active. By similar means, it was found that the gametocidal anti-malarials derived from 8-amino-6-methoxyquinoline, such as primaquine (*3.36*), act only after de-methylation and oxidation to the corresponding 5,6-quinone (*3.37*) (cf. Smith, 1956). Cycloguanil (the drug) has completely replaced proguanil (the pro-drug) as an anti-malarial. It is noteworthy that the much used anti-malarials (chloroquine, pyrimethamine, and mepacrine) were introduced as the true drugs.

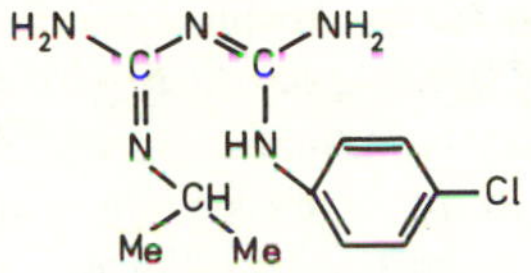

Proguanil
(3.34)

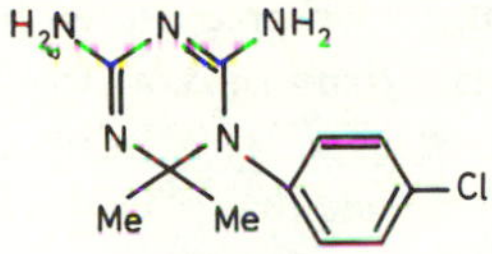

Cycloguanil
(3.35)

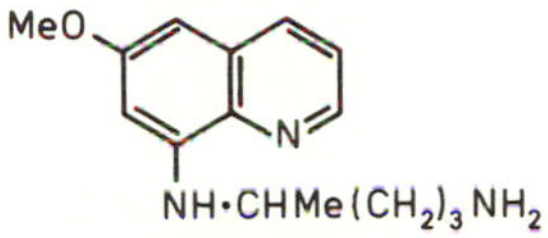

NH·CHMe(CH₂)₃NH₂
Primaquine
(3.36)

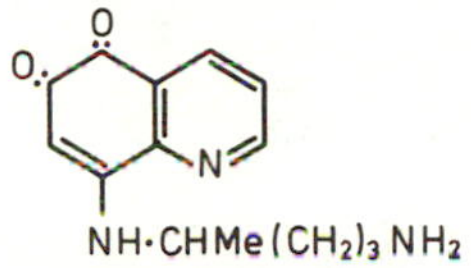

NH·CHMe(CH₂)₃NH₂
Quinone from (3.36)
(3.37)

3.6.3 Pro-drugs of the antibiotics

Esters of the 3-carboxylic acid group of ampicillin (Section 13.1) are used as pro-drugs, e.g. bacampicillin, talampicillin, and pivampicillin. Their more lipophilic properties ensure better oral absorption. The active drug is liberated by the non-specific esterases of the bloodstream. Esters of erythromycin (4.47), such as the stearate or ethyl succinate, survive the stomach's acidity, which erythromycin does not, but are hydrolysed back to the true drug in the duodenum. Children, who intensely dislike the bitter taste of chloramphenicol, are given the succinate ester. In both cases, hydrolysis liberates the true drug.

3.6.4 Pro-drugs in cancer therapy

Cyclophosphamide (3.38), the most selective anti-cancer drug of the nitrogen-mustard (or alkylating) class (Section 13.4), was designed as a pro-drug. This substance, much used in the clinic, is inert until metabolically converted, in several steps, to the active agent which is the aziridinium derivative of (3.39) (Connors et al., 1974; Foster et al., 1981). Tegafur (ftorafur) (3.40), the Russian-discovered anti-cancer drug which is also made and used in Japan, is an N-tetrahydrofuryl derivative of 5-fluorouracil (3.3) to which it is slowly broken down in muscle and liver. Given intravenously, it has the same spectrum of clinical usefulness as 5-fluorouracil (cancer of the breast, rectum, and colon) but is much better tolerated (Blokhina, Vozny and Garin, 1972; Valdivieso et al., 1976).

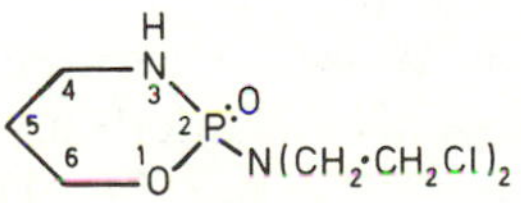

Cyclophosphamide
(3.38)

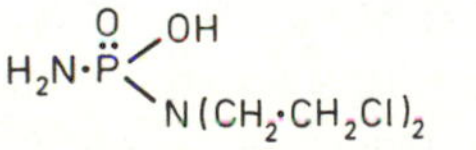

Phosphoramide mustard
(3.39)

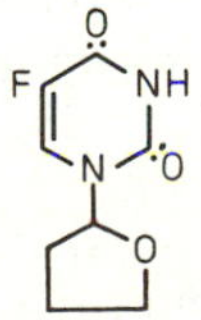

Tegafur
(3.40)

Experiments are in progress with anti-cancer pro-drugs which hydrolyse and liberate the true drug only at the lower pH (often about 6) which obtains in tumour cells (Bicker, 1974). In a completely different approach, nitrogen-mustard anti-cancer drugs were coupled to an antiserum against lymphoma cells. Powerful anti-cancer properties were evinced in the mice and it was suggested that the antibody operated a homing mechanism for the drug so that it could act at shorter range and in higher local concentration (Rowland, O'Neill and Davies, 1975).

3.6.5 *Miscellaneous examples*

6-Mercaptopurine (*3.14*) strongly suppresses the immune reaction in the human body and hence could be useful to secure survival donor grafts, but it is eliminated too fast for a sustained effect. Hence less easily eliminated derivatives were sought. The best of these was found to be azathioprine (6-1'-methyl-4'-nitroimidazol-5'-ylthiopurine) (*3.41*), which is slowly cleaved (non-enzymatically) in the body to 6-mercaptopurine and hence serves as a depot. The electron-attracting properties of the nitro-group give the desired lability to the $C-S$ bond (Elion, 1967).

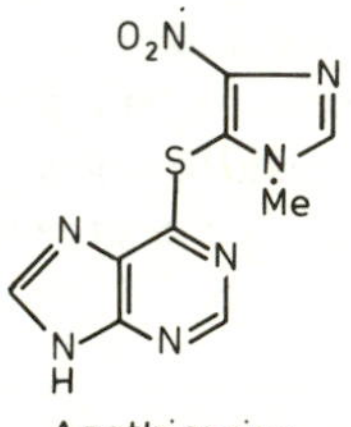

Azathioprine
(*3.41*)

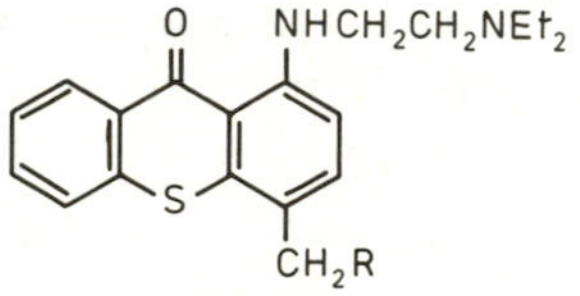

(a) Lucanthone (R = H)
(b) Hycanthone (R = OH)
(*3.42*)

Methyldopa (*12.50*), which is very much used for treating patients with very high blood-pressure, is decarboxylated and hydroxylated, in the body, to the true drug, α-methylnorepinephrine (*12.52*) (Iversen, 1967).

Benorylate, the aspirin ester of paracetamol (acetaminophen), is absorbed slowly from the gastrointestinal tract, and is rapidly hydrolysed to its two active constituents in the bloodstream. It is claimed that the prolonged absorption leads to prolonged analgesia.

An ingenious use of masking to overcome a problem in transport is the administration of 6-azauridine (see Section 4.0, p. 140) as its liposoluble 2',3',5'-triacetyl derivative, for oral medication. Whereas 6-azauridine is not taken up from the intestines, the triacetyl-derivative is absorbed, and then deacetylated to the true drug in the bloodstream (Welch, 1961). This drug, azaribine, provides one of the best known treatments for psoriasis (Crutchler and Moschella, 1975).

Metrifonate (*6.28*), an organic phosphate much used in treating schistosomiasis, is a pro-drug from which dichlorvos (13.29) is liberated as the active
form (Section 6.3.3). Lucanthone (*3.42a*), introduced in the 1940s for the
treatment of schistosomiasis, is without effect on the worms, *in vitro*. This led to
its replacement by hycanthone (*3.42b*), the true drug formed by metabolic
hydroxylation (see Section 4.0, p. 141).

Parkinson's disease is characterized by a deficiency of dopamine (*3.43a*) in the
brain. Although administration of this neurotransmitter is useless, because it
does not penetrate the blood–brain barrier, a very successful clinical procedure
is to give oral doses of the corresponding amino acid, levodopa (*3.43b*) (Cotzias,
Van Woert and Schiffer, 1967). The acid enters the brain on a specific carrier
and is decarboxylated there.

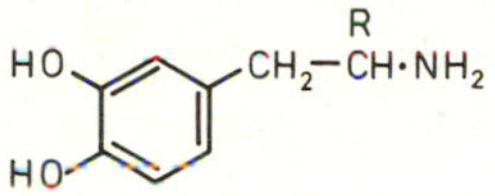

(a) Dopamine (R = H)
(b) Levodopa (R = CO_2H)
(3.43)

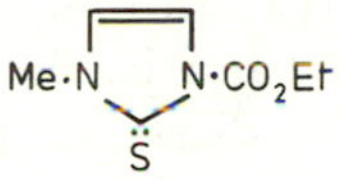

Carbimazole
(3.44)

Methimazole, 2-mercapto-1-methyl-imidazole, useful in thyrotoxicosis, is
usually given as the 2-ethoxycarbonyl-derivative (*3.44*) (carbimazole, 'Neo-
Mercazole') which forms a depot.

Pargylene (*3.45*) is used in the aversion therapy of alcoholism to inhibit
aldehyde dehydrogenase and so cause a highly unpleasant concentration of
acetaldehyde to build up in the bloodstream. Pargylene does not have this effect
directly, but only after the liver *e.r.* has metabolized it to propiolaldehyde (*3.46*)
which is the true drug (Shirota, DeMaster and Nagasawa, 1979).

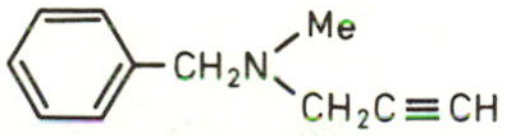

Pargylene
(3.45)

$OHC-C\equiv CH$
Metabolite of (3.45)
(3.46)

Several of the most-used benzodiazepine tranquillizers seem to act as the
metabolic product, nordazepam (*3.47*), which has the full therapeutic action of
the starting materials, and acts faster. Thus diazepam (*12.95*) ('Valium') loses
1-Me, chlordiazepoxide ('Librium') loses 4-*O* and hydrolyses 2-MeNH- to=*O*,
whereas clorazepate ('Tranxene') loses 4-CO_2, and in each case nordazepam is
the product. While the latter is available for prescribing, clinical experience
with the pro-drugs is greater; moreover they have very different pharmacokinetics, diazepam being absorbed quickly and chlordiazepoxide slowly from
the gastrointestinal tract. See further: Section 12.7; Garattini, Mussini and
Randall (1973), Nicholson *et al.* (1976), and Hollister (1978).

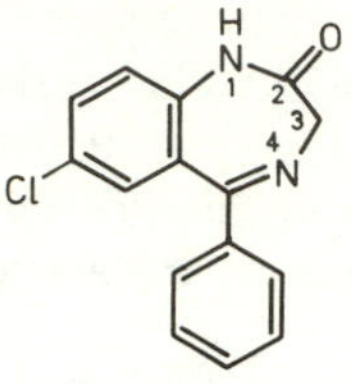

Nordazepam
(3.47)

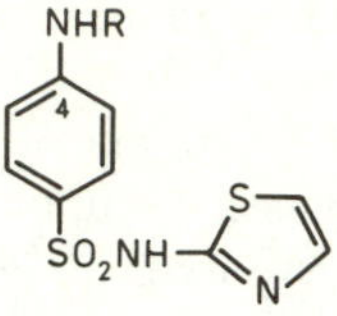

(a) Phthalyl-sulfathiazole R= $C_6H_4(CO_2H)CO-$
(b) Succinyl-sulfathiazole R= $HO_2C \cdot (CH_2)_2CO-$
(3.48)

N-4-acyl-derivatives of sulfathiazole have been much used in bacterial dysentery. Unabsorbed from the stomach and small intestine, they are hydrolysed by benign bacteria in the colon, liberating the anti-bacterial, sulfathiazole. The phthalyl-derivative (3.48a) proved more effective than the succinyl-derivative (3.48b) but, because of the widespread occurrence of sulfonamide-resistant intestinal bacteria, the following anti-bacterials are coming to be preferred for treating diarrhoea: trimethoprim, ampicillin, or tetracycline (for aerobes) and metronidazole (for the increasingly commoner anaerobes).

3.6.6 *Current trends in designing pro-drugs*

In recent years, imagination has been allowed to range more freely in planning pro-drugs, and new territories have opened up. A new era in the treatment of virus disease has begun with acyclovir (3.49), 9-[(2-hydroxyethoxy)-methyl]guanine, a pro-drug converted to the true drug at the level of the affected cell (rather than at that of the organ). This drug, effective against the whole family of herpes viruses, enters the parasitized cells where it becomes triphosphorylated by a kinase that the virus itself specifies. The phosphorylated product is highly toxic to the virus, but has no effect on healthy cells (Section 4.0) (Elion *et al.*, 1977).

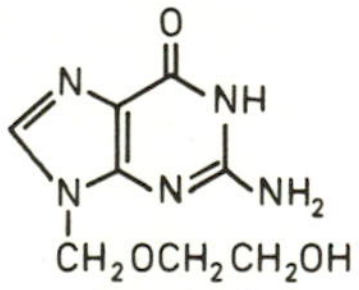

$CH_2OCH_2CH_2OH$
Acyclovir
(3.49)

$H_2NCH CO-NHCH-P=O$

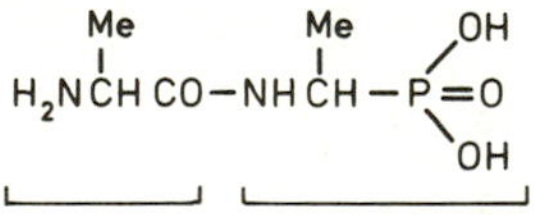

L-Alanine   L-Aminoethylphosphonic acid
Alafosfalin
(3.50)

Because of their extra external membrane, Gram-negative bacteria are accessible to fewer anti-bacterials than Gram-positive organisms. A new attempt to remedy this situation has produced alafosfalin (3.50) which makes use of a peptide permease in this outer membrane to transport the drug into the cell where it is hydrolysed to L-aminoethylphosphonic acid, a potent inhibitor of L-alanine racemase. The normal product of this enzyme, D-alanine, is an essential component of murein, the supporting framework in bacterial cell-wall (Section 5.3) (Allen *et al.*, 1978; Allen and Lees, 1980).

Another new, and successful, development is the 'IMBI' or irreversible mechanism-based inhibitor. This type of pro-drug uses the normal mechanism of the enzyme to become changed to a permanent inhibitor of that enzyme (Abeles and Maycock, 1976). IMBIs have a group that sufficiently resembles that of the normal substrate to be concentrated on the enzyme surface. The enzyme produces a double bond, across which a nucleophilic group (e.g. -SH) of the enzyme protein becomes added, thus permanently deactivating that enzyme. For more details of this type of drug, see Section 9.7b. Perhaps the most successful IMBI is 2-difluoromethylornithine (*3.51*) which irreversibly inhibits ornithine decarboxylase, the enzyme that decarboxylates ornithine to putrescine. The enzyme removes a fluorine atom, giving the double bond seen in (*3.52*), and the rest follows as described above (Prakash *et al.*, 1980).

$$
\begin{array}{ccc}
HCF_2 & & HCF \\
| & \longrightarrow & \| \\
H_2N\text{-}C\text{-}CO_2H & & H_2N\text{-}C \\
| & & | \\
(CH_2)_3 & & (CH_2)_3 \\
| & & | \\
NH_2 & & NH_2
\end{array}
$$

2-Difluoromethylornithine   Intermediate for Michael
(*3.51*)   acceptor
                                 (*3.52*)

3.6.7 *Self-limiting pro-drugs*

The self-limiting drugs discussed on p. 96 have their analogues among pro-drugs of which azathioprine (*3.41*) was the prototype (p. 102). Another early example was the propionyl ester ('estolate') of erythromycin which, unlike erythromycin itself (*4.47*), is stable to acid and rapidly absorbed through the stomach wall. This pro-drug is hydrolysed in neutral aqueous buffer with a $t_{0.5}$ of about 45 minutes. Hence the regeneration of erythromycin in the blood-stream does not depend on any enzyme (Tardew, Mao and Kenney, 1969).

A new approach to self-limiting pro-drugs has been made by N. Bodor and his associates. The true drug is quaternized with an alkylating agent that has as much lability built into it as the situation requires. If the quaternizing groups are bulky, the pro-drug usually has little physiological effect of its own. The usual quaternizing agents are esters of chloromethyl alcohol (e.g. *3.53*). Quaternization of tertiary amines proceeds rapidly at room temperature, but it should be noted that alcohols, and primary and secondary amines, also react (Renshaw and Ware, 1925). The products require only water to become dequaternized slowly (Bodor, 1982).

When it is desired to have the pro-drug absorbed on a biological surface, it is useful to use higher homologues of (*3.53*). For example, pilocarpine (*12.81*), after quaternization with chloromethyl palmitate, gave the derivatives shown as partial structure (*3.54*). This adduct produces a miosis (contraction of the pupil) that is ten times more potent, and more durable too, than pilocarpine can effect. Hence it has value in treating glaucoma through the slow release of pilocarpine (Bodor, 1981).

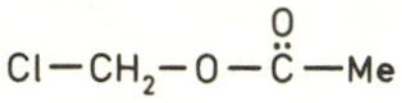

$$Cl-CH_2-O-\overset{O}{\overset{\|}{C}}-Me$$

Chloromethyl acetate

(*3.53*)

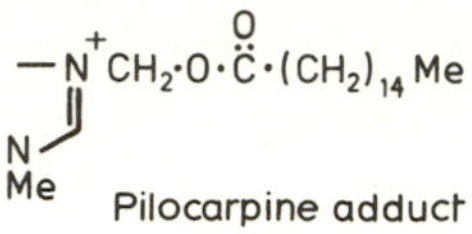

$$-\overset{+}{N}CH_2 \cdot O \cdot \overset{O}{\overset{\|}{C}} \cdot (CH_2)_{14} Me$$

Pilocarpine adduct

(*3.54*)

3.6.8 *Harmful metabolic changes*

In the present section and the previous one, we have discussed many examples of metabolic change that are beneficial to Man. We can now deal with the few, but highly important, examples where the so-called detoxifying enzymes of the body actually convert a harmless to a harmful substance. Thus, the *e.r.* converts dimethylnitrosamine to a substance which methylates the guanine of RNA to 7-methylguanine, a reaction that leads to acute liver necrosis (Magee, 1964). The toxic action of methanol, leading eventually to blindness, is caused by its transformation in the body into formaldehyde (Kini and Cooper, 1962).

Other cases are described in Section 13.5 under the heading 'Lethal examples'. Thus fluoroacetic acid is transformed to fluorocitric acid which blocks the enzyme aconitase. Several amines, such as benzidine and 2-aminonaphthalene are converted to carcinogenic hydroxylamines; and polycyclic hydrocarbons such as benzopyrene are converted to carcinogenic epoxides. The liver converts carbon tetrachloride to a free radical which causes liver necrosis (Slater, 1966).

Substances that have terminal double or triple bonds, if unconjugated to other multiple bonds, are oxidized by *P*-450 in the *e.r.* to toxic substances which attack this porphyrin and deactivate it. Examples are: ethylene, acetylene, vinyl chloride and the hypnotic ethchlorvynol which is 1-chloro-3-ethylpent-1-en-4-yn-3-ol. Thus, ethylene leads to the *N*-2-hydroxyethyl-derivative of *P*-450. A further adverse effect is that other drugs, if given at the same time, escape the usual metabolic transformation and so build up in the patient (Ortiz de Montellano, Beilan and Matthews, 1982).

3.6.9 *The activation of agrochemicals*

Many of the most successful organophosphorus insecticides, such as malathion and diazinon (*13.28*) have been designed to be non-toxic until metabolized by the insect into the true agent, the pathway used by the insect being one that is absent in vertebrates. This has been a very rewarding development of otherwise dangerous materials, and a full account will be found in Section 13.3.

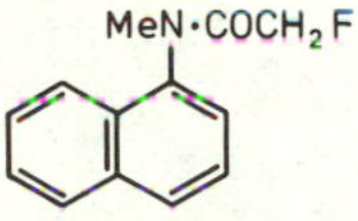

MeN·COCH$_2$F

Me$_2$N·C·SS·C·NMe$_2$
      S   S

Me$_2$N·C·SH
      S

Nissol
(3.55)

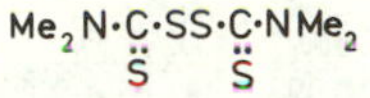

Tetramethylthiuram disulfide
(3.56)

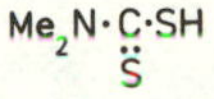

Dimethyldithiocarbamic acid
(3.57)

—NH·C·OMe
   O

   H S H
N·C·N·CO$_2$Me

N·C·N·CO$_2$Me
   H S H

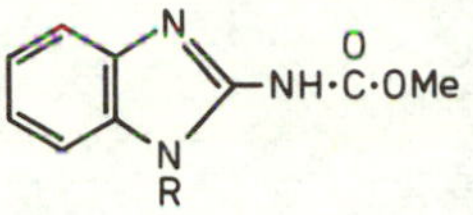

(a) R = –C(O)·NHBu (benomyl)
(b) R = –H   (BCM)
(3.58)

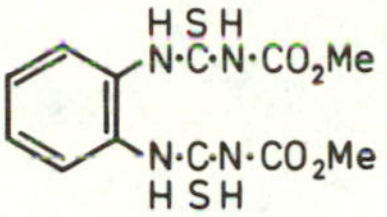

Methyl thiophanate

(3.59)

The highly selective N-methyl-N-(1-naphthyl)fluoroacetamide (3.55) ('Nissol'), is lethal to mites because they liberate fluoroacetic acid from it (see Section 13.5), but has little toxicity to mammals because they do not degrade it in this way (Hashimoto *et al.*, 1968).

Tetramethylthiuram disulfide (3.56), which is used to check the growth of fungi in seeds and turf, acts by reduction to dimethyldithiocarbamic acid (3.57) which is a widely used chelating fungicide (see Section 11.7.3).

When the fungicide, benomyl (3.58a), was introduced in 1966, it proved to be more active than any known agent by several powers of ten. This substance (methyl 1-butylcarbamoyl-2-benzimidazolecarbamate) easily loses the butyl-carbamyl-group (which may assist penetration into the plant) to give methyl 2-benzimidazolecarbamate (BCM) (3.58b) which has the same degree and range of activity (Clemons and Sisler, 1969). It has been quite clearly established that BCM is the only fungitoxic substance in plants treated with benomyl (Peterson and Edgington, 1969). Moreover another commercial fungicide, methyl thiophanate (1,2-*bis*-3'-methoxycarbonylthioureidobenzene) (3.59), similarly generates BCM when wet, and this seems to be the sole form in which it is active (Vonk and Sijpesteijn, 1971).

A most ingenious way has been found to increase the selectivity of herbicides in the phenoxyacetic acid series. Many weeds can degrade the sidechain of the intrinsically harmless ω-2,4-dichlorophenoxy-aliphatic acids until the lethal 2,4-dichlorophenoxyacetic acid is produced, whereas many economic crops lack the β-aliphatic oxidase necessary for this degradation (Wain, 1955, 1964) (see Fig. 3.6). The following observation led to this ingenious masking. In a homologous series of seven ω-2,4-dichlorophenoxyalkylcarboxylic acids, an alternation of growth-accelerating activity was found to occur as the series was ascended; but only the members with an *odd* number of methylene-groups had any activity. This led to the suggestion of β-aliphatic oxidation (first demonstrated in mammals by Knoop in 1904). Wain's masked herbicides are used commercially, e.g. MCP (γ-2-methyl-4-chlorophenoxybutyric acid), which efficiently kills the weeds in a legume crop, or flax in a field of clover.

2-Chloroethanephosphonic acid is used as a masked source of ethylene (a

Fig. 3.6 Selective latency in herbicides. Effect of spraying 0.2% solutions of ω-(2,4-dichlorophenoxy)-acetic (A), -propionic (P), -butyric (B), -valeric (V), -caproic (C) and -heptanoic (H) acids on charlock (top row) and clover (bottom row). Control plants are on the right. Photographed after 2 weeks. (Wain, 1964).

natural growth inducer in plants) to force flowering, and the ripening and abscission of fruit (Edgerton and Blanpied, 1968).

3.6.10 *Concluding remarks*

Although several substances originally thought to be agents have turned out to be only pro-agents, there is abundant evidence that the majority of agents act on their receptors in the same chemical form as that in which they are administered or applied. A detailed knowledge of permeability and enzymes can assist a skilful designer in finding useful pro-agents, but he must bear in mind that he is adding more complications to the long list of distribution problems that lie between administration and arrival at the receptor. In short: he has two distinct sets of pharmacokinetic problems to deal with instead of only one. The administration of pro-drugs is an artifice that is most used when the true drug is excreted or metabolized too fast for effective medication. In most cases, medical practitioners prefer to give the true drug in order to control the dosage according to the patient's response.

For further reading on metabolic activation and pro-drugs see Higuchi and Stella (1975).

3.7 Quantitative aspects of distribution. Pharmacokinetics. Sustained release

The earlier parts of this Chapter described the selective processes that regulate the access of a drug to its receptor. This was indicated qualitatively in Fig. 3.3. To transfer this thinking to the *quantitative* level, it is necessary to perform experiments to find the constants which govern each unit of the total process. Two quite distinct types of constant are relevant, (a) kinetic constant (also called rate constants, or velocity constants), and (b) equilibrium constants. The former give information on the speed of a process, the latter on the percentage composition of the mixture remaining when the equilibrium has been reached. Of the two, the kinetic constants are usually easier to measure in the living organism and are sufficient to provide a scientific basis for formulating dosage schedules. This concept of affixing constants to units of drug distribution is shown, in barest outline, in Fig. 3.7. Each pair of constants (one for each forward, and one for each backward direction) is usually determined first as a single overall constant which can be split into microscopic (i.e. constituent) constants if the nature of the work calls for it.

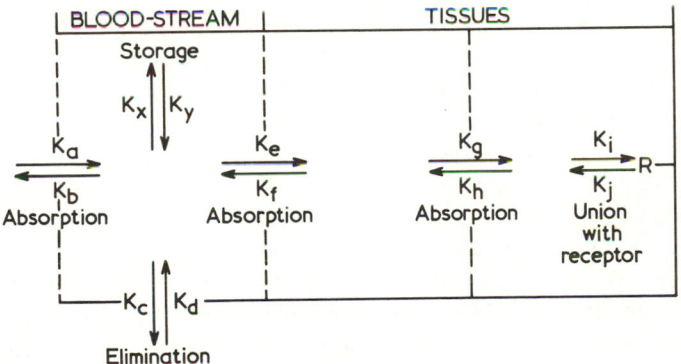

Fig. 3.7 Quantitative aspects of distribution.

3.7.1 *Pharmacokinetics*

This is the name given to the study of the kinetics of drug absorption, distribution, metabolism, and excretion, all of which are rate-controlled. The earliest studies, which were concerned with inhaled anaesthetics (Widmark, 1920; Dominguez, 1933), were not suited for general application. The fundamental equations were introduced by T. Teorell (1937) in his studies of insulin action. He provided simple kinetic formulae to monitor the concentration of drug at the site of administration, then in the blood and tissues, and finally after the inactivation or elimination. These equations were perfected and extended by Krüger-Thiemer (1960) in Germany, and Nelson (1961) in the United States, both of whom had the required mathematical ability and made use of the newer analytical techniques. Today pharmacokinetics is seen as one of the most important divisions of drug studies. It can determine the best dose for the patient and the right timing intervals between doses. It monitors the drug and metabolites, at every phase of distribution, and can detect areas where selectivity needs to be improved.

Even between substances that are chemically closely related, large kinetic differences operate at the various stages of distribution shown in Fig. 3.3, and these differences contribute very much to the selectivity of drugs. Thus the fate of each drug is ordained by the sum of the constants with which its chemical constitution has endowed it. (It should be borne in mind that these figures are constant only for the animal species in which they were determined.)

Without guidance from calculations based on these experimentally deter-mined constants, the physician finds himself perched on a tight-rope stretched between two extremes: ineffectiveness if the dose is too low, and poisoning the patient if it is too high. He needs to be given a dosage schedule which he can use to control two factors, (a) the degree of the drug's action and (b) the duration of this action. The rates of absorption and distribution govern the time of onset of the drug's action; the rates of metabolism and excretion govern the duration; the size of the dose, in combination with these effects, governs the intensity. This information is no more than vaguely indicated by animal experiments, and so the help of human volunteers is needed.

The usual difference in times of onset which occur when a dose is given intravenously or orally is shown in Fig. 3.8. It is evident that injection instantly gives the highest attainable blood concentration, whereas oral dosage gives a more delayed and less intense effect; however, the intensity of both effects declines at about the same rate.

The rate of transfer of a drug across a membrane can be best described by the differential dS/dt, where dS is the microscopic amount transferred in dt (a very short time); see Section 3.2. The amount (S) of drug on the outside of the membrane determines the rate of transfer across the membrane, as follows,

$$-dS/dt = kS$$

where k is the permeability constant. The rate of transfer will be faster the greater S is, in fact doubling the dose will double the rate. Similar differential equations govern distribution (which tends to be faster than absorption and elimination). S_0, the amount of unabsorbed drug at zero time is, of course, the dose (see also Section 3.2, p. 67).

The apparent volume of distribution (V_D) (see Section 3.1), discloses whether the drug remains in one compartment (usually the bloodstream) or is shared with a second compartment (usually the tissues). Chemotherapeutic agents

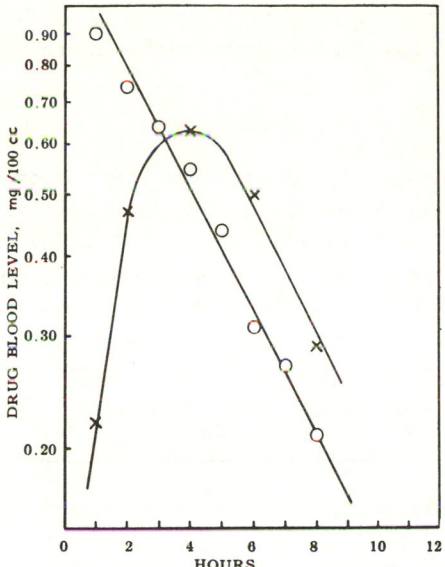

Fig. 3.8 Average blood levels of theophylline for human subjects given the same dose intravenously (circles) or orally (crosses). (Swintosky, 1956.)

work best if shared, because infection is commoner in the tissues than in blood. A two-compartment distribution can be confirmed by the biphasic character of the first-order plot of concentration against time. The symbol k_{12} relates to any transfer from compartment 1 to compartment 2, whereas k_{21} reverses the direction (examples of microscopic constants). Kinetics for two-compartment absorption were worked out by Loo and Riegelman (1968). The cardiac glycoside digoxin is typical of a two-compartment drug. With barbiturates, the second compartment is the tissue lipoid and not the tissue water.

Unlike the permeability constant, which is linked to the dose, all other constants must be linked to concentrations, regardless of whether they refer to transfers between compartments or to the final elimination. They require no great change in the foregoing equation, which now becomes

$$-dC/dt = \beta C$$

where C is the concentration in the compartment in front of the relevant membrane, β is the transfer or elimination constant, and $-dC/dt$ is the 'disappearance rate'. The elimination constant is usually taken as the sum of the metabolism and the excretion constants. Although first-order constants are most often encountered, a few of 'zero order' have been found. Zero-order constants are independent of the amount (or concentration) of drug present (Nelson and O'Reilly, 1960, 1961).

To obtain blood-levels as constant as possible during therapy, the best interval between doses is derived from the half-life of the drug in the blood-stream. This half-life ($t_{0.5}$) is calculated from the elimination constant (β) using the equation

$$t_{0.5} = 0.693/\beta$$

and it represents the time taken for half of the drug to disappear from the blood. Table 3.5 lists the half-lives of several common drugs. For a larger table of $t_{0.5}$ values, see Gilman, Goodman and Gilman (1980), p. 1675.

The ideal dose interval (τ) is $3.32\, t_{0.5} \cdot \log(1 + C_0/C_{min})$ where C_0 is the initial blood concentration and C_{min} is the lowest blood concentration that is therapeutically effective. Unfortunately, this equation often produces dose intervals

Table 3.5 Half-lives (in hours) of various drugs in the human body

Aspirin	0.3
Alprenolol	3
Ampicillin	1.3
Cephalexin	0.9
Chlorthiazide	1.5
Chlorpromazine	30
Cimitidine	2
Clonidine	9
Diazepam	50
Digitoxin	7
Digoxin	42
Erythromycin	1
Ethanol	0.2
Gentamicin	2
Hydralazine	2
Imipramine	13
Indometacin	2
Isoniazid	2
Methotrexate	8
Methyldopa	1.8
Morphine	3
Phenobarbital*	86
Prednisolone	2
Propranolol	4
Quinidine	6
Rifampicin (Rifampin)	2
Sulfamethoxazole	9
Tetracycline	10
Theophylline	9
Trimethoprim	11
Tubacurarine	2
Warfarin	37

*This drug has replaced the bromide ion (half-life 7 days).

that are too difficult to observe, such as 19 hours. It has become more usual to decide on a convenient dose interval (4, 8, 24, or 48 hours) and substitute it, as τ, into the following equation to find the best dose (D) (Wagner, 1967):

$$D = \tau . V_D . C_{av}/1.44t_{0.5}F$$

where C_{av} is the average required blood level, F is the fraction absorbed (ideally 1.0), and the other symbols have their previous significance.

The ideal dosing intervals can also be determined by drawing a line parallel to the base line of curves, constructed as in Fig. 3.8, at the minimal effective blood level of the drug, i.e. the level below which no beneficial effect on the patient can be observed.

In oral therapy, a *priming dose* may be needed for the drug to reach an effective blood level as quickly as possible. If the drug is one with a long half-life, a priming dose is essential, because it takes five half-lives for a drug to reach its plateau. Further doses are determined as in the foregoing.

Ideally, the size of doses should be determined for each patient by blood analyses performed at the beginning of treatment. This procedure builds the patient's idiosyncrasies (of distribution or metabolism) into the dose. When it is not possible to do this, it is usual to take published average figures, derived from a group of patients.

The pioneer studies of Krüger-Thiemer have established dose intervals and ratios of initial to maintenance doses for a large number of sulfonamide antibacterial drugs of which Table 3.6 presents a selection (Krüger-Thiemer and Bünger, 1961, 1965). In the early days of treatment with sulfonamide drugs, the wide span of their half-lives was not realized, and examples with long half-lives were often discredited because unwitting overdosage harmed the patients.

Table 3.6 Recommended dose intervals. Recommended ratios of priming dose (D^*) to maintenance dose (D) for obtaining steady blood-levels of drug

	Average half-life (hours)	Dose interval (hours)	D^*/D
Sulfathiazole	3.5	4	1.8
Sulfisoxazole	6.1	6	2.0
Sulfanilamide	8.8	8	2.1
Acetylsulfisoxazole	13.1	12	2.1
Sulfadiazine	23.5	24	3.0
Sulfamerazine	23.5	24	3.0
Sulfadimethoxine	41.0	24	3.0

One sometimes reads that a drug is 'cumulative', but accumulation in the body is often a function of the dosage pattern and not of the drug. Any drug, given too often or in too large a dose, will accumulate; the problem is most evident with drugs that have a long $t_{0.5}$.

Once the principles of these studies have been grasped, the distribution of a drug can be investigated in more detail. Dose schedules for sulfonamides like those in Table 3.6 were obtained as follows. The drug was given orally to healthy human volunteers, and specimens of blood and urine were taken at frequent intervals and analysed. The rate constants for the metabolism of various sulfonamides were obtained; also those for their excretion (see Fig. 3.9). Further work enabled the composite excretion constant β_1 to be split into two microscopic constants describing respectively the secretion of the drug by the kidney glomerulus, and the resorption from the kidney tubules into the blood-stream.

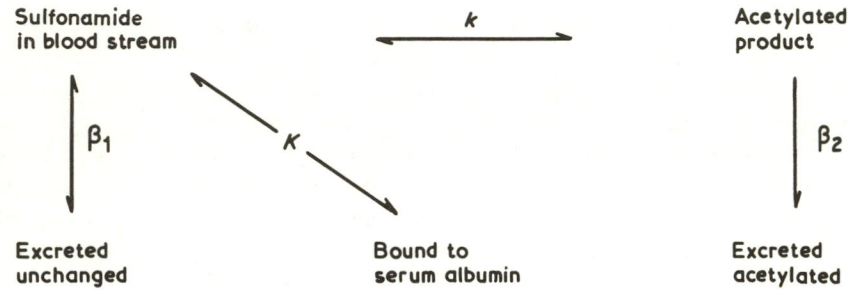

Fig. 3.9 Kinetics of the metabolism and excretion of sulfonamide drugs in the human body. (Nelson and O'Reilly, 1960.)

Studies such as these have shown that most of the host-related pharmaco-kinetic properties of the sulfonamide anti-bacterials are favoured by lipo-philicity of the drug. Thus the rate of absorption rises with increasing lipophilicity and so does the volume of distribution although, in the latter case, ionization plays a small part. Elimination requires lipophilicity *and* ionization about equally. Protein binding increases with lipophilicity except that *ortho*-substitution, while increasing lipophilicity, decreases protein binding (a steric effect). Metabolism (e.g. acetylation) is diminished by increase in lipophilicity. In contrast with much of the foregoing, the *in vitro* anti-bacterial action of sulfonamides is little influenced by lipophilicity but is strongly dependent on electronic effects, particularly ionization (see Section 10.5), and it is increased by *ortho*-substitution (Seydel, 1981). These studies, and others made with many distinct families of drugs have made valuable contributions to selectivity, for they showed that drugs had many unsuspected *independently variable* constants. Hence the pattern of drug distribution can be varied at will by making minute changes in the molecular structure. Two decades ago, this knowledge opened a new door for the drug designer to improve selectivity by controlling distribution, and at once led to improved anti-bacterial sulfon-amides, many with specialized uses in therapy (see further Section 9.3).

The use of a pro-drug in place of the drug requires a complete set of constants to be determined for both (Martin, 1967). When the elimination rate of the precursor is less than that of the drug, a blood level (of the drug) is obtained which declines more slowly than that obtained when the drug is given as such. The rate of conversion of precursor to drug also governs this result. When the drug-designer perseveres until he finds a precursor with favourable values for both these rates, he has on hand a substance capable of extending the dosage interval and maintaining a more uniform level of the drug between doses.

The absorption of orally administered *solids* is slow, and exponential. The solution rate is proportional to the drug's surface area, and the availability for absorption decreases in the order: solutions, suspensions, capsules, compressed tablets, coated tablets. Sodium salts of poorly soluble weak acids usually produce higher blood-levels than the free acids, because the hydrochloric acid of the gastric juice liberates the weak acid from its salt in a much finer form than any in which it can be marketed. The kinetics of dissolving drugs is reviewed by Wagner (1961).

The *bioavailability* of different specimens of a poorly soluble drug, when given orally and measured by the rate of the rise of plasma levels, varies enormously because of differing particle sizes. This test has revealed important differences between different brands of aspirin, diphenylhydantoin, cardiac glycosides (especially digoxin), tetracyclines, chloramphenicol, and dicumarol. The remedy is, usually: supersonically 'micronize' the drug.

3.7.2 *Sustained release*

For some therapeutic uses, slow absorption of drugs is a nuisance, but in other cases it can represent the ideal state, so long as it is regular. In Section 3.6, we have considered the use of pro-drugs to ensure slow release of the true drug. In sustained release, however, the drug is not chemically altered, but is surrounded by a physical barrier to its unrestrained diffusion. The kinetics of diffusion through this barrier are much studied with a view to achieving a tight control over drug release.

At first the desired restraint was sought by encasing each dose in a coating of shellac, waxes, or cellulose esters to protect the drug from attack by the acidity of gastric juice, while encouraging its release in the mild alkalinity of the small intestine. The antibiotic, erythromycin (which is acid-sensitive) is often prescribed in this way, but alternatively it is given as an ester (i.e. as a pro-drug) which resists acid but is hydrolysed later.

These enteric coatings sometimes evolved into concentric coatings designed to provide an initial dose followed by one or more delayed doses. Alternatively, tiny pellets, with different coatings timed to dissolve after a series of intervals, were incorporated in a single capsule. Although these devices work well for some patients, much evidence of individual variability has been met. The kinetics of slow dissolution of pellets of steroids, implanted under the skin, have been worked out (Ballard and Nelson, 1962). Other examples of slow release from a depot are provided by benzylpenicillin (penicillin G, ordinary peni-cillin). This is often given by deep intramuscular injection as its insoluble procaine salt, in aqueous suspension, to provide a tissue depot from which the drug is slowly released over 12 hours or more. For still slower release (several days), the procaine is replaced by another base, benzathine, which is *NN'*-dibenzylethylenediamine.

A combination of pro-drug and sustained release is provided by testosterone propionate injected in peanut oil, and by the anabolic esters used in carcinoma of the breast, also injected in oily depots (e.g. nandrolone decanoate USP).

Imaginative developments are advancing the subject of delayed action. The use of microencapsulation, an idea derived from modern copying papers, provides capsules as small as a few μm in diameter filled with micronized drug particles or ultrafine droplets. A coacervate of gelatin and gum acacia has been much used to form these microcapsules (Luzzi, 1970). Alternatively, liposomes which are oil drops of radius about 20 nm, are filled with hydrophilic or lipophilic drugs and injected into the bloodstream (Gregoriadis, 1977). Lipo-somes have special affinity for the spleen and liver into whose cells they discharge their contents by endocytosis or by fusion with the plasma membrane. Insertion of organic antimonials into liposomes increases potency 300-fold in treatment of visceral leishmaniasis, because both parasite and drug become concentrated in the same liver cells (Alving *et al.*, 1978). Attempts have been made to label the surface of liposomes with 'homing substances' to dispatch them to other organs.

The advent of silicone rubber ('Silastic Polymer') has further widened the possibilities for ensuring sustained release (Zaffaroni, 1974). When this sub-stance is prepared by polymerization in the presence of the drug, the latter can be relied upon to diffuse from it at a slow, steady rate, in any moist situation (for equations of diffusion, see Roseman and Higuchi, 1970). Thus a silicone rubber membrane, impregnated with pilocarpine, is available for the treatment of glaucoma when implanted in the conjunctival sac but many patients cannot retain this device. Light-weight pumps that can be worn on the backs of laboratory animals and deliver small steady increments of a drug are now in use. Research continues into micro-pumps, sensitive to feedback from a relevant blood constituent, to be implanted in the human body for the release, on biochemical demand, of the drug with which they have been charged.

For further reading on the modification of a drug's structure to improve the pharmacokinetics, see Notari (1973). For the theory and calculations of kinetic studies, see Krüger-Thiemer (1966); for the design of dose schedules, see Gilman, Goodman and Gilman (1980, p. 1675); for books on pharmacokinetics, see Gibaldi and Perrier (1975) and Notari (1980), also a concise review in Gilman, Goodman and Gilman (1980, p. 1). For a book on controlled drug delivery and sustained release, see Bruck (1983). See also the *Journal of Microencapsulation* (London: Taylor and Francis).

Box 3.2 Answers to questions in Box 3.1

(1) At pH 2.0 in the stomach, (a), (d), and (e) are ionized and hence will penetrate poorly, whereas (b) and (c) are non-ionized and hence will be the best absorbed.

(2) At pH 6 in the small intestine, only (c) is non-ionized and hence must be the best absorbed.

(3) The glomerulus allows all substances of low molecular weight to pass into the glomerular urine, regardless of ionization. The non-ionized substances are then usually resorbed by the tubules. At pH 4.2, (b) is half-ionized, (c) is non-ionized and the others are all ionized. Hence (c) will be best resorbed and hence have the lowest rate of excretion.

(4) At pH 7.3, (d) is 11% non-ionized, but (c) is 67% non-ionized and hence will penetrate even better than (d). The others are too strongly ionized to penetrate.

4

Comparative biochemistry: the second principle of selectivity

Of the three approaches to selectivity outlined in Section 1.2, Comparative Biochemistry has so far proved the most successful. In retrospect, it is surprising how slowly this subject developed, but the widespread misunderstanding among biochemists that all living cells had a common ground-plan of metabolism (see Section 1.2) hardly left the subject room to exist. Baldwin's book *An Introduction to Comparative Biochemistry* first appeared in 1937 and opened many people's eyes to the interest and possibilities of the subject, but it is only in the last three decades that comparative biochemistry has supported a substantial number of senior research workers. Today, however, it is a thriving and successful subject. In fact comparative biochemistry has undergone such a rapid growth in recent years that even the seven volumes of Florkin and Mason (1960–4) do not hold more than the outline of it. The following account of selectivity through comparative biochemistry will begin with the nucleic acids, because of their dominant position in the life of the cell.

4.0 Nucleic acids

The biological importance of nucleic acids began to emerge with the discovery by Oswald Avery, and his colleagues at the Rockefeller Institute, that the deoxyribonucleic acid of pneumococci carried heritable information, so that a specimen from one strain of this bacterium could confer its properties on a different strain, a process called transformation (Avery, MacLeod and McCarty, 1944).

(a) Deoxyribonucleic acid (DNA), the most important of the nucleic acids, occurs in mitochondria and chloroplasts, but most of it is in the nucleus. It is the carrier of all the cell's genetic information, the appropriate portion of which is placed in service instantly to meet changing circumstances in the cell. The information stored in DNA is encoded by the nature and order of the pyrimidine bases, thymine (*4.1b*) and cytosine (*4.2*), and the purine bases adenine (*4.3*) and guanine (*4.4*). Each strand of DNA has a deoxyribose-phosphoric acid back-bone to which these bases are attached. Usually DNA has two such strands wound around one another in a double-helix and held together by H-bonds between each opposing pair of bases (Watson and Crick, 1953; for bond lengths and angles, see Wing *et al.*, 1980).

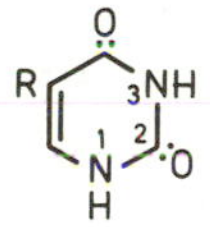

(a) R = H (uracil)
(b) R = Me (thymine)
(4.1)

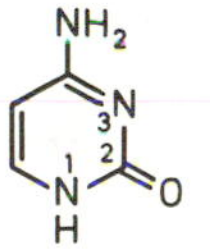

Cytosine
(4.2)

Adenine
(4.3)

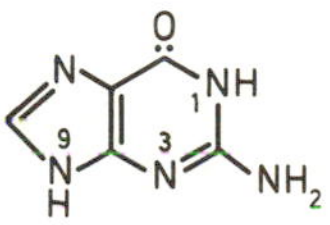

Guanine
(4.4)

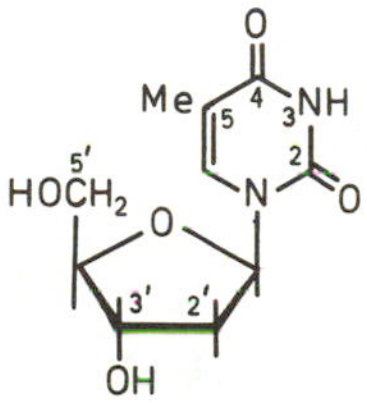

Thymidine
(4.5)

DNA, from many sources including mammalian and bacterial, has a molecular weight between 10 and 100 million. In the vertebrate nucleus, the spirals usually form rods of about 3 μm in length, and 18 Å diameter. However, non-nuclear DNA (as in mitochondria and bacteria) is often circular and in the latter, single-stranded. The vertical distance between the layers of bases is about 3.3 Å (measured from centre to centre of the molecules) so that there is no free space between these layers (Jordan, 1968). The purine and pyrimidine bases are planar, and the paired bases are in the same plane with one another and with the C-1' and C-4' atoms of the sugars to which they are attached. However, the planes of the sugars are nearly at right angles to those of the bases.

The structure of the DNA molecule is intimately related to its two primary roles: replication (gene duplication by synthesis of more DNA) and transcription (gene expression by synthesis of RNA) (see Fig. 4.1).

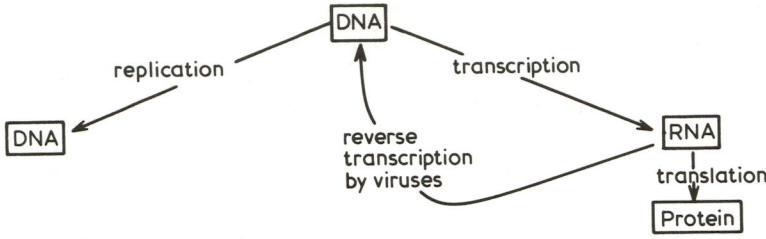

Fig. 4.1 Roles of nucleic acids in the living cell.

In the nucleus, DNA is synthesized from mononucleotides (such as deoxy-adenosine triphosphate) by a polymerase. Some preformed DNA is required as a template. Several substances are known which inhibit this synthesis by combining with the template and making it unavailable (for example, the acridines). Many tumour-producing viruses, which have RNA as their sole nucleic acid, contain reverse transcriptase, i.e. a polymerase that forms new DNA as a copy of viral RNA (Temin and Mizutani, 1970). Agents to inhibit this enzyme selectively are being sought.

The four bases mentioned account for most of those found in the DNA of higher forms of life. 5-Methylcytosine forms the principal exception; 25% of the cytosine in the DNA of plants is in this form, but animal DNA has much less, and bacterial DNA has only 0–2% (Vanyushin *et al.*, 1968). Bacterial and viral DNAs sometimes contain other methylated bases, such as 6'-methyladenine, 2-methyladenine, or 5-hydroxymethyluracil. In some phages, all cytosine is replaced by 5-hydroxymethylcytosine, made by a virus-induced enzyme in the bacterial host (Cohen, 1963).

The ratio of bases in bacterial DNA differs greatly, from one bacterial species to another, in that the sum of the two amphoteric bases (G+C), when divided by the sum of the two monofunctional bases (T+A), ranges between 0.45 and 2.80. In higher plants and animals, on the contrary, this ratio is confined between 0.6 and 0.9 (Belozersky and Spirin, 1958).

(b) Chromatin. DNA does not exist free in the nucleus, but is organized into a continuous rope of flat, wedge-shaped *nucleosomes*. Each nucleosome contains a double-helix of 150 base-pairs, superwound around a core of 8 molecules of basic proteins (histones). Apparently the outside of each nucleosome is available for union with RNA polymerase and the information-masking acidic proteins. During mitosis, this 'rope' can be folded into chromosomes. This example is from rat liver, typical for eukaryotes. The rather different structure of bacterial chromatin is also described (Finch, *et al.*, 1977).

(c) Ribonucleic acids (RNAs). Besides reproducing itself, DNA acts as a template for the synthesis of ribonucleic acids, of which the three principal kinds are messenger RNA (mRNA), transfer RNA (tRNA) sometimes called 'soluble RNA', and ribosomal RNA (rRNA). The RNAs closely resemble DNA in general structure, but only a portion of each molecule is in the helical form, and they have ribose in place of deoxyribose, and uracil in place of thymine. All three kinds of RNA have some methylated bases. All types of RNA have their part to play in the synthesis of proteins, but the details of protein synthesis differ little, whether taking place in bacteria or in the most highly evolved organisms. The molecular weights of all RNAs are lower than that of the parent DNA; that of a typical mRNA is often about one million, and of a tRNA about 25 000.

Messenger RNA (mRNA) is actually a whole family of RNAs, upon each member of which the DNA has transcribed a message which the mRNA has to translate into synthesis of a protein that is chosen according to the needs of the cell at that time (see Fig. 4.1). Each mRNA travels rapidly to the ribosomes, selects the tRNA-bound amino acids (see below) in the order required to make the needed protein, leaves the ribosome, and is destroyed. Each triplet 'codon' of consecutive mRNA bases specifies a particular amino acid, thus providing a genetic code, identical for all forms of life.

Transfer RNAs (tRNA) occur in many varieties, each of them specific for one amino acid to which they become esterified. These esters are attracted to the ribosomes in an order predetermined by the messenger RNA that is passing through the ribosome at that time (see Fig. 5.7). The ribosome is able, by expending the energy of guanosine triphosphate, to add each new amino acid residue to the terminal —CO_2H of the growing polypeptide chain, at the same time casting the tRNAs back into the cytoplasm to make a new supply of amino acid esters.

In detail, the 3'-hydroxy-group of the ribose (in the terminal adenosine residue of each tRNA molecule) is acylated, enzymatically, by an activated form of that amino acid for which this tRNA is specific. Amide-ester exchange of these esters with peptidyl-tRNA on the ribosomes then occurs under the direction of mRNAs. In the latter step, the growing peptide is transferred from the tRNA bond on the 'donor' site of the ribosomes to an aminoacyl-tRNA bond on the 'acceptor' site of the same ribosome. This reaction is catalysed by the enzyme peptidyl transferase. The peptide, lengthened in this way, is then entirely shifted to the donor site with the help of a translocation enzyme. The next aminoacyl-tRNA then becomes bound to the vacant acceptor site, and the process is repeated. The terminal groups of the tRNAs are always —XCCA, where C is a cytidylic, and A an adenylic, acid residue.

Several of the nucleotides in tRNAs have bases unusual for RNA. For example, the alanine-specifying tRNA has hypoxanthine, 1-methylhypoxanthine, 1-methylguanine, 2-dimethylguanine, thymine (ribose-bound), dihydrouracil (three times), and pseudouracil (ψ), i.e. uracil with ribose attached to C-5 instead of to nitrogen (twice). Other unusual bases often found in tRNAs include: 5-methylcytosine, 7-methylguanine, 2-methylguanine, and 1-methyl-, 2-methyl-, and 6-methyl-adenine. The precursors of all tRNAs have the usual purine and pyrimidine bases which are changed *in situ* by a methylase that transfers the methyl-group from methionine. This methylation increases specificity. In mammals, methylation occurs only in the nucleus, and is quite specific in pattern for each tissue. Malignant tumours perform such methylations at an abnormally high rate, producing some tRNAs different from those of the tissue of origin (Kuchino and Borek, 1978).

Ribosomal RNA (rRNA) is associated with about 50 different proteins in the ribosomal sub-units (see Section 5.4). A precursor of the rRNA is synthesized in the nucleoli and is then partly methylated there.

For further reading on the chemistry of nucleic acids, see Davidson (1976) and Neidle (1982).

Many of the most successful antibacterial, anti-protozoal, antiviral, and anti-cancer drugs act by inhibiting the synthesis of DNA, whether directly or indirectly. Understandably enough, many of these substances became established clinically before it was realized that their mode of action was interference with the cell's genetic material, the prime governor of its present health and future inheritance. To interfere deliberately in this way with the uneconomic cell has long been a cherished dream, but one that was held back by fear of damage to the genes of the economic cell. Consensus has now been reached that a drug is acceptable even if it interferes slightly with the synthesis of DNA in the host, provided that it does not damage (or even covalently combine with) the DNA of the host. The application of this rule safeguards the host's cells from the possibility of any mutation, while retaining many of the most selective drugs that Man has yet discovered.

It will be convenient to divide these drugs into five classes.

4.0.1 *Drugs that inhibit early stages in DNA synthesis*

Outstanding among drugs which inhibit the production of DNA from several stages back in the biosynthetic pathway are the sulfonamides and the 2,4-diaminopyrimidines used as antibacterials and anti-malarials. All of the chemotherapeutic *sulfonamides*, whether simple sulfanilamide (*4.6a*) or its more complex heterocyclic derivatives (*4.6b*) including sulfadiazine, competitively inhibit the enzyme dihydrofolate synthetase which produces dihydrofolic acid (*2.14*) (see p. 31). The basis of this inhibition, as outlined in Section 2.1 (p. 31), is the similarity in the steric and electronic properties of *p*-aminobenzoic acid (*2.12*) (which the enzyme is ready to build into new molecules of dihydrofolic acid) and the sulfonamides (*4.6*) which, when taken up by the enzyme, merely block it. The basis of the *selectivity* of these antibacterial sulfonamides depends on two factors, which reinforce one another: (i) mammals lack the enzymes necessary for the synthesis of dihydrofolic acid, and hence they tolerate these sulfonamides very well; (ii) pathogenic bacteria lack the permease with the aid of which mammals absorb dihydrofolic acid from the diet. Further relevant data will be found in Section 9.3. Dihydrofolic acid is only two steps away from the coenzyme required for biosynthesis of thymine and all the purine bases. Deprived of the substrates, especially thymine, bacteria soon die because they can make no new DNA.

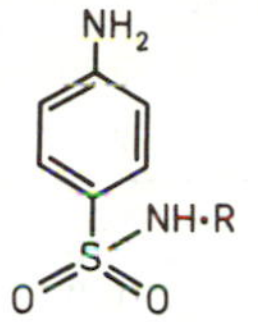

Sulfanilamide, and
derivatives
(a) R = H
(b) R = heterocyclic ring
(4.6)

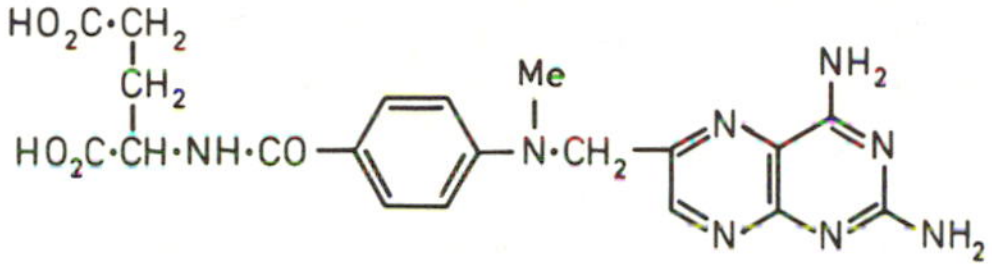

Methotrexate (amethopterin)
(4.7)

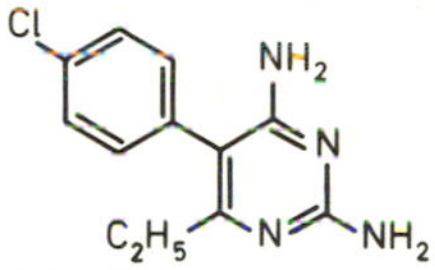

Pyrimethamine
(4.8)

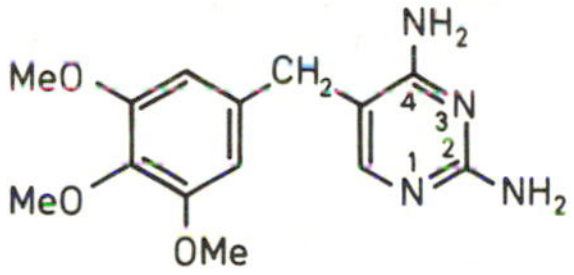

Trimethoprim
(4.9)

Another kind of much used drugs, interfering with the synthesis of DNA, acts by inhibiting dihydrofolate hydrogenase. This is the enzyme that reduces dihydrofolic acid to 5,6,7,8-tetrahydrofolic acid, a substance only one step short of the coenzymes for synthesis of the purine bases and thymine. Analogues of folic acid, based on 2,4-diaminopteridine such as *methotrexate* (4.7), are useful anti-cancer drugs (see Section 9.3.3) but are not absorbable by micro-organisms. To overcome this defect, Hitchings pared away at the molecule of methotrexate until he found that the power to inhibit dihydrofolate hydro-genase resided in 2,4-diaminopyrimidine. The addition of some lipoidal groups furnished the powerful anti-malarial drug, *pyrimethamine* ('Daraprim') (4.8) (Falco *et al.*, 1951). This drug (2,4-diamino-6-ethyl-5-(*p*-chlorophenyl)-pyrimidine) has become the most widely used of all prophylactics against malaria. The lipoidal groups in the molecule favour its uptake by the tissues and the malarial parasite (Hitchings, 1952), and these groups also increase the adsorption of the drug to the reductase by van der Waals forces (Baker and Shapiro, 1966).

The selectivity of this type of drug depends on the existence of *analogous enzymes*, namely dissimilar enzymes carrying out apparently identical functions in dissimilar organisms (more about these in Section 4.2).

As can be seen from Table 4.1, the plasmodial enzyme is inhibited by pyrimethamine at a concentration about 2000 times lower than that inhibiting the analogous mammalian enzymes. The concentration that inhibited the plasmodial enzyme corresponded to that achieved in the tissues after the usual prophylactic dose. These data established that the selective action of pyrimethamine in malaria is due to the extraordinary sensitivity of the enzyme in the parasite compared to that of the analogous enzyme in the host (Burchall and Hitchings, 1965).

Table 4.1 Concentrations ($\times 10^8$ M) of anti-folic drugs needed for 50% inhibition of dihydrofolate reductase, isolated from six sources

Substance		Human liver	Rat liver	Mouse erythrocyte	Pl. berghei	Tryp. equiperdum	E. coli
Pyrimethamine	(4.8)	180	70	100	0.05	20	2500
Trimethoprim	(4.9)	30 000	26 000	100 000	7.0	100	0.5
Methotrexate	(4.7)	9	0.2	(not done)	0.07	0.02	0.1

(Burchall and Hitchings, 1965; Ferone, Burchall and Hitchings, 1969; Jaffe and McCormack, 1967.)

Cycloguanil (3.35), the triazine anti-malarial formed in the human body from the inert pro-drug proguanil, also acts by inhibiting dihydrofolate hydrogenase (Wood and Hitchings, 1959).

The degree of inhibition of dihydrofolate hydrogenase by pyrimethamine was decreased very little when the chlorophenyl-group was changed to a butyl-group (Baker and Shapiro, 1966). However, quite a different substitution pattern was required to persuade the 2,4-diaminopyrimidine series to show effective anti-bacterial action. Eventually excellent results were obtained by providing the lipophilic substituent with a somewhat hydrophilic periphery, as in *trimethoprim (4.9)*, a very successful systemic anti-bacterial (Roth, Falco and Hitchings, 1962). Table 4.1 shows that, whereas pyrimethamine is the most selective (of representatives of the three classes of anti-dihydrofolate hydrogenase) against the malarial parasite enzyme, trimethoprim is most selective against the bacterial enzyme. It will be observed that methotrexate, although no longer kept apart from the target enzyme by a permeability barrier, remains a poorly selective drug against micro-organisms.

Table 4.2 Effect of trimethoprim (4.9) on isolated dihydrofolate reductase concentration (nM) causing 50% inhibition

Source: *Mammalian liver*		*Bacteria*		*Ratio*
Rat	1200	*Strept. faecalis*	0.96	1250
Ox	2400	*E. coli*	1.3	1840

(Roth and Cheng, 1982.)

Table 4.2 demonstrates the high selectivity that trimethoprim exerts against the dihydrofolate reductase of bacteria, both Gram-positive and Gram-negative types, while leaving the analogous mammalian enzyme unharmed. For clinical uses of this drug see Section 9.6.

Hydroxyurea (*4.10*) is a derivative of hydroxylamine that is well tolerated by the patient in the large doses required for the management of chronic granulocytic leukaemia resistant to other drugs. The molecular basis for its selectivity lies in its sequestering the iron required by the iron-dependent enzyme *ribonucleoside diphosphate reductase*, essential for the conversion of ribo- to deoxyribonucleosides, a vital stage in DNA synthesis (Krakoff, Brown and Reichard, 1968). It does not affect RNA synthesis. The same enzyme is inhibited by two other drugs that show clinical promise: guanazole (*4.11*), 3,5-diamino-1,2,4-triazole (Levi and Wiernik, 1976), and 5-hydroxypicolinic aldehyde thiosemicarbazone (*4.12*) (Agrawal and Sartorelli, 1975). All these substances are specific for the S phase of the cell cycle (see Section 5.1). Other thiosemicarbazones with anti-cancer properties are discussed at the end of Section 11.9.

$$H_2N \cdot C(:O) \cdot NH \cdot OH$$

Hydroxyurea
(*4.10*)

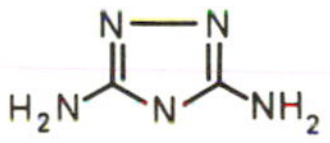

Guanazole
(*4.11*)

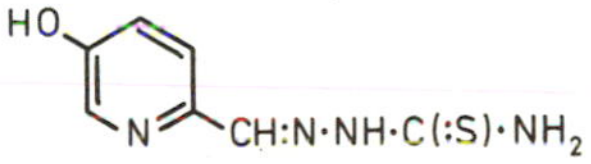

5-Hydroxypicolinic aldehyde
thiosemicarbazone
(*4.12*)

4.0.2 *Drugs that directly affect biosynthesis of DNA*

The pyrimidine and purine analogues are a class of drugs which inhibit the synthesis of DNA at a late stage. These anti-metabolite drugs are administered as the free bases or as nucleosides, but not as nucleotides which cannot penetrate into a cell. Several members of this class have earned a valued place in the treatment of cancer, and others are the mainstay of anti-viral therapy. The free bases are usually changed, in the cell, to deoxyribosides or deoxyribotides which compete, respectively, with the normal nucleosides and nucleotides (Langen, 1975). The principal ribotide-forming enzymes are: adenosine kinase, adenine PRPP transferase, hypoxanthine–guanine PRPP transferase, uridine phosphorylase, uridine/cytidine kinase, thymidine/deoxyuridine kinase, deoxycytidine kinase, and orotate PRPP transferase (Hakala, 1973). The reduction of ribonucleotides to deoxyribonucleotides (by ribonucleotide diphosphate reductase) provides another potential source for interference. For each base and riboside there exists an independent uptake mechanism that is also independent of any subsequent phosphorylation reaction.

A well-known anti-cancer analogue is *5-fluorouracil* (*3.3*) which is given intravenously in treating cancers of the breast, colon, and rectum. It alleviates these conditions but does not establish a lasting cure. We have already noted its outstanding selectivity in treating skin cancers (Section 3.0), where much of the selection depends on its pattern of distribution. However, when selective absorption is bypassed by parenteral injection, such selectivity as exists is biochemical, as follows. 5-Fluorouracil becomes an active cytostatic drug only after conversion to its nucleotide (5-fluorodeoxyuridylic acid) by a sequence of tissue enzymes. This nucleotide has an affinity for thymidylate synthetase that is several thousand times greater than that of the enzyme's natural substrate, deoxyuridylic acid. Thus it is able to keep the substrate off the enzyme, and no more DNA is synthesized (Reyes and Heidelberger, 1965). (This is the enzyme that normally transfers a methyl-group from methylenetetrahydrofolic acid to deoxyuridylic acid, which is the key step in passing from the uracil series to the thymine series.) Further selectivity was found in leukaemia cells (mouse) which had an analogous enzyme in place of normal phosphoribosyltransferase. This analogous enzyme had almost no affinity for uracil, but a very strong one for 5-fluorouracil (Kessel, Hall and Reyes, 1969). 5-Fluorouracil also interferes significantly with RNA synthesis.

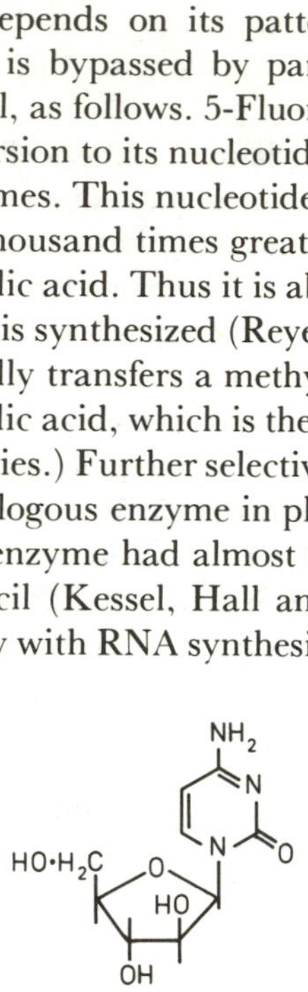

Cytosine arabinoside
(4.13)

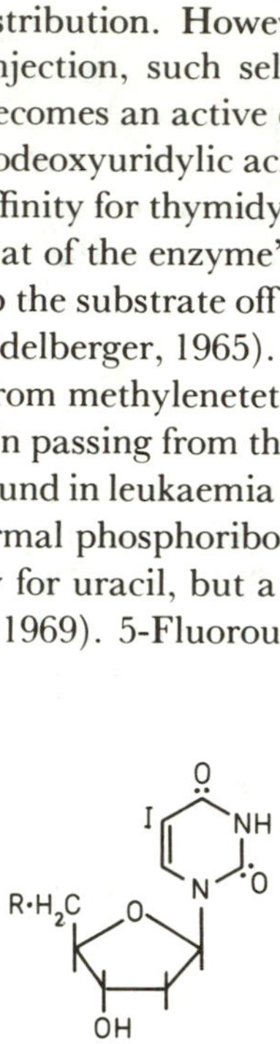

(a) Idoxuridine (R = OH)
(b) AIU   (R = NH$_2$)
(4.14)

Cytarabine (4.13) is an isomer of the RNA intermediate, cytidine whose 2'-hydroxyl-group has been swung around to become *trans* to the 3'-hydroxyl-group of the natural product. It is also known as Ara-C, cytosine arabinoside, and 1,β-D-arabinofuranosylcytosine. Only slightly incorporated into DNA, the principal action of this drug, which follows its conversion to the triphosphate by healthy cells, appears to be inhibition of DNA polymerase. Cytarabine is classed as a specific inhibitor of the S phase (Section 5.1) in the cell cycle (Skipper and Schabel, 1973). Its repressive effect on bone-marrow elements must be monitored throughout treatment. Because cytarabine is readily deaminated by human cytidine deaminase, it is usually given by slow intra-venous infusion. Cytarabine is valued for its ability to induce remission of acute leukaemia in chidren and adults, after which methotrexate, 6-mercaptopurine and their adjuvants are substituted to effect the cure. It seems to be the only effective drug in acute granulocytic leukaemia of adults (Kremer, 1975). For anti-viral therapy, vidarabine (see below) is preferred.

6-Mercaptopurine (3.14), first made by Elion, Burgi and Hitchings (1952) has a well-established place in the oral treatment of the acute leukaemias (Brulé *et al.*,

1973). Its action is very complex, and a little of it, after bio-transformation to thioguanidylic acid, is incorporated into cellular DNA (Tidd and Paterson, 1974; Parks *et al.*, 1975). However, it is not established that this is a therapeutically important reaction. Much of it is converted, in the cell, into 6-thioinosine 5′-phosphate (TIP) (Brockman, 1963). TIP inhibits conversion of inosine 5′-phosphate to adenosine 5′-phosphate, thus bringing neogenesis of purines to a halt (Salser and Balis, 1965). It also exerts feedback inhibition of the biosynthesis of phosphoribosylamine, a carbohydrate involved in the earliest steps in purine biosynthesis (Bennett *et al.*, 1963).

Azathioprine (*3.41*; p. 102) is the S-(1-methyl-4-nitroimidazol-5-yl) derivative of 6-mercaptopurine which it slowly releases in the body. It is much used as an immunosuppressant to prevent rejection of organ grafts, particularly those of the kidney (Elion, 1967; Elion and Hitchings, 1975). The 2-amino derivative of 6-mercaptopurine, known as thioguanine, is used to a smaller extent in cancer therapy and as an immunosuppressant.

More information on combating cancer is in Sections 1.1 (p. 15), 5.1, 6.3.6, 15d 13.4.

Turning, now, to anti-viral drugs, we must begin with 5-iododeoxyuridine (*4.14a*) (*idoxuridine*, IUdR) which made history by being the first drug to cure a viral disease, namely herpetic keratitis of the eye. Idoxuridine can quickly terminate what used to be a long-lasting and painful illness, sometimes ending in blindness (Kaufman, 1962). This drug acts by preventing the utilization of thymidine in cells by becoming converted to the 5′-monophosphate, which interferes with the use of thymidine. Unfortunately, it is also incorporated into DNA (in place of thymine), not only in typical DNA-containing viruses such as vaccinia and herpes simplex (Welch and Prusoff, 1966), but also into the DNA of mammals. Therefore care is taken not to use it systemically, for fear of mutagenesis. In the conjunctival sac, sealed off from the tissues and not easily absorbed, it has proved to be completely safe.

An insight into the mode of action of this drug is possibly given by the following work. When pseudo-rabies virus was exposed to IUdR, most of the thymine in the viral DNA was replaced by 5-iodouracil. The virus, when allowed to infect cultured mammalian kidney cells, continued to produce DNA, but this did not become protein-coated and hence did not generate new virus particles. That the new DNA was potentially infective was then shown by exposing it to thymidine: when much of the iodouracil had been replaced by thymine, infectivity was restored (Kaplan and Ben-Porat, 1966).

Later, a more soluble analogue of idoxuridine was introduced, namely *trifluridine* (5-trifluoromethyldeoxyuridine). It is used for the same purpose, acts by a very similar biochemical mechanism, and seems to be more effective (Pavan-Langston and Langston, 1975).

Although the only difference between idoxuridine (*4.14a*) and *aminoidoxuridine* or AIU (*4.14b*) is the exchange of a hydroxy- for an amino-group in the 5-position of the sugar residue, the newer amino-compound is metabolized so differently that a completely new territory of anti-viral compounds, with hitherto unattained high therapeutic indices, was realized (Chen and Prusoff, 1979). This substance, 5-iodo-5′-amino-2′,5′-dideoxyuridine, is converted to its cytostatic triphosphate by the kinase of herpes virus, but not by kinases in uninfected human, murine, or simian cells. It inhibits herpes simplex infection with no sign of toxicity to the host. However, when a mammalian cell was infected with herpes virus, the drug was incorporated into the DNA of both host and virus (Chen, Ward and Prusoff, 1976). This incorporation into host DNA does not matter, provided that all infected cells die and the replication of the virus is halted. Further development of this drug was possibly curtailed by the launching of the similarly acting acyclovir (see below).

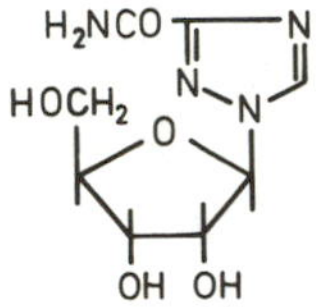

Ribavirin
(4.15)

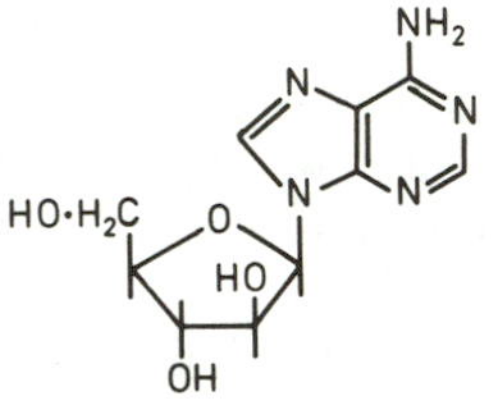

Vidarabine
(4.16)

Ribavirin (4.15) is of special interest because of its broad spectrum of attack. It is active against both DNA and RNA viruses. Also known as tribavirin and 'Virazole', this substance is 1-β-D-ribofuranosyl-1,2,4-triazole-3-carboxamide. *In vivo*, it is changed to the 5′-monophosphate which inhibits inosinate dehydrogenase, and hence prevents the formation of guanylic acid on which synthesis of both DNA and RNA depend. It is being used clinically for the treatment of viral pneumonia and measles although the results are not quite clear-cut (Smith and Kirkpatrick, 1980).

Three anti-cancer drugs that inhibit inosate dehydrogenase can appropriately be mentioned here 2-β-D-Ribofuranosylthiazole-4-carboxamide, chemically closely related to ribavirin (*4.15*), is in clinical trial against cancer of the lung (Plowman and Paull, 1982). The corresponding selenazole shows a high therapeutic index in cancer of mice (Srivastava and Robins, 1983). 3-Deazaguanine has been ordered for clinical trial because of good therapeutic results against solid tumours (Khwaja, 1982; Allen *et al.*, 1977).

Vidarabine (4.16) differs from adenosine in the same way that cytarabine differs from cytidine, namely the natural sugar has been replaced by arabinose. However, the anti-viral properties of vidarabine are more to the fore than any anti-cancer power, the exact opposite of cytarabine's properties. Vidarabine, also known as Ara A, is 9-β-D-arabinofuranosyladenine and it has a reasonably high therapeutic index (Müller *et al.*, 1977). Its most successful and spectacular application has been in herpes virus encephalitis, of which several thousands of cases break out in the United States each year, killing 70% of the sufferers

(Whitley *et al.*, 1977). In herpetic keratitis of the eye, it has been clinically found to be as effective as idoxuridine and less irritating. Vidarabine lacks the general usefulness of acyclovir (see below) in other herpetic infections.

Vidarabine is converted *in vivo* to the triphosphate (Cohen, 1976) which represses DNA polymerase more strongly in the virus than in the host cells, but the reason for this is not known. In higher than therapeutic doses it is incorporated into mammalian DNA. It is given intravenously.

The principal drawback to treatment with vidarabine is the ready deamination by adenosine deaminase, an enzyme widely present in human tissues. Improved therapy is being attempted by simultaneously giving an inhibitor of this enzyme such as *erythro*-9-(2-hydroxynon-3-yl)adenine (*4.17*) which has a little anti-viral action of its own. A more powerful inhibitor of this enzyme, pentostatin (deoxy-coformycin) (*4.18*) is also under investigation (Agarwal, Spector and Parks, 1977). It remains to be seen whether increasing the survival of vidarabine in this way can improve its therapeutic index which seems to be less than that of acyclovir.

9-(2-hydroxynon-3-yl) adenine
(4.17)

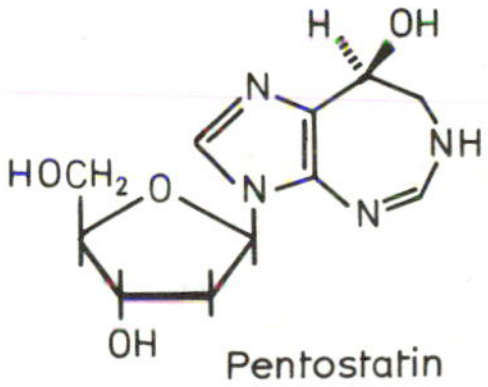

Pentostatin
(4.18)

Acyclovir (*4.19*) is undoubtedly the most interesting of the known anti-viral drugs because of its high therapeutic index. It is, for example, 3000 times more toxic to herpes simplex virus than to mammalian cells. Somewhat like amino-idoxuridine (*4.14b*), acyclovir is monophosphorylated by a virus-specified thymidine kinase and then converted to the triphosphoryl derivative which injures the virus by competing, against deoxyguanosine triphosphate, for the virus-specified DNA polymerase in the infected cells (Fyfe *et al.*, 1978). This inhibition puts an end to all DNA synthesis in these cells. Its phosphorylation does not take place in healthy cells, which are thereby spared, and this accounts for most of the selectivity. For the rest: the DNA polymerase of infected cells is more sensitive to acyclovir triphosphate than is the DNA polymerase of healthy mammalian cells which, in any case, receive very little of this product (Elion, 1980).

Chemically, acyclovir ('Zovirax', formerly called acycloguanosine) is 9-(2-hydroxyethoxymethyl)guanine and is made by condensing 2,6-dichloropurine with 2-benzoyloxyethoxymethyl chloride and exposing the product, in turn, to ammonia, nitrous acid, and ammonia again (Schaeffer *et al.*, 1978). It is the slightly indirect descendant of inhibitors of adenosine deaminase, such as (*4.17*) (Schaeffer and Schwender, 1974). That some of these substances, particularly those with 'opened-up ribose' in the 9-position, had anti-viral properties was not realized at first because of the high specificity imposed by thymidine kinase on its substrates. The anti-viral properties of acyclovir were first announced in 1977 (Elion *et al.*, 1977). As for structural specificity, virus-specified thymidine kinase reacts only feebly with adenine and pyrimidine analogues of acyclovir, nor does it phosphorylate deoxyguanosine (Elion, 1980).

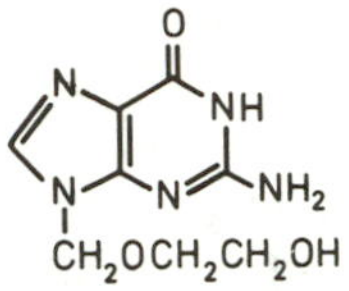

Acyclovir
(4.19)

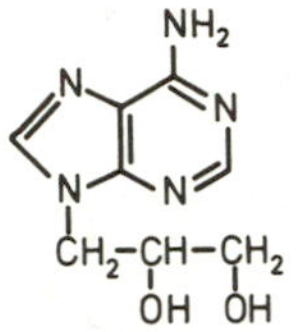

9-(2,3-dihydroxypropyl) adenine
(4.20)

Clinically, acyclovir has been found effective against all types of herpes infection, but not against other species of virus. It rapidly changes the prevalent herpes-2 type of genital infection to an inactive state without always eliminating this virus. Usually it is given parenterally, but also as tablets and ointment. It is also used prophylactically in patients receiving immunosuppressant drugs, whether for cancer therapy or prior to an organ graft, circumstances in which a quiescent herpes infection is apt to light up.

Meanwhile, somewhat *parallel* developments have been taking place in other countries. Holý in Czechoslovakia synthesized many analogues of (*4.19*) including dihydroxy analogues such as (*4.20*): the anti-viral testing was carried out in Belgium by de Clercq. Although a broader spectrum of anti-viral activity has been claimed for some of these products (de Clercq *et al.*, 1978; de Clercq and Holý, 1979), no new forms of treatment seem yet to have emerged.

BVDU (*4.21*) was synthesized in the University of Birmingham (England). Chemically, it is *trans*-5-(2-bromovinyl)-2'-deoxyuridine and has a biochemistry of action rather similar to that of acyclovir. It is more potent than the latter against varicella zoster virus and is highly active when given orally. It suffers in comparison with acyclovir by being almost inactive against herpes-2 infections. For laboratory studies, see de Clercq *et al.* (1979); Barr *et al.* (1981), and for clinical studies, de Clercq *et al.* (1980); de Clercq (1983).

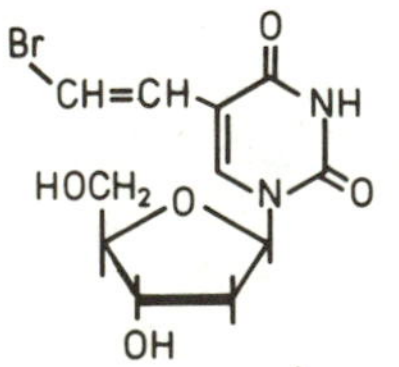

5-Bromovinyl-2'-deoxyuridine
(4.21)

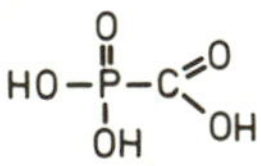

Foscarnet
(4.22)

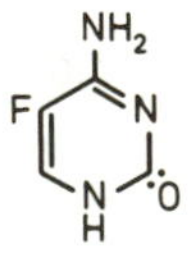

Flucytosine
(4.23)

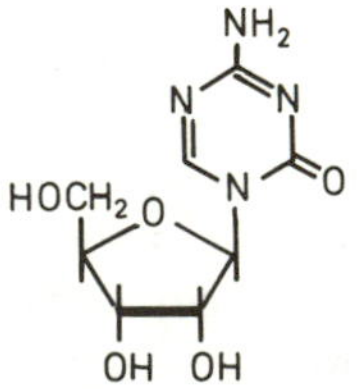

5-Azacitidine
(4.24)

Foscarnet (phosphonoformic acid, *4.22*) is an anti-viral which acts by binding to the pyrophosphate-combining site of DNA polymerase, but only on the virus-specified enzyme. It is usually applied topically to the lesions of herpes (Larsson and Öberg, 1981). Sodium phosphonoacetate acts similarly (Overby, Duff and Mao, 1977).

More information on combating viral infections will be found in Sections 1.1 (p. 11), 5.5, and 6.3.2.

Flucytosine (*4.23*) (5-fluorocytosine, 'Ancobon') has proved clinically successful as an orally active fungicide for treating such systemic diseases as candidosis and cryptococcal meningitis (Bennett, 1977). Its selectivity depends on the fact that mammals secrete it unchanged (just as they do cytosine), whereas fungi convert it to 5-fluorouracil and then elaborate this to its cytostatic nucleotide.

5-Azacitidine (*4.24*), isolated from *Streptoverticillium ladakans*, but now made by synthesis, was originally used as an anti-leukaemic agent in Czechoslovakia. It recently sprang into prominence through an ability to activate genes which can produce γ-globin but which are usually dormant in adults. This helps sufferers from β-thalassaemia, an anaemia in which the β-globin chain of haemoglobin is missing (Ley, 1982).

See Langen (1975) for the metabolic basis of the medicinal use of purine and pyrimidine analogues.

4.0.3 *Drugs that physically restrain DNA*

Many substances are known which, by inhibiting unwinding of the DNA helix, suspend both replication and transcription. Probably the most selective of these are the *intercalating agents*, which were briefly referred to in Section 2.2 and are discussed at length in Section 10.3. These drugs become stacked between the layers of purines and pyrimidines to which they cling by van der Waals forces, supplemented by stronger ionic bonds to the phosphate ions of the DNA backbone (Lerman, 1961). Typical examples are: the anti-cancer drugs, amsacrine, ellipticine, anthramycin, actinomycin D (dactinomycin, see Section 4.0.6), daunorubicin, and doxorubicin (adriamycin) (see Section 4.0.5); the anti-malarials mepacrine (quinacrine, 'Atebrin') and chloroquine; and the anti-trypanosomal drugs ethidium and quinapyramine. For more on intercalation, see pp. 132, 134 and Section 10.3.2.

A less selective DNA-restraining drug is *chlormethine* (*4.25*) which was not only the first of the nitrogen-mustard anti-cancer agents but actually the first anti-cancer drug to win clinical approval (Gilman and Phillips, 1946; Goodman *et al.*, 1946). These mustards, which have very unusual chemical properties (see Section 13.4), act by locking together two strands of DNA by the guanine residues (Goldacre, Loveless and Ross, 1949; Lawley and Brookes, 1967). In general, the selectivity of the 'mustards' is not high; cyclophosphamide (*3.38*), which has the best therapeutic index, is understandably the most used.

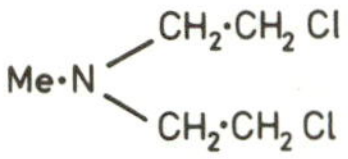

$$Me \cdot N \underset{CH_2 \cdot CH_2 \, Cl}{\overset{CH_2 \cdot CH_2 \, Cl}{<}}$$

Chlormethine
(4.25)

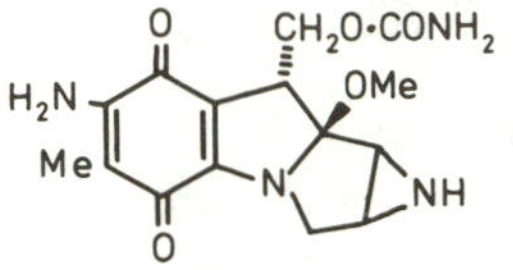

Mitomycin
(4.26)

Mitomycin (*4.26*), formerly mitomycin-C, is a violet-coloured substance based on a pyrrolizidine nucleus, one side of which is annelated to an aziridine ring, the other to benzoquinone. The strongly hypoxic cancer cells are thought to reduce the quinone function so that the product is an alkylating cross-linker of DNA (Kennedy, Rockwell and Sartorelli, 1980). Used in the palliative treatment of terminal gastric adenocarcinoma, it is known as a poorly selective drug, being mutagenic in Man and carcinogenic in rats (Crooke and Bradner, 1976).

Cisplatin (*11.48*) which is *cis*-diamminedichloroplatinum II, is an anti-cancer drug which is used regularly in treating solid tumours. It is thought to interlink two guanine bases from opposite strands of the DNA helix (Section 11.10).

4.0.4 *Drugs that destroy DNA*

Doxorubicin (*4.27*), formerly called *adriamycin*, is an anthracycline antibiotic that is much used for treating carcinomas and sarcomas. That it is an intercalating agent is indicated by the rise in T_m, decrease in buoyant density, decrease in sedimentation constant, and increased viscosity of native double-stranded DNA when adriamycin is added. Moreover, the visible spectrum, fluorescence, and ease of reduction of adriamycin were altered by DNA in the directions expected for an intercalated substance. In a set of derivatives, carcinostatic activity fell off with decreasing intercalation, and also with decrease in basic strength. From these data, it was concluded that the highly basic amino-group in the sugar-residue forms an ionic bond with a phosphate group of the DNA (just as can be seen in the X-ray diagram obtained for daunomycin), and the flat anthraquinone skeleton lies between a layer of purine and pyrimidine bases (di Marco and Arcamone, 1975). Adriamycin inhibits DNA-directed RNA polymerase and DNA polymerase; the latter is inhibited 50% at 7.4 μg/ml.

In addition to this intercalating property, doxorubicin has a destructive action on DNA which some would attribute to its activation of superoxide dismutase (Oberley and Buettner, 1979). Although it has been placed here, alongside other drugs that undoubtedly act by destroying DNA, its action is not fully understood. An opposing view is that its action is on the cell surface (Tritton and Yee, 1982; Israel and Potti, 1982). The chief drawback of this drug is its toxic action on the heart. For clinical results, see Blum and Carter (1974) and Gottleib and Hill (1974).

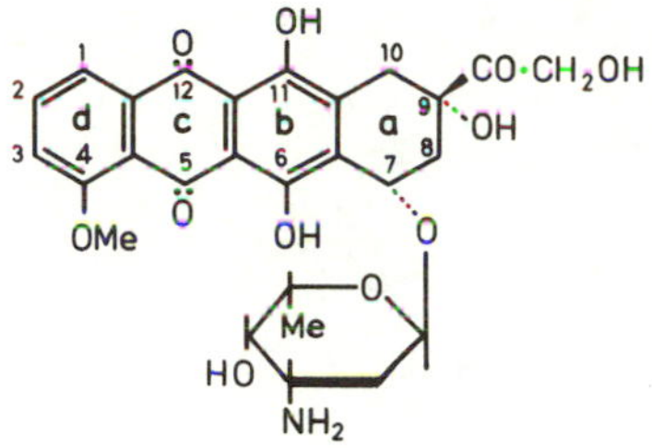

Doxorubicin (adriamycin)
(4.27)

Doxorubicin is isolated from *Streptomyces peucetius* by separation from the less valuable *daunorubicin* (daunomycin) which has the group —COMe in place of —CO.CH$_2$OH and is used principally in treating leukaemias (Arcamone *et al.*, 1969). The strongly intercalating properties of this antibiotic have been most carefully confirmed. X-ray crystallography shows that the amino-sugar portion projects into the minor groove, the methoxy-group projects beyond the base-pair area into the major groove, and the 9-hydroxy-group of the anthracycline ring-system is hydrogen-bonded to an adjacent guanine (Quigley *et al.*, 1980). This interpretation is supported by n.m.r. (Jain, Kozlowski and Rice, 1981), which indicates that rings c and b overlap a base-pair.

Many analogues of doxorubicin are being tested to find a less cardiotoxic, but equally carcinolytic, example. Of these, some like 4′-tetrahydropyranyl-doxorubicin are derivatives, whereas others, such as the Russian *carminomycin* and the Japanese *aclacinomycin A* have been isolated from related species (Carter, Sakurai and Umezawa, 1981). For reviews of the action of doxorubicin, see Schwartz (1983); Neidle and Sanderson (1983) and El Khadem (1982).

Bleomycins are a family of antibiotics which were isolated from *Streptomyces verticillus* by Umezawa (1973) in Japan. Pharmacopoeial bleomycin is a mixture of bleomycin A$_2$ (*4.28*) and bleomycin B$_2$ which differs from A$_2$ only in that the trimethylenesulfonium group (bottom right in formula) is replaced by a tetra-methyleneguanidinium group [—NH(CH$_2$)$_4$NH.C($:$NH)NH$_2$] (Oppenheimer, Rodrigues and Hecht, 1979). By adding different amines to the fermentation broth, about 400 other bleomycins have been made, of which pepleo-mycin (more resistant to metabolic hydrolysis; less toxic to lung tissue) seems the most promising.

Bleomycin is much used in the clinic because of its rapid attack on solid forms of cancer. It is one of the very few anti-cancer drugs that does not attack the bone-marrow. Its characteristic toxicity (fibrosis of the lung, and some effect on skin) is not shared with the few other drugs that are effective against solid tumours, such as doxorubicin and cisplatin. Hence, bleomycin is given (parenterally) in conjunction with one of these, so that the anti-cancer effect is additive whereas the toxicity to healthy cells in minimized (Carter and Blum, 1976; Carter *et al.*, 1976).

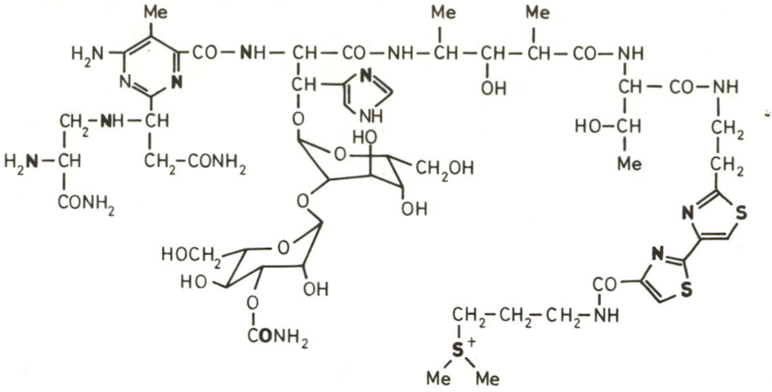

Bleomycin A$_2$
(4.28)

The characteristic effect of bleomycin is to break single strands of DNA in the tumour and prevent their repair; it has no effect on RNA (Nagai *et al.*, 1969).

With a molecular weight of about 1500, the structure of bleomycin at first looks rather complicated, but the mode of action is fairly well understood at the molecular level. Two main regions are concerned in the action, an intercalating region on the right of (4.28) and a chelating–oxidizing region on the left. Just as was found for the aminoacridines in the 1940s (see Section 2.2), the molecule of bleomycin is attracted to the phosphoric anions of DNA by the highly basic (sulfonium or guanidinium) group of the drug. As soon as the two molecules are close enough, the flat bi-thiazolyl pair of rings in the drug slip between the base-pairs of the DNA, an act of intercalation that provides the firm attachment required for the operation of the remainder of the molecule (Takita *et al.*, 1978; Chien, Grollman and Horowitz, 1977).

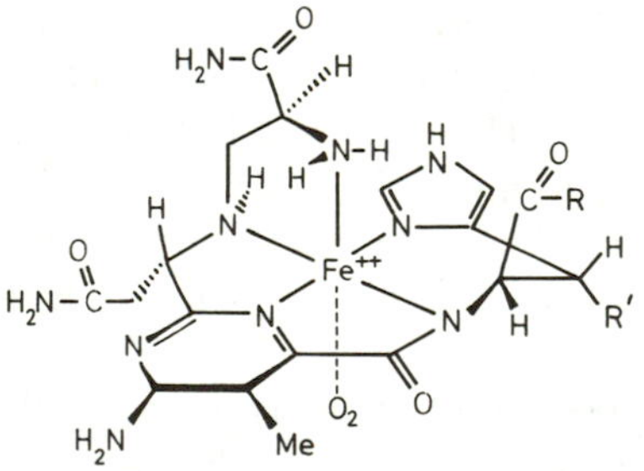

Co-ordination of Fe^{2+} and O$_2$ by bleomycin   Simplified analogue of (4.29)
(4.29)   (4.30)

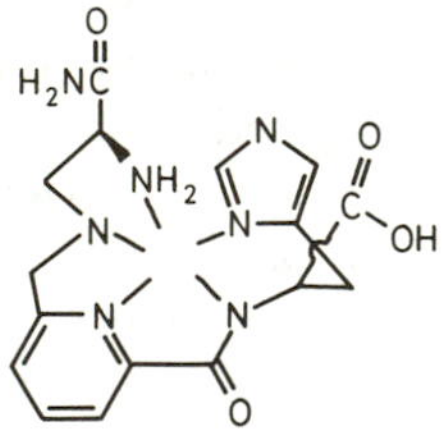

The bleomycins are purifiable as their well-crystalline copper salts. In X-ray crystallography, the six atoms nearest to the copper are shown in heavy type on the left side of (4.28). Neither bleomycin nor its iron salts have yet been obtained crystalline enough for X-ray work. In action, bleomycin combines with ferrous iron, which is usually, if only accidentally, present. This binding uses the five nitrogen atoms shown in (4.29). The sixth iron valence picks up atmospheric oxygen which it converts to the hydroxy-anion* radical ($^{\cdot}O^{-}$): this is retained in the iron-binding cage (Sausville, Peisach and Horowitz, 1978).

This activated iron–oxygen complex then oxidizes the 4'-hydrogen atom of the deoxyribose moiety in the DNA of cancer cells (Grollman and Takeshita, 1980). This attack causes the release of four propenal-base molecules, of which thymine-propenal (4.31) is typical. This scission is an oxidative cleavage of the bond between C-3 and C-4 atoms in the deoxyribose ring (Giloni et al., 1981).

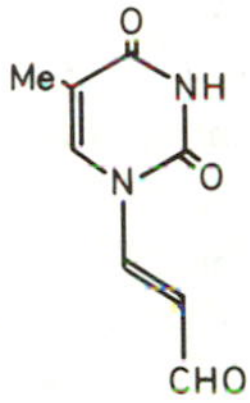

Thymine-3-propenal
(4.31)

Umezawa's group made a simplified molecule (4.30) with the iron-binding and oxygen-activating properties of (4.29); but it lacks an intercalating group and hence lacks the DNA-breaking action. However, molecules in which an iron-binding site is covalently linked to an intercalating moiety, are able to break DNA strands, apparently as effectively as bleomycin. Examples of such molecules are: haemin, linked to 9-aminoacridine (Lown and Joshua, 1982), and EDTA (Section 11.6), linked to an aminophenanthridine (Hertzberg and Dervan, 1982).

Thymine-propenal (4.31) is profoundly cytotoxic. It is not yet known whether the anti-cancer effect of bleomycin is largely due to the base-propenals to which it degrades the DNA of tumours.

The phleomycins, isolated from the same *Streptomyces*, differ from the bleomycins only by the presence of two extra hydrogen atoms in the thiazole rings, which remain flat. Oxidation yields the corresponding bleomycins (Umezawa, 1973). Their anti-cancer properties are only beginning to be explored.

Powerful 'amplifiers' (potentiators) for both phleomycin and bleomycin have been discovered in Australia (Brown and Grigg, 1982). The potentiators are nitrogen heterocycles of which the most potent seems to be 2-(5',7'-dimethyl-*s*-triazolo[1,5-*c*]pyrimidin-2'-ylthio)acetamide (Allen *et al.*, 1983). Part of the action of the potentiators seems to be unwinding the DNA helix sufficiently to allow better access of the antibiotic (Grigg, Edwards and Brown, 1971). There is some evidence that they may also stop mitochondrial energy being turned off before destruction of DNA is complete.

*The peroxy-anion radical ($^{\cdot}OO^{-}$), originally postulated, is less reactive.

Improvements have been made in the molecule of bleomycin by exposing the fermenting organism to foreign amines during its labours. When the sulfonium or guanidinium group was replaced by a weaker amine, the ratio of tumour damage to lung damage increased greatly. One such derivative, *pepleomycin*, differs from (*4.28*) only in having the —$NH(CH_2)_3S^+Me_2$ group replaced by —$NH(CH_2)_3$—$NH.CH(Me)Ph$. It is undergoing intense clinical study in Japan (Carter, Sakurai and Umezawa, 1981). In the USA, bleomycin-BAPP, which has —$NH(CH_2)_3NH(CH_2)_3NH(CH_2)_3NH$ as the replacing amine is being closely examined, as is *tallysomycin* which is a similarly modified phleomycin with an extra amino-sugar substituent.

Zinostatin (neocarzinostatin) and *auromomycin* are two other DNA-cleaving antibiotics which, however, have not achieved wide clinical use.

For more on molecular aspects of anti-cancer therapy, see alkylating agents (Section 13.4), and Neidle and Waring (1983).

Camptothecin, an alkaloid from the Tibetan shrub *Camptotheca acuminata*, cleaves DNA in a way that the cell ligases can repair (Horowitz and Brayton, 1970). The molecule has five fused rings of which the active portion is a pyridine lactone (*4.32*). It is principally used as a biochemical reagent for inhibiting biosynthesis of ribosomal and messenger RNAs while permitting formation of mitochondrial RNA (Abelson and Penman, 1972). For its anti-leukaemic properties, see Wall and Wani (1977).

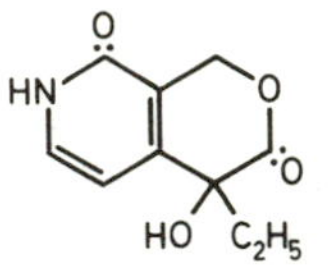

Active portion of camptothecin   Nalidixic acid
(4.32)   (4.33)

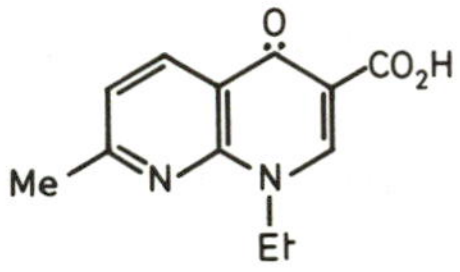

Caffeine, which is consumed daily in vast quantities by millions of people (without harm), increases the rate of spontaneous breakdown of DNA in *E. coli* and it is highly mutagenic in this bacterium (Grigg, 1970). It is a sobering thought that, under legislation current in many advanced countries, these adverse effects of caffeine on DNA could hold up the introduction of tea and coffee were they not already in common use. In assessing this, and other possible mutagens, it has to be borne in mind that most mutated cells are not viable, and a small percentage of mutations are actually beneficial.

4.0.5 *Miscellaneous anti-DNA drugs*

A urinary antiseptic with high selectivity, *nalidixic acid* (*4.33*) ('Negram', 1-ethyl-7-methyl-4-oxo-1,8-naphthyridine-3-carboxylic acid), has high potency against Gram-negative, and inertness toward Gram-positive, bacteria (a reversal of the usual trend in anti-bacterials). It has proved very successful in the clinic for curing urinary infections caused by rod-shaped bacteria which are usually difficult to treat. Nalidixic acid inhibits the synthesis of DNA in bacteria (but not in Man) without affecting either the use of existing DNA or the synthesis of RNA and protein (Goss, Deitz and Cook, 1965). The presence of an ionizing carboxylic acid group guarantees elimination of this drug in the urine. There is some evidence that chelation of heavy metals, by the oxo and carboxylic groups, is necessary for close contact between the drug and DNA (Crumplin, Midgley and Smith, 1980, a review).

Procarbazine (*4.34*), a derivative of *N*-methylhydrazine, injures DNA apparently by the conversion of the *N*-methyl group to formaldehyde. It is administered in Hodgkin's disease because it has no cross-resistance with other drugs given at the same time (Kreis, 1977). *N*-Methylformamide (*4.35*) also acts on DNA, by inserting a one-carbon fragment (Sartorelli and Le Page, 1958). Its metabolism is being studied (Gescher *et al.*, 1982), and it is undergoing Phase 2 clinical trial, (Professor M. Stevens, personal communication). The ethyl analogues of procarbazine and methylformamide lack anti-cancer activity, as does hexamethylmelamine. Because *N*-methylformamide shows no action on bone-marrow, it is likely to prove suitable for combination therapy with cyclophosphamide, cisplatin, or other myelosuppressants in treating lung and colon tumours.

Hexamethylmelamine (*4.36*), which is metabolized to pentamethylmelamine, also may function by releasing a one-carbon fragment to DNA. It is used in treating ovarian carcinoma (Wharton, 1979).

$$Me-NH-NH-CH_2-\langle\bigcirc\rangle-CO-NH-Pr^i$$

Procarbazine
(*4.34*)

MeNH—CHO

N-Methylformamide
(*4.35*)

Hexamethylmelamine
(*4.36*)

In concluding this section, two precautions necessary for responsible use of newly introduced anti-DNA drugs must be emphasized. Regular blood counts should be made, and the drugs should be withheld from women in the critical months of pregnancy.

4.0.6 *Drugs that inhibit synthesis of RNA*

The *rifamycins* are antibiotics isolated in Italy from *Streptomyces mediterranei* and then chemically altered to give more selective products. Rifamycin SV (*4.37a*) is the original substance but its derivative rifampicin (rifampin in the USA) (*4.37b*) is now the established drug in this series. It is much used, in conjunction with isoniazid (Section 11.9), for the cure of severe cases of tuberculosis but is still too expensive for general use (American Thoracic Society, 1980). Because it is one of the most selective of all known anti-bacterial drugs, it could profitably be used a great deal more in other infections, but is generally reserved for staphylococcal infections resistant to penicillin.

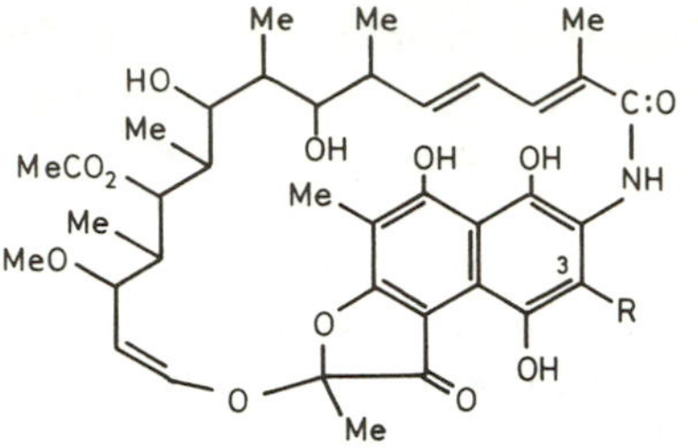

(a) Rifamycin SV, R = H   CH₂ ——————— CH₂
(b) Rifampicin R = —CH : N · N · CH₂ · CH₂NMe
(4.37)

Chemically, the rifamycins belong to a group of natural products known as ansamycins, from the Latin *ansa*, a handle. The name ansa-compound was coined by Lüttringhaus and Gralheer for substances containing an aromatic nucleus (naphthalene in the present case) around which is wrapped the 'handle' in the form of an aliphatic chain, joined in two places (Maggi *et al.*, 1966, 1968). At least three of the hydroxyl-groups are necessary for their action. The structure, which has not been synthesized, lends itself to chemical alteration, but principally only in the 3-position. Hydrogenation of the double bonds in the aliphatic ring progressively diminishes the activity in proportion as rigidity of the molecule is lost.

Rifamycins act on the β-subunit of the protein in bacterial DNA-dependent RNA polymerase. Combination is tight, though not covalent, and occurs in a 1:1 molecular ratio (this union does not take place in rifampicin-resistant organisms). As a result, no more RNA is synthesized, but neither DNA nor protein syntheses are affected. There is no action whatsoever on the RNA-synthesizing mechanism of mammals, which lacks a β-subunit, and hence the selectivity experienced by the patient is of the very highest degree (Tocchini-Valenti, Marino and Colvill, 1968).

Other ansamycins are found in Nature, e.g. streptovaricin (from another *Streptomyces*) which acts in the same way as rifamycin but is not so selective; amanitin (from the fungus *Amanita phalloides*) which is counter-selective, inhibiting eukaryotic but not prokaryotic DNA-dependent RNA polymerase; and maytansine (from the bark of an African flowering plant) which has anti-cancer properties but is not established in the clinic.

Dactinomycin (4.38), a bright red antibiotic isolated by Waksman and Woodruff in 1940, has an aminophenoxazine nucleus which bears two identical cyclic side-chains, each of which has one ester and five peptide linkages (4.39). The residues in each side-chain are: *N*-methylvaline, sarcosine, proline, valine, and threonine (Brockmann, 1960). Evidence from n.m.r. indicates that the —NH of valine is strongly hydrogen-bonded to the CO of sarcosine.

Dactinomycin is remarkably specific in inhibiting the synthesis of ribosomal RNA (DNA-primed) without any effect on the synthesis of DNA (Reich *et al.*, 1962). Use of radioactive dactinomycin showed that it was covalently bound to the guanine of the DNA starter, and did not combine with any other cell component at concentrations that block RNA synthesis. The phenoxazine nucleus intercalates into DNA near to a G-C pair, and the peptide portions project into the minor groove (Müller and Crothers, 1968) (for more on intercalation see Section 10.3.2).

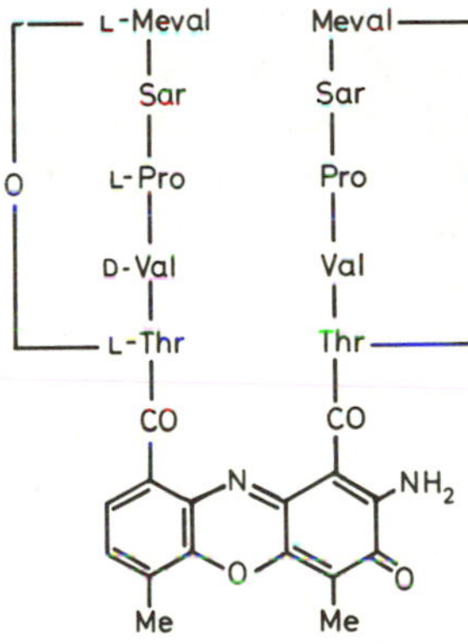

Dactinomycin
(4.38)

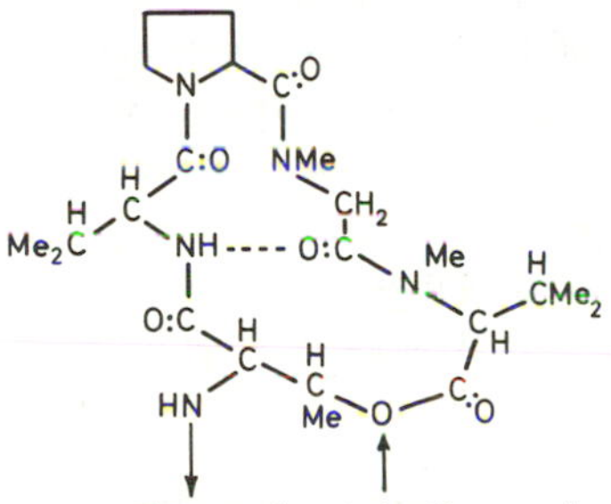

Side-chain of dactinomycin
↓ indicates point of attachment to nucleus
↑ indicates ester linkage
(4.39)

Dactinomycin, or actinomycin D as it was called when first introduced, has shown striking curative properties in Wilms's tumour of the kidney which forms a high proportion of all malignant tumours in children. Under its influence, even pulmonary metastases caused by this tumour regress (Farber and Mitus, 1968). Forms of cancer requiring longer treatment are not suitable for this drug, which has only moderate selectivity.

8-Azaguanine (*4.40*) was first prepared by synthesis but the antibiotic, pathocidin, isolated later from *Streptomyces albus* (Anzai and Suzuki, 1961), is identical. 8-Azaguanine is used to treat cancer of the brain, kidney, and liver because these organs are rich in guanase which deaminates it to harmless 8-azaxanthine, whereas tumours common in these organs lack this enzyme (Levine, Hall and Harris, 1963). A strong inhibitor of protein synthesis, 8-azaguanine is not incorporated into DNA in *E. coli*, T_2 bacteriophage (Smith and Matthews, 1957), cancer cells in culture, or in the intact mouse (Nelson *et al.*, 1975). Instead, it is converted by guanosine 5′-phosphate pyrophosphorylase into the nucleotide which is incorporated into messenger RNA, causing polysomes to break down into monomeric ribosomes: this puts a stop to protein synthesis (Kwan and Webb, 1967). In addition, 8-azaguanosine 5′-phosphate inhibits phosphoribosylpyrophosphate amidotransferase at an early step in purine bio-synthesis (McCollister *et al.*, 1964).

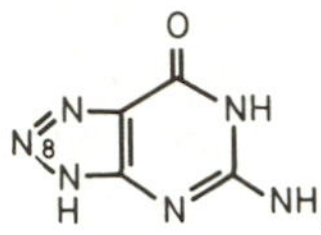

8-Azaguanine
(4.40)

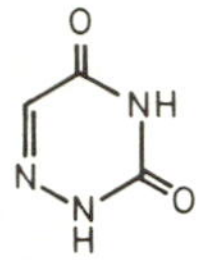

6-Azauracil
(4.41)

6-Azauracil (4.41) is used in agriculture as a fungicide to inhibit powdery mildew (e.g. in cucumbers). It is converted in the fungal cell to the ribotide which, being an analogue of orotidylic acid, blocks orotidylic decarboxylase (Dekker, 1968). Although a side effect on the central nervous system precluded its use in Man, the triacetyl-derivative of this ribotide, azaribine (see Section 3.6, p. 102), has proved valuable in the oral treatment of psoriasis (Calabresi and Turner, 1966).

Mithramycin, which has a tetrahydroanthracene nucleus with two chains of pyranose rings attached, is obtained from several species of *Streptomyces*. It inhibits synthesis of RNA while that of DNA continues. It can cause hypo-calcaemia and, in general, is not very selective, but has proved useful in treating testicular tumours (Hill *et al.*, 1972).

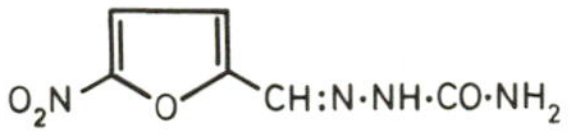

Nitrofurazone
(4.42)

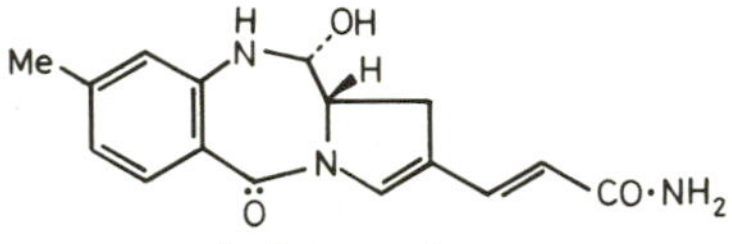

Anthramycin
(covalently hydrated form)
(4.43)

Nitrofurazone (4.42) (5-nitrofurfural semicarbazone, 'Furacin') is used top-ically with great success as a broad-spectrum anti-bacterial. It is also kept in reserve for arsenic-resistant late stages of trypanosomiasis. In dilute solution, nitrofurazone inhibits the formation of all types of RNA in *E. coli* (Tu and McCalla, 1976). Mutants that lack 'nitrofuran reductase' are not attacked, and hence reduction must precede activity of this drug. The active form is either the hydroxylamino analogue of this nitro-compound, or one differing from this by one electron. Nitrofurazone, with its reduction potential of −0.425 volts (versus the standard calomel electrode at pH 7) is an electron acceptor at a highly negative potential, just on the lower border of the biological range. The cor-responding derivative of benzene, *p*-nitrobenzaldehyde semicarbazone, has a reduction potential of −0.580 V, considered to render it biologically un-reducible (Sasaki, 1954). Of the nitrofurans, those with a conjugated side-chain are most easily reduced because the radical anion thus formed is stabilized by resonance (Lindberg, 1970). See, further, Section 5.4.3 (hydrogenosomes).

Related drugs include nitrofurantoin (*6.15*) (a urinary antiseptic) and metronidazole (*6.22*) (a nitroimidazole much used in amoebiasis and trichomoniasis).

Anthramycin (*4.43*), which readily undergoes covalent hydration (Section 2.5, p. 47), inhibits the synthesis of RNA by binding to the DNA template in RNA polymerase, which could account for its anti-tumour and anti-microbial action (Horwitz and Grollman, 1968; Kohn and Spears, 1970).

Hycanthone (*3.42b*), 1-(2-diethylaminoethylamino)-4-hydroxymethylthioxanthene-9-one, is used to cure schistosomiasis in a single intramuscular dose. It prevents the incorporation of uridine into the RNA of host cells in both susceptible and resistant worms. This effect is quickly reversed in host cells and resistant worms, a likely reason for its selectivity (Mattoccia, Lelli and Cioli, 1981). Its use has some drawbacks, particularly nausea, and there is evidence of carcinogenicity and mutagenicity in model experiments. Hence it is not advised for children, or those who risk being repeatedly infected.

Flavan (*4.44*), which is the parent of innumerable colouring substances in fruits and flowers, is slightly active against 20 strains of common-cold virus in plaque tests. The highly lipophilic 4′,6-dichloro-derivative is much more active. How to administer such a water-insoluble substance effectively to human volunteers presented a problem; but it does inhibit RNA synthesis in the parasitized cell, and the therapeutic index (mouse) is exceptionally high (Bauer *et al.*, 1981).

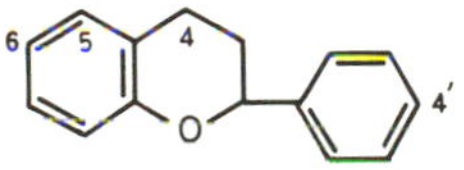

Flavan
(*4.44*)

4.1 Proteins

In every living cell, proteins are assembled on ribosomes by the co-polymerization of about 20 kinds of amino acids, all with the same L-optical configuration (see Fig. 5.7). The molecular weight of proteins varies from about 6000 to well over a million, but all have the common structure shown in Fig. 4.2, where R represents the familiar side-chains, such as methyl (for alanine) and *p*-hydroxybenzyl (for tyrosine). The *Atlas of Protein Sequence and Structure* (National Biomedical Research Foundation, Georgetown University, Washington, D.C.), published at intervals, contains the amino acid sequences of many hundreds of proteins (for other lists, see Section 17.4).

The sequence of the amino acid residues is determined by the genetic information in the cell's DNA, and expressed through the combined operation of mRNA and tRNA. This sequence is known as 'primary structure'. Such a

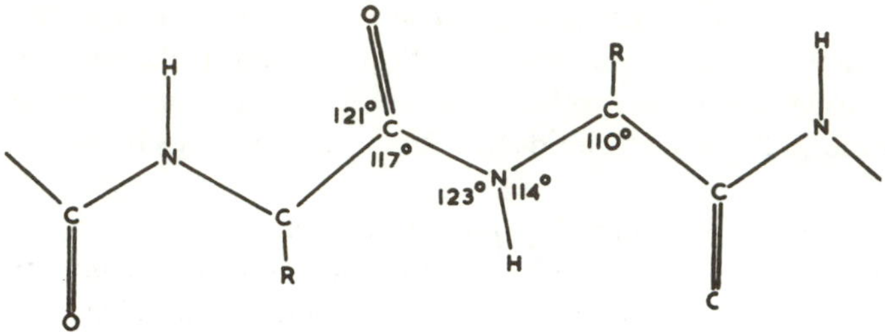

Fig. 4.2 Portion of primary polypeptide chain.

polypeptide chain is always folded into a secondary structure which is retained by multiple hydrogen-bonding between the —CO groups in one strand and the NH— groups in the next. The reversal of the chain's direction, to give strands, is forced wherever proline residues occur in it. This produces a sheet which has a repeat distance of 7.2 Å between *alternate* amino acid residues. As it occurs in nature, the sheet is slightly pleated to make room for small-to-medium side-chains, and the repeat distance is about 7.0 Å, as in fibroin, the protein of silk.

When the average size of the side-chain is large, a *helix* is formed instead of a sheet. These helices (also called α-helices, but the α is unnecessary) are all right-handed and have 3.7 amino acid residues in each turn. As Pauling and Corey showed, this is the structure of a wool fibre. These spirals have a repeat distance of 1.5 Å between amino acid residues measured along the *axis*; neighbouring amino acids unite the loops of the spiral by CO . . . HN bonding. Apart from X-ray crystallography at a resolution of 2 or 3 Å, knowledge of this secondary structure has been acquired by investigating the optical rotatory dispersion, and by measuring the number of hydrogen atoms which exchange *slowly* with the deuterium of D_2O (i.e. those forming interhelical bonds).

These helices are long strands which, in globular (i.e. non-fibrous) proteins, including enzymes, are folded back on themselves to give the irregular loops characteristics of tertiary structure. These loops are usually set in a form pre-determined by the order of the amino acids (Perutz, Kendrew and Watson, 1965). The setting is effected by covalent (disulfide), ionic, van der Waals, or hydrogen bonds. The molecule is compact but with a few water molecules inside. Almost all the polar groups (e.g. —OH, —NH_2, —CO_2H) are on the outside of the molecule. Because of this, globular proteins dissolve in water. The S—S bonds, which give ribonuclease (for example) the conformation on which its enzymatic action is dependent, unite the following pairs of amino acid residues, counting from the NH_2 end of the chain: 28—84, 65—72, 40—95, 58—110 (there are 124 residues in this protein).

Globular proteins have been found (by X-ray diffraction examination of their crystals) to be partly helical, partly pleated, and partly random. Randomness seems to be encouraged by residues of asparagine, aspartic acid, and phenyl-alanine. It also appears that the acidic constituents occur preferentially near the

NH_2 end of the chain, whereas the basic constituents (lysine, arginine, histidine) are found more often towards the CO_2H end (Cook, 1967).

No RNA is needed, in plants or animals, to synthesize the tripeptide, glutathione, nor in bacteria to synthesize polypeptides with up to 20 amino acid residues.

For a discussion of enzymes, see Section 9.0. For further information on the structure of proteins, see Neurath and Hill (1975), and the periodical, *Advances in Protein Chemistry*. Models of proteins may be visualized in three dimensions by using the Stereo Supplement to the book *Structure and Action of Proteins* (Dickerson and Geis, 1969).

4.1.1 *Drugs that act by inhibiting the synthesis of proteins*

Structural differences between bacterial (70S) and mammalian (80S) ribosomes may suggest that the discussion of the inhibitors of protein synthesis should be deferred to Section 5.4, where the different kinds of ribosomes are compared. However, sufficient biochemical differences are emerging to indicate that the protein-synthesis-inhibiting drugs should first be presented in this chapter.

Chloramphenicol (*4.45*), isolated from *Streptomyces venezuelae* in 1947 but now obtained entirely by synthesis, is the D-(-)*threo* isomer of 2,2-dichloro-*N*-[2-hydroxy-1-(hydroxymethyl)-2-(4-nitrophenyl)ethyl] acetamide. Its selective antibacterial effect, when given orally, depends on its inhibition of protein synthesis on the ribosomes of bacteria without affecting that taking place on mammalian ribosomes, even when it has free access to the latter (Rendi and Ochoa, 1962). The other three stereoisomers have no effect on protein synthesis.

X-ray crystallography shows that the two hydroxy-groups lie close together (the amide-group points away from these), and the whole aliphatic portion is in a plane roughly at right angles to the benzene ring (Dunitz, 1952). Many think that the amide-linkage in the chloramphenicol molecule makes it act as a stereospecific analogue of a natural dipeptide (Das, Goldstein and Kanner, 1966).

The antibacterial properties of chloramphenicol are unimpaired when the nitro-group is replaced by another strongly electron-attracting group, such as methylthio- (MeS—), azido (—N_3), or methylsulfonyl (MeSO$_2$—), hence the role of the nitro-group is simply electron-attracting. Of these analogues, the methylsulfonyl-compound, known as thiamphenicol, is used clinically in Britain, Italy, and Japan (unlike chloramphenicol, it is excreted unaltered through the kidneys, thus extending the usefulness of this series). The D-*threo* configuration has proved as essential for these altered molecules as it is for chloramphenicol itself (Freeman, 1970). The aliphatic portion of the molecule is more sensitive to change, e.g. the substitution of any hydrogen atom by a methyl-group leads to complete inactivation.

Bacteria quickly absorb chloramphenicol which is rapidly accumulated on

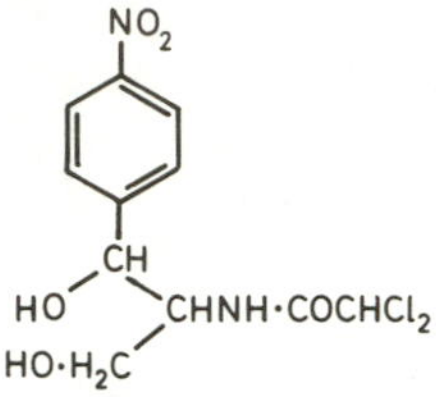

Chloramphenicol
(4.45)

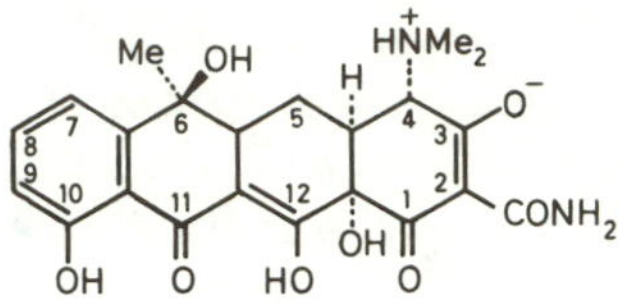

Tetracycline
(4.46)

the 50S subunit of the 70S ribosomes typical of bacteria, and protein synthesis is rapidly halted (learnt from studies with [^{14}C]chloramphenicol followed by ultrasonic rupture of cell walls). On the ribosome, chloramphenicol prevents formation of the peptide bonds by inhibiting peptidyltransferase at the A site (Jacoby and Gorini, 1967). By working with ribosomal cores that had been depleted of specific proteins, it was shown that chloramphenicol most specifically binds to protein L-16 (Roth and Nierhaus, 1975).

Chloramphenicol was the first broad-spectrum antibiotic to be used in medicine, but it came under a cloud when long-continued administration produced many cases of aplastic anaemia, which can be life endangering. Its use is now restricted to diseases where it is the most active known remedy, and which are likely to be cured quickly, within the safe period of the drug. Hence it is used to cure typhoid fever, bacterial meningitis, and anaerobic infections of the brain such as those caused by *B. fragilis*. It is the only common antibiotic to pass freely into the cerebrospinal fluid and to cross the blood–brain barrier. It also serves as a useful alternative to the tetracyclines in cholera and the rickettsial diseases such as typhus or Rocky Mountain spotted fever.

Tetracycline (4.46) and its derivatives are the most used of all 'broad-spectrum antibiotics'. Their selectivity depends on their preferential accumulation by bacteria, as was outlined in Section 3.0. Chelation of magnesium also plays an important part in their action, and this is discussed in Section 11.8. Tetracycline was prepared by the dechlorination of its 7-chloro-derivative ('Aureomycin'), the first medicinal tetracycline, isolated in 1947 from *Streptomyces aureofaciens*. It is a dimethylaminopentahydroxydioxo-octahydro*napthacene*carboxamide.

By using [^{3}H]tetracyclines it was shown that these antibiotics were bound to the 30S ribosome subunit (e.g. Connemacher and Mandel, 1965), where they inhibited the *m*RNA-directed binding of aminoacyl-tRNA (Sarkar and Thach, 1968). In this way they efficiently inhibit bacterial protein synthesis. Orally administered, tetracyclines cure infections caused by many species of Gram-positive and -negative bacteria, spirochaetes, and rickettsiae, with very few side effects, none of them serious. The earlier types, which needed large and frequent doses, have been supplemented by (a) two long-acting examples: demeclocycline [as (4.46) but with Cl in 7-position, and no Me in 6], and methacycline [gain of OH in 5-position, and dehydration of 6 to $=CH_2$], and (b) two much longer-acting examples which have the further advantage of being better absorbed: doxycycline [OH moved from 6 to 5 position], and minocycline [loss of Me and OH from 6-position, gain of NMe_2 in 7]. (Barza and Scheife, 1977). Many strains of bacteria have become resistant to the tetracyclines.

Erythromycin (4.47) is an orally active macrolide antibiotic isolated in 1952, from *Streptomyces erythreus* in soil from the Philippines. Studies with [^{14}C]erythromycin show that it inhibits protein synthesis in 70S (but not in mammalian 80S) ribosomes. In detail, it prevents peptidyl-tRNA transferring from the A to the D site on the 50S ribosomal subunit during translocation (Mao, Putterman and Wiegand, 1970).

Erythromycin has much the same spectrum of antibacterial activity as penicillin. Because erythromycin-resistant strains of the commoner bacteria abound, the use of this antibiotic is becoming confined to three diseases: mycoplasmal pneumonia, diphtheria, and Legionnaires' disease.

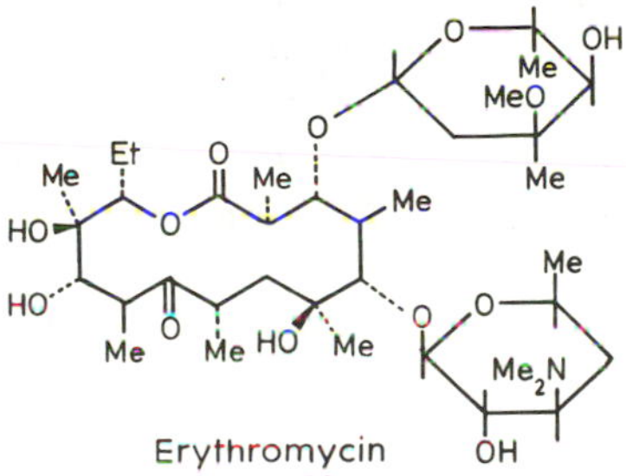

Erythromycin
(4.47)

Of the *aminoglycoside antibiotics*, the first to be discovered was *streptomycin* (Schatz, Bugie and Waksman, 1944) which was used for many years against tuberculosis and for infections with Gram-negative bacteria. In the long run, its record of causing kidney damage and incurable deafness have led to disuse. Its place has been taken by other aminoglycoside antibiotics. These antibiotics consist of amino-sugars joined to a central cyclitol ring. They are highly basic substances, unabsorbable orally and hence given intramuscularly.

The aminoglycosides penetrate the bacterial plasma membrane in a complex and individual way, described in Section 14.3. They act by inhibiting protein synthesis at an early stage. While harmless to the mammalian (80S) ribosome, they bind to the 30S subunit of the bacterial ribosome (Le Goffic *et al.*, 1979) and are actually bactericidal, whereas other antibiotics that interfere with protein synthesis are only bacteriostatic.

Gentamicin, a mixture of *(4.48)* with two of its methyl homologues, was isolated in 1963 from *Micromonospora purpurea* which is an actinomycete (Table 1.3). Although highly active against all common Gram-positive and -negative bacteria, gentamicin is reserved for infections of the urinary tract with Gram-negative rods. Although more selective than streptomycin (Milanesi and

Ciferri, 1966), it still carries the risk of kidney damage (often reversible) and deafness (irreversible), and is not a drug to be prescribed lightly.

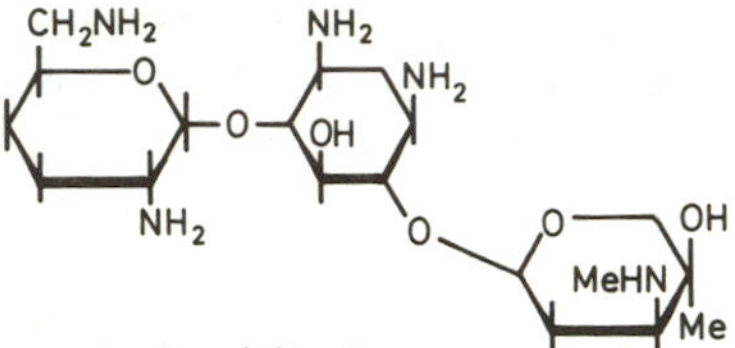

Gentamicin C₁ₐ
(4.48)

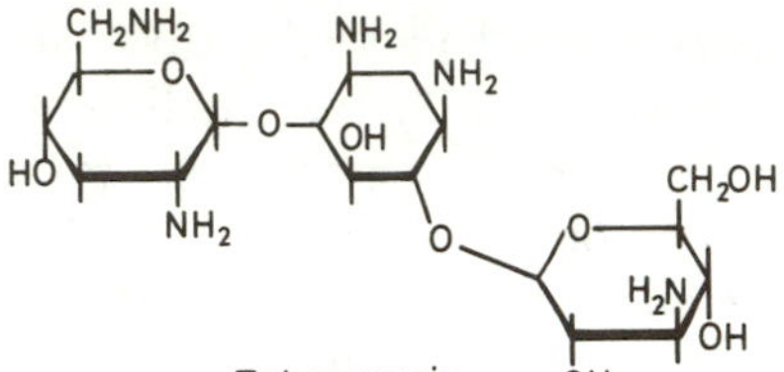

Tobramycin
(4.49)

Tobramycin (4.49) closely resembles gentamicin both in activity and in untoward effects, but has greater activity against *Pseudomonas* infections. *Netilmicin* (semi-synthetic) is coming into greater use because of its lowered ototoxicity (Lerner, 1983). *Amikacin,* which is semi-synthetic, finds special use in hospitals where gentamicin-resistant organisms are infecting patients (noso-comial infections); *paromomycin* is used in amoebic dysentery for its direct effect on the protozoon (Woolfe, 1965); *kanamycin* has given way to tobramycin; and *neomycin,* too little selective for internal use, is a common constituent of antibiotic ointments. For a review on aminoglycoside antibiotics, see Umezawa and Hooper (1982).

Clindamycin, a substituted pyrrolidine, is used only for intestinal abscesses caused by anaerobes like *Bacteroides,* and the related lincomycin has passed out of use.

Fusidic acid, a sterol carboxylic acid isolated from a fungus, is used for pneumonia and other coccal infections that are resistant to the commoner antibiotics, particularly in children. It is equally inhibitory to both bacterial and mammalian ribosomes, but is selective because it cannot penetrate the mammalian cytoplasmic membrane (Franklin and Snow, 1981).

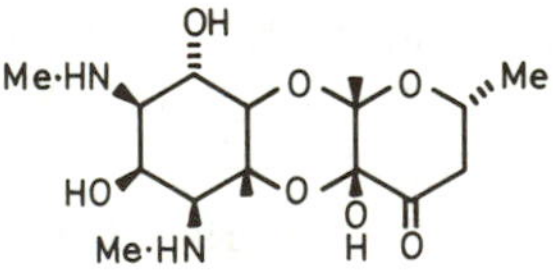

Spectinomycin
(4.50)

Spectinomycin (4.50) (decahydro-4a,7,9-trihydroxy-2-methyl-6,8-*bis*methyl-aminopyrano[2,3-*b*]benzodioxin-4-one, 'Trobicin') inhibits protein synthesis by binding to the 30*S* ribosomes. It is selective, but weak, and is reserved for treating, intramuscularly, cases of gonorrhoea resistant to penicillin and the tetracyclines (Reyn *et al.,* 1973).

Tunicamycin, a biochemist's tool not used in medicine, inhibits the gly-cosylation (usually on the amino-group of asparagine) of the proteins that line the ion-channels in plasma membranes; this renders the pores non-functional (Olden, Pratt and Yamada, 1978).

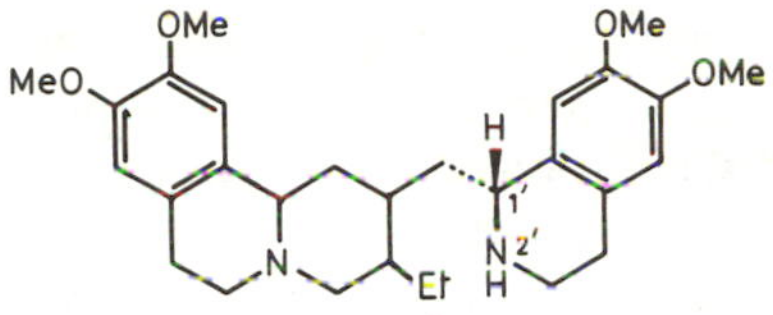

Emetine
(4.51)

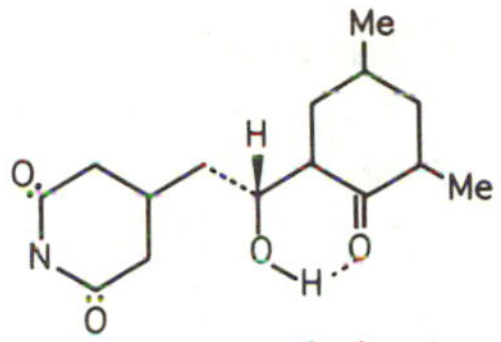

Cycloheximide
(4.52)

The alkaloid *emetine* (*4.51*), the first of all drugs against amoebiasis (see Section 6.0), is now losing ground against such non-nauseating but highly effective remedies as metronidazole and chloroquine. However, it is still considered a useful adjuvant. Emetine acts on both parasite and host by suppressing protein synthesis namely by preventing the translocation of peptidyl-tRNA from the acceptor site to the donor site in the ribosomes (Huang and Grollman, 1970). The likely basis of its selectivity is that the amoebae, unlike mammalian liver cells, cannot quickly enter into a recovery phase between courses of the drug (Grollman, 1966).

Cycloheximide (*4.52*), (obtained from *Streptomyces griseus* along with strepto-mycin) is counter-selective, for it inhibits protein synthesis in mammalian, but not in bacterial, cells. It is widely used by biochemists to inhibit protein synthesis whenever it is desired to establish if a given biochemical effect depends on the synthesis of new protein. It also has limited use as an agricultural fungicide. There are configurational and conformational similarities between emetine and cycloheximide. In emetine, the *R* (rectus) configuration at C-1' and the secondary nitrogen atom at N-2' are essential for activity, as inspection of a series of analogues, active and inactive, showed. Cycloheximide, which has a hydrogen-bonded nucleus (Siegel, Sisler and Johnson, 1966), shares much of emetine's essential structure (Grollman, 1966, 1968).

For a review on the design of drugs intended to inhibit protein synthesis, see Grollman (1971).

4.2 Analogous enzymes and coenzymes

Many examples are known where the enzymes carrying out apparently identical functions, in dissimilar cells or dissimilar genera, are themselves dissimilar. When the chemical differences between the two enzymes is small, we speak of isoenzymes but when it is much larger, they are called analogous enzymes.

(a) *Isoenzymes.* Many apparently pure enzymes have been separated by electrophoresis into a small number of pure proteins; each of these isoenzymes has a specificity similar to that of the crude enzyme, yet differing subtly in physical properties. One of the best-known examples, lactic dehydrogenase (LDH), exists in animal tissues as five isoenzymes. These are tetramers, formed by the association of two polypeptides A and B, and have the following composition: B_4, AB_3, A_2B_2, A_3B, and A_4. Creatine kinase and arginine kinase similarly have several isoenzymes. Isoenzymic patterns tend to remain constant for a given tissue in various mammalian species, but to differ from tissue to tissue in any one species. For selective inhibition of a particular isoenzyme, see Sections 9.4.6 (monoamine oxidase) and 9.7.1 (lactic dehydrogenase). Thymidine kinase, which catalyses the phosphorylation of thymidine, has two isoenzymes, one associated with the cytoplasmic and one with the mitochondrial fractions of mammalian cells. The mitochondrial form preponderates in the tissues of normal human adults whereas the cytoplasmic isoenzyme is the principal form found in tumours (Lee and Cheng, 1976).

(b) *Analogous enzymes.* Whereas isoenzymes seem to differ from one another mainly by an increment of electric charge, analogous enzymes (which perform the same function in *different* organisms) differ in a more fundamental way. Alternative names are isofunctional enzymes, equifunctional enzymes and (in the older literature only) homologous enzymes. The magnitude of their differences provides good conditions for deploying selectively toxic drugs. Analogous enzymes can be demonstrated by differences in kinetics, electrophoresis, or specificity (for substrates, coenzymes, or inhibitors). Here are examples of some of the chemical differences that exist between such enzymes. Numerous hexokinases have a high specificity for a single hexose, whereas others catalyse the phosphorylation of several carbohydrates. Differences in requirements for coenzymes, particularly metals, can be found. Thus, crystalline aldolases from yeasts and moulds require iron (Warburg and Christian, 1943), and so does that from the bacterium *Clostridium perfringens* (Bard and Gunsalus, 1950), whereas those from mammals, plants, and trypanosomes do not (Taylor, Green and Cori, 1948). Again, the superoxide dismutases, a family of enzymes found in both prokaryotes and eukaryotes, differ in that some have manganese but others have copper at the functioning site; moreover some have an isoelectric point near 4.5, others near 8 (Oberley, 1982).

There are two distinct classes of the enzymes known as FDP (fructose 1,6-diphosphate) aldolases, which carry out an important and early stage in glycolysis: (a), the varieties found in animals and higher plants cleave FDP by way of a Schiff base, whereas (b), those which occur in bacteria and fungi require a metal (usually Zn^{2+}) bound to the carbonyl group in the enzyme–substrate complex; special inhibitors exist for the second variety (Lewis and Lowe, 1973). More examples of differing metal requirements will be found in Section 11.1: see under *Helminths* in Section 4.4 for some other differences in analogous enzymes.

Table 4.6 clearly shows how antimonial drugs can distinguish between a pair of analogous enzymes, one in a parasitic worm, the other in its mammalian host. This is the selectively toxic basis of the classical treatment of schistosomiasis.

The ability of various diaminopyrimidines to distinguish between analogous forms of the enzyme dihydrofolate reductase is the basis of some of the best contemporary anti-malarial and anti-bacterial therapy (see Section 4.0, p. 123, Tables 4.1 and 4.2, and Section 9.3.3 and 9.6). Let us first look at the differences that exist between various vertebrate types of the enzyme, none of which is much inhibited by trimethoprim (4.9), and then proceed to invertebrate types, which are highly susceptible to this drug. The enzyme from chicken liver has only 75% identity of amino acid sequence with that from ox liver. Moreover, methylmercuric hydroxide activates the avian type twelvefold whereas it in-activates the bovine type. The avian type is much richer in basic amino acids and has an isoelectric point of 8.4 compared to 6.8 for the bovine type. This result is achieved in the avian type by the presence of lysine at positions 32, 106, and 154, whereas the bovine type has glycine, threonine, and glutamic acid, respectively, in these positions (Kumar et al., 1980).

No matter what the source of dihydrofolate reductase (DHFR), there is an acidic group in (or near) the 27th residue; for instance, in E. coli this is Asp-27. In fact it is always aspartic acid for prokaryotic forms, but glutamic acid for eukaryotic forms. This seems an important difference, because it is the aspartic acid residue in the prokaryotic enzyme that binds to the $2-NH_2$ of 2,4-diamino-pyrimidine inhibitors (Bolin et al., 1982).

Again, DFR from vertebrate sources has the residue Tyr-31 in place of the less less bulky Leu-27 of the bacterial enzymes; these residues line the pocket in which the pteridine nucleus has to fit in each case. Vertebrate enzymes, which have about 185 residues, are larger than those of bacteria with about 165 residues. How this difference comes about is seen in chicken liver enzyme which has three extra loops on the edge of the pleated sheet, all of them free from normal interchain hydrogen-bonding (Volz et al., 1982). Unlike bacterial DFR, mammalian DFR can reduce folate as well as dihydrofolate.

The DFR of the malarial parasites has a much higher molecular weight (103 000–210 000) compared to those of bacterial (17 000) and vertebrate (21 000) types, and performs a second enzymic function as thymidylate synthe-tase (methylation of a uracil ring).

See Volz et al. (1982) for a tabulation of sequences (aligned) for amino acid residues in seven DFR molecules: three bacterial, three liver (chicken, pig, ox), and one cancer (murine lymphoma) in origin. Only 25 residues are conserved as between bacterial and mammalian examples.

The different locations and bindings of methotrexate and the diamino-pyrim-idine and -triazine inhibitors, in these analogous enzymes, will be described in Section 9.3.3, with stereo-drawings.

The acetylcholinesterase in the worm Haemonchus contortus, which parasitizes the sheep's gut, is irreversibly inhibited by the organophosphorus drug, haloxon

(*13.36*), a much-used vermifuge. Yet the acetylcholinesterase of the sheep's gut is only temporarily affected and rapidly recovers. Worms not affected by haloxon have been shown not to have this *Haemonchus* variant of the enzyme (Lee and Hodsden, 1963). Pure glutamate dehydrogenase from *Trypanosoma cruzi* has a structure very different from that of the mammalian enzyme (Juan *et al.*, 1978).

Parallel enzymes in *two mammalian species* can be analogous instead of (as expected) identical. For instance, rabbit muscle adenylate kinase is inactivated by 0.8 mM N^6-iodoacetamidohexyladenosine 5'-phosphate, whereas the corresponding enzyme in pig muscle was quite unaffected even by a 2.8 mM solution (Hampton, 1976).

Apart from such inter-species differences, many differences have been found in the proportions of enzymes in the different tissues of a *single organism*, even such very closely related tissues as heart and skeletal muscle. Table 4.3 provides examples.

Table 4.3 Proportions of enzymes in muscle (rat)

	Heart	*Skeletal*
Citrate synthase	12	1
Enolase	1	8
Fructose biphosphatase	45	1
NADH dehydrogenase	35	1
Pyruvate kinase	1	8

(Dixon and Webb, 1979.)

The distribution of many other enzymes in mammals is limited to particular organs as exemplified in Table 4.4. The more specialized the mammalian tissue, the more individual its enzymes. The stomach's ATPase is uniquely blocked by omeprazole (Gustavsson, 1983).

Table 4.4 Enzymes that occur mainly in special organs of mammals (rat)

Enzyme	*Organ*
Arginase	Liver
Acid phosphatase	Prostate, kidney, liver, heart
Alkaline phosphatase	Kidney
Carboxyesterase	Pancreas
β-Glucuronidase	Spleen
Glucosaminephosphate isomerase	Intestines
Glutamine synthetase	Brain, liver
Mannosidase	Epididymis
Ribonuclease	Pancreas

(Dixon and Webb, 1979.)

Some information is available on the chemical differences between members of a pair of analogous enzymes. Thus the enzyme that hydrolyses cyclic adenosine monophosphate (cAMP) has been shown to have different molecular composition when specimens from different mammalian tissues were compared (Weiss and Fertel, 1977). Relevantly, pyruvate kinase from healthy liver (L), kidney (K), and muscle (M) of the rat gave inhibitory ratios (L:K:M) of 1:7.6:6 with 3'-methoxy-ADP, of 1:1.2:7.1 with 8-ethylamino-ADP, and 3:2:1 with methyl-(N-acetyl-ω-methylaminobutyl)-ADP (Hai, Abo and Hampton, 1982). Similarly, small changes in the substituents inserted into pyrazolo-[1,5-a]-1,3,5-triazine (*4.53*) bring about an inhibition of cAMP phosphodiester-ase in *different* tissues from this list: bovine brain, bovine heart, or rabbit lung (Senga *et al.*, 1982).

A pyrazolotriazine
(*4.53*)

Some very interesting differences can exist between an enzyme *in cancerous tissue* and the analogous enzyme in normal tissues. For instance, the specific activity of purine phosphoribosyl transferase in mouse tumour cells (Ehrlich ascites) had between 15 and 60 times the activity of that in liver, brain, spleen, heart, or kidneys of the same animal (Murray, 1966). Again, the adenylate kinase of a rat hepatoma was 22 times more strongly inhibited than the analogous enzyme from healthy rat muscle using an ATP analogue: P^1-P^5-*bis*-[8-(ethylthio)adenosine-5'-]pentaphosphate (Kappler *et al.*, 1982).

The organ-specific character of so many enzymes suggests possibilities for devising selective pro-drugs, masked with a group that the target tissue can specifically remove. An example is fosfestrol USP (diethylstilbestrol diphosphate), which remains biologically inert until hydrolysed by the acid phosphatase which is abundantly present in the prostate gland when carcinoma is present there. The liberated diethylstilbestrol brings about regression of the growth and this treatment is considered life-saving (Lambley and Ware, 1967).

An enzyme that is present in two communicating tissues may be active in only one of them because of a difference in pH. For example, the pH of cancer cells, provided glucose is available, tends to be lower (often pH 6) than that of the surrounding healthy cells (pH 7.3) (Rauen, 1964; Schloerb *et al.*, 1965). As a result of this difference, the enzyme β-glucuronidase (which has a pH optimum of 5.2) is much more active in tumours than in normal cells (Bicker, 1974). Some attempts have been made to adopt this phenomenon for designing anti-cancer drugs, and probably more will follow (see Section 10.2). That some human cancer tissues can have an extraordinarily high content of β-glucuronidase relative to normal tissue was first demonstrated by Fishman and Anlyan (1947); it is useful knowledge to combine with the pH difference.

A different kind of enzymatic individuality in cancer cells is shown by three liver tumours (from mouse, rat, and human) which had lost feedback control of cholesterol synthesis while the surrounding healthy liver tissue was exercising normal restraint on this synthesis. The site of this failure of control was traced to HMG-reductase, the enzyme that converts β-hydroxy-β-methylglutaric acid to mevalonic acid, on the way to squalene (Siperstein and Fagan, 1964).

The various intracellular organelles also have their characteristic enzymes, examples of which are given in Table 4.5.

Table 4.5 Characteristic enzymes of intracellular structures (rat liver)

Plasma membrane	ATPase; adenylate cyclase
Nucleus	Nicotinamide−mononucleotide adenylyl transferase
Mitochondria	Amine oxidase (outer membrane),
	Sulfite oxidase (intermembrane compartment),
	Succinate oxidase and cytochrome-*c* oxidase (inner membrane),
	Glutamate dehydrogenase and pyruvate dehydrogenase.
Endoplasmic reticulum	Glucose 6-phosphatase; NADPH−cytochrome reductase; Fructose diphosphate aldolase
Golgi apparatus	Galactosyltransferase
Peroxisomes	Acid phosphatase; catalase; ureate oxidase; D-amino acid oxidase
Lysosomes	Acid phosphatase

(Dixon and Webb, 1979.)

For more information on enzymes, see Section 9.0.

(*c*) *Coenzymes*. It is convenient to continue the foregoing discussion at the level of coenzymes, even though the differences observed must often depend on differences in the apoenzymes.

Many remarkable species differences have been found among the coenzymes. Most plants and animals synthesize their own ascorbic acid which is (among other tasks) essential for the hydroxylation of proline and lysine in the biosynthesis of collagen. However, Man, other primates, and the guinea pig are notable exceptions, so that for them, and for them alone, it is a vitamin, and must be taken in with food.

Differences between bacteria and Man in the absorption and the biosynthesis of dihydrofolic acid (*2.14*) and its derivatives are so great that the whole system of sulfonamide chemotherapy rests on it (Section 9.3.1). In brief, pathogenic bacteria can synthesize their requirements of folic acid, but cannot absorb preformed folic acid in their nutriment. Man, on the other hand, cannot synthesize this coenzyme, but has no difficulty in absorbing it from food.

Tumour cells (ascites hepatoma) of the rat liver can take up (i) pyridoxine phosphate and (ii) pyridoxal phosphate without eliminating the phosphate group, but the liver cells of normal rats always completely dephosphorylate these forms of vitamin B_6 (Ito, Nakahara and Sakamoto, 1964).

Early work on members of the cytochrome family established that the pattern of distribution varied more widely among bacteria than among eukaryotes. The extreme position is occupied by the anaerobic genus *Clostridium*, members of which have no cytochromes whatsoever (Keilin, 1933). The parasitic worms *Ascaris lumbricoides* var. *suum* (a nematode) and *Moniezia expansa* (a cestode) lack cytochrome *P*-450. This suggests that helminthicidal drugs can be designed which will persist in these parasites but be destroyed by the *P*-450-linked oxidases of their hosts (Douch, 1976).

The sideramines, iron-containing substances found only in bacteria, and thought to be bacterial equivalents of the cytochromes, are dealt with in Section 11.1, along with the ferredoxins, non-haem iron-containing proteins confined to bacteria and plants.

4.3 Nitrogen and phosphorus metabolism

This Section and the next one deal with catabolic processes. Because these are known to follow fairly uniform patterns in most living cells, the expectation of variations that could be exploited selectively may not seem very high. In spite of that, some variations do exist, and a selection of them is recorded here together with indications of their use.

Nitrogen metabolism has end-products more varied than those of fat or carbohydrate metabolisms, ranging in complexity from ammonia to the

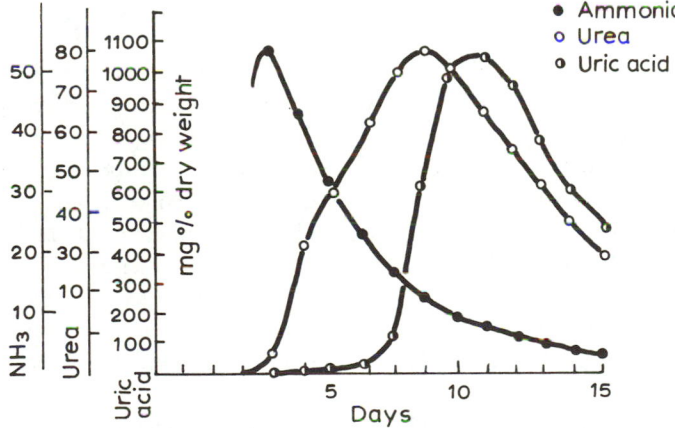

Fig. 4.3 Nitrogen excretion of chick embryo (after Needham).

Maximum of	Days after fertilization
Ammonia	4
Urea	9
Uric acid	11
(Baldwin, 1948a.)	

alkaloids. What is still more surprising, the end-products of nitrogen metabolism can vary within a single species, depending on the stage of development. For example, Fig. 4.3 shows that a chicken, in the egg, passes in turn through the stages of excreting ammonia (as though it were a fish), urea (as though it were a frog), and uric acid (typical of birds), all within a few days. This sequence may shed light on the well-known necessity for using different selectively toxic agents to deal with parasites at different stages. For example, the sexual form of the parasite of malignant tertian malaria can be attacked only by 8-aminoquinolines, e.g. primaquine (*3.36*); whereas the asexual form, although coexisting with the other in human erythrocytes, is injured by drugs of quite different molecular structure such as chloroquine (*10.31*) and mepacrine (*10.30*).

Some unusual amino acids occur in the cell walls of bacteria (Section 5.3). Several unusual biosyntheses of common amino acids have been recorded. Thus plants and bacteria make lysine only by decarboxylating diaminopimelic acid, but the lysine of fungi and mammals is made from 2-aminoadipic acid (Vogel, 1959). No amino acid of the D-series has been found in an active cell constituent of any creature higher on the evolutionary scale than the earthworm (*Lumbricus terrestris*), which contains the phosphagen lombricine (*O*-phosphodiester of guanidinoethanol and D-serine) (Ennor *et al.*, 1960).

The aromatic amino acids, phenylalanine and tryptophan, are made in bacteria and plants from shikimic acid (*4.54*) (3,4,5-trihydroxycyclohex-1-ene-1-carboxylic acid); but mammals cannot form the benzene ring and hence have to obtain these two amino acids from the diet (Gibson, 1964). From shikimic acid, plants elaborate gallotannins, and insects harbour bacteria which make protocatechuic acid (3,4-dihydroxybenzoic acid) which tans the proteins of their integument.

Because shikimic acid does not enter into mammalian metabolism, its synthesis and use are clear targets at which to aim selective toxicity. In bacteria, shikimic acid arises by cyclization of the carbohydrate 3-deoxy-2-oxo-D-*arabino*heptulosonic acid 7-phosphate, which is formed by the condensation of erythrose 4-phosphate and phosphoenolpyruvic acid. Shikimic acid undergoes biosynthesis to chorismic acid (*4.55*) which is the enolpyruvic ether of *trans*-3,4-dihydroxycyclohexa-1,5-diene-1-carboxylic acid. As its name indicates, this acid sits at a metabolic fork, the branches of which lead to prephenic acid, to phenylalanine (and hence to tyrosine), to anthranilic acid (and hence tryptophan), to ubiquinone, vitamin K, and *p*-aminobenzoic acid (and hence folic acid).

Glyphosate (*4.56*) which hinders steps in the biosynthesis of chorismic acid is a successful herbicide; devising other agents by inhibiting shikimic, prephenic, and chorismic acids is open for further exploration.

Adenosine triphosphate (ATP) is employed by all multicellular organisms for energy storage and transfer, but some bacteria (e.g. *Rhodospirillum rubrum* and *Proprionibacterium shermanii*) use inorganic pyrophosphate, and so does the dysentery-causing protozoon *Entamoeba histolytica*. The work of ATP is assisted

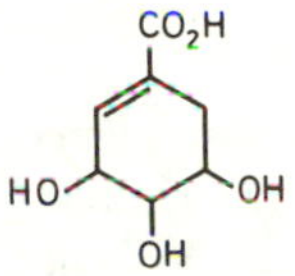

Shikimic acid
(4.54)

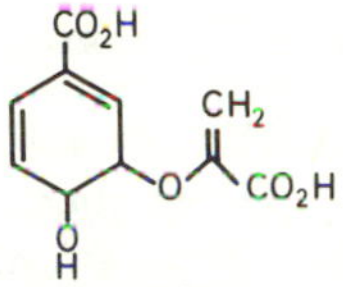

Chorismic acid
(4.55)

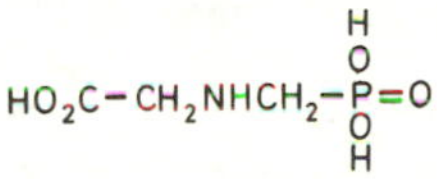

Glyphosate
(4.56)

by phosphagens which differ, interestingly, among the different forms of animal life. Phosphocreatine, which is the sole phosphagen for vertebrates, occurs in a few invertebrate species, but phosphoarginine is more common, and a few invertebrate species have both. Within any given phylum, the distribution of these two phosphagens often differs between closely related species, and even between the tissues of a single animal (Florkin and Mason, 1960). Some rarer phosphagens, which have a distribution restricted to a few invertebrates, are phosphoguanidoacetic acid, phosphoguanidotaurine, and lombricine (see above).

(a) *Purine and pteridine metabolism in protozoa.* Protozoa, for the most part, have no *de novo* synthesis of purines, and hence they have to obtain them from their hosts. For salvage of their own spent nucleosides, protozoa possess a wealth of enzymes, many of which have unusual substrate specifications. As a result, opportunities exist for the design of selective chemotherapeutic agents.

For example, allopurinol (9.51) inhibits the growth of several species of *Leishmania* (p. 9). It is converted in the parasite to allopurinol ribonucleoside which is stable in the human host but is rapidly converted by the parasite to the ribonucleotide by a parasite-specific phosphotransferase. The parasite then transforms this product to an analogue of adenylic acid by replacing $=O$ by $-NH_2$. The product is incorporated into the parasite's RNA, with lethal effect (Nelson *et al.*, 1979). *Trypanosoma cruzi* reacts similarly.

Growth of the Leishmanias is also inhibited by the antibiotic, formycin B, which is an isomer of allopurinol riboside from which it differs only in being a [4,3-*d*] instead of a [3,4-*d*] pyrazolopyrimidine. This antibiotic is curative in hamsters infected with *L. donovani* infections. Its curative activity depends on phosphorylation by the protozoa, a reaction which mammalian cells do not carry out (Carson and Chang, 1981).

Although the plasmodia that cause malaria synthesize pyrimidines *de novo*, few other protozoa do so. Instead they take uracil or uridine from the host, and convert some of it to thymidine (Kidder, 1967). Plasmodia synthesize folic acid and cannot use exogenous folate, but flagellates and ciliates require exogenous folate (Kidder, 1967).

For further information on purine metabolism in protozoa, see Hitchings (1982).

Schistosome worms require preformed purines from which they make their own nucleotides (Senft, 1970).

4.4 Carbohydrate and lipid metabolism

(*a*) *Carbohydrate metabolism* has at its core the chemical sequence known as glycolysis, i.e. the conversion of glucose (or glycogen) to lactic acid. This takes place, in all living cells, by the pathway uncovered by Meyerhof, Embden, Parnas, and Cori. Oxygen is not utilized, but nicotinamide–adenine dinucleotide (NAD) is reduced in proportion as glyceraldehyde 3-phosphate (anion) is oxidized to phosphoenolpyruvate. The NADH produced in this way is then used to reduce pyruvate to lactate. Glycolysis takes at least 11 steps, each with its own enzyme. The net balance of energy storage is that two (and three) molecules of adenosine triphosphate (ATP) are made for each molecule of glucose (and glycogen, respectively) utilized.

Such anaerobic metabolism can sustain life in animal cells only for short periods. However, if there is free access of oxygen, pyruvic acid passes from the above Meyerhof sequence to the tricarboxylic acid cycle (see Section 4.5), where it is completely oxidized to carbon dioxide and water. Alternative pathways exists for degrading glucose, through pentose phosphate in vertebrates, the choice of reactions varying with different tissues (see under *bacteria* below). All of the cell's ribose, so important in the synthesis of nucleic acids, comes from the pentose type of glycolysis.

Investigation of the comparative biochemistry of normal and cancerous cells has revealed few significant differences in glycolysis. Moreover, the enzymes used seem identical, even immunologically. However, Warburg (1927) showed that cancer cells have less oxidative phosphorylation than normal cells and hence more anaerobic glycolysis leading to enhanced production of lactic acid. In fact, rapidly dividing cancer cells have been found to lack glycerol 3-phosphate dehydrogenase (NAD-linked type) in the cytoplasm, an important difference from normal cells (Criss, 1973). The core of solid malignant tumours usually consists of hypoxic cells which newer therapy with nitroimidazoles is attacking (Section 5.1).

The following paragraphs summarize variations of glycolysis in invertebrates.

(*i*) *Insect* muscle utilizes the Meyerhof sequence only as far as pyruvate, and the NADH produced during triose phosphate oxidation seems to be reoxidized by the reduction of dihydroxyacetone phosphate to glycerophosphate (Chance and Sacktor, 1958). The major sugar in the plasma is α,α-trehalose, a disaccharide of glucose, and it plays a major part in the glucose-transport system of insects (Wyatt and Kalf, 1957). For a review of insect biochemistry, see Candy (1975).

(*ii*) *Helminths* are characterized by a high rate of carbohydrate metabolism associated with incomplete substrate oxidation. This is the case whether they live anaerobically (as intestinal worms do) or aerobically (like schistosomes). Quite unlike their hosts, many nematodes (e.g. *Ascaris*) and cestodes (e.g. *Hymenolepis*) terminate their carbohydrate metabolism by the formation of succinate, which they derive, via fumarate, from pyruvate. This process is coupled to the formation of ATP (Scheibel and Saz, 1966). Trehalose (see above) plays an important part in helminth carbohydrate metabolism. For a review of helminth biochemistry, see von Brand (1974).

Kinetic studies of the lactic dehydrogenase of schistosome worms and of rabbit muscle showed a large rate difference (Mansour and Bueding, 1953), also the pH optimum for the schistosome enzyme was significantly lower. The dissociation constant for the enzyme–pyruvate complex was 6–12 times higher for the worm enzyme than for the mammalian. An even more strikingly different pair of analogous enzymes in these worms is displayed in Table 4.6. In short, the phosphofructokinase of schistosome worms (the enzyme which converts fructose 6-phosphate to the diphosphate) is much more sensitive to the inhibitory effect of antimonials than is mammalian phosphofructokinase. The therapeutic success of antimonials in treating schistosomiasis depends on blocking this enzyme selectively (see Section 13.0 for more on antimonials). Because glycolysis is the main source of energy for the parasite, and this enzyme is a pacemaker (as defined in Section 2.4, p. 39), its substrate (fructose 6-phosphate) accumulates (Mansour and Bueding, 1954; Bueding and Fisher, 1966). (For other treatments of schistosomiasis, see Section 6.3.5.)

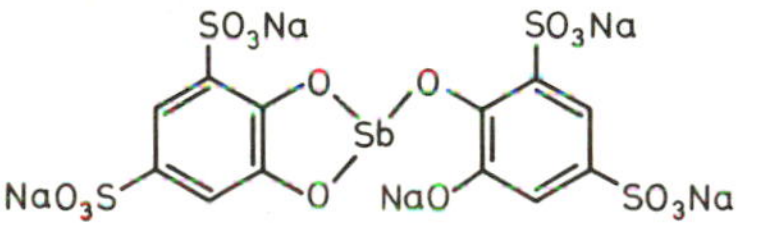

Stibophen
(4.57)

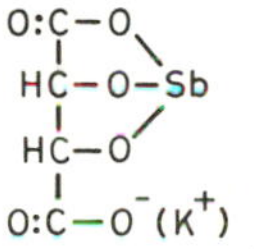

Antimony potassium
tartrate
(4.58)

Table 4.6 Percentage inhibition of phosphofructokinase activity by antimonials

Concn. of antimonial (M)	Antimony potassium tartrate (4.58)		Stibophen (4.57)	
	(A) Enzyme from Schistosoma Mansoni	(B) Enzyme from rat brain	as (A)	as (B)
1×10^{-3}	100	32	100	0
5×10^{-4}	100	4	100	0
3×10^{-4}	100	0	85	0
1×10^{-4}	70	0	44	0
3×10^{-5}	32	0	0	0
1×10^{-5}	2	0	0	0

(Mansour and Bueding, 1954.)

(*iii*) *Protozoa*, in general, can utilize exogenous acids but not carbohydrates for energy. They lack hexokinase, and accumulate starch as an energy reserve. However, the bloodstream forms of African trypanosomes are completely dependent on glycolysis (of the host's glucose) for their energy supply. Indeed, the intense aerobic glycolysis of the (highly motile) bloodstream forms of trypanosomes depends on a singular respiratory system, one that lacks cytochromes, has no citric acid cycle, and does not generate ATP. Instead, these parasites oxidize the NADH formed in glycolysis, and generate glycerol 3-phospate from dihydroxyacetone phosphate. The ultimate product of glycolysis, pyruvic acid, is attained through two enzymes, α-glycerophosphate dehydrogenase (located in a microbody) and α-glycerophosphate oxidase located in the kinetoplast (a solitary mitochondrion) (Fairlamb and Bowman, 1975). The former enzyme seems to require spermidine as coenzyme, and should offer a suitable selective site for chemotherapy (Bacchi, 1981) because these enzymes are absent from the host. (See Section 9.7 for chemotherapy achieved through suppression of spermidine synthesis.)

Another example of deviant biochemistry in protozoa is the utilization, by *Entamoeba histolytica*, of pyrophosphate in place of the usual ATP, as a coenzyme for phosphofructokinase (Deeves, Serrano and South, 1976). The major path of glucose utilization in this parasite is formation of pyruvate from phosphogluconate and from 2-keto-3-deoxy-6-phosphogluconate (Florkin and Mason, 1960).

(*iv*) *Plants* follow the Meyerhof pattern, but no less than three different glyceraldehyde 3-phosphate dehydrogenases exist: one reacts with NAD, two with NADP, processes linked to photosynthesis. In older plant tissues, the pentose phosphate pathway is also utilized and up to 50% of all carbohydrate may be metabolized by this route.

(*v*) *Bacteria* often use the Meyerhof sequence, but at least two alternative pathways have been found. In one of these (the pentose phosphate pathway) glucose 6-phosphate is oxidized to ribulose 5-phosphate, two molecules of which are quickly changed to one molecule each of glyceraldehyde 3-phosphate and sedoheptulose 7-phosphate. These phosphates then yield fructose 6-phosphate and a tetrose phosphate. In the other (the 2-keto-3-deoxy-6-phosphogluconate pathway), glucose is oxidized to gluconate by a primitive route without prior phosphorylation. The non-Meyerhof routes are favoured by those bacteria, such as the pseudomonads and aerobacters, which cannot utilize glucose 6-phosphate.

When nutrients are present in excess, many bacteria accumulate special energy reserves such as glycogen or β-hydroxybutyric acid (Wilkinson, 1966). Yeasts, and some bacteria, convert pyruvic acid to ethanol (instead of lactic acid), an idiosyncrasy which Man has carefully exploited.

(*b*) *Lipid metabolism.* Lipids are divided broadly into fats (triglycerides, esters of glycerol with three moles of fatty acids), steroids, and phospholipids such as phosphatidylcholine (lecithin) (4.59). All natural phospholipids have the L-3-glycerophosphoric structure. Each phospholipid is a family rather than a pure substance. The 2-position of lecithin is usually esterified by an unsaturated, and the 1-position by a saturated, fatty acid of the C_{16} or C_{18} series. Double unsaturation in the 2-position is common (esterification by linoleic acid). Phosphatidylethanolamine and phosphatidylserine (both found in the 'cephalin' of brain) are variations on the lecithin structure. Inositol phosphates, e.g. (4.60), occur in fungi, higher plants and animals. About 35% of the phospholipids of mammalian brain and nerves is in the form of plasmalogens; these resemble phosphatidylethanolamine, but the 1-chain is not saturated but an α,β-unsaturated ester, a structure which yields an aldehyde on hydrolysis. Many bacteria have phospholipids which are esterified by amino acids (CO of acid to OH of glycerol) (Macfarlane, 1962). Branched chains and the cyclopropane ring are often found in the fatty side-chains of bacterial phospholipids.

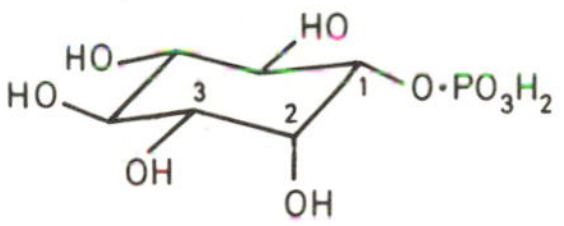

Phospholipids in mammals are synthesized (from smaller molecules supplied by the bloodstream) mainly in the tissues which use them, and they are broken down by a series of phospholipases secreted by the pancreas and gut. Flatworms cannot synthesize the fatty acids required to make their triglycerides and phospholipids, but they can alter their host's lipids to suit their requirements. Cholesterol, their only sterol, comes directly from the host (Meyer *et al.*, 1970).

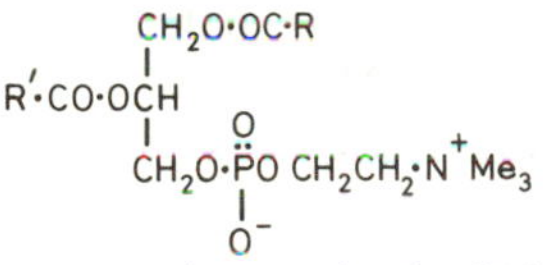

Phosphatidylcholine (lecithin)   Myoinositol phosphate
(4.59)   (4.60)

Like the flatworms, insects cannot synthesize steroids and so they degrade the plant steroids, stigmasterol and sitosterol, to obtain their essential cholesterol. The special enzymes used for this purpose are suitable targets for selective attack, because most other forms of life synthesize their steroids *de novo*.

The steroids of fungi are plant, not animal, in type and selective chemotherapy makes use of this difference (Section 14.3).

In the oxidation of fats, the classical β-oxidation to acetyl-coenzyme A occurs in the mitochondria of all kinds of cells. Some cells have one or two subsidiary mechanisms as well. Because the metabolism of fat produces twice as much water as that of either carbohydrate or protein, cells which have to encounter sudden dehydrating conditions usually have a high fat metabolism. Parasitic nematode worms are a striking example of this (Baldwin, 1948b).

4.5 The tricarboxylic acid cycle, and electron transport

After the glycolysis of carbohydrates and the β-oxidation of fats has taken place, the greater part of the energy of these nutrients still remains to be liberated, but only well-aerated cells can do this. These aerobic cells utilize acetyl-coenzyme A for this purpose. They obtain it not only from lipid metabolism but from the pyruvic acid which is the final product of glycolysis if conditions are aerobic. This acetyl-coenzyme A is utilized as fuel by being passed through the tricarboxylic acid cycle (synonyms: citric acid cycle, Krebs cycle), which converts it to carbon dioxide, water, and energy. Bacteria convert the pyruvic acid to acetyl-coenzyme A through acetyl phosphate, whereas vertebrate animals use adenyl acetate as the intermediate and a different set of enzymes.

Contrary to what was formerly believed, all bacteria and yeasts use the tricarboxylic acid cycle as the major pathway for terminal oxidation. A small shunt in the cycle is made by some bacteria (e.g. *E. coli* and *Pseudomonas aeruginosa*) and fungi (e.g. *Aspergillus* and yeasts) as follows. *Iso*citrate is dismutated to succinate and glyoxylate; the latter is then dimerized to malate which, like succinate, is a normal constituent of the cycle. The bloodstream form of many parasitic trypanosomes, which lack mitochondria, have no tricarboxylic acid cycle.

The operation of the cycle produces large quantities of the reduced forms of the nucleotides of adenine with nicotinamide (NAD and NADP) and riboflavine (FP). The regeneration of these coenzymes is effected by a transfer of electrons from the reduced forms to the oxygen of the atmosphere. In almost every kind of living cell, this transfer is mediated by some or all of the cytochrome respiratory chain (Section 5.4.3). Most of the organisms that lack all cytochromes have insignificant aerobic metabolism. Few enzymatic differences in the cycle have been demonstrated in mammals, but in the rat there is six times more aconitate hydratase in the heart than in skeletal muscle (Dixon and Webb, 1979).

(a) Inhibitors of the respiratory chain. This chain terminates in cytochrome oxidase for which the cyanide ion is a powerful and specific inhibitor.

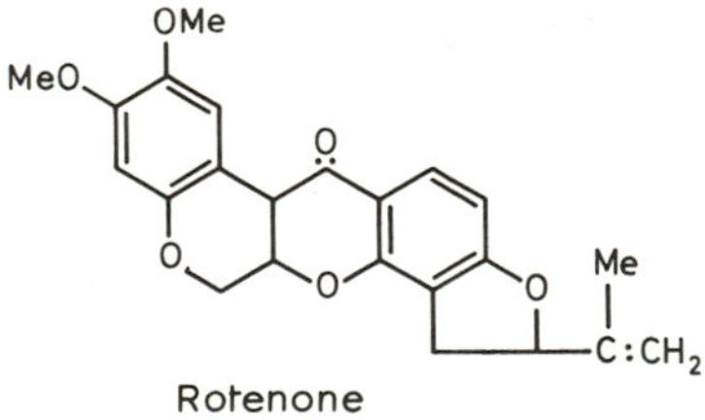

Rotenone
(4.61)

Rotenone *(4.61)*, an insecticide of vegetable origin, blocks the dehydrogenation of NADH in the respiratory chain, at a dilution of 10^{-8} M, by displacing ubiquinone from NADH dehydrogenase (Gutman *et al.*, 1971). It thus prevents the oxidation of pyruvate and glutamate (but not succinate). Fish, but not mammals, are highly susceptible to rotenone: mammals are selectively protected by rapid metabolic oxidation, but their *isolated* mitochondria are very susceptible (Ernster *et al.*, 1963). For a review on rotenone, see Yamamoto (1970).

Good selectivity is shown by a soil fungicide, sodium *p*-dimethylamino-benzenediazosulfonate ('Dexon'), which inhibits mitochondrial oxidation of NADH in the fungus *Pythium ultimum*. Sugar beets, which this fungus infects, have an enzyme in the mitochondria which decomposes this fungicide (Tolmsoff, 1962).

Many simpler molecules can uncouple oxidation from phosphorylation, so that the energy obtained by the combustion of nutrients ceases to be stored as ATP. Three classes of uncoupling agents are recognized: liposoluble weak acids, alkylating agents, and liposoluble strong bases. Phenols and other weak acids, by far the most numerous of the uncouplers, act by transporting hydrogen ions across the inner mitochondrial membrane until its resting potential is discharged (Scherrer and Howard, 1977). Artificial membranes made from mitochondrial lipoprotein similarly lose their potential (Büchel and Schäfer, 1970). The following examples of the selective use of uncoupling agents could be supplemented by the anti-fungal substances in Section 10.5, especially 2,4-dinitrophenol.

The point of attack of trialkyltin salts (R_3Sn^+), which are highly active fungicides, is believed to lie in the oxidative phosphorylation of mitochondria (Aldridge, 1958). The tributyl homologue is the most used because it is the least toxic for mammals.

Because many anti-inflammatory drugs, such as salicylates, are strong uncouplers of oxidative phosphorylation in the mitochondria, it had been thought that the anti-inflammatory action might be the result of such un-coupling. However, 2,4-dinitrophenol, a strong uncoupler, is not anti-inflam-matory, and the anti-inflammatory action of salicylates is an anti-prostaglandin effect (see Section 4.7).

The terminal respiratory systems of some bacteria (Dolin, 1961), protozoa (trypanosomes) (Grant, Sargent and Ryley, 1961), and helminths (*Ascaris*) (Kmetec and Bueding, 1961) are insensitive to concentrations of cyanides and antimycin A that are lethal to mammalian cells. It is evident that they can dispense with the whole respiratory chain after cytochrome *b*. However, they are particularly sensitive to vitamin K analogues, e.g. 2-hydroxy-3-2'-methyloctyl-1,4-naphthoquinone.

4.6 Photosynthesis

All green plants and a very few bacteria (non-pathogens) can utilize sunlight to split water into hydrogen and oxygen. The plant releases this oxygen to the atmosphere, thus helping to reconstitute the air that we breathe. More im-portant for the plant's nutrition is its ability to combine the hydrogen with carbon dioxide from the air to produce carbohydrates, proteins, and even fats, by an exceedingly complex cycle of reactions. Photosynthesis commences with the absorption of visible light by quantasomes (aggregates of about 200 molecules of chlorophyll) in the chloroplasts described in Section 5.4. This brings about the photolysis of water, which provides hydrogen radicals for reductive processes and electrons for replenishment of the chlorophyll. This part of the process is called the Hill reaction.

At the same time the light-excited chlorophyll molecules reduce the fer-redoxin in the chloroplast to an unusually low potential (about $-400\,mV$). This electron-transfer from water is used to reduce NADP and to phosphorylate ADP to ATP. These co-factors then carry the sun's energy to an elaborate synthetic cycle, in which ribulose and sedoheptulose are important intermediates, and most paths lead to glyceraldehyde 3-phosphate (and hence to starch or pyruvic acid). For further reading on photosynthesis, see Govindjee (1982). In the red algae (and the 'blue–green algae' now classified with bacteria) the primary absorbers of light are not chlorophyll but phyco-erythrin and -cyanin (re-spectively). The pyrrole pigments seem to be the evolutionary precursors of phytochrome (see Section 4.7).

(a) Herbicides. In recent years, many herbicides have been found that attack only the Hill reaction and hence are harmless to all animals. The most used of these herbicides are triazines such as simazine (*4.62*), which is non-toxic to mammals, no matter how great the dose. The action of the triazines can be demonstrated on isolated chloroplasts at the same concentration (10^{-7} M) at which the substances exhibit herbicidal action. Cereal crops absorb simazine, but detoxify it by hydrolysing the chloro-substituent to a hydroxy-group (Gysin, 1962). One of the largest uses of simazine is to kill weeds in maize crops. Although most vegetables are adversely affected by the triazine herbicides, the berry fruits are resistant, also rose bushes and garden shrubs; it has now become a common sight to see plants chemically weeded by triazines. These herbicides (also monuron, see below) are not destroyed by bacteria but remain in the soil for many years. Herbicides that block the Hill reaction in photosynthesis bind reversibly to a site near the chloroplast membrane where they block electron transport. This occurs in the vicinity of the secondary electron acceptor, Quinone B (Tischer and Strotmann, 1977).

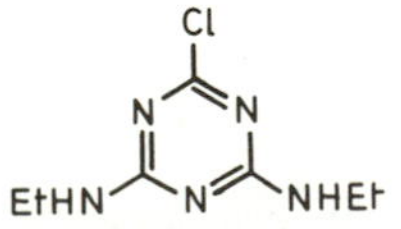

Simazine
(2-chloro-4,6-*bis*ethyl-
amino-1,3,5-triazine)
(4.62)

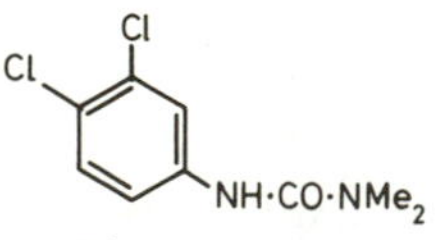

Diuron
(3,4 - dichlorophenyl-dimethylurea)
(4.63)

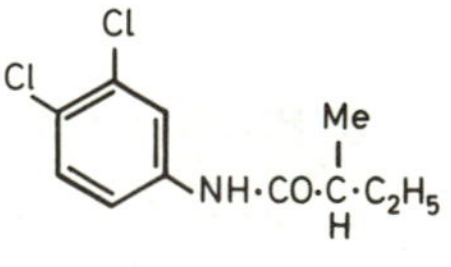

N-3,4-Dichlorophenyl-2-
methylbutyramide
(4.64)

Other much-used herbicides which depend on inhibition of the Hill reaction are phenylureas, acylanilides, and uracils such as bromacil. The most potent member of the urea family is diuron, 3-(3,4-dichlorophenyl)-1,1-dimethylurea (*4.63*), but monuron (the earlier and slightly less active 4-monochloro analogue) is still in use. The most potent amide is 3,4-dichlorophenyl-2-methyl-butyramide (*4.64*). For activity, these substances require a free imino-group, a carbonyl-group, a side-chain of rather critical length, and 4- (preferably 3,4-) substitution in the benzene ring. The best substituents are halogens, or methoxy- or methyl-groups (Good, 1961; Verloop, Hoogenstraaten and Tipker, 1976).

The bipyridylium herbicides, used as a substitute for ploughing in erosion-prone country, are contact herbicides. Because they are not appreciably trans-located, weeding with them produces a razor-sharp boundary. The most potent of these is paraquat (*4.65*), 1,1′-dimethyl-4,-4′-bipyridylium cation (used as the dichloride). This substance has long been in use in another connection, namely as an indicator for low reduction potentials. This colourless substance, known as methyl viologen, functions by forming a violet-coloured stable free radical (*4.66*) when the potential falls to -446 mV (Michaelis and Hill, 1933).

It was quickly noted that the 3,3′-bipyridyl analogue of paraquat was not herbicidal; what was needed was a structure that permitted formation and stabilization of a free radical upon reduction. Stabilization requires a co-planar molecule, so that in (*4.66*) the unpaired electron (shown as a large dot) can exchange rings with the positive charge and so obtain stabilization by the resonance of similar canonical forms (Brian, 1965).

It was next found that substances with too low a reduction potential formed too little of the free radical form in the plant; and when the results were recalculated on the basis of the amount of free radical formed, all members of the series were equitoxic (Homer, Mees and Tomlinson, 1960). High herbicidal activity is confined to analogues with E_0 between -300 and -500 mV. This makes it likely that they are reduced to free radicals by ferredoxin.

After application of these herbicides, rapid uptake into the chloroplast follows, and death occurs when the ratio of one molecule of paraquat to 100 molecules of chlorophyll is reached (Baldwin, Clarke and Wilson, 1968). Plants kept in the dark are unaffected by this concentration, but they rapidly die when light is admitted. Also there is no injury if oxygen is excluded or the amount of chlorophyll is deficient.

The bipyridyl free radicals are not a direct cause of death, but just a stage in a rapid cycle of reduction and re-oxidation. In the course of this cycle much hydrogen peroxide is formed, and some superoxide free radical which is considered to be the true toxic agent (Fridovich, 1975). When monuron (see above) is used to inhibit photosynthesis, diquat loses most of its effect. Although paraquat is completely safe in normal usage, large oral doses cause fibrosis of the lungs, and death may follow. For a book on bipyridylium herbicides, see Summers (1980).

3-Amino-1,2,4-triazole (*4.67*) (amitrole) is a selective herbicide that acts by inhibiting the synthesis of chlorophyll (see also Section 9.4.6).

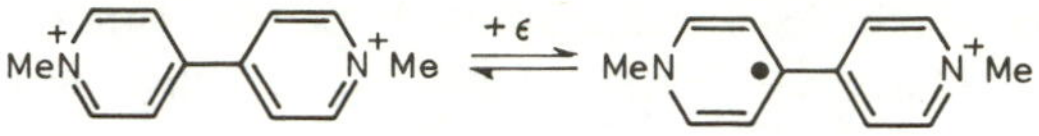

Paraquat (cation)
(4.65)

(radical)
(4.66)

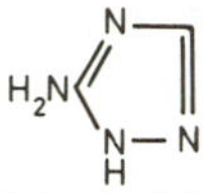

3-Aminotriazole
(4.67)

4.7 Hormones and pheromones

4.7.1 *Vertebrate hormones*

It is known that vertebrate hormones influence invertebrates, and vice-versa. Substances that show the pharmacological action of adrenaline (epinephrine) on the frog's heart have been extracted from protozoa, annelid worms, molluscs, and arthropods (cf. Wense, 1939). Conversely, pure adrenaline causes increased muscle-tone in annelida, molluscs, and arthropods. Again, substances with estrogenic action on vertebrates have been extracted from protozoa, coelenterates, annelid worms, molluscs, and echinoderms (Steidle, 1930). It seems to be Nature's way, to use chemicals already present in the environment (arising by degradation of nutrients, for example) as messengers that become more and more specialized as the evolutionary tree is ascended. Thus glycine has no hormonal function in plants, and may have none in worms, but in Man it is an important transmitter in the central nervous system. *Schistosoma mansoni* adults contain a high concentration of 5-hydroxytryptamine which they use as a neurotransmitter after obtaining it from their hosts (Bennett and Bueding, 1973).

Currently, much interest is being taken in selective possibilities of the eicosanoids (prostaglandins), a family of secondary messengers discovered by von Euler in 1934. Their occurrence is widespread in Nature. In mammals their usual task is to translate a flux of a hormone or neurotransmitter into the appropriate physiological result. Their action, which can be violent, is usually short-lived; specific dehydrogenases exist to degrade them rapidly after release. Some are stimulant, even irritant, whereas others have healing properties.

The starting point of their biosynthesis is linoleic acid, an essential dietary constituent for Man. This is converted to arachidonic acid (*4.68*) (a 20-carbon aliphatic acid with four double bonds) which is stored as a phospholipid in cell membranes. From these, it is liberated, on demand, by phospholipase A_2. Arachidonic acid is further metabolized by two pathways. In the first of these, it is acted on by *cyclo-oxygenase* to give an endoperoxide (*4.69*). From this key intermediate, prostaglandins of the important E and F series are formed readily by *prostaglandin synthetase*. Thus, prostaglandin E_2 (*4.70*) arises by simple isomerization. Chemically, it is (8R,11R,12R,15S)-11,15-dihydroxy-9-oxoprosta-5,13-dienoic acid.

The E series of prostaglandins is distinguished by a hydroxyl-group in the 11 position and an oxo-substituent at C-9. The F series has two hydroxyl-groups (at C-9 and C-11) and both series have a hydroxyl-group at C-15. Prostaglandins with the subscript '2' have the extra double bond between C-5 and C-6.

$$Me(CH_2)_4 \cdot (CH=CH\ CH_2)_3 \cdot CH=CH\ (CH_2)_3 \cdot CO_2H$$

Arachidonic acid
(Eicosatetraenoic acid)
(*4.68*)

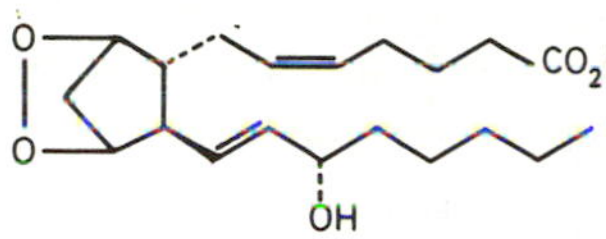

Endoperoxide PGG₂
(*4.69*)

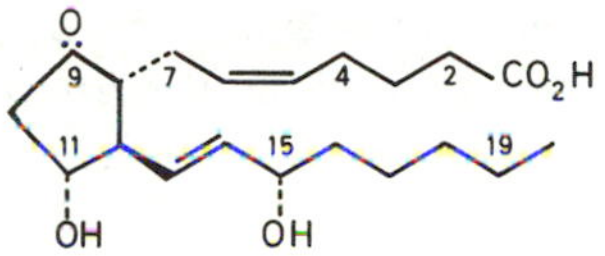

Prostaglandin E₂ (dinoprostone)
(*4.70*)

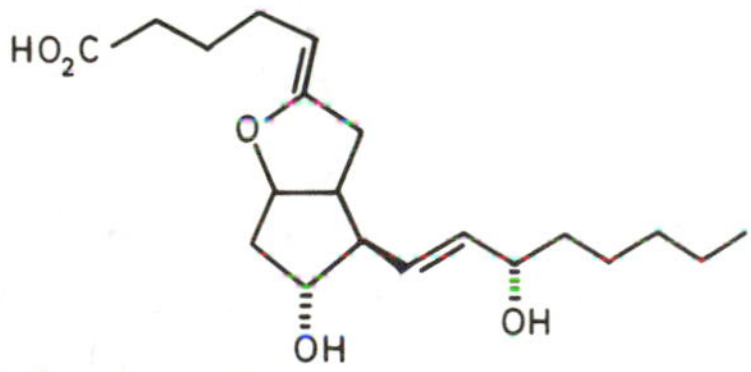

Epoprostenol (prostacycline)
(*4.71*)

The cyclic endoperoxide (*4.69*) gives, by the action of another enzyme (thromboxane synthetase), thromboxane A_2 which is responsible for the formation of thrombi (clots) in blood-vessels. Yet another enzyme (prostacyclin synthetase), acting on the same endoperoxide, gives epoprostenol (prostacyclin) (*4.71*), a substance which protects blood-platelets and prevents clotting. Epoprostenol has a half-life of only 3 minutes in water at 37°C. Chemically it is (5Z,13E)-(8R,9S,11R,12R,15S)-6,9-epoxy-11,15-dihydroxyprosta-5,13-dienoic acid.

In the second pathway by which arachidonic acid is metabolized, a battery of *lipoxygenases* converts it to a family of about four leukotrienes of which (*4.72*) is typical (Murphy, Hammarström and Samuelsson, 1979). These cause a much more long-lasting inflammation than that brought about by histamine and are held responsible for most of the distressing symptoms of asthma. Before the chemical constitution of leukotriene C was known, it was referred to as SRS (Slowly Reacting Substance).

Leukotriene C
(4.72)

The short half-lives of prostaglandins, and the fact that each one of them seems capable of both useful and unpleasant effects, depending on the location, have made this a difficult and provoking series in which to find useful new drugs. Longer life is being conferred on synthetic analogues by such devices as insertion of a methyl-group in the 15-position to protect the allylic hydroxy-group from 15-hydroxyprostaglandin dehydrogenase. Before mentioning the accepted clinical uses of prostaglandins E_1, E_2, and F_2, it must be said that in some situations they cause erythrema, oedema, and pain and, in addition, they sensitize pain receptors to other hyperalgesic substances such as histamine, bradykinin, and leukotriene C (Ferreira and Vane, 1974).

Alprostadil (PGE_1) is given intravenously to secure dilitation of blood-vessels before surgery for congenital heart disease. Its other use for improving peripheral vascular disease (ischaemia, Raynaud's syndrome) has largely been taken over by epoprostenol (Szczeklik et al., 1979). Dinoprostone (PGE_2) and dinoprost (PGF_{2a}) are used to induce labour, as alternatives to the traditional drug, oxytocin. They are also used as abortifacients in mid-trimester. Epoprostenol and its longer-lasting synthetic analogues look like becoming important in the management of thrombosis (Moncada and Vane, 1978). Epoprostenol is already used in the processing and storage of blood-platelets.

Several synthetic prostaglandin derivatives, such as doxaprost, are in clinical trials as bronchodilators in asthma. Several analogues of epoprostenol are undergoing trials for inhibiting gastric secretion and curing ulceration. The following synthetic prostaglandin derivatives are used as luteinizing agents in veterinary medicine: cloprostenol, delprostinate, and fluprostenol.

Up to the present, the biggest impact that prostaglandins have made in therapy has come from substances of very different chemical constitution, but capable of dispersing inflammation, and all are inhibitors of enzymes which metabolize arachidonic acid. This field was opened up by Vane's amazing discovery that the anti-inflammatory action of aspirin (4.73) (acetylsalicylic acid) is due to its inhibition of the biosynthesis of PGE_2 (4.70) (Vane, 1971). It is still the most-used drug for treating arthritis and rheumatism. The inhibition takes place at a concentration far below that which causes uncoupling of phosphorylation. Other non-steroidal, anti-inflammatory drugs act similarly, e.g. indometacin (4.74), ibuprofen (4.75), and mefenamic acid (Ferreira and Vane, 1974). Ibuprofen is one of a long series of 2-phenylpropionic acid drugs, cf. the similarly acting naproxen which is a naphthylpropionic acid. All of these drugs are more powerful than aspirin, but their side effects are also greater.

Because of the remissions common in the rheumatoid diseases, an objective comparison has proved difficult.

The site of action of these drugs is prostaglandin synthetase (described above) in the affected tissue. The seemingly inevitable gastric bleeding, seen with high doses, is thought to have the same origin. This is part of the prostaglandin paradox: these substances injure the stomach, but make for good health in rheumatic tissues (Flower, 1974). The action of these inhibitors is irreversible, so that dosage does not have to be continuous. Aspirin is also given prophylactically to decrease blood clotting and hence coronary infarction: the idea is to inhibit formation of thromboxane (see above).

The influence of ionization on the activity of non-steroid anti-inflammatory drugs is mentioned in Section 10.3.6. Possibly differently acting from these drugs is piroxicam (4.76), said to inhibit cyclo-oxygenase (see above), and to reduce migration of leucocytes into joint spaces. Chemically, it is 4-hydroxy-2-methyl-2H-1,2-benzothiazine-(2'-pyridyl)-3-carboxamide (Carty $et\ al.$, 1980).

The mild analgesics, phenacetin and paracetamol do not interfere with prostaglandin synthesis, nor does aspirin do so in headache-relieving doses.

Biosynthesis of prostaglandins is also inhibited by corticosteroids such as hydrocortisone, but by a different mechanism. They act on DNA to induce synthesis of a polypeptide known as macrocortin, about 15000 daltons in size, and this inhibits the phospholipase that releases arachidonic acid from membranes, as described above (Blackwell $et\ al.$, 1980).

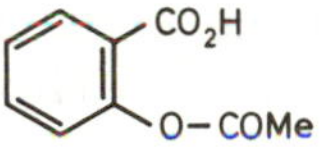

Aspirin (4.73), Indomethacin (4.74), Ibuprofen (4.75), Piroxicam (4.76)

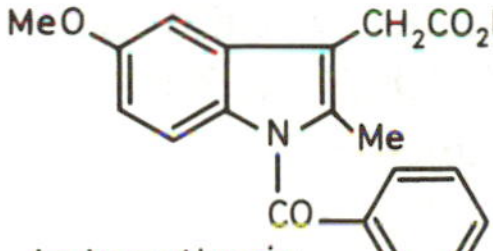

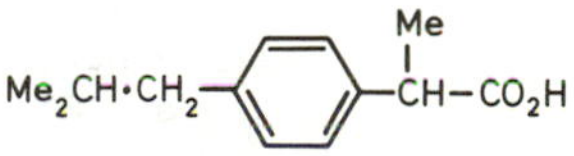

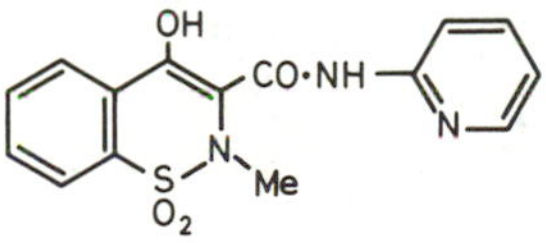

Other vertebrate hormones are dealt with under metabolite analogues (Sections 9.1 and 9.4.2), and steric factors (Sections 12.2 and 12.4).

4.7.2 $Insect\ hormones\ and\ pheromones$

In insects a high degree of control of physiological action is exerted by insect-specific hormones. These are being studied with two goals of interest for selective toxicity. One goal is to learn the nature of the hormones, to synthesize them inexpensively, and then to apply them in excessive amounts to confuse insect metabolism. The other goal is to synthesize antagonistic analogues (Section 9.1) for use as insecticides.

The main types of insect-specific hormone are: α-ecdysone, from the prothoracic glands, which causes moulting; the brain hormone that stimulates the prothoracic glands; and the juvenile hormone, in the corpora allata, which causes metamorphosis. Further, the corpus cardiacum releases a hormone that increases the amplitude of the muscles of heart and gut, and also an adipokinetic hormone which regulates the use of lipids as a source of energy in flight (Stone *et al.*, 1976).

Ecdysone (*4.77*) has the following features not found in mammalian steroid hormones: rings A and B are *cis*-fused, there is a hydrophilic substituent at C-2, the α-edge (which runs from C-3 to C-15) has hydrophilic substituents, and the side-chain on C-17 is very long (see Section 12.2 for more on steroids). The keto-group at C-6 is essential for action. Crustecdysone, the moulting hormone of crustacea (also present in some insects), is 20-hydroxyecdysone. Members of the ecdysone family can be prepared in quantity from plant products; but although many ecdysone analogues are insecticidal when injected, they are not absorbed through the insect cuticle and no practical use has yet been found. Because insects cannot synthesize this, or any other steroid, they have to absorb steroid intermediates from the diet (Lasser, 1966), a good point for selective attack.

A juvenile hormone (*4.78*) is a simple aliphatic substance: e.g. methyl 10-epoxy-7-ethyl-3,11-dimethyl-2,6-tridecadienoate in which both double bonds are *trans* (Röller *et al.*, 1967). Activity is largely determined by the configuration at the C-1 end of the molecule.

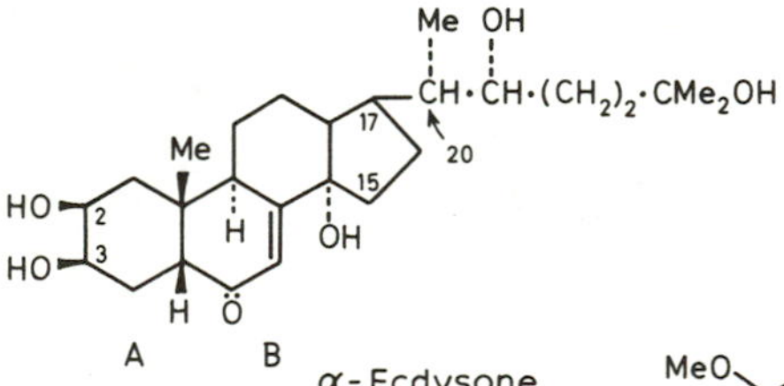

α-Ecdysone (4.77)

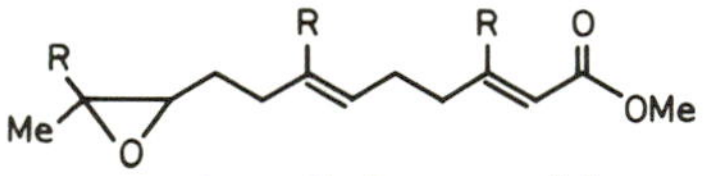

Juvenile hormone(s)
R is Me or Et)
(4.78)

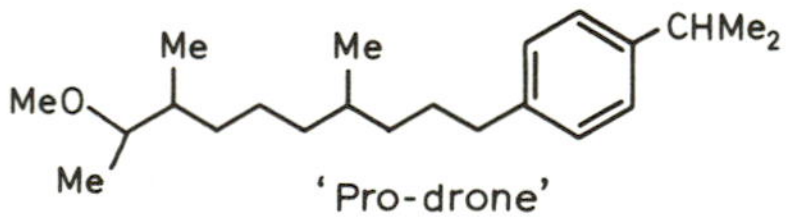

'Pro-drone' (4.79)

In an attempt to prepare mimics (agonists) of the juvenile hormone, the epoxide ring was recognized as a source of instability and therefore deleted. Typical mimics often have a terminal ester group that is conjugated with a double bond held *trans* to a long alkyl chain. Such a substance is methoprene, which is isopropyl (2*E*,4*E*)-11-methoxy-3,7,11-trimethyl-2,4-dodecadienoate. It has little toxicity to Man, birds, or fish and has been used with some success against mosquitoes (in slow-release polyamide microspheres) and manure-breeding flies. The main problem with juvenile hormone agonists is that they keep the insect at the larval stage where it usually does the most harm. Hence these mimics do best against those few categories of insects (mainly flies and mosquitoes) where the adults do the most harm. A subtle twist to this relationship was given by Meyer Schwarz of the Agricultural Environmental Quality Institute, Beltsville, Maryland, who eliminated the aliphatic double bonds, and inserted a benzene ring, to furnish 'Pro-drone' (*4.79*) (Schwarz and Miller, 1979). First tried on the housefly, this agent is proving particularly useful against fire ants. It is eaten by the ants and persists for several months in their fore-stomach. It causes more drones (specialized mating types) to be born at the expense of workers and this leads to the ants dying of starvation. No adverse effects were found on Man, birds, or fish.

Typical of juvenile hormone *antagonists* are the precocenes, e.g. (*4.80*) which are chromones occurring in garden plants. Even in low concentrations, they produce precocious moulting to give a sterile adult (Bowers *et al.*, 1976). Their site of action is the gland that synthesizes juvenile hormone, which activates the precocene, apparently by epoxidation, and is destroyed by the product. (Brooks, Pratt and Jennings, 1979).

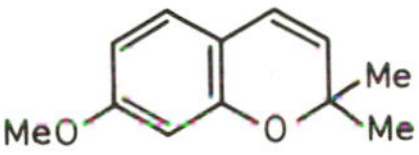

Precocene I
(*4.80*)

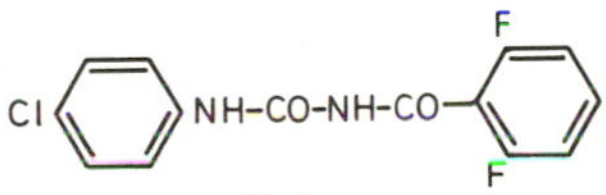

Diflubenzuron
(*4.81*)

One of the most useful of the insect hormone antagonists is diflubenzuron (*4.81*), which is 1-(4-chlorophenyl)-3-(2,6-difluorobenzoyl)urea. It prevents moulting in flies and mosquitoes by inhibiting formation of the enzyme that converts UDP-acetylchitosamine to chitin, an inhibition that leads to slow death. It also inhibits egg hatching and, for larviporous insects such as the African tsetse fly, no viable offspring appears. It is harmless to Man (Jordan and Trevern, 1978).

For an account of the physiology of insect hormones, see Novak (1975), and for insect biochemistry, see Candy and Kilby (1975).

Insect pheromones, which serve to convey messages between insects, have been isolated from many species and found to be both highly specific, and effective in mere traces. They are often aliphatic compounds, such as the termites' trail-following substance, dodecatrien-1-ol, but that secreted by the town ant is methyl 4-methylpyrrole-2-carboxylate, and several ant and beetle pheromones are derivatives of pyrazine (Barlin, 1982).

Other, sex-attractant, pheromones are secreted by the female. That from the silkworm moth is hexadeca-10-*trans*-12-*cis*-dien-1-ol, that from the gipsy moth (*Porthetria dispar*) is *cis*-7,8-epoxy-2-methyloctadecane; and the attractant released by the female bollworm moth is 10-propyl-5,9-tridecanol.

Insect pheromones are used in two ways: as a lure to poisoned baits, or to disrupt an insect's mating pattern. However, the practical application of pheromones has, so far, been small, largely because of the high specificity for a particular insect species. Their use seems to be confined to protecting crops of unusually high value. For further reading, see Rothschild (1981) and Kydonieus and Beroza (1982).

The slow maturation of desert locusts (males) is accelerated when they are crowded together, which suggests that a volatile stimulant is emitted.

The 'queen substance', which inhibits conversion of larvae to queens, and is secreted by the mandibular gland in the queen bee's head, is 9-oxodec-*trans*-2-enoic acid:

$$CH_3CO(CH_2)_5CH:CH \cdot CO_2H$$

4.7.3 *Plant pheromones and hormones*

The ova of the common brown seaweeds (*Fucus* spp.) release into sea water small amounts of *n*-hexane which attracts motile spermatozoa and so favours fertilization (Hlubucek *et al.*, 1970). Hexane is active at a dilution of 1 part in 10 million and exemplifies the principle that, no matter how commonplace a chemical may be, as soon as it gives an advantageous signal in a biological system, its specificity as a messenger can be perpetuated by natural selection.

The fungus *Achlya* (male strain) secretes a hormone oogoniol which causes the female strain to form egg-bearing branches. This hormone is a steroid, with oxygen atoms attached to the 7- and 15-positions, (McMorris *et al.*, 1975).

Higher plants have numerous hormones which, being confined to the vegetable kingdom, offer us special opportunities for the exercise of selective toxicity. The most important of these hormones is usually considered to be indolyl-3-acetic acid (*4.82*), discovered in 1934. It, and some related substances, known collectively as 'auxins', bring about the lengthening of plant cells, the setting of fruit, and other expressions of growth. For stereochemical data on auxins, see Section 12.2. Indolylacetic acid acts by derepressing the synthesis of specific RNA molecules, so that certain enzymes are more abundantly formed (Armstrong, 1966).

The *phenoxyacetic acids*, which are cheaper to synthesize and have a more intense and prolonged action, are much used in agriculture to mimic the effect of the auxins. They were introduced (Zimmerman, 1942), and are still much used, for the rooting of cuttings, to prevent premature dropping of fruit, and to promote setting of fruit in the absence of pollination. When they were being used in high concentrations for this purpose, they were found to act as herbicides (Templeman and Sexton, 1946). At these higher concentrations they become exaggerated agonists, causing unrestrained growth which brings about death through exhaustion of reserves (uncoupling of phosphorylation). This effect is highly selective and has led to universal employment of such phenoxyacetic acids as (*4.83*) for killing dicotyledon weeds in cereal crops (see Section 6.4.3).

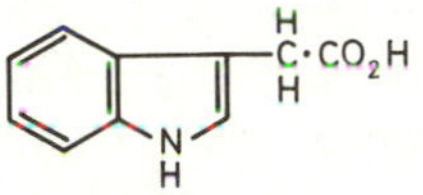

Indolyl-3-acetic acid
(4.82)

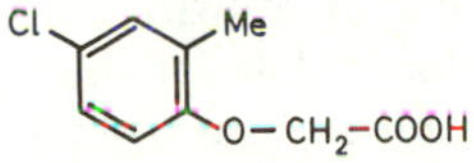

MCPA
Methylchlorophenoxyacetic acid
(4.83)

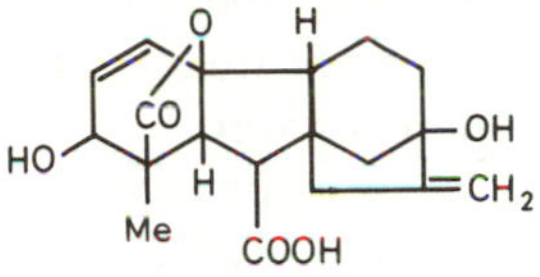

Gibberellic acid, GA_3
(4.84)

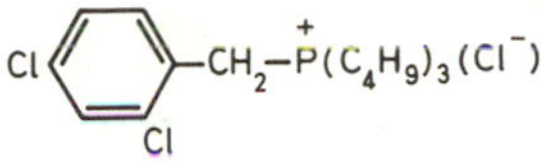

Chlorphonium chloride
(4.85)

The *gibberellins*, e.g. (*4.84*) discovered in 1926, are a family of about 30 related diterpene acids which, like the auxins, are synthesized in the growing tips of plants. In nature, they work in conjunction with the auxins. In agriculture they are used to break the dormancy of seed, delay the ripening of citrus fruits, increase the height of sugar cane (and hence increase yield per acre), and improve the malting properties of barley. They act by inducing DNA to make specific RNAs, one of which produces amylase (Varner and Chandra, 1964; Matthyse and Abrams, 1970). For a book on gibberellins, see Krishnamoorthy (1975).

Many simple synthetic compounds are used in agriculture to retard the action of gibberellins, for example chlormequat (2-chloroethyltrimethyl-ammonium chloride) which is used to dwarf cereal plants, *NN*-dimethyl-aminosuccinamic acid, and chlorphonium chloride (*4.85*). These inhibit the biosynthesis of gibberellic acid by fungi, and some of their effects on plants can be reversed by gibberellins (Lang, 1970). These inhibitors are also used to coax fruit trees into a more compact growth, leading to a higher yield of fruit in a given area.

The process of cell division requires purine-based hormones known as *cyto-kinins*, which are formed in plant roots. Zeatin (*4.86*) is a typical member, isolated from maize (Letham, Shannon and McDonald, 1964). For a review, see Letham and Palni (1983).

Abscisins, e.g. (*4.87*), are sesquiterpenes widely distributed among plants and chemically related to vitamin A. They cause dormant phenomena such as the suspension of growth and the fall of leaves and fruits. The first of them was synthesized by Cornforth, Milborrow and Ryback (1965).

The related substance *xanthoxin* (*4.88*) clearly plays a fundamental part in normal plant regulation by antagonizing indolylacetic acid, gibberellic acid, and kinetin. It is readily oxidized to abscisic acid, of which it is probably the natural precursor and with which it shares the *trans,trans* configuration (Taylor and Burden, 1972).

Ethylene is a very simple plant hormone, widespread in occurrence, which promotes or retards growth according to circumstances. It is much used in commerce to force the ripening of cold-stored fruit. For the use of chloro-ethanephosphonic acid, as a masked source of *ethylene*, see Section 3.6 (p. 107).

In plants, many aspects of morphogenesis are regulated by light, e.g. control of budding and flowering, leaf and stem expansion, seed germination, and the biosynthesis of many pigments. The chief mediator of these changes are the *phytochromes*, blue metal-free complexes of a tetrapyrrolemethine (similar to a bile pigment in structure) with proteins. Confined to green plants, phyto-chromes absorb at 660 nm, but the red rays of sunlight transform them to another active form absorbing at 730 nm. Located in semi-permeable mem-branes, the phytochromes are thought to react to illumination by regulating membrane potential (Roux and Yguerabide, 1973). Like everything else that has a specialized biochemistry, they offer the opportunity for selectively toxic interference.

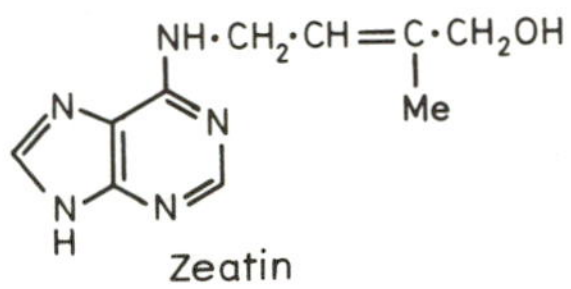

Zeatin
(4.86)

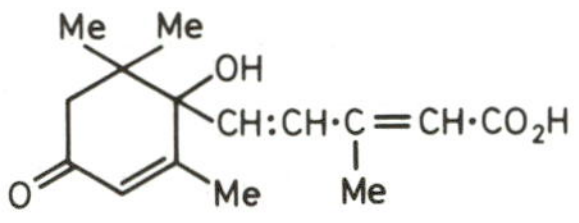

Abscisin II
(4.87)

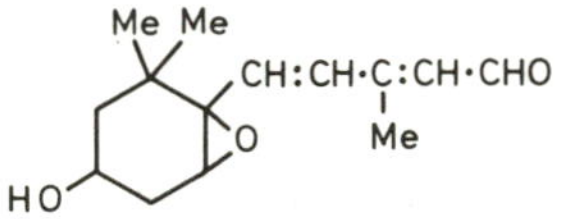

Xanthoxin
(4.88)

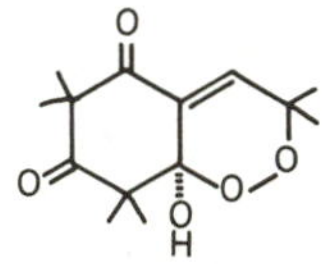

Substance G3
(Eucalyptus peroxide)
(4.89)

A family of cyclic peroxides, such as G3 (*4.89*), has been isolated from *Eucalyptus grandis* (Crow, Nicholls and Sterns, 1971; Sterns, 1971). They increase membrane permeability and, by stomatal closure, diminish respiration.

For further reading on plant growth modifiers, see Stutte (1977), Mandava (1979), and Letham, Goodwin and Higgins (1978).

4.8   Metabolism of foreign substances

Closely related genera often handle foreign substances in surprisingly different ways. The highly selective nature of many of the most useful of the organic phosphorus insecticides is due to two metabolic changes. The first of these (in the insect) makes the substance more toxic for insects, the second change (in the mammal) makes the substance less toxic for mammals. A large sector of the contemporary use of insecticides depends on this. For details see Section 13.3

Some other examples of selective alteration will now be given. Whereas vertebrates convert phenols to β-glucuronides, most insects form phenyl-β-glucosides instead. In a patient whose normal lung tissue converted 1-naphthol to 1-naphthyl sulfate, squamous carcinoma tissue made 1-naphthyl glucuronide from the same phenol (see Section 3.5, p. 87 for the biochemistry of these alternative routes) (Cohen, Gibby and Mehta, 1981).

The amino acid tryptophan is degraded by the higher plants to indolyl-3-acetic acid which is an important growth hormone for them, whereas bacteria usually degrade it to tryptamine, and mammals to 3-hydroxyanthranilic acid and thence to nicotinic acid which is an indispensable metabolite.

The most cautionary examples of selective metabolisms are those in which Man behaves differently from most other mammals, for herein lies one of the dangers in transferring results from laboratory animals to Man. Here are some examples. The only animals that dehydrogenate quinic acid to benzoic acid are Man and the Old World primates, for not even the New World primates do so. The anti-bacterial sulfadimethoxine is excreted by Man and the primates as the N'-glucuronide, whereas the common laboratory animals excrete it as the N-acetyl-derivative (Adamson, Bridges and Williams, 1966). Other aromatic amines such as aniline and sulfanilamide are acetylated in Man, and many other mammals, as well as in most species of birds, amphibia, reptiles, and fish; nevertheless dogs, frogs, and turtles do not perform acetylation. Further examples of different metabolic paths followed by Man on the one hand and mammals on the other have been traced for amphetamine, phenylacetic acid, and 6-propylthiopurine in Section 3.5 (p. 92).

The most selective known poison for mammals is norbormide (*4.90*), a rat-killer. It is 5-(α-hydroxy-α-2-pyridylbenzyl)-7-(α-2-pyridylbenzylidene)-5-norbornene-2,3-dicarboximide. Only the genus *Rattus* is affected, and death follows its powerful and irritant local vasoconstrictor action; this leads to ischaemia of most of the vital organs which then cease to function. It is non-lethal to over 30 species of other mammals (including mice and other rodents), birds, and fish. It is suggested that all animals other than rats can detoxify norbormide (Roszkowski, 1965). Norbormide has not proved very successful in rat extermination because the vermin soon recognize its odour and learn to avoid it.

For further reading on the selective metabolism of foreign substance, see Williams (1959), Jakoby (1980), Jakoby, Bend and Caldwell (1982), and the Journal *Xenobiotica*.

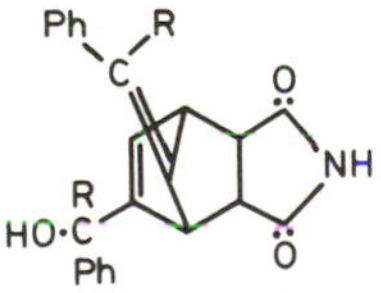

Norbormide
(R = 2-pyridyl)
(4.90)

4.9 Quantitative aspects of comparative biochemistry

So far, the differences discussed in the biochemistry of species have been mainly qualitative. But even where similar metabolic pathways are used by two species, *quantitative* differences become apparent. There are quantitative differences of accumulation (see Section 3.6), and quantitative differences in metabolism. An example of the latter is afforded by pathogenic trypanosomes which can be shown to utilize glucose two thousand times faster than their hosts, as follows. One million specimens of *T. rhodesiense*, weighing 0.0078 mg and consuming 0.031 mg of glucose in five hours, would consume 20 times their weight in 24 hours; but a man consumes only one-hundredth of his weight in that period. This intense carbohydrate metabolism of parasites is all the more vulnerable because usually so little energy is stored.

These quantitative differences do not always tell in favour of the larger creature. Man, for example, is 15 times more sensitive to atropine than the rabbit. However, he can safely take a dose of strychnine which would kill more than his own weight of rabbits, and he is unaffected by a concentration of hydrocyanic acid that is instantly fatal to dogs.

Remarkable differences can also be found within a single species. For example, the glutamine synthetase from rat kidney acts on its substrate ten times faster than does the analogous enzyme from rat muscle (Iqbal and Ottaway, 1970). Again, cancer cells maintain a more rapid cell cycle than normal cells: hence they are more sensitive to drugs which interfere with the synthesis of nucleotides.

For further reading on the selective metabolism of foreign substances, see Lehninger (1982) and Stryer (1981); and for plant biochemistry, see Bonner and Varner (1976). A simple account, for revising the basic facts, can be found in Campbell and Kilby (1975), and a related work in pictorial form, suitable for projection in Campbell and Smith (1982).

5

Comparative cytology: the third principle of selectivity

It is an everyday observation that the various forms of life differ greatly in size. The frontispiece of this book shows pictorially how mammals, insects, and bacteria form a series in which the size of average members decreases a thousandfold, then a thousandfold again, whereas molecules (of the size of drugs, vitamins, and coenzymes) are a thousandfold smaller still.

5.0 The variations of cell architecture

The sizes of cells do not differ so greatly. The average diameter of bacteria (1–4 μm) may be compared with that of a roughly spherical parenchymatous cell (15–70 μm) of which the largest part of plant tissues is composed. A few kinds of plant cells are much larger, notably the pulp cells of fleshy fruits (up to 1 mm), and the fibre cells [usually 1–2 mm (occasionally 100 mm) in length but of normal thickness]. Plant cells tend to remain small until the final division, after which they increase greatly in size as the vacuole fills with water.

Animal cells, which have no large vacuole, do not usually grow after division and tend to be slightly smaller than plant cells. Thus a typical epithelial cell may be $10 \times 10 \times 50$ μm, and an erythrocyte 8 μm wide. A few kinds of animal cells are much larger, notably nerve cells [in mammals 1–20 μm in diameter (i.e. the long axon, but the body of the cell may be 100 μm), and from 20 mm to 1 m long]; the fertilized ovum is also unusually large.

Plants differ from animals in many ways: absence of a nervous system, of muscles (and hence of locomotion), and of an efficient circulatory system;

presence (except in fungi) of a photosynthetic mechanism. Animals, on the other hand lack two features very characteristic of plants, namely chloroplasts, and cell walls. Organophosphorus compounds, which kill insects by preventing transmission of the nervous impulse, cause no harm in a plant's sap because of the complete absence of a nervous system. On this basis rests the very successful practice of plant chemotherapy against insects (Section 13.3). The unique structure of chloroplasts (Section 5.4) offers the reverse opportunity, for example, to kill weeds without harming bees. In the structure of viruses, quite extraordinary features may lend themselves to selective interference (Section 5.5).

The organization of plant and animal cells into a variety of tissues and organs makes a valuable division of labour possible in *multi*cellular creatures. A further division of labour occurs at the cellular level by differentiation into nerve cell, muscle cell, epithelial cell, and so on.

(a) *The width of synapses.* Many differences found by pharmacologists in the response of various adrenergically innervated smooth muscles has been traced to the width of the synaptic gap (Burnstock and Costa, 1975). Across the wide gaps (80–400 nm) characteristic of blood-vessels, nerve stimulation releases a concentration of norepinephrine which, even at its peak value, tends to be close to the threshold concentration. In such tissues, decreasing the amount of transmitter, either by depleting the vesicles (e.g. with reserpine) or by interfering with its release (using, e.g., bretylium), can easily inhibit neurotransmission to the involuntary muscle. On the other hand, these two types of drug have little effect on tissues in which this gap is small because the peak concentration of norepinephrine may normally be 100 times the threshold concentration. Examples of such tissues are the nictitating membrane, the vas deferens, and the sphincter papillae, all of which have a gap of about 20 nm only. The lack of inhibitory action of α-blocking agents in the 'narrow gap' tissues, follows from this: the usual blood levels of antagonists are not high enough to compete with the high concentrations of transmitter. The ability of cocaine to potentiate sympathetic postganglionic neurotransmission in some tissues but not in others can also be traced back to the gap size.

It is evident that selectivity through comparative cytology is the youngest, and least investigated, of the three principles of selectivity. One valuable source of recruitment is the series of postdoctoral fellowships (2 years in length) for finding correlations between morphology and pharmacology. These have been offered annually since 1964 by the Pharmaceutical Manufacturers Association Foundation, 1100 Fifteenth Street N.W., Washington, D.C. 20005.

5.1 Cytological aspects of cancer therapy

The characteristic organs of the mammalian body are able to perform their specialized functions through being made up of *highly differentiated cells,* each kind adapted for its particular work. Some examples are shown diagramatically in Fig. 5.1.

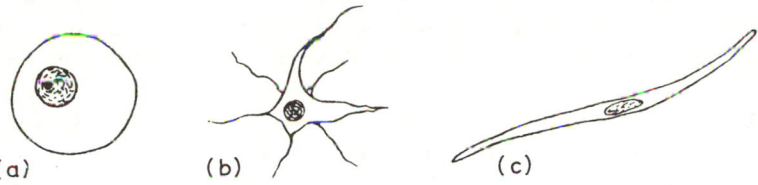

Fig. 5.1 Variations in the form of cells. (a) Ovum; (b) nerve cell; (c) muscle cell (Sinclair, 1966).

Much of the most effective treatment of cancer with drugs depends on comparative cytology. Cancer cells are normally highly specialized cells which have lost some or all of their differentiation. In short, they have regressed to a much simpler, more primitive type of cell* which (unlike the normal parent) divides continuously though inefficiently. Because a much higher proportion of cancer cells are actively dividing, they are more vulnerable than normal cells to drugs that are DNA anti-metabolites. By giving the patient well-spaced courses of those anti-metabolites that interfere with the synthesis (or incorporation into DNA) of purines or pyrimidines, a selective action against the cancer cells can be obtained. (Normal cells divide more quickly than most cancer cells, although fewer are in a state of division at any one moment, and hence make a good recovery in the unmedicated intervals.) For examples of these drugs, see Sections 4.0 and 13.4. For the theory and practice of applying them, see Brulé *et al.* (1973).

This leads to a discussion of what is known as the 'cell cycle'. All proliferating cells go through a series of phases when synthesizing new DNA. After mitosis (M phase), there is a resting phase (G_1) followed by the DNA synthesizing period (S), then a second resting phase (G_2) occurs before the cell enters mitosis. A third resting phase (G_0) is seen when the cell cycle is dormant, and this is so refractory to drugs that only the nitrosoureas (Section 13.4) can attack it.

The success of chemotherapy depends on (a) the tumour cells' susceptibility to the drug, and (b) the drug reaching the tumour in an effective concentration (C) and remaining there long enough (T) to kill the cancer cells. The optimal product ($C \times T$) has to be related to the ruling phase of the cancer cell's cycle. Generally, with appropriate suspensions of therapy to allow all normal cells to catch up with their mitoses, chemotherapy must be continued for many weeks, or even months. For more information on the cell cycle, see Shall (1981).

Another form of cancer treatment makes use of hormones (especially androgens and estrogens) to induce the cancer cells to re-acquire some differentiation,

---

*They can even be viewed as cells which have devolved from a metazoan life (with its genetic instructions for mutual adhesion and shared growth-information) to a protozoan life (with complete independence for each cell) (Goldacre, 1977).

and hence behave more like the healthy cells of the tissue in which the malignancy occurs. Post-menopausal breast cancers yield splendidly to treatment with estrogens, whereas a common type of pre-menopausal breast cancer is actually stimulated by them (Brulé *et al.*, 1973). However, about one-third of pre-menopausal breast cancers are dependent on a pituitary hormone, prolactin.

The retinoids, which are relatives of vitamin A, have attracted attention, in the last decade, as redifferentiating agents. For example, 13-*cis*-retinoic acid (*5.1*) not only prevents established epithelial tumours from progressing (in mice) but can sometimes bring about their retrogression (Bollag, 1972). Similarly, retinyl acetate, given orally, abolished carcinogen-induced mammary cancer in mice (Moon *et al.*, 1977). Although more than 100 retinoids have been tested in mice, an analogue with a sufficient therapeutic index has yet to be found for clinical use in cancer. Meanwhile the retinoids are performing well in dermatology, particularly in the treatment of acne and psoriasis.

Carcinostatic agents with redifferentiating properties include 5-bromodeoxyuridine, which initiates differentiation in neuroblastoma cells (Schubert and Jacob, 1970), and cytarabine (see p. 126), which stimulates *in vitro* maturation of human myeloblastic leukaemia cells to become normal macrophages or granulocytes (Takeda *et al.*, 1981).

For further reading on cell differentiation, see Bownes (1981), and on differentiation in malignant disease, see Sherbert (1974, 1981). For the growth kinetics of tumours, see Steel (1978).

Me Me

$-CH:CH\cdot CMe:CH\cdot CH:CH\cdot CMe:CH\cdot CO_2H$

Me

Isotretinoin
(13-*cis*-Retinoic acid)
(*5.1*)

$-N-CH_2CH(OH)-CH_2OMe$

N NO_2

Misonidazole
(*5.2*)

(a) Transformation, and lectins. In experimental cytology, viruses are used to convert normal cells, e.g. fibroblasts in culture, to what are known as 'transformed cells'. Although not malignant, these are partly de-differentiated and have, *inter alia*, the following properties: decreased self-cohesion, altered appearance due to surface changes, increased rate of growth, loss of the enzyme cyclic AMP phosphodiesterase, and decreased synthesis of sulfated mucopolysaccharides. All these trends can be reversed by adding dibutyryl cyclic adenosine monophosphate which, for example, can restore normal membrane function to tumour cells *in vitro*, as a result of which contact inhibition returns and multiplication becomes normal (Puck, 1977). For more on the transformed cell, see Cameron and Pool (1981).

Lectins, of which concanavalin A (obtained from jack beans) is an example, are carbohydrate-binding proteins of vegetable origin. They agglutinate malignant cells, although normal cells are unaffected. Agglutinability tends to increase as malignancy increases (Inbar, Ben-Bassat and Sachs, 1972). Usually agglutination leads to cell death without harm to normal cells present (Culp and Black, 1972), but a trypsin digest of concanavalin A restores malignant cells to the growth pattern of normal cells (Burger and Noonan, 1970). Just how to adapt these very striking results to clinical use has, so far, proved elusive.

(*b*) *Hypoxic cells*. Malignant cells in solid tumours usually have a core of hypoxic (almost anaerobic) cells, resistant to both chemotherapy and radiation. The nitroimidazoles, particularly misonidazole (*5.2*) have a small but definite cyto-static effect on such cells. What is more important, they sensitize these cells to radiation. It appears that the nitro-group is reduced by the hypoxic cell to a nitroxyl radical anion ($^\bullet NO_2^-$). The surrounding aerobic cells do not attack the misonidazole, and are unchanged by it. For clinical trial, see Walker and Strike (1980).

More selective examples are being sought, and the one-electron reduction potential seems to provide a good indication of success. Other hypoxic tissues such as nerves, skin, and cartilage showed damage when lipophilicity was increased, in this series, in order to improve access to the cores. For further reading, see Brady (1980).

5.2 Cytological aspects of immunotherapy

Man's immune system is a complex *recognition* device composed of well-separated cells which exchange both specific and non-specific signals with one another. This system consists of B- (bone-marrow generated) and T- (thymus-generated) lymphocytes, and there are about a million million (10^{12}) of each. Over this potentially chaotic scene, some other cells, the macrophages, exert a measure of control.

The B-lymphocytes are mainly engaged in producing antibodies, whereas the T-lymphocytes are more specialized and can be subdivided into helpers, killers, and suppressors. Both B- and T-lymphocytes have been found to be populations of clones, each of which has its particular specificity. Their surface receptors are capable of recognizing many millions of antigens. Normally, functioning of the immune system is kept at a low level by circulating catecholamines. Recipro-cally, when the immune system has been aroused to function fully, it strongly suppresses the amount of catecholamines stored in lymphoid tissues, such as the spleen.

In recent years, the hope has arisen that drugs of fairly low molecular weight might be found, drugs which could stimulate an impaired immune system and help patients to overcome the immunodeficiency found in chronic infections. It is also hoped that these drugs will help to restore health in the autoimmune diseases such as arthritis, goitre, sarcoidosis, myasthenia gravis, and lupus erethymatosus. It is even foreseen that this form of therapy might benefit cancer as well as restrain premature ageing. Needless to say, this is a tall order, but in the next few paragraphs we can trace what has already been accomplished.

Much interest was aroused by the work done by Edgar Lederer's group in France on muramyl dipeptide, MDP (*5.3*). This substance, *N*-acetylmuramyl-L-alanyl-D-isoglutamine, is liberated from bacterial cell walls by the lysozyme manufactured by the human body. By this clever ruse, the infecting organism is forced to supply a remedy to combat the infection which it is causing. This has been acclaimed as 'taking a hair out of the dog that bit you': proverbial advice. In sober detail, MDP vigorously stimulates the production of circulating antibodies, as well as intensifying *specific* tissue immunities. Thus it elicits both humoral and cellular responses (Ellouz *et al.*, 1974; Lederer, 1980).

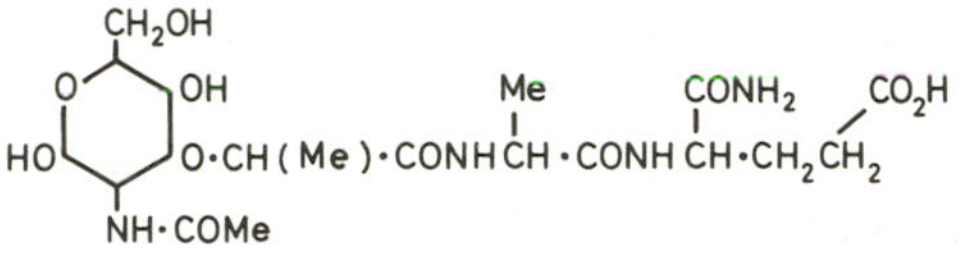

Lederer's muramyl dipeptide (MDP)
(*5.3*)

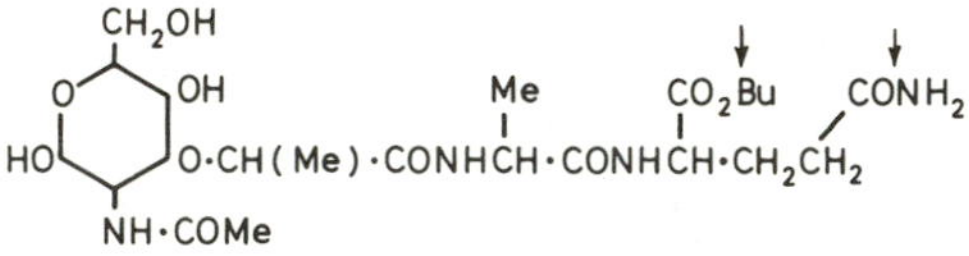

Lederer's improved dipeptide
(*5.4*)

The complexity of this phenomenon is almost overwhelming. The MDP acts on the host's macrophages which are stimulated to release soluble kinins, which regulate lymphocytes. These kinins then activate both the B- and the T-cells to produce the immunoglobin antibodies (IgG) (see Pick, 1980).

Two factors prevented MDP from assuming clinical importance: it had too short a half-life in the body and was highly pyrogenic. Many were the attempts, in France and elsewhere, to overcome these defects by ringing the changes on the amino acids and even inserting some very unbiological components. Eventually Lederer's group produced an analogue that was acceptable for clinical trials. This was the butyl ester of *N*-acetylmuramyl-L-alanyl-D-glutamic acid (*5.4*). The two important differences from MDP are shown at the arrow points of (*5.4*), namely conversion of the former carboxy group into an acid amide, and conversion of the former acid amide group into an ester (Lefrancier *et al.*, 1982). The type of tests that led to this happy result were: seeking a strong antibody response to an albumin antigen, and to a challenge with the bacterium *Klebsiella pneumoniae* (both tests done in mice).

Lederer's work began with the knowledge that mycobacteria, the bacteria that cause tuberculosis, have long been known to enhance immune response, and extracts of these germs have been widely used as immuno-adjuvants in laboratory tests. The best known of these laboratory aids is Freund's Adjuvant

which Lederer found could be replaced by MDP in some of the standard tests.
However, mycobacteria furnish yet other immuno-stimulants. For instance
mice, infected with tuberculosis, show regression of transplanted tumours
(murine fibrosarcoma). This effect is traceable to branched lipids, particularly
one known as the 'cord factor' which, chemically, is trehalose 6,6'-dimycolate*.
Trehalose is a simple dimer of glucose, and mycolic acid is a much-branched
aliphatic carboxylic acid containing 32 carbon atoms. Thus, the cord factor is
6,6'-di-O-(2-tetradecyl-3-hydroxyoctadecanoyl)-$\alpha\alpha$-trehalose. This substance,
when ultrasonically emulsified, and injected into the tumour mass, was carcinolytic in mice. The same property was shown, though to a lesser degree,
when glucose replaced trehalose which halved the molecular weight (to 638),
and also when palmitic acid replaced mycolic acid. However, the intensely
waxy character of these substances, e.g. (5.5), has retarded clinical development
(Lederer, 1980).

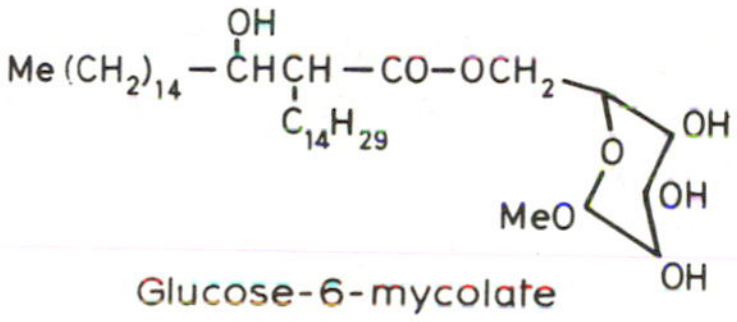

$$\text{Me (CH}_2)_{14} - \overset{\overset{\displaystyle OH}{|}}{C}H\overset{\overset{\displaystyle }{|}}{C}H - CO-OCH_2$$

Glucose-6-mycolate

(5.5)

Other immuno-stimulants were isolated by Umezawa and his colleagues in
Tokyo from microbial sources. These substances, which strongly inhibited
peptidases, could also activate T-cell immune responses in mammals. One of
them, bestatin, has found clinical use in those types of cancer where immune-
repression can be demonstrated. Bestatin is 3-amino-4-phenyl-2-hydroxy-
butyryl-leucine, and is non-toxic to Man (Carter, Sakurai and Umezawa,
1981).

Many attempts have been made to find immuno-stimulators among existing
synthetic substances. Levamisole (5.6), one of the best of these, acts as a
thymomimetic in mice. Thus, it can restore poorly functioning phagocytes and
T-lymphocytes and induce T-cell differentiation, but it does not increase T-cell
immune responses above their normal level. Moreover, it has no effect on
B-lymphocytes. It readily breaks down in the body to 2-oxo-1-mercaptoethyl-4-
phenylimidazole (5.7) which has all the properties of levamisole. Thus
levamisole seems to be the pro-drug for OMPI (5.7) which acts directly on
lymphocytes. OMPI also stimulates secretion of a factor, thought to be thymus
hormone, which, too, stimulates lymphocytes (Symoens *et al.*, 1979).

As can happen with immuno-stimulants, levamisole is immuno-suppressive
in larger doses. The latter property suggested the clinical trials, which are
taking place in Belgium, for the treatment of arthritis.

*Trehalose dimycolate is now being marketed for the treatment of cancer in horses, by Ribi
Immunochem Research, Hamilton, Montana.

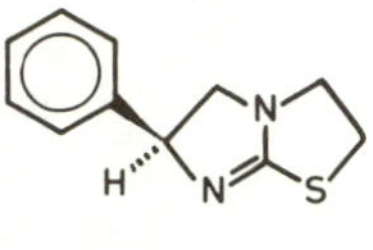

Levamisole
(5.6)

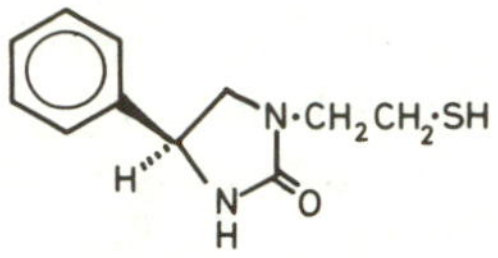

'OMPI'
(5.7)

N·CH$_2$CH$_2$SH

The production of T-lymphocytes can be promoted by as simple a substance as sodium diethyldithiocarbamate (5.8). It has little toxicity to Man and is undergoing clinical trials in France as a restorer of T-cells after general anaesthesia and surgery (Renoux and Renoux, 1979).

Glucan P, a 1,3-polymerized glucose from yeast, is a natural product with strong immuno-stimulant properties. Being particulate, it lodges in macrophages when injected into the veins of experimental animals. It stimulates the macrophages to multiply and to secrete lysozyme, and also to degrade tumours. It greatly diminishes mortality in protozoal, bacterial, and viral diseases. The end product of its metabolism is harmless: glucose. So far, no clinical trials in Man have been reported (DiLuzio, 1983).

The three principal human interferons and other lymphokines, which have large molecular weights, are being produced by hybridoma (gene-splicing) techniques, as is interleukin II, a polypeptide of molecular weight 35 000 which (inter alia) encourages growth of cytotoxic T-cells which can attack cancer cells. There are some indications that interleukin can be replaced by smaller peptides for this purpose (Anon., 1983). The lymphokines regulate immune functions rather than act directly. For further reading on the interferons, see Stebbing (1983).

$$Et_2N-\overset{\overset{S}{\parallel}}{C}-SNa$$

Sodium diethyldithiocarbamate
(5.8)

Powerful immuno-suppressants are regularly used during organ grafting, particularly azathioprine ('Imuran') (3.41) and cyclophosphamide (3.38) (see Section 3.6). More recently, a new immuno-suppressant, cyclosporin, has been introduced. This is a cyclic undecapeptide, obtained from fungi. Only one amino acid is in the D-configuration, but eight of the peptide nitrogen atoms are methylated. All side-chains are lipophilic. Its main target seems to be the T-lymphocytes but it does not affect leucocytes. Many successful transplants of bone-marrow and kidneys have been made under its protective influence (Powles et al., 1980). It also functions as an anti-schistosomal agent by suppressing the capacity of the parasite to adapt its surface antigens to resemble those of the host (Bueding, Hawkins and Cha, 1981). There are other immuno-depressants that lower the metabolism of T- and B-lymphocytes by inhibiting adenosine deaminase. Examples are pentostatin (4.18) (also known as deoxy-coformycin) and 9-(2-hydroxy-3-nonyl)adenine (4.17).

5.3 The cell wall

A cell wall, when present, provides external support but does not regulate permeability which is the function of the plasma membrane, a totally different structure (see next Section). A total lack of cell walls is one of the outstanding features of the animal kingdom. Only in a small proportion of unicellular animals can cell walls be found, as in *Eimeria* and *Chlamydomonas* which have walls of protein and cellulose, respectively. Yet the majority of the protozoa, including *Amoeba* and *Trypanosoma*, lack this outer cage. Cell walls are, however, a special feature of plants (including fungi) and bacteria.

(a) Multicellular plants have cell walls composed of microfibrils of cellulose, 10–20 nm wide and of variable length. These are embedded in an amorphous matrix of hemicelluloses and pectins. The latter are polymers of galacturonic acid, some esterified with methyl-groups, the others present as calcium salts. The cell wall is largely synthesized by enzymes present in the plasma membrane. For further reading on plant cell walls, see Siegel (1962).

(b) Fungi have cell walls consisting of a mosaic of carbohydrates plus a little lipid and protein. The principal carbohydrate is chitin, a polymeric *N*-acetyl-glucosamine, although this is absent from yeasts. (Incidentally, chitin makes up a large proportion of the exoskeleton of insects and crustaceans.)

Yeasts are single-cell fungi with cell walls consisting of two interlocking structures, either of which is sufficient to keep the characteristic shape. One structure is made entirely of glucan (a poly-anhydride of glucose), and the other is a mannan–protein complex bound to disulfide linkages. Only when both structures are broken is the cytoplasm extruded (Bacon *et al.,* 1965).

Unlike higher organisms, fungi are under high internal pressure and they burst if the cell-wall chitin is removed (e.g. with chitinase, obtainable from snails), unless the osmotic pressure of the medium is increased. Pentachloro-nitrobenzene (quintozene), a commercial agricultural fungicide, produces mycelial walls deficient in chitin (Macris and Georgopoulis, 1969).

(c) Bacteria also are under a high internal turgor (osmotic) pressure, particularly Gram-positive types (Brown, 1964; Mitchell and Moyle, 1957). Bursting is prevented by a thick cell wall which often constitutes 25% of the dry weight. When the protoplast (i.e. all of the cell inside the wall) grows, additional quantities of cell wall must be synthesized. Several antibiotics owe their place in therapeutics to their ability to prevent this synthesis (see the following). In every case the bacterium ruptures under its own osmotic pressure, but only if it is growing and therefore in need of new cell wall. Lederberg (1957), who first reported this phenomenon (actually for penicillin), noticed that lysis can be prevented by making the culture medium hypertonic.

Bacterial cell walls have an effective pore diameter of about 1 nm (Mitchell and Moyle, 1956). The wall of Gram-positive bacteria is 15–50 nm in thickness, and about half of it consists of murein*, a polysaccharide–polypeptide polymer which gives it its strength. About 25 layers of murein, lying on top of one another, reinforce this strength. Most of the rest of the wall consists of teichoic acids (to be described later). Murein is the specific substrate of lysozyme, an anti-bacterial enzyme found in tears and other body secretions, and in egg white; it is also used by phage viruses to gain access to bacteria.

The wall of Gram-negative bacteria is more complicated. Freeze etching, in combination with electron microscopy, reveals concentric layers. The murein layer, here only 2 nm thick and forming only 5–20% of the wall mass, is firmly attached, on its inner side, to the plasma membrane whereas its outer side (and this is unique to Gram-negative bacteria) is covered with another semi-permeable membrane consisting of lipoprotein and lipopolysaccharide. Teichoic acid is absent. The integrity of these layers depends on the presence of magnesium and calcium. It is thought that the exterior membrane has evolved, through natural selection, in order to protect the murein from digestion by lysozyme. In confirmation, it is noteworthy that Gram-negative murein is far less varied in details of its composition than Gram-positive murein, the latter apparently responding, under continuous selection pressure, to become less easily digestible to enzymes of predators (Nikaido, 1979). For drug designers, this structure in Gram-negative bacteria presents the difficulty of one more barrier to be crossed, but the reward of greater uniformity on the far side of this barrier.

The structure of murein (also known as peptidoglycan) will now be briefly described; then its biosynthesis will be outlined step-by-step. It is a cross-linked polymer of no fixed size, and it seems that one single molecule of it forms a sack around each bacterium. The latter can supply hydrolytic enzymes from the cytoplasm to help reshape this bag of murein when it is being enlarged during growth. In one direction, the molecule of murein consists of linear chains of disaccharides which are cross-linked, in the other direction, by side-chains of amino acid residues (shown diagrammatically in Fig. 5.2). Not all of the amino acid side-chains are used for cross-linking.

(d) The biosynthesis of murein. The first stage is biosynthesis of the characteristic monosaccharide. To begin with, N-acetylglucosamine-1-phosphate and uridine triphosphate are united, in the cytoplasm, to give uridine N-acetylglucosamine. This nucleotide then forms the 3-lactyl ether (in two steps), giving uridine diphospho-N-acetylmuramic acid [acetylmuramic acid is 3-O-D-lactyl-N-acetylglucosamine (5.9), a sugar found nowhere else than in prokaryote cell walls]. Five amino-acid residues are then added, the terminal two going on as the pair: D-alanyl-D-alanine. The composition of this pentapeptide varies

---

*The name 'murein' was bestowed by J.T. Park (1966), the first chemical investigator of the mode of action of penicillin. For Christian Gram and his stain, see footnote on p. 20.

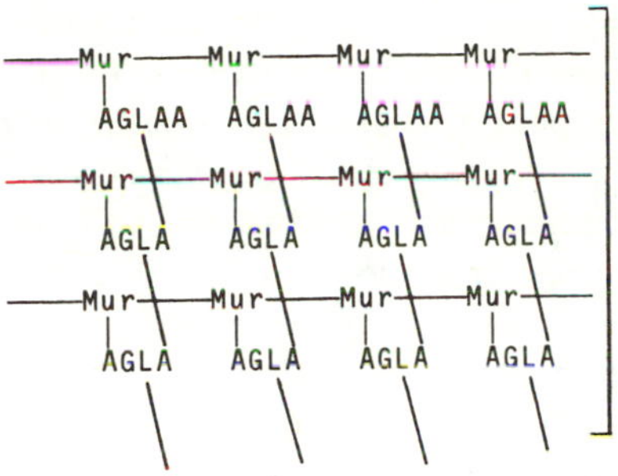

Fig. 5.2 Fragment of murein showing cross-linking of muropeptide units through their amino acids. Mur: the disaccharide. AGLAA: the five amino acids.

slightly with the species; that for *Staphylococcus aureus* is shown here (*5.10*, R = NH$_2$).

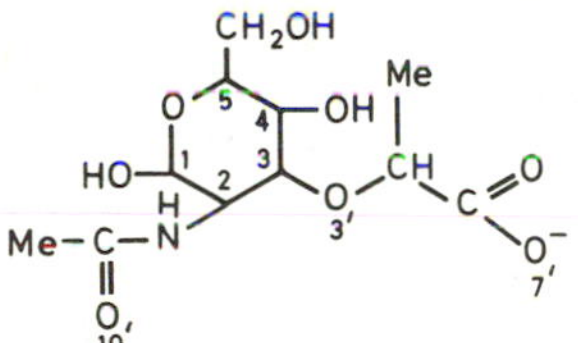

Acetylmuramic acid (anion)
(5.9)

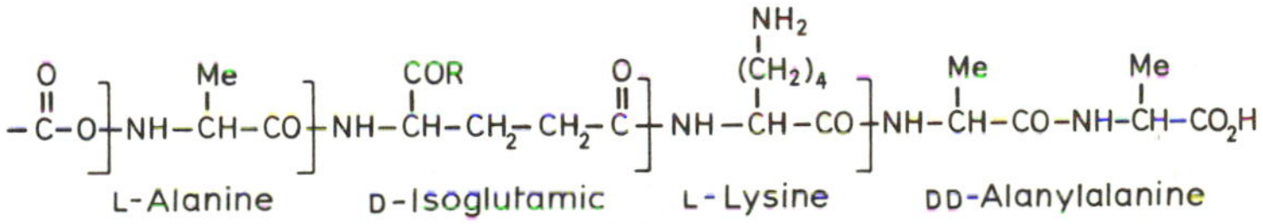

L-Alanine   D-Isoglutamic   L-Lysine   DD-Alanylalanine

Typical pentapeptide
R = OH or NH$_2$
(5.10)

This monosaccharide is transformed to the characteristic disaccharide as follows. It is first covalently united to a C$_{55}$ isoprenoid alcohol in the plasma membrane, releasing uridine monophosphate (Higashi, Strominger and Sweeley, 1967). A molecule of N-acetylglucosamine is then incorporated, and five glycine residues are added to the lysine residue (this is done in the reverse direction to normal peptide synthesis, and without recourse to ribosomes). The product is usually referred to as the 'disaccharide decapeptide'.

The next stage is the polymerization. Breaking loose from the C$_{55}$ group in the membrane, the disaccharide decapeptide forms a bond *from* the 1-position of the N-acetylmuramic residue *to* the 4-hydroxy-group of a terminal N-acetylgluco-samine residue of another disaccharide unit. By repetition of this step (some-thing like 50 times), the polysaccharide chain grows (Strominger *et al.*, 1967).

The final stage is the cross-linking. This is a transpeptidation reaction, needing no external supply of energy, between the terminal amino-group in the pentaglycine side-chain and the carbonyl moiety of the penultimate D-alanine residue of another pentapeptide chain. In this way, one molecule of D-alanine is eliminated and a new peptide bond is formed (Wise and Park, 1965; Tipper and Stromiger, 1965). By repetition of this process, the murein layer is completed.

The murein of Gram-negative organisms is constructed a little differently. No pentaglycine chain is present, and cross-linking takes place between the terminal amino-group of *meso*-diaminopimelic acid (which takes the place of lysine) and the penultimate D-alanine. The diaminopimelic acid is also linked covalently to the lipoprotein (Hofschneider and Martin, 1968). In spite of their lower content of murein, Gram-negative bacteria depend for their rigidity on murein, and burst if it is removed (Mandelstam, 1962).

Teichoic acids are polymers found on the C_{55}-site of the plasma membrane in Gram-positive bacterial species. They are composed of (polymerized) glycerol phosphate or ribitol phosphate; this polymer backbone carries species-variable substituents such as D-alanine, glucose, galactose, or amino-sugars. Teichoic acids take part in ion-exchange in the region between cell wall and plasma membrane and they concentrate magnesium ions. They also confer group-antigenic properties on the cell walls. Teichoic acids are not found elsewhere in Nature (Baddiley, Hancock and Sherwood, 1973).

Bacterial spore walls are like cell walls, but surrounded by a calcium dipicolinate complex, exterior to which is a protein shell, rich in disulfide bonds (Gould and Hitchins, 1963).

For further reading on bacterial cell walls, see Gale *et al.* (1981), and Franklin and Snow (1981).

Several antibiotics are known which act through injuring different stages of the biosynthesis of bacterial cell wall. Cycloserine (*(5.11)*, a structural analogue of D-alanine, has been found to inhibit two relevant enzymes, (a) one that racemizes L-alanine to D-alanine (*5.12*), and (b) the enzyme that synthesizes D-alanyl-D-alanine from D-alanine (Stromiger, Threnn and Scott, 1959). Cycloserine is held about 100 times more tightly by the latter enzyme than is its normal substrate, although each is in free equilibrium. Cycloserine is used in resistant cases of tuberculosis (see, further, Section 9.4.3). Other analogue-antagonists of alanine that have had some clinical trial include 3-fluoro-D-alanine and its 2-deutero-derivative (Kollonitsch *et al.*, 1973), also alafosfalin (*3.50*), described in Section 3.6, p. 104.

Benzylpenicillin (*5.13*), the first, and still the most used, member of the penicillin family, kills bacteria by combining covalently with a 'peptidoglycan transpeptidase', which normally brings about cross-linking, the final stage in murein biosynthesis (Izaki, Matsuhashi and Stromiger, 1968). Deprived of the opportunity to make new cell wall during a period of growth, the bacteria burst and die. This action of penicillin, which is discussed more fully in Section 13.1, is due to the acylating action of the 4-membered lactam ring. The action of the cephalosporin family (which is essentially similar) is described in Section 13.2.

Some less used antibiotics that attack quite different stages in cell wall synthesis are mentioned in the latter Section.

Cycloserine
(5.11)

D-Alanine
(5.12)

Benzylpenicillin
(5.13)

5.4 Sub-cellular architecture

Apart from the division of labour brought about by cell differentiation, and the grouping of similar cells into organs, a further division exists at the sub-cellular level. In this way, the many conflicting chemical reactions that take place simultaneously inside cells, achieve the required isolation in specialized compartments formed of selectively permeable membranes. The importance of these membranes is indicated by the fact that they comprise about 80% of the dry weight of an animal cell (O'Brien, 1967). Our present understanding of sub-cellular structure stems from the adaptation of the electron microscope to this task by Albert Claude in 1940.

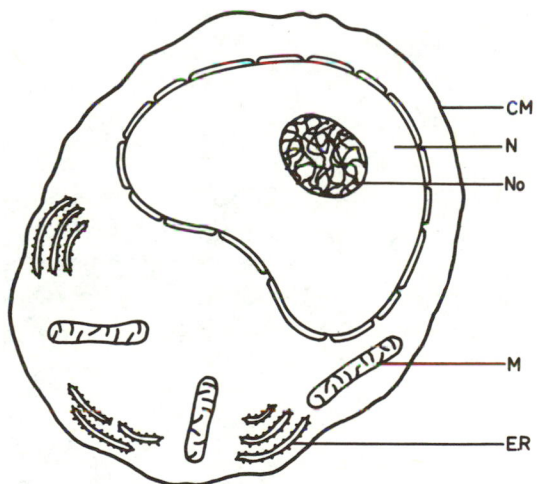

Fig. 5.3 Ultrastructure of a typical cell. CM, plasma membrane; N, nucleus; No, nucleolus; M, mitochondrion; ER, endoplasmic reticulum to which ribosomes are attached.

Figure 5.3 is a diagram of a typical cell showing the principal features seen under an electron microscope. Surrounding the whole cell is the plasma membrane, very thin, and highly selective of what it permits to enter. If a cell wall is present, it is external to this and is far thicker and more porous. Within the plasma membrane lies the cytoplasm, which is an ever-moving suspension of functional structures (called organelles) in the aqueous cytosol. The largest organelle inside the cell is the nucleus with its characteristic perforated membrane, and in the nuclear sap a nucleolus is shown. Also depicted in the cytosol (but in far fewer than their natural numbers, to assist clarity) are some of the other organelles that fill the cell: mitochondria, endoplasmic reticula, and ribosomes. These organelles provide compartments, interfaces, and membranes in which the chemical reactions of living matter take place in a specific order and facilitated by enzymes.

Such differences in comparative cytology can provide a basis for the exercise of selectivity in drugs and other agents. Nowhere is this more evident than in bacteria whose small size, compared to eukaryotic cells, leaves no space for a nucleus or even one mitochondrion. In place of a nucleus, the DNA is gathered into a single chromosome, laced through the plasma membrane; in place of

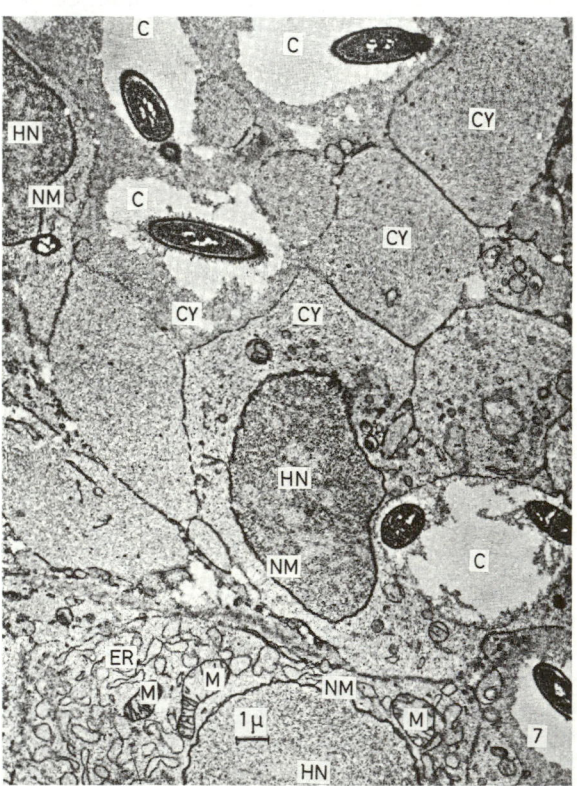

Fig. 5.4 Relative size of bacteria and host's organelles. Mouse spleen infected with *B. anthracis*. Bacteria seen as dark, oval objects. Host nuclei marked HN; host mitochondria marked M. (Roth, Lewis and Williams, 1960.)

mitochondria, the whole plasma membrane has to function as a mitochondrion, breaking down nutrients and storing the energy. Thus the exposed position of nuclear and mitochondrial functions in bacteria stands in contrast to the membrane-protected situation of these functions in the cells of higher organisms. Figure 5.4 helps one compare the relative sizes of a typical bacterium, nucleus, and mitochondrion.

Even in mammals, selectivity between the various tissues is possible because of the great specialization in shape and structure of the cells, and this is true also of the cells' ultrastructure. Nuclei from the various human tissues differ so much in appearance that forensic pathologists have long based an entire system of tissue identification on these differences. Mammalian mitochondria, too, differ from organ to organ (Section 5.4.3). Such cytological differences (i.e. between two tissues in the one organism) could be used in therapy.

The remainder of this chapter outlines the comparative morphology and function of organelles. For further reading on the division of labour inside cells, see Bourne (1970); for a descriptive atlas of the structure of cells and tissues, see Porter and Bonneville (1973).

5.4.1 *The plasma membrane*

All cells have a thin, fragile, lipoprotein membrane which regulates permeability between the cytoplasm and the outer environment. It is usually 50–100 Å thick, and the electron microscope shows three parallel layers (after fixation with permanganate): dense lines, each 20–30 Å wide, separated by 25–40 Å. A similar structure surrounds the organelles present in the cytoplasm, except that the membranes of nucleus, mitochondria, and chloroplasts are double. Our present conception of this membrane owes much to such techniques as freeze-etching (which sharpens the image of the outer surface), freeze-fracture (which reveals planes of cleavage within the membrane; both freeze techniques are used in conjunction with electron microscopy), low-angle X-ray diffraction, optical rotatory dispersion in polarized light, and fluorescence and ultraviolet spectrometry.

The current opinion, widely held, is that all biological membranes, including mammalian plasma membranes, have as a structural framework a phospholipid bilayer of which the characteristic feature is a parallel array of hydrocarbon chains, averaging 16 carbon atoms in length. This bilayer has some of the properties of a two-dimensional fluid in which individual lipid molecules can diffuse rapidly in the plane of their own monolayer, but cannot easily pass into the other monolayer. This lipid matrix provides the basic structure of the membrane. Whereas some protein molecules cover part of the membrane, particularly its outer surface, other protein strands penetrate the lipid layer, every here and there, and some of these strands are bunched together to form water-filled tubes or 'pores' (Wallach and Zahler, 1966). These proteins are responsible for most of the membrane's functions, e.g. receiving and transducing chemical signallers such as hormones, neurotransmitters, growth factors, and antigens, and also forming the three main types of cell junction, namely: tight, small gap, and synaptic. Moreover, these proteins are concerned in the transport of ions and molecules.

This fluid-mosaic membrane model allows proteins their freedom to diffuse in the plane of the lipid membrane and hence to become distributed over the cell surface in a pattern that is sometimes random and sometimes homogeneous (Singer and Nicolson, 1972).

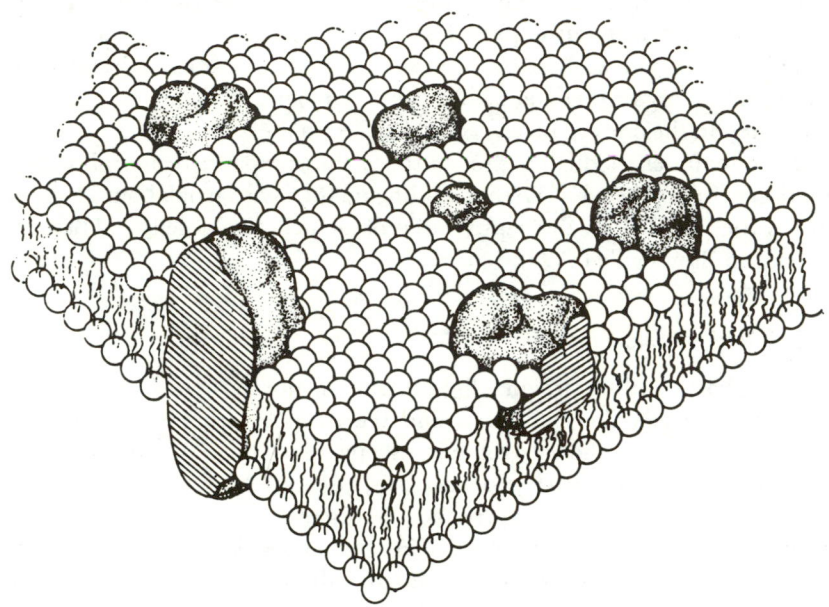

Fig. 5.5 Contemporary representation of plasma membrane (Singer and Nicolson, 1972).

Some of the proteins that span the bilayer seem to be coiled rods, as detected by freeze-fracture, whereas others appear to be globular (see Fig. 5.5). Most, possibly all, of these penetrating proteins have an attached carbohydrate molecule, which remains on the cytoplasmic side, apparently serving as a hydrophilic anchor for the protein.

Where two cells in a tissue are touching, the membranes contain an organelle called a *connexon* which regulates the passage of ions and small molecules between the cells. The connexon is a cylinder composed of six identical protein subunits, which can move so as to create or eliminate a central pore (Unwin and Zampighi, 1980).

Some important details of the composition of plasma membranes will now be discussed. Calcium, abundantly present, plays an important part in membrane stabilization, and in controlling the pores. The stability of biological membranes is considerable as the following experiment shows. Membranes from erythrocytes, mitochondria, and endoplasmic reticulum, when treated with phospholipase, lost about 70% of their phosphatidylcholine (*4.59*). This loss changed the area of the membrane, but the proteins remained bound to it and their conformation (as measured by circular dichroism) was little changed (Trump *et al.*, 1970).

The lipids consist of lecithin (phosphatidylcholine), triglycerides (ordinary fats), fatty acids, and cholesterol (see Section 4.4 for lipid chemistry). The phospholipid seems to predominate. Phospholipid bilayers have a characteristic 'melting point', namely a phase transition where they pass abruptly from a rigid to a fluid condition. This temperature depends on the nature of the head-groups as well as on the length, and degree of unsaturation, of the hydrocarbon chain. The highest melting point is found for lipids with long, unsaturated hydrocarbon chains.

In many, possibly all, biological membranes, the lipids are distributed asymmetrically. The outer half of the bilayer consists mainly of neutral lipids, whereas the inner half contains the negatively charged examples, particularly phosphatidylserine. The interior of such a membrane can be 300 mV more positive than the solution that bathes the outside. Such differences in potential can be measured by the potassium nonactin probe (Latorre and Hall, 1976) (see Section 14.2 for nonactin). Such potential differences indicate the source of some typical membrane properties such as the gating potentials of nerves.

When cholesterol is present, its plate-like steroid rings intercalate between the long chains of the phospholipid molecules, restricting their motion, and hence increasing the rigidity of the membrane as well as raising its melting temperature (Coleman, 1973). Both n.m.r. and spin-label e.s.r. studies in artificial phospholipid membranes confirm these conclusions (Gent and Prestegard, 1974).

The proteins of membranes often include enzymes (e.g. adenosine triphosphatase), and permeases (Section 2.1) are commonly found in membranes. It is logical to suppose that the protruding parts of the membrane proteins are made mainly of amino acids with *polar* side-chains, whereas the embedded parts are rich in amino acids with *non-polar* side-chains.

The proportion of lipid to proteins in different kinds of membranes varies greatly. Myelin membrane (which shields the nerve fibre) is at one extreme with a molar ratio of lipid to protein of 9:1, whereas the mitochondrial membrane lies at the other extreme with a 1:1 ratio. Myelin membrane lipid is composed of cholesterol, phosphatidylethanolamine, and cerebrosides (the latter are phosphorus-free condensation products of ethanolamine, fatty acids, and a hexose), whereas mitochondrial membrane lipid consists mainly of phosphatidylethanolamine, lecithin, and cardiolipin (diphosphatidylglycerol).

(a) *Animal membranes.* Mammalian plasma membranes are particularly rich in the phospholipids, namely phosphatidyl-choline, -serine, and -ethanolamine. Each membrane enzyme seems to need a specific phospholipid in order to function (Coleman, 1973). The principal residue in the 2-position of the glycerol moiety is that of arachidonic acid. Cholesterol is also present. The highly excitable membranes of nerve axons are discussed in Section 7.5.1.

(b) Plant membranes. The plasma membrane of plant cells is called the plasmalemma and, in spite of earlier doubts, is well established as the osmotic barrier of the cell. Plant cells are unique in containing large vacuoles which often consist largely of water isotonic with the cytoplasm, and are generally used as a repository for waste. However, at some phases of the cell's growth, the vacuole can become rich in enzymes. The tonoplast which surrounds each vacuole seems to be very like the plasma membrane in composition and properties.

A myoinositol-containing lipid is present in all the membranes of the fungus *Neurospora crassa*, and in an inositol-less mutant of this mould all the membranes are degenerate. Myoinositol is essential for balanced growth in yeast, and analogues of this inositol antagonize growth (Shatkin and Tatum, 1961). L-Myoinositol I-phosphate, which appears to be the form in which myoinositol exists in the membrane lipid, has the structure *(4.60)*.

Fungi, including yeast, have an absolute requirement for ergosterol in the plasma membrane. The polyene antibiotics, such as amphotericin *(5.14)*, used clinically, make this membrane leaky, or even rupture, it, by entering alongside the sterol, and spoiling the continuity (see, further, Section 14.3).

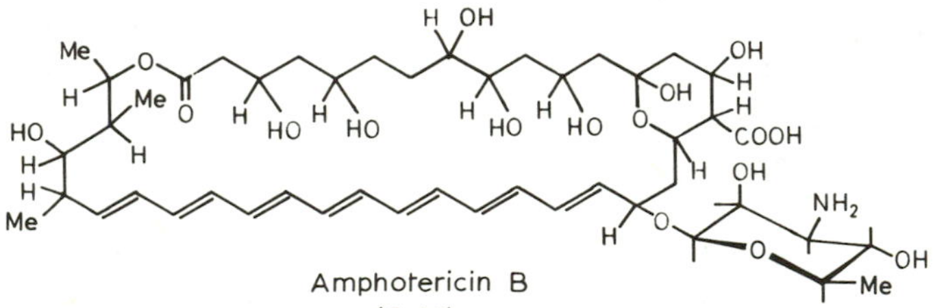

Amphotericin B
(5.14)

(c) Bacterial membranes. The cytoplasmic membrane of bacteria has both usual and unusual features. When the cell wall is completely hydrolysed by lysozyme, this membrane becomes the outer layer. It is 60–100 Å thick and sometimes extends into the cytoplasm as a few simple protrusions (Hughes, 1962). It forms about 10% of the dry weight of the cell and has a lipid content of about 25%. There is usually little lipid elsewhere in the cell. An analysis of the lipid of *M. lysodeikticus* shows that 80% is phospholipid, which is mainly diphosphatidylglycerol, but some phosphatidylinositol is also present. The diphosphatidylglycerol (a GPGPG-lipid) has C-15 branched, aliphatic groups (Macfarlane, 1961). Sterols are completely absent from bacteria.

The proteins have all the common (and no uncommon) amino acids. By chromatography on a lipophilic surface the plasma membranes of various bacteria yield many protein fractions, to one of which the DNA of the solitary bacterial chromosome is firmly attached (Daniels, 1971). Ribonucleic acid, too, is a normal component of the membrane (Hughes, 1962).

Plasma membranes of bacteria also contain permeases and the enzymes which synthesize cell wall (Crathorn and Hunter, 1958). Because of their small size, bacteria have no room for mitochondria; hence many mitochondrial enzyme are incorporated in the bacterial plasma membrane. The presence of so many vital enzymes in such an exposed position makes bacteria particularly vulnerable to selectively toxic agents (see further under mitochondria in Section 5.4.3). For further reading on bacterial membranes, see Gale *et al.* (1981).

Although bacteria have no sterols, another class of prokaryotes, the mycoplasmas (which lack all walls), have plasma membranes that depend for their integrity on the presence of a sterol, usually ergosterol. They cause disease in plants and animals, but are susceptible to polyene antibiotics and the tetracyclines.

(d) Cancer cell membranes. When a healthy cell undergoes a cancerous change, its surface structure becomes recognizably different. It has been shown, by intracellular electro-techniques, that cells form an orderly tissue by being able to recognize the presence of one another and to exchange 'messages' between their plasma membranes (Loewenstein and Kanno, 1967). This regulatory mechanism, which prevents unrestrained growth, is lacking in malignant cells. See also under 'transformation' in Section 5.1.

The increased content of sialic acid (*5.15*) in the surface layer of malignant cells (e.g. human leukaemic and lymphosarcoma cells) not only changes the glycoprotein composition, but makes a visible structural change (Van Beek, Smets and Emmelot, 1975).

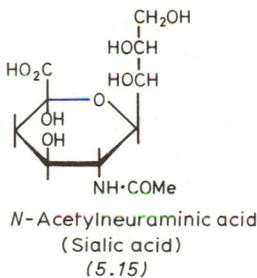

N–Acetylneuraminic acid
(Sialic acid)
(5.15)

(e) Artificial membranes. Much work on these is being carried out, some of it to obtain excitable membranes (e.g. Mueller and Rudin, 1967), some to study the permeability of ions. A favoured procedure is to ultrasonicate a mixture of lecithin and water, which produces a suspension of vesicles with a bilayer structure (Huang, 1969). These vesicles lend themselves to studies of transport across natural membranes.

(f) General reading on biological membranes. For the interactions and movements of proteins and lipids in membranes, and for membrane dynamics, see Houslay and Stanley (1982). For general reading on structure and membrane function, see Gomperts (1976), Chapman (1968–1982, 4 vols), and (for a shorter account) Finean, Coleman and Michell (1978); also the periodical, *Current Topics in Membranes and Transport*. Surface chemistry is discussed in Chapter 14.

5.4.2 *The nucleus*

Except for bacteria, all cells have a large, often spherical, nucleus in which one or several nucleoli can often be seen under the optical microscope. The nucleus is surrounded by an inner and an outer membrane each about 80 Å thick but several hundred Å apart. This membrane structure is perforated by easily deformable pores which (as measured on the oocyte of the amphibian, *Xenopus*) have an internal diameter of 800 Å, and are partly blocked by a plug of diameter 370 Å which has eight spokes. Molecules of diameter up to 90 Å can easily pass through these pores which, on occasion, are modified to allow egress of ribosomal particles of 200 Å in diameter (Unwin and Milligan, 1982).

The nucleolus is a web of threads with no surrounding membrane, and is rich in RNA. The nuclear sap contains chromatin threads which, at mitosis, become more densely organized and are seen as chromosomes (DNA+proteins). The total length of DNA in a nucleus is about 1 m, made up of about 3×10^9 nucleotides each with a molecular weight of about 350. F. Sanger has compared the amount of information stored in this quantity of DNA to that available in a large library.

The 23 pairs of chromosomes in a human nucleus contain, together, about 200 000 genes; at the other end of the scale, a virus may have as few as ten. [A gene is a strip of DNA that completely specifies a particular RNA, which itself completely specifies one protein (mRNA) or one amino acid (tRNA)].

Instead of a nucleus, bacteria store their genetic material as a single chromosome which is seen as a loop of DNA, lying unprotected in the cytoplasm, but attached to the plasma membrane. Typically, radioautography of *E. coli* that had been pulsed with [³H]thymidine showed that the DNA formed a two-stranded circle, about 800 μm long (Cairns, 1963). A strand of this length has about three million base-pairs, enough to form about 10 thousand genes. The replication points of the bacterial chromosome are bound to the cytoplasmic membrane (Smith and Hanawalt, 1967; Sueoka and Quinn, 1968). Bacterial DNA is more vulnerable to selective agents than that of higher forms of life because it is unprotected by a nuclear membrane, or even by histone molecules (Zubay and Watson, 1959). The highly selective action of the aminoacridines against bacteria in wounds provides an example (see Sections 2.2 and 10.3.1). See Fig. 5.4 for the relative sizes of a bacterium and a mammalian nucleus.

A *plasmid* is an extrachromosomal genetic structure, found in most of the bacterial species, but not constantly present. It varies in size from 1000 to 400000 nucleotide pairs, the latter size corresponding to about 600 genes. Plasmids can pass from bacterium to bacterium, even between members of different species. They are frequently responsible for introducing resistance to drugs (Section 6.5). For a review of plasmids, see Stuttard and Rozee (1979).

Some plant-infecting bacteria secrete plasmids which induce plants to manufacture *opines*, substances which help the bacterium flourish in the plant. For example, *Agrobacterium tumefaciens*, which induces tumours in dicotyledons, injects circular DNA programmed to make agropine, which is the 1,2′-lactone of *N*-1′-deoxymannitol-1′-ylglutamine (Tate *et al.*, 1982). Most opines are *N*-substituted amino acids. For a defensive response of plants to invasion, see phytoalexins, Section 6.4.2.

5.4.3 *Mitochondria*

Mitochondria, the energy generators-and-storers of the living cell, are present in all cells except bacteria. Mitochondria, which are the site of all oxidative phosphorylation, form rods or nearly spherical cylinders, from 0.2 to 3.0 μm in diameter. Often as many as a thousand mitochondria are present in a cell.

Under aerobic conditions, in which most cells grow, mitochondria are the site of (i) the tricarboxylic acid cycle which transforms (to carbon dioxide, water, and energy) the acetyl-CoA which is produced by the metabolism of both carbohydrates and fatty acids; (ii) the enzymes that oxidize and convert fatty acids to acetyl-CoA; (iii) the respiratory-chain enzymes which transmit, to atmospheric oxygen, the electrons removed from all the various metabolic substrates, and store part of the energy, obtained in this way, as adenosine triphosphate. The enzymes of carbohydrate glycolysis (the Meyerhof sequence) are in the cytoplasm.

Mitochondria are surrounded by two lipoprotein membranes, together about 180 Å thick. The inner membrane is folded into the cell as a series of invaginations known as cristae. About one-quarter of the protein part of the cristae consists of oxysomes (respiratory assemblies), i.e. ordered arrangements of riboflavine-protein, coenzyme Q, cytochromes b, c_1, c, a, and a_3 (in that sequence) together with their specific proteins. Ferredoxins (Section 11.0) also play an important part. The tricarboxylic acid cycle ensures the reduction of the first two members of the above chain, and each member is oxidized by the member on its right (in the above list), and so on to the end of the chain at cytochrome a_3 which is in equilibrium with atmospheric oxygen.

Mitochondria contain about 20 soluble and 20 insoluble proteins, most of them enzymes. Many of these proteins are made in the cytoplasmic ribosomes, but others are synthesized by the mitochondrial ones. Both DNA and RNA are present, the former as single-stranded rings with molecular weight of about 10 million (Sinclair and Stevens, 1966). Mitochondria reproduce by forming buds which break free.

Mitochondria undergo cycles of swelling and contraction. Swelling, corresponding to an energy-discharging phase, is initiated by calcium ions, also by thyroxine and other hormones; contraction is brought about only by adenosine triphosphate. Swelling corresponds to the conversion of chemical energy (from electron transport) to mechanical work, but excessive swelling corresponds to damage. The cristae are studded with patches of the enzyme adenosine triphosphatase, similar to that which constitutes the myosin fibrils of muscle. Although normal mitochondria take up calcium ions to the point where oxidative phosphorylation becomes uncoupled, ascites tumour cells do not react in this way, and hence it seems that the calcium-controlled uncoupling mechanism does not exist in these cancer cells (Thorne and Bygrave, 1974).

Large differences in structure can be seen between the mitochondria of various organisms (Palade, 1952). But even within a single organism, there are notable differences in the structure. In any mammal, mitochondria from different organs have different numbers of cristae: thus the mitochondria of the heart and kidney (rapidly respiring tissues) have numerous cristae whereas those of the liver are much fewer. Mammalian brain contains at least two populations of mitochondria, each with its own series of functional enzymes (Blokhuis and Veldstra, 1970). All mitochondria in the rapidly growing hepatoma 3924A (in contrast to those from the host cells: rat liver) were found to have deleted an important enzyme β-hydroxybutyrate dehydrogenase, also the matrices had lost affinity for stains although both membranes were intact (Pedersen et al., 1970). Liver mitochondria of both rat and chick have been separated into two populations of differing densities; the denser preponderates in embryos, the lighter in adults (Pollack and Woog, 1971).

Tissue distinctions can also be made on the basis of swelling. Brain tissue, confined in the rigid skull, has mitochondria which, through internal cross-bracing, can undergo no more than a 1% increase in volume in response to swelling agents, whereas the mitochondria of liver and kidney can expand to two or three times their normal volume. Thyroxine makes isolated rat liver mitochondria swell greatly whereas, under the same conditions, heart mitochondria are little affected (Tapley and Cooper, 1956). Chlorpromazine inhibits oxidative phosphorylation in intact mitochondria of brain, but not of liver (Berger, 1957).

The various structural differences mentioned above provide scope for the discovery of selective drugs, particularly when assisted by selective organ permeability.

Mitochondria and unicellular organisms. Bacteria have no mitochondria. Being of mitochondrial size, the bacterium has to function as its own mitochondrion; its plasma membrane, although it lacks cristae, has to attempt to carry out as many as it can of the complex activities of eukaryotic mitochondria. Hence many of the typical enzymes of eukaryotic mitochondria are located in the bacterial plasma membrane (De Ley and Docky, 1960). In particular, the enzymes of the tricarboxylic acid cycle are found there. More than 90% of the cell's succinic, malic, lactic, and formic dehydrogenases, as well as the cytochrome oxidase, are present in the plasma membrane of typical bacteria, e.g. *Staphylococcus aureus* and *Micrococcus lysodeikticus* (Mitchell, 1963).

The exposure, on the bacterial plasma membrane, of so many enzymes that, in the host, are well protected behind mitochondrial membranes, makes bacteria particularly susceptible to selectively toxic agents. The case of oxine, and other chelating drugs that act similarly, was discussed in Section 2.3. For the distribution of enzymes in mitochondria see Table 4.5.

In some protozoa the mitochondria are few, in other numerous, and more often filled with tubules than with cristae. The solitary kinetoplast of trypanosomes, which contains DNA, arises by division of an earlier kinetoplast, and contains all the information needed to produce mitochondria (and a cytochrome system) when these parasites infect their second host, the insect in which they reproduce as crithidia (amastigotes) (Vickerman, 1962).

The following anti-trypanosomal drugs bind irreversibly to, and disintegrate, the kinetoplast of *Trypanosoma rhodesiense* (in the mouse) without any effect on the cell's nucleus: diminazine, ethidium, pentamidine, hydroxystilbamidine, and trypaflavine (Macadam and Williamson, 1972). The lack of a histone covering over the kinetoplast makes it more vulnerable than the nucleus to these drugs (Steinert, 1965). In place of mitochondria, some protozoa have hydrogenosomes in which ferredoxin-activated enzymes, which convert pyruvate to acetyl-CoA, are selectively vulnerable to metronidazole.

For further reading on the structure and properties of mitochondria, see Tzagoloff (1982) and also Prebble (1981) who compares mitochondria, chloroplasts, and bacterial membranes.

5.4.4 *Chloroplasts*

The green photosynthetic organelles of plants resemble mitochondria. Chloroplasts are about 0.2 μm in diameter and surrounded by a double-layered lipoprotein membrane about 100 Å thick. The two major components of chloroplasts are the membrane system and the stroma. The membrane system contains grana and interconnecting lamellae which carry the chlorophyll molecules, whereas the stroma contains ribosomes (70S) and both RNA and DNA. In spite of the presence of DNA, chloroplasts (like mitochondria) are not autonomous but depend for many of their proteins on nuclear DNA and cytoplasmic ribosomes. Liposoluble spheres containing carotinoids are also present (Mercer, 1960). See Section 4.6 for the many herbicides whose selectivity is exerted on the chloroplasts.

For further reading on chloroplasts, see Goodwin (1966).

5.4.5 *The endoplasmic reticulum and 'microsomes'*

Most cells have a highly convoluted internal membrane, which forms an extensive network of tubules that seems almost to fill the cytoplasm. This membrane, called the endoplasmic reticulum (or *e.r.* for short), is a lipoprotein mosaic very much like the plasma membrane.

There are two kinds of *e.r.*, a smooth-surfaced form and one that appears rough because of the large number of ribosomes attached to it. The two varieties of *e.r.* can be separated, although only in a somewhat degraded form, by homogenization followed by differential ultracentrifugation. This treatment converts the smooth *e.r.* to spherical artifacts known as 'microsomes', which are used in experiments because they behave like native *e.r.* The *e.r.* is a storage organelle. In mammalian lymphocytes it stores the antibodies, whereas in the liver its membrane is the site of a host of enzymes which degrade foreign liposoluble substances (Section 3.5). Although in bacteria the *e.r.* is lacking, it is abundantly present in fungi.

See section 3.5 for other aspects of the endoplasmic reticulum.

5.4.6 *Ribosomes*

The ribosomes, which are the site of protein synthesis, consist of particles of 100–200 Å in diameter, built up from several kinds of proteins and RNA. Whereas the ribosomes of higher organisms form a sediment at 80S, those of bacteria do so at 70S. The latter are rich in magnesium, upon withdrawal of which (e.g. with EDTA), they split at once into two particles, one of 30S and one of 50S. Eukaryotic (80S) ribosomes have much less magnesium, and are much harder to split. These properties argue for a structural difference between prokaryotic and eukaryotic ribosomes, even though the process of protein synthesis follows a similar course with both (Vazquez, 1964).

About 60% of the 70S (bacterial) ribosome consists of RNA, and most of the remainder are proteins, of which *E. coli* has 55 kinds but only one molecule of each (Gale *et al.*, 1981). Bacterial ribosomes account for about one-third of the dry mass of rapidly growing *E. coli* cells.

Many drugs that inhibit protein synthesis are able to distinguish between bacterial and eukaryotic ribosomes, even in intact cells. This is illustrated in Fig. 5.6. The aminoglycoside antibiotics, such as streptomycin, bind to the 30S subunit exclusively, whereas chloramphenicol and erythromycin bind only to the 50S subunit. Through having no affinity for the host's 80S ribosomes, these

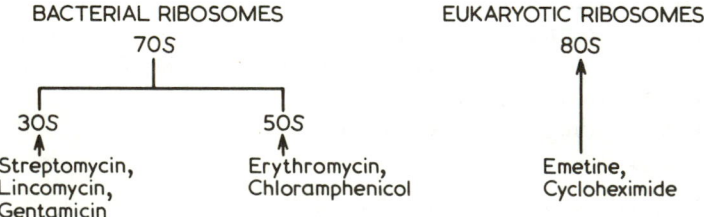

Fig. 5.6 Selective binding to ribosomes of several inhibitors of protein synthesis.

antibiotics exert great selectivity. The tetracyclines, which bind equally to $30S$ and $80S$ material in isolated ribosomal fractions, do not reach the $80S$ ribosomes under therapeutic conditions, on account of a highly selective distribution effect (see Section 3.0). Finally, counter-selective properties are shown by emetine and cycloheximide which bind to eukaryotic, but not to prokaryotic, ribosomes. (See Section 4.1 for more on drugs that act by inhibiting protein synthesis.)

The synthesis of proteins on ribosomes is shown diagrammatically in Fig. 5.7. The messenger RNA becomes bound to the smaller subunit, most likely in the cleft between the two subunits. The transfer RNA makes contact with the mRNA and, as a result, a polypeptide chain is biosynthesized on the larger subunit.

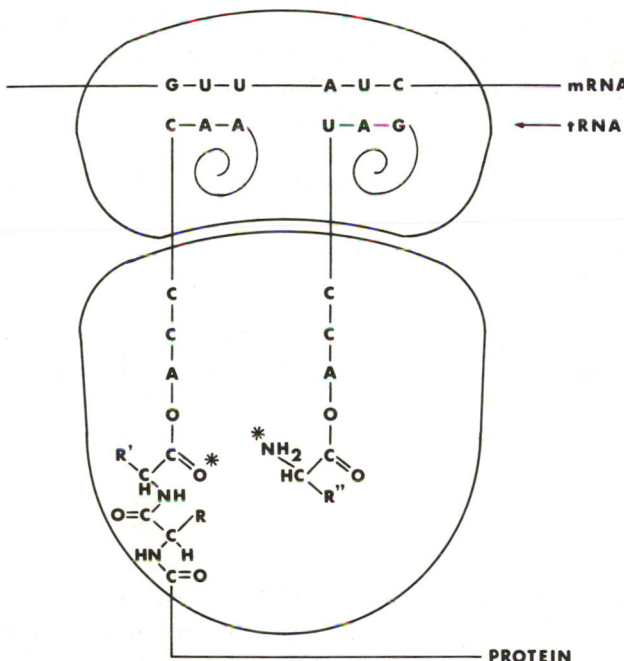

Fig. 5.7 Diagrammatic representation of protein synthesis taking place on the surface of a pair of ribosome sub-units. The peptide bond is being formed between two groups marked with an asterisk.

The mitochondrial ribosomes of mammals are small (about $55S$) and insensitive to erythromycin and lincomycin, although these antibiotics penetrate freely into the mitochondria. These ribosomes are sensitive to chloramphenicol which, fortunately, does not normally penetrate into mammalian mitochondria. Chloroplasts have ribosomes that are sensitive to most of the protein synthesis inhibitors that injure bacteria (Küntzel and Noll, 1967).

Viruses have no ribosomes and are completely dependent for reproduction on those of the host.

For further reading on ribosomes, see Nomura, Tissières and Lengyel (1974).

5.4.7 *Microtubules*

These organelles occur in all animal and plant cells. For example, human brain consists largely of microtubules (neurotubules) that sustain nerve function in that organ. They also play an important part in moving chromosomes (together and apart) during mitosis.

Colchicine is an alkaloid which specifically and reversibly combines with tubulin, a protein which normally polymerizes to form these tubules (Borisy and Taylor, 1967). The ketonic group of colchicine (*5.16*) provides the binding site. The powerful action of colchicine in blocking mitosis at metaphase, by arresting formation of the mitotic spindle, is due to its selectively binding tubulin there. Its rapid anti-inflammatory action in the treatment of gout is thought to be due to a repressive effect on the microtubules of leucocytes. Because it is a drug of low selectivity, after achieving this rapid initial effect in gout, it is quickly replaced by allopurinol (Section 9.4.4), an inhibitor of xanthine oxidase.

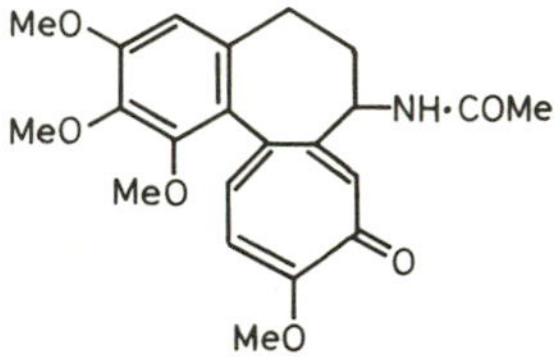

Colchicine
(5.16)

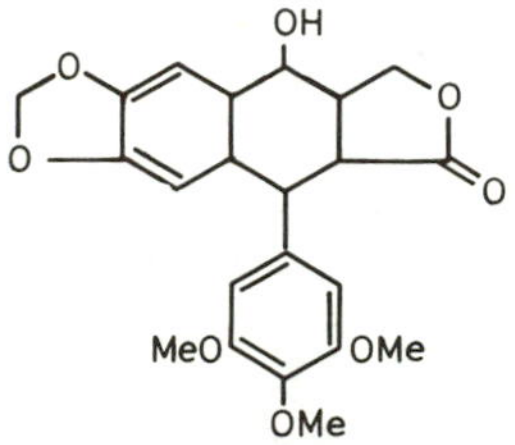

Podophyllotoxin
(5.17)

The benzimidazole anthelmintics, which enjoy widespread veterinary use (Section 6.3.5), act by powerfully binding to tubulin and hence interfering with the assembly of microtubules (Kelly *et al.*, 1977). Many of these anthelmintics carry a methoxycarbonylamino- (urethane) group, and there is some evidence that as simple a urethane as 2-methoxycarbonylaminobenzylamine acts at the same site as colchicine, namely at the growing tip of the microtubules which they inhibit. The selectivity of the *benzimidazole anthelmintics* is much higher than that of colchicine and one of them, thiabendazole, is used in Man for eliminating roundworms. See Dawson, Gutheridge and Gull, 1983.

The *benzimidazole fungicides*, such as benomyl (Section 6.3.4), are thought to act similarly to the benzimidazole anthelmintics.

Maytansine, an ansa-molecule related to rifamycin (p. 138), obtained from an African plant, is another inhibitor of the polymerization of microtubules, and has received clinical trial in cancer (Rebhun, 1975).

Podophyllotoxin (*5.17*), from *Podophyllum peltatum*, the tropical 'squirting cucumber', strongly inhibits the polymerization of tubulin and is used in dermatology to destroy warts and other skin neoplasms. Deoxy-podophyllo-toxin, from the berries of the juniper tree, has anti-viral activity. *Etoposide*, a semi-synthetic derivative of podophyllotoxin, is used clinically for the treat-ment of otherwise fatal tumours. Chemically it is demethyl-1-*O*-(4,6-*O*-ethylidene-β-D-glucopyranosyl)epipodophyllotoxin. It does not seem to pre-vent the synthesis of microtubules, but a decrease in DNA synthesis is noted (Loike, 1984).

The *Vinca rosea* ('periwinkle') alkaloids inhibit mitosis in metaphase by binding to tubulin, but *not* at the colchicine-binding site (Marantz and Shelanski, 1970). *Vincristine* (*5.18a*) is generally used in the first days of treat-ment of acute leukaemia in children. It gives a powerful start to the treatment, after which its place is taken by more selective drugs such as methotrexate and 6-mercaptopurine. It is also valued for treating acute lymphocytic leukaemia of adults. *Vinblastine* (*5.18b*), used in conjunction with cisplatin and bleomycin, can cure metastatic testicular tumours. Both alkaloids have proved useful in Hodgkin's disease and other lymphomas. It is remarkable that there is no cross-resistance between two substances of such similar chemical constitution. Research is in progress to obtain analogues of these alkaloids lacking the neurotoxic side-effects.

Apart from all these agents that destabilize microtubules, there are other substances, such as *taxol*, that stabilize them. Yet both have anti-cancer proper-ties, suggesting that the progress of malignancy requires an equilibrium between both processes. Taxol is a complex terpene, bearing four ester groups, fused to an oxetane ring. It occurs in gymnosperms related to the common yew tree (Horwitz *et al.*, 1982).

For further reading on microtubules, see Roberts and Hyams (1980) and Sakai, Borisy and Mohri (1982).

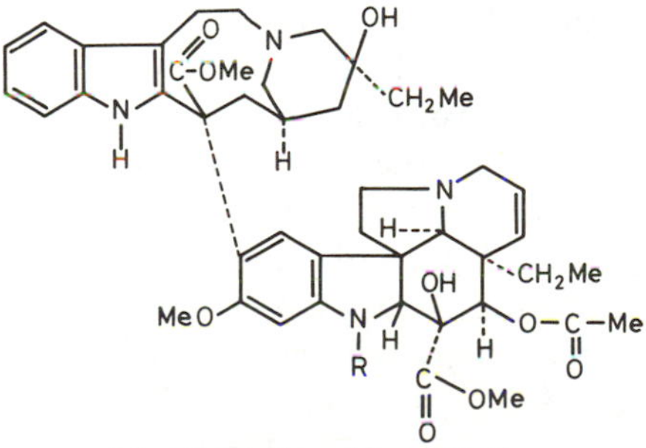

(a) Vincristine (R = —CHO)
(b) Vinblastine (R = Me)
(5.18)

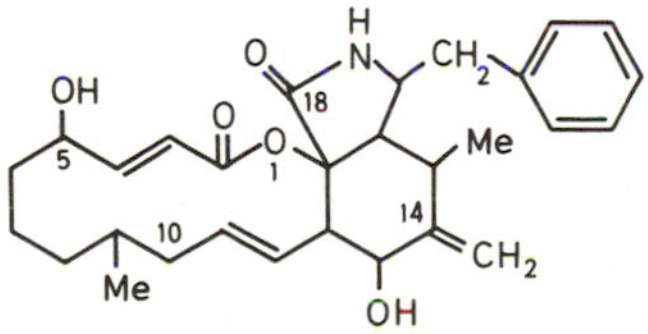

Cytochalasin B
(5.19)

5.4.8 *Other intracellular organelles*

(*a*) *Microfibrils*. Actin and myosin, which were once thought to be present only in muscle cells, also occur in microfibril organelles in many types of cell, including vertebrate and protozoan. For example, microfibrils, about 50 Å in diameter, play the leading role in such common cellular phenomena as protoplasmic streaming, cytokinesis, ruffling of membranes, nerve outgrowths, phagocytosis, and pinocytosis. This knowledge has accrued through the use of fungal products, cytochalasins, which are specific inhibitors of microfilament movement (Carter, 1972). Cytochalasin B (*5.19*) has been the most studied, biologically.

The non-muscle myosins share with muscle myosin the ability to hydrolyse the ATP and to form cross-links between actin filaments, but there is considerable variation in their other physical, chemical, and enzymatic properties. This variation in the myosins may explain the diversity of movements which different cells exhibit.

(*b*) *Dictyosomes*. These organelles, collectively termed the 'Golgi apparatus', are sack-shaped bodies of variable outline and cohesiveness, and are surrounded by a lipoprotein membrane. The known functions of dictyosomes are the partial synthesis and the assembly of glyco- and lipoproteins, the formation and packaging of secretory granules, the synthesis of lysosomal enzymes, and the assembly of membrane components. For a review, see Dauwalder, Whaley and Kephart (1972).

(*c*) *Lysosomes*. These are organelles, 0.2 to 3.0 μm in diameter, which have an acidic cytosol containing about 40 different kinds of hydrolytic enzymes. Among these are ribonuclease, deoxyribonuclease, several proteases, acid phosphatase, and an enzyme for splitting the sugar out of glycolipids. Lysosomes are formed by the endoplasmic reticulum. They can penetrate the cytoplasmic membrane and hence leave the cell, whereupon their membrane ruptures and liberates the enzymes. In some one-celled organisms, lysosomes help nutrition by providing extracellular digestion. In higher organisms they assist in the reorganization of tissues by the destruction of aged or merely superfluous cells. Lysosomes are labilized by vitamin A, an effect reversed by chlorpromazine (Guth *et al.*, 1965). Because they are more exposed than other organelles, they are thought of as promising sites of action for membrane-stabilizing and -destabilizing drugs. In inflammation and degenerative diseases, the therapeutic aim is to stabilize or repair the containing membrane of the lysosome (see Section 14.4). However, to attack tumours, lysosome rupture is being sought with selective agents that make use of the acidity of these organelles (see Section 14.3).

For further reading, see Dean and Barrett (1976) and Segal and Doyle (1978).

(*d*) *Peroxysomes*, in which the mammalian liver cells are rich, contain both D- and

L-amino acid oxidases and a crystalline mass of urate oxidase, as well as enough catalase to decompose the hydrogen peroxide generated by these oxidases.

(*e*) *Pre-synaptic vesicles* are exemplified by the minute *vesicles* in which acetyl-choline, after its synthesis in motor nerve terminals, is stored. These rupture when the arrival of a nervous impulse releases calcium ions. *Synaptosomes,* which are much larger, are formed from the snapped-off presynaptic ends of nerves, and are isolated by centrifugation. Depending on their source, they may contain norepinephrine, dopamine, serotonin, the corticotrophic-releasing factor, and other transmitters, but never acetylcholine. Catecholinergic synaptosomes actually synthesize catecholamines (Patrick and Barchas, 1974). The rupture of presynaptic vesicles in the insect is thought to play an important part in the insecticidal action of DDT (see Section 7.6.5).

(*f*) *Liposomes* arise physiologically after a fatty meal, when they are seen in the blood as finely divided fatty particles. The use of artificially produced liposomes as carriers of therapeutic material was discussed in Section 3.7.

5.5 Viruses

Viruses are non-cellular forms of life, structurally simpler than bacteria. They vary in size from rounded particles as small as 200 Å, to others that form long rods ($10\,000 \times 100$ Å).

The *virion* (the complete infective virus, such as exists extracellularly) consists of a core of either DNA or RNA (but not both) surrounded by a protective capsid (= shell) of one, or sometimes two, proteins. These two components are arranged in a highly ordered fashion that is characteristic of the particular kind of virus. Most of the viruses that infect animals are icosahedral (20-sided) and hence roughly spherical. But the viruses of measles and influenza are spirals. In the latter virus two kinds of protein (neuraminidase and haemagglutinin) form the capsid which is embedded in lipid, and this lipoprotein envelope encloses a coiled ribonucleoprotein tube. Poxviruses (which include the herpesviruses) are the largest known, brick-shaped and of very complicated structure. Many types of virus have a protein core around which the nucleic acid is arranged.

Phages have a special shape, consisting of a head and a tail. The head of T2 coliphage, for example, contains a single molecule of DNA (mol. wt. = 10^8), which weighs 2×10^{-10} μg and contains 2×10^5 nucleotide pairs. This molecule, neatly folded into an approximately spherical mass and surrounded by neatly packed (non-genetic) protein molecules, constitutes the 'head'. To this is attached a 'tail' built of a series of five structures: (1) the outer sheath consisting of a contractile, myosin-like protein with about 110 molecules of adenosine triphosphate, (2) a solid core, (3) a tip consisting of a spiked plate, (4) a few molecules of endolysine, a lysozyme-like enzyme, (5) a series of fibres which are wound around the distal end. A typical T-even coliphage, such as T2 or T4, is shown diagrammatically in Fig. 5.8.

Chemical components. Individual members of the various classes of virus contain from 1 to 20 kinds of proteins, some of which are enzymes (see below).

The molecular weight of nucleic acids, in various types of virus, varies from 2 to 160 million. RNA and the much larger DNA types are about equally common in viruses infecting Man. Typical DNA viruses are the adenoviruses (causing infections of the respiratory tract), the poxviruses, and the herpesviruses. Typical RNA viruses are the myxoviruses (including influenza), paramyxoviruses (mumps and measles), and those causing yellow fever and encephalitis. The nucleic acids are double-stranded in some types, but single-stranded in others.

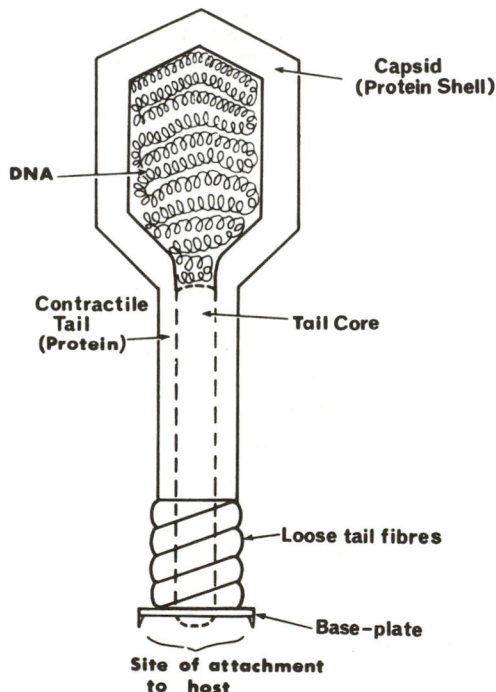

Fig. 5.8 Diagram of T-even coliphage. (After Kozloff, Lute and Henderson, 1957.)

Phages (viruses that infect bacteria) contain aliphatic polyamines such as putrescine $[H_2N(CH_2)_4NH_2]$ and spermidine $[H_2N(CH_2)_3NH(CH_2)_4NH_2]$, in quantities sufficient to neutralize from 30 to 50% of the DNA present. These amines seem to be stabilizers of the folded state of DNA, but they are non-specific cations as they can be replaced by an excess of magnesium ions without disturbing the functioning (Ames and Dublin, 1960). Many other viruses, e.g.

poliovirus and tobacco mosaic virus, lack these organic bases. All T-even phages infecting *E. coli* contain a folic acid derivative, dihydropteroyl-penta-glutamic acid, localized near the tail-plate. These phages also induce formation of this acid in the bacterium after infection (Kozloff *et al.*, 1970).

Although lipid barriers are common outside the capsids of viruses (and inside as well for poxviruses), no lipids are associated with adenoviruses or reoviruses. Enzymes are frequently present. Neuraminidase, in myxoviruses, releases sialic acid (*5.15*) from a glycopeptide widely present in mucus. Vaccinia virus (a poxvirus) contains the enzyme RNA polymerase which enters the host cell at the time of infection. A herpesvirus has adenosine triphosphatase. Lysozyme is common in phages. Carbohydrates, other than the pentose that forms part of the nucleic acids, have often been found. Analysis of purified herpes simplex virus showed proteins (70%), phospholipids (22%), DNA (6%), and carbohydrates (1.6%).

In viruses, no device for producing energy is present, nor is there any cell wall. Nevertheless each type of virion has considerable individuality and complexity, and it is hard to recall that 30 years ago each virus was thought to consist only of one molecule of nucleic acid and one of protein.

When a virus particle infects a cell, the nucleic acid (which is the sole infectious part) passes into the host and not only reorganizes the host's supply of intermediates to produce fresh viruses, but can also command the synthesis of new intermediates which it may require. A typical RNA virus such as *Vaccinia* enters the host cell's cytoplasm and, by means of reverse-transcriptase, it brings about the synthesis of complementary DNA and the host cell's DNA is repressed. As soon as the new DNA makes viral RNA and the host's ribosomes begin to translate the latter into viral proteins, the virus particles multiply fast in the host cell.

Coliphage attacks *E. coli* in a series of steps. First the long tail fibres come to rest on the surface of the cell, then the spiked base-plate is brought into contact with this surface. The lysozyme-like enzyme in the plate then depolymerizes a small area in the murein of the bacteria cell wall. The myosin-like sheath of the virus then contracts and the solid core pierces the bacterial membrane; finally the viral DNA is injected into the cytoplasm (Lwoff, 1961).

During the short period of its life cycle when a virus is extracellular, it is susceptible to chemicals, even to dilute soap solution. However, no useful attack can be made on virus diseases unless the viruses can be killed in the parasitized cell without harm to uninfected host cells.

For chemotherapy with the new anti-viral drugs see Sections 4.0 (p. 129) and 6.3.2. For an account of the biology of animal viruses, see Fenner *et al.* (1974), and for medical aspects of virology, see Fenner and White (1976).

6

Chemotherapy: history and principles

Discussion, in the last three chapters, of the three principles that govern selectivity provides the scientific basis for a review (in this chapter) of the history and principles of chemotherapy and (in the next chapter) of pharmacodynamics. The study of history grows in interest the more we realize that our best drugs are losing their value, due to drug resistance and metabolic destruction, grim phenomena that continue to plague the drug designer's best efforts. History helps us go back in time to the period just before a discovery was made so that we can see how it was achieved, sometimes in a most unfavourable climate of opinion. Thus we may hope, by reviewing the circumstances in which remarkable new discoveries were made, to learn better how to make others.

6.0 The early history of chemotherapy

Micro-organisms were discovered by Antonie van Leeuwenhoek in Delft (Holland) in 1676, but their role in causing infections, plagues, and epidemics was established only two centuries later, by Robert Koch through his work on anthrax. Koch, and his four postulates for ensuring that a suspected microbe actually caused a given disease made it possible for biologists to study infectious diseases experimentally for the first time. This set the scene for the discovery of chemotherapy, but it was slow to arrive as we shall see.

In 1891, Romanovsky made a most significant observation in St Petersburg (Russia). By the use of his special microscopic stain (eosin-methylene blue) he showed that the malarial parasite was damaged in the blood of patients undergoing treatment with quinine. The greatest effect was on the asexual intra-corpuscular forms, whose nuclei quickly disintegrated. After two days, no parasites could be seen in the blood. This led Romanovsky to state that quinine cured malaria *by damaging the parasite more than the host*. This conclusion had great historical importance, because no one had previously thought that a drug could act in this way. Rather it was thought that drugs strengthened the body's defences, or even released new energy.

The protozoon which caused malaria had been discovered by Laveran in 1880. Romanovsky predicted that specifics for other diseases would eventually be found, substances that would cause minimal damage to the tissues of the host, and maximal damage to the parasites (Romanovsky, 1891). This line of thought was so little in tune with the intellectual climate of the times that it was not further pursued, until Ehrlich resumed it with energy and insight, and discovered chemotherapy.

The term *chemotherapy* was coined by Paul Ehrlich (1854–1915), who defined it as *the use of drugs to injure an invading organism without injury to the host*. As his pupil Carl Browning wrote in 1929:

'Chemotherapy is a term coined by Ehrlich to indicate the treatment of infections by compounds, of known chemical constitution, which effect cure by leading to destruction of the pathogenic organisms or their products. Thus a contrast is implied between chemotherapeutic substances and the class of antibodies, which represent highly complex, specific products of the biological reaction to the infecting organisms.'

The essence of chemotherapy is the achievement of a differential effect whereby the host, in his struggle with a parasite, gains some advantage through the introduction of a drug. This leaves the majority of the struggle to be done by the host's leucocytes and other natural defensive forces, but a useful drug can tip the balance in the host's favour.

The progress of chemotherapy was greatly speeded by the non-identity of the economic and the uneconomic species (see Section 1.0), for this gave the possibility of examining (and even culturing) the parasite independently of the host. As a result, the governing principles of chemotherapy have been brought to light much sooner than those of pharmacodynamics (Chapter 7). The present chapter is concerned with the nature of these principles and the historical background which led to their discovery.

A few chemotherapeutic agents were known before Ehrlich's time. These were cinchona bark and ipecacuanha rhizome, for the cure of malaria and amoebic dysentery respectively, and mercury for alleviation of the symptoms of syphilis. Mercury began to be used in this way in the sixteenth century, cinchona and ipecacuanha in the seventeenth. Santonin and male fern have been used as anthelmintics since classical times. In view of the many centuries during which medication had been practised, this list is remarkable for its brevity.

6.1 Ehrlich's fundamental contributions

It is customary to take 1899 as the start of Ehrlich's interest in chemotherapy. At that time he was 45 years old, and had just been appointed Director of the Königliches Institut für experimentelle Therapie, in Frankfurt (Germany). Up to that date, his work had consisted of applications of chemistry to the furthering of biological knowledge, and selectivity was his guiding light. His earliest work, on the distribution of lead in the body, revealed preferential accumulation in the central nervous system. Next came his discoveries of the differential staining of tissues by dyes. His technique and his division of the stained leucocytes into acidophil, basophil, neutrophil, and non-granular, is used by clinical pathologists to this day. Belonging to this period is his discovery of vital staining (using methylene blue and neutral red), and his ranking of the different tissues of the body by their comparative oxygen requirements.

All of this work he capped by outstanding discoveries in immunochemistry. He showed, by the first test-tube experiments to be done in the field of immunity, that the neutralization of toxin by anti-toxin, and antigen by antibody, are direct reactions which do not require the presence of a living mammal, as had been thought. He then went on to state his side-chain theory, a chemical interpretation of the immune process, as follows. An antigen has two active areas, namely the *haptophore* (anchorer) and the *toxophile* (poisoner). Mammalian cells, he believed, have 'side-chains' which contain *receptors*, i.e. groups or regions that are complementary to the haptophores, and hence anchor them. This combination, he taught, was in itself harmless but brought the toxophile close enough to the cell to poison it. He believed that the normal function of the receptors was to anchor nutrient molecules. (For more on antibodies, which have a mol. wt. of about 150000, see Nisonoff, Hopper and Spring, 1975.)

His experimental investigations on immunity, begun in 1893, led to discovery of an accurate method for standardizing diphtheria anti-toxin. This prompted the State to create the Royal Institute for Experimental Therapy*, for the statutory control of sera and vaccines. When Ehrlich assumed the Directorship in 1899, his reputation as a chemical pathologist had already been established internationally. He could well have rested on his laurels, but the chemist in him made him eager to launch out in an entirely new direction: the cure of infectious diseases with *small* molecules, specific enough to be 'magic bullets' that could harm only the invader.

In visits to the Hoechst works near Frankfurt, Ehrlich was brought face to face with the German synthetic chemical industry and saw the manufacture of a profusion of synthetic analgesics, antipyretics, and anaesthetics (see Section

*Originally in Berlin, later moved to Frankfurt at that city's request.

7.1). It seemed logical to him that, because the factories were turning out simple substances that differentiated between various tissues in Man, it should be possible to synthesize other small molecules which would differentiate between Man and his parasites. The emphasis on low molecular weight illustrates the contrast which Ehrlich made between immunotherapy and chemotherapy. He realized that immunotherapy was a matter of strenghtening the defence forces of the body, but he conceived of chemotherapy as a direct attack upon the parasite. The problem, as he saw it, was to find chemicals with very much stronger affinities for the parasites than for the tissues of the host.

Many of Ehrlich's contemporaries did not think his new line of research was reasonable or likely to succeed. The State, in particular, had expected him to confine his research to immunology in the new institute, and noted that chemo-therapeutic research would be very expensive because of the large number of experimental animals required. A compromise was reached: that Ehrlich might investigate the therapy of cancer with small molecules. He took up this cancer work late in 1901, pursued it for a few years and made some interesting observations, but realized that in many ways the time was not ripe for this kind of work, and dropped it.

From 1904 onwards Ehrlich concerned himself chiefly with chemotherapy, selecting trypanosomiasis in the mouse as his principal model. Conditions for this work were poor in his Institute, and became steadily worse. However, in 1906 his gloom was dispersed by the splendid action of a benefactor, Frau Franziska Speyer who had come to regard Ehrlich's work with the greatest

Fig. 6.1 The Georg Speyer Haus, Frankfurt, Germany.

affection. In memory of her late husband, she purchased a large house near the University, and named it the Georg Speyer Haus which she equipped and handsomely endowed for Ehrlich to pursue his research into chemotherapy. These chemical and biological laboratories became the Mecca for the cream of Europe's young drug scientists. Ehrlich thus became, virtually, Director of two Institutes. Fig. 6.1 portrays the Georg Speyer Haus.

Ehrlich's first chemotherapeutic experiments were performed with dyes. Three series gave good results in the chemotherapy of trypanosomiasis in the mouse, namely the acridines, the triphenylmethanes, and the azo dyes, but none was outstanding. True, in 1904 Ehrlich and Shiga cured trypanosome-infected mice with trypan red (6.1) (a polyazo dye), which thereby became the first man-made chemotherapeutic agent. This aroused interest, but unfortunately the drug was inactive in Man, and his thoughts turned to organic arsenical compounds.

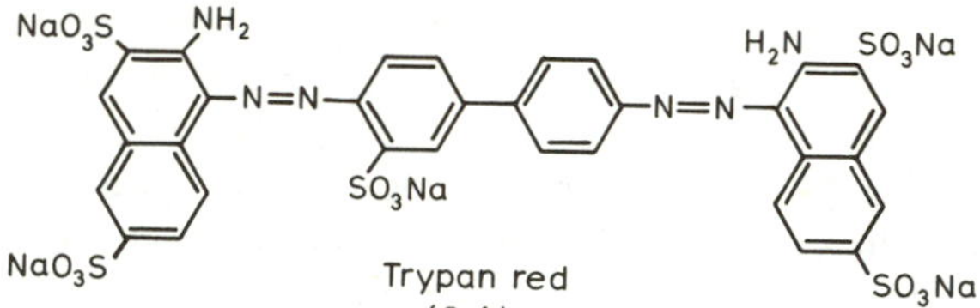

Trypan red
(6.1)

The status of arsenicals in infectious diseases at the beginning of the present century can be judged from the following. In 1902, Laveran and Mesnil injected arsenious acid ($HO \cdot As:O$) into mice infected with trypanosomiasis. All the mice died, but they 'died cured' which was considered a great advance at the time. In 1905, Thomas and Breinl (in Liverpool) had shown that an arsenical drug, 'Atoxyl' (6.2), had a slight, favourable action on human trypano-somiasis.* This discovery influenced Ehrlich to commence prolonged experi-mentation with aromatic arsenicals.

In 1908, a Nobel Prize in Medicine was awarded Ehrlich, 'in recognition of his work on Immunity'. Chemotherapy was not mentioned. Even in Germany, many of his colleagues thought that his new interpretations were wild, and had completely outrun the evidence. It could not be denied that Ehrlich had been working intensely on his self-chosen, seemingly improbable subject for ten years without producing even one result of use in human medicine.

Yet this run of bad luck was suddenly to change, for in 1910 a discovery of utmost significance was made, namely the anti-syphilitic drug arsphenamine (6.3), whose laboratory code number had been '606' (Ehrlich and Hata, 1910). It happened as follows. In 1905 Schaudinn had discovered that the causative organism of syphilis was a highly motile bacterium (a spirochaete) which he

*The rather optimistically named 'Atoxyl' (p-aminophenylarsonic acid) was discovered in 1860 by Béchamp, whose wrongly assigned constitution was corrected later by Ehrlich and Bertheim (1907).

named *Treponema pallidum*. Hata, working in the Kitasato Institute for Infectious Diseases in Tokyo, followed by discovering how to produce this disease in rabbits. Hence, for the first time, an experimental model was available for laboratory work in this disease. (Parenthetically it must be noted that, over and over again, progress in chemotherapy was to await discovery of a suitable laboratory model, in one disease after another.) Hearing of Hata's discovery, Ehrlich lost no time in bringing him to the Georg Speyer Haus (see Fig. 6.2).

Fig. 6.2 Paul Ehrlich (1854–1915) and Sahachiro Hata (1873–1938), who together discovered the curative action of arsphenamine ('Salvarsan') in syphilis.

It may seem a vast leap for Ehrlich to transfer his interest from a protozoon (*Trypanosoma*) to a bacterium (*Treponema*), but there was this in common: both were incessantly motile organisms, precariously dependent on a wildly racing

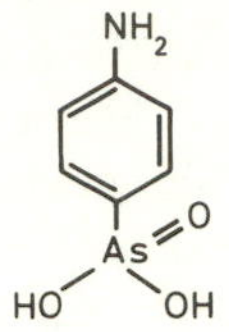

Atoxyl
(6.2)

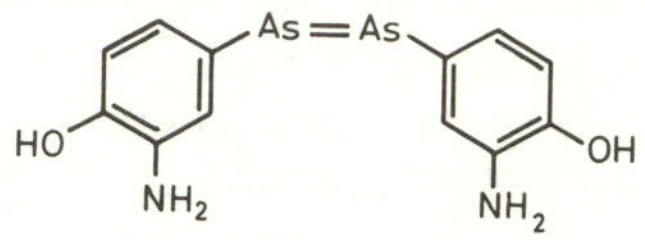

Arsphenamine (monomer)
(6.3)

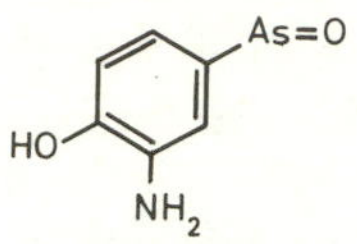

Oxophenarsine
(6.4)

metabolism. Ehrlich had an intuition* that *Treponema* would yield to his new arsenicals. Of these, he now had examples with increased selectivity, thanks to the collaboration of Professor A. Bertheim and Drs Kahn and Schmitz, chemists who had learnt to synthesize a wide variety of organic arsenical compounds, difficult though this task was (Ehrlich, 1909).

The synthesis of arsphenamine was reported by Ehrlich and Bertheim (1912) and it was patented and then manufactured by the firm, Hoechst (DRP 224 953), and sold by them under the name 'Salvarsan'. The initial announcement of a cure for syphilis was taken up by the newspapers, and Ehrlich became a world celebrity overnight. This fame gave him little pleasure, for he had worries that arose from the discovery. For example, arsphenamine was oxidized in the air to a more toxic product, now known to be oxophenarsine (*6.4*) a much more selective and desirable drug, and one that would in time replace arsphenamine (see Section 6.2). However, at that time, the uncontrolled oxidations were causing deaths, so that Ehrlich decided that arsphenamine must be issued only as single doses, and in sealed tubes from which all oxygen had been removed. He also issued directions for the preparation of solutions in sterile distilled water, neutralization, and intravenous injection without delay. These directions were often departed from, and the resulting disasters attracted unfortunate publicity. Ehrlich later introduced neoarsphenamine ('914'), a more soluble derivative.

Ehrlich always had entertained hopes of his 'magic bullets' curing with a single dose, but arsphenamine was not selective enough for a sufficiently large dose to be given. Hence treatment with it had to be spread over several months (Ehrlich, 1911).

From the observation point of today, there can be no doubt that arsphenamine, with all its faults, was the opening event in the chemotherapeutic revolution which transformed all treatment of infectious diseases. This happy result owes much to Ehrlich's quantitative approach.

In seeking drugs which would have a great affinity for the invader and little for the host, Ehrlich introduced the 'Chemotherapeutic Index' which he defined as the ratio:

$$\frac{\text{minimal curative dose}}{\text{maximal tolerated dose}}$$

*H. Uhlenhuth had noticed a slight, favourable effect of atoxyl in human syphilis, but increasing the dose caused blindness (Ehrlich, 1909).

Thus a substance which was curative of trypanosomiasis in mice at 2 mg/kg, and did not kill below 50 mg/kg, would have an Index of 1/25 (Ehrlich, 1911). Thus the idea of selective toxicity† was provided with a yardstick for measuring the degree of selectivity. How this measure was later developed and refined will be seen in Section 6.2.

When Ehrlich had transferred his interests from large molecules to small ones, he drew on part of his 'side-chain' (immunological) hypothesis to explain the mode of action of chemotherapeutic agents. They too, he supposed, had distinct haptophoric and toxophilic groups (he demonstrated this for arsenicals; see the end of this section), and they combined with cellular receptors whose normal function was to take part in cellular nutrition or respiration (Ehrlich, 1908).

Ehrlich insisted that drugs acted upon cells by ordinary chemical reactions. This is substantially the present-day view, although the phrase 'ordinary chemical reactions' conjures up a much wider variety of bonds than was known in Ehrlich's time (see Section 8.0). Ehrlich's highly original concepts, of chemically reactive groups on drugs and of chemically reactive receptors for them on cells, made an important advance in biological thought. His thesis that the most effective drugs would have a fairly low molecular weight has been well substantiated. Chemotherapeutic drugs have molecular weights ranging from about 140 to 1400, the latter being unusually high. On the other hand, γ-globulin, of which antibodies are fashioned, has a molecular weight of 150 000.

The work done by Ehrlich and his band of collaborators in the Georg Speyer Haus did not stand high in the esteem of all contemporary scientists. Many people of this period thought that his interpretations had outrun the experimental evidence. Chief among Ehrlich's critics was Uhlenhuth, an influential and respected pathologist. Uhlenhuth contended that drugs had no direct action on the parasite, but worked by stimulating the natural defences of the host. He made the most of the fact that 'Atoxyl' and trypan red can cure trypanosomiasis in the animals and yet do not attack trypanosomes in the test-tube (Uhlenhuth, 1907). This objection was taken seriously because these substances were two of Ehrlich's most important experimental agents. But in 1909 Ehrlich discovered that *tri*valent arsenicals were trypanocidal in the test-tube and he suggested that pentavalent arsenicals would also be found active if it were possible to keep trypanosomes alive in culture until they had time to reduce the drug. That this was actually so was not proved until (15 years after his death) a better culture method become available. However, Ehrlich did show that pentavalent arsenicals became active *in vitro* if they were first incubated with a reducing tissue (e.g. liver), and this result served to explain their activity *in vivo*.

The discovery of drug-resistance (discussed in Section 6.5) in Ehrlich's laboratories gave welcome support to his chemical hypothesis of absorption and combination. Ehrlich observed that those trypanosomes which were resistant to

†The term 'Selective Toxicity' was born, long after Ehrlich's time, in 1940.

trypan red did not absorb it, whereas susceptible trypanosomes were stained red by the dye. Thus the resistant parasites, by selective breeding, had lost the chemical group that took up the dye. Later, Ehrlich found two strains of trypanosomes, each resistant to a different kind of organic (aromatic) arsenical. There was no cross-resistance between these strains. Because no trypanosomes are resistant to inorganic arsenic (e.g. HO·As:O), Ehrlich concluded (i) that the resistance was directed against certain substituents in the benzene ring, (ii) that these substituents (e.g. —NH$_2$) were responsible for the uptake of the aromatic arsenicals by the parasite, and (iii) that the arsenoxide group (—As:O) was responsible for the death of the parasites. These results indicated the chemical composition of both haptophoric- and toxophilic-groups in those arsenicals which have selectivity (only *aromatic* arsenicals are selective, see Section 12.0).

For an account of Ehrlich's aims, and a sampling of his accomplishments up to 1908, see his address to the German Chemical Society (Ehrlich, 1909). For more biographical material, see Muir (1921); Browning (1955); Himmelweit (1956); also the 'Paul Ehrlich Centennial' issue (1954) of the *Annals of the New York Academy of Sciences*, Vol. 59, which includes assessments by various authors of Ehrlich's work and its present-day developments.

6.2 Chemotherapeutic drugs available before 1935.
The chemotherapeutic index

Although the discovery of anti-bacterial sulfonamides in 1935 immensely quickened the pace of chemotherapeutic research, the principles that govern chemotherapeutic action had been brought to light before that date. Before Ehrlich died in 1915, the following advances bore witness to the faith in chemotherapy that he had created. Antimony, in the form of tartar emetic, was shown to cure trypanosomiasis in mice (Plimmer and Thompson, 1907) and to influence the disease favourably, but without cure, in Man (Manson, 1908). In Ehrlich's Institute 'Optochin', a simple derivative of quinine, cured pneumonia in the mouse, but was ineffective in Man (Morgenroth and Levy, 1911). Next, tartar emetic (*4.58*) was introduced as a clinical treatment for the protozoal disease leishmaniasis by Vianna (1912) in Brazil. About this time, Rogers found that emetine (*4.51*) was the most active principle in ipecacuanha, and used it to cure advanced amoebiasis in Man (Rogers, 1912). Ehrlich had just shown that euflavine ('Trypaflavin') (*6.5*), which Benda had synthesized for him at the Hoechst works, could cure trypanosomiasis in mice, but it was useless in Man (Ehrlich, 1912). Thus the only clinically useful drugs discovered by Ehrlich were arsphenamine ('Salvarsan') and its solubilized derivatives. His influence on the development of chemotherapy came not so much from these discoveries as from his clear delineation of the principles involved in treating infectious diseases with chemicals of low molecular weight.

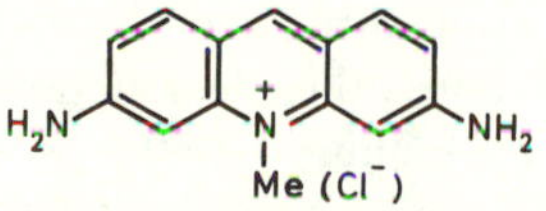

Euflavine ('Trypaflavine')
(6.5)

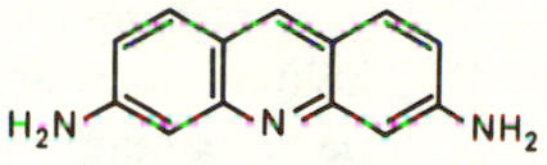

Proflavine
(3,6-diaminoacridine)
(6.6)

In 1913, Browning, a former pupil of Ehrlich, discovered the remarkably selective anti-bacterials, acriflavine and proflavine (6.6), which were chemically allied to euflavine (Browning and Gilmour, 1913). Apart from an extensive use of these acridines in wounds, the First World War passed without any notable advance in chemotherapy. The control of bacterial infection in the bloodstream was still an unrealized dream, and the feeling was widespread that chemotherapy had promised more than it was likely to give. This pessimism was premature, particularly as research on protozoal diseases was prospering. To begin with, Christopherson worked out the dose schedules that made anti-monials, for many long years, the drugs of choice in schistosomiasis (Christopherson, 1918), and other important remedies quickly followed.

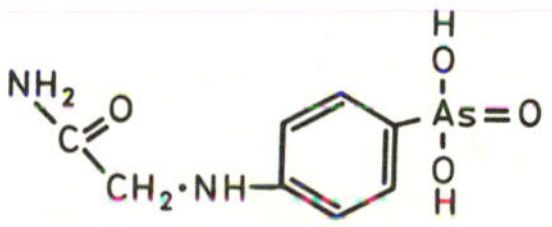

Tryparsamide
(6.7)

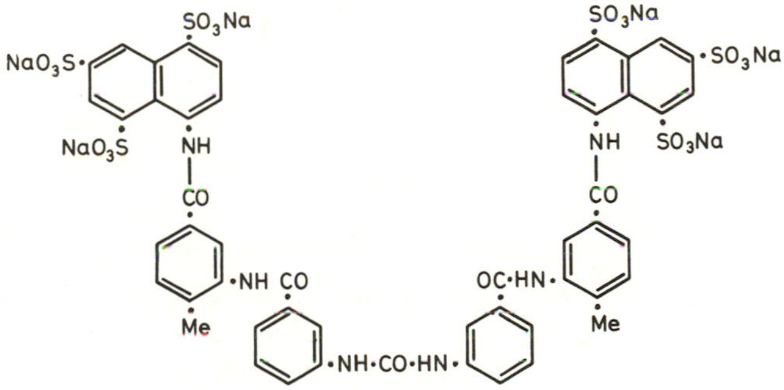

Suramin
(6.8)

In 1920, two drugs were discovered which became the basis of a successful treatment of sleeping sickness. These were (a) tryparsamide (6.7), a pentavalent arsenical derived from, but less toxic than, 'Atoxyl' (Jacobs and Heidelberger, 1919; Brown and Pearce, 1919), and (b) suramin ('Bayer 205', 'Germanin'), a colourless and highly effective analogue of trypan red (Heymann et al., 1917). Suramin (6.8) was later used for prophylaxis after it was found that a single dose gave immunity for three months (Roehl, 1920; Heymann, 1924). The next notable advance was the introduction of the first synthetic anti-malarials, pamaquine ('Plasmoquine') (6.9) (Roehl, 1926a; Schulemann, Schönhöfer and Wingler, 1932), and mepacrine (6.10) ('Atebrin') (Kikuth, 1932; Mauss and Mietzsch, 1933). They were inspired by the (weak) anti-malarial action that Guttmann and Ehrlich (1891) had found in methylene blue (6.11); but systematic work had to await discovery of a test object, and Roehl found this in malaria-infected finches.

In 1930, Leake introduced carbarsone (6.12) for the treatment of amoebiasis, a less nauseating drug than emetine which was henceforth reserved for the severer cases. Interestingly, this organic arsenical (4-ureidobenzenearsonic acid) had been synthesized by Bertheim in 1907, and found unsuccessful by Ehrlich against trypanosomiasis and syphilis (Leake, Koch and Anderson, 1930). Until the discovery by Laidlaw, Dobell and Bishop (1928) of a method of cultivating entamoebae *in vitro*, the experimental study of amoebicides had been severely handicapped.

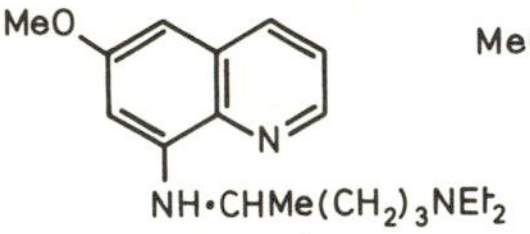

MeO— [structure] NH·CHMe(CH₂)₃NEt₂
Pamaquine
(6.9)

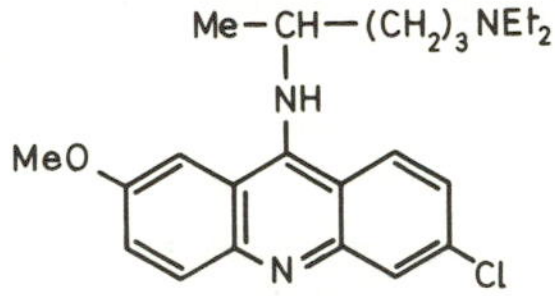

Me—CH—(CH₂)₃NEt₂ / NH
MeO— [structure] —Cl
Mepacrine
(Quinacrine, 'Atebrin')
(6.10)

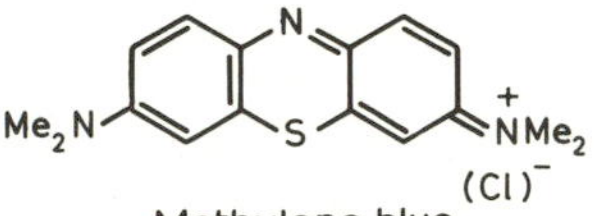

Me₂N— [structure] —⁺NMe₂
(Cl)⁻
Methylene blue
(6.11)

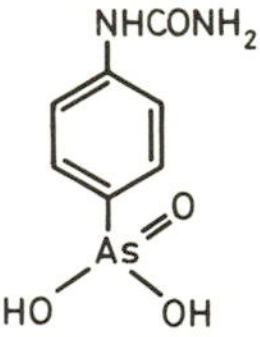

NHCONH₂
[structure]
As
HO　OH
Carbarsone
(6.12)

All these discoveries were made by pursuing clues which Ehrlich and his school had revealed. Indicative of the impact that chemotherapy was now making was the German State's ban on any disclosure of the constitution of suramin and mepacrine, because of the potential they lent to tropical warfare. This was a complete swing of the pendulum from the State's apathy in 1899. (Parenthetically, the secrecy was short-lived: the synthesis of suramin was solved by Fourneau *et al.* in 1924 at the Pasteur Institute in Paris; Maghidson and Grigorovski, in Moscow, published that of mepacrine in 1933.)

The year 1932 was notable not only for the introduction of mepacrine, but for the discovery by Tatum and Cooper of a much safer arsenical drug for use in syphilis, namely oxophenarsine (6.4) ('Mapharside') (Tatum and Cooper, 1934). This drug had been discovered by Ehrlich and his colleagues and undervalued by them in experimental work, so that the clinical application was not foreseen (see Section 13.0).

Yorke, Adams and Murgatroyd (1929) had found how to keep trypanosomes alive and unmetamorphosed in the test-tube for two days, whereas Ehrlich had only moribund parasites *in vitro*. With the help of this new technique, Yorke, Murgatroyd and Hawking (1931) showed that when normal trypanosomes were treated with a solution of a trivalent arsenical, they were rapidly killed and the solution would not kill fresh trypanosomes. On the other hand, when arsenic-resistant trypanosomes were treated with a trivalent arsenical, they were not killed, and the residual solution would still kill susceptible trypanosomes. It was then shown that the bodies of the killed parasites contained measurable amounts of arsenic and that the resistant parasites contained none (Reiner, Leonard and Chao, 1932). Advances in technique made it possible for others to show that fixation of drugs by the parasite could take place while the latter was actually circulating in the host's bloodstream. Thus von Jancsó (1932) used the fluorescent microscope to show that 'Trypaflavin' was taken up in this way. These discoveries confirmed the idea that accumulation of the drug by the parasite was a necessary step in the drug's lethal action, thus fulfilling one of Ehrlich's main predictions, one that had been controversial in his lifetime.

Another prediction confirmed in this period was the co-operation between drug and the host's defence forces. Ehrlich had held the view that the function of a drug was to disorganize the parasite's metabolism, after which the natural defence forces of the body would complete the actual destruction of the invader. This was confirmed when Kritschewsky (1928) showed that the host's reticulo-endothelial system provided such co-operation, because when this was artificially damaged, the efficiency of chemotherapeutic drugs was lowered greatly.

That a parasite need not be harmed merely by taking up a foreign substance, followed from Ehrlich's teaching that the haptophoric (anchoring) and toxophilic (poisoning) portions of a substance were distinct entities. Techniques of vital staining have now made this concept familiar.

As momentous as any discovery in the period we are considering, was the first demonstration of the chemical nature of the union between a drug and a parasite. This was accomplished by Voegtlin and his colleagues in the United States Public Health Services, who showed that the toxic action of arsenicals was due to the formation of As—S bonds with essential thiol-groups in the parasites (see Section 13.0). This, too, had been predicted by Ehrlich (1909).

Thus, at the end of this period, the scientific principles of chemotherapy, as foreshadowed by Ehrlich, had become established. It was now generally agreed that substances of relatively low molecular weight make the most satisfactory chemotherapeutic drugs, that their action is directly on the parasite, and that their reaction with the parasite is chemical in nature. Ehrlich's division of a drug into haptophoric and toxophilic portions still reminds us to look closely at the structure of every agent and to ask what each group is contributing to the total effect. But today we know that the properties of a molecule also depend on the electronic interaction of its component parts which, according to well-known laws of chemistry, augment or diminish one another's influence in fairly predictable ways (see Section 17.2).

6.2.1 *The chemotherapeutic index*

Ehrlich's Index (Section 6.1) has undergone some changes. In the course of time, it came to be used as a reciprocal (i.e. 25 instead of 1/25). It was found later that more repeatable results can be obtained if the LD_{50} (dose killing 50% of the test animals) was used instead of the maximal tolerated dose, and if CD_{50} (dose curing 50% of the animals) replaced the minimal curative dose. The chemotherapeutic index makes it clear that it is no advantage to find a new substance with half the toxicity to the host unless the new substance is *more* than half as active. With this idea to guide them, chemists and biologists have discovered drugs of tremendously improved chemotherapeutic index in the last 50 years. The toxicity of the best sulfonamide anti-bacterials to mammals is so low that it is difficult to measure accurately; that of penicillin is almost negligible.

The adoption of the LD_{50} was due to the English mathematician J.W. Trevan, who showed that there is a Gaussian (normal) distribution of variations in the intensity of cellular responses to *a given dose* of a drug. Fig. 6.3 makes this clear. When these intensities were summed for *increasing dosage* in accordance with the percentage of cells responding, a sigmoid relationship was obtained (Fig. 6.4; Trevan, 1927); this is called cumulative normal distribution. The reproducibility of biological tests is usually poor, relative to those made in physics and chemistry. Of the several factors that make for a spread of values are the many rates, and even kinds, of metabolic transformations due to genetic traits which differ from individual to individual, even in inbred strains of laboratory animals. In addition, many biological end-points are significantly affected in the laboratory by variations in feed, lighting, bedding, degree of

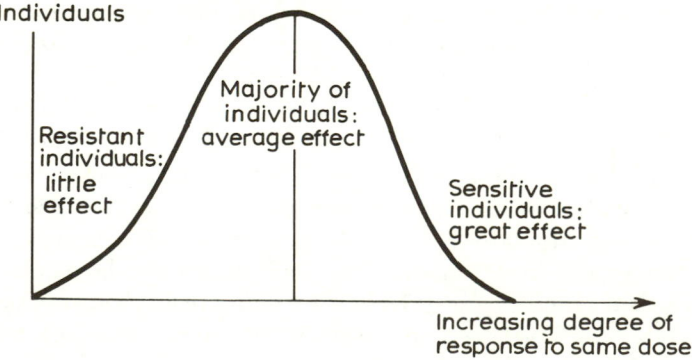

Fig. 6.3 A normal distribution curve.

crowding, and handling by the investigator. These adverse effects are countered by using as large a number of animals (or organs, or cells) as is economically possible for each experiment, and submitting the results to statistical analysis.

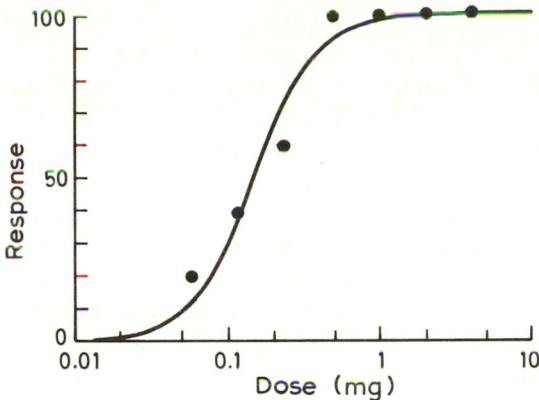

Fig. 6.4 A sigmoid dose–response curve (dose plotted logarithmically.)

Reverting to Fig. 6.4, the *response* may be the percentage cured, or a physiological response such as muscular contraction, or a measure of lethality (percentage killed). So long as the dose is plotted logarithmically, the portion of the sigmoid correlation curve that lies between 20 and 80% response is likely to be a straight line, a convenient starting point for comparing two specimens of the same agent (e.g. histamine liberated from a tissue).

Many aspects of the relationships between dose and effect, particularly as regards time and concentration, were analysed by Clark (1933, 1937). Further statistical analysis showed that the safety of a drug is better established by first determining the dose that causes evident toxicity but allows more than 90% survival: this dose is then progressively reduced. Judged by these high standards, several drugs now in use, particularly the cardiac glycosides, are seen to require a substitute with a better chemotherapeutic index. For more on the analysis of dose–response curves, see Waud (1981).

6.3 1935 and afterwards

The year 1935 forms a natural division in the history of chemotherapy because of Domagk's discovery of the chemotherapeutic properties of 'Prontosil', the first of the anti-bacterial sulfonamides. This discovery refuted a prejudice that chemotherapy applied mainly to protozoal diseases, and it opened a new era in medical and surgical treatment. Woods (1940) then discovered that sulfonamides blocked the use by bacteria of one of their essential constituents (*p*-aminobenzoic acid). This was in accord with Ehrlich's postulate that drugs combined with receptors in such a way as to prevent normal nutrients from reaching them (see Section 9.3.1).

6.3.1 *Antibacterials*

(*a*) *Antibacterial sulfonamides.* The discovery of sulfachrysoidine (*3.30*) p. 100 ('Prontosil'), followed a strange course. Gerhard Domagk, the discoverer, was born in Berlin in 1895, qualified in medicine in 1921, then followed a university career in chemical pathology at Greifswald and Münster. In 1927, he was invited to the Bayer Company in Elberfeld to take charge of the bacteriological and pathological research laboratories there, a position that he held for the rest of his life. While observing the phagocytosis of streptococci by the Kupffer cells of the liver, he decided to weaken the pathogenicity of the bacteria. He recalled a technique of Ehrlich's and asked for a selection of azo-dyes. Among those given him was 'Prontosil' which Mietzsch and Klarer had recently synthesized for other work. Noting how well his animals survived injection of the dye-treated streptococci, Domagk tried the effect of 'Prontosil' on established streptococcal infection in mice. The result was dramatically successful (Domagk, 1935). However his employers, reflecting the current climate of opinion, moved very slowly: they had seen many candidate drugs, including azo-dyes, fail as systemic antibacterials.

Domagk was immensely disappointed, but fate played a remarkable trick. On the fourth of December 1935, his little daughter who was making a Christmas present, punctured her hand with a needle while running down stairs for her mother to rethread it. Although she was rushed to hospital, and received the best treatment available in those times, severe inflammation and fever soon set in. Four days later, she seemed likely to die from streptococcal septicaemia, as was not at all uncommon in those days. Domagk begged for permission to administer 'Prontosil': the speed of her recovery confounded the physicians (Domagk, 1936).* Almost ignored in Germany, the first large and controlled clinical application of 'Prontosil' took place in London, where Colebrook and Kenny (1936) demonstrated its power in saving moribund mothers from puerperal sepsis. The drug was now truly launched.

A discovery of great importance for perfecting Domagk's discovery was made late in 1935 by Tréfouël and his co-workers in the Institut Pasteur, who showed that the inactivity of 'Prontosil' towards bacteria in test-tube experiments could be overcome by adding a reducing substance. In devising this experiment, they were guided by Ehrlich's parallel activation of pentavalent arsenicals (Section 6.1). Later, they were able to show that not only did the mammalian body

---

*Awarded the Nobel Prize in Medicine for this work, Domagk was gaoled by the German (Nazi) Government until he renounced it. He reaccepted it in 1947 after the Second World War. The rest of his working life was devoted to research on the chemotherapy of tuberculosis (see Section 11.9). He died in 1964. Biographies: *Deutsch. med. Woch.* 1948, **73,** 350; *Münch. med. Woch.* 1965, **15,** 739; *Chemiker Ztg.* 1964, **88,** 395.

reduce 'Prontosil' to give sulfanilamide* (*2.13*), but that this substance was equally active (and helpfully much more soluble) in the treatment of bacteraemia. Had the French workers not made this discovery, the evolution of sulfonamides would very likely have continued on 'hit-and-miss' lines, with immense resources squandered in the search for new types of *coloured* sulfonamides (Tréfouël *et al.*, 1935).

An important scientific advance was made when Marshall (1937) showed that the effect of sulfanilamide was proportional to the concentration reached in the blood of the patient being treated, and that, for a given dose, this varied from patient to patient. This was the beginning of the analytical control of blood-levels during chemotherapy, although it was soon found that the activity of some substances (e.g. mepacrine) is not proportional to their concentration in blood if greater concentrations exist, loosely bound, in the tissues.

The mode of action of sulfonamides was greatly clarified by Woods in 1940. It had been shown that tissue extracts, pus, bacteria, and particularly yeast extract contained a heat-stable substance of low molecular weight which would inhibit the action of sulfonamides on bacteria (Stamp, 1939). Woods, recalling that enzymes are inhibited by substances which chemically and sterically resemble their substrates (see Section 9.3.1), adopted this hypothesis: that the inhibitory substance in yeast is the substrate of an enzyme widely distributed in nature, and that it resembles sulfanilamide chemically. He found activity was concentrated in an alkali-soluble fraction of yeast, and that it ran parallel to a colour test for an aromatic amino-group. Activity was lost on esterification or acetylation, recovered on hydrolysis, and lost again on treatment with nitrous acid (Woods, 1940). Thus he made it clear that the active substance was an aromatic amino-acid. Because *p*-aminobenzoic acid (*p*AB) (*2.12*, p. 31) is the aromatic amino acid that most resembles sulfanilamide (*2.13*) he tried it as an inhibitor of bacteriostasis, and found that one molecule could prevent 5000 to 25 000 molecules of sulfanilamide from functioning.

Next, *p*AB was isolated (as the benzoyl-derivative) from yeast by Rubbo and Gillespie (1940), who showed that it was a growth-factor for a bacterium (*Clostridium acetobutylicum*), i.e. it was essential for this bacterium in the way that vitamins are for human beings. Since 1940, it has been found to be essential for many families and genera of micro-organisms, including the malarial parasite; most micro-organisms do not require an *external* source of *p*AB because they synthesize it. That the antagonistic action of *p*AB was exerted also *in vivo* was shown by Selbie (1940), who found that simultaneous administration of this acid prevented sulfanilamide from curing mice infected with streptococci.

Knowledge of this antagonism between sulfonamides and *p*AB, of Voegtlin's work on the mode of action of arsenic (see Section 13.0), and of his own similar work with mercurials, led Fildes (1940a) to appeal for a more rational approach

---

*Sulfanilamide, synthesized by Gelmo in 1908, was not known to have any biological activity. Yet, because 'Prontosil' was made from it, its antibacterial nature could easily have been observed in Elberfeld.

to chemotherapy. He pointed out that the action of a drug on a parasite had much in common with the inhibition of an enzyme by a foreign substance, and that many of the most successful enzyme inhibitors were analogues of substrates. [This relationship was already well established in pharmacodynamics (see Sections 2.1 and 9.2).] Fildes discussed two main types of enzyme inactivation as models for new drugs: (a) the inhibitor combines with the enzyme, and displaces a similarly constituted substrate or coenzyme, and (b) the inhibitor combines with the substrate or coenzyme.

This visualization of some of Ehrlich's receptors as enzymes provided an incentive for biochemists to study the enzymes of parasites, to discover the nature of the substrates, and to synthesize molecules that would be like enough to these substrates to be accepted by the bacterium, and yet different enough to interrupt its vital processes. By developing Fischer's analogy that enzymes and their substrates have a lock-and-key relationship, a chemotherapeutic agent could be seen as a key that fits the lock only well enough to jam it (see Section 9.1). This thought was received with too much enthusiasm! Countless analogues of natural metabolites were synthesized, but the majority of these proved to be as toxic to the host as to the parasite. Why these misfortunes occurred, and how they were overcome, is discussed in Chapter 9.

(b) *Antibiotics.* Seven years after the sulfonamides appeared on the scene, penicillin, the first chemotherapeutic antibiotic, began to be used in the clinic. Antibiotics are toxic substances, of low molecular weight, secreted by a few species of bacteria, and by a few fungi, most of them the lower fungi known as moulds. Most antibiotics are insufficiently selective to be used in medicine, but some twenty highly selective kinds are right in the forefront of current prescribing.

The name 'antibiotic' is a misnomer because it implies that organisms secrete these substances to fight their natural enemies. However, if penicillin were of survival value to *Penicillium notatum*, the few strains which manufacture this substance would be the commonest examples of this species whereas, in nature, they are actually the most uncommon. The fact is that bacteria and fungi produce what we call antibiotics at the time when they have *completed* the fast, logarithmic phase of growth, and stasis has set in. To get the highest yields, manufacturers of antibiotics make full use of this static phase, hastening its commencement and prolonging it artificially. Woodruff (1966) thought that stasis allowed existing enzymes to act on substrates which would normally have been used for growth and whose production, in higher forms of life, would be suppressed at this stage by feedback mechanisms. This, he suggested, was the origin of most of the antibiotics (all of which are polypeptides), and this production occurs only at the end of logarithmic growth (Schaeffer, 1969). It has been suggested that antibiotics are made by the bacteria to switch off normal metabolism, thus preparing the cell for sporulation (Schazschneider *et al.*, 1974). These authors note, for example, that tryocidin combines with DNA, in a way that inactivates many of the genes responsible for vegetative growth.

Before mentioning the principal types of antibiotics now available, the discovery and application of these substances will be briefly described. In 1929, Fleming discovered that a particular strain of the mould *Penicillium notatum*, which landed in, and lysed, a culture of staphylococci in his London laboratory, secreted a substance of low molecular weight that was highly toxic to Gram-positive bacteria, and had a low toxicity to mammals. However, he found it too unstable to isolate and did not persevere further.

In 1938, it occurred to Florey that much more potent antibiotics should be obtainable from bacteria and fungi than the known but feebly active examples: hydrogen peroxide, alcohol, pyocyanase, and acetic acid. Systematic investigations by Florey, Chain and their colleagues at Oxford led them back to penicillin and they discovered how to isolate it in a pure state (Florey *et al.*, 1940, 1941, 1949; Chain, 1948). They also demonstrated its remarkable clinical usefulness. In 1945, Fleming, Florey and Chain jointly received the Nobel Prize for Medicine. At first penicillin was used as the mixture of naturally occurring homologues; later it was found that the best of these, benzylpenicillin (see Section 13.1), could be produced exclusively if a related mould (*Penicillium chrysogenum*) was grown in the presence of phenylacetic acid (Moyer and Coghill, 1947). Industrial improvements in the fermentation and extraction procedures eventually made penicillin one of the least expensive of chemotherapeutic remedies.

The outstanding success of penicillin, in both local and systemic bacterial infections, followed from its high selectivity, the reason for which is discussed in Section 13.1. The mammalian toxicity is so low that it has come to be regarded as the most innocuous of all antibacterial drugs (however, a small proportion of patients become sensitized to it). In 1948, Brotzu found (in a sewage outfall on the Sardinian coast) another mould, *Cephalosporium*, which gave rise to the related cephalosporin series of antibiotics (see Section 13.2).

The world-wide search for antibiotics led to the discovery of many other anti-bacterials in cultures of lower organisms. The majority of these proved to be too toxic to mammals. Most of those which have achieved widespread clinical use have been derived not from moulds but from the actinomycetales, a distinct class of prokaryote organisms, often found in soil. From the nature of their cell walls, they are true bacteria. From various species of *Streptomyces*, a genus of the actinomycetales, the following clinically useful antibiotics were obtained: (a) *chloramphenicol*, isolated by Bartz (1948) from a culture obtained from a mulched field in Venezuela, (b) chlorotetracycline ('Aureomycin'), the first of the tetracyclines (Duggar, 1948), in a culture from a Missouri farmyard and (c) *streptomycin* by Waksman, in a culture from the gizzard of a chicken, and later from soils (Schatz, Bugie and Waksman, 1944). Later, many other examples were derived from yet other actinomycetes, such as adriamycin, amphoteracin, the bleomycins, cycloserine, erythromycin, gentamicin, kanamycin, lincomycin, neomycin, nystatin, oleandomycin, paromomycin, the rifamycins, spectinomycin, vancomycin, and viomycin. From the true bacteria (the eubacteriales) came bacitracin, colistin, the gramicidins and tyrocidins (Dubos, 1939), and the polymixins. A fungus (*Penicillium griseofulvum*) yielded griseofulvin (Oxford, Raistrick and Simonart, 1939) which, like amphotericin and nysatin, has purely anti-fungal properties (all three are used medicinally). Several antibiotics are in use as anti-cancer agents (see Section 4.0). Antibiotics act in many ways, but principally:

(a) by interfering with the synthesis of nucleic acid or protein (e.g. actino-mycin, the rifamycins, chloramphenicol, streptomycin, the tetracyclines; see Sections 4.0 and 4.1);

(b) by interfering with mucopeptide synthesis and hence the formation of new bacterial cell wall (e.g. cycloserine, penicillin, cephalothin; see Sections 5.3, 13.1 and 13.2); or

(c) by injury to the cytoplasmic membrane (e.g. amphotericin, polymixin; see Section 14.3).

The few that act by interfering with oxidative phosphorylation, or energy-releasing metabolic steps, e.g. antimycin, oligomycin (see Section 4.4) are unselective and lack clinical value, as does tunicamycin, a glycoside of uracil, which inhibits the glycosylation of proteins (Mahoney and Duksin, 1979).

For further reading on the mode of action of antibiotics, see Franklin and Snow (1981), Gale *et al.* (1981), the six-volume work edited by Hahn, completed in 1983, the ten-volume set edited by Bérdy, completed in 1982, the encyclo-paedia edited by Glasby (1979), and the journals *Antibiotics and Chemotherapy* and *Journal of Antibiotics*.

(*c*) *New synthetic antibacterials.* The success of antibiotics and sulfonamides as antibacterials encouraged new exploration of simple synthetic compounds. Isoniazid (*6.13*), a very simple substance (isonicotinic hydrazide), has played the leading part in reducing human tuberculosis from a dreaded scourge to an easily treated disease. It was discovered simultaneously by three laboratories namely Domagk's (Offe, Siefken and Domagk, 1952), where the inspiration had been his less effective drug, thiacetazone (Section 11.9), and those of Grunberg and Schnitzer (1953) and Fox (1953).

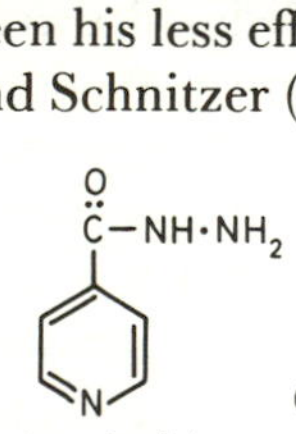

Isoniazid
(6.13)

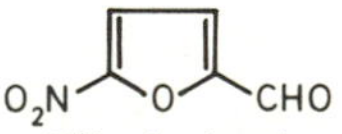

Nitrofurfural
(6.14)

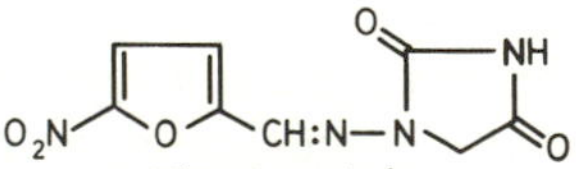

Nitrofurantoin
(6.15)

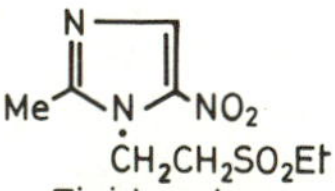

Tinidazole
(6.16)

Antiseptics active against Gram-negative bacterial rods are uncommon, but two excellent examples have been introduced. Of these, nalidixic acid (*4.33*) (about 1965), is highly active against Gram-negative bacteria in the urinary tract, but is inactive against Gram-positive species. How it acts by suppressing the synthesis of DNA was outlined in Section 4.0. The other example is a nitro-heterocycle. Until about 1945 it was thought that a nitro-group was too dangerous a substituent to introduce into a drug. This prejudice arose from the many cases of methaemoglobinaemia seen in munitions works where substances, such as trinitrotoluene, were absorbed through the skin and became reduced to derivatives of aniline in the body. However, non-benzenoid nitro-derivatives do not give this reaction. The introduction of nitrofuran drugs, all of them derivatives of nitrofurfural (*6.14*), by Dodd in 1946 opened up a rich vein in chemotherapy. Nitrofurantoin (*6.15*) ('Furadantin'), given orally, has proved to be an excellent urinary antiseptic, active against both Gram-positive and -negative organisms.

Nitroimidazoles were then found to have excellent anti-microbial properties. For example, tinidazole (*6.16*) is valued for the prophylaxis of anaerobic bacterial infection of the bowel, before operation, particularly with species of *Bacteroides*, also for the treatment of such infections (Hunt *et al.*, 1979). It is also effective in the protozoal diseases, amoebiasis, giardiasis, and trichomoniasis. For the anti-protozoal action of nitro-heterocycles, see Section 6.3.3.

Following the demonstration by Cowdry and Ruangsiri (1941) that dapsone (*9.17*) was effective in rat leprosy, this drug has become the standard treatment for human leprosy which it promptly alleviates, and slowly cures. To prevent the outgrowth of drug-resistant organisms, a supplementary drug is also used, either rifampicin (*4.37*) or the 2,3-diaminophenazine known as clofazimine.

6.3.2 *Antiviral agents*

Antiviral chemotherapy has recently reached the stage which antibacterial chemotherapy had attained in 1940 when, following the discovery of the sulfonamides, only a few effective drugs were available, and yet clues were on hand that were to lead to the discovery of a whole wealth of effective remedies.

When a virus infects a cell, several events occur that are specific for the invader and hence offer opportunities for selective attack. First of all, there is the contact with the cell, then the penetration of the cell's plasma membrane and the (often simultaneous) rejection of the viral coating-protein. If the virus is of the RNA type, reverse transcriptase is soon in manufacture, but in any case the synthesis of nucleic acid polymerases dominates this early stage of invasion. Next follows the synthesis of viral nucleic acids, structural proteins, and yet more enzymes, followed by the assembly of these components to form the complete virus. Finally, some thousands of these virions are liberated from each cell. Apart from the possibilities for finding selective inhibitors for each of these stages, the patient could also be helped by other drugs to control the secondary (non-viral) symptoms, which are often of an inflammatory or anaphylactic character.

The following antiviral agents have proved successful in human medicine.

First of all, two prophylactic agents: amantadine for the prevention of influenza and methisazone for preventing smallpox. These were intended for use during epidemics when it is too late to vaccinate a population that is at risk. Next come the successful therapeutic agents, idoxuridine (*4.14a*), and trifluridine for herpes infections of the eye, whereas the following drugs are used systemically against the various forms of herpes affecting the body. The principal drugs of this kind are vidarabine (*4.16*) (also known as Ara-A or adenine arabinoside), and the more selective drugs, acyclovir (*4.19*) and 5-bromovinyldeoxyuridine or BVDU (*4.21*). These drugs are inactive against RNA viruses, but ribavirin (*4.15*) acts on both RNA and DNA types and is used clinically in virus pneumonia and measles. All of these are discussed more fully in Section 4.0.2, leaving us free to review amantadine and methisazone.

Chemically, amantadine (*6.17a*) ('Symmetrel') is 1-aminoadamantane. Adamantane is tricyclodecane ($C_{10}H_{16}$), a cage-like hydrocarbon and the 1-amino-derivative is a strong base (pK_a 10.1). Prophylactic treatment with amantadine gives immediate protection as effective against influenza during epidemics as that which would have been provided by immunization, which must be begun two months earlier. The prophylactic dose (0.2 g) is taken orally each day for up to 3 months, and it is excreted unchanged in the urine. Unluckily, it acts on only the A strains of influenza virus, although epidemics with these strains are common, and can be dangerous. In one experiment, 238 volunteers in matched groups were given this drug, in a double-blind test. When challenged with a live vaccine of Asian influenza virus, the protected individuals showed a 60% lower incidence (Jackson, Muldoon and Akers, 1963). In a similar experiment on 850 volunteers, this substance was found to be a highly effectively preventive and also to decrease symptoms in infected subjects (Wendel, 1964; Little *et al.*, 1978).

Amantadine, at least in fowl-plague virus infections, exerts its effect by inhibiting the uncoating of the virus that has to accompany penetration into host cells (Kato and Eggers, 1969). Non-basic analogues of amantadine (especially alcohols, ketones, esters, and nitriles) are also antiviral and many of these, when tested against Newcastle disease virus on a monolayer culture of chick embryo fibroblasts, showed better selectivity (Aigami *et al.*, 1975). The analogue, rimantadine (*6.17b*) has also given good results in the prophylaxis and treatment of Influenza A (Krylov, 1976; La Montagne and Galasso, 1978). Influenza B virus has not yet been conquered by a drug.

Derivatives of a related cage-nucleus, 4-homoisotwistane (*6.18*), also showed good *in vitro* antiviral properties (Aigami *et al.*, 1976). For more on the chemo-prophylaxis of influenza, see Oxford (1977).

Methisazone (*6.19*) (1-methylisatin-3-thiosemicarbazone, 'Marboran') inhibits the multiplication of vaccinia virus in experimental animals. The anti-viral action is extraordinarily high. Mice infected intracerebrally with 1000 mean lethal doses (LD_{50}) of vaccinia (cowpox) virus required only 0.5 mg/kg for protection, and only 10 mg/kg was needed for protection against variola major

(a) 1-Aminoadamantane (R=−NH₂)
(b) Rimantadine (R=−CH(Me)NH₂)
(6.17)

Homoisotwistane
(6.18)

Methisazone
(6.19)

(smallpox) virus. The protective effect is immediate, whereas that of vacci-nation take 7−10 days to develop. During an outbreak in Madras, the drug was given to 1100 people who had been in intimate contact with cases of smallpox. Only three cases, all mild, occurred amongst them, but, in a similar number of untreated people, 78 contracted smallpox and 12 died of it. The drug gave protection even when administered too late in the incubation period for vac-cination to be of any use (Bauer *et al.*, 1963). However, it could not cure established infections. Now, following the recent elimination of smallpox from the world, methisazone is little used.

Methisazone does not interfere with the replication of viral DNA, the syn-thesis of viral mRNA, or the functioning of early viral mRNA. However, by disorganizing the formation of 'late' proteins, it prevents assembly of complete, infectious virions (Prusoff, 1967). The antiviral spectrum of this drug is wide, *in vitro*. It inhibits multiplication of all pox-viruses, and other DNA viruses too (e.g. the adenoviruses and varicella), as well as several groups of RNA viruses (e.g. poliomyelitis, common cold, influenza A and B, and some arboviruses). However, in Man, no benefit can be demonstrated in these diseases at any dose that does not cause intense nausea and vomiting (Turner, Bauer and Nimmo-Smith, 1962).

When both of the hydrogen atoms at the end of the side-chain of this drug are replaced by methyl-groups, the action against vaccinia is completely lost, but higher activity against another poxvirus, ectromelia, is found. Similarly, 'type 2' poliovirus, in tissue-culture, was selectively inactivated by 1-methyl-4',4'-dibutylisatin thiosemicarbazone (Bauer and Sadler, 1961).

Striking examples were given in Section 4.0.2 where extra selectivity is derived from the difference between enzymes responsible for the replication of virally nominated nucleic acids and those being used by the uninfected cell. Apart from these examples of the control of DNA viruses, possibilities exist for RNA viruses because the following enzymes, specified by these viruses, are quite different from those used by the uninfected cell: RNA transcriptases of influenza viruses, RNA replicases of the enteroviruses and rhinoviruses, and the reverse transcriptases of RNA tumour viruses.

Synthetic anti-viral agents discovered so far, tend to be specific for a limited range of viruses; even so, they are usually wider in spectrum than available vaccines, and less expensive. Moreover their ability to treat an *established* viral infection injects new hope into the very area where viral immunotherapy fails. Antiviral action has rarely been found among antibiotics.

A quite different approach to antiviral therapy is to seek small molecules which can stimulate the formation of interferon (a small antiviral protein) in the host. Such a substance is tilorone, which is 2,7-bis-(2-diethylamino-ethoxy)fluren-9-one (Krueger and Mayer, 1970). So far, this approach has not produced any result. The unpleasant side effects of interferon must also be taken into account.

For the rational designing of antiviral agents, see Grollman and Horwitz (1971) and Gauri (1981). For antiviral chemotherapy, see Came and Caliguiri (1982).

6.3.3 Antiprotozoal drugs

(a) *Diseases caused by sporozoa.* For the prevention and treatment of malaria, quinine came to be replaced, during the Second World War, by mepacrine (*6.10*). After the war, the latter was replaced for *prophylaxis* by the more effective pyrimethamine (*4.8*). For *treatment* it was replaced by chloroquine (*10.31*), at least among Caucasian people, many of whom disliked the temporary yellow coloration of their skin during mepacrine treatment. Chloroquine was dis-covered (Schönhöfer, 1938) in the Elberfeld (Germany) laboratories of the Bayer Company, who had been responsible for the first synthetic anti-malarials: mepacrine ('Atebrin') and pamaquine; but it came to light independently in the USA during the large-scale 'Antimalarial Survey' of 1941–1945 (Wiselogle, 1946).

The less host-toxic primaquine (*3.36*) replaced pamaquine (*6.9*) as a gameto-cide, used as an auxiliary drug to prevent the patient from infecting new mosquitoes (Elderfield, 1946). The first potent anti-malarial prophylactic proguanil (*3.34*) ('Paludrine') was discovered in England toward the end of the 1939–1945 war (Curd, Davey and Rose, 1945), and was tested on volunteers in North Queensland, Australia (Fairley, 1946). It has now largely been replaced by cycloguanil (*3.55*) and another selective prophylactic, pyrimethamine (*4.8*) ('Daraprim'), discovered by a combined American and British team (Falco *et al.*, 1951). Quinine is now reserved for cases of malaria resistant to these drugs. The emergence of resistant strains of *Plasmodium falciparum* has brought about a renewal of synthetic work designed to find new anti-malarials (see Section 10.3.5). For other data on malaria, and its management, see Sections 1.1, 9.6 and 10.3.5.

Coccidiosis in poultry and cattle, a group of diseases caused by various species of *Eimeria* (a sporozoon), has been a cause of tremendous economic loss in the poultry industry. The chances of infection have been greatly increased by contemporary battery methods of raising the birds, namely in a house of up to 30 000 birds with an earth floor covered by wood shavings. Excellent prophylaxis, and even cure, of the disease can be obtained with the following drugs mixed with the food: (a) sulfonamides, especially sulfaquinoxaline, (b) amprolium (*9.37*), a metabolite analogue of thiamine, or (c) nitrofurazone (*4.42*).

(*b*) *Diseases caused by flagellates.* Notable advances have occurred since 1935 in managing diseases caused by flagellates. Pentamidine (*10.27*) has proved the best of a series of aromatic diamidines for both prophylaxis and treatment of African sleeping sickness, although suramin (*6.8*) still finds a use for both purposes. The diamidines were synthesized by Ewins *et al.* in the late 1930s, but because of the Second World War, not published until 1942. The biological investigation was by King, Lourie and Yorke (1938), and Adler and Tchernomoretz (1942). Diminazene, another diamidine (*10.28*), has since become established as a valuable drug for treatment. In advanced cases, where the flagellates had invaded the brain, it was customary to prescribe tryparsamide (*6.7*), but this has now been replaced by the more selective French drug, melarsoprol (*13.3*) (Friedheim, 1949), but even this has some serious side effects.

Currently, attention is concentrated on devising drugs which relate to the unusual biochemistry of *Trypanosoma brucei,* the various sub-species of which cause African sleeping-sickness in both Man and farm animals. This organism metabolizes glucose by converting it to glycerol and pyruvate by anaerobic glycolysis. To this end, the parasite uses a characteristic enzyme, α-glycerophosphate dehydrogenase, which needs spermidine as a coenzyme. This enzyme is required to regenerate NAD^+ from the NADH formed by glycolysis. The new drug, 2-difluoromethylornithine (DFMO) (*3.51*) cures *T. brucei* infections in mice by preventing the synthesis of putrescine which is a necessary precursor of spermidine (Bacchi, 1981). This drug is a typical IMBI (irreversible mechanism-based inhibitor, see Section 9.7.2). It specifically inhibits the enzyme that converts ornithine to putrescine (ornithine decarboxylase). Even large doses of DFMO are non-toxic to mammals, but the drug has only a short half-life in the human body.

Chagas' disease, the South American form of trypanosomiasis, is unaffected by any of the above drugs, but both acute and chronic cases respond well to a nitrofuran, nifurtimox (*6.20*) ('Lampit') (Bock *et al.*, 1972). If irreversible organic lesions have not yet occurred, the cure rate is high.

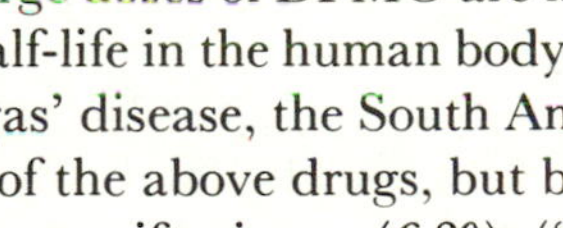

Nifurtimox
(6.20)

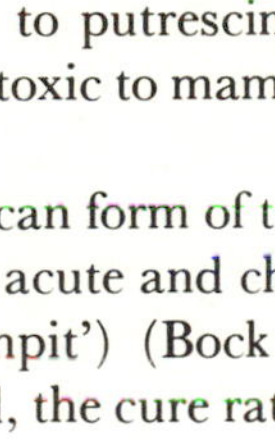

Sodium stibogluconate
(6.21)

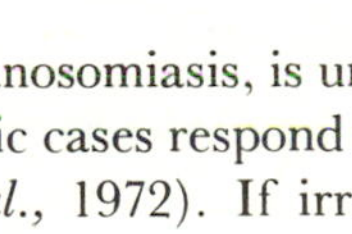

Metronidazole
(6.22)

For the trypanosomiasis of quadrapeds, that causes such large economic loss in Central Africa, a single injection of an aminophenanthridinium salt can effect a cure, in horses, cattle, and camels (Browning *et al.*, 1938; Carmichael and Bell, 1944). [See the review by Walls (1951) who did the first syntheses, and maintained interest in phenanthridines during a long period of neglect.] Homidium (*10.23*), better known under its trade-name 'Ethidium', seems the best of these phenanthridine trypanocides. Quinapyramine (*10.29*) is used as a prophylactic (Wilson, 1949). For more on trypanosomiasis, see Sections 1.1 (p. 9) and 10.3.5.

Leishmaniasis is successfully treated with the pentavalent antimonial, sodium stibogluconate (*6.21*). Given intramuscularly, it has displaced all earlier forms of antimony treatment. Intramuscular pentamidine (*10.27*) provides an alternative treatment.

In temperate climates, the most troublesome flagellate disease in humans is vaginal trichomoniasis, cured by the oral administration of metronidazole (*6.22*) ('Flagyl'), which is free from side effects (Keighley, 1962). By its inhibition of hydrogen production by *Trichomonas vaginalis*, metronidazole was shown to interfere with a ferredoxin system of E_0 about -400 mV (Edwards, 1982; O'Brien and Morris, 1972), ending in DNA breakage. See also nitrofurazone in Section 4.0 (p. 140) and hydrogenosomes in Section 5.4.3.

(*c*) *Diseases caused by amoebae.* Many useful new amoebicidal drugs have been found which exert their action by contact with *Entamoeba* (including its trophozoites) in the colonic lumen. Metronidazole (*6.22*) is the most clinically used example (Adams and MacLeod, 1977). Trophozoites not accessible to contact (e.g. those in liver abscesses) require systemically acting amoebicides. Here the physician's choice is no longer restricted to nauseating emetine, because chloroquine (*10.31*) has proved clinically effective.

6.3.4 Fungicides

Most of the non-systemic fungal infections of Man (such as athlete's foot and ringworm) yield to the Belgian drug miconazole (*6.23*) which is an imidazolylmethyl derivative of 2,4-dichlorobenzyl ether. It is also successful against vulvovaginal candidiasis as a 2% cream. The related substance, clotrimazole (*6.24*), is about equally effective (Plempel *et al.*, 1969). Miconazole interferes with the incorporation of ergosterol into the cytoplasmic membrane of fungi (Arndt, Schulz-Harder and Schulz-Harder, 1982). It is thus a synthetic equivalent of the polyene antibiotics. An older, and slightly less effective, anti-fungal drug is tolnaftate (*6.25*), the first of the directly fungicidal drugs. Older remedies, such as salicylic acid and salicylanilide worked keratolytically, by peeling off the stratum corneum in which the fungus is often concentrated.

Superficial fungal infections can also be treated by *internal* medication, such as the antibiotic griseofulvin (*6.26*) which destroys the mitotic spindle (micro-

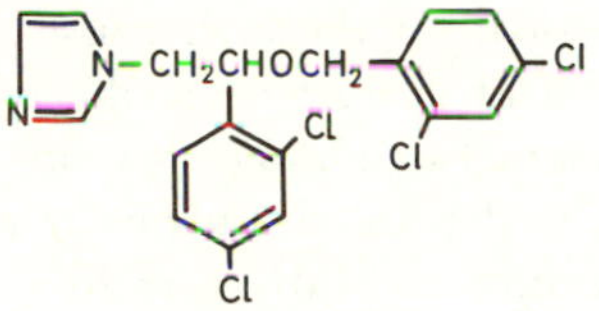

Miconazole
(6.23)

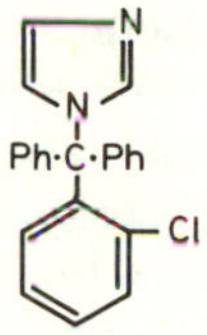

Clotrimazole
(6.24)

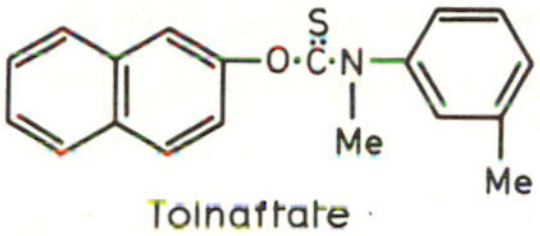

Tolnaftate
(6.25)

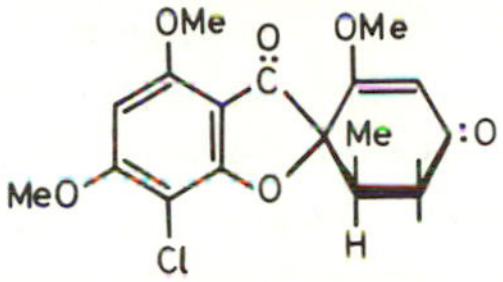

Griseofulvin
(6.26)

tubules) of fungi, but is remarkably selective towards Man (Samson, 1976). It is most used for fungal infections of the nails, where the treatment may take many months, but is probably the best available (Goldman, 1970). Griseofulvin does not affect bacteria, yeasts, and many kinds of hyphal fungi.

Nystatin (*14.19*) is a polyene antibiotic used specifically for infections of the skin, mucous membranes, or gut by *Candida albicans*.

Systemic fungal diseases (e.g. histoplasmosis and blastomycosis) are uncommon but, if untreated, often fatal. They usually take the form of lung infections or meningitis. The best treatment is still the polyene antibiotic, amphotericin B (*5.14*) whose mode of action is described in Section 5.4.1 (p. 192). It is usually administered intravenously. Flucytosine (*4.23*) is an excellent synergist (Section 4.0, p. 131), seldom given alone. Intravenous miconazole (*6.23*) provides alternative therapy, but there are frequent adverse reactions. An orally active analogue, ketoconazole ('Nizoral') was introduced in 1981, and looks promising.

For the treatment of fungal diseases by amphoteracin, see Bennett (1979); for the whole subject of anti-fungal chemotherapy, see the book by Speller (1980).

6.3.5 *Anthelmintics*

Worms that parasitize Man belong to widely separated zoological families (see Table 1.1, p. 18) and hence have very different anatomy, physiology, and susceptibility to drugs. The screening of various materials on worms was practised by Francesco Redi in 17th century Tuscany (Redi, 1684). In spite of this early start, very few substances now used against worms were employed before 1935, and those with the highest chemotherapeutic index have been discovered much more recently. Most of the parasitic worms, in an infected host, are found to be in the adult, non-growing stage of the parasite's life cycle. Hence they present a problem quite different from bacteria and protozoa which, because they are rapidly growing and multiplying, fall easy victims to drugs which interfere with nucleic acid or protein synthesis. Because in adult worms, these biosynthetic reactions are going at a slow pace, the most vulnerable biochemical pathways are those concerned with motor activity and the generation of metabolic energy. In some cases the selective toxicity shown by anthelmintics depends on a biochemical peculiarity of the parasite; in other cases a high concentration of a drug, unabsorbable by the host, is built up in the neighbourhood of the worms.

Beginning with the flat worms, we come to the most devastating of all worm diseases, schistosomiasis, whose natural history is sketched in Section 1.1 (p. 12). The traditional remedies, organic trivalent antimonials (see Table 4.6, p. 157; also Section 13.0), are still in use. Of these, intramuscular stibophen (*4.57*) is highly effective in *Schistosoma mansoni* and *S. haemotobium* infections, but causes vomiting. A single intramuscular injection of hycanthone (*3.42b*) usually cures infections by either of these two species but it is under suspicion as a mutagen and possible carcinogen.

The introduction of niridazole (*6.27*) in 1965 provided the first effective oral treatment for schistosomiasis, but only against *S. haematobium* infections, which it cured in about a week (Blair *et al.*, 1969). This drug, which is 1-(5-nitro-2-thiazolyl)-2-imidazolidinone, is concentrated in the embryonated eggs and the gonads of these worms. Selectivity resides in its ability to inhibit the natural inactivation of the worm's glycogen phosphorylase. Consequently, glycogenolysis increases, and malnutrition ensues (no equivalent reaction occurs in Man) (Bueding and Fisher, 1970).

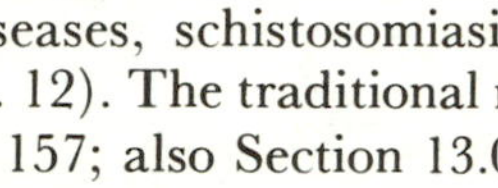

Niridazole
(6.27)

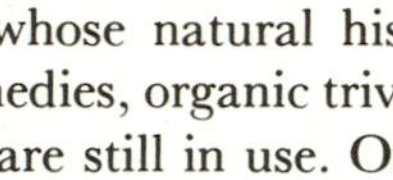

Metrifonate
(6.28)

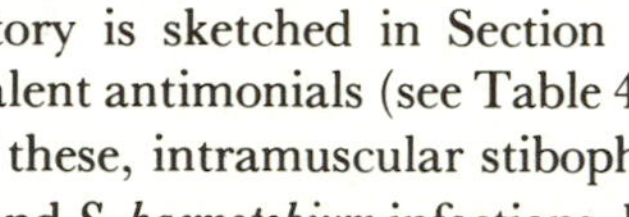

Oxamniquine
(6.29)

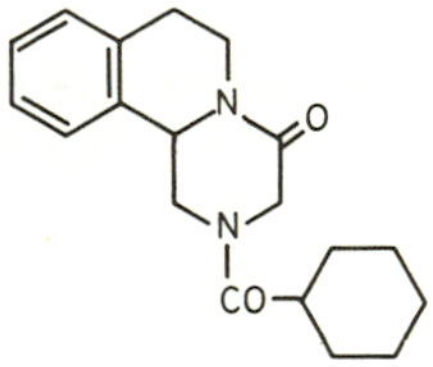

Praziquantel
(6.30)

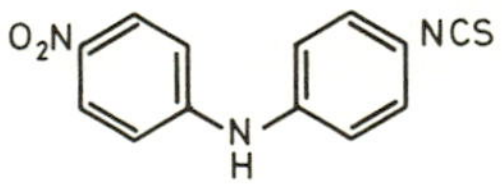

Amoscanate
(6.31)

Metrifonate (*6.28*) was the next remedy discovered for *S. haematobium* infections. This pro-drug (dimethyl 2,2,2-trichloro-1-hydroxyethylphosphonate) loses hydrogen chloride to give dichlorvos (*13.29*), which fatally combines with acetylcholinesterase (AChase) in the worms. Treatment usually consists of two well-spaced oral doses. In this assault on AChase, Man comes off astonishingly well. All the enzyme disappears from human serum during treatment, but it reappears later, and there are no symptoms of deprivation! In addition, this drug (the first to do so) shows worthwhile prophylactic properties (Jewsbury, Cooke and Weber, 1977).

To complement this attack on *S. haematobium*, oxamniquine ('Mansil') (*6.29*), was introduced as a single (oral) dose to cure *S. mansoni* infections. It proved to be well tolerated, even by children (Katz, Zicker and Pereira, 1977). Millions of patients have been treated with this drug in Africa and South America, and WHO have added it to their list of essential drugs. It is thought to act by inhibiting ornithine transaminase, an enzyme essential for supplying the proline that plays a critical part in the nitrogen balance of these worms, but not in Man (Goldberg *et al.*, 1980).

Two recently introduced drugs, praziquantel (*6.30*) and amoscanate (*6.31*), seem able to eliminate all three species of *Schistosoma* in a single oral dose, including the very hard-to-kill *S. japonica*. Praziquantel, which is a pyrazino-isoquinolinone, is thought to act by increasing permeability in the worm's plasma membrane. It is free from the suspicion of mutagenicity which hangs over most anti-schistosomal drugs (Wegner, 1981). Amoscanate, a simple derivative of diphenylamine, loses all activity if the nitro-group is deleted (Streibel, 1976). It has, unfortunately, shown some toxicity to the host's central nervous system. At present, praziquantel is first favourite for the treatment of schistosomiasis.

Lung flukes (*Paragonimus* spp.), common in the Far East and Latin America, yield to oral treatment with bithionol (*6.32*) which is 2,2'-thiobis(4,6-dichloro-phenol). Infestation with *Chlonorchis sinensis*, the Chinese liver fluke, usually treated with chloroquine (*10.31*), is still in search of the ideal remedy. The liver fluke *Fasciola hepatica*, common in herbivores throughout the World, occasion-ally infects Man and quickly yields to bithionol. *Fasciolopsis buski*, the giant intestinal fluke of Southeast Asia, can be eliminated with tetrachlorethylene.

Turning to the other division of flat worms, the cestodes or 'tapeworms', it is heartening to note that niclosamide (*6.33*) effects a prompt and painless cure for all four main types (Gönnert and Schraufstätter, 1960). A satisfactory alter-native is provided by dichlorophen (*6.34*). These two drugs owe their remark-able selectivity to their ability to inhibit the unusual anaerobic phosphorylation reaction, in the worms' mitochondria, whereby inorganic phosphate is incor-porated into ATP. At higher doses, these anthelminthics uncouple classical aerobic phosphorylation in Man, but the anaerobic system in cestodes is far more susceptible (Scheibel, Saz and Bueding, 1968). The hydatid-forming cestode, *Echinococcus granulosus*, can be eliminated from sheep dogs with bunamidine (*NN*-dibutyl-4-hexyloxy-1-naphthamidine), which is also an effective one-dose for tapeworm in dogs and cats (Gemmell and Shearer, 1968).

Bithionol
(*6.32*)

Niclosamide
(*6.33*)

Dichlorophen
(*6.34*)

The nematodes (round worms) are a large group, of which the various members need individual approaches. Filariasis, the mosquito-borne nematode infestation, is easily treated with diethylcarbamazine *(6.35b)* ('Hetrazan') (1-diethylcarbamoyl-4-methylpiperazine) which is remarkably selective (Hewitt *et al.*, 1947). This drug is inactive *in vitro* but sensitizes the filaria so that they become attacked by phagocytes. In the related disease caused by *Oncho-cerca*, which has a predilection for skin and eyes, diethylcarbamazine (oral) is effective for the juvenile state of the worms (microfilaria) but the adult state requires long treatment with intravenous suramin *(6.8)* which has unpleasant side effects. The Guinea worm, a larger skin-infester, yields to thiabendazole *(6.36)*, which is further discussed below, or metranidazole *(6.22)*.

(a) $R^1 = R^2 = H$ (piperazine)
(b) $R^1 = Me$; $R^2 = CO \cdot NEt_2$
(diethylcarbamazine)
(6.35)

Thiabendazole
(6.36)

In treating hookworm disease caused by *Necator* or *Ancylostoma*, the patient's nutrition must first be improved, after which either mebendazole *(6.37)* or pyrantel *(6.38)** rapidly eliminates the worms. Formerly employed, bephenium *(6.39)* and the chlorinated hydrocarbons are passing out of use. Thiabendazole *(6.36)* is used for mixed infestations, including *Ascaris*. The bendazole family of drugs are thought to act by selective inhibition of microtubule assembly (Section 5.4.7), whereas pyrantel acts as a selective nicotinic type ACh agonist at the ganglia. Pyrantel is therapeutically incompatible with piperazine *(6.35a)* which causes hyperpolarization at this junction, leading to flaccid paralysis of the worms which are then swept out, still alive, by peristalsis (Eyre, 1970; Desowitz *et al.*, 1970).

For infestation by roundworms (*Ascaris lumbricoides*), a single dose of pyrantel *(6.38)* is effective. Mebendazole *(6.37)* is almost as good, but piperazine *(6.35a)*, whose unprecedented high therapeutic index put it ahead of all existing vermifuges when it was introduced in 1953‡, has lost some ground to the marginally better pyrantel and mebendazole*. Piperazine has interest for the comparative toxicologist because it inhibits the normal contractile action of acetylcholine at the neuromuscular junction in nematodes without having any effect on the

---

*Pyrantel, which is 1,4,5,6-tetrahydro-1-methyl-2-[2-(2-thienyl)vinyl]pyrimidine, is usually given orally as a suspension of the poorly soluble pamoate.

‡It is not easy for us to recapture the elation of thirty years ago when, suddenly, it became possible to eliminate intestinal worms with a single dose of a remedy completely free from side effects. Until then, santonin and male fern, drugs which made the patient quite ill, had to be preceded and followed by vigorous purges; and often the whole cycle needed repeating.

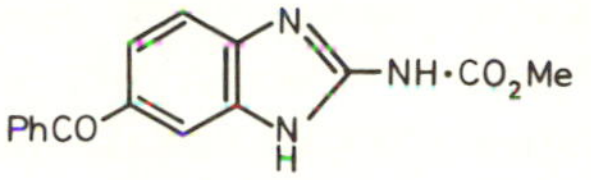

Mebendazole
(6.37)

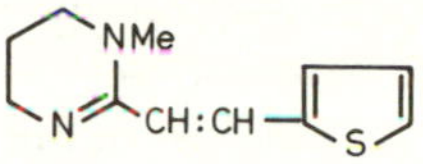

Pyrantel
(6.38)

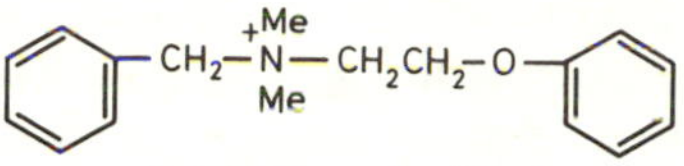

Bephenium (cation)
(6.39)

corresponding junction in mammals. Conversely, tubocurarine (2.6), which blocks this junction in mammals, has only a very weak action on this site in worms (del Castillo, Mello and Morales, 1964). Frog muscle, intermediate between that of worms and mammals, is blocked by both drugs (Bueding, 1962).

For the threadworm (*Strongyloides*), the best treatment is with thiabendazole (6.36) (2,4'-thiazoylbenzimidazole), a highly selective, broad-spectrum anthelmintic introduced by H.D. Brown and his colleagues in 1961. This drug is larvicidal at 10^{-10} M, surpassing by far the dilution at which penicillin is active against bacteria (Brown *et al.*, 1961). It acts by selectively suppressing assembly of microtubules in the worm which leads to inhibition of secretion of AChase, which is the worm's natural holdfast (Watts *et al.*, 1982).

The pinworm (*Oxyuris*), common among young children everywhere in the World, yields readily to pyrantel or mebendazole, and the latter drug is equally effective for whipworm (*Trichuris*) (Brugmans *et al.*, 1971).

For *Trichinella spiralis* infestations, so common in Europe, the United States, and Canada from eating undercooked pork foods, the most commonly used drugs are the corticoids as palliatives during the agonizing migration of the worm from gut to the host's muscle; thereafter thiabendazole is moderately effective.

Veterinary anthelmintics. Among the most used drugs are pyrantel (6.38), the organic phosphates [such as metrifonate (6.28) and dichlorvos (13.29)], and the benzimidazoles, particularly thiabendazole (6.36) which has a wide range of activity against both adult and immature worms in cattle, horses, pigs, and sheep. It is so selective, and so rapidly eliminated, that animals can safely be slaughtered, and milk from cows used, soon after administration.

However, by 1967 it was realized that thiabendazole was partly inactivated by metabolic hydroxylation in the 5-position, obviously a position which must be blocked. At about that time it was also found that anthelmintic action improved when a carbamato-group replaced the thiazoline ring in the 2-position. These discoveries led to a whole range of more active anthelmintics, namely mebendazole (6.37), and its variants in which the 5-benzoyl-group is replaced by a butyl (parbendazole), phenylthio (fenbendazole) or propylthio (albendazole) group, all highly selective variants of thiabendazole. Albendazole has a particularly broad spectrum of activity (Theodorides *et al.*, 1976).

Phenothiazine, the first of the selective anthelmintics for sheep, is now little used because of the large dose needed.

Tetramisole (and levamisole, the more selective laevorotatory isomer obtained by optical resolution) (*6.40*) ('Nilverm'), is 2,3,5,6-tetrahydro-6-phenylimidazo[2,1-*b*]thiazole. Much used for many kinds of nematode infections in cattle, pigs, and sheep (Raeymaekers *et al.*, 1966), in human medicine it has proved satisfactory only for ascariasis. Tetramisole acts by ganglionic (ACh) stimulation which leads to muscular paralysis of the worms which are then extruded by peristalsis. Its status as an immuno-regulator in human physiology was discussed in Section 5.2.

Ivermectin is described in Section 12.7.

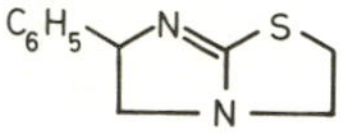

Tetramisole
(Levamisole)
(*6.40*)

6.3.6 *Drugs against cancer*

'Cancer' is a collective word which embraces more than 200 different diseases, each responding to therapy differently, but all characterized by unrestrained growth. The treatment of cancer by drugs is both pharmacodynamic (because both economic and uneconomic cells are part of the same organism) and chemotherapeutic (because it aims to exterminate the uneconomic cell). The phrase, 'The chemotherapy of cancer', introduced by Ehrlich, is now in general use. Discovery of a selective agent against cancer remained elusive until Alfred Gilman and his colleagues, about 1942 and in conditions of wartime secrecy, adapted the chemical-warfare vesicants known as nitrogen-mustards and produced chlormethine* (*4.25*), the first substance to win clinical approval for the treatment of cancer (Gilman and Philips, 1946; Goodman *et al.*, 1946). This launched the era of cancer chemotherapy.

Poorly selective though it was, chlormethine pointed the way to the discovery of the following nitrogen-mustard drugs, currently used in cancer therapy: cyclophosphamide (perhaps the most selective of them all), chlorambucil, melphalan, and uracil mustard; also the closely related thiotepa and busulfan; also the related nitrosoureas: lomustine, carmustine, semustine, and streptozocin (an antibiotic). All of these are alkylating agents, linking two strands of DNA (see Section 13.4 for details).

Next on the scene came the anti-metabolites of which the following analogues of purines and pyrimidines are in current clinical use: cytarabine, fluorouracil, azaribine, 6-mercaptopurine, and 6-thioguanine (see Section 4.0, p. 126); also

*This is the international name, agreed to by members of the World Health Organization. However, in Britain it is still called mustine and, in the United States, mechlorethamine!

the important pteridine (folic acid) analogue, methotrexate (see Section 9.3.3). Antibiotics that act on DNA and are currently used in cancer therapy include bleomycin, doxorubicin (adriamycin), and daunorubicin, all described in Section 4.0 (pp. 132–6). The principal other natural products in general use are the *Vinca* alkaloids, which work by inhibiting assembly of microtubules (Section 5.4.7).

Other anti-cancer drugs in clinical use include cisplatin (p. 132) which is also a linker of DNA strands, dacarbazine and hydroxyurea (both of which act on nucleic acid metabolism), dactinomycin, procarbazine, mitotane, the hormones and their antagonists, and two radioactive isotopes: iodine and phosphorus.

Although contemporary anti-cancer drugs are only moderately selective, their selectivity is increased in the clinic by enforcing a selective distribution, or by synchronizing them with the cell cycle (Section 5.1). At no time has interest in the chemotherapy of cancer stood higher than today, and this enthusiasm reflects current achievements (see Section 1.1, p. 15). This use of drugs enables the physician to function beyond what surgery and radiation can accomplish. Patients with widespread metastatic cancers can have these lesions reached and cured by chemotherapy even in situations where their presence is unsuspected and undetectable.

For reading on molecular targets for anti-cancer drugs, see the book by Sartorelli, Lazo and Bertino (1981).

6.4 Parallel developments in crop-protecting agents: Agrochemicals

Progress, parallel to that discussed for the chemotherapy of mammals, has also taken place in agriculture where the farmer must prevent, and even cure, the invasion of his crops by weeds, insects, and fungi. Interest in insecticides has a still broader base, because insects are also ectoparasites of Man and his domestic animals, and the vectors of infectious diseases. Insects have been wasting about 14% of the World's annual crop production, moreover plant diseases (caused by fungi or worms) waste another 12%, weeds another 9%, and rodents perhaps another 10%, an estimated total of 1800 million tons (see Cramer, 1967).

This Section (6.4) is divided into (6.4.1) insecticides, with further data on insect repellants and ectoparasites, (6.4.2) fungicides, (6.4.3) herbicides (weed killers), and (6.4.4) other pesticides, including defoliants, plant antiviral agents, anthelmintics, molluscicides, and rodenticides.

Appropriate reading is suggested after each subject, but works which cover all aspects of pesticides will be given now. For the chemistry of pesticides, uses and mode of action, see Hassall (1982) and Büchel (1983); for selectivity in pesticides, see Street (1975); for the principles of the distribution and physical behaviour of pesticides, see Hartley and Graham-Bryce (1980–1), and for a manual that includes all commonly used pesticides, see Worthing (1983).

Periodicals, with commencing dates, include *Pesticide Biochemistry and Physiology* (1970), *Pesticide Chemistry* (1972), *Advances in Pesticides* (1977), and *Progress in Pesticide Biochemistry* (1981). For a study of the action of pollutants on ecosystems (with special reference to chlorinated hydrocarbons) see Moriarty (1983).

6.4.1 *Insecticides*

Attempts to destroy insect pests on crops have early origins, but the results (using mainly lead arsenate, tobacco dust, soap, or petroleum) were mainly dispiriting. The turning point occurred about 1940 through Paul Müller's discovery (in the Swiss firm, Geigy) of the insecticidal action of DDT. With this event, the modern era of insecticides was born.

DDT (6.41) is 1,1,1,-trichloro-2,2-bis(4-chlorophenyl)ethane, and suffers from a plethora of synonyms. The international (WHO) name is clofenotane, but it is called dicophane in Britain and chlorophenothane in the United States.

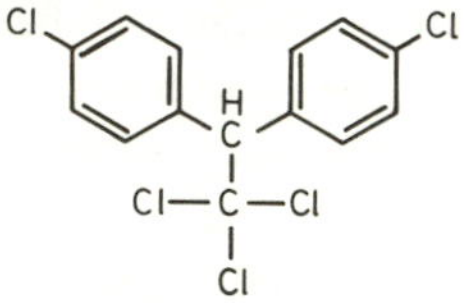

DDT
(Clofenotane)
(6.41)

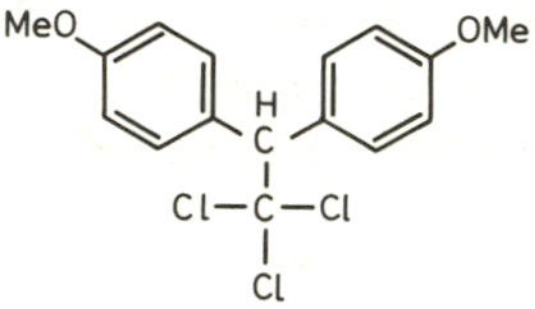

Methoxychlor
(6.42)

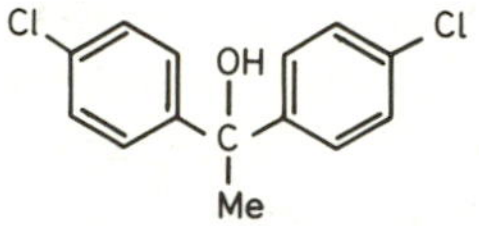

Dicofol, DMC, 'Kelthane'
(6.43)

For his discovery, Müller was awarded a Nobel prize in medicine, and the early uses of this agent (mainly by the Allied armies in the Second World War) were to control malaria and typhus during campaigns. Because of the war, very little was published about DDT until Läuger, Martin and Müller (1944) gave their historical account. It turned out that DDT had been known since 1874 when Zeidler first synthesized it, yet its biological properties remained unsuspected for more than six decades.

At the end of the Second World War, DDT was introduced as an agricultural and domestic insecticide, and it was distributed prolifically over farms and premises in the belief that it was completely selective. By increasing crop yields, it played an important part in alleviating the needs for food and clothing of an ever-expanding world population. In 1962, however, there was a sudden realization that, because of its poor biodegradability, DDT constituted an ecological risk, namely a potentially harmful effect of the residues on wildlife (mainly fish, large birds, and bees) and possible toxicological effects on Man through chronic overdosage. As a result, many countries banned the use of DDT in agriculture, or limited it to particular crops where no adequate substitute existed.

Special difficulties were encountered by the World Health Organization in its sustained battle against the mosquitoes that spread malaria in the tropics. The non-biodegradable nature of DDT offered the economic advantage that spraying was needed only once a year. However, biodegradable substitutes have come to the fore, such as methoxychlor (*6.42*), also the more persistent of the carbamates, such as propoxur (*13.40*), and of the organophosphates such as fenitrothion and chlorphoxim (Section 13.3).

The molecule of DDT permits a good deal of alteration, so as to introduce improvements, without loss of insecticidal properties. Thus, as we saw in Section 2.6, all five chlorine atoms can be replaced by methyl or methoxy with retention of a strong and selective insecticidal activity. As will be seen in Section 7.6.5, DDT (and the analogously acting pyrethrins) owe their insecticidal action to the overall *wedge shape* which props open the ion channels of cold-blooded creatures. The selectivity depends on this opening not occurring in warm-blooded creatures, because of heat-induced loss of electron-transfer binding between the benzene rings of the drug and oppositely charged flat surfaces near the mouth of the channel (Holan, 1971).

Holan (1969, 1971) recognized that the trichloromethyl group of DDT, and the dimethylcyclopropyl group of pyrethrins (*6.52*), are what keep the mouth of sodium channels open. He named these the 'Apex Groups' and found that they have an average diameter of 6.3 Å. This thinking led him to devise hybrid molecules, e.g. (*6.44*) with excellent insecticidal properties and improved biodegradation.

At the present time, the most used derivative of DDT, as a crop insecticide, is dicofol (*6.43*) whose excellent biodegradation properties have won it the approval of IPM (see p. 244).

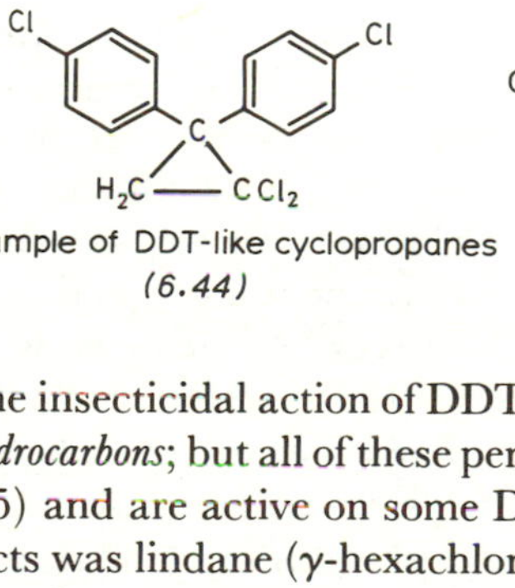

Example of DDT-like cyclopropanes
(*6.44*)

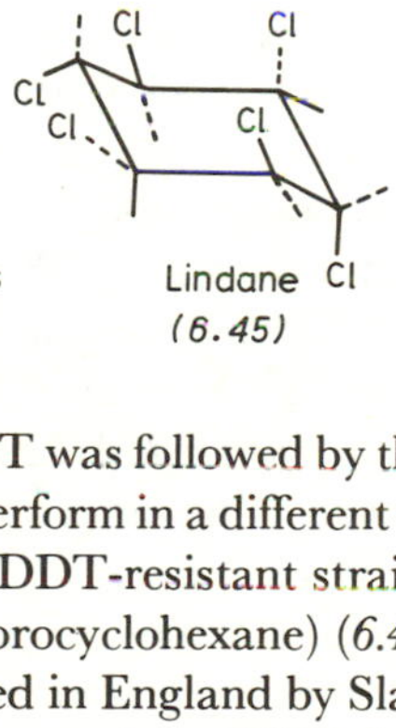

Lindane
(*6.45*)

Discovery of the insecticidal action of DDT was followed by the marketing of *other chlorinated hydrocarbons*; but all of these perform in a different way from DDT (see Section 7.6.5) and are active on some DDT-resistant strains. The first of these new products was lindane (γ-hexachlorocyclohexane) (*6.45*) once known as BHC or benzene hexachloride. Introduced in England by Slade (1945), this substance acts with greater speed and power than DDT, but being more volatile, it fairly soon evaporates. Its mode of action is described in Section 7.6.5. Lindane is the only chlorinated hydrocarbon accepted by the pharmacopoeias for application to the human body. It is one of eight stereoisomers. Of these, lindane is the one with three neighbouring chlorine atoms in the axial conformation, the other three being equatorial. Of each trio, the central member is described as *meso*, and the flanking members as *dl*. It is accepted that the insecticidal action is derived entirely from the shape of the molecule, but it does not act at sodium channels. The three commonest isomers (known as alpha-, beta-, and delta-) penetrate insect cuticle poorly (Armstrong, Bradbury and Standen, 1951). The alpha-isomer has an unpleasant odour and the beta-isomer lacks volatility. Hence stereochemical studies have been concentrated mainly on the gamma-isomer.

A study was made of 38 analogues of lindane in which one or more chlorine atom was replaced by H, F, Br, Me, OMe, SMe, or OH substituents. It was found that the radius of the substituent was critical when it was placed *meso*, whereas the volume was critical when it was placed *dl*. For activity against mosquitoes, it was necessary for there to be no biodegradation and for a certain lipophilicity to be exceeded. The *meso* bromo analogues (both mono- and di-) were more insecticidal than lindane, and the *meso* mono-ethoxy and mono-methylthio replacements gave products as active as lindane. Of the *dl* replacements, H and Me were the least disturbing, but no product as active as lindane emerged (Kiso *et al.*, 1978).

While Slade, in England, was launching lindane, Hyman (California) introduced the highly potent and persistent chlorinated cyclodiene series, which included *aldrin (6.46)*, *dieldrin, chlordane, heptachlor*, and *endrin* (Hyman, 1949). After a long period of widespread use, these unbiodegradable (and sometimes even carcinogenic) substances were considered in many countries to be insufficiently selective and their use has been limited or even suspended. It is puzzling that the chlorinated cyclodienes, which seem to have an insecticidal action identical to that of lindane, should be more toxic to vertebrates. It is true that each chlorinated cyclodiene has an activated double bond and dieldrin has one (and endrin, two) epoxy groups, all biologically hazardous and prone to biotransformations (e.g. van den Bercken and Narahashi, 1974).

After examination of 106 chlorinated cyclodiene insecticides on six selected types of insect, Soloway (1965) concluded that a strict geometry governs ability to react with insect receptors. Two electron-rich sites are required (Cl, O, N, S, or double bond) for electrostatic adsorption (dihydro-aldrin has only one such site, and has only a low toxicity to insects). Aldrin *(6.46)* is a typical example. Chemically it is *endo-endo*-1,2,3,4,10,10-hexachloro-1,4:5,8-dimethano-hexa-hydronaphthalene. The critical outline which these nearly spherical molecules require for activity is shown in *(6.47)*; it was produced by viewing molecular models along a line joining the bridgehead (methano) atoms. Lindane, which has a similar mode of action, has a similar outline. For further details of structure–action relationships, see Brooks (1974).

From the massive rejection of the chlorinated cyclodienes, two survivors emerge, both recommended by IPM (p. 244). The first of these is *endosulfan (6.48)* with inbuilt biodegradability. The second is *campheclor* (formerly called *toxaphene*) said to be a mixture of about 175 polychloro derivatives of cam-phene *(6.49)* with an overall empirical formula of $C_{10}H_{10}Cl_8$, and no double bond (Casida *et al.*, 1974). It, too, is biodegradable.

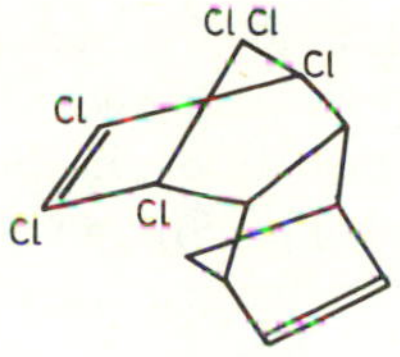

Aldrin
(6.46)

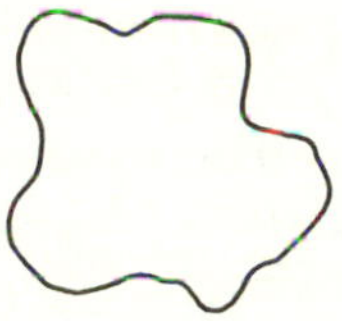

Critical outline for activity
in chlorinated insecticides
(6.47)

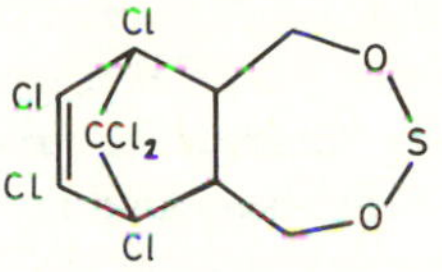

Endosulfan
(6.48)

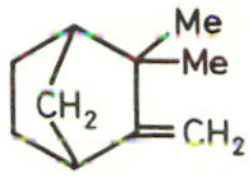

Camphene
(6.49)

At the present time there is much concern over the pollution of waters by polychlorobiphenyls (PCB) which are used as plasticizers, fire-retardants, high-temperature lubricants, and transformer oils; but these were little used as pesticides.

The *pyrethrins* are a pair of esters extracted by petroleum from Insect Flowers, a species of *Chrysanthemum* native to Dalmatia and much grown in Kenya. Pyrethrin I has insect-killing properties (including mode of action) that differ only in detail from those shown by DDT (see Section 7.6.5). It is also poisonous to fish, but is so readily metabolized, by both hydrolytic and oxidative pathways, in warm-blooded creatures that it is considered completely safe for household use, and still forms one of the mainstays of domestic fly sprays. Pyrethrin II, on the other hand, provides only what is termed 'knockdown', which is a rapid hypnotic effect with eventual recovery. Both pyrethrins are photolabile, and hence unsuitable for use on crops.

Chemically pyrethrin I consists of an ester formed from the alcohol (6.50) with chrysanthemic acid (6.51), whereas pyrethrin II has the same alcohol esterified with pyrethric acid (6.52). The extra ester group in pyrethrin II gives it a more limited distribution and half-life in the insect.

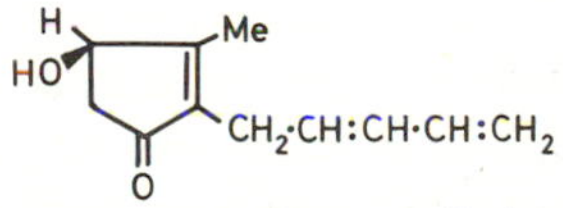

Alcohol of natural pyrethrins
(6.50)

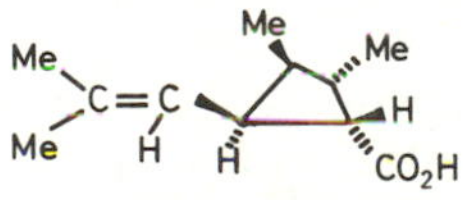

Chrysanthemic acid
(6.51)

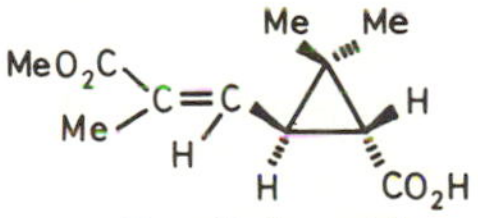

Pyrethric acid
(6.52)

A substitute for pyrethrin I was found in the more easily synthesized (1949) allethrin, which differed only in that the conjugated side-chain was contracted to $-CH_2 \cdot CH=CH_2$. It turned to have only 1% of the killing power of pyrethrin I, but has useful knockdown action. It and the similar tetramethrin, introduced in 1967, were able to fill gaps in fly-spray formulations.

From 1967, the efforts of Michael Elliott at Rothamsted (England) produced 'pyrethroids' of ever increasing potency, suited for bulk production, and useful in agriculture. The first of these, bioresmethrin (Elliott *et al.*, 1967), in which chrysanthemic acid is esterified with a furan-derived alcohol (*6.53*), proved to be the first synthetic substance with the full potency of pyrethrum I against a wide range of insects. It is non-toxic to mammals, and is still much used in market gardens, particularly on carrots, cabbage, and lettuce.

Elliott's team then made two advances which produced pyrethroids of high photostability. First, the five-membered ring in the alcohols was replaced by 3-phenoxybenzyl alcohol (Elliott *et al.*, 1973), and then the two methyl-groups which terminate the vinyl chain in chrysanthemic acid (*6.51*) were replaced by two halogen atoms. The most powerfully insecticidal of these products, delta-methrin (formerly called decamethrin) (*6.54a*), has a LD_{50} against the housefly of only 2 μg/kg, when synergized as described in Section 3.5. This remarkable potency is about 50 times that of pyrethrin I. Deltamethrin is 10000 times as toxic to insects as to mammals. In the field, it has been found stable to light, and resistant to weather. It is ten times as active against insect pests as any available carbamate, phosphate, or chlorinated hydrocarbon, but it is rather expensive to manufacture. The more economic permethrin (*6.54b*) and much more easily synthesized fenvalerate (*6.55*) fill useful economic gaps (Elliott *et al.*, 1974).

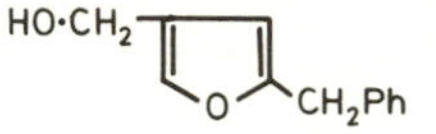

Alcohol of bioresmethrin
(6.53)

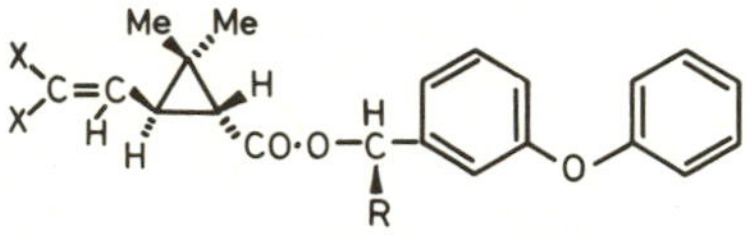

(a) Deltamethrin (decamethrin) X = Br ; R = CN
(b) Permethrin X = Cl ; R = H
(6.54)

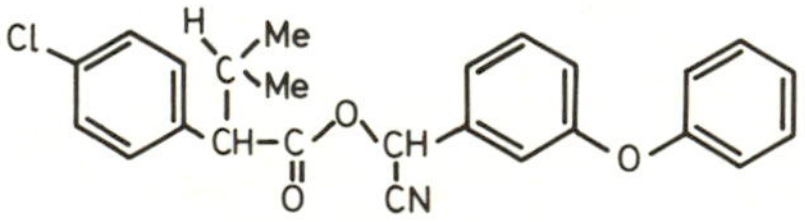

Fenvalerate
(6.55)

Insecticidal action in the pyrethroids depends on steric detail. So long as the cyclopropane ring is retained, the *gem*-methyl-groups must be free to help plug the receptor (as explained above, under DDT). This provision bans any isomer with the *1S* configuration. Fenvalerate (*6.55*) illustrates a new-found latitude in this region, but an isolated double bond (C=C) must be retained to get the necessary 'Apex' effect (p. 239). A *cis* relationship of the attachment of the vinyl to the *cyclo*propyl ring is helpful, but not essential (Shono, Unai and Casida, 1978).

In the alcohol moiety of pyrethroids, the unsaturated chain attached to the cyclopentane ring of pyrethrin I must be allowed, when modifying the molecule, to adopt a configuration that is *not* co-planar with that ring. This is no less true when the alcoholic moiety becomes a phenoxybenzyl alcohol, as in deltamethrin (*6.54a*). When a cyano-group is present in this alcohol, as in (*6.55*), the configuration *S* is essential for activity (Elliott *et al.*, 1974).

Most pyrethrins and pyrethroids are toxic to fish, but this is not the case with two hybrids of DDT and pyrethrin structures, namely cyclophenthrin and phencyclate. The former is α-cyano-3-phenoxybenzyl 1-(4-ethoxyphenyl)-2,2-dimethylcyclopropanecarboxylate, and phencyclate is the 2,2-dichloro analogue (Holan *et al.*, 1978, 1983; Wakita *et al.*, 1983). These are racemates, the inactive isomer of which prevents the fish receptor from taking up the active isomer whereas no such protection occurs at the insect receptor (G. Holan, personal communication).

For further reading on the chemistry of pyrethroids, see Elliott, Janes and Potter (1978).

The *organophosphate* insecticides, when introduced, about 1945, made a large and welcome addition to what was available. Because a fuller account is given in Section 13.3, it need be said here only that they are nerve poisons which initially act in a quite different way from the pyrethroids and chlorinated hydrocarbons, for they are inhibitors of acetylcholinesterase. The earliest examples, discovered in Germany by Schrader, were as toxic to the spraymen as to the insects. Little by little, adequate selectivity was built into the molecules, and control over their half-life in the field became part of the molecular design also.

When insects are poisoned by organophosphates, the free acetylcholine level rises sharply, causing great increase in spontaneous nerve activity (neuronal hyperexcitation), both autonomic and somatic. This brings about ion imbalance, and liberation of tissue toxins followed by paralysis, dehydration, and death, a sequence reminiscent of that caused by the chlorinated insecticides (Smallman and Fiske, 1958).

Unlike the chlorinated hydrocarbons and pyrethroids, the organophosphate insecticides tend towards hydrophilicity, and the more hydrophilic types are taken up by plants from the soil. These 'systemic insecticides' have made a great contribution to selectivity because they poison only those insects that bite the plant. The organophosphates were later joined by the similarly acting *carbamate* insecticides, less hydrophilic but also less persistent. They are described in Section 13.3.

Rotenone (*4.61*), an oxygen-heterocycle of plant origin, is another suitable agent for leaf-biting pests. Unfortunately it is destroyed by light. For the source of its action on the respiratory chain, and its selectivity, see Section 4.5. Attempts have been made to find selective inhibitors of α-glycerophosphate metabolism, of extreme physiological importance to the insect for flight, and controlled by enzymes which are no more than analogous to their mammalian equivalents (Marquardt and Brosemer, 1966).

Nicotine (*7.26*) and the tobacco dust of which it is the active constituent have long, and rather fruitlessly, been used on plants. Nicotine is a counter-selective substance, injurious to only a few types of insects, but highly poisonous to Man. Attempts to alter the molecule to find a more selective variant (Section 7.3) have failed.

Chlordimeform (*(2.36)*) was introduced in the early 1970s to control mites, ticks, and moths in the field. Chemically it is *N*-(2-methyl-4-chlorophenyl)*N'N'*-dimethylformamidine, and it has shown no cross-resistance with other insecticides. Its selectivity is attributed to its mimicry of octopamine (*2.35*), which is an important neurotransmitter in arthropods but not in vertebrates (Evans and Gee, 1980). Death occurs partly by exhaustion after overstimulation, partly by starvation. The only effect on mammals appears to be an inhibition of monoamine oxidase, 300 times weaker than that evoked by the much-used anti-depressant drug, tranylcypromine *(9.47)* (Aziz and Knowles, 1973). The suggestion that inhibition of this enzyme also contributes to the effect on insects and mites is not tenable (Neumann and Voss, 1977).

There are strong indications that chlordimeform is not the true agent, but rather a pro-agent, because its action is prevented by those methylenedioxy-compounds that inhibit those enzymes in the *e.r.* that degrade foreign substances (see p. 94). The true agent is considered to be the lower homologue, in which one methyl group has been lost from the —NMe₂ substituent. In tests with this homologue, the methylenedioxy substances actually synergized the acaricidal action, because they inhibited biological destruction of the agent (Knowles and Roulston, 1973).

Pheromones and analogues of insect *hormones* (both agonistic and antagonistic) continue to be investigated without, so far, producing substances of wide-ranging usefulness (see Section 4.7).

Ever-increasing and widespread *resistance* presents a gloomy outlook for the most-used insecticides today. Cross-resistance of insects, in many localities, to the chlorinated hydrocarbons (both kinds) and the organophosphates has highlighted early indications that the pyrethroids, too, will be trapped in this net of multiple cross-resistance. All of these apparently diverse types of insecticide release acetylcholine at an important phase of their action, and some part of the trouble may be due to that. That so much resistance has arisen in so few years is traceable to overapplication of the insecticide, neglect of helpful climatic factors, timing, and unselective destruction of useful insect predators (e.g. birds and certain kinds of mites).

From these considerations of evoked resistance, plus the poisoning of useful pollinating insects and the destruction of so many fish in rivers and lakes, a worldwide movement began in 1976 which culminated in the policy of *'Integrated Pest Management'* (IPM) adopted by the constituent members of the Food and Agricultural Organization (FAO). It was agreed to draw up lists of the most selective insecticides and use only those that spared predators; also to use environmental aids such as timing, temperature, and rainfall to supplement or even replace the use of insecticides; to use the insecticides only in minimal quantities; to use only those that are appropriate to the identified infestation; and closely to monitor residues (Metcalf, 1980).

The five most recommended insecticides, chosen with these aims in mind, were trichlorofon (13.31), malathion (13.27b), methoxychlor (6.42), carbaryl (13.39), and the inorganic substance, cryolite (sodium aluminofluoride).

A supplementary list was also drawn up for selective agents with particular rather than general uses: dimethoate (13.35), diazinon (13.28), dicofol (6.43), chlorpyrifos (13.32), endosulfan (6.48), lindane (6.45), phosphamidon (13.33), methomyl (13.41), camphechlor (see p. 240), and three pyrethroids: deca-methrin (6.54a), permethrin (6.54b), and fenvalerate (6.55) (Metcalf, 1980).

Among the new types of insecticides that are being sought are ones that would inhibit the tanning (sclerotization) of insect cuticle in which N-acetyldopamine is polymerized (cf. 7.48), others that would inhibit receptors for acetylcholine or for GABA (γ-aminobutyric acid) (12.91), and others to inhibit choline acetyl-transferase. A promising lead in another direction is 29-fluorostigmasterol, harmless to mammals and plants, but metabolized by insects to toxic fluoro-acetate (Prestwick, Gayen and Kline (1983)).

For further reading on the neurochemistry of arthropods, see Treherne (1966), and for insect biochemistry see Rees (1977), Rockstein (1978), Wilkinson (1976), and Hutson and Roberts (1982). For the various methods available for insect pest management, see Metcalf and Luckman (1982); for mode of action and selectivity, see Beeman (1982) and Coats (1983). The following handbooks give the composition of all insecticides: Worthing (1983), and Entomological Society of America (1981).

(a) Ectoparasites of animals. So far we have discussed the control of arthropods on plants only, but selective agents are available for treating infestations of animals also. Blowfly-strike in the crutch of sheep, which undermines their health, has for many years been successfully overcome, in Australia, by spraying with the organophosphate, diazinon (13.28), which is now being replaced by the still more effective triazine, cyromazine (vetrazine) (6.56). Another organophos-phate, metrifonate (6.28), is sprayed on horses to prevent the bot fly from laying eggs on the coat (otherwise the horse swallows the eggs, then voids live maggots in the dung, thus starting a new cycle).

A new and highly effective treatment for ectoparasites in farm animals is to give a dose, oral or subcutaneous, of the antibiotic, ivermectin (see Section 12.7).

In domestic life, crannies are sprayed with organophosphates to banish silverfish, termites, and cockroaches. Domestic pets are dusted with powders containing propoxur (13.40) or carbaryl (13.39); these carbamates, because of their short half-life in the human, are esteemed because medicated pets will be embraced by children.

In Man, scabies can be rapidly cured by applying either benzyl benzoate or lindane. Pediculosis (infestation with lice) is effectively treated with carbaryl, lindane, or isobornyl thiocyanate.

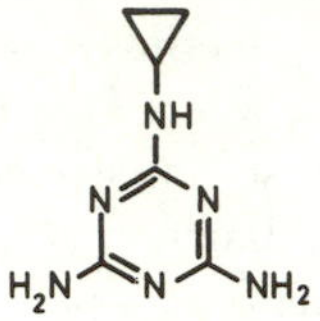

Cyromazine
(6.56)

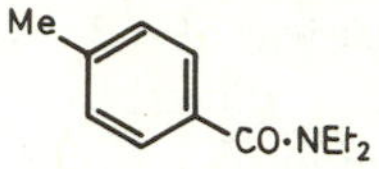

Diethyl-*m*-toluamide
(6.57)

(b) *Insect repellants*. The health of human beings is safeguarded by effective insect repellants such as *NN*-diethyl-*m*-toluamide (*6.57*), hexamethylenebenzamide and the weaker, but bland, dimethyl phthalate.

(c) *Fumigants*. These are non-selective, volatile substances used to kill all insects in a confined space after vacation by all higher forms of life. Typical examples are hydrocyanic acid, sulfur dioxide, ethylene oxide, and nicotine.

6.4.2 *Fungicides*

Fungi, far more than bacteria, viruses, or worms, are the main agents of plant disease. Pathogenic fungi cause the well-known blights, wilts, mildews, scabs, and rusts to which Man's economic crops are prone and they have, throughout history, played havoc with his food supplies. The economic and social devastation wrought by such fungal diseases as blight in potatoes, mildew on vines, and rust and smut in wheat, have been graphically recorded (Large, 1940). These depredations of fungi were traditionally fought by crop rotation, open planting, and the use of disease-free seeds and resistant varieties. Yet so adaptable are fungi that these methods do not suffice, and fungicides remain the most important method of controlling fungal diseases.

Until recently most available fungicides were biochemically non-specific, and depended for their selectivity on favourable distribution coefficients (Section 3.3). They were usually routed to the fungus (rather than to the plant) by providing them with extra lipophilic character, often by means of adding a hydrocarbon side-chain to the molecule. The basis of this selectivity was the protection given to the plant by its cuticle which does not readily permit the passage of either lipophilic substances or the cations of heavy metals, whereas fungi (particularly their conidia or sporulating structures) easily accumulate substances of these types. For example, apple leaves absorbed only 1 μg/g (dry weight) of captan (*6.58*) (*N*-trichloromethylthiotetrahydrophthalimide) from a 25 μM aqueous solution in 30 minutes, whereas the conidia of *Neurospora crassa* took up 7000 times as much under the same conditions (Richmond and Somers, 1962). Glyodin (2-heptadecylimidazoline) and dodine (*n*-dodecylguanidine) behaved similarly. In the fungus, such substances tend to be non-specific, attacking *many* enzymes or other vulnerable cell constituents. As a class they have little toxicity for vertebrates, and they do not incite fungi to produce resistant strains.

After use of these contact fugicides had become well established, more selective substances were found which penetrate into the plant and either prevent penetration of fungi or even kill those that have already penetrated.

In what follows, we shall discuss first the contact fungicides, then the systemic ones that effect 'plant chemotherapy'.

(a) Contact fungicides. These seek to kill the fungus before it has penetrated the plant. This kind of protection is still practised, often very successfully and with inexpensive, traditional inorganic agents. Basic cupric calcium sulfate (Bordeaux mixture), introduced by Millardet in 1885, is highly selective and still much employed, as is elemental sulfur which has been used for this purpose during some 2500 years. Sulfur, by contact with living tissues, forms two mildly anti-fungal substances: hydrogen sulfide and pentathionic acid. A great advance was made in 1934 when Tisdale and Williams reported the high anti-fungal action of dimethyldithiocarbamic acid (*3.57*) as its iron and zinc salts, ferbam and ziram. These have no toxicity for Man, cattle, or the higher plants, but show a wide spectrum of toxicity for fungi, and are made and sold in enormous quantities. These complexes with heavy metals are valued for their enhanced adhesive properties, but the metals have to be below copper on the affinity scale (Table 11.1). How these substances act by chelating copper, adventitiously present, is described in Section 11.7.3.

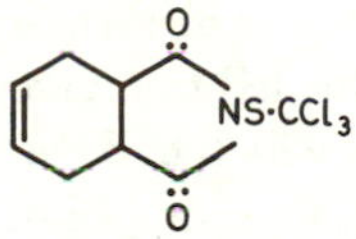

Captan
(6.58)

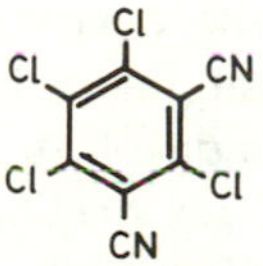

Chlorothalonil
(6.59)

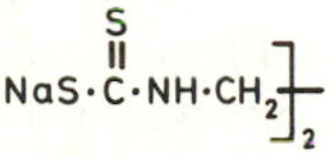

Nabam
(6.60)

One of the most used of all contact fungicides, particularly on fruit trees, is captan (*6.58*), introduced in 1952 by Kittleson. Inside the fungus, it liberates thiophosgene ($CSCl_2$) which reacts with the free hydroxy- and amino-groups of enzymes (Brown, 1978). Folpet is a related agent in which the cyclohexene ring has been replaced by a benzene ring. The following anti-fungals are thought to act upon a solitary mercapto-group: the mercurials, and the salts of ethylene-1,2-bisdithiocarbamic acid, e.g. nabam (*6.60*). A precise attack on the SH-group of cysteine-149 in glyceraldehyde 3-phosphate dehydrogenase is made by chlorothalonil (*6.59*) (2,4,5,6-tetrachloroisophthalonitrile) (Long and Siegel, 1975).

The modes of action of these and other sprayed-on anti-fungal agents have been classified by Sijpesteijn (1970). The paired mercapto-groups in dihydro-lipoic acid and dihydrolipoic dehydrogenase, which are important for the oxidation of pyruvate in fungi (Wren and Massey, 1965), as in other forms of life, are inactivated by dialkyldithiocarbamates, and 8-hydroxyquinoline, which require copper as a co-toxicant. Oxine is used to mould-proof wood, canvas, and outdoor equipment generally. Sijpesteijn then lists oxidative phos-phorylation as the target for trialkyltin salts and the nitrophenols. How azauracil (*4.41*), which is used for powdery mildew, acts is explained on p. 140.

(b) Systemic fungicides. 'Plant chemotherapy', the systemic use of fungicidal agents, is more difficult than animal chemotherapy because plants have no true circulation, no phagocytes to assist the agent, and no mechanism for the detoxication or excretion of the agent when its work is done. Nonetheless, a startling improvement in fugicides was heralded by the introduction in 1966 of benomyl (*3.58a*) [(methyl 1-butylcarbamoyl-2-benzimidazolecarbamate) 'Benlate'], after the careful report of Delp and Klopping (1968). This agent is usually placed in the soil, gains access to the transpiration stream through the root hairs, and ends in the leaves. In this way, deeper penetration and a more even application are possible than with spraying techniques. Benomyl exerts its anti-fungal effect at concentrations as low as 0.02 parts per million, and it is active against a very wide range of fungal diseases (Selling, Vonk and Sijpesteijn, 1970). As recounted in Section 3.6, benomyl is hydrolysed by water to methyl 2-benzimidazolecarbamate (*3.53b*), which is the actual agent; methyl thiophanate (*3.59*) is also used in the fields as a precursor of this carbamate (see Section 3.6, p. 107).

Unfortunately, resistance to benomyl soon set in, although resistance to fungicides was previously unknown. As a result, larger doses had to be used, and fewer species of fungus could be subjugated. This led to a vigorous search for other systemic fungicides. Thiabendazole (*6.36*), which had been introduced in 1961 as a veterinary vermifuge (see Section 6.35), was found to be a good systemic fungicide for plants. All these benzimidazole-based fungicides are thought to inhibit the assembly of microtubules (see Section 5.45).

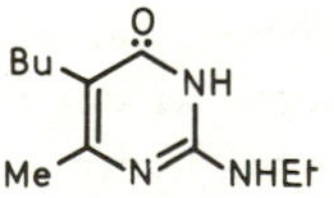

Ethirimol   Carboxin
(*6.61*)   (*6.62*)

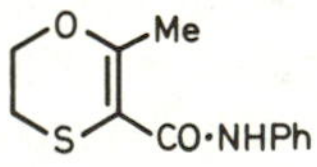

Another successful systemic agent, ethirimol (5-*n*-butyl-2-ethylamino-6-methylpyrimid-4-one) (*6.61*) ('Milstem') is much used against powdery mildew which formerly caused great loss in cereal crops (particularly barley) through-out Europe. It acts by inhibiting folic acid metabolism, somewhat similarly to pyrimethamine in Section 9.3.3 (Bent, 1970).

One systemic fungicide that has been outstandingly successful, although specialized to act only against basidiomycetes (such as smuts and rusts in barley and wheat), is carboxin (6.62) ('Vitavax') (5,6-dihydro-2-methyl-1,4-oxathiin-3-carboxanilide) (Edgington, Walton and Miller, 1966). The essential structure in carboxin is a methyl-group placed ortho to an anilide group; the nature of the nucleus is not of prime importance and a benzene ring will suffice (Carter, Huppatz and Wain, 1975). These anilides attack succinic dehydrogenase in the basidiomycete mitochondria (White and Thorn, 1975), and act systemically in the emergent plant. Even as simple a congener as 2-methylbenzanilide (mebenil) acts similarly.

Much of the current interest in fungicides is centred on those which interfere with the synthesis of ergosterol by the fungal cell. This sterol gives indispensable support and permeability control to fungal membranes. One of the many successful inhibitors is tridemorph (6.63) which acts systemically and controls powdery mildew. This substance, 2,6-dimethyl-N-tridecylmorpholine, inhibits C-14 reductase, and also the shift of a double bond, from the 8- to the 7-position, in the conversion of fecosterol to episterol as part of the synthesis of ergosterol. Selectivity is attributed to the fact that biosynthesis of ergosterol occurs via the lanosterol pathway in fungi but via the cyclosterol route in the higher plants. Similarly acting agents are marketed in which the N-tridecyl group is replaced by a 3-phenylbutyl group (Kerkenaar, Uchiyama and Versluis, 1981; Kerkenaar, 1983).

Ergosterol synthesis is inhibited differently by another class of fungicides, namely by preventing demethylation at C-14 during biosynthesis. As a result, useless sterols accumulate such as 14-methylfecosterol. Examples of this class are: fenarimol (6.64), which is 1-(2,4'-dichlorodiphenyl)-1-pyrimidin-5-yl-methanol, and triadimefon (6.65) which is 1-(4-chlorophenoxy-3,3-dimethyl-1-(1,2,4-triazol-1-yl)butan-2-one. These are somewhat wedge-shaped molecules with alternating lipophilic and hydrophilic areas. The heterocyclic fragment can be varied to be pyridine, piperazine, or imidazole as in buthiobate, triforine, or imazilil, respectively (Kato, 1983).

All ergosterol inhibitors act at low dosage, and no field resistance has been encountered.

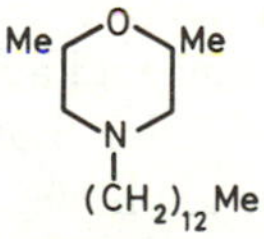

Tridemorph
(6.63)

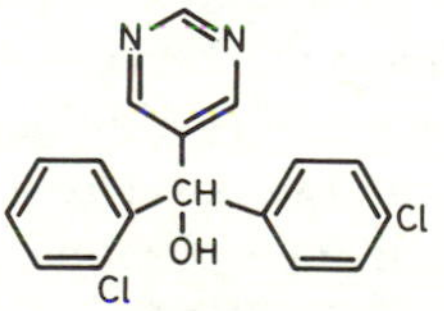

Fenarimol
(6.64)

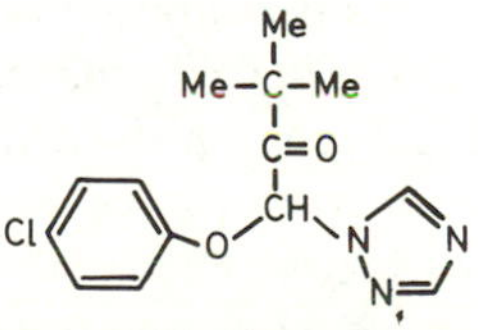

Triadimefon
(6.65)

Of the phosphorus-containing systemic fungicides, typical examples are IBP ('Kitazin P') (6.66), which is di-isopropyl S-benzyl phosphorothioate, and edifenphos, which is ethyl SS-diphenyl phosphorodithioate. These injure fungi by inhibiting the transmethylation process that produces lecithin in membranes (Kodama, Yamashita and Akatsuka, 1980).

In one of the *indirect* approaches to plant health, long advocated by Horsfall (1972), probenazole (*6.67*), which has no anti-fungal properties, is used to prevent infection of rice plants. It does this by preventing the mature hardening of piercing organelles on the fungus so that they cannot penetrate even cellulose film. This substance, 3-allyloxy-1,2-benzoisothiazole-1,1-dioxide, and penta-chlorobenzyl alcohol seem to act by the same mechanism (Sekizawa *et al.*, 1982).

Finally, a systemic fungicide that has been very successful against *Phytophora* and other soil-borne pathogens is metalaxyl. This substance, methyl *N*-(methoxyacetyl)-*N*-2,6-xylylalaninate (*6.68*), acts by preventing the incorporation of uridine into RNA in fungi (Davidse *et al.*, 1982).

IBP ('Kitazin P')
(6.66)

Probenazole
(6.67)

Metalaxyl
(6.68)

Horsfall (1972) suggested other ways in which the biochemical interactions between plants and fungi might be disturbed so as to favour the plant. (1) The basic metabolism of the plant may be so altered that it becomes a less attractive host; for example, when plant growth substances lower the sugar level in foliage, resistance to fungi usually increases. (2) Systemic agents may inactivate fungal toxins and tissue-destroying enzymes. (3) The agent can protect the plant by lignifying the point of attack, as phenylthiourea seems to do. Certainly much of the damage caused by fungi arises from fungal toxins such as tentoxin, a cyclic tetrapeptide secreted into fruits and grains by *Alternaria tenuis* which inhibits photophosphorylation (Steele *et al.*, 1976).

(*c*) *Other aspects of fungicides*. The disinfection of stored seeds prior to planting has traditionally been accomplished with mercurials and, in recent times, with hexachlorobenzene. Unfortunately if, as does happen, the seeds are mistakenly eaten, these chemicals can cause severe illness. Hence there has been a search for substances that injure fungal spores but are selective enough not to hinder germination or harm humans. Benomyl meets these criteria.

For killing fungi in the soil, salts of methyldithiocarbamic acid and various chloronitrobenzenes have been found suitable; for soil fumigation more volatile substances such as chloropicrin and methyl isothiocyanate are preferred.

In spraying crops, many of the affected tissues cannot be reached directly, but depend on receiving fungicide by slow *translocation*. As Rich (1954) has shown, the particles of copper-containing fungicides are positively charged and cling more readily to the negatively charged leaf than to one another. As the wind and the rain dislodge particles from each residual clump, they move to aspects of the leaves not covered by the original spraying.

(d) Phytoalexins. In practice, it is found that every species of plant can protect itself from all but a very few species of fungi, most of which are highly specific for particular varieties of plant. Many higher plants protect themselves against not only fungi, but also bacteria, viruses, and nematodes. They do this by producing phytoalexins, substances absent from healthy plants. Thus *Leguminosae* synthesize isoflavonoid protectants, the *Solanaceae*, diterpenes, and the *Compositae*, polyacetylenes (Cruickshank, 1963; Bailey and Mansfield, 1982).

For further reading on fungicides, see Hutson and Roberts (1982). The handbook edited by Worthing (1983) gives the composition of all fungicides.

6.4.3 Herbicides (weed killers)

Only in recent times have farmers become receptive to the idea of selectivity in eliminating uneconomic plants (i.e. weeds). In 1895–7 it was demonstrated independently by Bonnet in France, by Bolley in America, and by Schultz in Germany that a solution of cupric sulfate killed the ubiquitous charlock (*Sinapis arvensis*) without injury to the crop in which it was a weed. While continuing to use their traditional fungicides and insecticides, farmers turned a deaf ear to this news of the World's first selective herbicide. In 1911 a Frenchman showed that a solution of sulfuric acid could be safely used on crops to destroy the weeds (Rabaté, 1927). So slowly did thinking about herbicides move, that not until 1932 were systemic trials of Rabaté's method made in England and the USA (see Fig. 3.1, p. 57). These trials established the high selectivity of this simple acid, but problems arose from corrosion of equipment, and the increased acidity of the soil. Soon two more Frenchmen discovered the selective weed-killing properties of dinitro-*o*-cresol, a substance which had been known since 1866 (Truffaut and Pastac, 1932, 1944), and this enjoyed a few years of cautious use.

By the middle 1930s, a good deal of the prejudice against selectively toxic weed killers had been overcome, and the agricultural industry became receptive to the phenoxyacetic acids. These substances were designed to resemble the natural auxin, indolylacetic acid (*4.82*) but, unlike the latter, to resist destruction by the plant. At first they were used only for their hormone-like action, to promote the setting of unfertilized fruits, and to promote the growth of roots (Zimmerman, 1942). When using them in excess as leaf sprays, it was found that they acted as herbicides (Templeman and Sexton, 1946). This discovery led eventually to the use of these substances in colossal amounts throughout the world, killing weeds as the result of inducing excessive growth. The two most used herbicides in this class are 2-methyl-4-chlorophenoxyacetic acid (*4.83*) and 2,4-dichlorophenoxyacetic acid. [See, further, Sections 4.7 (p. 170) and 12.2.]

These two phenoxyacetic acids, the former preferred in Britain, the latter in the United States, have proved outstandingly successful for killing weeds in cereal crops. It is not known why cereals are relatively unaffected. When the cereals were, in an experiment, made to absorb as much of the herbicides as the weeds do (normally cereals absorb less), they remained unharmed (Wood, Wolfe and Irving, 1947). On the whole dicotyledons are killed and monocotyledons survive, but there are exceptions to this rule. In general, phenoxyacetic acids are harmless to Man who excretes them rapidly, largely unchanged.

To kill monocotyledonous weeds, such as wild oats in a wheatfield, derivatives of diphenyl ether are used, such as diclofop, which is 2[4-(2,4-dichlorophenoxy)phenoxy]propionic acid (6.69). To deal with woody plants, 2,4,5-trichlorophenoxyacetic acid was introduced and proved effective. Unfortunately, it has often been contaminated, in manufacture, with dioxin (p. 42).

Diclofop-methyl
(6.69)

Picloram
(6.70)

Ioxynil
(6.71)

An ingenious method for increasing the selectivity of herbicides in the phenoxyacetic acid series, by β-degradation of intrinsically harmless homologues, was recounted in Section 3.6 (p. 108) and illustrated in Fig. 3.6.

It was later found that many carboxylic acids, provided that they were ortho-substituted, behaved similarly to the phenoxyacetic acids and many of these were marketed, offering small advantages in particular circumstances. A typical example is picloram (6.70) which was introduced in 1963 for controlling annual weeds and deep-rooted perennials. It is more potent than the phenoxyacetic acids and can destroy weeds in a cereal crop that are resistant to the phenoxyacetic acids (Kefford, 1966).

The cyanophenol herbicides arrived as an intelligent replacement for Truffaut's dinitro-o-cresol. A typical example is ioxynil (6.71) 2,6-di-iodo-4-cyanophenol (Wain, 1963). It is a contact herbicide which powerfully uncouples oxidative phosphorylation (Section 4.5), in fact much more strongly than 2,4-dinitrophenol does (Kerr and Wain, 1964). It and the less expensive, if slightly slower acting, 2,6-dibromo-4-cyanophenol (bromoxynil), are used against young dicotyledonous weeds in cereal crops. They are also available as the octoate esters, for better penetration.

Useful specificity is shown by 4,6-dinitro-2-secbutylphenol (dinoseb), much used to eliminate dicotyledonous weeds from crops which are themselves dicotyledons, particularly peas and lucerne (alfalfa) (Roberts, 1954). The ionization and penetration of nitrophenols is discussed in Section 10.5.

Some other substances, whose action is quite the contrary of MCPA (methylchlorophenoxyacetic acid), can eliminate monocotyledonous weeds from a dicotyledon crop. For this purpose 2,2-dichloropropionic acid (dalapon, 'Dowpon') is much used. Its mode of action is discussed in Section 9.4.1. The carbamates, which are mitotic poisons, introduced by Templeman and Sexton (1945), are also available for this purpose. The most selective example is barban (6.72) (4-chlorobut-2-ynyl N,3'-chlorophenylcarbamate), which suppresses the

growth of wild oats in cereal crops (Crafts, 1964). The carbamates act by
inhibiting the synthesis of RNA and protein (Kobayashi and Ishizuka, 1978).

Cl——〈 〉—NH·CO·O·CH₂·C≡C·CH₂Cl
Barban
(6.72)

Me—〈triazine with OMe〉—NH·CO·NH·SO₂—〈 〉Cl
Chlorsulfuron
(6.73)

The triazines, such as simazine (*4.62*), are the most used of all soil-applied
herbicides; small changes in the substituents give excellent control over selectivity and persistence. These substances [and the phenylureas, such as diuron
(*4.63*), which are also soil-types] act by interfering with the Hill reaction in
photosynthesis (see Section 4.6). Both of these parent substances, simazine and
diuron, have undergone extensive modification for particular purposes, but
each is still extensively used. A newcomer, chlorsulfuron (*6.73*), combines
molecular features from both parents, but is notable for the greater dilution at
which it acts, and also for the speed (all cell division is halted within 1 hour).
This substance, 1-(2-chlorophenylsulfonyl)-3-(4-methoxy-6-methyl-1,3,5-tri-
azin-2-yl)urea, is taken up by wheat, oats, and barley plants which rapidly
inactivate it. It is harmless to Man (Campion and Tichon, 1981).

Whereas some herbicides, such as dalapon (above), need to be sprayed on the
foliage, others may be applied to the soil (the phenoxyacetic acids work both
ways). Soil-applied herbicides have to be strongly adsorbable on soil particles,
and readily taken up by the root hairs of plants. They are selective against
quick-growing weeds which usually have the most copious surface roots, and
they are more sparing of economic crops because the latter usually have roots
that are slow-growing, tougher, and deeper. Many are available for use in the
soil at seed-sowing time, to exert 'pre-emergence control'. A class apart are
those agents used for 'chemical ploughing'. These herbicides are sprayed on the
soil, and kill all forms of plant life, exactly where they fall, spreading neither to
right nor left. They become inactivated, moreover, by exposure in 1–3 days. Of
these agents, the bipyridylium herbicides (e.g. diquat and paraquat), which act
by liberating free radicals, are described in Section 4.6. Another much-used
example is glyphosate (*4.56*). It is absorbed by leaves but, through translocation, becomes shared with the plants' roots. The primary site of action is on
protein synthesis (Tymonko and Foy, 1978), presumably by inhibiting choris-
mate mutase (Section 4.3) (Davis and Harvey, 1979). Glyphosate, which is
N-phosphonomethylglycine, can kill woody or fleshy plants that are little
affected by other herbicides (Barrett, 1974).

The growth of plants can be retarded by inhibiting the transport of the auxin,
indol-3-ylacetic acid. Schneider (1964) prepared a series of fluorene derivatives
to produce dwarf varieties of economic plants. Another use of these morphactins, as they are called, is to retard the growth of dicotyledonous weeds in a

grain crop until, overgrown by the crop, these weeds are finally destroyed by lack of light, meanwhile serving to protect the soil from drying (Ziegler, 1970). One of the best of the morphactins is chlorflurecol-methyl (*6.74*).

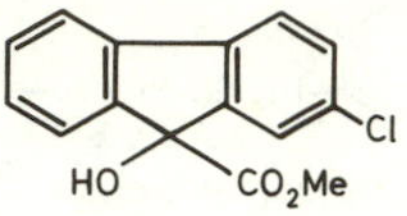

Chlorflurecol-methyl
(*6.74*)

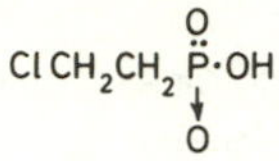

Chloroethane-phosphonic acid
(*6.75*)

(*a*) '*Plantidotes*'. Plants suffering from toxicity inflicted by an insufficiently selective agent, can be revived by an appropriate antidote. In fact, the antidote and the herbicide are often mixed together and sprayed at the outset. Thus the thiocarbamate herbicides become safer if 1,8-naphthalic anhydride or a dichloroacetamide derivative is used with them, e.g. on a maize crop. For a book on this subject, see Pallos and Casida (1977). A typical combination is *NN*-diallyl-2,2-dichloroacetamide with the herbicide, butylate (*S*-ethyl *NN*-di-isobutylthiocarbamate).

(*b*) *Defoliants*. The aerial spraying of defoliants was introduced in East Africa to eliminate tsetse flies, which carry the trypanosomes of sleeping sickness (Blackman, 1954). The most used substances are aliphatic arsenicals (e.g. sodium cacodylate), 3-amino-1,2,4-triazole, and the phenoxyacetic acids and their *n*-butyl esters. When a defoliation campaign moves near to agricultural land, ominous ecological problems, such as soil erosion, arise. An unsuspected hazard of defoliation by 2,4,5-trichlorophenoxyacetic acid (2,4,5-T) in the Vietnam war, was the presence of a small percentage of 2,3,7,8-tetrachloro-dibenzo-*p*-dioxin (*2.28a*), a by-product of synthesis (for the mode of action of dioxin, see p. 42). Conifers are defoliated with picloram (*6.70*).

A defoliant is essentially a substance that causes rapid senescence, and there is no more effective way to do this than to increase the level of ethylene in the leaf. 2-Chloroethanephosphonic acid (*6.75*), which liberates ethylene in the cell, effects this and is also being used (at lower concentrations) to accelerate ripening of fruits. For a review, see Osborne (1968).

For further reading on herbicides, see Fletcher and Kirkwood (1982), Audus (1976), and Ashton and Crafts (1981). For a handbook giving the composition of all herbicides, see Worthing (1983). For a specialized work on the phenoxy-alkane herbicides, see Hee and Sutherland (1981).

6.4.4 *Other pesticides*

(*a*) *Plant antiviral agents*. These are still uncommon. Low concentrations of the antibiotic blasticidin S selectively inhibit virus multiplication in bean and tobacco plants (Hirai *et al.*, 1966). For a book on plant viruses, see K. Smith (1977).

(*b*) *Agricultural anthelmintics*. Nematodes can impoverish soil and attack roots. 1,2-Dibromo-3-chloropropane (DBCP) ('Nemagon') and 1,2-dibromoethane are volatile nematocide formerly used to rid soil of these undesirable worms. Plants are remarkably tolerant to them but they are toxic to humans.

Anthelmintics for farm animals were dealt with in Section 6.3.5.

(*c*) *Molluscicides*. By killing the worm-carrying snails in the watercourses of Egypt and other countries where bilharzia in Man is widespread (see Section 1.1), an essential stage in the life cycle of this parasite can be broken. The traditional use of copper sulfate for this purpose has been replaced by the more efficient amides such as 2,5-dichloro-(4-nitrosalicyl)anilide ('Bayluscid'). N-Tritylmorpholine (trifenmorph), another potent molluscicide, is very selective; other forms of aquatic life are little affected by it and mammals unharmed (Boyce, Jones and van Tongeren, 1967). In a field test in Brazil, bistributyltin oxide ($Bu_3Sn-O-SnBu_3$) (hexabutyldistannoxane), in an asphalt base, retained a strong molluscicidal activity when immersed in a stream for a year, and was not inactivated either by immersing in mud or drying in the sun (Gilbert *et al.*, 1973). The Ethiopian water plant *Phytolacca dodecandra* is useful to grow in infested streams because it secretes a molluscicidal glycoside.

(*d*) *Rodenticides*. Annually huge amounts of crops and food stacks are spoilt by rats, and many people die from such rodent-borne diseases as bubonic plague, rickettsiosis, leishmaniasis, spirochaetosis, leptospirosis, and (through rat-borne lice) typhus. Because of the rat's habit of fighting his brethren to the point of drawing blood, anti-coagulants were introduced as (chronic) rodenticides, namely the indanediones and hydroxycoumarins. Of the latter, warfarin (*9.40*) has given outstanding service, although resistance to its action has developed among many scattered populations of the common rat (*Rattus norvegicus*). This resistance has been partly overcome by incorporating calciferol (vitamin D) with it in the bait. Of the acute rodenticides, norbormide (*4.90*) is absolutely specific but rats soon become 'bait-shy' because of its characteristic odour. See Section 4.8 for more about this agent, and a related development. N-3-Pyridylmethyl-*N'*-*p*-nitrophenylurea (pyrinuron) kills both rats and mice, after a single dose, by inhibiting their nicotinamide metabolism. The rats die from paralysis and pulmonary arrest. Mice are not so easy to poison as rats, for the mice eat a little here and there, and not much from any single bait. The weak point of a mouse's defence is the habit of compulsive self-grooming, which can be utilized by strewing a selectively medicated powder to cling to its paws and fur. Such traditionally used rodenticides as arsenic, thallium, zinc phosphide, and naphthylthiourea (antu) are now little esteemed.

Since the early 1900s, biological control of the rat and field mouse has been attempted, on a large scale, from one end of Europe to another. Usually a species of *Salmonella* bacteria (sometimes misbranded 'rat virus') was used, and false claims were made that the organism was specific for the rat. Actually these bacteria are pathogenic to Man and several epidemics have been started in this way (Wodzicki, 1973). A joint WHO/FAO expert committee strongly condemned this form of 'biological control' in 1967.

Biological control of rabbits, using *Myxoma* virus, has been successful in Australia, where it was first undertaken. Although a moderate degree of resistance has now set in, the survivors are easily exterminated with sodium fluoroacetate (non-selective).

For a book on the ecology of pesticides, see A. Brown (1978); for a list of 1500 pesticides with molecular formulae and properties, see Büchel (1983).

6.5 Resistance to drugs and other agents

In 1905 Franke and Roehl, while working with Ehrlich, discovered the phenomenon of *drug resistance* in the following way. Mice, suffering from trypanosomiasis, were treated with drugs in doses that were too small to cure; inevitably a relapse took place but, surprisingly, the renewal of treatment with normally curative doses of the drug also failed. The trypanosomes had developed a resistance which was hereditary, and usually irreversible. As Yorke later observed, resistant trypanosome strains may require up to 250 times the normal concentration of drug before they are injured. Such massive dosage is usually more than the host can tolerate.

Ehrlich (1909) thought that resistance was caused by stepwise withdrawal, or masking, of the receptor by the organisms. However, this was not the case, and it remained for Yorke *et al.* (1931) to show that, at least in trypanosomes, resistance is caused by diminution of uptake by the parasite. With highly resistant parasites, all the drug remains in the medium and can be used to kill susceptible parasites. These workers had the advantage of being able to culture the trypanosomes *in vitro,* a technique unknown in Ehrlich's day.

Following Franke and Roehl's discovery, Ehrlich demonstrated three distinct types of resistance in trypanosomes. Parasites which had become resistant to trypan red *(6.1)* resisted all other azo-dyes. Some other strains, which were resistant to 'Atoxyl' (*p*-aminophenylarsonic acid) *(6.2)*, resisted all other phenylarsonic acids, and a third type, resistant to parafuchsin *(10.5)*, resisted all other triphenylmethanes. Yet a strain resistant to one of these three classes of drugs did not resist the other two classes unless specially trained to do so.

Later, Ehrlich recognized two classes of arsenical agents between which there was no cross-resistance. The first of these, typified by *(6.4)*, is distinguished by the presence of water-attracting substituents, which may be —OH, —CONH$_2$, or —SO$_2$NH$_2$ (in place of NH$_2$, or in addition to it), but only such as are not ionized (as anions) at pH 7. Nearly all the arsenical drugs that have achieved clinical use belong to this sub-class, members of which often have a favourable therapeutic index. It was also found that the acridine, trypaflavine *(6.5)* belonged to this sub-class because trypanosomes made resistant to it were automatically resistant to arsenicals of type (*6.4*). The second class of resistance-prone arsenicals lacked hydrophilic substituents.

In these arsenicals, the initial state of oxidation of the arsenic is of no consequence. The eventual toxic action of the drug follows on biological conversion to the arsenoxide group (—As : O), whether in an organic arsenical or in arsenious acid (HO · As : O) (see Section 13.0). These facts, taken in conjunction with the knowledge that no resistance to arsenious acid could be achieved (although carefully sought), show that the non-arsenical part of the molecule is responsible for the uptake of the organic drug and that the resistant parasite is able to block this uptake. On the other hand, a susceptible trypanosome can accumulate in its interior a 500 times greater concentration of arsenic than exists in the external medium (Eagle, 1945).

Today a rough distinction is made between natural and acquired resistance, as illustrated by the following three cases. The *Mycobacterium* that causes human tuberculosis has natural resistance to penicillin; *Staphylococcus aureus* (the common yellow pus organism) has many strains that are susceptible to penicillin but can easily acquire resistance to it; *Streptococci* lack natural resistance to penicillin and do not acquire it either. To extend the last example: the spirochaete that causes syphilis, although abundantly exposed to arsenicals and later to penicillin, never developed resistance to either drug.

Resistance arises mainly by *natural selection*, namely by the outgrowth of a naturally resistant strain after the drug has killed all the susceptible strains. Resistance seldom arises by drug-provoked mutation, because the use of mutagens as drugs and agrochemicals is avoided. In addition to natural selection, drug resistance can be conferred by gene-transfer as in the mating of insects, the conjugation of Gram-negative bacteria, and the transformation of pneumococci.

Resistance by gene-amplification is another possibility. This has been demonstrated in aphids, the female of which can produce offspring without fertilization by the male, thus creating genetic duplicates (clones) of herself. These aphids, when exposed to increasing concentrations of parathion (*13.24*), became 15 times less sensitive to it in 15 generations, and it was found that they had built up their production of a parathion-destroying esterase by induction, steadily switching dormant esterase-manufacturing genes into production. See below for other examples.

In the commonest form of drug resistance, namely that arising by selection of a naturally arisen mutant, the situation is as follows. A culture of about ten million bacteria may contain only one organism resistant to a particular drug. The chance of finding an organism resistant against two different drugs is often put at 1 in 10^{14}, and against three drugs 1 in 10^{21}. Hence the great success of using several drugs simultaneously when the uneconomic cell is known to be able to acquire resistance. On the other hand, attempted elimination of a susceptible organism by one solitary drug gives the resistant cells an opportunity to proliferate (see Tuberculosis, Section 11.9).

The ingenious technique of replica-plating was devised to show that organisms resistant to a drug need have had no previous contact with it (Lederberg and Lederberg, 1952). Bacteria (*E. coli*) were grown on an agar plate, and replicas of the colonies were transferred to several other agar plates by printing with velveteen. One such plate contained streptomycin. When (after incubation) the position of a colony resistant to streptomycin was found in the antibiotic-containing plate, colonies were harvested from identical positions in the streptomycin-free plates. This process was repeated, and finally an entire plate was obtained of streptomycin-resistant organisms that had never been in contact with streptomycin. Another proof of the pre-existence of resistant organisms was provided by the bacteriologist's habit of freeze-drying and storing cultures, from which viable specimens can be drawn after the passage of whole decades. Many bacterial strains that were stored, long before currently used antibacterials were developed, have been found to be immediately resistant to them. The explanation of these results is that an agent usually attacks just one metabolic step, and that a resistant organism does not use this step but has an alternative pathway (which is not always advantageous to it).

Resistance is a widespread, but not universal, phenomenon. Its possibilities, although frightening in special cases (insects to most known insecticides; the staphylococcus to most of the formerly useful antibiotics), are not unlimited. Replica-plating discloses that *E. coli* can be made to achieve only a threefold increase in resistance to chloramphenicol (Cavalli-Sforza and Lederberg, 1956), and this seems to be representative of what natural selection usually accomplishes.

The four main types of resistance (arising by natural selection or by gene-transfer) will now be described.

6.5.1 *Type 1 resistance: Exclusion from site*

Historically, this was the first kind of resistance to be recognized. Trypanosomes resist organic arsenicals by altering the plasma membrane in a way which prevents uptake of the drugs (Yorke *et al.*, 1931). Similarly, *Staphylococcus aureus* can become resistant to tetracycline by modifying the plasma membrane which normally facilitates uptake of this drug (Section 3.0). The organism is thus protected from the drug's action even though the drug's target (the ribosome) remains as susceptible as ever (Sompolinsky *et al.*, 1970).

In much the same way, a cultured line of lymphoblasts, from lymphatic leukaemia in the mouse, became resistant to methotrexate through a decrease in the transport of this drug, effected by modifying the plasma membrane (Hill *et al.*, 1979). It was already known that, in mouse leukaemia, the survival time of the leukaemic cell suspensions is inversely proportional to the rate of uptake of methotrexate (*4.7*), but this rate begins to decline soon after treatment is instigated (Kessel *et al.*, 1965). This phenomenon has not been demonstrated in human leukaemia.

The physical basis for Type 1 resistance is suggested by the work of Haest *et al.* (1972) who showed that the cell membrane can adjust its net charge by varying the proportion of phosphatidylglycerol (anionic) to lysylphosphatidyl-glycerol (cationic). It is easy to see how, in this way, either anionic or cationic drugs can be excluded by setting the scene for the operation of Coulomb's law. In different cases, this can be effected by using genetic information already present, by selection, or by gene transfer.

6.5.2 *Type 2 resistance: Increased enzyme production. DNA amplification*

Quite the commonest way for an organism to achieve resistance is to let the susceptible clones be replaced by others that have an increased production of an enzyme, either the target (if that is an enzyme) or else a drug-destroying enzyme.

(*a*) *Drug-destroying enzyme is increased.* In this way, *Staphylococcus aureus* can become resistant to penicillin in the clinic, whether by clone-emergence or by capturing a helpful plasmid. Penicillin-resistant strains isolated from patients secrete the enzyme β-lactamase ('penicillinase'). This enzyme hydrolyses the drug to penicilloic acid, which is biologically inert (see Section 13.1). Penicillinase-producing staphylococci are inherently quite sensitive to penicillin. Hence *small* inocula can be inhibited by low concentrations of the antibiotic. It is, in effect, a race between the speed with which penicillin can kill the bacteria and the speed with which they can produce enough of the enzyme to destroy the penicillin (Knox, 1962). Actually penicillin can be made to induce some strains of *Staph. aureus* to produce penicillinase. No permanently resistant population of this bacterium has arisen in this way, and the organisms return fairly rapidly to the 'uninduced' susceptible state when the penicillin is withdrawn. Much of the detail of penicillinase-induction was first worked out in *Bacillus cereus* (Pollock and Perret, 1951).

In many hospitals, over 90% of the nursing staff harbour penicillinase-containing, penicillin-resistant *Staph. aureus* in the nose, a state of affairs that is rare outside of hospitals. This has been attributed to the continual inhalation of traces of penicillin, which destroys the sensitive strains and thus liberates the nostrils as ideal culture-areas for growth of resistant strains (Gould, 1957).

Other examples follow of resistance achieved by increased production of a destructive enzyme. The treatment of acute leukaemia with cytosine arabino-side (*4.13*) fails in proportion as malignant cells with a higher concentration of cytosine deaminase appear (Steuart and Burke, 1971). Acquisition of insensitivity to 6-mercaptopurine by acute lymphocytic leukaemia cells in Man, and by murine sarcoma 180/TG, was shown to be due to an increase in alkaline phosphatase which causes degradation of the tumour-inhibiting nucleotide to which the cell converts this pro-drug (Rosman *et al.*, 1974).

Insects usually become resistant by a Type 2 mechanism, most often through the acquisition of a mixed function oxidase (Casida, 1973), very similar in nature to the combined oxidases of the human endoplasmic reticulum (see Section 3.5). This genetic change brings with it resistance to all the main types of chlorinated hydrocarbons, to the phosphates and carbamates, and (though in reduced degree) to some of the newer pyrethroids and the pyrethrin synergists (Tsukamoto and Casida, 1967). Resistance to organophosphates can, alternatively, be effected by organophosphate esterase and, sometimes, by glutathione S-transferase (Casida, 1973).

In addition, two different types of resistance to chlorinated hydrocarbons occur in insects, particularly flies: one is to DDT and the other is to dieldrin, chlordane, and benzene hexachloride (BHC). DDT resistance in flies is commonly due to the increased rate of conversion of DDT (dichlorodiphenyltrichloroethane) (6.41) to DDE (dichlorodiphenyldichloroethylene), which is inactive. This is achieved by an increased production of the enzyme 'DDT-dehydrochlorinase' (Winteringham and Barnes, 1955). The normal function of this dehydrochlorinase is unknown in the mutant which the agent has caused to be selected. Resistance to BHC in houseflies has been traced to its conversion to water-soluble sulfur-containing derivatives, probably mercapturic acids (Bradbury and Standen, 1959).

(b) *Target enzyme is increased.* The widespread resistance of the malarial parasite to drugs acting on dihydrofolate reductase (DHFR) has this origin. The Uganda (Palo Alto) strain of pyrimethamine-resistant *Plasmodium falciparum* contains from 30 to 80 times as much DHFR as the susceptible strain. The enzyme was unaltered, for it showed no change in inhibitor-binding, or K_m, for the substrates (Kan and Siddiqui, 1979).

In human leukaemia, the malignant white blood cells readily develop resistance to methotrexate (4.7) when treated with this drug. At the same time, an excess of DHFR, the enzyme that methotrexate is employed to block, is found in the patient's leucocytes (Bertino et al., 1965). The 200-fold increase in this reductase found in murine sarcoma and lymphoma cells which became resistant through methotrexate treatment, has been traced to induction, by the drug, of extra copies of the gene (Alt et al., 1978). For more on such gene amplification, see Fox (1984), Borst (1984).

6.5.3 Type 3 resistance: Decreased enzyme production

Human neoplasm cells in culture show a different type of resistance to 6-mercaptopurine from that cited above, in that they delete the enzyme responsible for converting this pro-drug to the therapeutic nucleotide (6-thioinosine 5'-phosphate) (Brockman, 1963). The deleted enzyme is inosine 5'-phosphate pyrophosphorylase. Similarly, resistance to 8-azaguanine is accompanied by loss of the enzyme guanosine 5'-phosphate pyrophosphorylase in human epidermoid carcinoma cells (Brockman et al., 1961). Resistance to 5-fluorouracil also depends on the uneconomic cell's ceasing to convert this pro-drug to the nucleotide.

In experimental tumours, resistance to 6-mercaptopurine is usually caused by deletion of an enzyme required to convert this pro-drug into the active molecule. The enzyme is hypoxanthine–guanine phosphoribosyltransferase (Harrap, 1976). Resistance to this drug in leukaemic patients seems to follow the same course (Rosman and Williams, 1973).

A related kind of resistance, 'target withdrawal', falls somewhere in between Type 2 and Type 3. In an example, the clinical resistance of *Staphylococcus aureus* to erythromycin, the 50*S* ribosome subunits were found to have been methylated by an enzyme peculiar to the resistant strain (Lai and Weisblum, 1971). Also, in some examples of clinical resistance of pneumococci to sulfonamides, the target enzyme, dihydrofolate synthetase, has been found chemically altered (Ortiz, 1970).

The widespread resistance, in rats, to the anti-coagulant rodenticide warfarin (*9.40*), seems to be of this type. Ribosomes from resistant animals contain less warfarin than normal, due to replacement of the usual binding protein by one with less affinity (Martin, 1973).

Apart from the two types of methotrexate resistance outlined above, a third type has been encountered in the laboratory. In this type, the leukaemic cells change the structure of their dihydrofolate reductase so that it loses all affinity for methotrexate, while still performing its normal function in the cell. Presumably, this takes place by selection. For a review of the three types, see Harper and Kellems (1981). Another case of apparent 'enzyme redesigning' has been reported for some types of housefly resistance to organophosphates and carbamates (Plapp, 1976). In this, the enzyme (AChase) performs its acetylcholine-destroying function normally, but no longer takes up the inhibitors, which are known to depend on a feature in the uptake site different from that required by acetylcholine (Section 13.3; O'Brien, 1971). However, many more cases are known of insect resistance to organophosphates and carbamates where the insecticide is destroyed by increased production of mixed-function oxidase (in the *e.r.*), or (for organophosphates only) by increase in non-specific esterase, or glutathione-dependent alkyl- (or aryl-) transferase (Plapp, 1976).

It will be realized that in Type 3 resistance, and much of Type 2, the uneconomic cells have to make increased use of an alternative biochemical pathway.

6.5.4 *Type 4 resistance: Increased metabolite production*

Type 4 resistance takes the form of secretion by the organism of an excessive amount of the substance to which the drug is a metabolite-antagonist. Several workers, for instance, have shown that staphylococci, pneumococci, and gonococci become resistant to the usual concentrations of sulfonamide drugs by secreting extra quantities of *p*-aminobenzoic acid (cf. Landy and Gerstung, 1944). The *p*-aminobenzoic acid produced in this way has been isolated chromatographically (Moss and Lemberg, 1950).

6.5.5 *Gene-transfer.*

The following is the classical example. When sensitive pneumococci were cultivated in a medium to which an extract of penicillin-resistant pneumococci had been added, a proportion of the resulting organisms were penicillin-resistant and continued so on sub-culture. The transforming factor was found to be a fragment of DNA, a bacterial gene. In this way sensitive cells can be made resistant to penicillin without ever having been in contact with that drug (Hotchkiss, 1951).

Another aspect of gene-transfer, known as *infectious multiple drug resistance*, has, in some localities, undermined the usefulness of a wide range of drugs normally able to cope with serious intestinal infections. As was first suspected about 1960, many Gram-negative bacteria in the intestines carry a plasmid (i.e. a transferable particle of DNA), also called the 'R Factor' or episome. This can invade other bacteria if two species conjugate, and thus transfer drug resistance from the first to the second species. The bacteria in question are those that cause dysentery, cholera, typhoid fever, tularemia, and plague. This R Factor often confers simultaneous resistance to sulfadiazine, streptomycin, the tetracyclines, and analogues of all these, by chemical elaboration of the drug to give an inert substance. The following two examples are from Umezawa's Institute in Tokyo. *E. coli* plasmids were found to inactivate streptomycin by esterification with adenylic acid (Takasawa *et al.*, 1968). Plasmids from both *E. coli* and *Staph. aureus* inactivate chloramphenicol by acetylation, and kanamycin by phosphorylation (Doi *et al.*, 1968).

These R Factors are units of non-chromosomal genetic inheritance that exist independently of any exposure to drugs. One of their chief functions is to act as centres for detoxication of foreign substances, similar to the endoplasmic reticulum of mammals (Section 3.5). Examination of a historical bacteria-bank showed that conjugative plasmids were as common in the various enterobacteria before medical use of antibiotics began as they are in drug-sensitive strains today. However, very little drug resistance was seen in this old collection (Hughes and Datta, 1983).

Plasma-conveyed drug resistance has raised serious problems in the treatment of infections caused by the enterobacteria. Chief among such infections are shigellosis, often called bacillary dysentery, and salmonellosis, which is often referred to as food poisoning and may be caused by any of 1200 serotypes of *Salmonella*. Typhoid fever is also caused by *Salmonella* (*S. typhi*). The resistance problem was first recognized in Japan in 1959–60, when *Shigellae* were shown to have become resistant to tetracyclines, streptomycin, sulfonamides, and chloramphenicol. The Central American outbreak due to *Shigella dysenteriae I*, which was resistant to these drugs, caused over 12 000 deaths in Guatemala in 1968–69 and also spread to Mexico. In 1972, similarly resistant strains of *S. typhi* appeared in India, Mexico, and Vietnam (Anon., WHO, 1974).

For a book on plasmids and transposons, see Stuttard and Rozee (1979).

6.5.6 *Overcoming resistance*

In general, there is no known procedure for infallibly effecting this aim, but the following examples of achievement may prove helpful in a specific case.

In insects, any genetic mutation that exhibits resistance to insecticides usually has less Darwinian fitness to survive. As a result, reversion towards susceptibility has often been found after application of the insecticide is halted. An example is the recapture of susceptibility of anopheline mosquitoes in Maharashtra State (India) after a change of insecticide type (Raghavan, 1969). Again, increased susceptibility to pyrethroids was found in spider mites (which are serious pests of food crops) after they had become resistant to organo-phosphorus insecticides (Chapman and Penman, 1979). We have noted, above, how tumours that have become resistant to 6-mercaptopurine have deleted their hypoxanthine phosphoribosyltransferase. But this change prevents them from using hypoxanthine arising by salvage, and hence they become more susceptible to methotrexate which prevents the synthesis of inosine (Harrap, 1976).

Type 2 resistance to some agents can be overcome by blocking the destroying enzyme, when suitable inhibitors are found. Thus dicofol, 1,1-bis(p-chlorophenyl)ethanol, 'Kelthane' (*6.43*) which is structurally related to DDT, overcomes resistance in flies by blocking the enzyme which degrades DDT to to 1,1-bis(p-chlorophenyl)-2,2-dichloroethylene (DDE). Typical results are given in Table 6.1.

Table 6.1 Effect of dicofol (*6.43*) in overcoming the resistance of houseflies to DDT

Amount applied to fly in μg		Inhibition of enzymic conversion of DDT to DDE (%)	Mortality of flies (%)
DDT	Dicofol		
0.65	0.0	0	0
0.65	0.06	20	2
0.65	0.65	49	50
0.65	1.30	65	72
0.65	6.50	84	100

(Perry, Mattson and Buckner, 1953.)

In the laboratory, chelating agents have been used to overcome resistance in bacteria. For example, ethylenediaminetetra-acetic acid (EDTA) restored susceptibility in polymyxin-resistant strains of *Pseudomonas*, presumably by removing calcium and magnesium from the plasma membrane (Brown and Richards, 1965). EDTA also restored susceptibility to hospital-isolated strains of penicillin-resistant *Staphylococcus aureus* (Rawal, 1969). Further, oxytetracycline, a known chelating agent (Section 11.8), rendered penicillin-resistant *S. aureus* susceptible to penicillin (Michael, Michael and Massell, 1967).

Plasmids can be inactivated *in vitro* by short exposure to simple amino-acridines (Mitsuhashi *et al.*, 1961; Bouanchaud, Scavizzi and Chabbert, 1968). The chelating agent, 7-hydroxytropolone, inhibits the plasmid-caused adenylation of the aminocyclitol antibiotics and thus overcomes resistance in cultures of *E. coli* (Allen, 1981). Unfortunately, overcoming drug resistance in the clinic, a problem of increasing magnitude, can so far be met only by increasing the dose up to the patient-toxic limit, or switching to an alternative drug.

6.5.7 *General remarks on resistance in the clinic.*

A useful distinction can be made between resistance acquired (a) in the patient and (b) apart from him. Thus, in patients undergoing treatment for tuberculosis with streptomycin, isoniazid, or *p*-aminosalicylic acid, the causative organism (*Mycobacterium tuberculosis*) often becomes resistant to one or more of these drugs. This has been found to take place, stepwise, by a selection of resistant strains within the patient's body. Resistant staphylococci, on the other hand, are very often acquired from other infected people (Knox, 1962).

The discovery of multiple resistances in his experimental organism (the trypanosome), and the existence of multiple receptors which this finding indicated, led Ehrlich (1909) to suggest *multiple therapy*, the treatment of the sick patient with two or more drugs, simultaneously, each aimed at a different receptor. When applied at the start of the treatment, this strategy can greatly delay the onset of any kind of resistance. In current therapy, this approach is much used in tuberculosis, cancer, and anti-bacterial therapy. In practice, it usually effects a vast decrease in the parasite/host ratio before resistance has had time to set in.

Resistance is often a purely regional phenomenon, at least at first. When the malarial parasite *Plasmodium falciparum* had become insusceptible to chloroquin in Southeast Asia and South America too, it remained highly susceptible elsewhere. A population of inadequately dosed sufferers forms a dangerous reservoir of resistant organisms.

When one pauses to think how widely drugs and other selective agents are used, and how widespread the resistance phenomenon has become, it is gratifying to find that many organisms remain universally susceptible (see above, p. 258).

In pharmacodynamics, the parallel phenomenon to resistance is the loss of effect of a drug, either through the body having learnt to degrade it (as happens with barbiturates, Section 3.4) or by the sensitivity of the receptor apparently declining (as happens with nicotine and morphine).

For further reading on the biochemical basis of drug resistance, see Mitsuhashi (1982), Bryan (1982), and Mihich (1973). For more on resistance of tumors see Fox and Fox (1984). For pharmacodynamic tolerance, see Section 3.5.3.

6.6 Therapeutic interference

The term *therapeutic interference* was coined by Browning and Gulbransen (1922) to describe the following phenomenon which they discovered. Mice suffering from trypanosomiasis could not be cured by injection of the usual dose of euflavine (*6.5*) if they had previously been fed parafuchsin (*10.5*). Although parafuchsin is itself slightly trypanocidal, it was used here in an ineffective dose. The phenomenon has often been confirmed and a typical experiment is shown in Table 6.2. Organisms cannot pass the interference effect on to the next generation, a point of difference from drug resistance.

It can be seen from this Table that one part of parafuchsin can prevent 10 parts of euflavine from acting. Interference between these substances has been demonstrated *in vitro*, where the parafuchsin again prevents the euflavine from killing the organism (Von Jancsó, 1931). The same antagonism has also been demonstrated respirometrically (Scheff and Hasskó, 1936); both substances separately depress consumption of glucose, but a mixture does not.

Table 6.2 Therapeutic interference Injections into trypanosome-infected mice

Parafuchsin	Euflavine	Result
Nil	Nil	Died 5th to 7th day (parasites + + +)
0.05	Nil	Died 5th to 7th day (parasites + + +)
0.25	Nil	Died 5th to 7th day (parasites + + +)
Nil	0.5	Cured on 3rd day
0.05	0.5	Died 6th to 7th day (parasites + + +)
0.25	0.5	Died 7th day (parasites + + +)

(Schnitzer, 1926.)

The term therapeutic interference has been confined to cases where there is chemical similarity between the two substances and hence no likelihood of their reacting chemically with one another. The phenomenon brings to mind the often-met situation in pharmacodynamics where an agonist can be converted, by an increase in molecular weight, to an antagonist of low or zero efficacy, but with a greater affinity for the receptor (see Section 7.5.2, p. 294).

In the clinic, the current situation is that many pairs of interfering drugs are known. The wise physician prescribes as few drugs as possible at any one time, while warning the patient against self-medication while treatment is in progress. For a discussion of drug interactions, see Melmon and Gilman (1980).

7

Pharmacodynamics

The word pharmacodynamics, as used in this book, means the study of those examples of selective toxicity where the economic and uneconomic species are constituent cells *in the one organism*. Thus it stands in contrast to chemotherapy, where the uneconomic species is a different organism from the economic one.

The beginnings of pharmacodynamics as a science is traceable to the efforts of Rudolf Buchheim (1820–1879), born in Saxony, the son of a physician. Four years after graduating in medicine, he was promoted to the rank of full professor at the (Baltic) University of Dorpat, where he established the World's first pharmacological laboratory and attracted many brilliant young men to work in it. In 1867, he transferred to a comparable position in Giessen (Germany), where he remained until his death from a stroke.

Buchheim taught that the mode of action of drugs should be investigated by scientific means in order to introduce a more rational basis for therapy. This way of thinking was, at that time, revolutionary, yet he went on to write (Buchheim, 1872):

'If we translate our often obscure ideas about drug actions into an exact physiological language, this should without doubt be a considerable achievement. However, scientific recognition of the action of a given drug would imply our ability to deduce each of its actions from its chemical formula'.

Buchheim also pressed for the study of metabolism and of statistical methods to raise pharmacology to the level of an exact discipline, equal in status to chemistry and physiology. He placed special emphasis on experimentation with simple models.

Although no great discovery can be connected with Buchheim's name, he was a great teacher and introduced into pharmacodynamics the methods that were essential for its later achievements. A forerunner whose writings may have influenced his early thinking was the French physiologist François Magendie (1783–1855).

For further reading on Buchheim, see biographies by his disciple Schmiedeberg (1912), and by Habermann (1974).

7.0 Pharmacodynamics and chemotherapy compared

In pharmacodynamics at least three problems arise which have little or no counterpart in chemotherapy.

In the first place, pharmacodynamic results are usually required to be reversible. The patient who submits to an anaesthetic does not expect to be deprived of feeling permanently. In chemotherapy, on the other hand, the toxic action is most esteemed when it is most irreversible.

In the second place, pharmacodynamic drugs are expected to act with a *graded* response. In proportion to the severity of a spasm or excessive secretion, so should various doses of the remedy exactly neutralize what is morbid without inflicting upon the patient any further loss of function. The reverse of the graded response, namely the all-or-nothing effect so desirable in chemotherapy, must be avoided in pharmacodynamics.

In the third place, the worker in pharmacodynamics finds his biological testing material harder to isolate in quantity and in a uniform condition. Ideally he should begin with the simplest system possible, namely the selectively toxic agent and a uniform population of the cells on which it is to act. Once these fundamental relationships have been explored, the natural complicating factors may be added gradually so as to enable the work eventually to serve a utilitarian end. Thus he should progress from effector cells to the tissue in which they occur, from these to the organ and, eventually, to the entire plant or animal.

However, the student of pharmacodynamics finds difficulty in studying cells in isolation. It is often not practicable to obtain a uniform population of these, undamaged and functional. Moreover, he may be interested in the communication mechanism (e.g. a synapse) between two different kinds of cells. But it is a sad fact that most pharmacological test-objects contain a high percentage of entirely extraneous cells which are rich 'sites of loss' (as defined in Section 3.4). This makes it difficult to obtain results that are significantly quantitative. Moreover, the tradition has been to begin most exploratory work in perfused organs or even with an intact animal.

Fortunately, from time to time some progress has been made in the direction of A.J. Clark's ideal: single-cell pharmacology. Thus the intra- and extracellular response of single nerve or muscle cells can be recorded while iontophoretic administration of a drug is being made into the extraneuronal environment (del Castillo and Katz, 1957; Katz, 1966).

In general, pharmacodynamic drugs would be expected to have less specificity than chemotherapeutic ones, because of the greater difficulty of discriminating between receptors *in the one organism* (especially as many of them happen to be receptors for a single neurotransmitter, acetylcholine). As a consequence of this, pharmacodynamic drugs often fail to reverse the disease process, but merely arrest or retard its progress; when this is the case, the prolongation of therapy is likely to be accompanied by long-term side effects.

From these considerations, it might be concluded that the study of chemotherapy is immeasurably simpler than that of pharmacodynamics. However, those who study chemotherapy have to perform pharmacodynamic work also in order to follow the side effects of their drugs on the host. Even drugs, such as piperazine and penicillin, which have no side effects at all in the average patient, nevertheless need to be studied pharmacodynamically to discover their pattern of distribution in the body, and the origins of their favourable selectivity.

7.1 Early history of the use of synthesis to find new drugs

The use of pharmacodynamic remedies had its remote origins in tribal practices, but even the great civilizations of antiquity knew of only a few drugs acceptable today (notably opium, ergot, and belladonna). Early in the seventeenth century, the pragmatic flavour of Baconian philosophy led to the publication of pharmacopoeias, books which listed the most valued drugs and set standards for them. Among the very first of these works the *London Pharmacopoeia* of 1618 contained several remedies that are still in use, but also many quaint and futile items such as the fat of dogs, eels, storks, and hedgehogs, the excrement of various animals, and stones from a patient's bladder. The first isolation of an alkaloid (morphine, by Sertürner* in 1804) turned attention to the value of obtaining pure active substances from the vegetable or animal mass in which they were commonly dispersed.

Chemical synthesis did not seem, at first, a likely source of useful remedies. However, after Wöhler's synthesis of urea, in 1828, a whole range of synthetic organic chemicals was introduced into medical practice. After several false starts, some relatively simple substances were adopted as inhalation anaesthetics, namely nitrous oxide, ether, and chloroform (1844–7) which, by giving the surgeon adequate time for his task, initiated a tremendous refinement and extension of the possibilities of surgery.

What is so odd, from our present-day point of view, is that these anaesthetics were first tried on Man, not on laboratory animals. Later it was found that the changes induced in Man could be reproduced in experimental animals and this altered the method of testing new substances as potential anaesthetics. For human trials, those compounds were selected which had shown maximal activity without obvious side effects in laboratory animals. This change of

*This bold young fellow, a repairer of pharmacists' stills and balances, describes flushing, vomiting, and coma in three men whom he persuaded to swallow 90 mg of his new crystals (Sertürner, 1806).

approach initiated a procedure which is now traversed in the search for all new remedies.

Between 1860 and 1905, organic syntheses were energetically pressed into the service of finding hypnotics, i.e. drugs to induce sleep in the sleepless patient. The first hypnotics were relatively simple substances, such as chloral hydrate (Liebreich, 1869) and paraldehyde (Cervello, 1882), and progressed in complexity to barbitone ('Veronal') (Fischer and von Mering, 1903). The immaturity of organic chemistry not only conditioned the choice of substances used as sedatives but, through them, it selected the type of sedation which came to be regarded as desirable (cf. McIlwain, 1957). How limited this type was, has been made clear by relatively recent discoveries of the kinds (and proportions) of sleep regarded as normal, and of drugs which sedate different regions of the brain quite selectively.

In the last quarter of the nineteenth century, a vigorous search for temperature-reducing drugs, with quinine as a vague model, brought to light a whole class of mild synthetic analgesics, such as acetanilide (Cahn and Hepp, 1887) which was followed in the next year by phenacetin (*3.24*). Sodium salicylate was introduced as a combined analgesic and anti-rheumatic in 1875 by Buss, and aspirin by Dreser in 1899.

The alkaloid cocaine, the first local anaesthetic, was introduced by Koller (with some help from Sigmund Freud) in 1884. Caught up in a contemporary urge to simplify natural products as described in Section 7.3, Einhorn(1905) produced the synthetic (and in many ways more useful) local anaesthetic, procaine.

Mention has been made (in Section 6.1) of how these pharmacodynamic discoveries led Ehrlich to seek simple synthetic remedies for infectious disease: this initiated his opening the subject of chemotherapy in 1899. For the next 50 years its rate of progress far outpaced that of pharmacodynamics, largely for the reasons given at the beginning of this section. Pharmacodynamics, in short, is much the more complex of the two divisions of pharmacology. Fortunately, in the last few decades, a more quantitative and molecular approach to pharmacodynamics has enabled it to forge ahead. This new progress is important in the more industrialized countries where most serious illnesses are non-infectious in nature (Section 1.1).

7.2 Some common molecular patterns in pharmacodynamic drugs

It is one of the curiosities of nature that many plants find it convenient to excrete waste nitrogen in molecules that have patterns which affect vertebrate nerve and muscle, although these two kinds of tissue are completely lacking in the plant kingdom. Alkaloids are typical waste products of plant metabolism, allowed to accumulate only in those parts of the plant which can easily be shed, such as the bark, leaves, and fruit. The majority of alkaloids are biologically inert in mammals. Of the 25 alkaloids of opium, only 4 show any effect on Man.

Inspection of the formulae of medicinal alkaloids shows that a certain pattern of atoms is repeated. This pattern consists of the tertiary amino-group connected by two or three (rarely four) saturated carbon atoms to another tertiary amino- (or to an alcoholic or ether) group, or to an unsaturated ring, for example (7.1), (7.2) and (7.3). This pattern occurs in lobeline, atropine, cocaine, quinine, strychnine, and morphine, to give a few of the many examples. These complex alkaloids have other groups and rings, to be sure, but the heart of their pharmacological action seems to reside in the above patterns which can, in whole or part (although this is immaterial) occur in a heteroaliphatic ring such as piperidine. The hydroxyl-group, when present, is often acylated. This pattern is somewhat reminiscent of those found in such potent mammalian nitrogenous agonists as acetylcholine (7.4), noradrenaline (norepinephrine) (7.5), adrenaline (epinephrine) and histamine (7.6).

$$R_2N \cdot CH_2 \cdot CH_2 \cdot NR_2 \qquad R_2N \cdot CH_2 \cdot CHR \cdot OH \qquad \text{(ring)}\text{---}CH_2 \cdot CH_2 \cdot NR_2$$

└── Pharmacologically active regions of alkaloids ──┘

(7.1)   (7.2)   (7.3)

$$\overset{O}{Me \cdot \overset{\|}{C} \cdot OCH_2 \cdot CH_2 \cdot N^+Me_3} \ Cl^-$$

Acetylcholine (chloride)
(7.4)

Noradrenaline (norepinephrine)
(7.5)

Histamine
(7.6)

Further, this pattern turns up often in contemporary pharmacodynamic agents, especially in local anaesthetics, anti-histamines, anti-emetics, tranquillizers, and in drugs intended to replace one or more uses of atropine. This is not a chance occurrence, because Fourneau and Bovet, working in Paris in the Institute Pasteur, began about 1935 to explore the pharmacodynamic effect of attaching this pattern of atoms to aromatic and heteroaromatic rings. The discovery of anti-histamines and neuroleptics. Two novel and highly valuable classes of drug, was an early reward for this inspired approach.

It may at first seem bold to suggest that all these substances, whether alkaloids, neurotransmitters, or the newer synthetic drugs, are acting on a few, chemically similar, receptors. The pattern itself allows of considerable variation without loss of a particular biological effect, e.g. local anaesthetics may be either type (7.1) or (7.2), the number of connecting carbon atoms may be two or three, and these may carry small substituents such as methyl-groups. This points to a conclusion that the receptors of vertebrates have not so great a specificity for their substrates as many enzymes have (Ariëns, 1960; Fastier, 1964).

In line with this conclusion is the fact that few pharmacodynamic drugs have *complete* biological specificity. They usually give results suggestive of combination with a single, fairly specific, receptor, but in addition show a number of side effects indicative of combination with other receptors. For example, it is reasonable to think that each of the following six properties of a drug would involve a different kind of receptor: local anaesthetic, anti-histaminic, spasmo-lytic, analgesic, anti-acetylcholine, and prolongation of the refractory period of the heart. Yet each of the seven following substances displays *all* of these properties (although each possess *one* of them in an enhanced degree): procaine, mepyramine, papaverine, pethidine (meperidine), atropine, quinidine, and sparteine. It has been suggested that the basis of these common properties is that such substances can antagonize acetylcholine, histamine, and possibly adrenaline too, in a number of tissues (Burn, 1950).

Although the specificity of receptors is not so strict as that of the most important anabolic and catabolic enzymes, it is at least as strict as the degrad-ative enzymes of microsomes (see Section 3.5). Thus at ganglia, nicotine (but not muscarine) can take the place of acetylcholine, whereas at postganglionic parasympathetic synapses, muscarine (but not nicotine) can take its place (see Table 7.1). Further specificity is shown in acetylcholine *antagonism*, for which tubocurarine is specific at the neuromuscular junction, hexamethonium at ganglia, and atropine at parasympathetic postganglionic synapses.

Although it seems that receptors have only a moderate specificity, it should still be possible to find specific drugs by (a) arranging their ionic (Chapter 10) and lipophilic (Sections 3.2 and 17.1) properties to give them the best possible access to the site, and (b) providing them with such blocking groups as sterically prevent their access to other sites (Fastier, 1964). At the same time, the structure should be kept as simple as possible to avoid side effects.

7.3 Simplification of the structure of natural products

Late in the last century, a beginning was made in the task of simplifying the molecular structure of natural products while retaining the therapeutic action. It was hoped in this way to obtain substances which could be more easily synthesized and which might be free from toxic side effects introduced by unwanted parts of the molecule. The introduction of salicylic acid into medicine in 1875 was one of the earliest results of this endeavour, and this acid and its salts and derivatives such as aspirin replaced the use of willow bark, and the glucoside of salicyl alcohol which this bark contained.

In the last century, many plant alkaloids were found to have powerful pharmacological actions. The simplification of the structure of cocaine (*7.7*) to give the local anaesthetic, procaine ('Novocain') (*7.8*), was the first notable success in modifying the structure of alkaloids to suit the combined needs of manufacturer and clinic (Einhorn, 1905). Procaine, unlike cocaine, does not penetrate mucous membranes but is non-addictive and lacks other side effects of cocaine. It at once achieved widespread use as a dental anaesthetic, and is still much used.

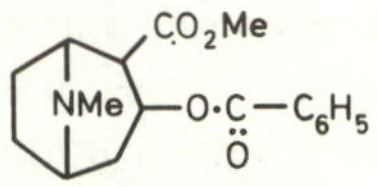

$$\text{Cocaine structure with } CO_2Me, NMe, O\cdot\overset{..}{\underset{..}{C}}-C_6H_5$$

Cocaine
(flattened formula)
(7.7)

$$Et_2N\cdot CH_2\cdot CH_2\cdot O\cdot \overset{..}{\underset{O}{C}}-C_6H_4(p-NH_2)$$

Procaine
(7.8)

7.3.1 Simplification of the molecule of cocaine

The guiding principle in these alkaloidal simplifications is that *saturated* heterocyclic rings are to be broken open (mentally, that is), a process which affects physical and chemical properties very little. Thus cocaine (7.7), which has a tertiary amine group in two saturated rings which bear an alcoholic group esterified with benzoic acid, was simplified to an isomer (7.9) of benzamine without loss of local anaesthetic properties. Next, the other saturated ring was opened to give 3-diethylaminopropyl benzoate (7.10). The basic strength, which had gone up through loss of the methyl ester-group of cocaine, was advantageously (see Section 17.2) lowered by deleting a methylene-group, thus restoring better penetration of the nerve plasma membrane. One final change led to procaine (7.8): an amino-group was inserted in the benzene ring to achieve an effective hydrophilic/lipophilic balance. It must not be supposed that the physical principles behind each of these changes could be clearly stated in 1905. Yet they must have been understood intuitively, because they were soon being applied to the molecules of other alkaloids.

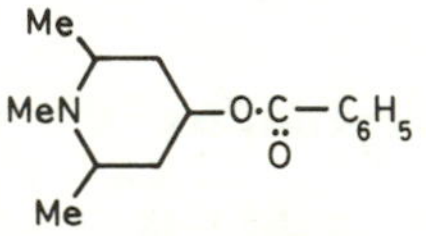

$$\text{Isobenzamine structure with } Me, MeN, Me, O\cdot\overset{..}{\underset{O}{C}}-C_6H_5$$

Isobenzamine
(7.9)

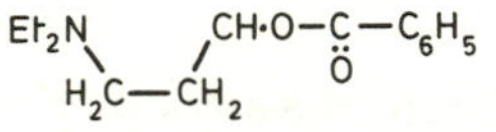

$$Et_2N, H_2C-CH_2, CH\cdot O-\overset{..}{\underset{O}{C}}-C_6H_5$$

3-Diethylaminopropyl benzoate
(7.10)

It is now known (Sinnema *et al.*, 1968) that the molecule of cocaine (7.11) has a very rigid three-dimensional conformation*, whereas that of procaine (7.8) is loosely jointed, which is no disadvantage, for it allows better adaptation to the receptor.

Later the structure of procaine was modified to obtain, for particular medical uses, a substance which could easily penetrate mucous membranes. A redistribution of lipophilic groups gave the much-used local anaesthetic, tetracaine (amethocaine, 'Decicain') (7.12) (these steps are reviewed by Barlow, 1968). It is now known that a tremendous number of chemicals have local anaesthetic

---

*A refresher on molecular structure: when every second bond is a double bond (e.g. benzene) the molecule is flat and rigid. A saturated ring (no double bonds), as in cocaine, is three-dimensional and fairly rigid. A saturated aliphatic molecule without a double bond (e.g. ethanol) is three-dimensional and flexible. For conformation, see Section 12.3.

activity. The ester-group can be replaced by an amide-group, or even a reversed amide-group as in lidocaine (*7.13*) (Lignocaine, 'Xylocain'). A feature of esters is that they are soon hydrolysed by serum esterases, so that the duration of the anaesthesia is self-limited, whereas the amide type is valued when prolonged action is required. In recent years, lidocaine has become the most used anaes-thetic for dental and mucous membrane work.

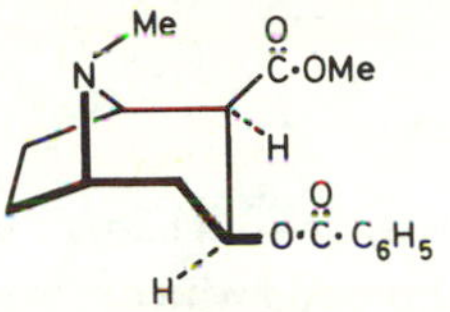

$$Me_2N \cdot CH_2 CH_2 \cdot O \cdot \overset{\text{O}}{\underset{\text{O}}{C}} - C_6H_4(p-NHBu)$$

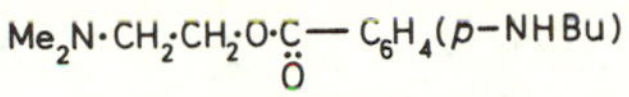

Tetracaine (amethocaine)
(*7.12*)

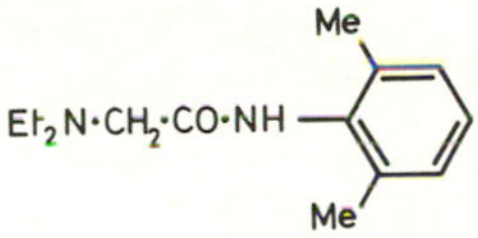

Cocaine
(conformational formula)
(*7.11*)

$$Et_2 N \cdot CH_2 \cdot CO \cdot NH -$$ [Me, Me benzene ring]

Lidocaine (lignocaine)
(*7.13*)

Quite early in the quest for a simplified cocaine, the simplest structure of all was discovered in benzocaine (ethyl *p*-aminobenzoate) (*7.14*). Too insoluble to be injected, it has found a permanent place in anaesthetic therapy as a constit-uent of dusting powders for inflamed areas. Although so insoluble in water, it is soluble in oils, and this has opened up a second area of application. The fact that it is incapable of ionization at physiological pH values, ensures that its anaes-thetic action is weak (see Section 10.5).

Another amide-type local anaesthetic (cf. lidocaine) is cinchocaine (dibu-caine, 'Nupercaine') (*7.15*) one of the most used spinal anaesthetics. It is 2-butoxy-*N*-[2-(diethylamino)ethyl]quinoline-4-carboxamide.

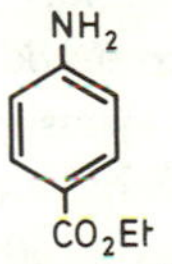

NH_2 ... CO_2Et

Benzocaine
(*7.14*)

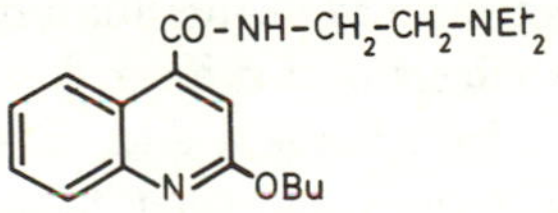

$CO - NH - CH_2 - CH_2 - NEt_2$ ... N ... OBu

Cinchocaine (dibucaine)
(*7.15*)

7.3.2 *Some general observations on simplification*

Before describing other simplifications, two important points must be made. The natural product being simplified is most often an antagonist. In that case, too great a simplification must be avoided, otherwise agonistic effects will appear. This follows from a general principle to be outlined in Section 7.5.2. A further caveat against undue simplification is that, shorn of sterically hindering groups, a drug that acted on only a couple of receptors could be freed to act on others also.

The greatest benefit conferred by simplification of the cocaine molecule was that every product was completely free from the euphoric property of cocaine and the addiction that this euphoria facilitates.

7.3.3 *The simplification of atropine*

Whereas in the simplification of cocaine the investigators had a difficulty [they did not know (and we still do not know) the receptor on which it was intended to act], no such problem existed for atropine which antagonizes the action of acetylcholine at the postganglionic receptors, but not at the ganglia, nor at the voluntary nerve—muscle junctions. In other words, it is a typical anti-muscarinic drug.

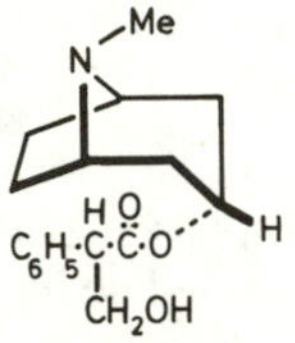

Atropine
(conformational formula)
(7.16)

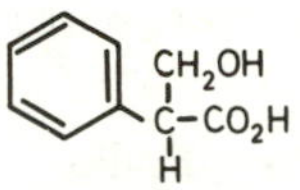

Tropic acid
(7.17)

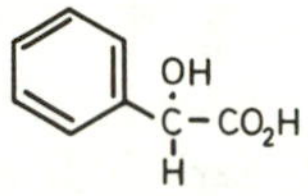

Mandelic acid
(7.18)

The simplification of atropine (DL-hyoscyamine) (7.16) was begun at the start of this century. Although both atropine and cocaine have a tropine ring system, the conformations are different, as can be seen by comparing (7.11) and (7.16) (Fodor, 1960). Many attempts were made to prepare a stripped-down analogue that would conserve the typical anti-muscarinic action while confining it to selected parts of the human body. It was hoped to overcome the unpleasant side effects of this drug: loss of visual accommodation, dryness of mouth, difficulty in urination, and, in susceptible subjects, CNS stimulation leading to sleeplessness, and even hallucinations or behavioural problems. It was realized that (as with cocaine) not just one simplified product was required, but a set of these, each with its own specialized use. Thus, the loss of visual accommodation, through cycloplegia, is exactly what the ophthalmologist requires, and dryness of the mouth becomes a goal in treating excessive salivation.

It was soon found that any simplification of atropine's esterifying acid (tropic acid) (7.17) diminished the activity. By changing this to mandelic acid (7.18), a cycloplegic agent (homatropine) was obtained with a shorter, more

practical duration of muscular paralysis which greatly helped both ophthalmologist and patient.

Next, the bicyclic ring system was broken open, as with cocaine, and
simplified to diethylaminoethanol which is the simple choline derivative that
forms part of the procaine molecule (7.8). At this stage, it was found that local
anaesthetic properties were appearing, a state of affairs that was remedied by
increasing the molecular weight of the esterifying acid. This gave us the
pharmacopoeial drug dicycloverine (diclomine) (7.19) much used to relieve
spasm and hypermotility of the colon. Here the esterifying acid is dicyclo-
hexanoic acid and similar use was made of diphenylacetic acid, phenylcyclo-
hexylacetic acid, and related substances. The action of dicycloverine on the gut
is not purely anti-muscarinic but also has a direct spasmolytic effect on the
muscle. To obtain a purely anti-muscarinic substitute for atropine, it was
necessary to make the aminoalcohol component a little more complex, as in the
pharmacopoeial drugs, piperidolate (7.20), which is 1-ethyl-3-piperidyl
diphenylacetate, and the related oxyphencyclimine (7.21). These are much used
as intestinal sedatives.

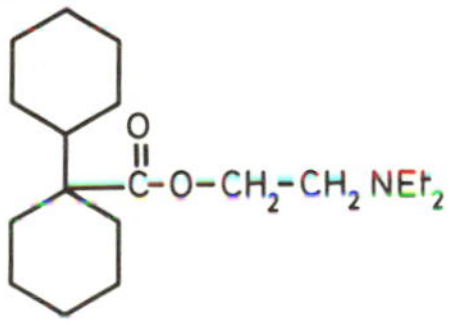

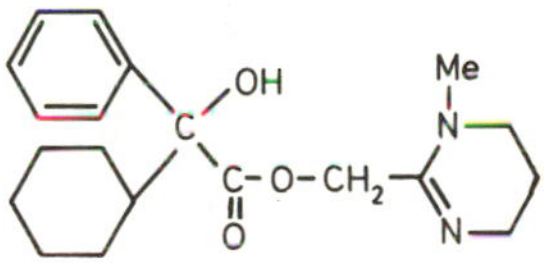

Dicycloverine (dicyclomine)   Piperidolate   Oxyphencyclimine
        (7.19)   (7.20)   (7.21)

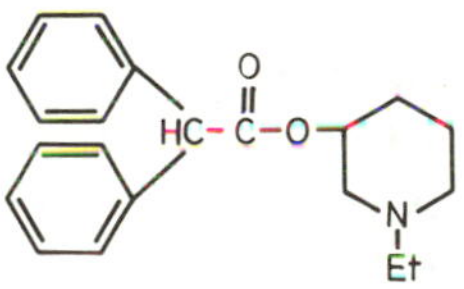

Two specialized products for producing mydriasis in the eye are tropicamide
(7.23) and cyclopentolate (7.22). They are weaker than homatropine and do not
give a full cycloplegia. Trihexyphenedyl (7.24) (benzhexol, 'Artane') is a
simplified atropine adapted for use in the central nervous system where it plays
a supportive role in the treatment of parkinsonism with levodopa. The ultimate
in the manipulation of the atropine molecule is quarternization of the ring-
nitrogen atom to obtain some ganglion-blocking action (in addition to the
anti-muscarinic action), an extra effect that is not prominent in low doses
(Barrett et al., 1953). Typical of this class of substance is propantheline (7.25)
which is di-isopropyl methyl 2-(xanthen-9-ylcarboxyloxy)ethyl ammonium
bromide, a powerful drug much used as a gastrointestinal sedative. Here, the
xanthene nucleus simply provides another kind of high-molecular-weight
carboxylic acid for the esterification. In fact, all these specialized, simplified
atropines are built on the principles described above, modified only to influence
distribution.

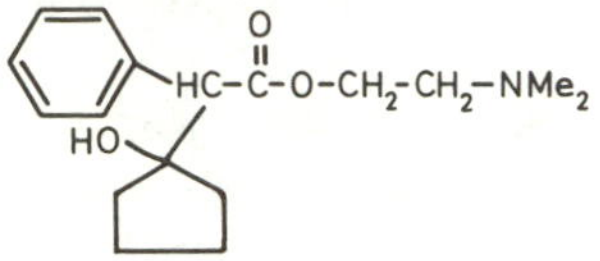

Cyclopentolate
(7.22)

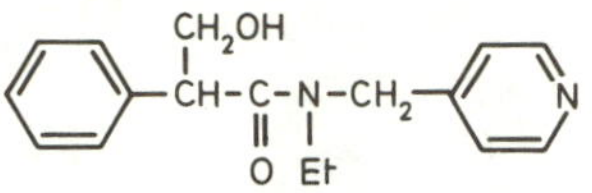

Tropicamide
(7.23)

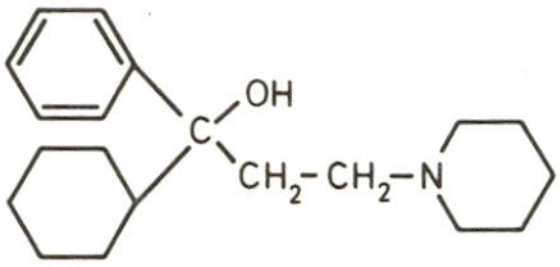

Trihexyphenedyl (benzhexol)
(7.24)

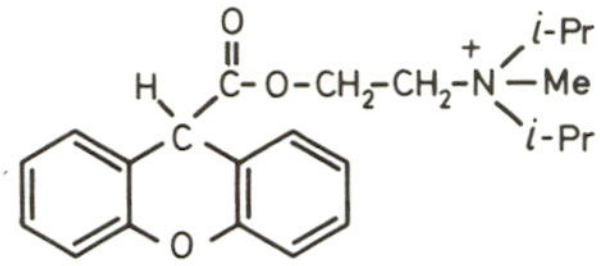

Propantheline
(7.25)

7.3.4 *The simplification of nicotine* (7.26)

This is one of the few alkaloids with agonistic properties. It imitates the neurotransmitter action of acetylcholine at both the ganglionic and the voluntary neuromuscular synapses (Table 7.1). It also has an action on the central nervous system that is both euphoric and addictive (as in cigarette smoking). For this reason, it is not used in human medicine, but its insecticidal action has been used in agriculture and the fumigation of flour mills against moths. The nature and potency of its action are hardly changed by quaternization (further methylation) of the nitrogen in the saturated (pyrrolidine) ring. When the quarternary salt was modified by opening the saturated ring, as was done for cocaine, the product, and its homologues, had typical nicotine-like activity, e.g. at the neuromuscular junction (rat and chick). Substance (*7.27, n* = 1) was as potent as nicotine, and substance (*7.27, n* = 2) was 2.6 times as potent (Barlow and Hamilton, 1962).

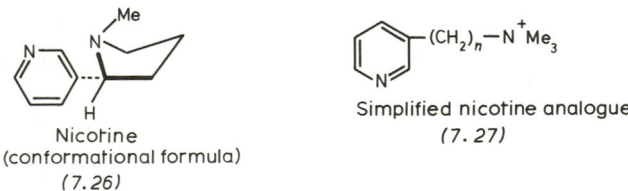

Nicotine
(conformational formula)
(7.26)

Simplified nicotine analogue
(7.27)

7.3.5 *Attempts to simplify the molecule of tubocurarine* (2.6)

Attempts to simplify tubocurarine were made while it was still thought to have two quarternized amino-groups, whereas it has in fact only one (Everett, Lowe and Wilkinson, 1970). In spite of this misconception, some very useful drugs were developed. It was related in Section 2.0 how Crum Brown and Fraser (1869) had shown that various alkaloids acquire curare-like properties by methylation to quaternary ammonium bases. Even simple aliphatic quaternary amines, so long as they possess at least one methyl-group, have this curare-like action, but they are far too weak to be useful adjuncts to surgical anaesthesia. Reviewing the evidence, Ing (1936) came to the conclusion that tubocurarine (*2.6*), and similarly acting quaternary amines, worked by blocking the action of acetylcholine on the muscle end-plate of the neuromuscular junction.

In 1948, many substances with two or more quaternary nitrogen atoms were synthesized and some were found to produce neuromuscular block, e.g. decamethonium (*7.28*) (decanebistrimethylammonium iodide) (Paton and Zaimis, 1948; Barlow and Ing, 1948). Close examination showed that this substance not only acts durably as an acetylcholine blocking agent (as tubo-curarine does), but first acts momentarily as an acetylcholine analogue, causing depolarization which is easly demonstrated with non-polarizable silver elec-trodes (Burns and Paton, 1951). Suxamethonium (*7.29*) also has this type of action (Bovet and Bovet-Nitti, 1949). After brief muscular contraction, pro-longed relaxation follows the use of these drugs because they are strongly adsorbed and hence prevent newly liberated acetylcholine from reaching its receptors. Decamethonium was used for some years to secure muscular relax-ation during light anaesthesia, but suxamethonium is preferred because its effect is self-terminating thanks to hydrolysis by esterases (in the serum of normal patients). There is no antidote for these depolarizing-curarizing agents whereas the action of tubocurarine or gallamine can be promptly terminated by neostigmine. The disulphide (*7.30*) is another ingeniously self-terminating agent, the disulfide bond being easily broken by cysteine (Khromov-Borisov, Gmiro and Magazanik, 1969).

$$I^- \{Me_3\overset{+}{N}-(CH_2)_{10}-\overset{+}{N}Me_3\}\, I^-$$

Decamethonium iodide
(*7.28*)

$$Me_3\overset{+}{N}-(CH_2)_2-O-\overset{O}{\overset{\|}{C}}-CH_2$$
$$Me_3\overset{+}{N}-(CH_2)_2-O-\underset{O}{\overset{\cdot\cdot}{C}}-CH_2$$

Suxamethonium (cation)
(*7.29*)

$$Me_3\overset{+}{N}-\langle\bigcirc\rangle-S-S-\langle\bigcirc\rangle-\overset{+}{N}Me_3$$

Muscle relaxant from USSR
(*7.30*)

O—CH₂·CH₂·N⁺Et₃
O—CH₂·CH₂·N⁺Et₃

O—CH₂·CH₂—N⁺Et₃

Gallamine (cation)
(*7.31*)

$$Me_3\overset{+}{N}-(CH_2)_6-^+NMe_3$$

Hexamethonium (cation)
(*7.32*)

The simplest clinically successful substance with a true tubocurare-like action is gallamine (*7.31*) (Bovet, 1947). Interestingly, it has no *N*-methyl group. For a muscle-relaxing drug, as used during general anaesthesia, current practice is favouring the self-terminating drug, atracurium (*3.19*), described in Section 3.6. Decamethonium now has only historical significance.

In the course of investigating the homologous series of bisonium compounds which led to the discovery of decamethonium, *hexamethonium* (*7.32*) was found to block the receptors for acetylcholine within autonomic ganglia, a type of activity which tubocurarine shows only weakly. At first these bisonium ganglion-blocking drugs were used clinically to treat hypertension, but drugs which block the sympathetic nerve endings, e.g. propranolol, give fewer side effects and are now preferred.

7.3.6 *The simplification of physostigmine*

This alkaloid (*7.33*), also known as eserine, blocks the enzyme acetylcholinesterase (Section 12.5) at nerve–nerve and nerve–muscle synapses (see Table 7.1). Stedman (1926) came to the conclusion that the methylcarbamate-group of physostigmine was responsible for its action, and he esterified simple derivatives of phenol with (mono- and di-)methylaminoformic acid. The presence of a basic group, ionized at pH 7, was found to be necessary also, and so he took the only possible way of increasing the low basic strength of an aromatic amino-group, namely by quaternization. The pharmacological results led to the introduction of neostigmine ('Prostigmine') (*7.34*) as a substitute for physostigmine (Aeschlimann and Reinert, 1931). It is free from some of the side effects of the latter and has, in addition, a small ACh agonist action. It is much used in the treatment of myasthenia gravis, a disease characterized by muscular weakness, and in overcoming postoperative bladder atony.

Many simpler carbamates, as well as organic phosphates, which have this property of blocking acetylcholinesterase, are used in treating glaucoma, also as anthelmintics and insecticides (see Section 13.3).

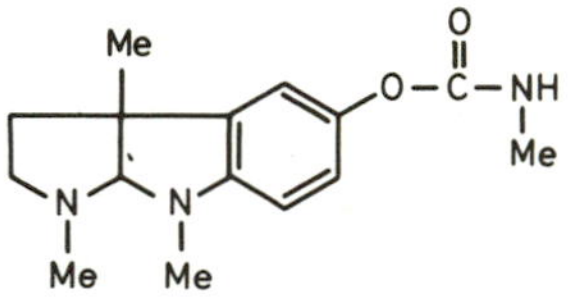

Physostigmine (eserine)
(*7.33*)

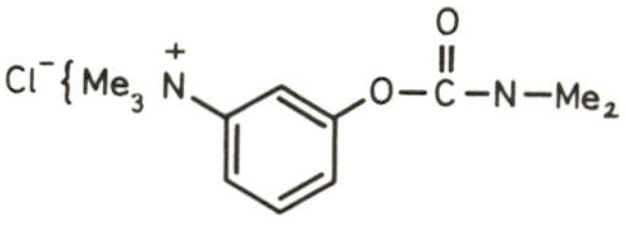

Neostigmine
(*7.34*)

7.3.7 *The simplification of morphine*

Morphine (*7.35a*) is an alkaloid that has the morphinan (*7.36*) skeleton. The molecule has the shape of a strap sandal, with the piperidine ring forming the strap. Formula (*7.36*) places the phenanthrene (three fused hexagons) ring system in the plane of the paper whilst the piperidine ring, denoted 9,14,13,15,16,17, rises towards the viewer. A simplified analogue of morphine was sought in vain until *pethidine* (*7.37*) was found by Eisleb and Schaumann (1938) while searching, in Germany, for an atropine-type anti-spasmodic. Commentators found two features in common between morphine and pethidine: the piperidine ring, and the quaternary carbon atom numbered 13 in the morphinan nucleus.

Pethidine (known as meperidine in the USA) has only one-tenth of the analgesic action of morphine but, unlike the latter, is well absorbed orally. It lacks the cough- and diarrhoea-suppressing actions of morphine; the addictive properties of the latter are still there. One of the principal contemporary uses of pethidine is in childbirth where it causes less respiratory depression at equi-analgesic doses.

This remarkable discovery made it clear that the molecule of morphine could be drastically simplified without qualitative loss of action. Tradition indicated that the saturated heterocyclic ring should be opened. During the Second World War, unnamed scientists at the Hoechst company in Germany discovered methadone (*7.38*), a non-heterocyclic analgesic, which is quantitatively as effective as morphine when given parenterally and more active than morphine orally. It lacks only morphine's ability to control diarrhoea. Methadone is much used, today, as an opioid analgesic, and also, although it is itself addictive, in treating addiction to heroin. The history of the discovery of the synthetic opioids is related by Bergel and Morrison (1948).

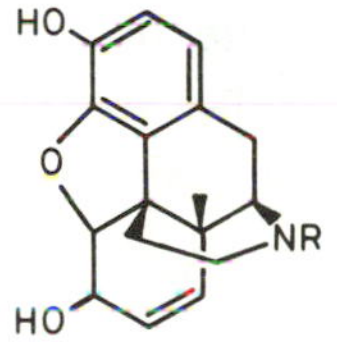

(a) Morphine (R = −Me)
(b) Nalorphine (R = −CH₂CH : CH₂)
    (*7.35*)

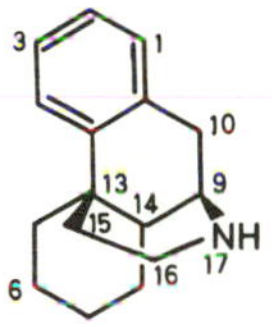

Morphinan
(*7.36*)

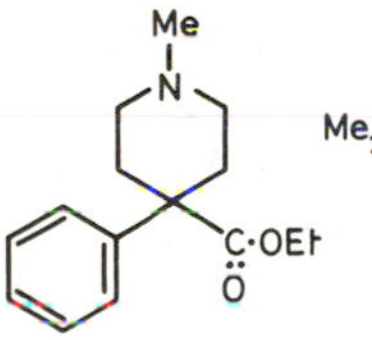

Pethidine (Meperidine)
(*7.37*)

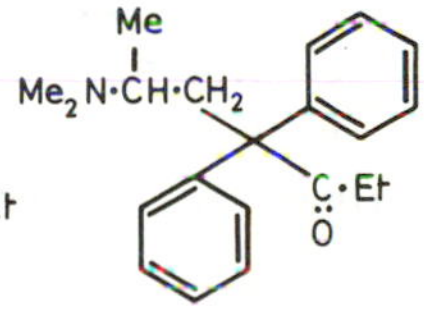

Methadone
(*7.38*)

The success of methadone sparked off a search for a simplified *codeine*. The latter, the 3-methyl ether of morphine, is a mild analgesic, midway in action between morphine and aspirin. It has only one-twelfth of the activity of morphine, given parenterally. It is usually administered orally to relieve cough or visceral pain. Propoxyphene (*7.39*) ('Davron') is much used in the United States as a codeine replacement.

Returning to methadone (*7.38*), this seems to have no more in common with the structure of morphine than the prominent quaternary carbon atom. But even this feature is not essential for opioid activity because a tertiary nitrogen atom (attached to a benzene ring) can serve the same purpose (presumably to create molecular congestion). An example is etonitazine (*7.40*) which closely imitates the analgesic action of morphine, but is 1500 times as potent (Gross and Turrian, 1957). It is also highly addictive.

An important step in the separation of analgesic from addictive properties in morphine was made by replacing the *N*-methyl group by an allyl group, giving nalorphine (*7.35b*), allylnormorphine. This turned out to be highly analgesic and non-addictive, but was of little clinical value because it stimulated the central nervous system, often producing delirium. Surprisingly, it acted as an antagonist in cases of morphine poisoning. This paradoxical agonist–antagonist action of nalorphine began to be understood in 1975 when it was found that morphine was an agonist, acting on bodily receptors normally reserved for some polypeptides (see, further, Section 12.8). Not long after this, it became evident that there are three kinds of these receptors, that morphine is an agonist at only two of them, and nalorphine at only two, but only one kind of receptor is shared for analgesic action.

Knowledge that euphoria (the prerequisite for addiction) could be suppressed by an allyl group on the ring-nitrogen atom led to the discovery of pentazocine ('Talwin') which makes use of a similar loading group attached to a simplified nucleus (Bellville and Forrest, 1968). This drug, N-methylbutenyl-5,9-dimethyl-2'-hydroxy-6,7-benzomorphan [benzomorphan is (7.41)], has parenterally about one-quarter of the analgesic effect of morphine. Its lowered addictive potentiality is bought at the risk of some central nervous system stimulation (e.g. uncontrollable, racing thoughts).

Pursuing these lines of work led to the introduction of butorphanol, which is 9-cyclobutylmethylmorphinan-3,14-diol [see (7.36) for morphinan]. It has about five times the analgesic action of morphine, when given parenterally, and the same duration of action. It is less addictive than morphine. It has adverse effects on the central nervous system like those produced by pentazocine, but these are less per equi-analgesic dose (Dobkin, 1975).

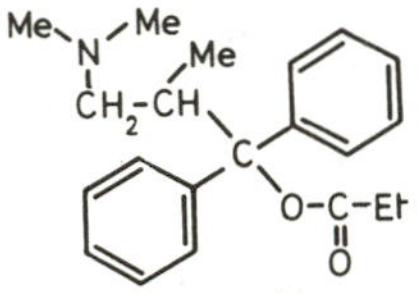

Propoxyphene
(7.39)

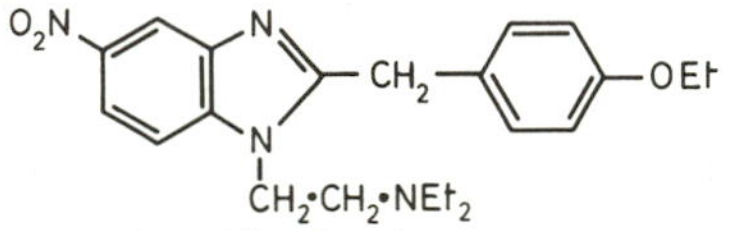

Etonitazine
(7.40)

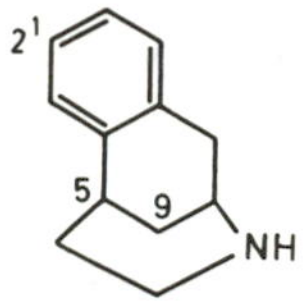

6,7-Benzomorphan
(7.41)

Dextromethorphan, a simplified *codeine*, is a much-prescribed cough suppressant, with neither analgesic nor addictive properties. Chemically, it is (+)-3-methoxy-N-methylmorphinan. Its laevo stereoisomer has the analgesic and euphoric properties of morphine and (although not used in medicine) stands at the end of a line of work initiated by Grewe (1947) who was the first to show that the morphinan series can produce strong analgesics.

7.3.8 *Other examples of simplification*

It is interesting to consider the enkephalins as Nature's simplification, for certain purposes, of the endorphins (Section 12.8). For his part, Man has successfully shortened some of Nature's bioactive *polypeptides*, as in the deletion of the first three of its amino acids from the undecapeptide neurotransmitter, Substance P. Again the synthetic pentapeptide, pentagastrin, is used to stimulate gastric secretion in place of the natural heptadecapeptide, gastrin. Moreover, the potent blood-pressure-reducing monoamide, captopril, was devised by simplifying a natural nonapeptide (see Section 9.4 for all examples).

The convulsant alkaloid, *strychnine*, which has seven rings, is exactly imitated, in its ability to act as antagonist at the glycine receptor and hence cause convulsions, by the three-ringed fragment (*7.42*) (Hershenson *et al.*, 1977).

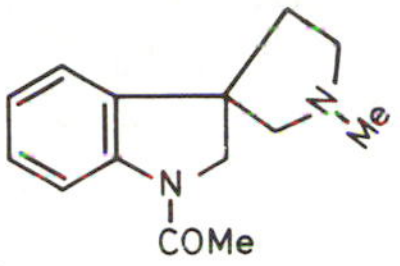

'Simplified strychnine'   Epinephrine (adrenaline)   Tuaminoheptane U.S.P.
(7.42)   (7.43)   (7.44)

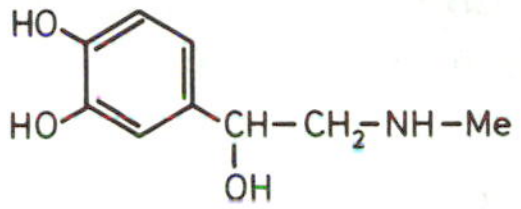

The anti-malarial alkaloid, *quinine*, when deprived of both its heterocyclic rings, furnishes molecules that retain high anti-plasmodial activity (see Section 10.3.5).

The four-ringed *cardiac glycosides*, such as digitoxin, have been deprived of the lactone ring without loss of activity (Section 14.1).

Simplification of the female *sex hormone*, the steroid estradiol (*12.32*), gave the widely employed diethylstilbestrol (*12.31*) (see Section 9.4.2).

It might hardly be expected that the benzene ring of *epinephrine* (*7.43*) could ever be discarded, but *tuaminoheptane* (*7.44*) is a sympathomimetic drug with strong alpha-adrenergic activity, free from stimulation of the central nervous system. It is used as a nasal decongestant.

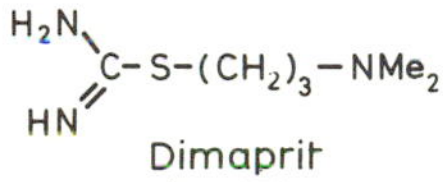

Dimaprit
(7.45)

The imidazole ring of *histamine* (*7.6*) was replaced by isothiourea to give *dimaprit* (*7.45*), which is a highly specific agonist for the H-2 receptor of the stomach on which it has 20% of the activity of histamine (Durant, Ganellin and Parsons, 1977). Higher and lower methylene homologues are almost inactive. It is used as a diagnostic agent to stimulate gastric secretion.

7.4 Recognition of the importance of measurement

The standardization and quantification of technique, and the rigorous application of statistics to the results has, in the last four decades, provided pharmacodynamics with a solid base such as it had not previously known. The use of experimental and control groups in tests on laboratory animals followed by similar tests on the human subject, have made it possible to evaluate the efficacy and side effects of drugs reliably, and also to make meaningful comparisons with established remedies.

No less revolutionary in the quantification of pharmacodynamics was the insistance on *size relationships* which Alfred Joseph Clark (London) introduced in his book, *The Mode of Action of Drugs on Cells* (Clark, 1933). The mnemonic diagram that forms the frontispiece of the present book will serve to introduce this aspect. In it we see a typical mammal (dog) followed by a typical insect (flea) which is a thousand times smaller, and this is followed by a typical microbe (streptococcus) which is yet a thousand times smaller, and finally a typical bioactive molecule (*p*-aminobenzoic acid) which is one thousand times smaller still. Thus we have familiar reference objects staked out over a size-range of 10^{12} (a million-millionfold).

To proceed further, a molecule of muscle protein, if unfolded, would be about 8 nm long. This could be compared with the length of a muscle cell (about 60 μm). If a model of some common pharmacodynamic drug, one with a molecular weight in the usual range of 150–300, is built from a molecular model kit (in which 0.1 nm is reproduced as 1 cm), the molecule of the drug will seem to be about 2 in (5 cm) long, the molecule of protein 10 yd (9 m) long, and the muscle cell 3½ miles (5.5 km) long by 200 yd (180 m) wide. These size relationships show that many useless collisions are likely to take place before a drug hits the desired receptor, and that the drug has many opportunities to land on a 'site of loss' (defined in Section 3.4). Nevertheless, given sufficient specificity between drug and receptor, the receptor is eventually reached. By no means alone among drugs, the natural agonists, epinephrine, histamine, and acetylcholine show strong pharmacological effects on test-organs at dilutions as great as 10^{-9} M (Clark, 1933).

Calculations of this kind help us realize what a prodigious number of molecules are present in one microgram of a drug (namely three thousand million million, assuming a molecular weight of about 200). It would need only four million molecules to cover a staphylococcus which has a surface area of about 2 μm². Moreover, to *cover* a cell is unnecessary for pharmacodynamic purposes. The amount of acetylcholine, epinephrine, and histamine taken up by frog heart, frog stomach, and rat uterus (respectively), when the organs are markedly inhibited by these hormones, corresponds to about 10^{14} molecules per gram of moist weight. This quantity of hormone can cover only 1 cm². The surface area of the cells in 1 g of frog's heart is about 6000 cm², and hence these hormones are active when only about 1/6000 of the surface is covered. Historically, this provided early evidence that they act by uniting with certain specific receptors, which form only a minute fraction of the total cell surface (Clark, 1933).

It is rewarding to apply these simple calculations to experimental data, in order to build a mental picture of the proportion of drug to target. For example, the lowest effective concentration for acetylcholine on the frog's heart was found to be a 5×10^{-19} M solution (Boyd and Pathak, 1965). This is equivalent to only 330 molecules per ml and must be approaching the very limit of physiological detectability. Actually Hahnemann, the founder of homeopathy, claimed therapeutic effects from various drugs at a dilution of 1 in 10^{60}. This claim strains credulity because it corresponds to one molecule in a sphere of circumference equal to the orbit of the planet Venus (Clark, 1933).

An example of how a test-preparation can be refined, so as to present fewer sites of loss for the drug, is afforded by the neuromuscular junction. Until recently, this had to be studied in an isolated 'nerve–muscle' preparation (such as the rat phrenic nerve diaphragm) which contained much extraneous material. But in 1957, it was found that drugs could be applied iontophoretically, from a point-source, right on to the motor end-plate of a frog's sartorius muscle. This close-range application of acetylcholine set up transient potential changes of several millivolts (del Castillo and Katz, 1957). This proved to be the best model system for obtaining correlations between structure and activity of drugs at the neuromuscular junction.

New aspects of structure–activity relationships came into view when new techniques enabled a further advance in measurement to be made. For example, Merritt and Putnam (1938) discovered the clinically useful anti-epileptic drug phenytoin (*15.4*), by first seeking a substance capable of suppressing electrically induced convulsions in laboratory mammals. This discovery was a great advance in anti-epileptic therapy because phenytoin is free from hypnotic action. Hitherto it had been thought that anti-epileptics must cloud the consciousness as phenobarbitone did. This work gave much encouragement to the use of physical methods of measurement in seeking new remedies.

How dose–effect relationships came to be accurately measured and judged is outlined in Section 6.2.

7.5 How agonists and antagonists act on receptors

The concept of a 'receptor' was explained in Section 2.1. For most chemotherapeutic agents it has been satisfactory to assume that the observed physiological response occurs as soon as the agent becomes adsorbed on the receptor, and that (short of death of the cell) the response continues so long as the agent remains adsorbed. This is usually satisfactory also in pharmacodynamics. Several pharmacodynamic drugs, e.g. organic phosphorus derivatives, are known to act by inhibiting identifiable enzymes, and often the very atom with which a drug combined covalently in such an enzyme has been identified (see Section 13.3). When, as is more usual, the combination is not covalent, combination follows the law of mass-action and can generally be described by the same relationships as are used in the Langmuir adsorption isotherm (see Section 8.1, also, for an early example: Gaddum, 1926). In all these cases, the response of a tissue seems to depend on the proportion of its specific receptor-groups occupied by the drug. This assumption can be referred to as *simple occupation theory*. The action of most *inhibitors* can be fairly well described by this theory.

However, drugs which *activate* receptors are more complex in their actions.

The structural requirements for these activators are much more specific than those for antagonists. Several hypotheses have been evolved which will be discussed a little later in this section. As there can be no action without expenditure of the chemical energy that is stored in Nature as complex phosphates, agonists work by liberating energy from energy stores, even if only indirectly.

7.5.1 *Chemical and electrical conduction compared*

At least 80% of known pharmacodynamic agents act exclusively on nerve or at nerve–muscle junctions. Hence it is desirable to preface further discussion with a brief account of the propagation of an impulse along nerves, and across synapses. Nerve cells (*neurons*) consist essentially of a central, roughly spherical portion, which is drawn out into fibres, one of which (the *axon*) is much longer than the others. At the junction with a muscle fibre or a second nerve, the axon is swollen into a synaptic *terminal* which lies about 20–400 nm from specialized receptors on the muscle or nerve with which it has to communicate. Thus, at the synapse, there is a gap (*synaptic cleft*) between pre- and post-synaptic structures. Neurons of similar function are often collected together and such groupings are called *ganglia*. Fig. 7.1 shows, diagrammatically, a nerve–nerve synapse and a nerve–muscle junction which are essentially similar. For a general article on synapses, see Eccles (1965), and for a book, Katz (1966).

The main cation inside cells is potassium (K^+), and that outside is sodium

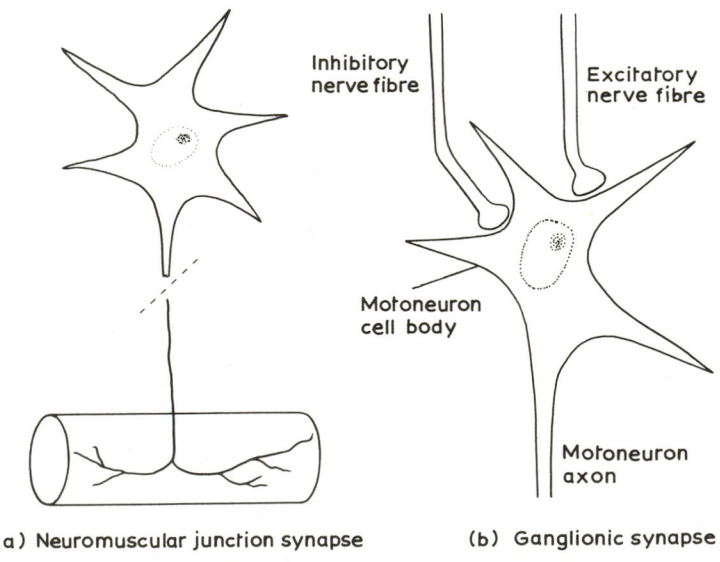

(a) Neuromuscular junction synapse   (b) Ganglionic synapse

Fig. 7.1 Junction between (a) a nerve and a muscle, and (b) three nerves

(Na^+). Because of the difference in ion concentrations across the cytoplasmic membrane of nerve and muscle cells, there is usually a resting negative potential of 50 to 100 mV (compared to the outside, defined as 0V). It is as though the membrane were a miniature battery with the negative terminal on its inside. When it loses this potential, it is said to be depolarized. This lipoprotein membrane permits *active* transport (Section 3.2) in both directions. It is generally accepted that the enzyme, Na^+/K^+-ATPase, is the energy transducer that facilitates the transport of Na^+ out of, and K^+ into, the cell against the electrochemical gradients of these ions.

The passage of a nerve impulse along a nerve fibre is a complicated physico-chemical event which follows upon the depolarization of the fibre at one end, either within the nervous system (for a motor fibre) or at the peripheral sense organ (for a sensory fibre). This change in potential brings about a transient increase in the membrane's permeability to sodium ions, which pass rapidly into the nerve fibre, thus reversing the resting level of membrane potential. This process is immediately followed by a transient increase in the permeability to potassium ions which pass out of the fibre, thus restoring the membrane potential. This sequence of changes is confined to a very small area, but the associated ionic currents depolarize the next portion of the membrane which goes through the same sequence of events. In this way the impulse, which is essentially this transient reversal of membrane potential, is propagated along the fibre at a rate that depends upon the fibre diameter. The conduction velocity varies from 0.1 to 100 m s^{-1} (Hodgkin, 1964).

More recently, it has been shown that cation channels in the squid's giant axon can be blocked by hydrogen ions. From the voltage needed to overcome this block, it was calculated that passage of Na^+ through its channels is controlled by two acids, of pK_a 4.6 and 5.8 respectively (Wanke, Carbone and Testa, 1980).

The rapid entry of sodium can be selectively prevented by tetrodotoxin (*7.46*), a spherical perhydroquinazoline molecule from which a highly basic guanidinium group protrudes like a tongue. It occurs naturally in various species of fish, amphibia, and molluscs. By excluding sodium ions, tetrodotoxin blocks the generation of a potential (and hence of transmission) in both nerve and muscle. Because transmission at synapses is unaffected, mammals die by a purely peripheral toxicity, usually by respiratory paralysis (Narahashi, Moore and Scott, 1964). Potassium transport is unaffected.

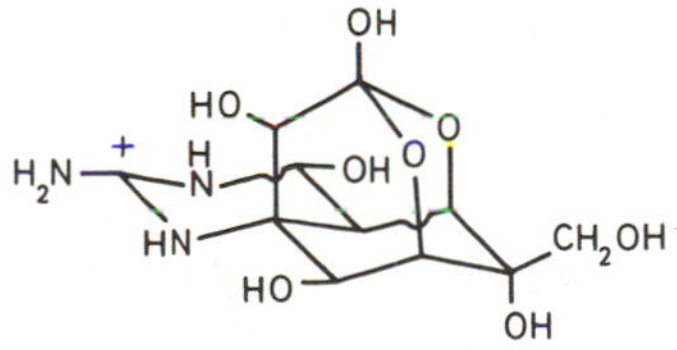

Tetrodotoxin
(7.46)

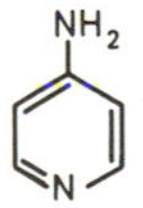

4-Aminopyridine
(7.47)

As guanidinium is one of the few cations that can replace sodium ions in the production of action potentials (Watanabe *et al.*, 1967), it is reasonable to suppose that the guanidinium group of tetrodotoxin enters a sodium channel which the rest of the molecule then blocks. Saxitoxin, a perhydropurine with two guanidinium groups (obtained from unicellular ocean dinoflagellates), acts almost identically. For a general review on tetrodotoxin (and the related saxitoxin), see Gage (1971), and for the chemistry, see Goto *et al.* (1965).

The rather strong heterocyclic base, 4-aminopyridine (*7.47*) similarly blocks potassium channels thereby enhancing the influx of calcium ions during the depolarization of nerve terminals (Thesleff, 1980). Substances that can block calcium channels in muscle (e.g. verapamil, nifedipine) were introduced during the last decade as vasodilatory drugs (see Section 14.2).

Returning now to the normal firing of a nerve, it should be noted that, inevitably, the passage of a single impulse leaves behind a small elevation of the intracellular $Na^+:K^+$ ratio, which is eventually restored to normal (without change in the resting potential) by a slow, energy-consuming process of ion transport called the 'sodium–potassium pump' (Hodgkin and Huxley, 1952). Of the separate channels for sodium and potassium ions, usually only one (or neither) of these is open. Either can be blocked independently of the other with foreign ions, and both can be kept open together with veratridine or DDT (Baker, 1968). In vertebrate cardiac muscle and vertebrate smooth muscle, Ca^{2+} replaces Na^+ for carrying the inward current (Reuter, 1973).

When a nerve impulse arrives at a synapse, it causes the release of a minute amount of a chemical substance called a *neurotransmitter* (or synaptic trans-mitter) which is a localized hormone. This substance diffuses across the gap and, on reaching the far side, it interacts with receptors (on the postsynaptic membrane of a muscle or nerve cell. As soon as it has acted, the neuro-transmitter is removed from the synaptic cleft, either by reabsorption into the presynaptic store (norepinephrine) or by enzymic destruction (acetylcholine). In this way, the synapse is rapidly restored to its resting state and is then ready to respond to another impulse.

That transmission of the nervous impulse across synapses was *chemical* occurred to T.R. Elliott while still an undergraduate at Cambridge. Published work suggested to him that sympathetic nerve impulses released minute amounts of an agent that diffused across the synapse and was taken up on the far side (Elliott, 1905). Experimental confirmation was slow to arrive. Henry Dale (1914) supposed that acetylcholine could be the parasympathetic transmitter and, because it could not be detected, he further assumed that an enzyme must be nearby to hydrolyse it. Not until 1921 was there a convincing experimental proof of chemical transmission across this synapse (Loewi, 1921), and in 1926 the first neurotransmitter was finally identified and found to be acetylcholine (Loewi and Navratil, 1926). Dale was shown to be right about acetylcholin-esterase, too.

(a) *The divisions of the nervous system.* The first division is into *central* (brain and spinal cord) and *peripheral* (all other nerves). The peripheral system is divided into *sensory* (collecting information) and *motor* (providing action). The latter is divided into *somatic* (acting on voluntary muscles) and *autonomic*. The latter is divided into *sympathetic* and *parasympathetic* which usually oppose one another's action and, in this way, exercise fine control at the involuntary (unconscious) level. These divisions are conveniently set out in Fig. 7.2.

Central
Peripheral ⌐ Sensory
    Motor ⌐ Somatic
        Autonomic ⌐ Sympathetic
            Parasympathetic

Fig. 7.2 Divisions of the nervous system.

(b) *The peripheral nervous system.* Acetylcholine (7.4) acts at more kinds of synapses than any other neurotransmitter, and norepinephrine (7.5) (noradrenaline) takes second place. Only in recent years has it become evident that there are other neurotransmitters, particularly in the gut, each playing some minor but apparently essential role. Some, at least seem to be evolutionary residuals because, in invertebrates, dopamine (7.48) and 5-hydroxytryptamine (serotonin) (7.49) can be observed playing the roles that belong to acetylcholine and norepinephrine in mammals (however, acetylcholine is also much used by invertebrates). Adenosine is present in some of the presynaptic adrenergic terminals of mammals where it seems to act as a modulator of norepinephrine release. The human gut receives neurotransmission from purinergic nerves, activated by adenosine triphosphate (Burnstock *et al.*, 1970), but normal peristalsis depends on acetylcholine. It may take several years before the role of minor peripheral neurotransmitters can be clarified. For more on purinergic transmission, see Burnstock (1981). For a discussion on such polypeptides as bradykinin, substance P, and the enkephalins as possible neurotransmitters (or are they only local hormones?) in the gut, see Regoli (1982).

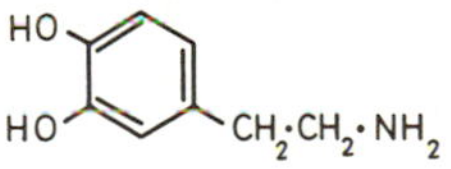

Dopamine
(7.48)

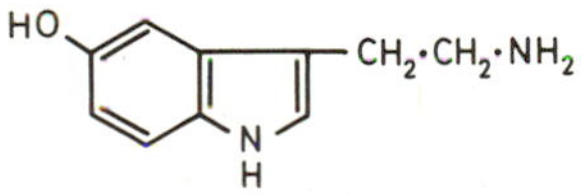

5-Hydroxytryptamine
(7.49)

In mammals, hormones are distinguished from neurotransmitters by being liberated into a whole tissue, or the bloodstream, and not confined to a point source in synapses. Histamine (7.6) and epinephrine (adrenaline) (7.43) are such hormones, but are not neurotransmitters.

Events happening at the neuromuscular junction (see Fig. 7.1) on the arrival of a nerve impulse will now be described. The topography of this junction, already explored with the light microscope, has become even better known through the electron microscope (Katz, 1962). In a typical example (the junction of voluntary nerve with the sartorius muscle of the frog), the nerve-tip gives off slender branches (total length about 1 mm) which lie in gutters formed in the tip of the muscle fibre. Nerve and muscle are separated by a *synaptic cleft* of about 100 Å, and it is only this part of the muscle (the end-plate) which is ordinarily sensitive to acetylcholine. This transmitter is stored in the pre-synaptic terminal within synaptic vesicles which are about 200–500 Å in diameter. In the frog there are about three million vesicles per junction.

At the neuromuscular junction, acetylcholine interacts with receptors on the muscle end-plate. The resultant depolarization triggers an action potential which is similar to a nervous impulse (described above). For about one milli-second, acetylcholine opens channels that allow sodium ions to flow inwards, there is a temporary loss of membrane potential, and the muscle begins to contract. The acetylcholine is then destroyed by acetylcholinesterase, which is located elsewhere in the end-plate. This process occupies only a few milli-seconds and can be reproduced by administering acetylcholine close to the end-plate region. However, injection directly into the interior of the muscle cell produces neither depolarization nor contraction.

The amount of acetylcholine released by a nerve terminal in one impulse is 1.5×10^{-10} μg (as found in both cat and frog). This is 5×10^6 molecules, or 200 times as much as is needed to depolarize the end-plate. However, enough acetylcholinesterase exists at the neuromuscular junction to split 10^9 molecules of the transmitter every millisecond. The dissociation constant of the complex formed by acetylcholine with its destructive enzyme is 2.6×10^{-4} (Nach-mansohn, 1959).

The muscle end-plate is thus a *chemically excitable* membrane, and stands in contrast to the *electrically excitable* membrane which covers the bulk of every nerve fibre as described above. The conductance (permeability to ions) of a chemically excitable membrane is changed only by the specific chemical messenger (synaptic transmitter). These changes in ionic conductance produce membrane potential changes that are proportional to the concentration of the transmitter.

Acetylcholine is also the transmitter at sites other than the neuromuscular junction (see Table 7.1), namely at all autonomic ganglia (whether sympathetic or parasympathetic), at parasympathetic end-organs, and some synapses in the central nervous system. A hypothesis that acetylcholine is concerned in trans-mission *within* all nerve cells (Nachmansohn, 1959) is not widely entertained.

Dale (1914) noted that some of the actions of acetylcholine can be imitated by nicotine (*7.26*) and others by muscarine (*7.50*), and the terms nicotinic and muscarinic have turned out very useful in classifying the sites at which this neurotransmitter acts.

Table 7.1 Drugs acting at peripheral synapses

| Synapses | Stimulants | | | Antagonists of natural transmitter |
	Natural transmitter	Mimic of natural transmitter	Protector of transmitter from destruction	
1. Voluntary muscle (neuro-muscular junction)	acetylcholine (7.4)	nicotine (7.26)	physostigmine (2.8)	tubocurarine (2.6)
2. Ganglionic (sympathetic and para-sympathetic)	acetylcholine	nicotine	physostigmine	hexamethonium (7.24) nicotine (if in excess)
3. Postganglionic nerve-endings (para-sympathetic)	acetylcholine	muscarine (7.50)	physostigmine	atropine (7.16)
4. Postganglionic nerve-endings (sympathetic)	norepinephrine (noradrenaline) (7.5)	phenylephrine (7.51)	(see text)	propranolol (12.56)

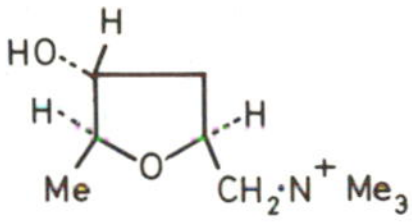

L (+) - Muscarine (cation) (7.50)

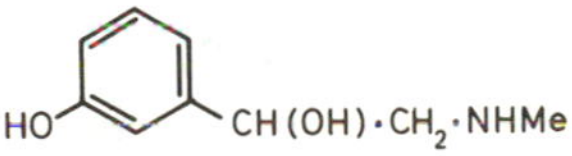

Phenylephrine (7.51)

$HO \cdot \text{---} \cdot CH(OH) \cdot CH_2 \cdot NHMe$

Norepinephrine (7.5) is the transmitter at sympathetically innervated end-organs (see Table 7.1). After initiating synaptic transmission, it is inactivated mainly by reabsorption into the synaptosomes of the presynaptic terminal which secreted it (Iversen, 1967). A little of the norepinephrine is metabolized, in the neurons by monoamine oxidase, and extraneuronally by catechol-O-methyltransferase which effects 3-O methylation. Norepinephrine can be saved from the destructive action of these two enzymes by, respectively, phenelzine (9.46) and pyrogallol. The four kinds of receptors for norepinephrine are described in Section 12.4.

Fig. 7.3 summarizes the anatomical features of the autonomic nervous system. This differs from the voluntary system (exemplified by the neuro-muscular junction in Fig. 7.1) in that one extra control point exists for each fibre, namely an extra synapse (situated in a ganglion). In sympathetic fibres, these ganglia are not far from the spinal cord: in parasympathetic fibres, the ganglia are near the end-organ, i.e. the one which the nerve is to influence. As was pointed out, all such ganglionic synapses are activated by acetylcholine. The postganglionic nerve endings (those which make contact with the end-organ) transmit each impulse to the organ by means of acetylcholine in parasympathetic fibres, but by norepinephrine in sympathetic fibres.

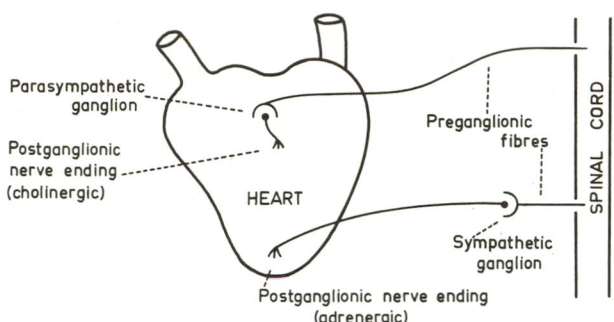

Fig. 7.3 Diagram showing examples of control by autonomic nervous system.

The distinction between *excitatory* and *inhibitory* transmitters will now be outlined. As Fig. 7.3 indicates, some nerves, when stimulated, can inhibit and others stimulate. There is nothing in the chemical nature of the neuro-transmitter that will indicate which of these functions its secretion will bring about: that depends entirely on matters of anatomy, which adds to the burden of our memories! For example, at the neuromuscular junction, and within auto-nomic ganglia, acetylcholine is an excitatory transmitter which depolarizes and hence fires neurons or causes muscle to contract. However, in the heart, acetylcholine is an inhibitory transmitter.

(*c*) *The central nervous system* (c.n.s). Five main regions are distinguished: (i) the cerebral cortex, which is the ultimate centre for processing information and operates abstract thought, memory, and consciousness, (ii) the limbic system, which lies beneath the cortical mantle and integrates the ruling emotional states with visceral and motor activities (it contains the hippocampus, amygdaloid complex, hypothalamus, olfactory lobes, thalamus, and other defined regions), (iii) the midbrain which connects the cerebral hemispheres to the spinal cord and contains the reticular activating system which regulates sleep, (iv) the cerebellum, small and located behind the cerebral hemispheres, controls posture, and (v) the spinal cord which co-ordinates sensory information from skin, viscera, and muscles with information from higher levels, and also transmits to those levels.

In the central nervous system both slow and fast changes of potential occur at synapses. The former, quite unknown in the peripheral system, are thought to be effected mainly by polypeptides, some of which are secreted at a synapse (even by a nerve that is also secreting a peripheral-type fast neurotransmitter), and others are secreted by cells (non-synaptic secretion of what some workers call 'neurohormones').

Acetylcholine seems the most widely distributed of the neurotransmitters in the c.n.s., and both muscarinic and nicotinic receptors have been recognized, the former being the more common (Kuhar, 1978). Three catecholamines (epinephrine, norepinephrine, and dopamine) also play important roles as neurotransmitters, each with its own system of nerves to operate on (Moore and Bloom, 1979). 5-Hydroxytryptamine (*7.49*) is an important transmitter in the midbrain and pons.

Several amino-acids are important neurotransmitters in the c.n.s. Glutamic and aspartic acids seem to be excitatory transmitters in the entire brain. γ-Aminobutyric acid (GABA) and glycine are important inhibitory transmitters, the former in supraspinal interneurons and the latter at spinal interneurons (Curtis and Johnston, 1970). Of the polypeptide neurotransmitters, the most studied have been the endorphins and enkephalins (see Section 12.8), Substance P (an undecapeptide that helps transmit the sense of pain (von Euler and Pernow, 1977, somatostatin, and gastrin, and cholecystokinin whose action in the gut has been well researched. For more on GABA, see Section 12.7.

There is evidence to suggest that inhibitors act by driving chloride ions into the cell while potassium ions escape (Boakes *et al.*, 1971).

The brain is a highly complex organ with many distinct regions. It may be a long time before all the functions and interplay of its neurotransmitters can be discovered. For a book on the cell biology of the brain, see Watson (1976).

For an overview of the chemical pharmacology of all kinds of synapses, see Triggle and Triggle (1976).

7.5.2 *The principal hypotheses of drug action*

The classical work of Clark (1937), based on the application of Langmuir's adsorption isotherm (see Section 8.1), assumed that the effect of a drug is proportional to the fraction of receptors occupied by drug molecules, and that a maximal effect is obtainable only when all receptors are occupied by the drug. As a statement of the action of inhibitors (so important for work in chemotherapy and agriculture) this *simple occupation theory* can hardly be bettered. However, it is inadequate to explain the kinetics of agents which elicit a positive response: substances like the synaptic transmitter acetylcholine and innumerable artificial agonists.

Clark realized this, and said: 'The action of acetylcholine depends on at least two separable factors, firstly the fixation of the drug by certain receptors, and secondly the power to produce its action after fixation' (Clark, 1937). Further experimental work in Holland established these concepts firmly under the names *affinity* and *intrinsic activity* respectively (Ariëns, 1954). An inhibitor (e.g. a chemotherapeutic drug, insecticide, or fungicide) requires only affinity; but many pharmacodynamic drugs must have intrinsic activity as well, namely the ability to produce a natural physiological response as soon as combination between drug and receptor has occurred. The affinity is a measure of the attraction between the stimulant and the receptor: numerically it is the reciprocal of the dissociation constant of the complex so formed. The intrinsic activity is a measure of the ability of the complex to evoke a positive biological response, and it is expressed as the amount of response recorded for the fraction of receptors occupied (Ariëns, 1954, 1964).

The equation defining the dose–response curve is first-order with respect to both drug and receptor, and there is often a close agreement between the theoretically and experimentally obtained curves (Ariëns, van Rossum and Simonis, 1957). These authors compared the intrinsic activities of various neuromuscular depolarizing drugs by plotting their elicited degree of muscular contraction against increasing concentrations of the drug. It was found that when some drugs reached their maximal effect (i.e. one which no higher dosage could increase) they produced much less contraction than did other drugs. The dose required to obtain the maximal effect was used as a measure of the affinity of the drug. (The curves obtained were usually hyperbolas, similar to that in Fig. 8.2.)

Traditionally, help in understanding drug–receptor interaction has been sought from (apparently) parallel phenomena in enzymology. It was recalled that enzymes have two types of specificity which are exerted independently of one another, (a) different substrates are *bound* with different affinities, and (b) different substrates *react* at different rates. To Ariëns in 1957, it seemed that intrinsic activity corresponded to the constant that governs the transformation of an enzyme–substrate complex into products [see Section 9.1, Equation (iii)]. Because the drug was not changed by the receptor, it was assumed that the drug catalyses the breakdown of a receptor–substrate complex, and that it is this breakdown which triggers the response. The drug was thus seen as an equivalent of the cofactor in enzyme chemistry (cf. Welsh, 1948). A somewhat related thought was entertained at this time: that the receptor is opened by the agonist and sealed by the antagonist, as in Levine's hypothesis of the action of insulin which he supposed to open membranes that normally bar the passage of glucose (Levine and Goldstein, 1955, p. 360). This was the beginning of all our contemporary thinking that agonists produce conformational changes.

The Ariëns *composite occupation theory* was soon supplemented by that of Stephenson (1956). Using new, precise data for the effect of acetylcholine and histamine on guinea-pig ileum, Stephenson found, from the slope of the concentration-versus-effect curve, that the activity was not proportional to the number of receptors occupied. He was led to conclude, (a) that a maximal effect can be produced by the occupation of only a small proportion (even an exceedingly small proportion) of available receptors, (b) that response was not *linearly* proportional to the number of receptors occupied, and (c) that different drugs have different capacities to initiate a response, and hence that they occupy different proportions of the receptors when producing equal responses. This property, Stephenson called *efficacy*. This was defined as the reciprocal of the fractional occupancy required to produce a response equal to 50% of which the tissue is capable. A drug had high efficacy if it occupied a very small proportion of available receptors when it gave the maximal response of which the test-organ was capable.

There is a general resemblance beteen the Ariëns concept of intrinsic activity and the Stephenson concept of efficacy, but the latter needs more data for its calculation and seems to be a more fundamental attribute. Dose, for example, is one of the parameters of efficacy because all reasonably efficacious members of a homologous series can elicit the maximal activity of the test-organ, but at different doses. Stephenson extended the old idea, that all drugs were either agonists or antagonists, to include a third class: *partial agonists*. These have only moderate efficacy coupled with high affinity. Hence they can elicit a feeble response from the test-organ, but they can also prevent an agonist from eliciting its full activity. Table 7.2 shows the distribution of efficacy and affinity in a homologous series of cholinergic (muscarinic) agonists. Unfortunately, the efficacy of many common pharmacodynamic drugs has yet to be determined. The much-used concept of *spare receptors* (receptors unoccupied when the concentration of a drug is just enough for it to exert its maximal effect) sprang naturally from Stephenson's work.

Table 7.2 Efficacy contrasted with affinity. Comparison of activity* of alkyl trimethyl-ammonium ions on guinea-pig ileum

Alkyl group	Me	Et	Pr	Bu	Pent	Hex	Hept	Oct	Non	Dec
Efficacy	94	31	4.3	200	200	21	2.2	1.4	1.0	0.6
Affinity $\times 10^{-3}$	0.2	0.6	1.6	3.8	8.5	19	41	63	110	190

*(cholinergic, muscarinic agonists). (Stephenson, 1956.)

Knowledge of the free energy of association of ions permits calculation of the equilibrium distance separating the two centres of charge. This is so because the work done in removing a unit charge from another charge to an infinitely distant position is equal to this free energy. The work done can be found from the Coulomb equation (Pressman *et al.*, 1946). Using this relationship, Burgen (1965) calculated the equilibrium distance of the quaternary nitrogen atom of acetylcholine, from the negatively charged group in the receptor, as 3.29 Å. This calculation was based on the difference of free energy of interaction of acetylcholine (*7.4*) and dimethylbutyl acetate (*12.66*) as calculated from the response of guinea-pig ileum. He showed similarly that the basic centre, of acetylcholine and similarly acting analogues, makes a closer fit on the receptor than does that of chemically related antagonists (presumably because the intense van der Waals bonding of the 'tail' dislocates the positioning of the 'head'). Hence the distinction between agonists and antagonists was suggested to lie in the agonists' greater intimacy of receptor-association and consequent ability to induce conformational changes. Similar calculations and conclusions were made by Belleau, Tani and Lie (1965).

The association of increasing molecular weight with increasing antagonistic power is well known. An antagonist is always bulkier than the corresponding agonist and it is obvious that the likelihood of forming extra van der Waals bonds with the receptor increases the chances of the bulkier molecule having a longer retention time. Because a molecule's kinetic energy of translation (which is an important factor in desorption) does not change with increase in molecular weight, any gain in size by the molecule increases its time of residence on the receptor.

One can recognize three stages of partial agonism, as one passes from a pure agonist to a pure antagonist. Starting with the sympathomimetic amines, one might select ephedrine (*12.12*) as an example of a pure agonist. Yet its tachyphylactic* properties reveal a hidden antagonistic trait. From there one can proceed to typical *partial agonists* such as ergotamine which shows strong sympathomimetic properties which are opposed by higher doses. Finally one arrives at the almost pure antagonists, such as phentolamine (*12.48*) whose initial burst of slight agonistic action is followed by sustained depression.

Again, with parasympathomimetic drugs examined at the neuromuscular junction, a similar sequence can be traced as molecular complexity increases from members with a strong stimulant but occasionally small antagonistic action, through the type that is first stimulant and then antagonistic (e.g. tridecamethonium), to tubocurarine (*2.6*) which is primarily antagonistic but retains the ability to excite the end-plate under special conditions. Another example is afforded by the dialkyl-glutaramides, e.g. bemegride (*15.3*) which can be analeptic and convulsant when the alkyl-group is ethyl, hypnotic and anticonvulsant when it is larger than propyl; but when it *is* propyl, the drug is convulsant in small doses and hypnotic in large ones (Laycock and Shulman, 1967). These graded series correspond to diminishing rates of dissociation from the receptor. Again, apparently sedative drugs like phenytoin (*15.4*) and 5-ethyl-5-(1,3-dimethylbutyl)barbituric acid exhibit either stimulant or depressant action on the central nervous system (intact mouse), depending on the dose (Shulman and Laycock, 1967).

In contrast to the occupancy hypothesis of drug action stands the *rate hypothesis* (Paton, 1961) which suggests that the biological effect produced by an agonist is proportional to the rate of combination of a drug with its receptor, but not to occupancy *per se*. This conclusion was drawn from some meticulous kinetic studies with histamine, acetylcholine, and related stimulants on the isolated guinea-pig ileum, which showed that each contraction declined immediately after it had reached its peak. Rate theory attempts to answer the often-asked question, why do repeated doses of a stimulant usually produce steadily diminishing responses? It suggests that when time has not been allowed

*Tachyphlaxis (desensitization) is a steadily dwindling response to the repeated application of the same dose of a drug.

for the release of all the drug first applied, insufficient receptors are available for subsequent doses, and stimulation is proportional to the rate of drug-receptor combination. The fading of response of guinea-pig gut to repeated doses of acetylcholine, histamine, and a homologous series of quaternary amines is cited in support of this argument (Paton, 1961). Similar results had been obtained with a homologous series of S-alkylisothioureas (Fastier and Reid, 1952), but the phenomenon seems to be rare, and is opposed to the common finding of 'spare receptors' (see above). An ideal system for investigating rates must not offer any barrier to diffusion of the drug, and it must not impose any delay between the application of the stimulus and the registration of the response. Few pharmacodynamic systems meet these requirements. The historic importance of the rate hypothesis is that it has made pharmacologists more conscious of kinetic aspects. Rate must always be a component in quantitative studies of drug action, but it seems usually to be a relatively small one.

(a) *Conformational and allosteric factors in the action of drugs.* Once it had been observed that many substrates change the conformation of their enzymes as a first step in their interaction (Section 9.0), it was natural to attempt to apply this idea to the reaction of a drug with its receptor. Karlin (1967) postulated that two conformational states of the receptor exist, in equilibrium with one another. It is supposed that one state (called T) can initiate no physiological action whereas the other (R) can do so. Drugs are presumed to act as antagonists or agonists, depending on whether they are so constituted as to bind to (T) or to (R) respectively. This hypothesis grew out of the Monod–Wyman–Changeux (1965) discovery of allosteric transitions in enzymes (Section 9.0). The strongest evidence in support of the idea of allosteric receptors is possibly the co-operative effect (i.e. a sigmoid, instead of the usual hyperbolic, response) at the motor end-plate (frog) after stable cholinergic drugs (e.g. carbachol) have been applied to it.

At the present time, it seems that classical occupation theory has been expanded to assume that agonists, at least, produce conformational changes, and that these can open an ion channel, thus initiating the physiological response. However, it assumes that different drugs produce different kinds of conformational change, some not so effective as others: this was done to explain why chemically similar drugs could differ in efficacy. That there can be more than two conformational states of the receptor was later shown by current measurements on four agonists at the frog's muscle end-plate (Colquhoun et al., 1975).

Some other conformation-related phenomena will be mentioned. Many test-objects, exposed to a large dose of an agonist, temporarily lose their sensitivity after the drug has been washed out. The rate of recovery of sensitivity is constant for the tissue, regardless of the drug used. In one explanation, it is suggested that desensitization depends on a change of the agonist–receptor complex (AR) to a different conformational state (AR') which then dissociates, leaving the altered receptor (R′) free to revert slowly to the original receptor (Katz and Thesleff, 1957). Study of the kinetics of development and recovery from desensitization by cholinergic drugs in chick and frog muscle indicated that the physiological effect of the drug was proportional to the fraction of receptors in the active conformation (Rang and Ritter, 1970). The 'metaphilic effect' is a related phenomenon: the affinity of receptors for cholinergic antagonists (tubocurarine, gallamine) was increased when a relevant agonist (carbachol) was applied shortly beforehand (Rang and Ritter, 1969). This was interpreted to mean that the agonist generates more (R′), the form with which it is supposed the antagonist preferentially binds.

For reading on conformational changes induced by drugs (e.g. chlorpromazine, streptomycin, cardiac glycosides), see Levitzki (1973). Although there is much circumstantial evidence for drugs imposing conformational change on their receptors, more direct evidence (such as the X-ray demonstrations available for enzymes) is awaited.

7.6 The natural divisions of pharmacodynamics

Various classifications of pharmacodynamic action are practised. Sometimes drugs are classified according to their main therapeutic use, and thence subdivided according to the pharmacological pathway by which this goal is reached. Thus drugs for use in hypertension are divided into (a) those with mainly central action (e.g. clonidine), (b) those which block sympathetic ganglia (e.g. pentolinium), (c) those which block sympathetic nerve-endings (e.g. propranolol), (d) those which relax the muscle of the blood-vessels (e.g. nitrites), and (e) those which attack the molecular cause by blocking the angiotensin-forming enzyme, e.g. captopril (Section 9.4.3). In a more fundamental approach, drugs are classified by their sites of action, and the therapeutic effect is traced from this. That approach will be followed here. Because most pharmacodynamic drugs act on the nervous system, we shall begin with the central nervous system and proceed to peripheral fibres and endings. This will take us to those few drugs that act on muscles, glands, and elsewhere. At the end, a special section on insecticides, which are mainly nerve poisons, will be found.

7.6.1 *Drugs acting on the central nervous system*

These are customarily divided into depressants and stimulants, according to their ultimate effect. It might be supposed that the depressant effects are achieved with antagonists and the stimulant effects with agonists. This is not necessarily so, for strychnine which acts as an antagonist for inhibitory nerve fibres, is a violent convulsant. Again, morphine, which is an agonist of the enkephalin receptor, is used clinically for its powerful depressant effects.

The *central nervous system depressants* can usefully be divided into the following classes:

(i) The *general anaesthetics*, either inhaled (ether, halothane) or intravenously injected (thiopentone) are further discussed in Section 15.0. These substances depress the higher centres first, but when the dose is increased, centres in the medulla become blocked and respiration ceases. The introduction of muscle-relaxing drugs, about 1940, abolished the need for *deep* anaesthesia during operations.

(ii) The *hypnotics*, such as chloral hydrate and barbiturates milder than those used as general anaesthetics, are given by mouth to induce sleep, but produce little analgesia. Ethanol has a more complex action: in some people it depresses the sensory system more than the motor system and produces restlessness.

(iii) The so-called *anxiolytics*, which have a strong hypnotic action but act on more selective areas of the brain. They are mainly benzodiazepines (Section 12.7) such as diazepam.

(iv) *Analgesics.* The strong analgesics such as morphine (Section 12.8) act at centres different from those activated by the mild analgesics such as acetaminophen. Morphine also has local effects, particularly on the bowels. Aspirin is both a mild analgesic and a useful anti-rheumatic, the latter effect being exerted locally.

(v) *Anti-convulsants*, such as phenytoin (*15.4*), which has no hypnotic effect, are used in epilepsy. Phenobarbitone has both anti-convulsant and hypnotic properties.

(vi) *Anti-tussives* are exemplified by dextromethorphan, which is free from analgesic activity, and codeine which has both actions. Coughs can also be controlled by drugs that act on peripheral endings or on muscle (Eddy, 1969).

(vii) *Appetite suppressants.* These are sympathomimetic drugs most of which (e.g. dextramphetamine) stimulate the c.n.s. Fenfluramine is one that lacks this unwelcome effect.

(viii) *Anti-Parkinsonism drugs.* These help regions of the brain which lack dopamine because of Parkinson's disease. The most used drug of this kind is L-dopa (*3.43b*).

(ix) *The neuroleptic drugs*, sometimes called 'major tranquillizers' or psycholeptics, e.g. chlorpromazine (*12.110*) and haloperidol (*12.112*), calm the disordered mind of the schizophrenic by blocking dopamine receptors in the brain's corpus striatum (see, further, Section 12.9).

(x) *Hallucinogens* have little medical use nowadays. Examples: lysergic acid diethylamide (LSD), and tetrahydrocannabinol which is the active principle of *Cannabis indica* or marihuana.

Central nervous system stimulants are less used in medicine than are the depressants. They are conveniently divided as follows:

(i) *Psychomotor stimulants* which produce euphoria, a sense of physical and mental well-being coupled with increased power of mental concentration and a lowering of the barriers against physical labour. This is effected at the expense of the body's reserves of mental and physical energies and hence leads to tiredness which is ameliorated by another dose of the stimulant, and so on. To drink a little tea or coffee each day is the mildest form of this drug-dependence, but abstinence nearly always brings on withdrawal symptoms, the commonest of which is severe headache. The stimulant action of caffeine seems to depend on antagonism of the adenosine receptor in the brain (Snyder *et al.*, 1981). Caffeine (*7.52a*) is a common constituent of headache remedies. Ephedrine (*12.12*) is a stronger psychomotor stimulant, but not widely abused because the higher doses, necessitated by tachyphylaxis (footnote, p. 294), frequently inhibit the passing of urine. A more dangerous psychomotor drug is amphetamine ('Benzedrine') (*9.44*) which has been shown to act by driving norepine-phrine and dopamine out of their stores. When the synthesis of these two neurotransmitters is blocked by a low dose of α-methyltyrosine, the central effects of amphetamine disappear (Sulser and Sanders-Bush, 1971). This suggests that these two transmitters become the true drugs, in doses higher than anything available through the patient's normal physiology. The dextro stereoisomer (dextramphetamine) has a higher proportion of central to peripheral effects and is employed in narcolepsy ('attacks of sleepiness') and (with caution) in the hyperactive child. Taken by members of the public to increase wakefulness or for euphoria, the amphetamines have produced many a physical and mental wreck. The insomnia produced by psychomotor stimulants, even caffeine, is welcomed by some (e.g. truck drivers) and deplored by others. Theophylline (*7.52b*), which is at least as strong a c.n.s. stimulant as caffeine, is mainly used to relax bronchial smooth muscle in asthma and obstructive chest conditions.

(ii) *Anti-depressants used in the treatment of psychoses.* The monoamine oxidase inhibitors, such as tranylcypromine (Section 9.4), allow unusual concen-trations of norepinephrine and 5-hydroxytryptamine to build up in relevant parts of the brain. The tricyclic anti-depressants such as imipramine (*12.114*) seem to control the same transmitters but in a different way (Section 12.9). Lithium carbonate, introduced to control the manic component of manic-depressant psychosis, seems to help in both phases. Electroconvulsive therapy seems to act by liberating dopamine and 5-HT (5-hydroxytryptamine) in the brain (Green, Heale and Grahame-Smith, 1977).

(iii) *Antidotes to drug-induced central inhibition.* These have almost passed out of use; barbiturate-poisoned patients, for example, are preferably ventilated mechanically. Typical of these analeptics (stimulants), which are con-vulsant in higher doses, is strychnine which blocks the inhibitory action of glycine on the spinal cord (Curtis and Johnston, 1970; Eccles, 1957), and bicuculline which, like picrotoxin, blocks the inhibitory action of γ-aminobutyric acid (Curtis *et al.*, 1971).

(iv) *The centrally acting blood-pressure-lowering drugs* such as clonidine and α-methyldopa (Section 9.4.2) seem to be selective c.n.s. stimulants of catecholamine metabolism.

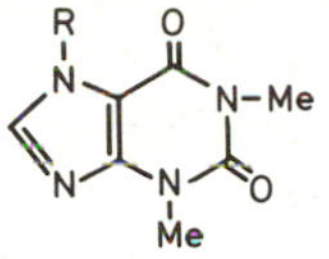

(a) Caffeine (R = Me)
(b) Theophylline (R = H)
(7.52)

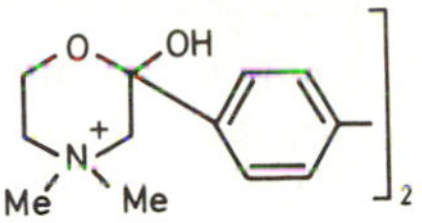

Hemicholinium cation
(7.53)

7.6.2 *Drugs acting on peripheral nerve fibres*

No stimulant drugs are known in this class which consists mainly of the local anaesthetics, much used in dental and spinal anaesthesia. These preferentially block the smaller fibres with the result that sensory nerves are affected more than motor nerves. What local anaesthetics do is to elevate the threshold for excitation and thus they block propagation of the nervous impulse without depolarizing the fibre.

Local anaesthetics lack any hypnotic or general anaesthetic action and are, if anything, more prone to excite the central nervous system than to depress it. Their action, as we know from studies of ionization constants, is on the inside of the nerve fibre (i.e. not on the outside, and not in the axoplasm). Simplification of the molecule of cocaine (Section 7.3) seemed to point to a receptor specific for an ethanolamine structure. While support is lacking for Nachmansohn's (1959) concept that local anaesthetics act by competition with acetylcholine, most contemporary workers think that the drug combines with a receptor in the sodium channel, which is thereby physically blocked (Narahashi and Frazier, 1971; Ritchie, 1975; Strichartz, 1976; and Hille, 1977). It was Straub (1956) who first pointed out that decreasing sodium permeability must block generation of the action potential. Thus implanted microelectrodes, in a single fibre of the frog sartorius muscle, showed that procaine competed with external sodium ions, and the sodium-derived conductance (which normally follows a stimulus) was suppressed (Inoue and Frank, 1962).

The fact that local anaesthetic action can be partially reversed by high external pressure (Kendig and Cohen, 1977) may point to an additional, non structurally specific, contribution to local anaesthetic action (Metcalfe and Burgen, 1968). This would not be surprising for at least one substance in ten, selected at random, has local anaesthetic action, however slight, usually accompanied by an unfavourable therapeutic index.

For a book on local anaesthetics, their mode of action, and their clinical application, see Covino and Vassallo (1976).

7.6.3 *Drugs acting on peripheral nerve-endings*

Stimulation of *sensory* nerve-endings occurs in tasting and smelling, and by electrical, thermal, and mechanical means. Veratrine and the amidines stimulate nerve-endings in heart and lungs, resulting in a reflex lowering of blood-pressure. Apart from this, little is known of the chemical stimulation of sensory nerves by drugs, and no therapeutic applications exist.

Chemical stimulation at *motor* nerve-endings, on the contrary, has an immense literature to which Section 7.5 may serve as an introduction. See also Table 7.1.

(*a*) *Cholinergic synapses*. Drugs with an acetylcholine-like action, but of higher selectivity and longer acting, and those which prolong the action of this neurotransmitter by blocking acetylcholinesterase, are used in treating disordered heart rhythms, glaucoma, myasthenia gravis, or to stimulate the bladder and gastrointestinal tract after operations. See Section 7.3 for notes on the discovery of some of these drugs, Section 12.6 for further discussion of the agonists, and Sections 13.3 and 12.5 for more information on the cholinesterase inhibitors.

As explained in Section 7.3, there are two different ways in which the action of acetylcholine can be blocked at the neuromuscular junction: with or without depolarization. Drugs with either kind of action are much used to secure muscular relaxation during anaesthesia. Similarly, there are two different types of ganglion-blocking drug: those (e.g. tetraethylammonium salts, hexamethonium) which block the receptors for acetylcholine but cause no depolarization, and other (e.g. tetramethylammonium salts) which block these receptors *and* cause prolonged depolarization (Paton and Perry, 1953). Clinical interest in ganglion-blocking drugs has waned because lack of selectivity between sympathetic and parasympathetic ganglia led to unpleasant side effects. For drugs that block postganglionic release of acetylcholine, see atropine in Section 7.3.

Other inhibitors depress the *synthesis* of acetylcholine in the motor nerve-terminals, as does hemicholinium (*7.53*) whose choline contribution to the structure can be seen in the span from the hydroxy-group to the nitrogen atom. These substances, which have found no medical use, interfere with choline transport and thereby deprive acetyltransferase of its substrate. Their action is reversed by choline (Birks and MacIntosh, 1957).

(*b*) *Adrenergic* synapses*. Since the pioneering work of Barger and Dale (1910), the many pharmacological effects of β-phenylethylamines† have become much better understood. As explained in Section 7.5.1, the arrival of an impulse at a

---

*Dale's original name 'adrenergic' is still used, although 'noradrenergic' or 'norepinephrinergic' would be more correct.

†This use of β for 'on the side-chain, in the position next to the benzene ring' and α for 'on the side-chain in the position next to the amino-group' has nothing to do with Ahlquist's division of all sympathetic receptors into α or β types (see Section 12.4).

presynaptic terminal releases norepinephrine, which transmits the impulse to the receptor. This action is terminated mainly by reabsorption of the norepinephrine into the presynaptic terminal, and relatively little part is played by the two destructive enzymes: monoamine oxidase (MAO) and catechol-O-methyltransferase (COMT). Drugs are known which prevent this uptake by becoming adsorbed on the terminal. The structural features optimal for this adsorption have been worked out by Burgen and Iversen (1965) and Trendelenburg (1972). It appears that a hydroxyl-group in position 3 or 4 increases uptake as does α-methylation, whereas O-methylation, N-substitution, and β-hydroxylation decrease it. That these specifications differ from those needed for combining with either an α- or a β-sympathetic receptor (defined in Section 12.4) points up the complexity of structure–activity studies in this area. Of the substances which prevent the uptake of norepinephrine by the presynaptic terminal, metaraminol (7.54) has four times the affinity of the natural substrate. Cocaine (7.11) and amphetamine (9.44) also show this effect, although it is only a part of their total spectrum of activities.

Metaraminol
(7.54)

Guanethidine
(7.55)

Another common activity for a substituted β-phenylethylamine is to accumulate in the presynaptic terminal and push the local stores of norepinephrine out into the synaptic gap. If this is done rapidly, the action must necessarily be of limited duration, as the stores eventually become depleted. Ephedrine (12.12) acts principally in this way (Schümann, 1961), and provides an excellent morning treatment for hay-fever, but its centrally stimulant effect causes disturbance of sleep if it is taken later in the day. The next morning, when norepinephrine stores have built up again in the adrenergic terminals, the patient can derive benefit from another dose. (Ephedrine, unlike norepinephrine, is not destroyed in the stomach and may be given orally.) Amphetamine (9.44) has this combination of peripheral and central actions but with the latter preponderating. Tyramine, on the other hand, has a purely peripheral action.

Comparing the two actions described in the last two paragraphs (prevention of re-uptake of norepinephrine by presynaptic terminal, and enforced release of norepinephrine from this terminal), it is notable that, although many substances like metaraminol, tyramine, and ephedrine show both effects, they exercise these properties quite independently. In other words, when put into ranking order for one effect, they are not in order for the other (Burgen and Iversen, 1965). However, both classes of drug give the patient a temporary overdose of his own norepinephrine (Burn and Rand, 1958).

Another class of substance that accumulates in the presynaptic terminal pushes the stored norepinephrine out no faster than it can be destroyed by monoamine oxidase. Examples are guanethidine (7.55), which is much used medicinally, in blood-pressure of moderate severity, to block, then slowly deplete these norepinephrine stores (Boura and Green, 1965), and the more dangerous alkaloid, reserpine, which is now little used.

Many β-phenylethylamines can not only compete with norepinephrine, in binding at the presynaptic terminal, but can themselves be liberated by each nervous impulse and activate the postsynaptic receptor, thus functioning as *false transmitters*. Usually the efficacy of a false transmitter is less than that of norepinephrine (this is the case for metaraminol and α-methyldopamine), but a few, notably α-methylepinephrine, have a higher efficacy than norepinephrine (Kopin, 1968).

Some false transmitters are known which are not taken up by the presynaptic terminal, but display a direct norepinephrine-like effect on the receptor. The most used of these is phenylephrine (7.51) ('Neo-synephrine') which is applied to inflamed mucous membranes. Whereas phenylephrine acts exclusively on α-receptors, isoprenaline (12.44) affects only β-receptors, and salbutamol (12.54) only one type of β-receptor (more about this in Section 12.4).

Some depressants of the sympathetic system act by interference with synthesis of norepinephrine, as α-methyltyrosine does (not used medicinally), whereas others, such as propranolol (12.56), compete with norepinephrine at the postsynaptic receptor, and are clinically highly successful for reducing high blood-pressure (see further Section 12.4).

From the above outline, it becomes clear that modes of action of adrenergic amines are highly varied, which allows for specialized applications in therapy. Some are used to raise blood-pressure, others to lower it, or else as central stimulants in narcolepsy, as nasal decongestants, for treating asthma or allergies, to curb appetite, and to control bleeding. For further reading on release, uptake and destruction of norepinephrine and its analogues, see Iversen (1975), and Schümann and Kroneberg (1970); for the stereochemistry of catecholamines see Section 12.1, and the review by Patel, Miller and Trendelenburg (1974).

In discussing the pharmacodynamics of postganglionic nerves, should the vague words 'sympathomimetic' and 'parasympathomimetic' be avoided? Because most parts of the body are innervated by both sympathetic and parasympathetic nerves, opposing one another, a 'parasympathomimetic' drug may be acting by stimulating the parasympathetic, or by blocking the sympathetic, receptors.

For further reading on chemical aspects of neuropharmacology, see Triggle and Triggle (1976) and Cooper and Bloom (1982).

7.6.4 *Drugs acting on muscles, glands, and elsewhere*

Whereas, as we have seen, most of the known pharmacodynamic drugs act on nerves, most of the remainder act on muscles. Let us first consider the natural agonists as they play their physiological roles. When a nervous impulse reaches a nerve—voluntary muscle junction (Section 7.5.1), the acetylcholine secreted by the nerve-ending causes the muscle end-plate to generate a current and transfer it to the outer membrane of the nearest muscle fibres. These transmit it to the interior of the muscle by way of the sarcoplasmic reticulum, which is a reservoir of calcium ions, retained there by a protein, calsequestrin. Passage of the current causes a sudden release of Ca^{2+} which stimulates muscle contraction by binding to the protein, TbC, a component of the troponin complex. This binding liberates myosin, and contraction ensues. Similar events follow the liberation of ACh or norepinephrine by postganglionic parasympathetic or sympathetic terminals, respectively.

Apart from this activation of muscle by neurotransmitters, the hormone epinephrine constantly secreted by the suprarenal gland, maintains the normal tone of involuntary muscle, relaxing lung muscle by activating the β_2 receptors and stimulating heart muscle by its action on β_1 receptors. Histamine (7.6), another tissue hormone, contracts muscles of the uterus and intestines and dilates the muscular fibres of blood-vessels. See Section 9.4.5 for agonists and antagonists of histamine receptors. 5-Hydroxytryptamine (7.49) which, in addition to its central neurotransmitter action, is a widespread tissue hormone, causes strong contractions in smooth muscle.

Some examples of botanical origin will now be mentioned. Ergometrine powerfully stimulates the smooth muscle of the uterus, and constricts the muscular fibres of blood-vessels. Papaverine, a poppy alkaloid, has a direct anti-spasmodic action on gut muscle. A number of simple aliphatic amidines, guanidines, isoureas, and isothioureas, all of which are strong bases, raise blood-pressure and contract the intestine by direct action (Fastier, 1949).

The principal cardiac glycosides (from digitalis, strophanthus, and squill) have a direct and selective action on heart muscle, increasing the force of the contraction (for details, see Section 14.1). Quinidine, which is an optical isomer of quinine (10..33), and procainamide (7.56) have a therapeutically useful depressant effect directly on the heart and are used to correct arrhythmias. This effect is shown by many other substances which have a relatively lipophilic aromatic ring linked (by an ester, ether, ketone, or carbinol bridge) to a basic group (Thomas, 1981).

Several drugs, used in reducing moderately high blood-pressure, directly relax the vascular muscles in the capillary bed throughout the body (examples: hydralazine (11.47), the organic nitrates, the inorganic nitrites, and prazosin (7.57) (Stokes and Weber, 1974). Drugs like verapamil and nifedipine, that block the slow calcium channel in smooth muscle and myocardium (see Section 14.2), are much used to reduce high blood-pressure, and also to prevent attacks of angina.

Drugs which relax skeletal muscle, whether acting through spinal reflexes, as do mephanesin, carisoprodol (3.11) and the benzodiazepines, or acting directly on muscle as dantrolene does (Davidoff, 1978), often handicap the patient more

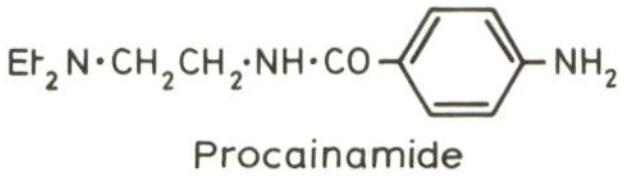

Et$_2$N·CH$_2$CH$_2$·NH·CO⟨benzene ring⟩NH$_2$

Procainamide
(7.56)

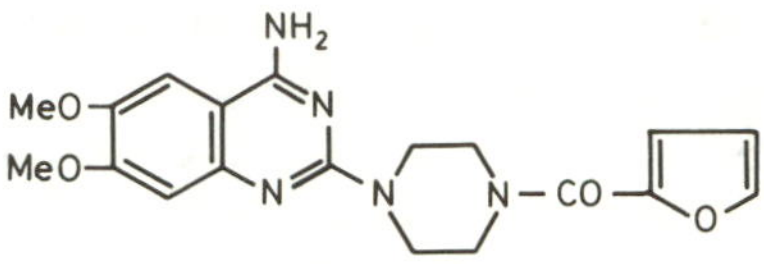

MeO / MeO — quinazoline ring with NH$_2$ — N—N piperazine — CO — furan ring

Prazosin
(7.57)

by the muscular weakness which they produce than did the spastic condition that they so effectively overcome. However, they have a useful place in treating acute spasm.

Dicycloverine (dicyclomine) (7.19), apart from its anti-muscarinic action on postganglionic nerve-endings, has a direct action on the muscle of the bladder and is used to treat bed-wetting (Awad, Downie and Kiruluta, 1979).

For further reading on the structure and function of muscle, see Smith (1972). Microfibrils, which are the primitive equivalents of muscle, were discussed in Section 5.4.8.

Several drugs act on exocrine and endocrine glands. Histamine brings about secretion of hydrochloric acid into the stomach, and the great success of the drug cimetidine (9.58) is that it relieves peptic ulcer by preventing this.

Some antagonists of the formation or secretion of hormones are known. For example, the thiocyanate ion (absorbed from excess intake of cabbage) inhibits the concentration of iodine in the thyroid gland. The thiourea-derived drugs, such as propylthiouracil and carbimazole (3.44), prevent the iodination of tyrosine in thyroglobulin, and are much used in treating hyperthyroidism.

Many drugs cause diuresis by influencing the secretory and resorptive properties of the kidneys which were discussed in Section 3.1. Several drugs are known which act directly on the haematopoietic system (e.g. phenindione, warfarin, sodium calcium edetate, amethopterin), on fibrous tissues (e.g. salicylates, prednisolone, cortisone), and on tumours.

For wider reading on systematic pharmacodynamics, see Gilman, Goodman and Gilman (1980).

7.6.5 Agents acting on insect nerves

Very few of the agents directed against insects behave like pharmacodynamic drugs with their reversible action and graded response. Most are, in fact, insecticides and have been dealt with, alongside chemotherapeutic agents, in Section 6.4.1. However there is an aspect of insecticidal action that is conveniently dealt with in the present chapter. Most insecticides in current use are nerve poisons, and the mode of action becomes more comprehensible if studied alongside human medication of the nervous system (Section 7.5.1). The following should be read in conjunction with Section 6.4.1. Of the most used insecticides, the chlorinated hydrocarbons and the pyrethroids bring about death through overstimulation of the nervous system, both centrally and peripherally.

The central nervous system of insects consists essentially of a double nerve cord situated ventrally and punctuated by segmental ganglia from which the peripheral nerves arise. The axon of such a nerve measures up to 10 μm in diameter and is enclosed in a thin non-myelinated lipoprotein sheath. These axons are bundled into nerves which are surrounded by dove-tailed layers of neuroglial cells, and the whole is enclosed by a protein lamella. The polarization of a resting nerve is very similar to that of vertebrate nerve (see Section 7.5.1). On electrical stimulation, successive spikes can be obtained at intervals of a millisecond; but the action potentials are propagated only at about 2 m s^{-1}, i.e. some 50 times slower than in the larger myelinated axons of vertebrates.

The chemical mediators in the central nervous system of insects are far from completely explored, but acetylcholine (ACh) and octopamine (2.35) play important roles. The neurotransmitter at insect ganglia is acetylcholine, but that at the neuromuscular junction is not acetylcholine but L-glutamic acid (Usherwood and Machili, 1968), and for this neurotransmitter, no selective antagonist has yet been found. The receptors for the neurotransmitters are well protected by selectively permeable membranes.

The principal action of DDT (6.41) is to stimulate the sensory, central, and motor axons (Roeder and Weiant, 1948). It does not inhibit acetylcholinesterase as the phosphorus insecticides do (Section 12.3). However, the net effect is a large increase in free acetylcholine which seems to come from a bound form (e.g. presynaptic vesicles) (Lewis, Waller and Fowler, 1960); it is not due to any increased rate of acetylcholine synthesis (Rothschild and Howden, 1961), but to a disorganization of the axonal membranes.

The insecticidal molecule becomes wedged into Na$^+$-conducting channels in such a way that sodium-propagated conductance is unimpeded because the channels cannot return to their closed, non-conducting conformation. As a result, the axon is kept in a sustained state of polarization which leads to repetitive firing, and eventual death. Calcium ions, which have a protective influence on the sodium ion channel, can prevent this action of DDT (Beeman, 1982). The potency of DDT decreases as the temperature is raised* and there is little effect above 30°C (Guthrie, 1950). The unblocking of sodium ion channels runs parallel to this decline in activity. Thus, warm-blooded animals are exempt from the toxic effects of DDT (which is just as well, because it is as poisonous to insects as hydrocyanic acid is to Man).

The onset of DDT poisoning (compared to that exerted by pyrethrin II) is slow, and the legs (but not wings) make useless movements. Death is not due to exhaustion from these movements, because anaesthetized insects die just as fast. Lindane and the dieldrin–aldrin family of chlorinated insecticides produce a still slower onset, and useless wing (but not leg) movements are seen.

The intoxication produced in insects by DDT is fatal if the dose is high

---

*This 'negative temperature coefficient' is attributed to a charge-transfer adsorption of the benzene rings (Holan, 1971).

enough. However, if death does not occur, because of under-dosage or resistance, a secondary effect occurs, namely a release of abnormal amounts of a
neurophysiologically active substance which can also be released by vibration
or electric shock. The ventral nerve cord seems to be the principal source of this
endotoxin which often eventually causes death (Sternburg, Chang and Kearns,
1959). The nature of the substance is not known, but it is not acetylcholine. It is
produced also by the organophosphate insecticides, but not by lindane.

The modes of action of lindane *(6.45)* and the chlorinated cyclodienes (the
dieldrin–aldrin family) seem to be identical, but quite different from that of
DDT. They act presynaptically at cholinergic junctions in the central nerve
system of insects, facilitating both spontaneous and evoked release of acetylcholine from an excess of which the insect appears to die. They do not block
acetylcholinesterase (distinction from the organophosphorus insecticides,
described in Section 13.3) (Beeman, 1982) and their action is not lessened by
heat, as is that of DDT.

The cyclodienes, particularly aldrin, dieldrin, and endrin, are considered
unsafe for any kind of use. Lindane, on the other hand, has high selectivity in
favour of Man. It is included in the World's principal pharmacopoeias as an
insecticide, acaricide, and larvicide. A 0.2% alcoholic solution is applied topically for head lice and a 1% emulsion is employed in the treatment of scabies, in
humans. The use of DDT in human medicine, for similar purposes, has greatly
diminished in the last decade. It must, however, be pointed out that it has a high
safety record in Man, and that government bans on its use are concerned with
its accumulation in the food chains of birds and fish where it interferes with the
calcium metabolism of shells and bones, respectively (see p. 96). In 1956,
Hayes, Durham and Cueto, of the USA Public Health Service, fed 35 mg of
DDT per man per day (i.e. 200 times the highest average dietary intake of that
time in the USA) to human volunteers for 18 months. None of the subjects
developed any symptom related to this chemical.

Workers employed in the formulation of DDT since 1945 have absorbed
doses many hundred times greater than the average public intake without
incurring ill-health. In a study of 35 men who had 11 to 19 years of exposure to
DDT in the Montrose plant in California (which has produced it continuously
since 1947), careful medical examination disclosed no ill-effects attributable to
this agent. The storage of DDT and its metabolites in the men's fat was found to
be 38 to 647 p.p.m., as compared to an average of 8 p.p.m. in the surrounding
population and slowly increasing storage in the liver was noted. It was
estimated that the average daily intake was 18 mg per day as compared to 0.04
mg in the public at large (i.e. at that time, for it is now much less). No sign of
cancer was found in any of the workers (Laws, Curley and Biros, 1967; Warnick
and Carter, 1972). Experiments on the CFI strain of mice, which have a 22%
incidence of spontaneous liver cell tumours, showed an increase when fed
2 mg/kg of DDT (Tomatis *et al.*, 1972). The Balb/C strain, which is not prone to

spontaneous tumours, showed no incidence at this dose, but some at a tenfold higher dose (Terracini *et al.*, 1973).

Pyrethrin I, the ester of *(6.50)* with *(6.51)*, is extracted from a variety of *Chrysanthemum*. Its action closely resembles that of DDT, both in propping open the sodium channel and in the steep negative temperature-dependence. It is highly lipophilic, easily penetrates the insect cuticle, and is translocated by the haemolymph to the peripheral and central ganglia on which it acts. Pyrethrin II, the corresponding ester with *(6.52)*, has more of a hypnotic action: it knocks flies to the ground quickly, but permits recovery. Although the oscilloscope tracings differ in detail, all evidence points to the target sites of DDT and the pyrethrins being closely related, and situated on nerve membranes. High concentrations of calcium ions, as with DDT, block the action (Beeman, 1982).

Pyrethrins are highly selective in favour of Man, thanks to the mammalian abundance of mixed-function oxidase (p. 89) and non-specific mixed esterases.

For more on all insecticides discussed in this Section, see Section 6.4.1.

8

The forces available for binding an agent. Chemical bonds. Adsorption

As soon as the forces of distribution (Chapter 3) have brought a drug molecule close to a receptor, the two are sure to be hurled into fleeting contact by thermal agitation. How durable and how selective that contact may be depends on the chemical nature of both drug and receptor, and on the type of bonds they can form.

In what follows, the various types of chemical bond will be discussed with reference to the part that each can play in drug–receptor unions. This outline will lead to a discussion of adsorption, which is much the same kind of phenomenon but described in different terms. The chapter will conclude with examples of highly selective binding found in non-biological settings, serving to remind us that selectivity flourishes independently of life.

8.0 Types of chemical bonds

In the past, it used to be claimed that some drugs acted by 'physical' and some by 'chemical' means. It was also suggested that the physical properties of a drug were responsible for getting it to the site of action, but there it reacted chemically. In these connotations, the word 'chemical' was strangely limited to the formation of covalent bonds.

Such a contrast between 'physical' and 'chemical' now seems unreal. Langmuir (1916, 1917) made it clear that physics and chemistry are interwoven: every substance has physical properties, and the physical properties which it has are the inevitable result of its particular chemical structure. Thus 'physical' and 'chemical' are not alternate sets of properties, but different aspects of the same set. For example, in raising a kettle of water to the boiling-point we may seem to be carrying out a purely physical process, but, as Langmuir pointed out, it is actually a depolymerization, involving the rupture of countless millions of hydrogen bonds: in short it is a chemical reaction.

The appearance of Pauling's book *The Nature of the Chemical Bond* in 1939 did much to codify knowledge of the various types of bonds between atoms. The four most important chemical bond types for the student of drugs to understand are: covalent bonds, electrostatic bonds, the hydrogen bond, and van der Waals forces. Within these main types, special varieties can be recognized. To make or break any of these bonds, whether covalent or otherwise, is a true chemical reaction with a substantial energy change. Modern knowledge of bonds rests on the fundamental basis of quantum mechanics, and the understanding of valence that has come from the application of quantum mechanics to atomic orbitals. Bonds strengths, usually experimentally determined, can be calculated too. The principal bond types will now be discussed in turn.

8.0.1 *The covalent bond*

Very few drugs form a covalent bond with the receptor: those that do so, such as penicillin and the organic phosphates, are dealt with in Chapter 13. Most drug effects are reversible by simply washing the agent away (see Fig. 2.2 and accompanying text). This quick reversal means that the biological activity does not depend on the formation of a covalent bond. The metabolism (biological degradation) of a drug, however, depends entirely on covalent bond changes.

The covalent bond is formed from two electrons, one contributed from each of the atoms forming the bond between them. They are the typical bonds of the organic chemist who spends so much of his time making and breaking them. They are usually much stronger than other types of bond, having bond-strengths lying between 70 and 200 kilocalories* per mole. Thus a single bond between two carbon atoms (C—C) has 83 and a double bond (C=C) has 146; a single bond between carbon and oxygen (C—O) has 86 and a double bond (C=O) has 178; a single bond between carbon and nitrogen (C—N) has 73 and a double bond (C=N) has 147 (kcal/mol). The greatest strength of bond that can be broken readily (i.e. non-enzymatically), in the 20–40°C range, is 10 kcal.

However, it must not be supposed that difficulty encountered in washing a drug out of a tissue necessarily implies covalent bond formation. The drug molecule may simply be *clathrate*, i.e. physically imprisoned in the folds of a macromolecule (biopolymer). The textile industry has long made use of this principle in dyeing cotton fabrics with long, narrow molecules of the type of Sky Blue (*8.8*). These are polyazo-dyes of low molecular weight, known as 'direct

*Multiplication by 4.18 will give kilojoules, the preferred SI unit. The standard literature is in kilocalories, and is slow to change.

dyes for cotton'. They require no mordant, and garments thus dyed retain their colour through many washings (Lapworth, 1940). The clathrate effect is, of course, greatly assisted by non-covalent bonds.

A variant of the covalent bond, although usually much weaker, is the *co-ordinate bond* that is formed when both electrons of the bond have to come from *one* atom, for reasons of valency. A familiar example is the union of a hydrogen ion with an anion to give the non-ionized form of the acid, or of a hydrogen ion with an amine to give an ammonium ion (e.g. $CH_3 . NH_2 + H^+ \rightleftharpoons CH_3 . {}^+NH_3$). This is the phenomenon known as ionization, which covers quite a range of bond-strengths and is discussed quantitatively in Chapter 10. In a related example of co-ordinate bond formations, a metal cation can replace the hydrogen cation in the previous examples. The metal complexes which arise in this way are discussed quantitatively in Chapter 11. Other examples of the co-ordinate bond are to be seen in nitrogen oxides $(R_3N \rightarrow O)$ and nitro-compounds $(R . NO_2)$.

In what are called *conjugate systems*, the concentrations of electrons, normally shown as double bonds, are delocalized (i.e. more diffused). By 'conjugate system' is meant a molecule (or portion thereof) in which every second bond would conventionally be written as a double bond (examples: benzene, butadiene, pyridine). The new electronic equilibrium is illustrated by benzene of which six of the valency electrons (one from each carbon atom) unite in forming the π-layer which completely covers both faces of the molecule. This condition is maintained by *resonance*. However, resonance can exist without conjugation, as in amides, which have a structure intermediate between *(8.1)* and *(8.2)*, but not otherwise representable. The double-headed arrow is the accepted sign for resonance. Note that in the canonical forms [such as *(8.1)* and *(8.2)*] which make up the resonance hybrid, no *atom* ever shifts its position: only electrons and charges are mobile.

Resonant systems are planar (flat), as has often been demonstrated by X-ray crystal analysis.

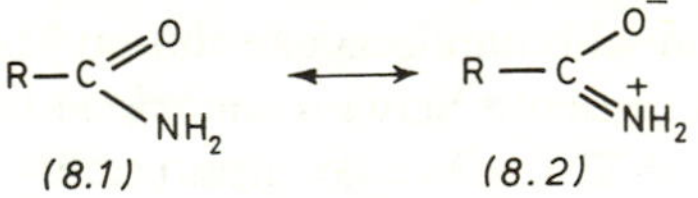

(8.1)   (8.2)

The amide group

8.0.2 *The electrostatic bond*

These bonds are thought to be very important for attracting agents to their receptors, just as they attract substrates to enzymes. This is because they are long-range forces that begin to act from far away. In this, they are unique among bonds. Their other characteristic feature is the readiness with which they interchange. In a biological environment, where there are so many competing ions, they may last only about 10^{-5} second, because of ion exchange. However, a drug–receptor bond that has been formed electrostatically, can persist for a long time if distances are right for shorter-range bonds (hydrogen and van der Waals) to take over.

Most often, the electrostatic bond is between two ions (the *ionic bond* or 'salt linkage'). There are variants in which the bond is formed between an ion and a dipole or between two dipoles. They are all maintained by purely electrostatic forces. The ionic bond has a strength of about 5 kcal/mol and declines by the second power of the distance between the opposite charges. Sodium chloride (Na^+Cl^-) is a typical example. In aqueous solution, each ion is able to move about freely so long as it does not leave the field of its counterion; in other words, the bond is non-directional and non-rigid.

The simplest example of an ionic bond being reinforced by another kind of bond is provided by the cations of all amines (except quaternary amines) which simultaneously form both hydrogen bonds and ionic bonds with the anions of carboxylic acids, as in (*8.3*) where the anion is shown in its mesomeric form (i.e. with the charge delocalized). These hydrogen-bonded salts are credited with 10 kcal/mol of bond-strength. Amidines form even more tightly bound salts (*8.4*). Again, two molecules can be held together at one point by an ionic bond and elsewhere by van der Waals bonds, as in the intercalation of 9-amino-acridine into DNA (Section 10.3.2). These reinforcements greatly increase the permanence of the bond.

Salt of amine
(8.3)

Salt of amidine
(8.4)

The stabilization of ion-pairs (i.e. pairs of oppositely charged ions) by short-range bonds is illustrated by the analytical method known as ion-extraction analysis. For example, picric acid (*8.5*), in aqueous solution, can be readily determined by titration with an aqueous solution of methylene blue (*8.6*) if a layer of chloroform is present. Neither methylene blue (a chloride of a very strong base) nor picric acid dissolves appreciably in chloroform, whereas methylene blue picrate is quite soluble in this solvent. The end-point is taken as the first appearance of a faint, permanent blue colour in the aqueous phase (Bolliger, 1939). Normally a salt has no properties other than those of the ions of which it is composed. But when large areas of the two ions can be brought into intimate contact, and both (*8.5*) and (*8.6*) have flat structures, the total bonding (ionic+secondary) is so strong that the water of hydration which all ions possess is squeezed out, and the salt becomes liposoluble. Similarly cationic organic drugs, like mepacrine (*6.10*), can be determined, in the presence of an immiscible solvent, with coloured sulfonic acids such as methyl orange (*8.7*) (Brodie, *et al.*, 1945) or bromothymol blue. Such ion-extraction is a model for the uptake of drug cations by receptors, and for the carrier-aided passage of agents through semi-permeable membranes (see Section 3.2).

Picric acid
(8.5)

Methylene blue
(8.6)

Methyl orange
(8.7)

One can best picture an *ion–dipole bond* by recalling that many non-ionized substances have quite large dipole moments and hence they carry a fractional positive charge on some of their constituent atoms and a fractional negative charge on others (see Section 17.2 for a list of substituent groups, classified by charge). Such a molecule can be attracted by an ion and form a bond between this ion and a region of the molecule that is rich in the charge that is opposite in sign to that which the ion carries. The bonds formed in this way are somewhat weaker than purely ionic bonds. The most familiar example of an ion–dipole bond is that which unites all ions to water (if in aqueous solution) and makes the properties of these hydrated ions so different from those of the anhydrous ion in a crystal.

The antibacterial sulfonamides have a very interesting relative, dapsone (*9.17*), the most used drug in the treatment of leprosy. Sulfadiazine, and the majority of antibacterial sulfonamides, have been shown to inhibit the relevant enzyme (dihydrofolate synthetase) best if in the form of the anion. However, dapsone is not capable of ionization, but forms an ion–dipole bond to the enzyme. The result is a milder agent with minimal side effects during the many long years that a cure of leprosy requires.

Finally there are the rather weak *dipole–dipole bonds* whose strength declines with the third power of the distance. In the Wilson-Bergmann model of acetyl-cholinesterase (Fig. 12.3), a dipole–dipole bond was postulated between the doubly bound nitrogen atom of the imidazole ring (in the enzyme) and the fractional positive charge of the carbon atom in the ester group of acetylcholine.

8.0.3 *The hydrogen bond*

Hydrogen bonds are both short in range and restricted in possible angles. Because of these stringencies, they possess a highly selective character. For this reason, hydrogen bonds are considered to be very important in drug–receptor interactions. Also, they play an essential role in maintaining folding in proteins.

The high concentration of positive charge in an uniquely small volume enables the hydrogen atom to act as a bond between two electronegative atoms (mainly O, N, and F): e.g. —O—H . . . N—. Hydrogen-bondable atoms must have complete octets of which at least one electron-pair is unshared. The atoms bound by hydrogen may be in the same or in different molecules. The bond-strength is usually 3–5 kcal/mol. To form a hydrogen bond the atoms must be free to take up a position along the line of the OH or NH axis, and at a particular distance (e.g. 2.7 ∞A for an O—H . . . O bond). They dissociate readily on warming.

The greater the difference in electron affinity of the two atoms linked by hydrogen (even if both atoms are of the same element, say, nitrogen) the stronger the hydrogen bond, whose nature lies roughly between the co-ordinate and the electrostatic bond. Hydrogen bonds involving sulfur are weak, those involving halogens (except fluorine) are weaker still, and those in which carbon participates are barely observable. Hydrogen-bonding is recognized by changes in the infrared spectra, ionization constants, melting-points, volatility, and solubility.

Hydrogen bonds are both short-range and angle-restricted. Because of the stringency of the conditions under which they can be formed, they have a highly selective character. For this reason, hydrogen bonds are considered to be important in drug–receptor interactions, and they are known to play a key role in maintaining the structure of proteins.

Ice, paper, and nylon are typical examples of solids whose mechanical properties are controlled mainly by their hydrogen bonds. For a review of the nature of hydrogen-bonding, see Kollman and Allen (1972).

8.0.4 *Van der Waals bonds*

Because this type of bond (or 'force' in another nomenclature) can be exerted only when the geometry of the two kinds of molecule concerned allows a very close fit of two atoms concerned in each bond, it is considered that bonds of this kind are of paramount importance in uniting a drug to its receptor. The presence of an ionized group in the drug can pull it into the close quarters required for van der Waals forces to operate. For an example of this sequence of bondings, see the aminoacridines in Section 10.3.

The attraction between an antigen and its antibody consists entirely of short-range forces of the van der Waals and hydrogen-bond type (Pardee and Pauling, 1949). Van der Waals bonds operate whenever any two atoms belonging to different molecules are brought sufficiently close together. These bonds arise from the fact that all molecules possess energy which leads to internal vibration. The temporary dipoles which this vibration creates in the constituent atoms, induce dipoles in neighbouring atoms of other molecules, a process which results in a net attraction. Such forces vary inversely as the seventh power of the distance, so that they are significant only over a short range. (Hence if two atoms are separated twice as far as they were previously, the attraction falls to 1/128 of its former value; cf. the attraction of an ionic bond which would fall to only 1/4 of it former value, and a dipole–dipole bond to 1/8.)

Van der Waals attraction increases with rise in atomic weight; it is negligible for hydrogen atoms, and about 0.5 kcal/mol between pairs of atoms of atomic weight 12–16 which are of greatest significance in drug–receptor unions. These forces add up to a significant attraction between any two molecules that can fit one another so well that many atoms in one molecule can touch those in the other molecule. In this way, strong bonding (e.g. 5 kcal/mol) may result from close juxtaposition of an agent and its receptor.

No *direct* way of measuring van der Waals forces in isolation is known, but their values can be obtained by orbital calculation or by subtracting other forces from the measured sum-total of all forces between two molecules or between a molecule and the whole of its environment. Of these total forces the following can be calculated reliably: ion–ion, ion–dipole, and dipole–dipole. (Here 'dipole' is used to signify a *permanent dipole* whose magnitude and orientation are known or can easily be determined, whereas van der Waals forces are between *oscillating dipoles*.)

Actually van der Waals forces are the resultant of four kinds of ultimate forces, namely the London attraction, the Debye attraction, the temperature-sensitive Keesom force, and the Born repulsion. The last-named arises as follows. Two atoms in any organic molecule are usually about 1.4 Å apart, but the atoms in *different* molecules cannot get as close as this. Two molecules begin to repel one another strongly as soon as their respective atoms come within about 3.0 Å (for C, O, or N atoms, but only 2.4 Å for two hydrogen atoms). These minimal distances are attributed to the addition of two van der Waals radii (1.2 Å for H, 1.55 Å for N).* These powerful repulsive forces come into play whenever two non-bonded atoms, even in the same molecule, approach to a distance equal to the sum of their van der Waals radii. This repulsion sets the upper limit to van der Waals attraction.

For a review on van der Waals bonds (forces) see Pitzer (1959).

8.0.5 *Miscellaneous aspects of bonding*

Electron-transfer complexes are a special electronic manifestation of some non-localized covalent bonds. As was outlined above, molecules with two or more conjugated double bonds have some highly delocalized electrons which form a π-electron layer covering the conjugated area. Through further delocalization, caused by substituents with powerful dipoles, this π-layer can have either a deficiency or an excess of π-electrons compared to the normal (two π-electrons contributed by each double-bond). Hence, particularly in ring systems, it is desirable to distinguish between π-deficient substances (e.g. nitrobenzene and pyridine) and 'π-excessive' substances (e.g. aniline and pyrrole) (Albert, 1968). A strongly π-deficient substance can form a very weak complex with a strongly π-excessive substance, seemingly sharing electrons almost as freely as though the two molecules were two adjacent rings in the one molecule. The thermo-dynamic basis for this effect is poorly understood, and the principal observable

*For other van der Waals radii, see Table 9.1, also Bondi (1964).

fact is the appearance of a new absorption peak in the ultraviolet or visible spectrum.

Interest in electron-transfer complexes began with spectroscopic studies, three decades ago (Benesi and Hilderbrand, 1948). It was then found that riboflavine forms such a complex with tryptophan, and also with 5-hydroxy-tryptamine (serotonin). Some think that this type of bonding is significant in the body's chemistry (Szent-Györgi, 1960). The alternative name 'charge-transfer complexes' originally referred to those electron-transfer processes effected by the action of light. For further reading on electron-transfer complexes, see Slifkin (1971).

Hydrophobic bonding. This term was invented by Kauzmann (1954) to describe the van der Waals attractions between atoms in the non-polar parts of two molecules immersed in water. Because van der Waals bonds operate over such short distances, there is no room for water molecules in the vicinity of the mutually bound paraffinic surfaces. Conversely the attraction of water molecules for one another, by hydrogen bonds (see end of Section 3.1), ensure that molecular regions lacking oxygen or nitrogen atoms (which are themselves hydrogen-bondable) tend to be squeezed out of water. Thus no new kind of bond is involved in the term 'hydrophobic bond', and its use served only to remind a reader that the region (of a molecule) under discussion is one free from oxygen and nitrogen atoms. The term lacks the thermodynamic validity of the four principal types of bond discussed above.

That the concept of 'hydrophobic bonding' does not square with the facts, is pointed out by Hildebrand (1979) who maintained that the very concept of a hydrophobic effect is unreal. No hydrophobic substance has yet been discovered: every substance is hydrophilic, although hydrocarbons are very little so. In support of the hydrophilicity of hydrocarbons, Hildebrand instanced that the energy required to evaporate a mole of butane from its aqueous solution (at 1 atm and 25°C) is 0.65 kcal greater than from its own pure liquid. He also pointed out that if you pour some octane on ice, you can see that the ice is instantly wet by it (Hildebrand, 1979).

Tanford (1979), a regular user and defender of the term 'hydrophobic bond', admitted that the work of adhesion (in erg/cm² at 25°C) for hexane to hexane is -35.8, for hexane to water is -39.5, and for water to water is -144. Thus the attraction of hexane molecules for water is about 10% greater than the attraction of hexane molecules for one another. However, the attraction of water molecules for one another is 3.6 times as great as that of hexane for water. This is why hexane is pushed out of water.

For two books devoted to the 'hydrophobic effect' see Tanford (1980) and Ben-Naim (1980). For useful information on the hydrophilic and lipophilic regions of molecules, see Section 17.1.

For further reading on bonds formed between molecules, see Pauling (1967); for information on valency, see Speakman (1968).

8.1 Adsorption

Many direct and indirect references to adsorption have been made in this narrative. Now that the main types of chemical bond have been reviewed, it is possible to discuss this property at a fundamental level, namely in terms of bonds. Let it be clear that adsorption is nothing more than the sum of all the different kinds of bonding that are taking place. There is no component of adsorption other than the types of bonding discussed above.

A substance is said to be adsorbed when it is concentrated reversibly at a surface. It was not a well-understood phenomenon until Langmuir (1916, 1917, 1918) clarified the subject in three papers. Adsorption involves exactly the same types of bond (especially van der Waals, hydrogen, and ionic bonds) as are involved in chemical reactions in the bulk phase. Formerly, workers tried to differentiate between 'chemical' and 'physical' adsorption, but it became evident that, because all bonds are chemical bonds, all adsorption is chemical.

A surface has two special features which can make reactions taking place there quantitatively different from analogous reactions taking place in solution. Firstly, a surface presents a 100 % concentration of the substance involved. As the substance is sparingly soluble (if it were soluble, it would not be present as a surface), this concentration enormously increases the opportunities for the reaction to take place. For example, a crystal of silver chloride has a surface concentration of 7 M, as one can quickly verify from the molecular weight. On the other hand, a saturated solution of this substance (1×10^{-5} M) contains practically no silver chloride.

The other special feature about a surface is that it is apt to contain unsatisfied valencies, which, elsewhere in the solid, are used to bind the constituent atoms to one another. This is evident from the representation of a piece of carbon in Fig. 8.1. It is obvious that the finer the carbon is ground, the more residual valencies there will be, and the more active an adsorbent it will become.

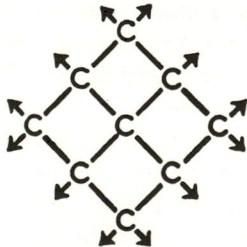

Fig. 8.1 Residual affinity of fragment of carbon.

Except in the rare cases where covalent bonds are made, adsorption is a reversible process, and equilibrium is established according to the mass-action law. In 1918 Langmuir derived the following equation from this law to permit a more accurate quantitative treatment of adsorption than had been possible.

$$\frac{x}{m} = \frac{abc}{1+ac}$$

This equation states that, if the temperature is kept constant, (i.e. isothermal conditions), the weight (x) of a substance absorbed per weight (m) of adsorbant is proportional to the term on the right where c is the concentration of *un*adsorbed substance, and a and b are constants. The equation expresses the fact that the adsorbant becomes saturated at high values of c, (because, when c is much larger than 1, the term on the right becomes simply b). This 'isotherm', as it is called, is represented graphically as a hyperbola (Fig. 8.2). It is evident that this equation is a general treatment of a subject which has some familiar special cases, e.g. the union of a hydrogen ion with an ammonia molecule (to give an ammonium cation) which follows the same hyperbolic curve and is one of the simplest examples of adsorption by ionic bonds.

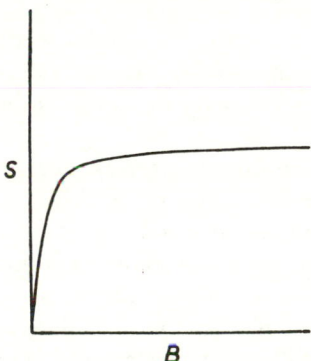

Fig. 8.2 Typical Langmuir isotherm. *S*, concentration of substance adsorbed (x/m). *B*, total concentration of substance.

In the adsorption of drugs on their specific receptors, it is very often found that doubling the dose produces smaller and smaller additional responses, and that this falling off corresponds to the shape of a hyperbola (Clark, 1933). This relationship is sometimes complicated by metabolic destruction of the drug (see 'Sites of loss' in Section 3.4).

In general chemistry, Langmuir's isotherm fits the experimental data in a very large number of cases where the adsorbed layer is unimolecular. However, some other patterns of adsorption are known and a convenient general classification has been made, as follows, by Giles *et al.* (1960).

(i) L-curves, normal Langmuir isotherms (e.g. Fig. 8.2) indicative of molecules adsorbed *flat* on the surface. The more solute is adsorbed, the harder it is for additional amounts to become fixed.

(ii) S-curves, indicative of *vertical* orientation of adsorbed molecules. In the initial part of the sigmoid curve, the more solute is already adsorbed, the easier it is for additional amounts to become fixed. This is called the co-operative effect.

(iii) H-curves, high-affinity curves which commence at a high value of the 'concentration on solid' axis; often given by solids adsorbed as micelles, and by high-affinity ions that exchange with low-affinity ions.

(iv) C-curves, constant-partition linear curves given by substances which penetrate into the adsorbant more readily than the solvent can.

It is important to bear in mind that 'dissolved' proteins and other large molecules are in the colloidal state. Such suspensions although transparent to the eye, present an enormous surface for adsorption. For example, a cubic centimetre of human serum contains 100 m² of protein surface.

8.2 Non-biological examples of selectivity

Chemical selectivity, of a degree comparable with the specificity shown by drugs, can be found in simple non-living material. Thus it is not necessary to assume any special biological forces to explain the action of biologically active agents.

Examples will be chosen from two topics, *dyeing* and *flotation*, which require chemicals that can differentiate between rather similar substances. Seemingly irrelevant to the study of selective toxicity, these parallel investigations can throw light on some of biological selectivity's knottiest problems.

In industry, two chemically related fibres, such as cellulose and cellulose acetate, are often woven into an invisible design and dyed in contrasting colours, simultaneously. Thus a dyebath containing both 'Chlorazol Sky Blue FFS' (*8.8*) (in USA: 'Pontamine Sky Blue 6BX') and 'Dispersol Yellow 3G' (*8.9*) (in USA: 'Acetamine Yellow CG') will dye the cellulose fibres pure blue and the cellulose acetate fibres pure yellow, giving a fabric with, for example, a yellow design on a blue ground. In short, each fibre has combined very strongly with its selective dye and has rejected the opposite type. The structure of the cellulose fibre consists of long, rather flat molecules, packed together in sheets. Each molecule is liberally studded with hydrogen-bonding groups. Hence it is not surprising that the majority of water-soluble dyes which have an affinity for unmordanted cellulose are also long and flat and are liberally studded with hydrogen-bonding groups (Lapworth, 1940; Ruggli, 1934). 'Chlorazol Sky Blue' is such a substance. In cellulose acetate, however, five out of every six hydroxyl-groups have been blocked by acetylation and the molecule has taken on the general characteristics of a lipophilic ester. Hence it is not surprising to find that the majority of dyes which have an affinity for cellulose acetate are very soluble in esters. Dyeing of this kind is simply a partitioning effect, one not requiring any hydrogen-bonding groups. In fact the possession of these is disadvantageous (the introduction of a single sulfonic-group has been found to destroy all affinity for cellulose acetate). For further information on the connexion between constitution and selectivity in these dyes, see Lapworth (1940) and Green (1937).

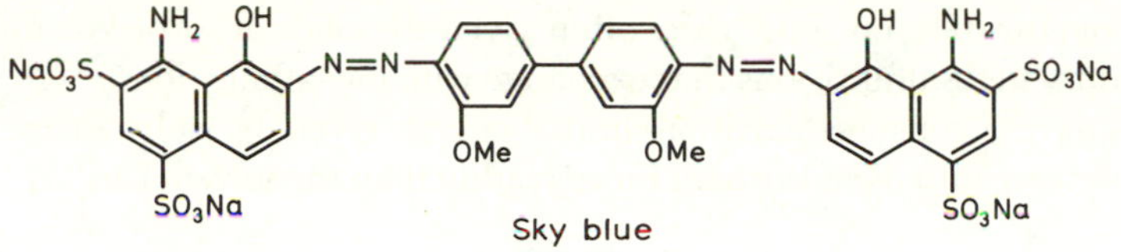

Sky blue
(8.8)

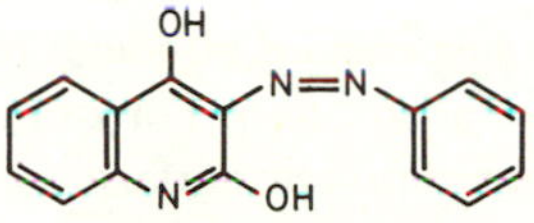

Dispersol yellow
(8.9)

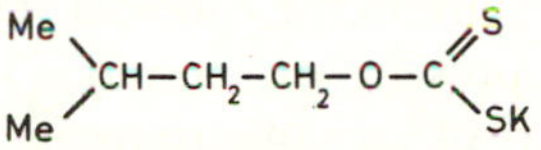

Potassium isoamyl xanthate
(8.10)

In mining technology, flotation agents are commonly used to separate the constituents of a mixed ore. The crushed ore is placed in a tank of water through which a stream of air bubbles is rising. When the specific 'collector' is added, the mineral which it selects will rise to the surface, off which it is mechanically scraped. Thus to float mercuric sulfide, enough potassium isoamyl xanthate (8.10) is added to give a 1 in 100000 solution. Similarly sodium stearate will cause copper silicate to float, and in both examples all quartz (sand) will remain at the bottom of the tank. This and similar processes are familiar to every mining engineer and are used commercially on a immense scale. For further reading on flotation, see Sutherland and Wark (1955).

Details for carrying out lecture demonstrations of (a) dyeing a mixed fabric blue and yellow simultaneously from the same bath, and (b) separating, by flotation, red mercuric sulfide, followed by green cupric silicate from their mixture with sand, will be found in the first four editions of this book. The degree of selectivity shown, in chromatography, by cellulose, alumina, and silica also lends itself to demonstrations.

PART TWO

Studies, in depth, of topics from part one

9

Anti-metabolites: antagonistic analogues of coenzymes and enzyme substrates

The smallest change in the chemical constitution of a selectively toxic agent often makes an enormous change in its biological activity, and many examples of this have been given in Chapter 2. The present chapter deals in greater detail with one cause of high specificity, namely a close resemblance between (a) the normal substrate (or coenzyme) of an *enzyme* and (b) an agent which inhibits it. In previous editions, these agents were called 'metabolite analogues', but the word 'metabolite' has become broader in meaning. To the biochemist, 'metabolites' are the intermediates of metabolism, such as the pyruvate ion, fatty acids, oxygen, and the nucleotides. Many pharmacologists, however, use 'metabolite' for the degradation product of a xenobiotic (foreign substance). Here we shall speak of anti-metabolites, meaning antagonistic analogues of coenzymes and the substrates of enzymes. Such substances have also been called 'metabolic inhibitors'.

9.0 Enzymes

The process of growth and division involves cells in ceaseless chemical activity. For the most part this activity takes the form of chemical reactions between enzymes and substrates, whereby the enzymes remain unchanged, and the substrates are transformed into other substances by the breaking or making of covalent bonds. Even the organic coenzymes may undergo a covalent change in the course of their functioning, e.g. nicotinamide adenine dinucleotide receives, and then loses, a hydrogen atom on the carbon atom in position 4. Similarly, diphosphothiamine receives, and then loses, an acetyl-group on C-2.

Some coenzymes ('co-factors') are inorganic cations, many others are simple organic molecules, but many enzymes need no coenzyme. The organic coenzymes are either synthesized by the organism, or by another organism and taken in with food. The protein portion of an enzyme that has a coenzyme is called the apoenzyme. The substrates are either (a) food or (b) simple molecules formed from the food by the action of other enzymes (both degradative and synthetic). More than 2000 different enzymes are known, many of them obtained pure by electrophoresis and (most effectively of all) by affinity chromatography (Dixon and Webb, 1979). If isoenzymes (Section 4.2) were considered, the number would be much greater. Some enzymes may be present in cells only to the extent of a few molecules, or even only as 'information' in the DNA which can cause the enzyme to be synthesized when it is needed; but other enzymes may constitute at least 50% of the dry weight of the cell, as myosin (an adenosine triphosphatase) does in muscle cells.

Until recently, the conventional purification of enzymes, e.g. by electrophoresis, was slow and the yields poor. The recent introduction of affinity chromatography enables an enzyme to be obtained pure and in high yield. In this technique, a specific inhibitor of the enzyme is attached, covalently and by a flexible link such as —$CH_2.CH_2$—, to a polymeric adsorbant. The enzyme collects on the inhibitor, from which it can usually be freed by a change in pH (Cuatrecasas, Wilcheck and Anfinsen, 1968).

The first enzyme to be completely synthesized was ribonuclease. This was done both in solution, by the classical chemical methods (Hirschmann *et al.*, 1969), and by Merrifield's solid-support technique (Gutte and Merrifield, 1969).

The most successful method for disclosing enzyme structure is the X-ray diffraction study of a suitable crystal. For definitive results, a resolution of 2 Å is eventually attempted, but this is extremely time-consuming, even when every modern computational aid is available. Nevertheless the structures have been established for about a hundred enzymes, showing not only every amino acid residue but also their disposition in space.

9.0.1 *Conformational adaptation of enzyme to substrate*

Koshland (1964) coined the term *induced fit* to express his idea that substrates and coenzymes can change the conformation of enzymes and vice versa. This has since been confirmed abundantly by X-ray diffraction studies of enzymes, examined first alone and then with a slow or reluctant substrate, or simply a coenzyme. For example, the apoenzyme of lactate dehydrogenase was seen to be greatly deformed in the presence of its coenzyme, NAD; also this coenzyme, which normally is closely folded, was stretched right out by the apoenzyme (Adams *et al.*, 1970). Even earlier, the conformation of an acetylglucosamine residue was seen to be distorted from a 'chair' to a 'half-chair' configuration when bound to the active site of lysozyme (Blake *et al.*, 1967). Today, it is known that the active site of an enzyme is usually composed of two or more amino acid residues that would lie far apart if the chain were extended. For example, the action of the enzyme lysozyme (from white of egg) on its substrate in the bacterial cell wall (murein; Fig. 5.2) requires the proximity of residues 35 and 52 (glutamic and aspartic acid, respectively) from two strands of the enzyme (Phillips, 1966). Lysozyme has predominantly a hydrophobic interior and hydrophilic exterior in its tertiary structure, an arrangement quite common in enzymes.

Similar X-ray studies have shown that yeast hexokinase consists of two lobes which form a bay into which the substrate, glucose, enters. The two lobes then close tightly around the substrate through an 8 Å movement of the polypeptide backbone, and only the 6-OHCH$_2$-group of the glucose remains exposed (Steitz, Shoham and Bennett, 1981).

The tightness with which an enzyme can enfold its substrates is often revealed by its behaviour with inhibitors. For example, dihydrofolate reductase*, which is the receptor for the drugs methotrexate (*4.7*), trimethoprim (*4.9*), and pyrimethamine (*4.8*), is easily hydrolysed by proteolytic enzymes, but not when one of these drugs is present (Burchall, 1966; Hakala, 1966).

Further examples of conformational adaptation† will be given, each one quite individual in its mechanism, yet collectively supplying a picture of considerable uniformity. Ribonuclease forms a compact structure, hydrophilic outside, lipophilic inside, with a slot to receive the substrate. The active site makes use of histidine residues (numbers 12 and 119) that would otherwise be remote from one another (Kartha, Bello and Harker, 1967). The dimensions of the folded ribonuclease molecule are about $30 \times 30 \times 38$ Å (mol. wt. 15 000). When charged with a substrate, e.g. cytidine phosphate, one histidine residue binds the phosphate group, the other the sugar. The tertiary structure of α-chymotrypsin has been similarly worked out: the active site depends on the closeness of two amino acids that are on different strands, namely serine-195 and histidine-57

---

*Data on the amino acid composition of the active site of this enzyme are in Section 9.3.
†For a diagram of the conformational change that takes place when haemoglobin forms a link with oxygen, see Fig. 12.2.

(Matthews *et al.*, 1967). The hydroxyl group of the serine effects alcoholysis of the peptide bond to give an ester which is then hydrolysed in a reaction catalysed by the imidazole ring of the histidine residue.

With the help of an inhibitor, human carbonic dehydrogenase has been shown, in X-ray crystallography, to sport its active site in a cavity, 12 Å deep, at the bottom of which lies a zinc atom, bonded to three histidine residues (nos. 94, 96, and 119) and one water molecule. It is here that carbon dioxide, or a sulfonamide inhibitor, becomes bound (Kannan *et al.*, 1975).

Carboxypeptidase A, an important enzyme of the pancreas which hydrolyses only peptides with hydrophobic side-chains, was examined similarly at 2 Å resolution. The zinc atom, a necessary co-factor, was found to lie in a shallow depression in the surface of the molecule, adjacent to a deep lipophilic cavity. The zinc was bound by the following three residues: two histidines (69) and (196), and glutamic acid (72). The enzyme had 307 amino acid residues, and the mol. wt. was 34 500. Comparison of the contour maps for the enzyme with and without a reluctantly hydrolysing substrate (glycyltyrosine) shows that the phenyl-group fits into the deep cavity and thus forces the peptide carbonyl oxygen atom (of the substrate) against the zinc, which consequently loses a co-ordinated molecule of water. The free carboxyl-group of the substrate forms an ionic bond with arginine (145). This causes the arginine to move through 2 Å, thus disrupting nearby hydrogen bonds. This disturbance causes the free hydroxyl-group of tyrosine (248) to rotate through 120°, so that it enters the shallow depression close to the peptide bond of the substrate, and probably protonates the nitrogen atom in the susceptible peptide bond. Glutamic acid

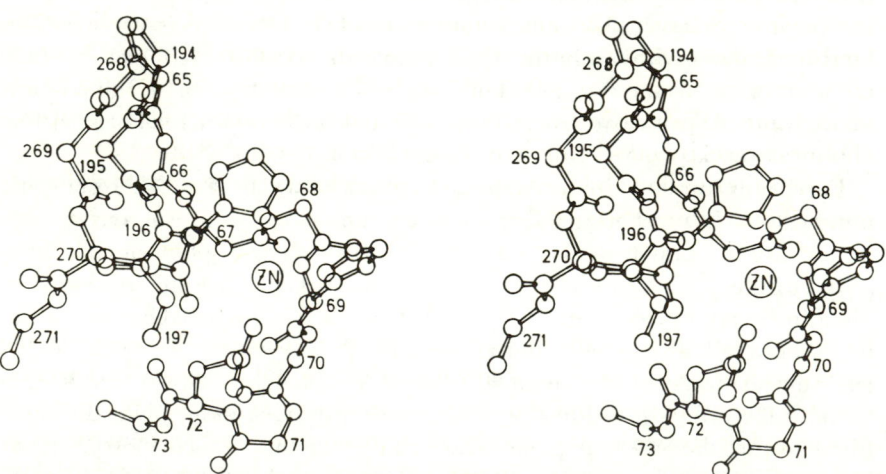

Fig. 9.1 Stereo-diagram of the active site in carboxypeptidase A. The substrate (not shown) is bound by Zn^{2+} and three amino acid residues: His-69, Glu-72, and His-196. Glu-270, which will be the final participant in the hydrolysis, is shown in its position before Arg-145 makes its 2 Å shift (Lipscomb, 1970.)

(270) then uses its anion to complete the hydrolysis (Lipscomb *et al.*, 1969; Lipscomb 1970; Breslow and Wernick, 1977). This situation is shown in a stereo-diagram (Fig. 9.1). For the structure of a *glyco*-protein, actually the neuraminidase of influenza virus, as found by X-ray crystal analysis, see Varghese, Laver and Colman (1983).

The more precise information which X-ray crystallography has provided about active sites has expunged those speculative sketches which, until recently, decorated the literature of enzymology. It seems a safe prediction that the similar sketches of drug receptors (e.g. Fig. 12.3) will go the same way.

9.0.2 *Sequencing*

The *order* of amino acids in an enzyme can seldom be completely established from X-ray data, because of localized blurring, whereas sequencing of enzyme hydrolysates gives it faster and quite exactly. Although sequencing gives no information about the active site, some valuable clues have been obtained in those cases where a reagent can form a covalent bond with a group in the active site. The organic phosphates (Section 13.3) have often proved useful substrates for studying proteolytic enzymes. Thus di-isopropyl phosphorofluoridate (*13.26*), taken up by chymotrypsin as a pseudo-substrate, alkylphosphorylates the enzyme at the active centre, thus making this enzyme unavailable for true substrates. It was found that one molecule of the inhibitor reacts with one active centre in each enzyme molecule. Hydrolytic degradation of the inhibited enzyme produced di-isopropylphosphorylated serine, and from this it was concluded that a serine residue played an important part in the active centre of the enzyme (Schaffer, May and Summerson, 1954). Degradative studies of this kind have shown that the sequence Asp-Ser-Gly occurs at the active centre of chymotrypsin (which has 230 amino acids in each molecule, of which 28 are serine) and trypsin, but Glu-Ser-Ala is found at the active centre of acetyl-cholinesterase, pseudocholinesterase, and liver aliesterase (Sanger, 1963).

Other information about enzyme active sites has been obtained from the use of molecular probes (Section 2.4).

9.0.3 *Enzyme kinetics*

Enzymes play their vitally important part in metabolism by speeding the numerous chemical reactions occurring within the cell. They act by lowering the free energy barrier and thus increase the speed at which a reaction will proceed. For example, the energy of activation for the decomposition of hydrogen peroxide is lowered, from 18 to 2 kcal/mol, by the enzyme catalase which accelerates the reaction ($H_2O_2 \rightarrow H_2O + O$) by a factor of 1.6×10^{11}.

This remarkable catalytic activity of enzymes depends only on ordinary principles of physical and organic chemistry: (a) The substrate molecule and the enzyme interact, changing the conformation of one or both, particularly in any way that favours holding the transition state more strongly than the substrate; (b) the enzyme can then initiate the scission of a covalent bond in the substrate thanks to the proximity of the reacting groups; (c) this is assisted by an electric field being supplied at very close range in an unusually anhydrous environment i.e. one of low dielectric constant; and (d) the amino acid side-chains, lining the pocket of the active site, can often supply new sites for temporary bonding, as a step towards completing the reaction. In bimolecular reactions, the enzyme can bring the two substrates together in higher concentration, more closely, and in better orientation than would be possible in solution.

Enzymes have two types of specificity: (a) various degrees of affinity for the substrate, and (b) various rates at which different substrates undergo reaction after binding has taken place. These are called, respectively, binding specificity and kinetic specificity and each is exerted independently of the other (Koshland and Neet, 1968).

The first to be isolated of those complexes which enzymes form with their substrates was the complex of D-alanine, D-amino acid oxidase, and its co-factor FAD (riboflavine adenine dinucleotide). It consisted of hexagonal purple crystals and was stable in the absence of air. Electron spin resonance measurements showed that the co-factor, present in these crystals, was in its free-radical monohydro-form (FADH) (Yagi, 1965).

9.0.4 *Simple models of enzymes*

Although all known enzymes are proteins, it has proved instructive to devise non-protein models in order to learn how enzymes accelerate common reactions to such a high degree while, at the same time, conferring extra specificity on both substrates and products. A starting point was found in histidine, known to be a protein component participating in many active sites. Here the essential group proved to be the imidazole ring (*9.1*) whose pK_a (about 7) enables it to act as both a proton donor and a proton acceptor at physiological pH values. Imidazole, and particularly 4'-imidazol-4-ylbutyric acid (*9.2*), greatly accelerate the hydrolysis of esters through general acid–base catalysis (Bruice and Benkovic, 1966).

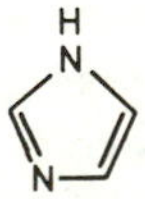

Imidazole　　　　4'-Imidazol-4-ylbutyric acid
(*9.1*)　　　　　　　　　(*9.2*)

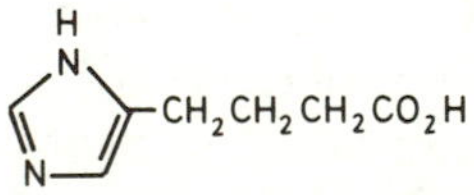

Behaviour that is much closer to that of enzymes has been obtained with the cyclodextrins (cycloamyloses) obtained by the degradation of starch (Cramer, 1956). These act as efficient esterases for phenyl acetate and its ring-substituted derivatives. X-ray diffraction analysis shows a torus (ring) consisting of six α-D-glucose units in the usual conformation; hydroxyl-groups form crowns around top and bottom of the torus, which has an internal diameter of about 5 Å. The carbonyl-group of the acetate 'guest' presses against the secondary hydroxyl-groups of the 'host': the closer the approach, the faster the hydrolysis. The 'guest' ester acylates a hydroxy-group in the 'host', thus releasing phenol. The non-acylated 'host' is then regenerated by intramolecular catalysis, and the cycle can continue indefinitely. The cycle further resembles an enzyme reaction by obeying Michaelis–Menten kinetics (Section 9.1), and by the ease with which it is inhibited by analogues of the normal 'guests' (Bender and Komiyama, 1978).

By the covalent addition of two imidazole groups to the cyclodextrin torus, artificial ribonucleases have been formed, and these have notable specificity with respect to both substrates and products (Breslow, 1983). Similar accelerations and specificities were found by coupling pyridoxamine covalently to a cyclodextrin to give an artificial transaminase. The accelerations produced by these models are only about 200-fold, but work in progress holds out hope for obtaining far greater velocities (Breslow, 1983).

From these models, as well as from X-ray studies of the natural enzymes, it is clear that hydrolysis requires the close approximation of a nucleophilic group (the anion of glutamic acid, aspartic acid, or imidazole) to the electrophilic centre (the carbon atom of the C–O group of an amide linkage) and the simultaneous presentation of a proton to the nitrogen atom of the same linkage. Cyclodextrin models have shown that enzyme-like action can be obtained with a molecule of mol. wt. just under 1000, and a structure of great simplicity and symmetry. For further reading on such model enzymes, see Jencks (1969), and Bender and Komiyama (1978). A recently reported macrocycle has accelerated by 10^{11}-fold the acetylation of amino acids (Cram, 1983).

Further light on enzyme action is shed by micelles of amphiphilic molecules, which catalyse many chemical reactions and are often regarded as very simple models for enzymes ('amphiphilic' and 'micelle' are defined in Section 14.0). The disposition of hydrophobic (inside) and hydrophilic (outside) groups in a micelle resembles that of enzymes. Like enzymes, too, micelles are denatured by heat or urea, and they show specificity towards substrates (Jencks, 1969). Reactions that liberate anions, e.g. the hydrolysis of esters, are catalysed by cationic micelles, e.g. those of cetyltrimethylammonium bromide (*14.16*) (Menger and Portnoy, 1967). The substrate is concentrated in the micelle. That van der Waals bonding then occurs is shown by the increasing efficacy of catalysis as the size of the substrate molecule is increased. The charged head groups of the micelle provide a charged environment (that may stabilize a transition state of the substrate) and are thought of as the actual catalysts. The low dielectric constant of the rest of the micelle can destabilize any charge on the substrate, its transition state, and its products.

What a micellar catalyst can achieve is exemplified by sodium dodecyl sulfate which accelerates 500000-fold the binding of Cu^{2+} by porphyrins (Lowe and Phillips, 1961). For general reading on catalysis by micelles, see Cordes (1973) and Fendler and Fendler (1975).

9.0.5 *Biological regulation of enzymes*

Enzymes functionally related to one another in a metabolic sequence are organized either in particles, or embedded in membranes. In these systems the enzymes are coupled to one another chemically, thermodynamically, and by physical location. Hence, as soon as a reaction product appears from the first enzyme it can become the substrate for the next one, and so on. Such sequences of enzymes are often controlled by an *allosteric effect*, as follows. Enzymes near the beginning of a chain of reactions may have a second receptor (at some distance) which can combine with a natural inhibitor. (This inhibitor is often the normal product of an enzyme further down in the chain of reactions.) The combination causes a conformational change in the enzyme and as a result its normal substrate cannot get to the active site. This feedback mechanism plays an important part in the self-regulation of the cell's metabolism (Monod, Changeux and Jacob, 1963). It has been shown that the presence of natural purines in a living cell can prevent further purine synthesis until the concentration falls below a certain level. Similar feedback control has been demonstrated with both natural and unnatural pyrimidines (Bresnick and Hitchings, 1961). The control of catecholamine synthesis by feedback mechanisms is shown in Fig. 9.2. For more on allosteric regulation, see Cohen (1980).

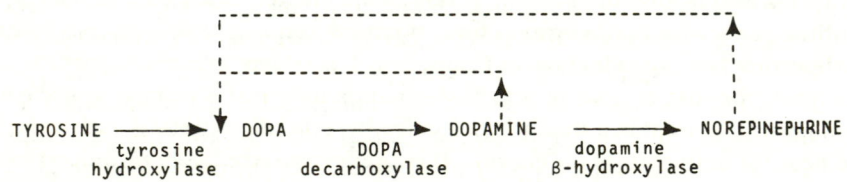

Fig. 9.2 Control of catecholamine synthesis.

9.0.6 *Enzymes and drugs*

Many selective agents owe their useful properties to inhibiting particular enzymes. Before any such relationship can be established, the following requirements must be met (Hunter and Lowry, 1956): (a) the enzyme should be inhibited in the intact cell, as well as in the isolated state; and (b) enzyme inhibition must not require concentrations of the agent greater than are commonly used to elicit the pharmacological effect on the cell.

By these standards, organic phosphates, and also urethanes, have been shown to exert their biological effects by inhibiting acetylcholinesterase; and oximes exert their antidotal effect by reversing this inhibition (see Section 13.3). Many of the most interesting and useful examples of agents acting by inhibiting enzymes are those which show a close resemblance, both in structure and in electron-distribution, to a substrate or coenzyme. Such agents are called 'anti-metabolites' and most of the rest of this chapter will be devoted to them.

For a simple introduction to the chemistry of enzyme action, see Williams (1969), for a list of enzymes with discussion, see Dixon and Webb (1979), and for a multivolume work on enzymes, see Boyer (1970–1983). Isoenzymes and isofunctional enzymes were discussed in Section 4.2. For the classification of enzymes, see International Union of Biochemistry (1978).

9.1 Anti-metabolites (antagonistic analogues): definition, derivation, and mode of action

Those metabolites (substrates or coenzymes) that are present in only small amounts in a cell or tissue can be antagonized by *anti-metabolites* (analogues). The molecules of each such analogue have a region which is similar to that region of the metabolite which makes contact with the enzyme protein. To be effective, this similarity must be not only in dimensions, but also in electron-distribution, because most of the active sites on enzymes are highly polar. Each analogue exerts its antagonism by occupying and blocking the enzyme site used by the metabolite (see Section 9.2).

Although a metabolite can be turned into an analogue by effecting a chemical alteration in the molecule, the change must not be too large or the normal action is lost without any antagonistic action being created. Various additions and subtractions of a methyl-group from thiamine seriously diminish its vitamin properties (see introduction to Chapter 2) but do not create antagonists. In general, the loss or gain of a methyl-group is too great a change to make an antagonist, at least in a smallish molecule where the affected site is likely to be in or near the active site of the enzyme. It is safer to conserve the steric properties of the metabolite (to ensure attachment to the enzyme) but vary the electronics (to make the drug unsuitable as a substrate). We shall soon encounter some antagonists formed by exchanging hydrogen for fluorine, and methyl for chlorine, in conformity with these guidelines.

Some antagonists are of such a chemically simple nature that their relevance as metabolite analogues is often overlooked. For example, inorganic cations are in competition with other inorganic cations (Sections 9.2 and 11.0). Even the hydrogen ion, one of the most important of all metabolites according to Mitchell (1979)*, enters into competition with cations both organic and inorganic (Section 10.3.1). Ethanol in beverages acts partly by competing with water

*Mitchell's chemiosmotic concept states that a proton gradient is the sole coupler of electron transport to ATP synthesis.

(for distribution at least). Ethanol can also antagonize the toxic effects of methanol, by displacing it from the enzyme that oxidizes it (Röe, 1955).

Several examples of competition between simple anions have been observed. Thus the perchlorate and the thiocyanate anion inhibit concentration of the iodide anion by the thyroid gland, without affecting the oxidative incorporation of the iodide anion into thyroxine (Stanbury and Wyngaarden, 1952). Likewise, the organism *Nitrobacter,* which oxidizes nitrite to nitrate, is inhibited by cyanate or chlorate (anions), an effect easily reversed by washing (Lees and Simpson, 1957).

Some physiological processes are regulated by pairs of analogous metabolites, for example the polyene sex hormones of algae (Kuhn, 1940), and the mammalian sex hormones. Of two prostaglandins present in human lungs, PGE_2 relaxes, and PGF_{2_a} contracts, bronchial muscle. Similarly prostaglandin D_1 inhibits the increase of vascular permeability produced in the rat skin by prostaglandins E_1, E_2, and D_2 (Flower and Kingston, 1975). Another pair of chemically similar regulators is prostacyclin and thromboxane, in the human body. The former is the most potent natural inhibitor of platelet aggregation and a powerful vasodilator, whereas the latter has exactly the reverse properties. Yet both substances arise from the same precursor, prostaglandin endoperoxide (Section 4.7).

Very few examples are known of antagonists made by rearranging the order of groups at an asymmetric carbon atom. One such antagonist is D-histidine which inhibits the enzyme histidase which normally opens the imidazole ring of L-histidine (Edlbacher, Baur and Becker, 1940). This method is seldom effective in such small molecules, because the space-relationships required for adsorption on the enzyme are the very ones altered by the rearrangement. What is more often found is that, if the (+)isomer has a strong physiological action, the (−)isomer has the same kind of action, but much weaker. In other words, it is seldom an antagonist. For this reason, a mixture of two optical antipodes (or the racemized substance) usually has the averaged potency of the two constituents, as happens with thyroxine, atropine, and DL-epinephrine. An ingenious use of unnatural optical isomers is in the design of polypeptide drugs, to make them indigestible by the patient's gastric juices (see enkephalin analogues in Section 12.8).

A reliable method for procuring an antagonist for a large molecule is to use a smaller molecule which resembles a repeating unit in the large one. Thus starch is hydrolysed by amylase to maltose which is an anhydro-dimer of glucose. This enzyme is strongly inhibited by glucose, although that sugar does not occur in the hydrolysis. Galactose, which is a stereoisomer of glucose, is less inhibitory, and fructose, which is an isomer of glucose but not a stereoisomer, is not inhibitory at all (Wohl and Glimm, 1910).

Small molecules can have near-homologues for antagonists, e.g. malonic acid (*9.3*) antagonizes the oxidation of succinic acid (*9.4*) by succinic dehydrogenase (Quastel and Wooldridge, 1927).

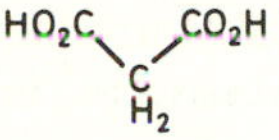

Malonic acid
(9.3)

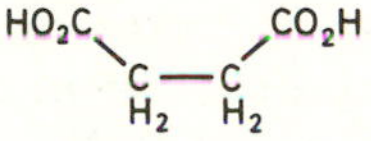

Succinic acid
(9.4)

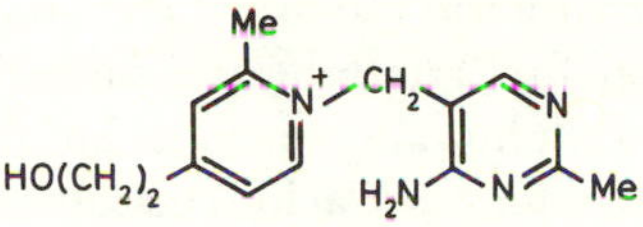

Pyrithiamine
(9.5)

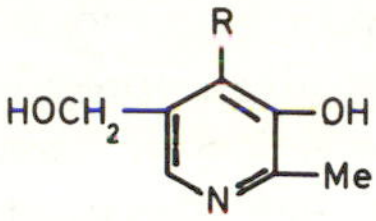

Pyridoxin (R = —CH₂OH)
(9.6)

Another way to form antagonistic analogues is to make a small change in the atoms that form a ring. When the sulfur atom in thiamine (2.1) was replaced by an ethylene group, the product, pyrithiamine (9.5), produced characteristic symptoms of thiamine-deficiency in mice (Woolley, 1950). It seemed only logical to replace the thiazole by the pyrimidine ring, because these are isosteres, just as thiophene and benzene are (Hartough, 1952). It was found that pyrithiamine displaced thiamine from the enzyme that normally phosphorylates it to give thiamine pyrophosphate, the coenzyme of pyruvate dehydrogenase. Unfortunately, pyrithiamine did not prove to be a useful drug, but modification of its structure led to amprolium (9.37), successfully used in treating a protozoal disease of poultry.

Similarly, it is valid to make the reverse change, namely to replace an ethylene moiety by sulfur, e.g. converting phenylalanine to thienylalanine, which is a strong antagonist for this amino acid in micro-organisms (Dittmer, 1949). In many other cases, such changing of the benzene nucleus into thiophene has been effective. For example, α-thienylalkylamines resemble the corresponding phenylalkylamines in hypertensive activity, and 2-thenoic esters of alkylamines resemble the corresponding benzoic esters as local anaesthetics (Hartough, 1952).

One of the best general methods for obtaining antagonists is to substitute one electron-attracting group for another. Thus —COOH may be replaced by —COCH₃, —SO₂OH or —SO₂NH₂. When designing replacements of this sort, it is important not to alter the ionization of any basic group present in the molecule. Thus the amino-group in p-aminobenzoic acid (anion) (9.7) is not ionized, and hence it is not admissible to replace —COOH by SO₃H because the basic group would then become ionized and the new substance too dissimilar to be a good analogue (this has been experimentally verified). On the other hand, the sulfanilamide anion (9.8) is satisfactory as an antagonist for the p-aminobenzoic acid anion (9.7), as will be further discussed in Section 9.3.

Effective antagonists have been made by substituting fluorine for hydrogen, as in the man-made agonists* of androgens and corticosteroids (see Gilman,

*In these agonists, the presence of fluorine brings about reversible blockade of the degrading e.r. enzymes.

Goodman and Gilman, 1980); and also in the anti-cancer drug, fluorouracil (Section 4.0.2) and the anti-fungal agent, flucytosine (*4.23*). Many such drugs are used in human medicine, but (a note of caution) the substitution of fluoro-citric acid for citric acid is disastrous (Section 13.5).

Similar replacement of a methyl-group by chlorine is often effective, as in the riboflavine antagonists (Kuhn, Weygand and Möller, 1943). Table 9.1 shows why these changes are sterically appropriate, and why the substitution of chlorine or methyl for hydrogen does not usually give effective antagonists. The unifying feature is that these analogues must be so similar to the substrate that the enzyme is deceived into taking up the foreign molecule in place of the substrate. Yet the analogue must be dissimilar enough to be incapable of functioning as the substrate does. That is to say, it must either fail to undergo the very next chemical reaction normal for the substrate, or if it does undergo this reaction, the product must be unacceptable to an enzyme later in the sequence of reactions.

Table 9.1 Some van der Waals radii relevant to metabolite antagonism.(Speakman, 1968.)

Substituent	Radius (Å)	Substituent	Radius (Å)	Substituent	Radius (Å)
H	1.2	N	1.55	C_6H_5	~1.8
F	1.35	Cl	1.8	CH_3	2.0
C	1.4	S	1.85	I	2.15
O	1.4				

Antagonists can be made by omitting some small group that normally undergoes a covalent change during normal metabolism. Thus desoxy-pyridoxine [(*9.6*), R = —CH$_3$] produces signs of vitamin deficiency in Man, rapidly reversed by this vitamin [(*9.6*), R = —CH$_2$OH] (Mueller and Vilter, 1950) Desoxypyridoxine becomes biologically active only after it is phos-phorylated in the body to a derivative that competes with pyridoxal phosphate, the coenzyme of the amino acid decarboxylases.

Apparently no molecule is too large to have an antagonist: acetylation of the thyroid-stimulating hormone (TSH), a protein of the pituitary gland, gives an analogue which accumulates in the thyroid gland and reduces hyperthyroidism by blocking the action of TSH (Sonenberg and Money, 1957). Again, very small chemical alterations of other polypeptide pituitary hormones, such as oxytocin and vasopressin, produce antagonists to the relevant hormones (Dyckes *et al.*, 1974; Manning *et al.*, 1977). Thus, the replacement of the 1-amino group in oxytocin by penicillamine gave a potent antagonist to that hormone.

Nature has evolved many examples of antagonism at the polypeptide level. A polypeptide (mol. wt. 6512) called aprotinin ('Trasylol'), abundantly present in mammalian tissues, inhibits the hypotensive kinin of the pancreas (kalikrein) as well as the proteolytic action of the enzymes trypsin, chymotrypsin, and plasmin. It has been used with success to treat pancreatitis (Trapnell, 1977). Again, melanostatin and somatostatin polypeptides from the hypothalamus, inhibit release, respectively, of melanocyte-stimulating hormone (a poly-peptide) and somatotrophin (growth hormone, a protein) by the anterior pituitary gland.

After Woods discovered the anti-metabolite nature of the action of sulfon-amides on bacteria (Section 9.3.1), an intense search began for other antagon-ists that would be useful drugs. This turned out to be an unexpectedly difficult project because economic and uneconomic cells share many common bio-chemical pathways. For some time the principal successes were confined to anti-folic acid drugs, commencing with the sulfonamides (Section 9.3.1, 9.3.3). However, little by little many other useful metabolite antagonists were dis-covered, some by chance but others by design, and these are reviewed in Section 9.4. The position today is that comparative biochemistry must provide strong evidence of a reaction *uniquely* carried out by the uneconomic cell for it to be worth while to make analogues of the metabolite involved. Fortunately some of the innumerable *unselective* agents, synthesized during this long period of misunderstanding, have been welcomed by biochemists as specific reagents for blocking various metabolic processes *in vitro*.

For reviews of metabolite antagonists in the widest sense, i.e. without special reference to selective toxicity, see Hochster and Quastel (1963–1973).

(*a*) *Quantitative aspects*. The relationship between metabolites and anti-metabolites is usually competitive. That is, if x molecules of metabolite are antagonized by y molecules of analogue, then $10x$ molecules of metabolite require $10y$ molecules of analogue to give the same biological end-point, and so on. As such competitive reactions are freely reversible, the antagonism of x molecules of metabolite by y molecules of analogue can be abolished by another x molecules of metabolite, and so on. Malonic acid and succinic acid have a competitive relationship of this kind; sulfanilamide and *p*-aminobenzoic acid provide another example.

For each pair of substances there will be a unique *index of inhibition*, defined as the ratio of the number of molecules of analogue (to those of metabolite) required to give 50% inhibition. This ratio will vary with the biological species, but is always the same for any one species. It is obviously an expression of the relative affinity of analogue and metabolite for a receptor-group, but it also includes a term for differences in the penetration of the two substances when the site of action is not exposed. The amount of inhibition which any analogue can produce therefore depends on two things: firstly, its affinity for the receptor relative to that of the metabolite, and, secondly, the relative amounts of analogue and metabolite available at the site of action.

The affinity of a substrate for an enzyme is represented by Equation (i):

$$[E]+[S] \rightleftharpoons [ES] \tag{i}$$

where $[E]$ is the concentration of the enzyme, $[S]$ is that of the substrate, and $[ES]$ that of the complex which they form. The affinity of a drug, when it is an inhibitor (I) of an enzyme, is similarly formulated. Hence the dissociation constant for an inhibitor (K_i) is:

$$K_i = \frac{[E][I]}{[EI]} \qquad \text{(ii)}$$

Unlike an inhibitor, a substrate is changed by the enzyme into a product (P), and the sequence of events becomes:

$$[E]+[S] \underset{k''}{\overset{k'}{\rightleftharpoons}} [ES] \overset{k'''}{\to} [E]+[P] \qquad \text{(iii)}$$

The ratio k''/k' equals K_m, the Michaelis constant, which records the dissociation of the enzyme–substrate complex [ES] into its components [E] and (S) (Michaelis and Menten, 1913). It pertains to the Equilibrium (iv) which is formally analogous to (ii):

$$K_m = \frac{[E][S]}{[ES]} \qquad \text{(iv)}$$

The *index of inhibition* of an inhibitor is the ratio K_i/K_m; the smaller the index, the more efficient the inhibitor.

The Michaelis constant is obviously inversely proportional to the affinity of the enzyme for the substrate, and is numerically equal to the substrate concentration when the reaction has reached half its maximal velocity. The dimensions of K_i and K_m are recorded as g mol per litre. However, they are not true equilibrium constants but ratios of velocity constant for the forward and reverse reactions. A suitable determination of K_m is by the graphical method of Lineweaver and Burk (1934), where the initial rate of formation of ES is plotted against substrate concentration, both as reciprocals. This should give a straight line and, if it does, the value of K_m is shown at the intersection of slope and abscissa.

Because the busy metabolic traffic of cells produces a *steady state* more often than an equilibrium, much use has been made of an equation devised by Briggs and Haldane (1925) who showed that it was unnecessary to assume an equilibrium between [E] and [S]. They derived Equation (v), formally similar to the Michaelis–Menten equation but free from this assumption and suitable for steady-state conditions.

$$K_{bh} = \frac{k''+k'''}{k'} \qquad \text{(v)}$$

This constant, more general than the earlier one, is a measure of the simultaneous dissociation of ES in two opposite directions. Where K_{bh} replaces K_m, the index of inhibition becomes the ratio K_i/K_{bh}. All of these calculations assume that the inhibitor is not being partly segregated (or even destroyed) by other biological material which may be present.

The molar inhibitory index of malonic acid is 1/3 (Thron, 1953). Even more

economical inhibition is shown by carbon monoxide when it replaces oxygen on
haemoglobin (1/210). Values below unity, such as these, are uncommon,
because it is rare to find an analogue that has a greater affinity for a natural
receptor than the normal substrate has. Methotrexate (4.7) may owe its
unprecedentedly high index of 1/10000 to the fact that the most basic nitrogen
atom is N-1 in this inhibitor, whereas it is N-5 in dihydrofolic acid which does
not make for so good a fit (see Sections 4.2 and 9.3.3). This lower-than-unity
inhibitory index (1/10000) of methotrexate stands at one extreme of what is
commonly found for a useful drug, and the greater-than-unity index (300/1) of
sulfanilamide versus p-aminobenzoic acid stands at the other extreme beyond
which little therapeutic gain might be expected.

Why is it considered to be desirable for the inhibitory index of a drug to be
low? It is because most enzymes have evolved, under the selection pressure of
Nature, to handle their natural substrates efficiently. But another reason for the
inefficiency of many competitive inhibitors is the *steady-state* condition of the
living cell. The target enzyme is continuously supplied with substrate (from the
previous enzyme in the chain) and its product is continuously removed by the
next enzyme. Hence target enzymes usually turn out to be *under*saturated with
substrate and so have great reserve capacity. It is true that the addition of a
competitive inhibitor to a steady-state system initially inhibits the target
enzyme; but accumulation of more of the normal substrate of the inhibited
enzyme soon overcomes this inhibition. There then exists a new steady-state
situation in which the throughput of the pathway is unaltered because the
concentration of the substrate in the target enzyme is maintained at a higher
level in order to counterbalance the effect of the inhibitor (Cleland, 1970). This
exemplifies the difference between the chemistry of the model systems that are
studied in the laboratory and more complex systems that have evolved in the
living cell.

9.2 History of analogue antagonism prior to 1940

Studies of analogue antagonism were foreshadowed by the London work of
Sidney Ringer (1883) who found, from a helper's error which he was quick to
interpret, that the sodium (cations) in a solution of sodium chloride could not
maintain the beat of an isolated heart unless balanced by calcium and
potassium. As a result of this work, the physiologically balanced solutions,
named after Ringer, Locke and Tyrode, were developed.

The next relevant discovery was made in 1910, when it was noted that some
enzymes are blocked by substances whose molecular structure resembles that of
the normal substrates. Thus amylase, which normally hydrolyses starch, is
inhibited by glucose (see Section 9.1). Again, malonic acid (9.3) competitively
inactivates the enzyme succinic dehydrogenase by displacing the normal substrate, succinic acid (9.4), from the enzyme (Quastel and Wooldridge, 1927). A
similar phenomenon in physiology is the toxic action of carbon monoxide

(C—O), which is due to its displacing a similarly shaped molecule, oxygen (O—O), from combination with haemoglobin (Douglas, Haldane and Haldane, 1912).

Also, as has been related in Section 2.1, it was shown how the alkaloid physostigmine (*2.8*) contracted the pupil of the eye by blocking acetylcholin-esterase, thus allowing local ACh (*2.7*) to stimulate the muscle (Stedman, 1926). It was soon found that the portion of the molecule which inhibited the esterase was the methylcarbamoyloxy-group (*2.9*) (Stedman and Stedman, 1931). In 1932, it was discovered that carbachol (*2.11*), which has a carbamoyloxy-group, has much of the biological effect of acetylcholine (*2.7*). This made it clear for the first time that the reason why physostigmine blocked acetylcholinesterase was because the group (*2.9*) caused that enzyme to adsorb the physostigmine instead of its normal substrate, acetylcholine.

Although this story proved a little too complex to be widely appreciated, and few were inclined to generalize from the evidence, the topic acquired a new facet when Ing (1936) published his much-quoted review. In this he pointed out that the alkaloid tubocurarine (*2.6*), which blocks neuromuscular transmission (see Section 2.1), must do so by competing with acetylcholine for a receptor in voluntary muscle (both substances are quaternary amines). In addition, Ing postulated that the somewhat weaker curariform action of innumerable quaternary ammonium, phosphonium, arsonium, stibonium, and sulfonium salts arose from their competition with acetylcholine (Ing, 1936). (For further reading on the physiological functions of acetylcholine and how they are antagonized by drugs, see Section 7.3, 7.6.3, 12.5 and 12.6.)

Soon after this, the first anti-vitamin was discovered by a happy accident. Woolley *et al.* (1938) prepared two analogues of nicotinic acid, namely 3-acetylpyridine and pyridine-3-sulfonic acid. Believing that the analogues would have, at least qualitatively, the biological action of the vitamin, they fed them to dogs suffering from nicotinic acid deficiency. To their surprise, the condition was worsened. A clear picture of what was happening emerged as soon as Woods (1940) demonstrated the reversal of the antibacterial action of sulfanilamide by *p*-aminobenzoic acid, and pointed out that this reversal depends on the structural similarity of the two substances.

9.3 The folic acid antagonists

There are antagonists of the biosynthesis of dihydrofolic acid, and antagonists of its utilization. The history of the discovery of the antibacterial sulfonamides, typical antagonists of biosynthesis, was given in Sections 2.1 and 6.3.1. In 1940, Woods showed that the anti-bacterial action of sulfanilamide depended on its competition with *p*-aminobenzoic acid (*9.7*), which is a natural metabolite (Woods, 1940). Later this competition was shown to take place at the site on the enzyme dihydrofolate synthetase, which uses *p*-aminobenzoic acid to build up the molecule of dihydrofolic acid (*2.14*) (G.M. Brown, 1962).

The enzyme accepts sulfanilamide in place of its normal substrate because of their close resemblance, electronically and sterically. *p*-Aminobenzoic acid has an acidic pK of 4.9 and is not a zwitterion like glycine. The anion (*9.7*) appears to be the biologically active form. Sulfanilamide is a weaker acid (pK 10.3), little ionized at a physiological pH. The primary amino-group in both substances is weakly basic (pK_a 2.5 and 2.6 respectively) and not ionized at a physiological pH. The dimensions of *p*-aminobenzoic acid (anion) (*2.12*) and of sulfanilamide (molecule) (*2.13*) are similar (see formulae). They have similar length, and width; both molecules are planar, and the amino-group must be both primary and situated *para* to the other group for there to be any biological activity in each case. The given dimensions are little changed by ionization (Bell and Roblin, 1942).

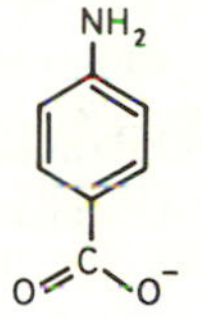

p-Aminobenzoic acid (anion)   Sulfanilamide (anion) (R = H)
        (*9.7*)   (*9.8*)

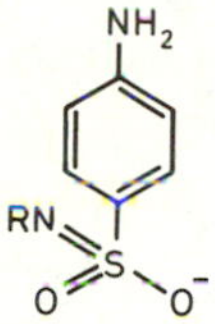

As soon as the clinical value of sulfanilamide was established, many modifications of the molecule were attempted in the hope of finding more active analogues. It was soon found that the most valuable sulfonamides were those in which the 'R' in (*9.8*) was a heterocyclic ring. As Bell and Roblin showed (1942), this increased the ionization (as an acid) of the sulfonamide; and those sulfonamides that were completely ionized at pH 7 and hence resembled *p*-aminobenzoic acid still more closely, were found to be the most strongly antibacterial of all (see, further, Section 10.5). Although those of the sulfonamides which cannot ionize (as acids) can show useful antibacterial action (examples: dapsone, sulfaguanidine), it is always much weaker than that given by well-ionizing sulfonamides. Also, the minimal inhibitory concentration of sulfadiazine (2-*p*-aminophenylsulfonamidopyrimidine) against *E. coli* is 1.02 μmol/litre whereas sulfanilamide is more than 100 times weaker (Krüger-Thiemer and Bünger, 1965). This correlated with the enhanced ionization of sulfadiazine (pK_a 6.5) which yields 75% of anion, at equilibrium (at pH 7). In all these *N*-substituted sulfonamides (*9.8*), the 'R' is out of the plane of the rest of the molecule, and hence it cannot interfere with the adsorption on the receptor normally occupied by (*9.7*).

The selectivity of the antibacterial sulfonamides depends on the non-utilization of *p*-aminobenzoic acid by mammals, which do not make their own dihydrofolic acid, but obtain it in food. Pathogenic bacteria, on the other hand, cannot absorb preformed dihydrofolic acid (Wood, Ferone and Hitchings, 1961), and hence are vulnerable to the sulfonamides which prevent them from synthesizing it.

Sulfapyridine (Ewins and Phillips, 1939), the first heterocyclic-substituent-type sulfonamide, was soon superseded by sulfathiazole, which was in turn displaced by the three more selective sulfapyrimidines shown in Table 2.5; these, by 1942, were widely accepted as the most useful and innocuous of the sulfonamides for oral use in a wide variety of severe bacterial infections.

At the present time, the largest use of sulfonamide antibacterials is as urinary antiseptics, e.g. those caused by *E. coli* and *Proteus mirabilis*. They are also used in treating nocardiosis (of lungs or foot), trachoma (eye), lymphogranuloma venereum, dermatitis herpatiformis, and valued for the prophylaxis of strepto-coccal infections in those prone to them, and in preventing recurrence of rheumatic fever.

The antibacterial sulfonamides fall mainly into two classes, (a) those that are eliminated fairly rapidly and (b) those that maintain a high blood-concen-tration for a long time. Of class (a), the most used are:

(i) Sulfadiazine, which is *N'*-(pyrimidin-2-yl)sulfanilamide (*9.9*), a well-established example, in fact the reference substance to which all others are compared. That it gives a therapeutic concentration in cerebrospinal fluid, enlarges its sphere of usefulness.

(ii) Sulfafurazole (sulfisoxazole in the USA) (*9.10*) is *N'*-(3,4-dimethyl-isoxazol-5-yl)sulfanilamide, another general-purpose example which achieves a higher concentration in the urine than sulfadiazine.

(iii) Sulfamethoxazole ('Gantanol') (*9.11*) has a rather long half-life for this class (12 hours) and has become the drug of choice for synergism with trimethoprim (see Section 9.6). It is often said that this synergic use with diaminopyrimidines has given the antibacterial sulfonamides a new lease of life (see Section 9.6).

(iv) Sulfacitine ('Renoquid') (*9.12*) and (v) sulfamethizole ('Thiosulfil'; 'Urolucosil') (*9.13*) are highly favoured as urinary antiseptics because of their short half-life in the bloodstream, and failure to accumulate elsewhere.

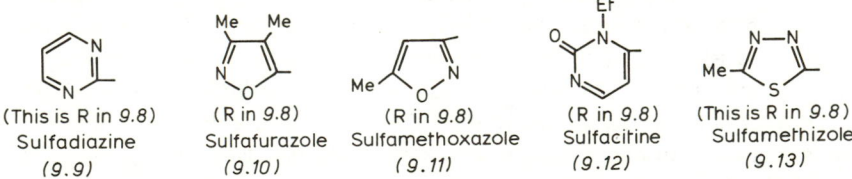

(This is R in 9.8)	(R in 9.8)	(R in 9.8)	(R in 9.8)	(This is R in 9.8)
Sulfadiazine	Sulfafurazole	Sulfamethoxazole	Sulfacitine	Sulfamethizole
(9.9)	*(9.10)*	*(9.11)*	*(9.12)*	*(9.13)*

Sulfonamides that are designed to be eliminated rapidly (class a) must have a high solubility in urine, and so must their acetyl-derivatives into which they are always partly converted. Failure to observe this precaution in the 1940s led to many deaths from sulfathiazole through kidney blockage. There is no problem of this kind with sulfonamides of class (b), namely those that maintain so high a blood-concentration that one dose per day often suffices. The principal disadvantage of prescribing these drugs is that, should the patient show an adverse reaction, as many as 3 days may elapse before the body is clear of them. The most feared adverse reaction is the Stevens-Johnson syndrome or erythrema multiforme which, though unusual, can be fatal. The most prescribed examples are:

(i) Sulfamethoxypyridazine ('Lederkyn') (9.14) which is N'-(6-methoxy-pyridazin-3-yl)sulfanilamide.

(ii) Sulfamethoxydiazine ('Bayrena'), N'-(5-methoxypyrimidin-2-yl)sulfanilamide.

(iii) Sulfametopyrazine (sulfalene, 'Kelfizine'), N'-(3-methoxypyrazin-2-yl)sulfanilamide (9.15).

(iv) Sulfadimethoxine ('Madribon'), N'-(3,6-dimethoxypyrimidin-4-yl)sulfanilamide.

(v) Sulfadoxine ('Fanasil'), N'-(5,6-dimethoxypyrimidin-4-yl)sulfanilamide. This, one of the safest, is much used in conjunction with a diamino-pyrimidine to produce sequential blocking (Section 9.6).

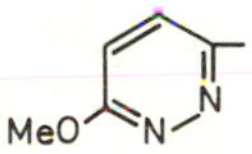

(R in 9.8)   (R in 9.8)   (R in 9.8)
Sulfamethoxypyridazine   Sulfametopyrazine   Sulfacetamide
(9.14)   (9.15)   (9.16)

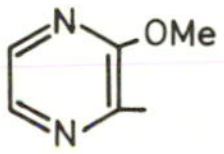

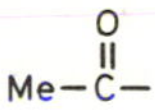

Also prescribed are some specialized sulfonamides for particular conditions: silver sulfadiazine (topical application to severe burns), sulfacetamide sodium (9.16) (in eye infections), sulfapyridine (in dermatitis herpetiformis), sulfasalazine (inflammation of the bowel), and phthalylsulfathiazole (pre-operative, to lower count of intestinal flora).

Factors which govern the distribution of sulfonamide drugs are discussed in Section 10.5.

It is important to note that many *non-sulfonamides* are known that can compete with *p*AB (*p*-aminobenzoic acid). The most used of these is dapsone (9.17), 4,4'-di(aminophenyl)sulfone, the sheet anchor of all leprosy therapy. Others contain no sulfur atom, but have the requisite steric and electronic resemblance to *p*AB. For example, the insertion of a chlorine atom into either the 2- or the 3-position of *p*AB makes an active anti-*p*AB substance (Wyss, Rubin and Strandskov, 1943). Diaminobenzil (2.15) is several times more active against bacteria than sulfanilamide, and is reversed by *p*AB (Kuhn, Weygand and Möller, 1943). Again, *p*-aminobenzenearsonic acid, atoxyl (6.2) has a typical sulfanilamide-like action (Albert, Falk and Rubbo, 1944). Although, in general, arsonic acids are not antibacterial, atoxyl forms an exception because it resembles *p*AB sufficiently, sterically and electronically, to compete with it.

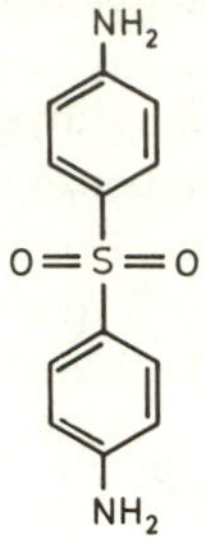

Dapsone
(9.17)

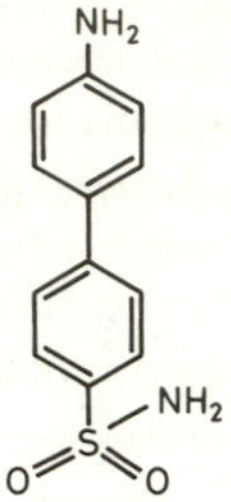

Spaced analogue
of sulfanilamide
(9.18)

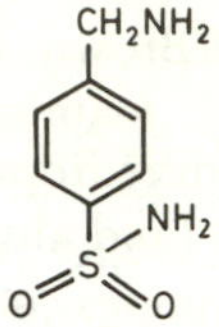

Mafenide
(9.19)

What features does a molecule require (other than the steric and electronic ones discussed in Section 2.1) to have an anti-*p*AB action? Two features are needed to make it acceptable to the enzyme, dihydrofolate synthetase, as a replacement for *p*AB. Firstly, it is essential that it should have a primary aromatic amino-group. This *N*-4-group must not be substituted in any way unless by a group which will readily break down in the body and liberate the primary amino-group. An azo- or anil-linkage can be depended on to break in this way, as in sulfachrysoidine (*3.30*) ('Prontosil'), but acyl or alkyl groups cannot (Northey, 1948). Secondly, an electronegatively charged group is required, placed *para* to the amino-group, and at the same distance as in *p*-aminobenzoic acid. The necessity for keeping the distances between the amino- and the electronegative-groups similar to that obtaining in *p*-aminobenzoic acid is illustrated by 4-amino-4'-sulfonamidodiphenyl (*9.18*), which has no anti-*p*AB effect.

Mafenide (4-aminomethylbenzenesulfonamide) (*9.19*) ('Marfanil'), which has only a 'paper-resemblance' to sulfanilamide, is a highly basic substance with special activity against the *Clostridia* which cause gas-gangrene. It is not antagonized by *p*-aminobenzoic acid (Jensen and Schmith, 1942) and appears to play no part in folic acid metabolism.

Many useful drugs contain the sulfonamido-group but are not anti-bacterial because they have not been designed as analogues of *p*AB. Some are diuretics (Section 9.4.7), and others anti-diabetics (Section 12.4).

9.3.1 *How sulfonamide drugs act by antagonizing synthesis of dihydrofolic acid*

The discovery of the folic acids led to a complete understanding of how sulfonamides antagonized *p*-aminobenzoic acid. This discovery took place as follows. A bright yellow substance, that had an anti-anaemic effect in vertebrates, had been isolated some time earlier from liver, yeast, green leaves, and bacteria. Degradation, followed by synthesis, established the constitution as in formula (*9.20*) minus the two hydrogen atoms in the 7- and 8-positions (which had been lost in the handling). This substance was named folic acid (pteroylglutamic acid) (Waller *et al.*, 1948). The molecule of dihydrofolic acid (*9.20*) consists of three main regions: these are the glutamic acid, the *p*-aminobenzoic acid, and the 2-amino-4-oxo-6-methylpteridine regions.

Bacteria use *p*-aminobenzoic acid only for conversion to 7.8-dihydrofolic acid (Griffin and Brown, 1964). Thus, *E. coli* condenses *p*-aminobenzoic acid (and, alternatively, *p*-aminobenzoylglutamic acid) with 2-amino-4-oxo-6-hydroxy-methyl-7,8-dihydropteridine (*9.21*) (as the 6-pyrophosphate) to give dihydro-pteroic acid (and alternatively, dihydrofolic acid) (Jaenicke and Chan, 1960). The sulfonamides competitively inhibit the isolated enzyme *dihydrofolate synthe-tase* which catalyses these steps (G. Brown, 1962). From *Lactobacillus plantarum* two enzymes responsible for this synthesis have been isolated in a pure state (Shiota *et al.*, 1969a). The first of these catalyses the esterification of the pteridine (*9.21*) to its pyrophosphoryl derivative. The second is Brown's dihydrofolate synthetase. This second enzyme has also been isolated from several strains of *Pneumococcus*, found to have a mol. wt. of 90 000, and to need ATP and Mg^{2+} as coenzymes (Ortiz, 1970).

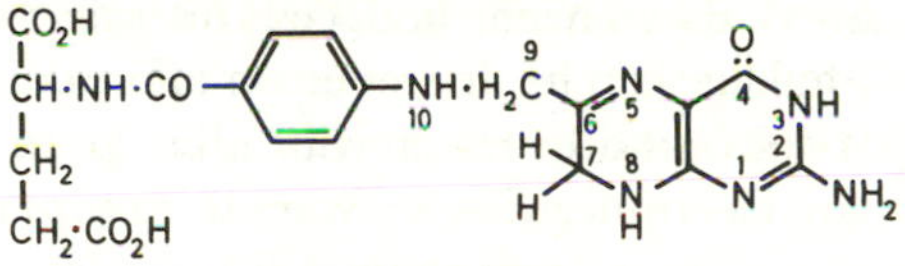

Dihydrofolic acid
(9.20)

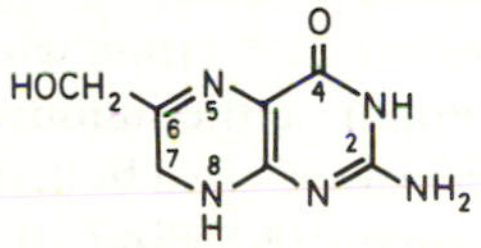

Pteridine intermediate in
folic acid biosynthesis
(9.21)

The effects of pteroic and folic acids (also their 7.8-dihydro-derivatives) cannot be antagonized by sulfonamides. This is most clearly demonstrated in the very few strains of bacteria which are capable of adsorbing folic acid (e.g. *Streptococcus faecalis*, and some *Lactobacilli*). (These bacteria are non-pathogenic to Man.) Table 9.2 demonstrates the exact proportionality between the amount of sulfadiazine required to inhibit Ralston's strain of *Streptococcus faecalis* and the amount of *p*AB required to reverse this inhibition. In contrast, the amount of folic acid required for reversal is constant, regardless of the amount of sulfa-diazine used. This indicates that the sulfadiazine interferes with the synthesis of folic acid from *p*-aminobenzoic acid, but does not interfere with the *utilization* of folic acid as, for instance, methotrexate does. Similarly, *L. arabinosus* (which requires an external source of *p*AB) is inhibited by sulfonamides, and this inhibition is reversed competitively by *p*AB and non-competitively by folic acid. The amount of folic acid (measured by disintegrating the bacteria, and titrating the homogenized culture against *L. casei* which responds to folic acid but not to *p*AB) produced by this organism is ordinarily proportional to the amount of *p*AB in the medium. However, if sulfanilamide is also present, it is decreased in proportion to the amount used, over a 10000-fold range of sulfanilamide concentrations (Nimmo-Smith, Lascalles and Woods, 1948). The production of folic acid is similarly inhibited in *E. coli* by sulfanilamide (Miller, 1944).

Table 9.2 Competitive and non-competitive inhibition

Substance	Amount needed for 50% antagonism of effect of sulfadiazine on Ralston's Strept. faecalis			
Sulfadiazine →	1	10	100	1000
p-Aminobenzoic acid	0.003	0.03	0.3	3.0
Pteroyglutamic acid (folic acid)	0.0003	0.0003	—	0.0003
Thymine (4.1b)	0.06	0.25	0.25	0.25

All values are μg per ml. (Lampen and Jones, 1946.)

The above explanation of the action of sulfonamides, namely enzyme inhibition, seems to account for most of the therapeutic action of these drugs. However, in some experiments the sulfonamide was made to unite with the pteridine intermediate (9.21). Thus, Brown's enzyme, in a cell-free system containing the sulfonamide, was incubated for 2 hours; the resulting inhibition was irreversible by pAB, an effect which does not occur in growing cells (G. M. Brown, 1962). In pursuit of this effect, folate-synthesizing enzymes from *E. coli* were made to convert sulfamethoxazole (9.11) to the pteroic acid analogue: N'-3-(5-methylisoxazolyl)-N^4-(7,8-dihydro-6-pterinylmethyl)sulfanilamide, which was isolated by chromatography and found identical with a synthetic specimen (Bock *et al.*, 1974).

Of all the reactions catalysed by folic acids, the synthesis of thymine is, in many cases, the most sensitive to a shortage of these acids (see Section 9.3.2), although, in some micro-organisms, purine synthesis is inhibited first. It will be observed in Table 9.2 that thymine can antagonize the action of sulfadiazine on the streptococcus and that it does this almost non-competitively. It is not surprising that so large an amount is needed, because thymine is not a catalyst (like pAB and folic acid) but a cell constituent, the demand for which increases as the cell continues to grow.

The action of anti-pAB drugs on most of the *pathogenic* bacteria and protozoa is not reversed by folic acid and its derivatives, because these do not penetrate into the organisms.

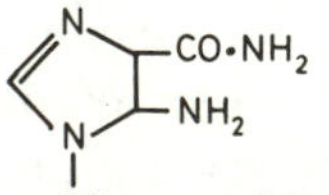

Aminoimidazole-
carboxamide ribotide
(9.22)

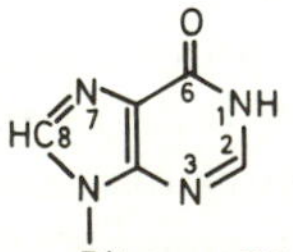

Ionosinic acid
(9.23)

9.3.2 *The role of folic acid and other pteridines in Nature*

As indicated above, derivatives of folic acid play a key role in the biosynthesis of purines and pyrimidines. These pteridines are the coenzymes responsible for inserting the carbon atoms into both positions 2 and 8 of purines, and they also insert the methyl-group into thymine (*4.1b*). When bacteria are treated with low concentrations of sulfonamides, 4-aminoimidazole-5-carboxamide ribotide (*9.22*) accumulates in the culture media. This substance is an intermediate in the biosynthesis of inosinic acid (*9.23*) from which all purines are derived.

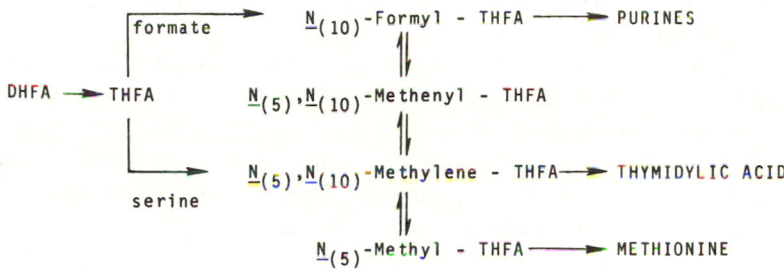

Fig. 9.3 Biosyntheses and functions of folic acid coenzymes. [DHFA and THFA = di-(and tetra-)hydrofolic acid, respectively.]

The coenzymes are formed from dihydrofolic acid (*9.20*) by the enzyme dihydrofolate reductase which gives tetrahydrofolic acid (Osborn, Freeman and Huennekens, 1958), and this is then modified by various one-carbon substituents. The coenzyme for the insertion of *C*-2 into purines, i.e. for the formylation of the ribotide (*9.22*) to give (*9.23*), is $N_{(10)}$-formyl-5,6,7,8-tetra-hydrofolic acid. For the earlier stage of the insertion of *C*-8, i.e. for the formyl-ation of glycinamide ribotide to form (eventually) (*9.22*), the same coenzyme is used and $N_{(5)},N_{(10)}$-methenyltetrahydrofolic acid (*9.24*) is no longer thought to be concerned. The coenzyme for inserting the methyl-group into uridylic acid to give thymidylic acid is $N_{(5)},N_{(10)}$-methylenetetrahydrofolic acid (*9.25*), the same coenzyme is responsible for the interconversion of two amino acids, namely serine and glycine. The uridylate→thymidylate methylation is effected by thymidylate synthetase (mol. wt. 67000). The synthesis of methionine and the catabolism of histidine are also effected by pteridine coenzymes, each with its specific apoenzyme. Folic acids with polypeptides formed from the glutamyl portion seem to have coenzymic functions as yet unexplained. The dihydrofolic acid of human plasma bears only one glutamyl residue but variants with from 3 to 7 glutamyl residues are common in Nature, particularly in bacteria (Kisliuk, 1981). Humans obtain most of their folic acid from vegetables, as polyglutamyl-folic acid. Liver and yeast are also good sources. Fig. 9.3 shows the biosynthesis and functions of the folic coenzymes.

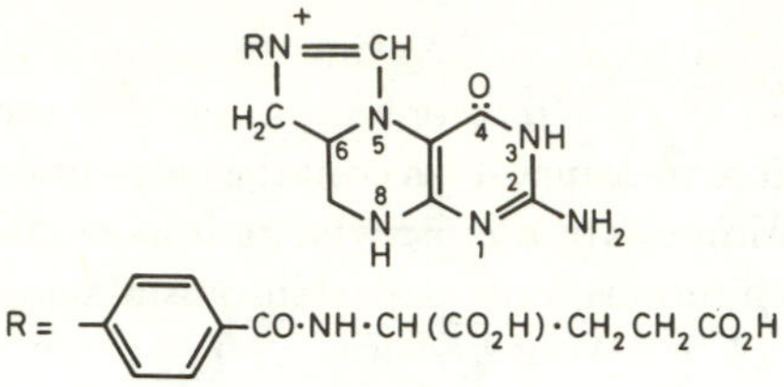

$$R = \langle\text{benzene ring}\rangle\text{-CO·NH·CH(CO}_2\text{H)·CH}_2\text{CH}_2\text{CO}_2\text{H}$$

Methenyltetrahydrofolic acid
(9.24)

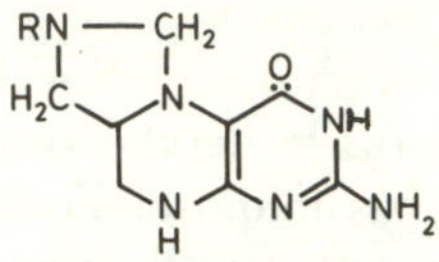

Methylenetetrahydrofolic acid
R as in 9.24
(9.25)

The biologically active pteridines are formed in nature by the degradation of the purine, guanine (4.4) (in the form of GTP), to give 2-amino-5-formamido-4-oxo-6-[(5-O-triphosphato-D-ribosyl-)amino]pyrimidine (Shiota *et al.*, 1969b). This pyrimidine (9.26) is cyclized by the enzyme dihydroneopterin triphosphate synthetase (obtainable from *Lactobacillus plantarum*), which forms a new bond between the exocyclic nitrogen atom and *C-2'* of the ribose moiety. From the dihydroneopterin [2-amino-4-oxo-6-(D-erythro-1',2',3'-trihydroxypropyl)-7,8-dihydropteridine], formed in this way, an enzyme called hydroneopterin aldolase (obtainable from *E. coli*) liberates glycollic aldehyde and leaves 2-amino-4-oxo-6-hydroxymethyl-7,8-dihydropteridine, which is readily built into the molecule of folic acid, as explained in Section 9.3.1.

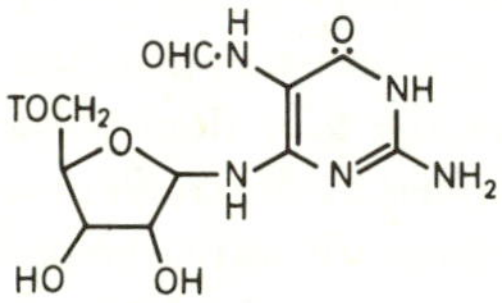

(T = triphosphate-group)
(9.26)

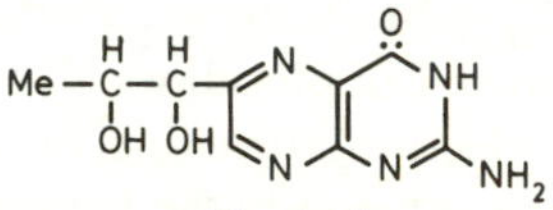

Biopterin
(9.27)

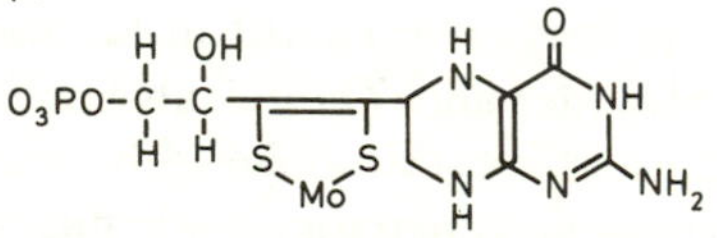

Molybdopterin
(9.28)

The participation of pteridines in metabolism is by no means restricted to the folic acid derivatives, nor is an amino-group in the 2-position essential for biological activity. The biosynthesis of the vitamin riboflavine takes place via the pteridine 6,7-dimethyl-8-ribityl-2,4-dioxopteridine. In this reaction, one molecule of this pteridine gives and another molecule accepts the four carbon atoms required to form the *o*-xylene ring of riboflavine (Plaut, 1964). The precursor of this pteridine is GTP of which the ribose-group becomes the ribityl-group of riboflavine.

Biopterin (9.27), an ivory-coloured pteridine, biosynthesized by animals and widely distributed in Nature, is L-erythro-2-amino-4-oxo-6-(1,2-dihydroxy-propyl)pteridine. Once again, GTP is the precursor. A reduced form, 5,6,7,8-tetrahydrobiopterin is the co-factor required for the set of enzymes that use molecular oxygen to oxidize phenylalanine to tyrosine (Kaufman, 1964), tyrosine to dopaminecarboxylic acid (*3.43b*), tryptophan to 5-hydroxytrypto-phan, and possibly in the formation of melanin. This co-factor uses molecular oxygen. It has found clinical use in one type of phenylketonuria in children (Cederbaum, 1979). Tetrahydrobiopterin improves parkinsonism symptoms (Narabayashi, 1982).

A pteridine is considered to be the primary receptor of the electrons liberated by light in *photosynthesis* (Fuller *et al.*, 1971); and a pteridine in the *mammalian eye* is presumed to be the agent which protects against the blinding effect of light (Cremer-Bartels, 1975). Cancerous cells break down biopterin to neopterin, the concentration of which in the urine may have diagnostic significance (Rokos *et al.*, 1980).

Another important naturally occurring pteridine is molybdopterin, appar-ently (*9.28*), which is the coenzyme of xanthine dehydrogenase, aldehyde oxidase, nitrate reductase, sulphite oxidase and presumably other enzymes that need both molybdenum and iron to function (Johnson and Rajagopolan, 1982). Xanthopterin, a co-lymphokine (p. 182), inhibits proliferation of lymphocytes (Ziegler *et al.*, 1983).

The biological importance of pteridines initiated, in 1952, a series of inter-national symposia on pteridine chemistry and biology, the seventh of which was held in 1982, in St. Andrews, Scotland [for the proceedings see Blair (1983)].

The chemistry of pteridines is in many ways unusual, principally because of the strong electron-attracting properties of the four doubly-bound nitrogen atoms, which antagonize the aromaticity implied by the conventional structural formula. The tendency of the pteridine ring to add a molecule of water covalently across a double bond, even at room temperature (as exemplified by xanthopterin) is further discussed in Section 2.5.1, and reviewed by Albert (1967, 1976). Stronger nucleophilic substances, such as acetone, keto acids, and mercaptans, are added even more readily. Another peculiarity of pteridines and some other nitrogenous heterocycles is the effect of hydrogen-bonding groups (e.g. $-NH_2$, or $-NH-C = O$) in *reducing* the solubility in water (Albert, Brown and Cheeseman, 1952). Strong chelating properties have been demon-strated and measured in naturally occurring pteridines (Albert, 1953).

For further reading on the biochemistry of folic acid and related pteridines, see Blakley (1969).

9.3.3 *Drugs that act by blocking dihydrofolate reductase*

Folic acid is an indispensable pro-vitamin for Man and other mammals, and lack of it quickly causes macrocytic anaemia and gastrointestinal disorders. Hence it was with some trepidation that chemists began to make metabolite antagonists based on the pteridine nucleus. Nevertheless, they were successful and many very valuable drugs have been obtained in this way.

Because pteridines play such an important part in catalysing the synthesis of purines and pyrimidines, various analogues of folic acid have been tested as tumour-inhibiting agents. High activity was shown by aminopterin, i.e. folic acid in which the 4-oxo-group was replaced by an amino-group. Clinical trials, from 1958 onwards, showed that methotrexate (9.29) was more selective and this has become an important anti-cancer drug. This substance, discovered by Seeger *et al.*, 1949, differs from folic acid in two details, the replacement of the 4-oxo-group by an amino-group, and the replacement of the 10-hydrogen atom by a methyl-group. In the human subject, methotrexate is partly converted to a form with two extra glutamyl residues (Covey, 1980).

It was soon learnt that methotrexate brings about a remarkable remission of symptoms in the acute lymphocytic leukaemia of young persons (Farber, 1952). Today it is considered to be the leading drug for the treatment of this disease. Unfortunately, leukaemic cells can develop resistance to the drug. This is countered by the following multiple drug therapy. First, a very short course is given of the powerful anti-tubule drug, vincristine (5.18a) (Section 5.4.7). Then prolonged treatment begins with methotrexate, often with a synergist that is usually 6-mercaptopurine (Skipper, Schabel and Wilcox, 1964). The happy result is the high percentage of cures in what had been, until a few years ago, a rapidly lethal disease.

In two other types of cancer, methotrexate brings about a lasting cure. Choriocarcinoma, a fast-growing tumour of pregnancy with normally a high death-rate, is quickly and completely cured by methotrexate (Ross *et al.*, 1965). Before this discovery, 5 out of every 6 women struck by this type of cancer had died within one year. A highly malignant lymphoma, discovered by Burkitt in African children living in hot, wet areas, usually begins in the jaw, rapidly spreads throughout the body, and kills the patient within six months. Methotrexate has a dramatic effect in Burkitt's lymphoma and often effects a complete cure (Bertino and Johns, 1967). Cautious as clinicians are about the use of the word 'cure' in cancer, it is agreed that these are true cures.

Methotrexate is a drug of only moderate selectivity. Prolonged medication with high doses has an adverse effect on the red blood cells, leading to macrocytic anaemia; later the production of white cells often diminishes. In spite of this, the therapeutic range and the effectiveness of methotrexate have been greatly improved by what is called a 'rescue programme'. In this therapy, the citrovorum factor, 5-formyltetrahydrofolic acid [see (9.20) for numbering], is injected one day later while the methotrexate treatment is suspended. For example, the osteogenic sarcoma of young people can be controlled by one hundred times the largest normally safe dose of methotrexate if citrovorum

Methotrexate
(9.29)

factor rescue periods are introduced. The administration of thymidine along with the methotrexate and citrovorum factor can protect the host from thymine depletion and allow the therapy to rest on purine depletion, which seems to have a greater selective advantage (Frenkel and Hitchings, 1957; Tattersall, Jaffé and Frei, 1975).

Another use for methotrexate is in the treatment of psoriasis, a common and rather intractable skin complaint (Weinstein, 1977). In this disease, because of an excessive production of dihydrofolate reductase, epidermal cells are shed every 3 days, compared to the normal 27 days. This drug must be given systemically; side effects restrict it to short-term therapy.

(a) *How methotrexate acts.* Methotrexate inhibits specimens of dihydrofolate reductase isolated from both mammalian and bacterial cells, although it cannot normally penetrate into the latter (Nichol and Welch, 1950; Werkheiser, 1963). The action of methotrexate is highly specific: 50% inhibition is effected by a 10^{-9} M concentration of the agent which has hardly any effect on any other enzyme. This inhibitor is bound to the enzyme about 10^4 times more tightly than the substrate, a most unusual effect (see Section 9.1), and hence is one of the least reversible of reversible inhibitors. The differences between its action on dihydrofolate inhibitors from different species is not large (contrast with the diaminopyrimidines, below) (Werkheiser, 1963). The failure of methotrexate to penetrate into one-celled organisms stems from their lack of a carrier system that is present in vertebrates. As methotrexate has the low log P of -1.85, it cannot diffuse passively across the plasma membrane.

Dihydrofolate reductases, depending on the species, can catalyse the reduction of both folate and dihydrofolate, although dihydrofolate reacts 4–10 times faster than folate with vertebrate enzymes. However, the rate of folate reduction is negligible for bacterial enzymes (Blakley, 1969).

X-ray diffraction analysis of dihydrofolate reductase (DHFR), co-crystallized with methotrexate, has shed much light on the action of this inhibitor. This work, one of the earliest visualizations of a drug interacting with its receptor (Matthews *et al.*, 1977), has since been refined to the remarkably clear resolution of 1.7 Å (Bolin *et al.*, 1982). A typical diagram of DHFR, its coenzyme (NADPH), and methotrexate is shown in Fig. 9.4. The enzyme depicted there is from *Lactobacillus casei* and the same authors also report on DHFR (with co-crystallized methotrexate) from the bacterium *E. coli*. However, they have not been able to co-crystallize methotrexate with DHFR from any vertebrate source.

Details of the binding of methotrexate to the *L casei*-derived enzyme are as follows. The Asp-26 anion of the enzyme is bound to the amidinium cation formed from the N-1 and 2-NH_2 atoms in the drug (the pK_a is reported in N-1 column of Table 9.3). The 4-NH_2 group of the drug is hydrogen-bonded to the carbonyl groups of both Leu-4 and Ala-97. The α-carboxylic group of the glutamic acid moiety of the drug forms a salt linkage (ionic bond) with the basic

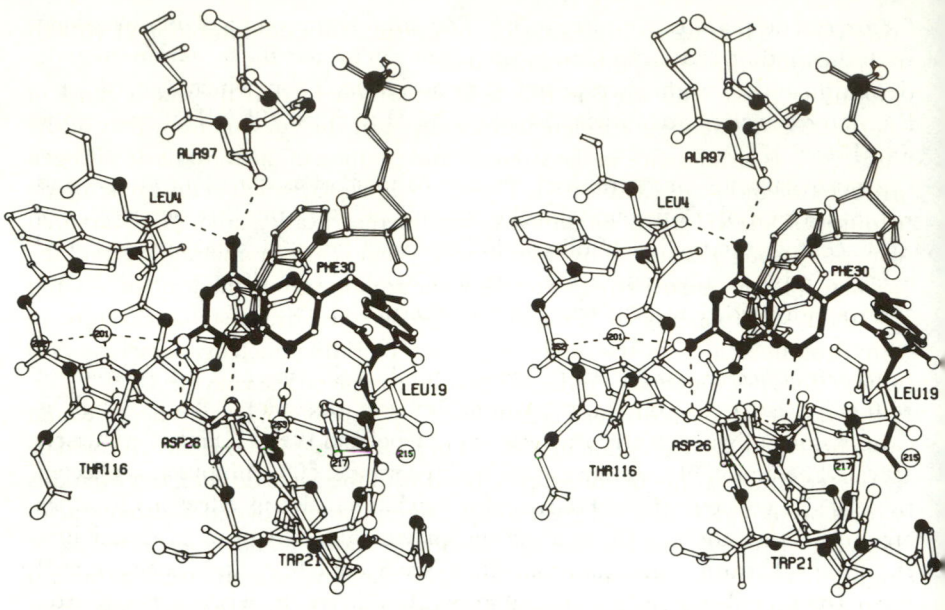

Fig. 9.4 Stereo-diagram of the pteridine-binding site of *L. casei's* dihydrofolate reductase. Methotrexate (an inhibitor) is indicated by solid bonds, protein by open bonds, and the NADPH molecule (only a portion shown) by striped bonds. Carbon atoms are represented by small (and oxygen by larger) open circles, nitrogen atoms by blackened circles. Large numbered circles represent molecules of water (fixed). (Bolin *et al.*, 1982.)

side-chain of Arg-57. The pteridine ring fits into a pocket lined by Leu-4, Ala-6, Leu-27, Phe-30, and Ala-97.

The *p*-aminobenzoyl moiety lies in a neighbouring pocket surrounded by the lipophilic side-chains of Leu-27 and Phe-30 (on one side) and of Phe-49, Pro-50, and Leu-54 (on the other).

The coenzyme, NADPH, is fully extended; the nicotinamide portion lies close against the pteridine ring in a position that would facilitate delivery of a hydride anion from the pyridine to the pteridine ring at C-6 (Filman *et al.*, 1982).

The enzyme derived from *E. coli* presented an almost identical picture (Bolin *et al.*, 1982).

Now we come to consider the most remarkable relationship ever encountered between a metabolite and its antagonistic analogue. The position occupied on the enzyme by methotrexate, as described above, *is the reverse* of that occupied by dihydrofolate, of which the Californian workers supply the following details (Bolin *et al.*, 1982). N-1 is unbound, 2-NH$_2$ is bound only to water, as is O-4. N-3 is bound by a hydrogen bond to Asp-26, N-5 is unbound, and N-8 has a van der Waals bond to Leu-4. Thus, compared to methotrexate, the normal substrate is very loosely bound, at least as regards the pteridine ring. The *p*-aminobenzoic and glutamic portions, however, are identically bound in both metabolite and antagonist. Here we see the basis for the unusual index of inhibition (1:10000) for methotrexate versus dihydrofolic acid.

Among those who work in this field, it is thought that the major cause of methotrexate's having a different orientation in the enzyme from that of the substrate, is a difference in the strength and position of the most basic nitrogen atom in the molecule (Table 9.3). Thanks to the replacement of the cyclic amide group (CO-4/NH-3) by an amino group, methotrexate is about one hundred times as strong a base as dihydrofolic acid. What is more remarkable is that the strongest basic centre in methotrexate is the amidine group made up of N-1 and C-2-NH_2 components in resonance, whereas the strongest basic centre in dihydrofolic acid is in the other ring of the pteridine nucleus, namely at N-5 which does not touch any centre of opposite charge on the enzyme.

Methotrexate's most basic centre has a pK_a of only 5.71, and hence would only be 2% ionized as cation in water at pH 7.3 (Table 17.1). Hence, it could be expected to form only an ion–dipole bond with the aspartic acid, but even this would be enough to give it a great competitive advantage over dihydrofolic acid. However, the pK_a, which was determined in water, could be expected to be higher in the lipid-lined cavity which the pteridine ring of methotrexate (but not that of dihydrofolate) fits so closely. This increase in basic strength follows from the decreased dielectric constant and absence of competing water molecules. This suggestion of increased basic strength in the methotrexate molecule when it is bound to the enzyme is confirmed by n.m.r. measurements taken on methotrexate in which C-2 had been 90% enriched by [13]C (Cocco et al., 1981). It was found that the very characteristic signal for C-2 in the cationic species persisted right up to pH 10, whereas in folic acid, similarly treated, the basic pK_a of 2.4 did not change. It would have been valuable to know what dihydrofolic acid did under these conditions, but apparently it was not sufficiently stable to permit a firm conclusion. (All this ^{13}C work was done on DHFR from *Streptococcus faecium*.)

Table 9.3 Ionization constants of folic acid and its derivatives (in water) A = by spectrometry; B = by n.m.r. of ^{13}C-enriched C-2

		N-1 (base)	N-5 (base)	CO-4/NH-3 (acid)
Folic acid	A	2.35	<−1.5	8.38
	B	2.40		8.25
Dihydrofolic acid	A	1.38	3.84	9.54
Methotrexate	A	5.71	<−1.5	(absent)
	B	5.73		

[Sources: (A) Poe, 1977; (B) Cocco et al., 1981.]

(*b*) *Simplification of the methotrexate molecule.* This anti-cancer drug has little toxicity for most bacteria and protozoa (Wood, Ferone and Hitchings, 1961) because it does not penetrate into unicellular organisms. It was at first thought worthwhile to improve its penetration by increasing the lipophilicity of the molecule. It was soon realized that most of the antagonism of dihydrofolate resided in the 2,4-diaminopteridine portion of the molecule. The following two results exemplify this endeavour.

2,4-Diamino-6,7-diphenylpteridine is as active as quinine in suppressing malaria due to *Plasmodium gallinaceum* in the chick, and 2,4,7-triamino-6-phenylpteridine (the diuretic, triamterene) suppresses parasitemia in mice and rats infected with *Pl. berghei* (Aviado, Brugler and Bellet, 1968). 2,4-Diamino-6,7-diethylpteridine is highly effective in laboratory tests against *Vibrio cholerae* (Collier and Waterhouse, 1950), but does not affect many other bacterial species. None of these results led to useful new drugs.

A more successful approach was the simplification of the methotrexate molecule, as initiated by G.H. Hitchings, who found that the omission of unessential features of (*9.29*) gave better penetrating analogues. These 'simplified amethopterins' were 2,4-diamino-1,3,8-triazanaphthalenes, of which (*9.30*) (2,4-diamino-5-methyl-6-*n*-butyl-pyrido[2,3-*d*]pyrimidine) is one of the most active (Robins and Hitchings, 1955). The action, on both Gram-positive and -negative bacteria, was assisted by the lipophilic groups in the 5- and 6-positions and also by the increase in basic strength that inevitably follows loss of doubly-bound nitrogen from the ring. When tested on DHFR isolated from several bacterial species, the inhibitory effect of these triazanaphthalenes was found to be proportional to their effect on intact bacteria (Burchall and Hitchings, 1965; Hitchings and Burchall, 1965).

Unfortunately these triazanaphthalenes introduced a new form of toxicity for the patient whose histamine-*N*-methyltransferase they inhibited, causing accumulation of unbearable concentrations of histamine. It was necessary to modify the 6-substituent, and this led to an analogue of (*9.30*), namely 2,4-diamino-6-(2,5-dimethoxybenzyl)-5-methylpyrido[2,3-*d*]pyrimidine. This proved to be as effective as methotrexate in inhibiting DHFR from human leukaemic cells, and it caused regression of leukaemias and solid tumours in rats and mice (Grivsky *et al.*, 1980; Duch *et al.*, 1980). Clinical results are awaited. Several quinazoline analogues are active against leukaemia, but nothing of clinical significance has yet been reported.

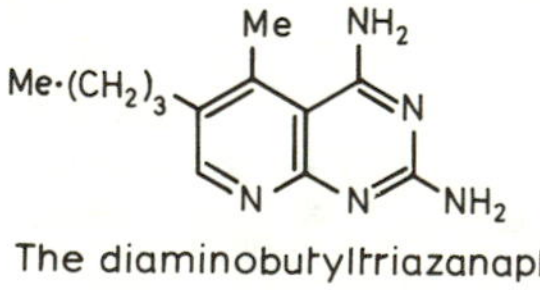

The diaminobutyltriazanaphthalene
(9.30)

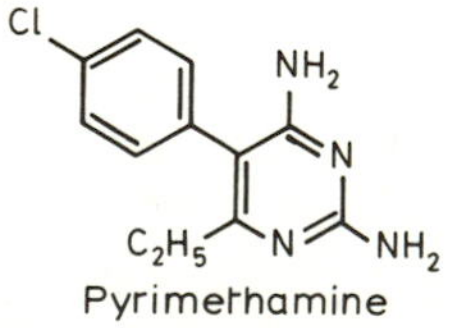

Pyrimethamine
(9.31)

So far, further simplification of the methotrexate molecule has not aided cancer research, but has proved to be very much worth while in the war against micro-organisms. As recounted in Section 4.0 (p. 123) Hitchings, continuing the dismemberment of the methotrexate molecule, found that quite simple 2,4-diaminopyrimidines have a powerful anti-folic acid effect on micro-organisms, and used this knowledge to develop pyrimethamine (*9.31*) ('Daraprim') (Falco *et al.*, 1951). This substance, 2,4-diamino-6-ethyl-5-*p*-chlorophenylpyrimidine, is one of the most powerful anti-malarials known, and has become the most widely used of all prophylactics against malaria. Pyrimethamine is a *causal* prophylactic, acting on the pre-erythrocytic form of the parasite, especially in the liver. In a weekly dose of 25 mg, it has proved to be a highly effective suppressant of falciparum malaria. It is not, however, suited for treating acute malaria. Cycloguanil (*3.35*), a dihydrotriazine counterpart of pyrimethamine, was discussed in Section 3.6. A large dose of it is sometimes injected as the insoluble pamoate salt to give a prolonged prophylactic action. For the combined action of pyrimethamine and sulfonamides, see sequential blocking in Section 9.6.

The action of pyrimethamine on DHFR, isolated from a variety of living species, and abstracted in Table 4.1, showed great selectivity against the enzyme from the malarial parasite. As Table 4.1 shows, the plasmodial enzyme is inhibited by pyrimethamine at a concentration about 2000 times lower than that inhibiting the analogous mammalian enzymes. These data established that the selective action of pyrimethamine in malaria is due to the extraordinary sensitivity of the enzyme in the parasite compared to that in the host. The plasmodial enzyme has demonstrable individuality for, when isolated from *Pl. berghei*, it has a mol. wt. of about 200 000, which is 10 times as large as the analogous mammalian and bacterial enzymes. (The malarial parasite, like bacteria, cannot absorb preformed folic acid.)

For the highest anti-malarial activity in vertebrates, the 2,4-diamino-pyrimidines require a strongly lipophilic substitution, such as a phenyl-group in the 5-position and an alkyl-group in the 6-position. The inhibition of isolated dihydrofolate hydrogenase is not greatly altered when the 5-phenyl-group in pyrimethamine is changed to a *n*-butyl-, or even a 4-phenylbutyl-group, but shortening of the butyl-group leads to progressive loss of affinity for this enzyme (Baker and Shapiro, 1966).

For the highest *anti-bacterial* activity in vertebrates, the 2,4-diaminopyri-midines require that the massive lipophilic substituent be modified with a slightly hydrophilic outer area. The best of these compounds proved to be trimethoprim (*9.32*) (Roth, Falco and Hitchings, 1962). It can be seen from Table 4.1 that whereas pyrimethamine is most selective against the malarial parasite enzyme, trimethoprim is selective against a bacterial enzyme as well. Then, as regards selectivity in favour of Man, these diaminopyrimidines have immensely higher therapeutic indices than methotrexate. Table 4.2 provides examples in which trimethoprim exerts a powerful action against the DHFR of both Gram-positive and -negative bacteria, leaving the analogous mammalian enzyme unharmed. Its index of inhibition for bacteria averages 1:1000 (Roth and Cheng, 1982).

The high degree of selectivity exhibited by trimethoprim for the bacterial over the mammalian enzyme is not shared widely among its analogues. First, it is required that all three positions (3-, 4-, and 5-) in the benzene ring be occupied. Thus the 3,5-dimethoxy analogue of (*9.32*) has a ratio of bovine liver to *E. coli* (inhibitory concentration on pure DHFR) of only 116 compared to 1840 for trimethoprim (Roth and Cheng, 1982), and the analogue with an unsubstituted benzene ring has almost no selectivity of this kind. However, excellent select- ivity and activity is shown by 3,5-dialkoxy analogues carrying, as a 4-substit- uent, isopropenyl, methylcarbonyl, or methoxycarbonyl group. Perhaps the best rivals of trimethoprim are the 3,5-diethyl-4-methoxy- and the 4-methyl- thio-3,5-dimethoxy- (metioprim) analogues. However, they have not yet had as much clinical exposure as trimethoprim. In this series, generally, increased lipophilicity favours action against Gram-positive at the expense of Gram- negative. Apart from the need to achieve a good hydrophilic/lipophilic balance in this benzyl substituent of the pyrimidine ring, a strong steric requirement is evident (Roth and Cheng, 1982).

Coccidiosis, a malaria-related protozoal disease of poultry, is better con- trolled with drugs of the trimethoprim than those of the pyrimethamine types of DHFR inhibitors. The most used remedy is diaveridine (*9.33*).

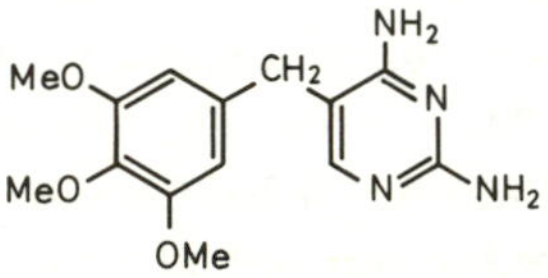

Trimethoprim
(9.32)

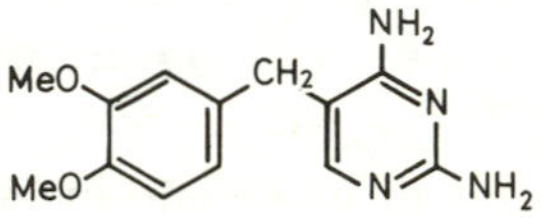

Diaveridine
(9.33)

It is frustrating that no X-ray crystal structure is known for DHFR contain- ing any diaminopyrimidine inhibitor, but the difficulties of this very tedious and exacting work are enormous. At least good results, at 2.9 Å resolution, are available for 2,4-diamino-5,6-dihydro-6,6-dimethyl-5-(4'-methoxy- phenyl)-1,3,5-triazine. This is a close analogue of the anti-malarial cyclo- guanil (*3.35*) from which it differs only in the replacement of a chlorine atom by a methoxy group (Volz *et al.*, 1982). This strong inhibitor binds to specimens of this enzyme from the malarial parasite and from vertebrates (in the present case: chicken liver). It does not bind appreciably to bacterial specimens. It was found that the triazine and the phenyl rings occupied positions analogous to those taken up by the pyrimidine and pyrazine rings, respectively, of methotrexate in the bacterial enzymes. Glu-30 and Tyr-31 played an important part in binding this inhibitor, and each moved 3Å when the inhibitor was inserted into the apoenzyme. The NADPH coenzyme lay in a long, shallow groove winding across the face of the apoenzyme and was fully extended (Volz *et al.*, 1982).

To the small extent to which trimethoprim binds to chicken liver DHFR, X-ray crystallography reveals it to be squeezed into a butterfly conformation, less energetically favourable than the position it occupies in the *E. coli* enzyme where the two rings are nearly perpendicular to one another (Matthews, 1981).

It is most gratifying that these useful 2,4-diaminopyrimidine drugs have been discovered mainly by the exercise of scientific reasoning. For biological activity, the maximal effect is obtained when the pK_a lies between 7 and 8 (Roth and Strelitz, 1969). This requirement indicates that a cell membrane has to be traversed (see Section 10.5).

Trimethoprim is most used in medical practice in conjunction with a sulfonamide, with the aim of achieving sequential blocking (Section 9.6). All anti-folic drugs are used with caution during pregnancy to avoid damage to the foetus.

For further reading on folic acid biochemistry, see Blakley (1969); for a review of structure–action relationships in 2,4-diamino-pyrimidines, -pteridines, and -quinazolines, as anti-bacterial, anti-cancer, or anti-malarial drugs, see Roth and Cheng (1982) and Roth, Bliss and Beddell (1983).

9.4 Other metabolite analogues of proven value in prophylaxis and therapy

9.4.1 *Introduction*

It is not difficult to make anti-metabolites, but it has proved very hard to discover ones which are selective, because it is rare to find a metabolite that is important in the uneconomic species and yet unimportant in the economic species. When a metabolite occurs in all living cells, as thiamine does, it must seem unlikely that any selective action of analogues could be achieved. Nevertheless, in favourable cases this selectivity has been demonstrated, and has led to useful and widely employed agents. Success often depends on the analogue being taken up unequally by the two species. Relatively small changes in a molecule can so alter cellular uptake as to make an agent either (a) inaccessible to the economic species, or (b) preferentially absorbed by the uneconomic species, or at least concentrated in the region where the uneconomic species is segregated.

Curiously enough, a metabolite analogue may be treated as a metabolite by one species and as an antagonist by a related species. For example, dethiobiotin

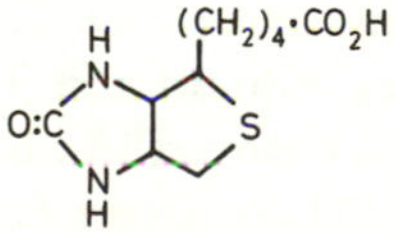

Biotin
(9.34)

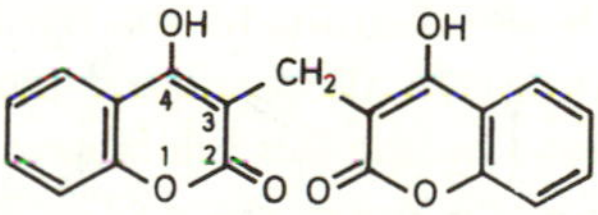

Dicoumarol
(9.35)

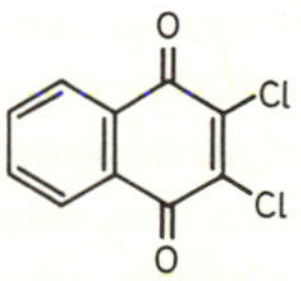

Dichlone
(9.36)

is able to replace biotin (*9.34*) as a growth-factor for yeast, yet it antagonizes the growth-promoting effect of biotin on *Lactobacillus casei*. It is known why this should be so: the yeast inserts the missing sulfur atom, but the bacterium is unable to do this (Dittmer and du Vigneaud, 1944). Likewise, the fungus *Endomyces*, when made resistant to pyrithiamine (*9.5*) by repeated sub-culture in its presence, was found to owe this resistance to its ability to break down the pyrithiamine and synthesize thiamine (*2.1*) from one of the large fragments (Woolley, 1944).

A natural agonist may have a function to perform in several different enzymes, each of which may require a different inhibitor. For example, the blood-clotting factor, vitamin K (*2.2*), can be specifically counteracted by dicoumarol (*9.35*) which is, however, non-toxic to micro-organisms (Goth, 1945). On the other hand, 2,3-dichloronaphthoquinone (*9.36*) (dichlone), which is harmless to mammals, exerts its outstanding anti-fungal properties by antagonizing vitamin K (Woolley, 1950b).

Anti-metabolites are usually more effective in injuring those cells which require an external source of the metabolite, and pyrithiamine (*9.5*) provides an example (Woolley and White, 1943). Examples will now be given of anti-metabolites which have become accepted as effective prophylactics or remedies.

9.4.2 *Analogues of vitamins and coenzymes*

Amprolium (*9.37*), an analogue of thiamine (*2.1*), has proved highly successful in coccidiosis of poultry. This drug acts by inhibiting the transport of thiamine, in *Eimeria*, 50 times more strongly than in cells of the chicken gut (James, 1980).

Sodium 2,2-dichloropropionate (dalapon, 'Dowpon') is much used for killing grass in dicotyledonous crops such as beet and lucerne. Competitive studies in bacteria (*E. coli*) point to its being an antagonist for the incorporation of pantoic acid [i.e. the left-hand side of (*9.38*)] into pantothenic acid. This is the main site of its action on grasses; its effect is diminished by external pantothenic acid (Hilton *et al.*, 1959).

Amprolium (cation)
(9.37)

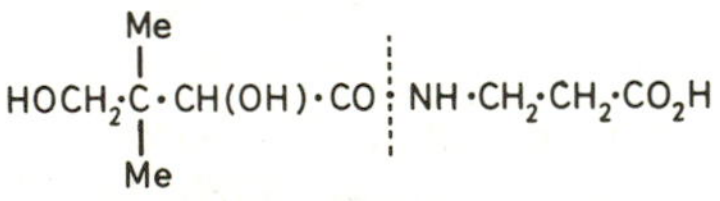

Pantothenic acid
(9.38)

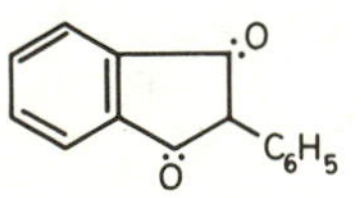

Phenindione
(9.39)

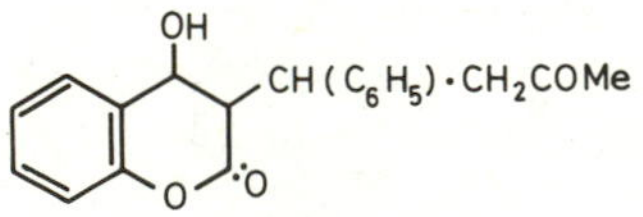

Warfarin
(9.40)

The first of the oral anti-coagulant drugs, dicoumarol (*9.35*), reached clinical trial in 1941. This substance [3,3'-methylene-bis(4-hydroxycoumarin)] had just been isolated from a spoilt batch of hay and identified as the factor causing bleeding in cattle eating the hay. Dicoumarol is much used as an anti-coagulant in the long-term prophylactic treatment of hypertensive patients at risk from coronary thrombosis and stroke. The minimal requirements for anti-coagulant activity in coumarins are a 4-hydroxycoumarin nucleus with either hydrogen or alkyl in the 3-position. Dicoumarol, and its more powerful analogue warfarin (*9.40*), exert their action by antagonizing regeneration of vitamin K (*2.2*) from its 2,3-epoxide (Zimmerman and Matschiner, 1974). The epoxide is formed when the vitamin is exerting its biological role of introducing an extra carboxyl group into ten glutamate residues in prothrombin, a step which enables this protein to chelate calcium more powerfully (Stenflo *et al.*, 1964).

Coumarin anti-coagulants are inactive *in vitro* and take 36 hours to develop their action in the body. Although the antagonism that exists between them and vitamin K is very clear (an excess of either neutralizes a dose of the other), it, too, requires the lapse of many hours. The indandione anti-coagulants, of which the most used member is phenindione (*9.39*), act a little more quickly and were devised in an attempt to find a molecule that resembled the structure of vitamin K more closely (presence of lipophilic group in 3-position; replacement of phenolic by ketonic group).

9.4.3 *Analogues of hormones and neurotransmitters*

In his book *Antihormones*, Agarwal reminds us that anti-metabolites are available to act against each of the five major classes of steroid hormones as well as against thyroxine, the polypeptide messengers, and the neurotransmitters (Agarwal, 1979). Some of these antagonists have so far achieved the status only of laboratory aids to the biochemist, others are effective drugs. In the present Section we shall keep to structural analogues that are accepted as valuable drugs and are prescribed regularly.

What may strike some readers as paradoxical is the occasional use of antagonistic analogues to achieve (indirectly) an agonistic effect, of which several examples will be discussed among the neurotransmitters. The reverse of this phenomenon is the use of agonistic analogues to achieve an antagonistic effect by attention to regularity of dosage and knowledge of the body's cycles. A well-known and possibly unique example of this is the use of an estrogen or a progestin, or (most usually) a mixture of the two as a daily dose ('the pill'), for the prevention of conception and taken (it is estimated) by 50 million women worldwide. In this rite, the most frequently used estrogens are ethinyl estradiol and its 3-methyl ether, mestranol, both of them simple agonistic analogues of the normal physiological female hormone, estradiol. The most used progestins are norgestrel and norethindrone which are simple agonistic analogues of women's natural progestin, progesterone. In all four contraceptives, the natural hormones have been slightly altered, mainly by inserting an ethinyl (—C⁝CH) group to make the drug active orally and to resist metabolic destruction. Both types prevent conception by inhibiting ovulation. The estrogen does this by inhibiting secretion of follicle-stimulating hormone (FSH), the progestin does it by inhibiting release of luteinizing hormone (LH); it is largely a matter of doing persistently what Nature does only intermittently.

Several true anti-estrogens are known, the most prescribed of which is tamoxifen (9.41), which is effective in treating those mammary cancers that are estrogen-dependent. It binds to the protein receptors for estradiol thus depleting the cytoplasm of available estrogen receptor. It is metabolized to the 4-hydroxy-derivative which is thought to be the active form (Borgna and Rochefort, 1981). Clomiphene and nafoxidine are similar products. The design of tamoxifen was devised, a little indirectly, from that of diethylstilbestrol (12.31) which is a more powerful estrogen than the natural hormone (12.32). For more on the steroid hormones, see Sections 2.4 and 12.2. For insect hormones and their antagonistic analogues, see Section 4.7.

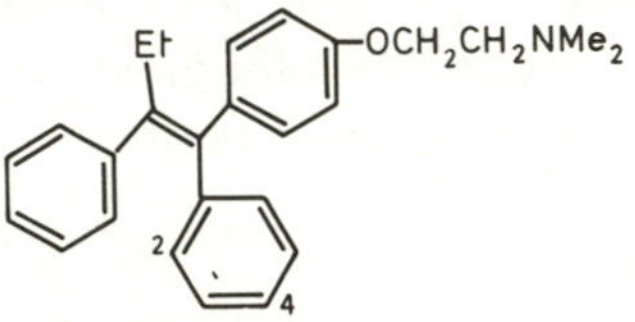

Tamoxifen
(9.41)

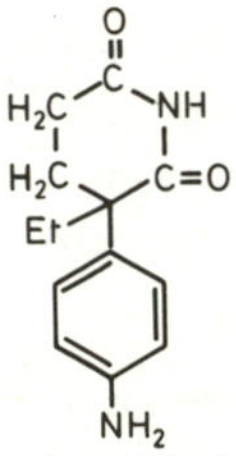

Aminoglutethimide
(9.42)

Aminoglutethimide (9.42), seemingly an inhibitor of the biosynthesis of all hormones derived from cholesterol, has proved useful in treating some types of hormone-dependent breast cancer (Henderson and Canellos, 1980).

Anti-metabolites based on the neurotransmitters will now be discussed. First, there are the analogues of acetylcholine (2.7) which block access of this neuro-transmitter to its receptors. Examples are the muscle relaxants used in general anaesthesia such as tubocurarine (2.6), gallamine (7.31), and suxamethonium (7.29); and also those based partly on atropine such as propantheline (7.25). Other drugs, such as physostigmine (7.33), by blocking the access of acetyl-choline to its destroying enzyme (AChase), enhance the action of this neuro-transmitter, as do other carbamate inhibitors of the same enzyme (Section 13.3). These pseudoagonists are used in glaucoma, myasthenia, and atony of the bowel and bladder. See Section 12.6 for further discussion of the agonists, and Sections 13.3 and 12.5 for the cholinesterase inhibitors. The characteristics which enable an agonist molecule to be modified to give an antagonist molecule have been discussed with special reference to acetylcholine in Section 7.5.2.

Adrenergic anti-metabolites that were conceived as analogues of nore-pinephrine (7.5) (noradrenaline) are playing a very important part in the treatment of disease. The following examples illustrate the wide range of their usefulness. Metaraminol (*7.54*), by preventing uptake of norepinephrine by the presynaptic terminal, is used to raise lowered blood-pressure and combat the state of shock. Ephedrine (*9.43*), by accumulating in the presynaptic terminal, and steadily pushing stores of norepinephrine out into the synaptic gap, gives valuable relief in hay fever and other allergic states. Some central stimulation can be a troublesome side effect. Amphetamine (*9.44*) ('Benzedrine') has both actions of ephedrine but the central stimulation is much the stronger. It is used to curb appetite for food and also to promote wakefulness but its addictive properties make for caution in prescribing.

The foregoing three drugs act by giving the patient an overdose of his own norepinephrine, but the following have more individual properties. Guanethidine (*7.55*) accumulates in the presynaptic terminals and steadily pushes out norepinephrine stored there, but at such an abnormally slow rate that it is destroyed as fast as it is liberated. This drug provides an alternative, if insufficiently selective, method for lowering moderate degrees of high blood-pressure. Bretylium (*12.4*), debrisoquine and bethanidine work somewhat similarly.

Other drugs modelled on catecholamine transmitters will be found in Section 12.4, e.g. phenylephrine (agonist) and methyldopa.

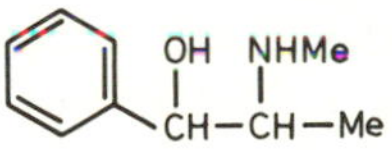

Ephedrine
(*9.43*)

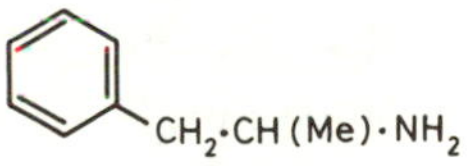

Amphetamine
(*9.44*)

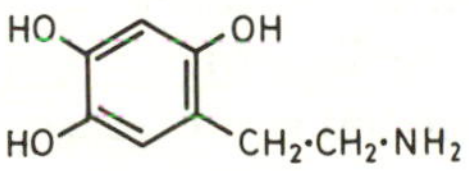

6-Hydroxydopamine
(*9.45*)

Propranolol (*12.56*), and similarly structured molecules, are effective antagonists of norepinephrine on postsynaptic receptors (Section 12.4). They are much used in the treatment of hypertension.

An analogue of norepinephrine that is selectively destructive, but valuable for that reason to the experimental physiologist, is 6-hydroxydopamine (*9.45*). When a small dose is given to laboratory animals, it is distributed to all sympathetic nerve-endings where it is oxidized to the corresponding *o*-quinone. The latter combines covalently with these endings and, by permanently deactivating them, assists in their mapping (Thoenen and Tranzer, 1968).

In the 1940s, it was found that hydrazines bearing a liposoluble substituent, such as phenelzine (*9.46*), gave relief from severe mental depression. It is generally held that these substances act by combining with monoamine oxidase (MAO) in the brain, and this inhibition causes accumulation of the amine neurotransmitters: norepinephrine, dopamine, and 5-hydroxytryptamine. However, there were serious side effects, so that these hydrazines came to be replaced by lipophilic primary amines which did not combine quite so strongly with the enzyme. Of these, tranylcypromine (*9.47*) is one of the most favoured at the present time. All these amines incorporate some steric hindrance to avoid being deaminated themselves.

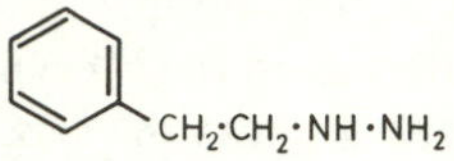

Phenelzine
(9.46)

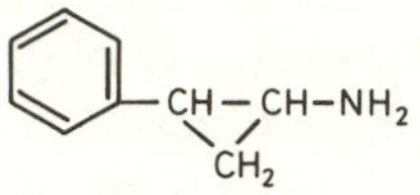

Tranylcypromine
(9.47)

Human MAOs, a small family of isoenzymes, are divided into Type A which preferably deaminates norepinephrine and 5-hydroxytryptamine, and Type B which seems to prefer dopamine and xenobiotic amines such as benzylamine (Cawthon and Breakefield, 1979). The primary amine type of MAO inhibitor still gives rise to dangerous side effects which, although rare, are difficult to reverse because the binding of these drugs to the enzyme is so tight. For this reason some physicians prefer to treat mental depression with the tricyclic anti-depressants (Section 12.9) (which, however, are unsuitable for cardiac patients), or else prescribe electroconvulsive therapy, itself a liberator of neuro-transmitter amines (Green, Heale and Grahame-Smith, 1977).

For more on monoamine oxidase and its modification, see Singer, von Korf and Murphy (1979). For a book dealing with the biochemical basis of neuro-pharmacology, see Cooper and Bloom (1982).

Chlordimeform (2.36), a much used insecticide which gently inhibits mammalian MAO, seems to owe its usefulness in the field to mimicry of the arthropod neurotransmitter, octopamine (2.35) (see Sections 2.6.1 and 6.4.1).

9.4.4 Analogues of amino acids

The antibacterial drug, cycloserine (5.11), is a structural analogue of D-alanine (5.12); the biochemical basis of its action was set out in Section 5.3. In short, it interferes with the formation of D-alanine and of the condensation of this to D–alanyl–D–alanine, so that there is none of this dipeptide to form the penta-peptide–bearing monomer from which all new cell wall must be made (Strom-inger, Ito and Threnn, 1960). A simpler analogue was later found in 3–fluoro–D–alanine, which is highly effective in vivo against all common Gram–positive and –negative bacterial infections, and whose activity can be antagonized by D–alanine (Section 5.3). A more fundamental attack on this system was made by L–alafosfalin (3.50), which inhibits L–alanine racemase, the enzyme which makes D–alanine. It is highly effective in vivo against Gram-negative bacterial infections (Section 3.6, p. 104). Up to the present, these analogues of the alanines have not achieved widespread clinical use.

The antibiotic azaserine (9.48), which is the ester of serine with diazoacetic acid, is a structural analogue of glutamine (9.49) and they act as mutual

$$N_2CH \cdot CO \cdot OCH_2 CH (NH_2) CO_2H$$
Azaserine
(9.48)

$$H_2N \cdot CO \cdot CH_2 CH_2 CH (NH_2) CO_2H$$
Glutamine
(9.49)

competitors in purine synthesis (Buchanan, 1957). For more details, see Section
9.7.2. It is used as a sequential blocking agent (Section 9.6) in cancer treatment.

(*a*) *Anti-metabolites of peptides*. The difficulties that have plagued attempts to
discover new polypeptide drugs are the ready digestibility of the drug by the
gastrointestinal tract and (even when this is avoided by, say, replacing some of
the L-amino acids by their D-enantiomers) the poor absorbability from that
tract. These difficulties can be overcome by replacing some of the peptide
linkages (—CO.NH—) by less-digestible polar linkages, usually ether or
ketone, or even by non-polar links such as the cyclohexane ring. The products
are called 'peptoids' (Farmer and Ariëns, 1982).

Substance P, an undecapeptide in the central nervous system, is thought to be
a neuromodulator for relief and, in higher doses, transmission of pain.
Antagonists have been obtained by omitting the first three amino acids and
substituting D-enantiomers for proline-4 and tryptophans-7 and -9 (Caranikas
et al., 1982). Clinical studies are not available.

Some effective analogues, both agonistic and antagonistic, of the natural
opioid oligopeptides (enkephalins) are reported in Section 12.8.

The pentapeptide, pentagastrin, has replaced natural gastrin (a heptadecapeptide) as a diagnostic agent that stimulates gastric secretion. This pentapeptide   is   *t*-butyloxycarbonyl-β-Ala-Trp-Met-Asp-Phe-NH$_2$.   It   can   be
administered as a snuff.

Inhibitors of the angiotensin-converting enzyme, also known as peptidyl
dipeptidase (PDP), were long viewed as likely sources of drugs for treating
hypertension. This enzyme appears to raise blood-pressure, physiologically, by
two mechanisms. First, it removes the terminal dipeptide from angiotensin I to
give the octopeptide, angiotensin II which causes powerful vasoconstriction.
Second, it also inactivates bradykinin (a vasodilating nonapeptide present in
the blood) by removing two dipeptides from it.

The venom of the snake, *Bothrope*, furnished terprotide, a nonapeptide
inhibitor of PDP,*pyro*Glu-Trp-Pro-Arg-Pro-Gln-Ile-Pro-Pro, and it was established that the terminal proline group was what prevented the enzyme from
hydrolysing the peptide. However, terprotide was ineffective orally because
peptidases in the human gut were less discriminating. It was then decided to
investigate peptoids, incorporating (for example) succinyl amino acids equal in
length to the dipeptide fragment. The first such product to achieve clinical use
was *captopril* (*9.50*). It is a highly potent drug for lowering abnormally high
blood-pressure, is effective orally, and has a high therapeutic index (Ondetti,
Rubin and Cushman 1977; Ondetti and Cushman, 1981). This work emphasized the necessity for a terminal pyrrolidine ring, a free carboxylic acid (or
isostere of that), a further carbonyl group, a group such as mercapto or phosphoryl to chelate the zinc in the enzyme, and a *S,S* configuration (*S* is defined in
Section 12.1).

Related work led to enalapril which is undergoing study in hyperpiesia. It is

N-(1-ethoxycarbonyl-3-phenylpropyl)-L-alanyl-L-proline. Both of these drugs also compete directly against angiotensin II and may owe most of their usefulness to this (Unger, Ganten and Lang, 1983).

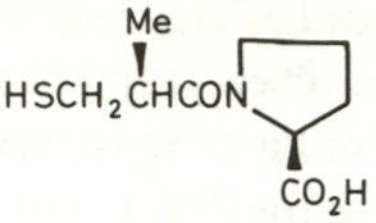

$$HSCH_2\overset{Me}{\underset{}{CHCON}}$$

CO₂H

Captopril
(9.50)

9.4.5 *Analogues of purines and pyrimidines*

Many uses of analogues of the natural purines and pyrimidines were discussed in Section 4.0. The following are used clinically in cancer chemotherapy: 5-fluorouracil (*3.3*), cytarabine (*4.13*), 6-mercaptopurine (*3.14*), and 8-azaguanine (*4.40*). Clinically useful anti-viral drugs include acyclovir (*4.19*) which is a guanine derivative, idoxuridine (*4.14a*), trifluridine (*4.14b*), and vidarabine (*4.16*) which is an analogue of adenosine. The immuno-suppressant, azathioprine (*3.41*), is a pro-drug for 6-mercaptopurine, but has a different use. Flucytosine (*4.23*) and 6-azauracil (*4.41*) are selective fungicides, the former used orally for systemic fungal infections in Man, the latter used agriculturally. Azaribine, the triacetyl derivative of 6-azauridine, is given orally for treating psoriasis.

Allopurinol (*9.51*) is an anti-metabolite of hypoxanthine (*9.52*). Unlike purines described in Section 4.0, it is not built into a nucleotide, indeed, its action does not involve the formation of any covalent bonds. Allopurinol, 4-oxopyrazolo[3,4-*d*]pyrimidine, is one of the most effective known drugs for reducing the uric acid load of patients with gout or diseases where uric acid is excreted excessively. This drug, which is given orally, is free from side effects. It acts by blocking the oxidation of hypoxanthine (to xanthine) by the enzyme xanthine oxidase. A small portion of the given dose is oxidized to the corresponding 2,4-dioxopyrazolopyrimidine (alloxanthine), which blocks the oxidation of xanthine to uric acid by the same enzyme (Elion *et al.*, 1966).

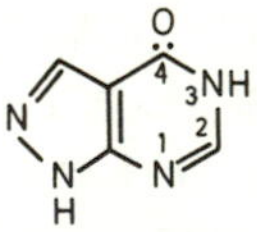

Allopurinol
(9.51)

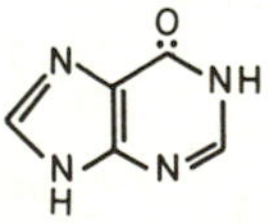

Hypoxanthine
(9.52)

9.4.6 *Analogues of histamine*

(*a*) *Antagonists of the H₁ receptor.* Anti-histamine drugs are those which occupy the receptors on which histamine (*9.53*) is normally free to act. The H₁ receptors are principally concerned in allergic reactions of the skin and mucous membranes. The most relevant test of a H_1-antagonist is its ability to abolish the wheal produced in Man by an intradermal injection of histamine (there are also several valid guinea-pig tests for early screening). This is normal, competitive antagonism between agonist and antagonist. Drugs of this kind give quick relief in coryza (hay fever), urticaria (hives), and pruritis (itching). In the mid-1930s, Bovet and his colleagues in the Pasteur Institute (Paris) started the search for this type of drug (Bovet and Staub, 1937), and 10 years later were able to review the relevant structure–action relationships (Bovet, 1947).

The first example with clinical value was antergan (*9.54*) (phenbenzamine) discovered by Halpern (1942). Its weak action was much improved by changing the weakly basic *N*-benzylaniline group to the stronger 2-benzylaminopyridine group, giving mepyramine (*9.55a*) (pyrilamine, 'Neoantergan', 'Anthisan') (Bovet, Horclois and Walthert, 1944), which is still in regular use. By now it was realized that many derivatives of *NN*-dimethylethylenediamine had H_1-antagonistic properties. Parallel work in the United States produced the closely related and equally valuable tripelennamine (*9.55b*) ('Pyribenzamine'). The pK_a values for these two drugs were 4.0 and 3.9 respectively for the weaker of the two basic groups (the more basic group has pK_a 9). Thus the weaker group had not reached the basic strength of histamine (pK_a 6.0, derived from both ring-nitrogen atoms acting together in mesomerism), although the stronger basic group in the anti-histamines would ionize just as well as the stronger group in histamine (pK_a 9.8).

It was soon found that excellent anti-histaminic drugs could be based on ethanolamine instead of ethylenediamine and a notable example from the United States was diphenhydramine (*9.56*) ('Benadryl'). Members of this class couple their anti-histaminic effect with sedative properties and are prescribed when a high degree of somnolence is advantageous. There is only one ionizing group (pK_a 9.0).

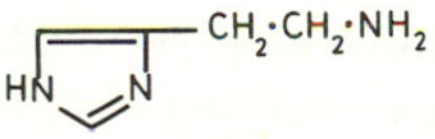

Histamine
(*9.53*)

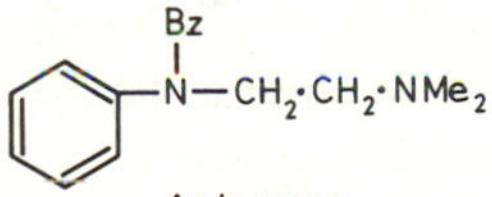

Antergan
(*9.54*)

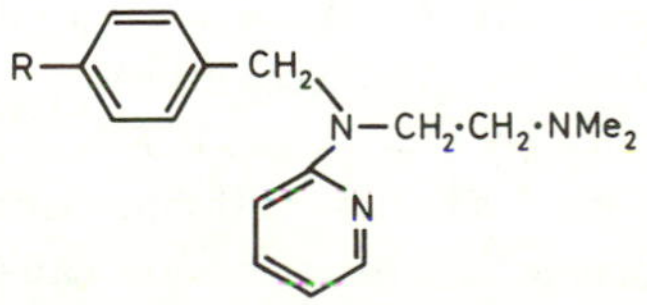

(a) Mepyramine (R = OMe)
(b) Tripelennamine (R = H)
(*9.55*)

The sedative property which is so highly expressed in diphenhydramine existed in a less measure in other antihistamines. In an endeavour to lessen it, to help those who need medication at times when they must be alert, chlorphenamine (*9.57*) (chlorpheniramine, 'Chlortrimeton') was introduced. In this type, the second polar atom (N or O) has been eliminated from the aliphatic chain. The pK_a values (4.0 and 9.2) resemble those of earlier compounds. Another type with a decreased incidence of drowsiness has the two nitrogen atoms of ethylenediamine joined by two saturated carbon atoms to give a piperazine ring. Chlorcyclizine (*9.56a*) ('Histantin', 'Diparalene') provides an example. The search for H_1 antagonists that could not cross the blood-brain barrier, and hence would be non-sedative, has produced the sterically-hindered astemizole ('Hismanol') which is 1-(*p*-fluorobenzyl)-2-[1-(1-*p*-methoxyphen-ethyl)-4-piperidylamino]benzimidazole, used for hay fever (Laduron *et al.*, 1982).

For a discussion of the steric contribution to the action of H_1 anti-histaminic drugs, see Casy and Ison (1970); and for a more general discussion, see Rocha e Silva (1978).

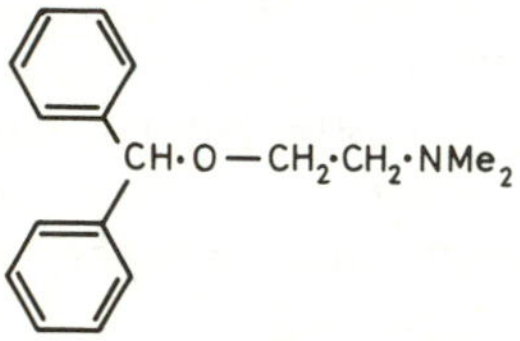

Diphenylhydramine
(*9.56*)

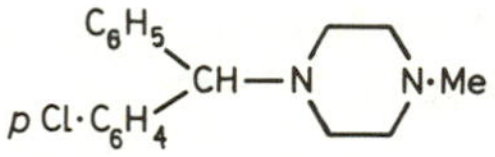

Chlorcyclizine
(*9.56a*)

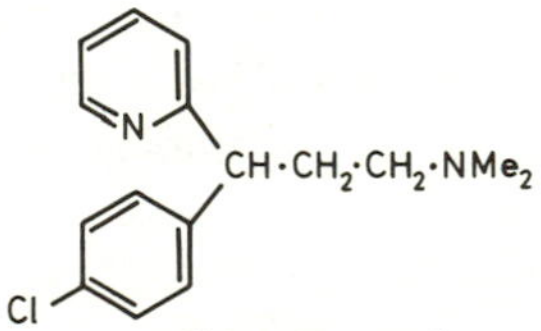

Chlorphenamine
(*9.57*)

(*b*) *Antagonists of the H_2 receptor*. Receptor H_2 is used by histamine for stimulation of the uterus, the cardiac atria, and the secretion of acid by the stomach. Search for an antagonistic drug was initiated by J.W. Black in the hope that such a drug could alleviate, and even cure, peptic ulcer. Although disputed at the outset, this thesis was vindicated by the discovery of cimetidine (*9.58*) ('Tagamet') which is much used clinically (Gray *et al.*, 1977).

An important early clue was that 5-methylhistamine (*9.59*) was an agonist for H_2, but not for H_1, receptors. This situation calls to mind the drug methacholine, which is the β-methyl derivative of acetylcholine and has all the latter's activity at muscarinic receptors but none at the nicotinic receptors (Section 12.6). It was found that strong H_2 agonist activity in derivatives of histamine required the possibility of tautomerism of the hydrogen atom (moving from one nitrogen atom to the other), whereas the absence of this possibility generated strong H_1 agonists such as 2-(2-aminoethyl)pyridine (Durant, Ganellin and Parsons, 1975).

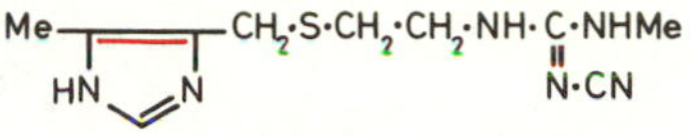

Cimetidine
(9.58)

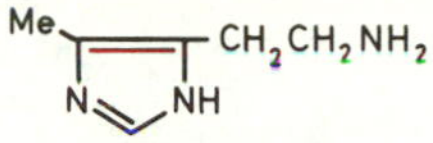

4-Methylhistamine
(9.59)

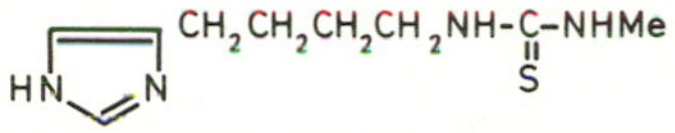

Burinamide
(9.60)

It was realized at the outset that the desired antagonistic molecule would have to be more complex in its structure than the agonist, as exemplified by the H_1 antagonists: this general principle was discussed in Section 7.5.2. Agonistic activity could be prevented only by having a non-ionizable side-chain; antagonistic activity increased as the side-chain was lengthened. Methylurea was recognized as a suitable termination for the side-chain, but several other flattish structures containing a double bond gave similar help in adsorption. These considerations led to the first H_2 antagonist, burinamide (9.60). It was shown to be a competitive and specific antagonist for histamine in guinea-pig atrium and rat uterus, receptors never influenced by H_1 antagonists. Unfortunately its activity proved too weak for clinical use (Black et al., 1972), and it was not well absorbed orally.

The same team developed this molecule by replacing one —CH_2— group of the side-chain by —S—. This exchange not only facilitated synthesis but lowered the pK_a of the nucleus from 7.3, which was thought to be too high, because that of histamine is only 6 (imidazole ring). Fine-tuning put a methyl-group in position 5 of the nucleus, giving the orally well-absorbed drug, metiamide (9.61), with a pK_a of 6.8. Thus, at pH 7.3, the cationic ionization of burinamide, metiamide, and histamine are 50, 24, and 5% respectively. Several hundred cases of peptic ulcer were successfully treated with this drug, but it was withdrawn when it was found to reduce the count of granulocytes in the blood of a small fraction of patients (Durant et al., 1977).

This side effect was traced to the thiourea portion of the molecule. Search for a non-ionizing isosteric group suggested the cyanoguanidine substituent. Making this exchange produced cimetidine (9.58) which is N-methyl-N'-cyano-N''-[(5-methylimidazol-4-yl)methylthioethyl]guanidine which, like metiamide, has a pK_a of 6.8 (Ganellin, 1981). A more recent arrival, with similar clinical properties to cimetidine, is ranitidine (9.62) ('Zantac'). The molecule contains the side-chain of cimetidine altered only in that $=CH.NO_2$ replaces the $=N.CN$ group with which it is isosteric. However, the imidazole ring has been replaced by a furan ring which is non-basic, and able to lower the basic strength of the trimethylamine side-chain. For its ability to decrease acid secretion in the stomach, see Bohman (1980). Replacement of the imidazole ring by isothiourea, to give dimaprit (7.45), is discussed at the end of Section 7.3.

For discussion of anti-histaminic drugs, see the volume edited by Rocha e Silva (1978).

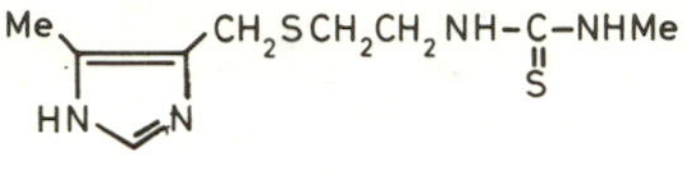

Metiamide
(9.61)

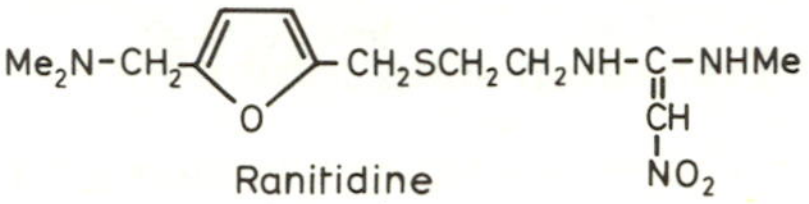

Ranitidine
(9.62)

9.4.7 Other analogues

The herbicide, amitrole (4.67) (3-amino-1,2,4-triazole), seems to be an anti-metabolite of some natural imidazoles. In yeast, it causes imidazoleglycero-phosphate to accumulate (Weyter and Broquist, 1960); also histidine can reverse its inhibition of the growth of yeast and of the alga Prototheca (Casselton, 1964).

The analogue of a quite simple entity, the bicarbonate ion, will now be described. The observation that sulfanilamide caused hens to lay eggs without any shell, coupled to the observation that this drug also produces an alkaline diuresis, led to the discovery that the sulfonamide anion (9.63) can competitively block the access of the bicarbonate anion (9.64) to the enzyme carbonic anhydrase. The O—O separation in a sulfonamide anion is 2.4 Å, and in carbonic acid 2.32 Å. This antagonism prevents dehydration to carbon dioxide. The high concentration of bicarbonate ion which builds up is excreted with an equivalent of sodium ions, causing strong diuresis. This blocking of the receptor cannot occur if the =NH-group is substituted, and hence most of the anti-bacterial sulfon-amides do not give this effect. These results suggested the insertion of hetero-cyclic groups, as R in (9.63), in order to obtain greater anionic ionization than in sulfanilamide, and substances 2000 times as active were found, the activity increasing with the ionization (Miller, Dessert and Roblin, 1950). The best of these substances, acetazolamide (9.65) ('Diamox'), found clinical use as a powerful orally administered diuretic, but it is tachyphylactic (self-limiting) and has largely been replaced by newer diuretics (Section 14.1). In acute glaucoma, however, an intravenous injection, by diminishing secretion of the aqueous humour, can save sight.

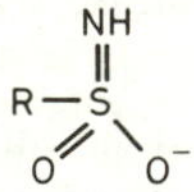

Sulfonamide ion
(9.63)

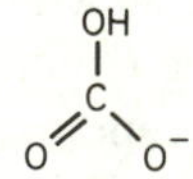

Bicarbonate ion
(9.64)

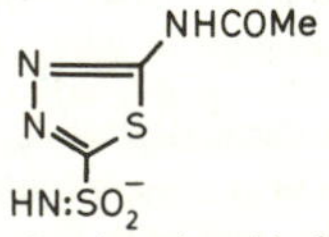

Acetazolamide ion
(9.65)

Thioglucose, in which the ring-oxygen atom of glucose is replaced by sulfur (giving 5-thio-D-glucose), is selectively active against hypoxic cancer cells and is being studied in conjunction with radiation therapy. It also protects oxic cells from radiation damage. It inhibits selectively the transport and cellular uptake of glucose. It is also under trial as a reversible male contraceptive (Zysk *et al.*, 1975).

9.5 'Transition-state' inhibitors

In this chapter we have encountered many analogues that interfere with enzyme action by (a) occupying the place of the normal substrate or coenzyme *and* (b) lacking the capacity to react with that enzyme. In this way, they block the normal biochemical process of that enzyme. In terms of Emil Fischer's lock-and-key symbolism of enzyme action, the inhibitor is the misfitting key that cannot turn the lock and prevents insertion of the right key. When, as is usual, the inhibitor readily dissociates from the enzyme, we do not know whether it had formed only a primary inhibitor–enzyme complex or if the latter had moved on to form an *unstable intermediate* by covalent bonding of substrate to enzyme*. X-ray crystal analysis can solve this problem, given a sufficient source of pure enzyme. At least it can be said of methotrexate, a drug very strongly held by dihydrofolate reductase, that co-crystallization of enzyme, coenzyme, and substrate does not lead to covalent bonding (Section 9.3.3). The high index of inhibition shown by this drug is related to its being able to make more effective contact with the active site than the substrate can.

Related to this phenomenon is the ability of some inhibitors to induce more conformational change in the enzyme than the substrate can. Thus the kinetics of the binding of allopurinol (*9.51*) [which is an anti-metabolite of hypoxanthine (*9.52*)] to xanthine oxidase demonstrate that binding is slow, but tight, and a steady state is not encountered (Cha, Agarwal and Parks, 1975). The authors think that this behaviour indicates a large conformational change.

Let us now turn to those enzyme inhibitors which are designed to resemble, not the substrate, but the transition state of that substrate. Interest in this derives from the widely held belief that the active site of a free enzyme is more nearly complementary to the transition state of the substrate than to the free substrate (Pauling, 1948). The aim of such work is to produce inhibitors that are much more tightly held.

It is usually not known what the transition state of a substrate is when it is on

---

*A *transition state* is defined as the most unstable species on the reaction pathway. It is to be found at the peak of a diagram in which the free energy of the reactants is plotted against time. In the transition state, chemical bonds are in the process of being broken. In those enzyme reactions that proceed through an *unstable intermediate*, this has newly formed bonds and occupies an energy trough between the two peaks of *two* transition states. Hammond (1955) postulated that a transition state must structurally resemble the unstable intermediate, for whose structure spectroscopy often supplies details. Hence, when a drug designer wishes to make an analogue of a transition state, he has the structure only of the unstable intermediate to guide him.

the enzyme. Hence the assumption is usually made that this state is the same as the one that has been established for the same reaction performed in a protein-free environment with the aid of an inorganic catalyst such as acid or alkali. Let us review some common types. When an amide (or peptide) bond is being hydrolysed, the first reaction to take place is the addition of a molecule of water across the C=O double bond to give a tetrahedral adduct. Thus an amide (9.66) gives the intermediate (9.67), which rapidly dissociates to the product (9.68) which is the corresponding carboxylic acid. In another example, a primary amine can condense with an aldehyde-group (9.69), as in pyridoxal, to give the unstable intermediate (9.70) which dissociates to give the azomethine (Schiff base) (9.71).

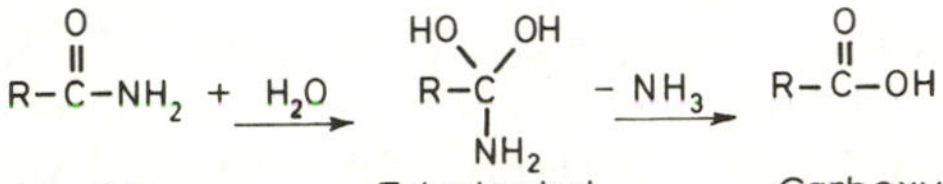

Amide (9.66) + H_2O → Tetrahedral intermediate (9.67) → −NH_3 → Carboxylic acid (9.68)

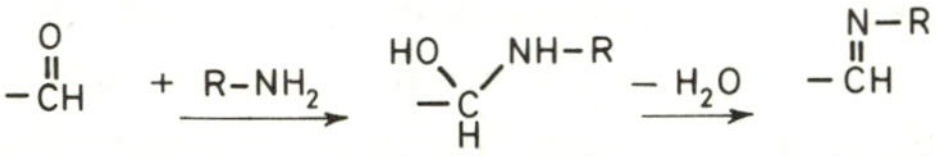

Aldehyde (9.69) + $R-NH_2$ → Tetrahedral intermediate (9.70) → −H_2O → Azomethine (9.71)

Pentostatin (deoxycoformycin) (4.18) and its ribose analogue, coformycin, strongly inhibit adenosine deaminase, the enzyme that converts adenosine (9.72) to inosine (9.74), presumably through the intermediate (9.73). This reaction is similar to the hydrolysis of an amidine to an amide, and it needs to be prevented when medicating patients with adenine-formulated drugs such as vidarabine (4.16). It can be seen that the structure of pentostatin resembles the

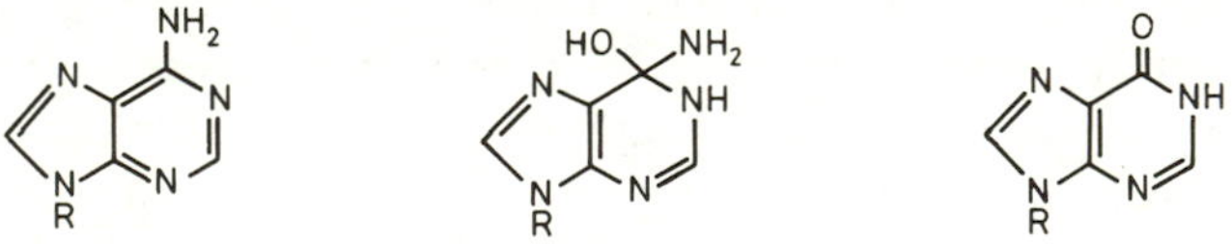

Adenosine (9.72)

Tetrahedral intermediate (R = ribose, in all cases) (9.73)

Inosine (9.74)

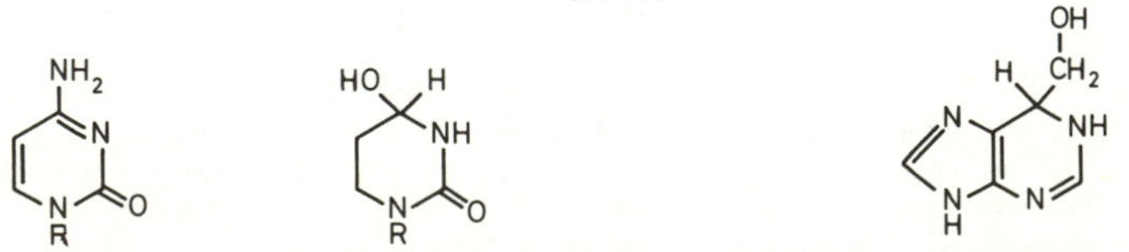

Cytidine (9.75)

3,4,5,6-Tetrahydrouridine (R = ribose) (9.76)

1,6-Dihydro-6-hydroxymethylpurine (9.77)

intermediate (9.73) *in the critical area*. It, and coformycin, are very strongly bound to the enzyme, apparently irreversibly (Cha, Agarwal and Parks, 1975).

Similarly, 3,4,5,6-tetrahydrouridine (9.76) is a potent and specific inhibitor for the deamination of cytidine (9.75) by cytidine deaminase (Carmenier, 1968). The tight binding of this inhibitor is attributed to its resemblance, in the critical position, to a tetrahedral intermediate analogous to (9.73). Similarly, adenosine deaminase is strongly inhibited by 1,6-dihydro-6-hydroxymethylpurine (9.77) (Evans and Wolfenden, 1970).

The boronic acids owe their powerful inhibition of peptidases to the stability of the tetrahedrally disposed hydroxyl groups, which makes them good analogues of the unstable intermediates formed in hydrolysing the peptide bond (9.67). The illustrated example (9.78) is specific for chymotrypsin which has a requirement for an aromatic side-chain on its substrate (Koehler and Lienhard, 1971).

Aldehydes, too, make good peptidase inhibitors. All aldehydes are in equilibrium with a tetrahedral hydrated structure, (9.79) for example, and the more electron-attracting its substituents, the higher is the proportion in the form of this methylene glycol. Some organisms produce aldehydes as specific inhibitors of proteolytic enzymes. For example, the leupeptins (acetyl-L-leucyl-L-leucylanginal is typical), produced by at least 17 species of *Streptomyces*, potently inhibit plasmin, trypsin, and papain, but not chymotrypsin. In mice, the leupeptins have an anti-inflammatory effect when given orally, prevent radiation injury, and inhibit blood coagulation and tumorigenesis (Umezawa, 1972).

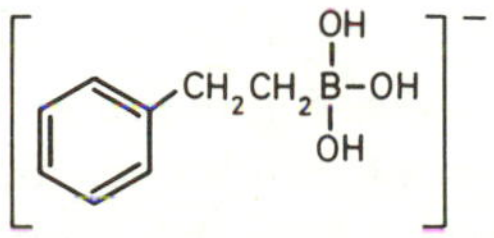

Phenylethane-2-boronic acid (anion of hydrate) (9.78)

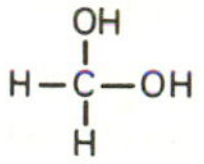

Formaldehyde (hydrated) (9.79)

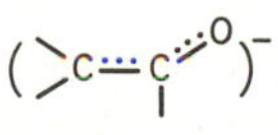

Enolate anion (9.80)

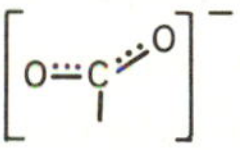

Carboxylate anion (9.81)

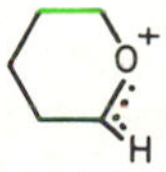

Transition state of a hexose (9.82)

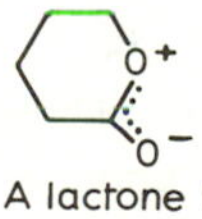

A lactone (canonical form) (9.83)

Other types of 'transition-state' inhibitors act by providing ions related to carbanions and carbenium cations. It has been found that several enzymic reactions in which the transition state probably resembles an enolate anion (9.80) are strongly inhibited by the analogous carboxylate anion (9.81). Similarly, most glycosyl-transferring enzymes are strongly inhibited by lactone structures, analogous to the hexose substrate. It is thought that the transition states for these reactions resemble alkoxycarbenium ions such as (9.82) which, like the lactones (9.83), have a half-chair conformation and a positive charge on the ring-oxygen atom.

It has been suggested, although the evidence is inconclusive, that penicillin provides transition state inhibition of peptidoglycan transpeptidase (Section 13.1) before this antibiotic acylates the enzyme (Lee, 1971). Several 1,6-dihydro-8-azapurines have shown high therapeutic indices in experimental cancer of mice (Albert, 1980). These candidate drugs were intended as 'transition-state' inhibitors of purine biosynthesis. In general, it would appear that the subject of 'transition-state' inhibitors may have a rosy future but not, so far, many achievements in therapy.

For further reading on 'transition-state' inhibitors, see Lindquist (1975) and Lienhard (1972). Some cases where there is a second site of bonding between the irreversible inhibitor and the enzyme are reserved for Section 9.7.

9.6 Sequential blocking

The therapeutic usefulness of anti-metabolites can be increased by selecting two such agents which, singly, will block two consecutive steps in the cell's metabolism (Hitchings, 1952). Natural agonists in cells are gradually built up from components moved along the enzymatic equivalent of a factory's production line, each stage of assembly being carried out by a different enzyme. The arithmetic of sequential blocking is this: if the first enzyme is blocked to the extent of 90%, then only 10% of the partly completed factor reaches the second enzyme. If one is fortunate enough to discover how to block the second enzyme also by 90%, then only 1% of the partly completed factor emerges, and that may be too little to sustain life in the parasite. It may be asked why it is not sufficient to use more of the first drug and block the first enzyme by 100%. The answer is: because of the usual shape of a dose–response curve: increasing the concentration of a drug beyond a certain point seldom leads to much further response. This is because adsorption from solution usually follows a hyperbolic curve (Langmuir's isotherm, see Section 8.1). Thus to increase the blockade of an enzyme from 90 to 100% would call for an amount of the drug so excessive that it may endanger the patient's life.

An example of sequential blocking is the use of a sulfadiazine with pyrimethamine (9.31) in toxoplasmosis, a protozoal disease (Wettingfeld, Rowe and Eyles, 1956). In this sequence, the sulfonamide blocks the incorporation of p-aminobenzoic acid into dihydrofolic acid, and the pyrimethamine prevents the reduction of this pteridine to tetrahydrofolic acid (Sections 9.3.2 and 9.3.3). In malaria, as early as 1959, Hurly made the observation that pyrimethamine and sulfadiazine potentiated one another to such a degree that the combination could actually cure *Pl. falciparum* infections. Thus, less than 0.1 m.e.d. (minimal effective dose) of pyrimethamine and 0.25 m.e.d. of sulfadiazine were, together, as effective as 1.0 m.e.d. of either drug separately. In current tropical medicine, 'Maloprim', a combination of pyrimethamine and dapsone (9.17) (the latter chosen because of its slow rate of excretion which matches that of pyrimethamine), forms an excellent replacement for chloroquine in cases of *Pl. falciparum* malaria resistant to this drug. This is fortunate because that is the most lethal species of the parasite. Another combination used for sequential blocking of this species is 'Fansidar' which is a mixture of pyrimethamine and sulfadoxine (*9.84*) (*N'*-5,6-dimethoxypyrimidin-4-ylsulfanilamide). In these combinations, the sulfur-containing drug must be a slowly excreted one, because of the long half-life of pyrimethamine, namely 4 days (much longer than that of cyclo-guanil). For more on the efficacy of these combinations in treating drug-resistant malaria, see WHO (1973a). Unfortunately, strains are now being encountered with resistance to both chloroquine and the sequential combi-nations. Such cases are being treated with quinine (rather host-toxic), and mefloquine is held in reserve.

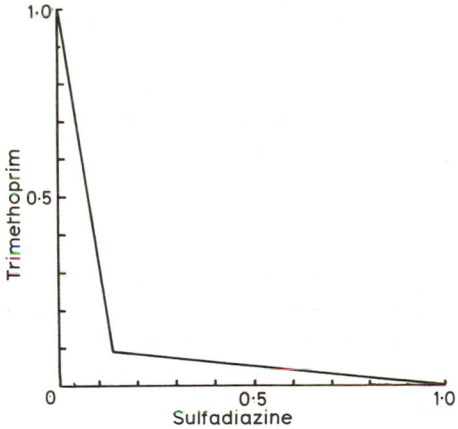

Sulfadoxine (anion)
(*9.84*)

Similar sequential combinations are used, very effectively, in treating poultry infected by the protozoon, *Eimeria*.

Fig. 9.5 Synergism of trimethoprim and sulfadiazine in treatment of mice infected with *Proteus vulgaris*. The units on the axes represent doses of each drug needed to give half the maximal percentage survival. Each point on the graph represents a combination of fractions of these doses. (The sulfadiazine unit was 0.14 mg, and that of trimethoprim 4 mg, per mouse.) (Hitchings and Burchall, 1965.)

Sequential blocking offers very practical advantages in *antibacterial* chemo-therapy also. The synergistic effect of trimethoprim (*9.32*) with sulfadiazine, shown in Fig. 9.5, is evident. In medical practice, the most striking and beneficial results have been obtained by selecting a sulfonamide with a similar pattern of distribution and persistence to those of this diaminopyrimidine, namely sulfamethoxazole (*9.11*), (3-*p*-aminobenzenesulfonamido-5-methylis-oxazole), under the name *co-trimoxazole* ('Septrin', 'Bactrim', 'Eusaprim'). This combination, which sequentially blocks first the synthesis and then the reduc-tion of dihydrofolic acid, is much used clinically in bacterial dysentery, bronchitis, and long-standing infections of the urinary tract by *E. coli*, *Proteus mirabilis*, *Neisseria gonorrhoeae*, *Klebsiella*, *Streptococcus*, and *Staphylococcus* (Cattell *et al.*, 1971).

For further reading on chemotherapy effected by sequential blocking of folate metabolism, see Hitchings (1983).

9.7 Analogues that form a covalent bond

9.7.1 *Active-site-directed irreversible inhibitors (ASDIIs)*

Several examples are known where a metabolite analogue is first attracted to the enzyme commonly used by the metabolite and then forms a covalent bond with this enzyme. Thus penicillin, as described in Section 13.1, may appear to be a metabolite antagonist for D-alanyl-D-alanine; but instead of these substances being freely competitive at the transpeptidase, the penicillin combines irrevers-ibly with this enzyme by forming a covalent bond. Again, the quaternary ammonium group of pyridine-2-aldoxime methiodide (*13.27*) causes this molecule to come to rest on acetylcholinesterase on the site usually fitted by the quaternary ammonium group of acetylcholine; if this enzyme is already inactivated by phosphorylation, the pyridine antidote brings about activation of the enzyme by transferring the phosphoryl-group to its own hydroxyl-group (see Section 13.3).

It seems reasonable that other useful agents could be made by designing them as metabolite analogues (to bring about concentration on the enzyme), and then giving the molecule one further group which could form a covalent bond with the enzyme which should thus become permanently inactivated. B. R. Baker, working with lactic dehydrogenase as a model, distinguished two kinds of inhibitory alkylating agents, those which alkylate the site which normally combines with the substrate, and those which alkylate adjacent sites. These two classes of inhibitors are called endo- and exo-alkylating agents, respectively. Baker addressed himself to the study of exo-types. Noting that lactic dehydro-genase, of which salicylic acid (*9.85*) sufficiently resembles lactic acid (*9.86*) to be an inhibitor (although a readily reversible one), he designed and made 4-iodoacetamidosalicylic acid (*9.87*) which irreversibly inhibited this enzyme (and L-glutamic dehydrogenase as well). Whereas iodoacetamide does not inhibit these enzymes, the iodoacetamide-group in this acid does so, because the similarity of the rest of the molecule to salicylic acid causes the inhibitor to be concentrated on the enzyme (Baker *et al.*, 1962). This inhibition is irreversible because hydrogen iodide is split off and a covalent bond is formed with the enzyme.

This inhibitor can even discriminate between isoenzymes. Thus, whereas it irreversibly inhibits the lactic dehydrogenase of skeletal muscle, it does not affect that of the heart. A related compound, 5-(phenoxycarbonylamino)sali-cylic acid, irreverisbly inhibits the heart (but not the skeletal) isoenzyme, whereas its 4-isomer irreversibly inhibits the skeletal (but not the heart) iso-enzyme (Baker and Patel, 1964; Baker, 1967).

For further reading on active-site-directed irreversible inhibitors, see Baker (1967). This is an idea which may lead to useful new agents, but has yet to produce a widely used drug.

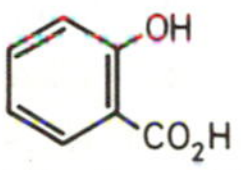

| Salicylic acid | Lactic acid | 4-Iodoacetamidosalicylic acid |
| (9.85) | (9.86) | (9.87) |

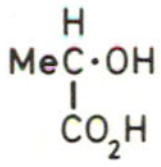

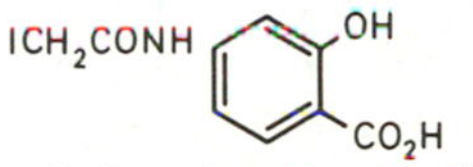

9.7.2 *Irreversible mechanism-based inhibitors (IMBIs)*

This is a newer type of inhibitor which requires not only that the medicament should covalently attack the enzyme, as in the ASDIIs, but that it should also be a pro-drug which only the enzyme can convert to the true drug (see Section 3.6 for pro-drugs). The name 'Irreversible Mechanism-Based Inhibitors' is usefully contracted to 'IMBIs'. Another valid name is 'Enzyme-Activated Irreversible Inhibitors'. Sometimes they are facetiously, but ever so inaccurately*, referred to as 'suicide inhibitors'. On the basis of an accidental observation by Konrad Bloch (Morisaki and Bloch, 1972), the subject gathered interest after appear-ance of the review by Abeles and Maycock (1976). Those who would design such a drug must have knowledge, from kinetic studies, of the mechanism by which the chosen enzyme works. Is it, for instance, of the kind that (after combining with the substrate) (a) strips off a hydrogen atom so that a con-jugated double bond is formed?, or (b) converts the substrate to an electrophile by protonation or by oxidation?

Up to the present, the commonest type of reaction between an enzyme and an IMBI takes the form of a Michael reaction. The pro-drug (IMBI) is so designed that the enzyme will generate a Michael acceptor in it. The latter (Michael, 1887) is essentially a double bond that is polarized (and hence activated) by its proximity to an electron-attracting substituent (Bergmann, Ginsburg and Pappo, 1959). As a rule, the IMBI carries the activating group, which may be a fluoromethyl, ketone, ester, or nitrile group. The enzyme must then exercise its normal function of creating a double bond. Naturally the IMBI has been

---

*Whose suicide is being inhibited? Is it suicide or murder?

designed also to have the characteristic group of the normal substrate without which it would never be taken up by the enzyme that it is designed to injure.

A successful example of an IMBI of this kind is 2-difluoromethylornithine (*9.88*), a drug which irreversibly inhibits ornithine decarboxylase. This is the enzyme which, with its pyridoxal coenzyme, normally decarboxylates ornithine to putrescine, a reaction of special importance to some protozoa and cancer cells. The pro-drug, so like ornithine in structure, is covalently bound by the pyridoxal coenzyme. Thus, the 2-amino group of the IMBI creates an azomethine (Schiff base) with the aldehyde group of the pyridoxal (*9.90*), just as happens with ornithine. At this stage, the enzyme normally removes a hydrogen atom from ornithine. It turns out that it can also remove fluorine, similar in size to hydrogen (Table 9.1) though much more electronegative. Thus, the enzyme removes a fluorine atom from the IMBI, converting it into the Michael acceptor (*9.89*). This operation converts the IMBI to a true drug, which attacks a Michael donor on the enzyme (thought to be a —SH group), thus irreversibly inactivating the enzyme (Prakash *et al.*, 1980).

Difluoromethylornithine (*9.88*) has a high therapeutic index and is proving curative in clinical trials of leukaemia and malaria. Unfortunately it has a very short half-life in the human body and that leads to difficulty in maintaining adequate doses.

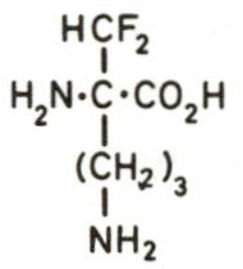

2-Difluoromethylornithine
(*9.88*)

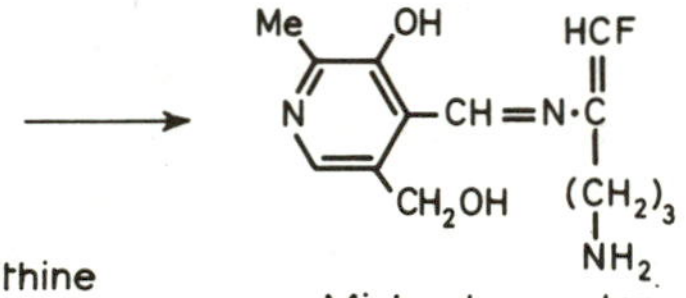

Michael acceptor
(*9.89*)

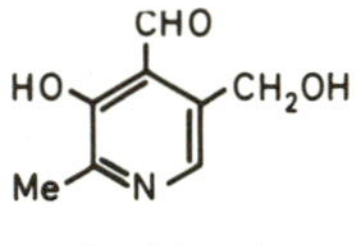

Pyridoxal
(*9.90*)

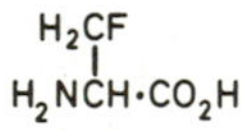

3-Fluoroalanine
(*9.91*)

A related procedure is seen at work in an anti-bacterial IMBI, D-3-fluoro-alanine (*9.91*), which is being clinically studied. This substance irreversibly inhibits alanine racemase, the enzyme that provides essential D-alanine for synthesizing the walls of bacteria (Wang and Walsh, 1978).

Michael's original (non-biological) donors were carbanions derived from an activated methylene-group as in diethyl malonate, but the most likely Michael donors in enzymes are the anions of either the NH— group in imidazole (as in Bloch's original discovery), or the hydroxy-group of serine, or the mercapto-group of cysteine.

Instead of using fluoromethyl-groups, activation can be achieved by a vinyl or ethinyl substituent. The aim here has been to increase concentrations of γ-aminobutyric acid (GABA) (*12.91*) in the brain. GABA is an inhibitory neurotransmitter, but injecting it into a convulsant patient does not help him, because it does not cross the blood–brain barrier. 4-Aminohex-5-enoic acid (*9.92*), which is the 4-vinyl derivative of GABA, when administered peripherally to mice increases the GABA content of brain threefold. It does this by inhibiting 4-aminobutyric acid transaminase (GABAase), apparently as follows. This IMBI combines with the pyridoxal coenzyme to form an azomethine, from which the enzyme abstracts a proton, allowing the double bonds in the IMBI-coenzyme adduct to become co-ordinated and hence a Michael receptor, leading to irreversible deactivation of the GABA-destroying enzyme (Lippert *et al.*, 1977). This IMBI does not act when the coenzyme is in the pyridoxamine state. It has considerable specificity because it does not work on a GABAase specimen from bacteria. Again, it is inactive with some other pyridoxal-requiring enzymes: mammalian glutamate decarboxylase and aspartate aminotransferase. However, the latter is inhibited by vinylglycine (2-aminobut-3-enoic acid) (Rando, 1974).

$$HC{=}CH_2$$
$$|$$
$$H_2N{-}CH{-}CH_2CH_2CO_2H$$

4-Aminohex-5-enoic acid
(*9.92*)

$$C{\equiv}CH$$
$$|$$
$$H_2N{-}CH{-}CH_2CH_2CO_2H$$

4-Aminohex-5-ynoic acid
(*9.93*)

The analogous 4-aminohex-5-ynoic acid (*9.93*) similarly elevates the GABA level in mouse brain (Jung and Metcalf, 1975). Moreover, 4-amino-5-fluoropentanoic acid, injected intraperitoneally into mice, raised the brain level of GABA 16-fold (Silverman *et al.*, 1981). The same mechanism is postulated except that the pyridoxamine form of the coenzyme participates (Silverman and Levy, 1981). In this case, fluorine was eliminated as the IMBI alkylated the coenzyme's amino-group, after which the enzyme removed a proton to give a Michael acceptor.

Some enzymes create a Michael acceptor in a different way, namely by oxidizing the IMBI. Thus an aminoacetylene can be oxidized to the corresponding iminoacetylene. For example, pargylene (*3.45*) is oxidized by monoamine oxidase (MAO), the first enzyme in the biosynthesis of catecholamines, to the imminium compound ($PhCH_2 \cdot N^+Me{=}CH{-}C{\equiv}CH$). Similarly, a hydroxyacetylene can be converted to a ketoacetylene. Pargylene's transformation product adds to the Michael donor, known to be the central ring-nitrogen atom of the riboflavine coenzyme (Abeles and Maycock, 1970). *N*-(1-Methylcyclopropyl)benzylamine, an IMBI related to the much-used antidepressant drug, tranylcypromine (*9.47*), is oxidized by monoamine oxidase to the highly reactive cyclopropiminium ion-radical which is thought to attack the riboflavine coenzyme (Silverman and Hoffman, 1981).

For enzymes that cannot effect Michael reactions, but tend to form electrophiles, IMBIs can be designed differently. Thus a diazoketone may be turned, by the enzyme, to a diazonium ion by protonation, and this should lead to alkylation of the active site of the enzyme. Thus the antibiotic, azaserine (*9.48*), is sufficiently like glutamine (*9.49*) for the enzyme formylglycinamide amino-transferase to take it up, and be inactivated by it. This prevents the change of the amide-group in the substrate (formylglycinamide ribonucleotide) to an amidino-group, an inhibition which terminates all purine biosynthesis. Note that the glutamine was a co-factor (ammonia supplier) in this reaction (Buchanan, 1978). Azaserine may be the IMBI with which physicians have had the longest experience because it has been used in cancer research and treatment since 1954.

Other enzymes can turn a pro-drug into an electrophile by inserting an oxygen atom. Thus the mixed function oxidases of the *e.r.*, described in Section 3.5, can oxidize difluoromethyl derivatives of the normal substrate to give an acyl fluoride which acylates the active site of the enzyme. Thus, aromatase is the enzyme normally concerned with the synthesis of estrone from androst-4-ene-3,17-dione. The 19,19-difluoro-derivative of this substrate acts as an IMBI which, in the manner described, becomes permanently attached to the enzyme in the 19-position (Marcotte and Robinson, 1982). These authors hope that this drug will prove useful in treating breast cancer.

Somewhat similarly, an α-cyano analogue of a substrate can be activated by some enzymes that can oxidize it to a hydroxyketenamine, thus:

$$NC—CH_2R \rightarrow HN{=}C{=}C(OH)R$$

For example, 4-hydroxybenzyl cyanide is an IMBI that can inhibit dopamine β-hydroxylase. Inhibition of this enzyme decreases production of epinephrine and it is hoped that this IMBI will turn out to be useful in reducing sympathetic tone (Baldoni, Villafranca and Mallettee, 1980).

For further reading on IMBIs, see Penning (1983) and Seiler, Jung and Koch-Weser (1978).

9.8 Special relationships between agonists and antagonists

Anti-metabolites need not be *analogues* of metabolites. In fact substances which do not resemble metabolites in any way can be used to combine with or destroy an apoenzyme, a coenzyme, or a substrate (Fildes, 1940). All that is necessary is a high affinity, and the combination is often effected by covalent bonds. These antagonists are not displaceable by the substrate. An example of such an antagonist is the porphyrin, haemin, which splits thiamine into two substances, one containing the pyrimidine ring and the other the thiazole ring. This causes thiamine-deprivation symptoms in animals fed on raw fish, in which this substance is abundant (Woolley, 1952).

The demonstration of competitive antagonism between two substances does not prove that one of them is a metabolite. Thus the severe contractions of the gut caused by barium chloride are competitively inhibited by sodium sulfate (which is not a normal gut constituent). In this case, there is a chemical reaction between the two substances, and the toxic barium ion is precipitated as the highly insoluble sulfate. Obviously, before a substance can be classed as a metabolite, its presence in Nature must be demonstrated, and also its origin and function. Morphine and allylnormorphine are examples of a pair of drugs which compete with one another without either of them being natural metabolites (Unna, 1943) (see Section 7.3). Similarly 2,4-dichlorophenoxyacetic acid, an artificial plant growth regulator, is strongly antagonized by the equally artificial analogue, 2,4-dichlorophenylthioisobutyric acid (Griffiths *et al.*, 1966). These examples may be classed with the classical 'Therapeutic Interference' effect of Section 6.6.

To assist the study of pairs of drugs that are antagonistic to one another without either of them necessarily being a metabolite, Schild (1947) introduced the very useful concept of pA_x. This is the negative logarithm of that (molar) concentration of an antagonist which reduces the effect of a multiple dose (x) of a drug to that of a single dose. Convenient values of x are 2 and 10. The concept runs like this: agonists that act on the same receptor are expected to produce the same pA_x with all competitive antagonists. This is a consequence of the mass-action law, and it applies whatever the efficacy and the affinity of the agonist may be.

The pA_x value can be used to classify drugs, because only those drugs that act on the same receptors are likely to give the same pA_x value with the same competitive antagonist. For example, the antagonistic effect of diphenyl-hydramine (*9.56*) was measured, on guinea-pig ileum, against two agonists that caused contractions. These agonists were histamine and 2-(pyrid-2-yl)ethyl-amine (*9.94*). The same pA_x was obtained, and this result strongly suggested that both agonists acted on the one receptor.

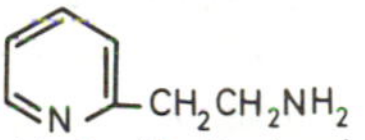

2-(Pyrid-2-yl) ethylamine
(9.94)

Moreover, a given drug/antagonist pair can be used to find if receptors in different tissues give the same pA_x values, a coincidence which would heighten the likelihood of these receptors being identical. Thus, similar pA_x values were obtained for the antagonistic effect of acetylcholine in such varied preparations as the frog heart, chick amnion, rat intestine, and guinea-pig lung (Arunlakshana and Schild, 1959).

9.9 Pharmacogenetics

In all work on enzyme inhibitors, it must be remembered that *individuals* do not necessarily possess the same content of enzymes as the population as a whole. Thus many people, particularly in Africa and the Middle East, have a deficiency of glucose 6-phosphate dehydrogenase, a biochemical lesion which does them no harm until they are prescribed the anti-malarial drug, primaquine (*3.36*). These patients quickly develop haemolytic anaemia, due to their biochemical lesion which is inherited on an incompletely dominant gene (Beutler, 1959; Fraser and Vesell, 1968). The anaemia is caused by a haemolytic quinone formed in the first stage of metabolism; the enzyme which should normally destroy the drug is missing.

Variable acetylation is a much studied pharmacogenetic phenomenon. It affects both hepatic and jejunal acetylation by *N*-acetyl transferase. Two phenotypes are distinguished in Man, the slow and the rapid acetylators, the relative proportions of which vary from country to country (Karim, Elfellah and Evans, 1981). About 90% of Orientals (Japanese and Chinese) are genetically rapid *N*-acetylators of isoniazid (*6.13*) whereas only 40% of the black or white citizens of the United States show this trait (Kalow, 1962). Slow acetylators become exposed to larger doses, and hence higher blood levels, of the drug than faster acetylators. However, isoniazid presents a worse situation because rapid acetylators produce acetylhydrazine which can cause liver damage (Mitchell, Thorgeirssen and Black, 1975).

The same inheritance controls the acetylation (deactivation) of the anti-bacterial sulfonamides, the anti-arrhythmic cardiac drug, procainamide (*7.56*) (Woosley *et al.*, 1978), the blood-pressure-lowering drug, hydralazine (*11.47*), and the amine derived by metabolism of the sedative, nitrazepam (*12.98*). In each case, rapid acetylation reduces the effect of the drug, but slow acetylators of hydralazine are unfortunate for they are prone to develop systemic lupus erythematosis, with arthritis-like symptoms (Batchelor *et al.*, 1980).

The enzyme alcohol dehydrogenase can be inherited in fast and slow types, the former commoner in Orientals who can be embarrassed by the prickling and flushing caused by sudden release of acetaldehyde (Propping, 1978). Debrisoquine (*9.95*), which is a blood-pressure-lowering drug which acts similarly to guanethidine (*7.55*), is converted to the inactive 4-hydroxy-derivative by a higher proportion of Caucasians than Orientals (Kalow *et al.*, 1980).

Another pharmacogenetic effect is the unwelcome rise in intraocular pressure when glucocorticoids are placed in the eye. A sampled Caucasian population showed 5% of responders. This effect was found to run in families, as does the presence of atypically weak *plasma* cholinesterase in those patients who find difficulty in hydrolysing suxamethonium (*7.29*) at the end of an operation (Kalow, 1962).

For further reading on pharmocogenetics, see Kalow (1962, 1980) and World Health Organization (1973b).

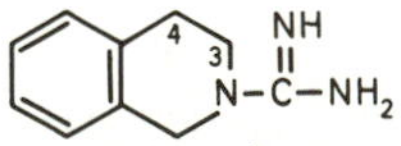

Debrisoquine
(*9.95*)

10

Ionization

Quaternary aliphatic amines, such as acetylcholine, must necessarily be completely ionized at all pH values. This follows from their structure: they have no proton to lose, and the nitrogen atom cannot exceed a covalence of 4. However, most biologically active substances are *weak* acids or bases, and hence the extent to which they are ionized is highly dependent on pH. Among such substances, the ionization constants (and hence the degree of ionization at a fixed pH) can be varied widely by making small alterations in the chemical structure. There are two ways of varying the degree of ionization, (a) to employ one substance at various pH values, and (b) to employ several related substances (of varying ionization constants) at a fixed pH. These controls are valuable to the investigator because ions behave differently from their corresponding uncharged forms. For example, ions undergo different chemical reactions, they penetrate membranes differently, and they become adsorbed to different types of substances.

10.0 The nature of ionization

Many substances are known which do not increase the electrical conductivity of water when dissolved in it. These are called non-electrolytes (chloroform and sucrose are examples) and they depress the freezing-point of water proportionally to their molar concentration. Acids, bases, and salts, on the other hand, increase the electrical conductivity of water. The majority of biologically active agents are acids, bases, or salts, and hence electrolytes. All electrolytes depress the freezing-point of water to a greater extent than would have been expected from their molar concentration. Dilute solutions of hydrochloric acid, sodium hydroxide, and sodium chloride give twice as much depression as would have been expected. This led the Swedish chemist Arrhenius to formulate the theory of ionization of electrolytes (1884–7). Thus, in solution, hydrochloric acid consists entirely of hydrogen cations and chloride anions (H^+ and Cl^-), sodium hydroxide of sodium cations and hydroxyl anions (Na^+ and OH^-), and sodium chloride of sodium cations and chloride anions (Na^+ and Cl^-). Sodium sulfate gives three times the expected depression, and it has been shown that, instead of a molecule of Na_2SO_4, three ions are present, namely, two sodium cations and one sulfate anion ($SO_4{}^{2-}$).

(*a*) *Salts.* In general, salts are completely ionized in dilute solution: of the few exceptions, the halides of mercury, cadmium, and lead are the most notable. Because salts are completely ionized, they have no biological properties other than those of the individual ions of which they are composed. Thus calcium chloride can have no conceivable physiological effects other than those peculiar to calcium ions and to chloride ions. This simple conception needs modification when a salt is derived from either a weak acid or a weak base, because some of the uncharged species is liberated by hydrolysis (see below), thus adding its own biological effect to those of the constituent ions of the salt.

In general the physiological properties of a well-ionizing salt can be neither more nor less than the sum of those of its ions. For example, Hata (1932) examined the toxicity for the mouse of 3,6-diamino-10-methylacridinium (*6.5*) chloride and the corresponding iodide, and found the iodide half as active, weight for weight. He then compared the ability of these substances to save mice infected with streptococci; once again the iodide was half as active. Now, as both of the anions are biologically inert at the dilutions tested, the biological activities must be proportional to the amount of cations in these substances which are completely ionized in neutral solutions (Albert, 1966). The formula-weights are respectively 260 and 351, so that the iodide would have 74% of the potency of the chloride. The biological results are in as good agreement with this calculation as could reasonably be expected from the serial dilutions that were used. Hence these experiments amount only to a biological comparison of the diaminomethylacridinium cation with itself. The anti-bacterial properties of a number of other acridine salts have been tested as their sulfates, nitrates, hydrochlorides, and hydriodides. As expected, no effect attributable to the anion was found (Browning *et al.*, 1922).

(*b*) *Acids and bases.* Unlike salts, acids and bases need not be completely ionized in solution. Strong acids (e.g. hydrochloric acid) and strong bases (e.g. sodium hydroxide) are completely ionized in the pH range of 0–14, but weak acids or bases show variable ionization within this pH range. Even small variations in pH on either side of the neutral point (pH 7) make considerable changes in the proportions of drug ionized in such cases as barbiturates, alkaloids, local anaesthetics, and anti-histaminics. Several examples will be given later.

Salts which are formed from a weak acid (or from a weak base) hydrolyse partly, in solution, to the acid (or base) from which they are derived and which are incompletely ionized. This situation is simpler than it may seem, because the degree of ionization in solution depends on only two factors, the pH and the pK_a. The latter (which will be defined below) is a constant for any acid or base. Hence, if the pH is controlled, the degree of ionization depends only on the nature of the acid (or base) added, *regardless of whether or not it has previously been neutralized*. Thus the same ratio of atropine ions to atropine molecules will result from the addition of atropine hydrochloride, atropine sulfate, or free atropine to a bath that has been buffered at pH 7. If the pH of the bath is raised, the proportion of atropine ions to atropine molecules will decrease, but the new ratio will again be independent of the form in which the atropine was added. Because it is confusing to speak of 'free' or 'non-ionized' acids and bases, the term 'molecule', or 'neutral species' is customarily used for all uncharged forms.

In 1944, some American bacteriologists (who shall be nameless, here) published a paper in which they found that 9-aminoacridine (*10.8*) was 64 times more bactericidal than its hydrochloride (against *Pneumococcus* type III in glucose-broth). The authors did not say whether their medium was buffered. If it was buffered, the same result should have been obtained in both cases, and hence the technique was at fault. If the medium was not buffered, the results could have no quantitative significance because of the greatly increased pH of a solution of a strong base (pK_a 10.0 in this case) compared to that of solutions of its salts. As it happened, the striking increase in the anti-bacterial action of acridines, when the pH rises, had long been on record (Graham-Smith, 1919; Browning, Gulbransen and Kennaway, 1919).

For further reading on acids, bases, and ionization, see Albert and Serjeant (1984) and Bell (1973).

10.1 The ionization constant (K_a)

An essential part of Arrhenius's theory of ionization was the application of the *law of mass-action* to describe the state of ionic equilibrium. Thus, acetic acid (CH_3COOH) is a weak acid which ionizes in water to give some hydrogen ions (H^+) and some acetate anions (CH_3COO^-). The product of the concentration of the ions (which is $[H^+] [CH_3COO^-]$) always bears a fixed ratio to the concentration of the neutral molecules $[CH_3COOH]$. This ratio is called the acidic ionization constant (K_a), or more simply the ionization constant. Thus:

$$K_a = \frac{[H^+][CH_3COO^-]}{[CH_3COOH]} \qquad (i)$$

and this has been found, experimentally, to be 1.75×10^{-5} (at 25°C).

Sometimes the expression 'dissociation constant' is used for ionization constant. The latter term is more precise because many complexes, such as enzyme systems, 'dissociate' into their components, and micelles into their monomers: the relevant equilibria are expressed as dissociation constants similarly derived from the law of mass-action. Nevertheless, such constants are not ionization constants.

The state of ionization of weak bases also can be described by acidic ionization constants. For example, ammonia is a weak base which can take up hydrogen ions to form ammonium ions. This is, of course, equivalent to thinking of the ammonium ion (NH_4^+) as a weak acid which ionizes in water to give some hydrogen ions (H^+) and some molecules of ammonia (NH_3). Thus:

$$K_a = \frac{[H^+][NH_3]}{[NH_4^+]} \tag{ii}$$

and this has been found, experimentally, to be 5.5×10^{-10} (at 25°C).

The use of acidic constants to describe the ionization of bases was introduced in 1929 by Brönsted in order to record the ionization of both bases and acids on the *same* scale, just as a single pH scale serves to measure both acidity and alkalinity.

Earlier workers had toyed with a separate scale for bases, writing:

$$K_b = \frac{[OH^-][NH_4^+]}{[NH_4OH]} \tag{iii}$$

and the value of K_b (the basic dissociation constant) was found to be 1.8×10^{-5} (at 25°C). Equation (iii) is unreal because no substance with pentacovalent nitrogen, such as NH_4OH, is capable of existence. Equations (i) and (ii) show that an acid produces hydrogen ions and a base receives them. Thus both acid and base can be related in terms of a single quantity, their affinity for the hydrogen ion. This relationship allows the use of the acidic ionization constant for both acids and bases.

(a) *The definition of* pK_a. Ionization constants are small and inconvenient, but their negative logarithms (known as pK_a values) are convenient to say and to write. Thus the pK_a of acetic acid is 4.76 and of ammonia, 9.26. When the older literature offers only pK_b values for bases (e.g. 4.74 for ammonia), these can be converted to pK_a values by subtraction from the negative logarithm of the ionic product of water (K_w) at the temperature of determination. (The value of pK_w is 14.16 at 20°C, 14.00 at 25°C, and 13.58 at 37°C.)

It is evident that pK_a values provide a very convenient way of comparing the strengths of acids (or of bases). The stronger an acid is, the lower its pK_a (the stronger a base is, the higher its pK_a). The pH at which an acid or base is half ionized is equal to its pK_a. When the pH is one unit below the pK_a, an acid is 9% ionized, and a base is 91% ionized (see Section 17.0).

Any acid or base, if partly neutralized, is an effective buffer within the range from one unit below the pK_a value to one unit above it. Biologists use initiative in selecting buffers suitable for particular experiments. They do not use anionic buffers (such as citrate and phosphate, which form complexes) if metal cations are essential to the system under study. To give just one example, phosphate buffers inhibit the action of isocitrate dehydrogenase (in pig's heart) by removing manganese (Lotspeich and Peters, 1951). In place of these anionic buffers, cationic buffers such as N-ethylmorpholine (pH range 7.0–8.2) and 'Tris' (aminotrishydroxymethylmethane) (7.5–8.7), and the zwitterionic buffers such as 'Hepes' and 'Tes' (6.9–8.1) (Good et al., 1966) are well established. For a useful book on buffers, see Perrin and Dempsey (1974).

(b) *How to locate a* pK_a *value.* Figures for about 500 common substances will be found in Albert and Serjeant (1984). For other values, see the comprehensive lists compiled for the International Union of Pure and Applied Chemistry: for *organic bases*, Perrin (1965b, and large supplement 1972), for *organic acids*, Kortüm, Vogel and Andrussow (1961) with a large supplement by Serjeant and Dempsey (1979), and for *inorganic acids and bases*, Perrin (1983). If your library has not catalogued these under the compilers' names, but under 'International', it will probably be quicker to ask the librarian to locate them. Table 10.1 gives the relative strengths of some common acids and bases; this table should be committed to memory because it provides a comparison for new pK_a values encountered in reading. Acids and bases of equivalent strength have been placed opposite one another.

A pK_a value of about 5 (cf. acetic acid) is typical of a great many mono-carboxylic acids, both aliphatic and aromatic. The value of 10 is typical of a phenol. Acids with pK_a values greater than 7 scarcely affect neutral indicator paper and, if greater than 10, do not even taste acidic.

The value of 11 for ethylamine is representative of aliphatic bases; that of 5 for aniline is typical of aromatic bases which are much weaker. As pK_a figures are logarithms, there is a difference of one millionfold (i.e. antilog 6) between the strengths of ethylamine and aniline. Many alkaloids and other biologically active bases have pK_a values about 8. Bases with pK_a values less than 7 scarcely affect neutral indicator paper.

Electron-releasing groups (e.g. $—CH_3$) strengthen bases and weaken acids, whereas electron-attracting groups (e.g. $—NO_2$) weaken bases and strengthen acids. The electronic properties of various substituents are summarized in Section 17.2. These influences are now so well understood, quantitatively, that the ionization constant of an acid or base can be calculated in advance of synthesis (Perrin, Dempsey and Serjeant, 1981). This saves a great deal of time which might be spent in synthesizing compounds of unfavourable ionization properties. Needless to say, it is essential to determine the actual constant, once the material becomes available.

For detailed practical instruction in the determination of ionization constants in the laboratory, see Albert and Serjeant (1984).

Table 10.1 Relative strengths of some common acids and bases

Acids	pK_a	Bases	pK_a
Hydrochloric acid	<0	Sodium hydroxide	>14
Phosphoric acid (first proton lost)	2		12
	3	Ethylamine	11
	4		10
Acetic acid	5	Ammonia	9
Carbonic acid	6	Quinine, strychnine	8
Phosphoric acid (second proton lost)	7		7
	8		6
Hydrocyanic acid, boric acid	9	Aniline, pyridine	5
Phenol	10		4
	11		3
Phosphoric acid (third proton lost)	12		2
Glucose	13	p-Nitroaniline	1

(c) *Calculation of the percentage ionized.* The degree of ionization of any base, in aqueous solution, can be calculated from the following equation, provided that two things are known, the pH of the solution and the pK_a of the substance:

$$\% \text{ ionized} = \frac{100}{1+\text{antilog}\,(\text{pH}-pK_a)} \qquad \text{(iv)}$$

This equation shows that the degree of ionization varies with pH, but that it does not bear a straight-line relationship to the extent of the change. On the contrary, a sigmoid curve is followed, as shown in Fig. 10.1.

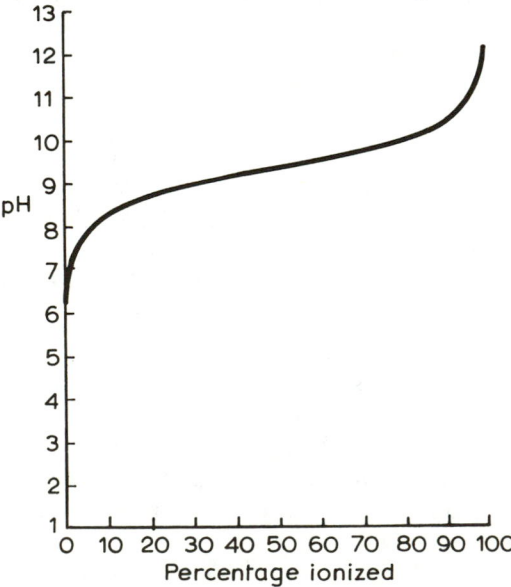

Fig. 10.1 Typical curve obtained in the potentiometric titration of an acid (boric acid, $pK_a = 9.21$ at 20°C.)

It will be seen from this figure that a small change in pH can make a large change in ionization, particularly if the pH of the solution is numerically close to the pK_a of the substance investigated. This is readily seen from Table 10.2. By way of example: if one were working at pH 7 with a nitrophenol of $pK_a = 7$, half of the substance would be in the ionized state. If the pH were allowed to rise to 8, the phenol would be nearly all ionized; but if it fell to 6, almost all of the phenol would be non-ionized.

Table 10.3 may be regarded as a magnification of the central part of Table 10.2.

Table 10.2* Calculation of the extent of ionization, given pK_a and pH

$pK_a - pH$	Per cent ionized (if anion)	Per cent ionized (if cation)
−4	99.99	0.01
−3	99.94	0.10
−2	99.01	0.99
−1	90.91	9.09
0	50.00	50.00
1	9.09	90.91
2	0.99	99.01
3	0.10	99.94
4	0.01	99.99

*An extended form of this table will be found in Section 17.0.

Table 10.3 Calculation of the extent of ionization where pK_a is close to pH

$pK_a - pH$	Per cent ionized (if anion)	Per cent ionized (if cation)
−0.5	75.97	24.03
−0.4	71.53	28.47
−0.3	66.61	33.39
−0.2	61.32	38.68
−0.1	55.73	44.27
0	50.00	50.00
+0.1	44.27	55.73
+0.2	38.68	61.32
+0.3	33.39	66.61
+0.4	28.47	71.53
+0.5	24.03	75.97

(*d*) *Polyelectrolytes*. These are molecules with a large number of ionizing groups which may have the same charge, as in polyacrylic acid, or both charges may be present, as in proteins. The titration curves of polyelectrolytes have a rather indefinite ('smeared-out') appearance. Contrary to what happens with mono- and di-ions, a polyion retains a large fraction of the oppositely charged counterions close to it. For further reading, see Albert and Serjeant (1984).

(*e*) *Zwitterions*. So far, no mention has been made of zwitterions (internal salts which carry both positive and negative charges). In discussing whether a substance is zwitterionic or not, the pH range in which the information is required must be specified, because a sufficiently alkaline solution will change the zwitterion to an anion, and a sufficiently acid solution will change it to a cation.

For example, the acidic group in glycine has a pK_a of 2.2, and the basic group has a pK_a of 9.9. It is evident from Table 10.2 that at pH 3.2, only 90% of the acidic groups (but *all* of the basic groups) will be ionized; hence at this pH, about 90% of the substance will be present as zwitterion and 10% as cation. Glycine is almost entirely zwitterionic between pH 3.2 and 8.9.

Not all substances with an acidic and a basic group are zwitterions (internal salts), because there may be no pH at which both groups are ionized. The pH of the solution is another factor. Thus a substance is 90% or more in the zwitterionic state when the pH is at least one unit above the acidic pK_a and at least one unit below the basic pK_a. If this rule is applied to *p*-aminobenzoic acid (acidic pK_a 4.8; basic pK_a 2.7), it will at once be seen that this substance, unlike glycine, is exclusively anionic in neutral and alkaline solutions. For other tests to distinguish between zwitterionic and merely amphoteric substances (e.g. glycine and *p*-aminobenzoic acid respectively), see Albert and Serjeant (1984).

In a zwitterion, each ionizing group has two constants, one for the fraction that has the other group ionized and one for the remaining fraction. These are called microscopic constants, and the macroscopic constants (as determined by spectrometry or potentiometric titration) are gross composites of these. Historically, tyrosine was one of the first zwitterions to be examined for microconstants (Edsall, Martin and Hollingworth, 1958) and the same methods are still used. See Section 12.4 for the pursuit of microconstants in the catecholamine series.

10.2 Differences in ionization that can bring about selectivity

Students often ask, 'How can the degree of ionization be relevant where biological action is concerned? Surely poorly ionizing substances produce at least a few ions and, as these are removed, fresh ones are generated from the neutral molecules according to the mass-action law, as in Equation (i)?'

This objection is intelligently argued, but it must be remembered that ions are not bound to their receptors by *covalent* bonds, and hence they can easily leave the receptors. In short, to keep a receptor saturated with a given ion, there must be an excess of this ion in the solution bathing the receptor. For example, crystal violet* is bacteriostatic to *E. coli* at a dilution of 1 in 10000, but not at 1 in 20000, although both solutions are intensely violet. The violet colour denotes the presence of the cation, which is the active form of this antibacterial drug (the neutral species is colourless).

Equations (v) and (vi) help to explain this phenomenon. Equation (v) shows that the magnitude of K_a (the ionization constant) governs the proportion of cation (BH^+) to neutral species (B) at any given hydrogen ion concentration $[H^+]$. Equation (vi) similarly deals with the magnitude of K_s which is the stability constant of the ion-pair (ABH) formed by combination of the cation (BH^+) of the drug with the vulnerable anionic group (A^-) of the bacterium. This complex is maintained by an ionic bond supplemented usually by a hydrogen bond or several van der Waals bonds.

$$K_a = \frac{[B][H^+]}{[BH^+]} \text{ (ionization constant)} \tag{v}$$

$$K_s = \frac{[ABH]}{[A^-][BH^+]} \text{ (stability constant)} \tag{vi}$$

If K_a and K_s are of comparable magnitudes, then any deficiency of cations in the solution will be replenished from the drug–bacterium complex (ABH) as well as from the non-ionized drug (B). Under these circumstances the ionization constant of the drug becomes the limiting factor governing its use: not only must ions be present, but they must be present abundantly. It is not to be supposed from this reasoning that the acids in receptors are necessarily weak. Rather, the dwell-time of an adsorbed cation, unless its ionic bond is reinforced by hydrogen bonding or van der Waals forces, is very short because of the inevitable exchange with sodium or potassium cations (from the saline solution that bathes the area).

Many bacteria are much more sensitive than *E. coli* to crystal violet. Thus *Streptococcus pyogenes* is inhibited by a dilution of 1 in 320000 (though not by 1 in 640000); and *Staphylococcus aureus* is killed by 1 in 2000000. Although these differences may be partly due to differences in sites of loss, there are different K_s values for different species and (in the hosts) for different tissues in the one species. This affords a basis for the selective action of cationic drugs.

Another way in which ionization can be enlisted in the service of selectivity is available when the target area has an unusual pH, as in gastric juice or urine. The abnormally low pH of malignant cells was mentioned in Section 4.2. Bicker (1974) forecast that the glucuronides of cytostatic phenols and amines would be selective against cancer cells because the optimal pH for β-glucuronidase activity is 5.2. However, utilization of the pH difference between normal and malignant cells need not depend on a pro-drug finding an enzyme to hydrolyse it. 4-Dodecylpyridine (*10.1*), which has a pK_a of 5.5, circulates harmlessly in the mouse's bloodstream as the neutral species, and is non-toxic to that host on

*Crystal violet is the NN′N″-hexamethyl derivative of (*10.5*).

intraperitoneal injection of 0.5 g kg^{-1}. However, the increased acidity of pathogenic cells trap it as the cation which, from the structure, is highly surface-active
and injures these uneconomic cells (Firestone, Pisano and Bonney, 1979).

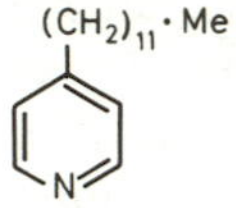

4-Dodecylpyridine
(10.1)

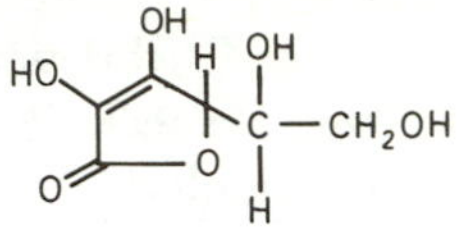

Ascorbic acid
(10.2)

Other types of selectivity depend on properties in which ions, as a class, differ
from molecules. These differences will now be discussed under three headings:
covalent reactivity (the making and breaking of covalent bonds), adsorption at
surfaces, and penetration of membranes.

(a) *Covalent reactivity.* Nitration of the neutral species of aniline occurs in the
ortho- and *para*-positions, whereas its cation becomes nitrated mainly in the
meta-position. Thus by changing the acidity of the nitration mixture, the proportions of the isomers can be controlled. An example of greater biochemical
interest is ascorbic acid (*10.2*), which is readily oxidized by air when present as
the di-anion, whereas the mono-anion and the molecule (either of which can be
made the principal species present, by a change in pH) are quite stable
(Weissberger and LuValle, 1944). This is shown in the top half of Table 10.4,
where a decrease in alkalinity is seen to slow the rate of oxidation in proportion
as the concentration of di-anion is lowered, even though the proportions of the

Table 10.4 Autoxidation of ascorbic acid

pH	Di-anion %	Mono-anion %	Molecule %	Velocity constant min^{-1}×10^3	
9.21	0.5	99.5	—	22	No metallic ions
8.71	0.2	99.8	—	11	,, ,, ,,
7.61	0.01	99.99	—	0.7	,, ,, ,,
5.80	—	97.8	2.2	0.1	,, ,, ,,
4.70	—	79.4	20.6	0.03	,, ,, ,,
9.31	0.6	99.4	—	101	Copper-catalysed
7.49	0.01	99.99	—	170	,,
6.12	—	99.0	1.0	134	,,
5.08	—	90.1	9.9	91	,,
3.87	—	36.0	64.0	26	,,
2.59	—	2.9	97.1	2	,,

(Weissberger and LuValle, 1944.)

other two species are thereby increased. In the presence of cupric ions, the autoxidation of ascorbic acid goes much faster; but the mechanism is obviously different, for here the rate is proportional not to the concentrations of the di-anion or of the molecule, but to that of mono-anion (Table 10.4, lower half).

The breaking of covalent bonds, by enzymes, greatly influences the metabolism of selectively toxic agents. Each may become changed to a more active or to an inert substance. However, because ion and molecule must react at very different rates, variations in pK_a among members of a series will produce marked differences.

(*b*) *Adsorption at surfaces*. There are two kinds of adsorption, the *indiscriminate* and the *specific*. Indiscriminate adsorption is shown by amphiphilic substances, i.e. those having a water-attracting end-group attached to a comparatively large residue that has little affinity for water. Ordinary soap is an example of this class of substances. It will be recalled that water molecules are extensively hydrogen-bonded to one another and that substances dissolve in water only by virtue of their ability to break some of these bonds and form new bonds with water molecules. Amphiphilic molecules, dissolved in water, are in a state of uneasy equilibrium, because the hydrocarbon portion is constantly being squeezed out by the water molecules in their endeavour to unite with one another.

This property of water gives rise to indiscriminate adsorption because such substances become deposited on any surface that presents itself, regardless of its chemical nature. Soap, for instance, is thus caused to accumulate not only at the air–water interface of a vessel in which its solution is stored, but also at the glass–water interface. Furthermore if various objects are immersed in the soapy solution, soap will be found to accumulate on them also. This is typical indiscriminate adsorption, in which molecules tend to be adsorbed more strongly than ions, e.g. oleic acid is more strongly adsorbed than (sodium) oleate. This happens because an ion is more strongly hydrated than the corresponding neutral species, so that the latter has the greater tendency to undergo expulsion.

Specific adsorption is more important in drug action. It is often shown by hydrophilic substances which tend to leave water when they can accumulate on a surface that has a chemically complementary character. Among the commonest cases of complementarity is that of an anion attracted to a positively charged portion of a surface or, conversely, a cation attracted to a negatively charged portion. In such cases, an ion is obviously adsorbed much more strongly than the corresponding molecule. Examples of this phenomenon among selectively toxic agents are developed later in this chapter (e.g. the aminoacridines). Perhaps even more striking examples are the phosphorylated vitamins B which, water-soluble as they are and occurring in food at immense dilution, nevertheless become concentrated on their complementary surfaces in enzyme systems, with astonishing speed and efficiency.

(*c*) *Penetration of membranes*. Every living cell is surrounded by a semi-permeable membrane which is two to four molecules in thickness. Such a membrane, usually lipoprotein in nature, is strongly charged, and its interior is almost inaccessible to ions. The difficulties in the way of the penetration of ions are (a) their relatively greater size, due to hydration, and (b) their charge, which is either *similar* to the portion of protein surface which they approach (resulting in repulsion) or is *opposite* (resulting in fixation).

However, neutral molecules, provided they have no more than three water-attracting groups and a molecular weight of not more than 150, usually penetrate membranes readily (see Section 3.2). A good example of the control that ionization exerts over penetration is illustrated in Fig. 10.11. Some new brain-imaging amines, containing ^{123}I, have been designed with pK_a values that could take advantage of the pH gradient that exists between blood (pH 7.4) and brain sap (pH 7.0). After diffusing into the brain as molecules, they encounter the lower pH of the cytoplasm which increases the percentage present as cations (non-diffusible) (Tramposch, Kung and Blau, 1983).

It must not be supposed, however, that no ions ever penetrate natural membranes. In the first place, specific mechanisms exist for the uptake of each ion for which the cell has a need. Thus cholinium ions are taken up by a 'Type 2' mechanism (see p. 68), and Na^+ and K^+ ions by an energy-consuming 'Type 3' mechanism (p. 69). The human gut is readily permeable to sodium and chloride ions, and fairly permeable to citrate and acetate ions; but for Ca^{2+}, Mg^{2+}, Fe^{2+}, and phosphate, it has evolved specific, easily saturated intake mechanisms, which reject all but the traces of these ions essential for the body's needs, and sulfate and tartrate anions pass through the intestines largely unabsorbed. In the second place (and this refers specially to synthetic agents) a non-penetrating ion can often be made penetrating by the addition of a lipophilic-group; the chloro- and other lipophilic-substituents of the anti-malarials mepacrine ('Atebrin') and chloroquine assist in the necessary penetration of red blood cells.

A further effect of ionization upon penetration must now be considered. The action of a drug obviously depends upon ability to reach its receptor. Now an agent with a pK_a between 6 and 8 is in the position that, at the physiologically interesting pH of 7, it is always in equilibrium with at least 10% of its more poorly represented ionic species (see Table 10.2). Such an agent will encounter membranes through which only the non-ionic species can pass. Yet this species, when it has penetrated, is likely to encounter an aqueous medium of similar pH value, and in this it is obliged to re-form ions until the same degree of ionization exists on both sides of the membrane. Quite a large number of agents have pK_a values between 6 and 8: but some have pK_a values which lie entirely outside this range and hence they have a different pattern of distribution and different types of action.

(*d*) *Pseudo-bases*. The time taken for ionic equilibria to occur, in solution, is so exceedingly small (e.g. 10^{-7} s) that ionic reactions may be regarded as instantaneous. However, there are a few kinds of cation which slowly combine *covalently* with hydroxyl ions to give non-ionized substances known as 'pseudo-bases' from which the original cation can be regenerated by acid. In different examples, any time from a minute to a week may be needed for equilibrium to be achieved. The two most common examples of pseudo-bases are heteroaromatic quaternary compounds and triphenylmethane dyes. Suitable methods for calculating the equilibrium ionization constants (expressed as pK_a^{Eq}), and the velocity constant at which equilibrium is attained, are available, e.g. Goldacre and Phillips (1949) for quaternary amines, and Cigén (1958) for the triphenylmethanes.

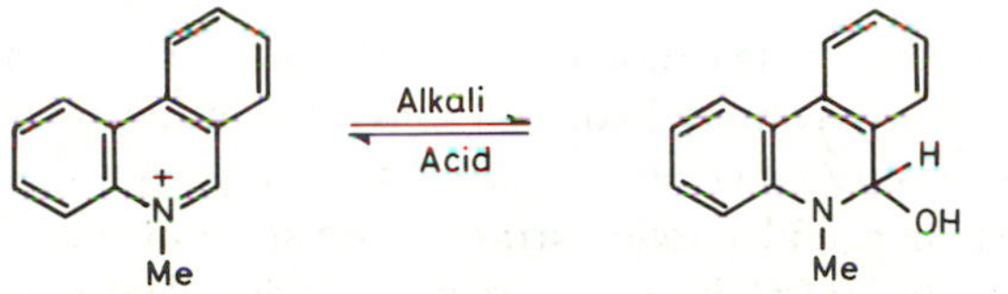

5-Methylphenanthridinium cation
(10.3)

Carbinol (or pseudo-) base
(10.4)

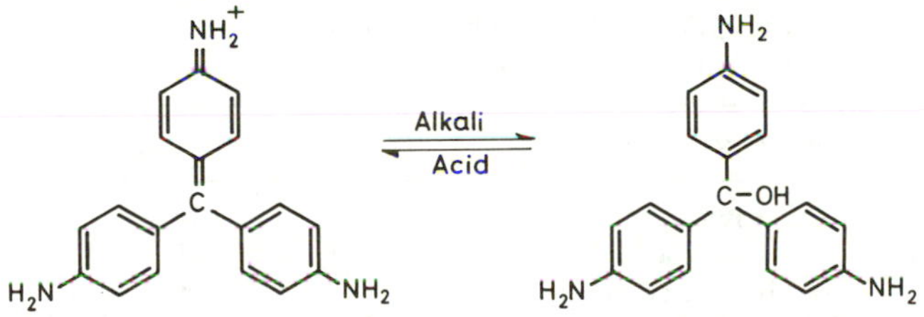

Parafuchsin
(10.5)

Carbinol (or pseudo-) base
(10.6)

Because pseudo-base formation can assist permeability, the relevant chemistry will now be discussed. Formula (*10.3*) shows the 5-methylphenanthridinium cation, such as exists in phenanthridine methochloride. This substance will serve to represent all quaternary amines which have a double bond attached to a ring-nitrogen atom (the cations of the related *tertiary* amines do not form pseudo-bases). Because tetracovalency, as in (*10.3*), is the maximal valency of nitrogen, and because a methyl-group is not mobile like a hydrogen ion, there can be *no* neutral species corresponding to structure (*10.3*). Thus one might suppose that it would remain ionized at every pH. However, this ion undergoes a slow, reversible reaction with a hydroxyl ion which, by the formation of a new covalent bond, produces the neutral molecule (*10.4*) (Magrath and Phillips, 1949). This secondary alcohol is the pseudo-base of (*10.3*). Pseudo-bases have a much greater liposolubility than the cations from which they are derived, and it is probably in the form of the pseudo-bases that these ions enter cells. No explanation of the action of the trypanocidal quaternary heterocycles (see Section 10.3.5) can be complete without considering both the equilibrium, e.g. (*10.3*) $\rightleftharpoons$ (*10.4*), and the time of half conversion ($t_{0.5}$).

This tendency to pseudo-base formation increases with the complexity of the heterocyclic nucleus. Thus, whereas pseudo-base formation does not occur for the 1-methylpyridinium cation, and only at a very high pH for the 1-methyl-quinolinium and 2-methylisoquinolinium ions, the reaction takes place much more readily upon further annelation (as in 10-methylacridinium). The inser-tion of an electron-attracting substituent (such as —NO_2), or a second doubly-bound ring-nitrogen atom, as in quinoxalinium and quinazolinium, has the same effect. Some examples are listed in Table 10.5, in which pK_{ROH} is the pH at which the pseudo-base and the quaternary cation are present in equal concen-trations. Formally pK_{ROH} is similar to a pK_a value, but the equilibrium is only slowly achieved.

Formula (*10.5*) shows the cation of parafuchsin, a typical triphenylmethane dye, which reacts slowly with hydroxyl ions to give the pseudo-base (*10.6*). Pseudo-bases of both types (*10.4*) and (*10.6*) react rapidly with alcohols to give colourless ethers; hence alcoholic extraction must be avoided in colorimetric estimation of the amount of dye taken up by a biological specimen.

Table 10.5 Equilibrium between quaternary cations and their pseudo-bases

Quaternary cations	pK_{ROH}
1-Methylquinolinium	16
2-Methylisoquinolinium	15
10-Methylacridinium	9.9
5-Methylphenanthridinium (*10.3*)	10.4
1-Methyl-3-nitroquinolinium	6.7
2-Methyl-4-nitroisoquinolinium	5.0
1-Methylquinoxalinium	8.6
3-Methylquinazolinium	<7

(Bunting and Meathrel, 1972.)

Zwitterions are usually rather inert pharmacologically. For example histidine, which is a zwitterion, has none of the marked physiological properties shown by the closely related cationic substance, histamine. Again the vinyl group of quinine can be oxidized to a carboxylic acid, thereby converting the cation to a zwitterion: concomitantly, the anti-malarial properties disappear. When this acid (quitenine) is esterified, the substance of necessity becomes cationic again and this ester is strongly anti-malarial (Goodson, Henry and MacFie, 1930). Similar examples have been found in the acridine series (see Section 10.3.1).

10.3 Substances that are more biologically active when ionized

It is now known that many kinds of organic cations are antibacterial, but this knowledge came only in the second quarter of this century. It had been found earlier that aliphatic amines (which exist mainly as cations at pH 7) were bactericidal, and aromatic amines (which exist mainly as molecules at pH 7) were not. However, no early worker sensed a connection between antisepsis and organic cations; antibacterial activity was supposed to be due 'to the presence of hydroxyl ions liberated through ionization of the alkylammonium hydroxides which are formed through the combination of the amines with water' (Morgan and Cooper, 1912). In other words, the effect was ascribed to alkalinity: yet the same bases, tested in the form of salts or in a neutral buffer solution, show the same bactericidal properties. The problem began to be better understood in 1924 when Stearn and Stearn suggested that triphenylmethane dyes owe their antibacterial activity to a reaction of the *cation* with some anionic groups of bacteria to give feebly dissociated complexes of the type discussed in Section 10.2 [cf. Equation (vi)]. Although the Stearns did not know the percentage ionization of their dyes,* they predicted that the salts of many strong bases would be found to be antibacterial because these would provide a sufficient supply of cations in the physiological pH range. They also showed that increasing the pH of the medium increased the antibacterial activity by bringing about increased (anionic) ionization of the receptors of the bacterium. They pointed out that this alkalization of the medium must not be carried to the point where it begins to suppress the ionization of the antiseptic itself.

The first rigorous proof of a positive correlation between ionization and biological action was made 17 years later (Albert, Rubbo and Goldacre, 1941). This work showed that a quantitative relationship existed between antibacterial action and the percentage ionized as cations in the aminoacridine series. This correlation was then confirmed and extended (Albert *et al.*, 1945).

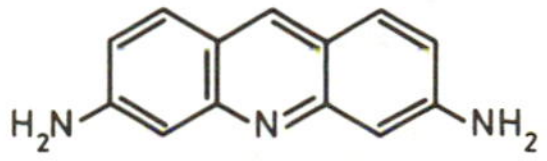

Proflavine
(3,6-diaminoacridine)
(10.7)

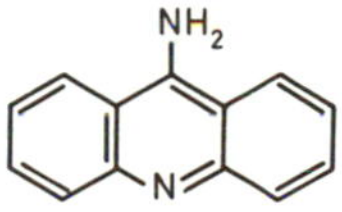

Aminacrine
(9-aminoacridine)
(10.8)

10.3.1 *The antibacterial aminoacridines*

Acridine (*2.16*) is a planar, feebly basic molecule. However, two of the five possible monoaminoacridines are rather strong bases, namely the 3- and 9-amino isomers; this strength comes from the resonance immanent in their cations as explained in Section 2.2 (p. 33). For reasons of valency there can be no extra resonance in 2- or 4-aminoacridine, and there is little in the 1-isomer because an *ortho*quinonoid disposition of bonds (which it would require) is energetically unfavoured (Albert and Goldacre, 1946; Albert, Goldacre and Phillips, 1948).

*The ionizations could not be measured at that time because of technical difficulties with pseudo-bases, since overcome.

The aminoacridines were introduced by the Scottish pathologist Carl Browning in 1913 as antibacterials for use in wounds. One of these, proflavine (*10.7*) (3,6-diaminoacridine), was shown to be toxic to a wide range of Gram-positive and -negative organisms without injurious effect on human tissues (Browning and Gilmour, 1913; Browning *et al.*, 1917; War Office, 1922). The quaternized acridines, such as euflavine, are more toxic to mammals yet no more active as anti-bacterials than the non-quaternized acridines, such as proflavine. Apart from these quaternized acridines, no source of pseudo-bases is encountered in the acridine series; thus the difficulties encountered in working with triphenylmethane dyes were easily avoided.

Preliminary experiments in 1939 had convinced my colleagues and me that

Table 10.6 Dependence of bacteriostasis on ionization in the acridine series

Acridine	Minimal bacteriostatic concentration for Strept. pyogenes (48 *hours* incubation in 10% serum broth at 37°C; pH 7.3)		Per cent ionized as cation (pH 7.3, 37°C)
(unsubstituted)	1 in	5000	1
4-Amino-		5000	< 1
2-Amino-		10 000	2
1-Amino-		10 000	2
4,5-Diamino-		< 5000	< 1
2,7-Diamino-		20 000	4
3-Amino-		80 000	72
9-Amino- (aminacrine) (*10.8*)		160 000	99
3,9-Diamino-		160 000	100
3,7-Diamino-		160 000	76
3,6-Diamino- (proflavine) (*10.7*)		160 000	99
2-Amino-9-methyl-		20 000	3
1-Amino-4-methyl-		20 000	1
4-Amino-5-methyl-		< 5000	< 1
2-Amino-6-chloro-		< 5000	< 1
3-Amino-9-chloro-		< 5000	11
3-Amino-7-chloro-		40 000	20
3-Amino-6-chloro-		40 000	24
9-Amino-1-methyl		160 000	99
9-Amino-2-methyl		160 000	99
9-Amino-3-methyl		160 000	99
9-Amino-4-methyl-		320 000	99
9-Amino-1-chloro-		160 000	86
9-Amino-2-chloro-		160 000	94
9-Amino-3-chloro-		160 000	98
9-Amino-4-chloro-		80 000	83

(Albert *et al.*, 1945.)

acridines became more antibacterial in proportion as they became more ionized, i.e. as cations (Albert, Rubbo and Goldacre, 1941). These results were extended to 101 acridines and 22 species of bacteria (Albert *et al.*, 1945; Albert and Goldacre, 1948). Some typical results are given in Table 10.6, which is an expansion of Table 2.2, in terms of derivatives tested. For comparison of 100 derivatives, see Albert (1966, p. 437).

Of the five possible monoaminoacridines, two are well ionized at pH 7, whereas the remainder are poorly ionized. (The chemical reason for this difference was given in Section 2.2, p. 33.) It can be seen from Table 10.6 that the two

Table 10.7 Comparative sensitivity of 22 species of bacteria toward aminoacridines
Highest dilutions completely preventing visible growth in 48 hours at 37°C
(Medium: 10% serum broth, pH 7.3)
All substances were completely ionized under the conditions of the test

Organism		A	B	C	D	E
Gram-positive species:						
Cl. welchii	1 in	320 000	640 000	160 000	1 280 000	320 000
Cl. sporogenes		160 000	320 000	160 000	640 000	160 000
B. subtilis		80 000	160 000	80 000	320 000	80 000
B. anthracis		160 000	320 000	80 000	640 000	80 000
Strep. pneumoniae		320 000	320 000	160 000	640 000	160 000
Strept. pyogenes A		160 000	320 000	160 000	640 000	160 000
Strept. viridans		160 000	640 000	160 000	640 000	160 000
Staph. aureus		40 000	80 000	40 000	160 000	40 000
Lactobacillus		320 000	320 000	160 000	640 000	160 000
C. diphtheriae		160 000	320 000	40 000	320 000	160 000
Mycobact. phlei		160 000	320 000	160 000	640 000	160 000
Average inhibitory dilution		160 000	300 000	110 000	540 000	140 000
Gram-negative species:						
E. coli		40 000	40 000	20 000	40 000	40 000
Proteus vulgaris		40 000	40 000	10 000	20 000	40 000
Ps. pyocyanea				(completely inactive)		
B. Friedlanderi		80 000	80 000	20 000	160 000	80 000
Eberthella typhosum		80 000	160 000	80 000	80 000	80 000
Shigella dysenteriae		320 000	320 000	160 000	320 000	320 000
Vibrio cholarae		160 000	160 000	160 000	320 000	160 000
Pastuerella		160 000	160 000	40 000	160 000	160 000
Brucella		640 000	640 000	320 000	1 280 000	320 000
Haem. influenzae		320 000	640 000	320 000	640 000	160 000
Neiss. meningitidis		320 000	640 000	320 000	2 560 000	320 000
Average inhibitory dilution		100 000	130 000	60 000	150 000	100 000

A 9-Aminoacridine (*10.8*) (aminacrine).   *B* 4-Methyl-9-aminoacridine.
C 3,6-Diaminoacridine (*10.7*) (proflavine).   *D* 3,6-Diamino-4,5-dimethylacridine.
E 3,7-Diaminoacridine.
(Albert *et al.*, 1945.)

isomers which are well ionized have a powerful antibacterial action, whereas
the three that are poorly ionized have only a feeble action. The same correlation
is seen in the remaining parts of Table 10.6 (compare the *well* ionized methylaminoacridines with the *poorly* ionized methylaminoacridines, and the well
ionized chloroaminoacridines with their poorly ionized isomers). In all cases
ionization has brought about a large increase in activity: usually eight- to
sixteen-fold. Plainly, structure is relatively unimportant in this series except in
so far as it influences ionization. This is true for a wide variety of bacterial
species (see Table 10.7 which extends Table 10.6 in terms of species, both
Gram-positive and -negative).

For a further extension of Table 10.6 to include one hundred acridines and
species of bacteria (anaerobes and aerobes, Gram-positive and -negative), all
supporting the correlation of bacteriostasis with ionization, see Albert (1966).

The ionization which governs the anti-bacterial action of acridines must be
cationic in character. By inserting acidic groups into the nucleus of a strongly
basic acridine, it is easy to make a substance which is zwitterionic, and hence no
longer cationic. Such a substance is 9-aminoacridine-2-carboxylic acid (Table
10.8). It is seen that this has lost its antibacterial action, but regains it upon
esterification which restores the cationic condition. By inserting acidic groups
into the nucleus of a weakly basic acridine, an anionic substance is formed. Such
a substance is acridine-9-carboxylic acid (Table 10.8). It is seen that this
substance, like its ester and like acridine itself (none of them cationic), has no
appreciable antibacterial action.

Table 10.8 Importance of cationic ionization for bacteriostasis in the acridine series

Substance	Min. bacteriostatic concentration for Strept. pyog. after 48 hours' incubation at 37°C (Medium 10% serum broth; pH = 7.3)	Per cent ionized (pH 7.3; 37°C)			
		Cation	Anion	Zwitter-ion	Neutral molecule
9-Aminoacridine (*10.8*)	1 in 160 000	99	0	0	0
9-Aminoacridine-2-carboxylic acid	<5000	0	0.2	99.8	0
Methyl ester of above	160 000	89	0	0	11
Acridine	5000	0.3	0	0	99.7
Acridine-9-carboxylic acid	<5000	0	99.3	0.7	0
Methyl ester of above	<5000	0	0	0	100

(Albert *et al.*, 1945.)

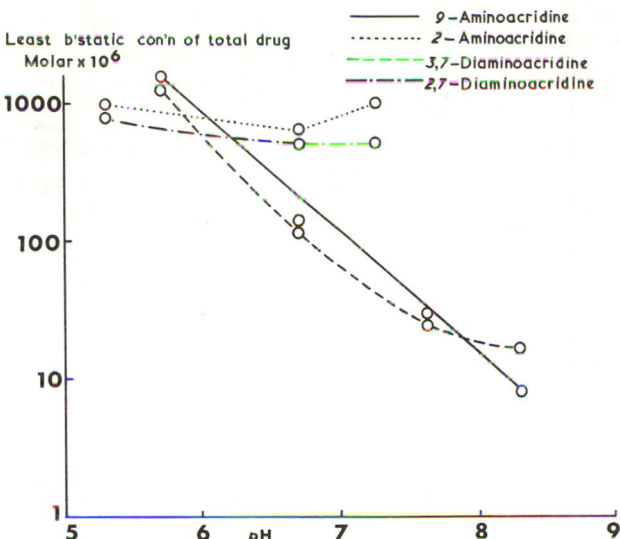

Fig. 10.2 Competition between hydrogen ions and acridines (cations + molecules). Organism: *E. coli*.

Much more has been learnt about the mode of action of acridines by using a test-organism that withstands both acid and alkaline media (*E. coli*). In Fig. 10.2, the logarithms of the minimal bacteriostatic concentrations of four acridines have been plotted against a range of pH values from 5.5 to 8.5. One of these acridines is a strong enough base to be ionized throughout this pH range, the others are not. It will be seen that the result is a rather disorderly set of curves. However, when the amount of actual cation (instead of the total amount of substance) is plotted on the ordinate, an orderly set is obtained (Fig. 10.3). These curves make a family of parallel straight lines, and they lead to some interesting conclusions.

Firstly, it is evident that the most important factor governing bacteriostasis at any pH is the amount of the acridine *present as cation* (this depends on both the pH of the medium and the pK_a of the acridine; see Table 10.2) and not the total amount of the acridine (cation+neutral species). Secondly, the slope of the curve shows that there is direct competition between acridine cations and hydrogen ions. The simplest interpretation is that acridine cations compete with hydrogen ions for a vitally important anionic group on the bacterium. The location of this site is discussed below. That cations of the same size and flatness as those under discussion can liberate hydrogen ions from bacteria is evident from Table 10.9.

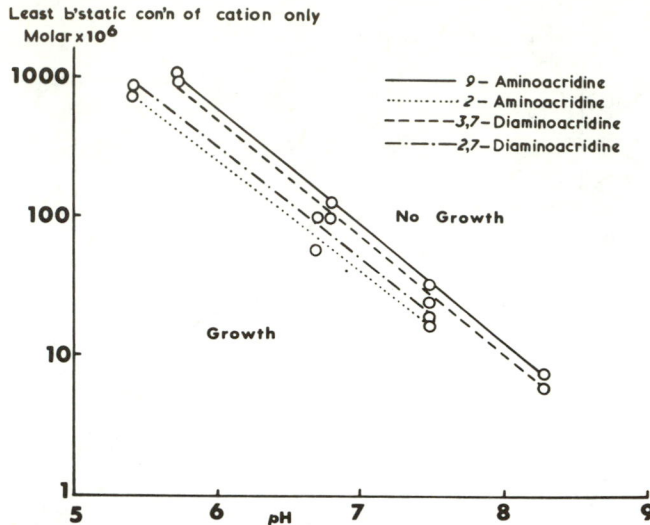

Fig. 10.3 Competition between hydrogen ions and acridine cations. Organism: *E. coli.*

Fig. 10.3 also suggested that a smaller amount of an aminoacridine would be effective in wounds if these were prevented from becoming acid. Some very interesting work along these lines was carried out during the Second World War by the Australian Army, who had adopted 9-aminoacridine hydrochloride as the best of the aminoacridines. They recommended that sodium bicarbonate lavage should precede treatment with this acridine (*10.8*). The excellent results of this therapy established it as a valuable measure for the prevention and treatment of sepsis in war wounds, even deep, ramifying, purulent, and neglected wounds. As a result of the laboratory tests, summarized in Tables 10.6 and

Table 10.9 Liberation of hydrogen ions from *Bacillus bellus,* caused by adsorption of crystal violet cations*

(1) *pH of bacterial suspension*	(2) *pH of crystal violet solution*	(3) *pH of mixture of 1 and 2*	*Fall in pH caused by adsorption*
8.5	8.5	6.6	1.9
6.5	6.3	4.8	1.7
5.5	5.5	3.8	1.7
5.0	5.1	3.5	1.6
4.1	4.2	3.0	1.1

(McCalla, 1941.)

*pH was measured with a glass electrode. Crystal violet is the *NN'N''*-hexamethyl-derivative of parafuchsin (*10.5*).

10.7, and the clinical success in treatment of severe war wounds (Poate, 1944; Turnbull, 1944), this acridine was made official in the British Pharmacopoeia, under the name Aminacrine.

The four steps that led to the discovery of the exact *site of action of aminoacridines* were, (i) the discovery that only *flat* molecules were effective (Albert, Rubbo and Burvill, 1949), (ii) finding that aminoacridines stain only the DNA and RNA of vertebrate cells (Strugger, 1940), (iii) discovery that aminoacridines are intercalated into DNA (Lerman, 1961), and (iv) finding that aminoacridines inhibited bacterial DNA polymerase by binding to the DNA template (Hurwitz *et al.*, 1962). Evidence relevant to (ii) and (iv) will be considered now, whilst that relevant to (i) and (iii) is deferred to Section 10.3.2 because evidence from larger and smaller nuclei makes an important contribution.

(*a*) *Vital staining*. It is firmly established that living mammalian cells, including those of the brain (Manifold, 1941) are not easily harmed by aminoacridines (Russell and Falconer, 1941, 1943; Selbie and McIntosh, 1943). In 1940, two independent laboratories showed that aminoacridines accumulate in the nucleic acids of living vertebrate cells without harming them in any way. The RNA in the nucleolus and cytoplasm was seen to fluoresce fiery red and the DNA in the nucleus green, when viewed under the fluorescence microscope (Strugger, 1940; Bukatsch and Haitinger, 1940). Even the delicate germ cells were unharmed by the application of these stains, and subsequent reproduction took place normally. In this way, the technique of vital staining, which Ehrlich had discovered, was refined further and a clear demonstration given that *nucleic acids selectively accumulate aminoacridines* in the living cell.

The following mammalian (rat, mouse, rabbit) cells have been observed to take up aminoacridines and concentrate them in the nucleus: lung, heart, liver, bone-marrow, spleen, tongue, kidney, and small intestines (De Bruyn, Robertson and Farr, 1950). That aminoacridines are not accumulated by the cell's protein is not surprising because their anti-bacterial action, *in vitro*, is not lowered by serum protein (Browning and Gilmour, 1913). Nucleic acids, on the other hand, strongly inhibit bacteriostasis by proflavine and acriflavine (McIlwain, 1941). The maximal amount of dye which dissolved DNA can bind follows an order for the monoaminoacridines that is the same as the order of their antibacterial action, namely; 9- > 3- > 1- > 2- > 4- (Jackson and Mason, 1971).

(*b*) *Inhibition of nucleic acid polymerases*. Two pieces of evidence point to the outside of the bacterial plasma membrane as the site of action of the aminoacridines. Firstly, there was no loss of antibacterial action when ionization (as cation) was increased from 70% to 100%, thus removing the more readily penetrating neutral species. Secondly, there was a great loss of activity when the aminoacridines were made more lipophilic (Albert *et al.*, 1945).

In bacteria, a single, ring-shaped chromosome contains most of the DNA of the cell, and it is externalized at one point of the cytoplasmic membrane (Ryter and Jacob, 1964). RNA is present in the cytoplasmic membrane (Yudkin and Davis, 1965) and in the ribosomes.

It has been shown, *in vitro*, that proflavine (30 μM) inhibits bacterial DNA polymerase by 85% and RNA polymerase by 30% (Hurwitz *et al.*, 1962). Some typical data for the RNA system are shown in Fig. 10.4. The inset graph (of the reciprocal of the velocity plotted against the reciprocal of the DNA-primer concentration) points to the site of action of proflavine as being mainly on this primer, thus inhibiting DNA and RNA synthesis in the living bacterium. The intercalating mechanism by which aminoacridines combine with DNA helices and impede strand separation is outlined in the next Section (10.3.2).

More detailed studies have since been made of the mode of action of amino-acridines on DNA-primed RNA, e.g. in *E. coli* (Nicholson and Peacocke, 1966), *Azotobacter* (Canellakis *et al.*, 1976), and a bacteriophage (Sarris, Niles and Canellakis, 1977). Inhibition of the corresponding enzyme in eukaryotes is uncommon, although 9-aminoacridine (500 μM) completely inhibited incorporation of UTP in an attempted transcription of calf thymus DNA by rat liver RNA polymerases (Zoncheddu *et al.*, 1980).

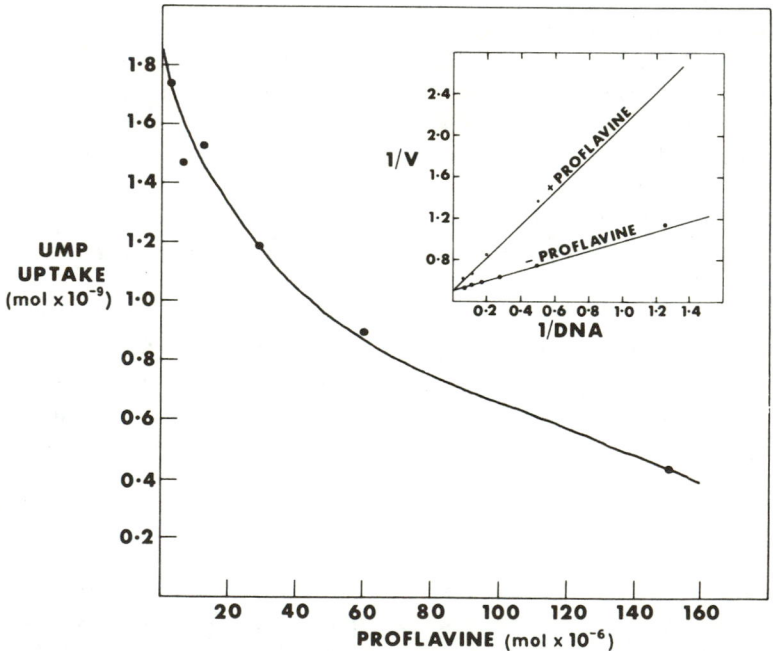

Fig. 10.4 Inhibition of RNA polymerase (*E. coli*) by proflavine as measured by decreased UMP uptake when DNA, ATP, GDP, and CTP are present in excess. *Inset:* Lineweaver–Burk plot of velocity of RNA synthesis in proportion to concentration of DNA template supplied (two straight lines: one with and one without proflavine) (Hurwitz *et al.*, 1962).

At least of historic interest, suggested modes of action, other than inter-calation into DNA, for the antibacterial aminoacridines will be briefly discussed. It has been considered whether acridines exert some of their anti-bacterial action by inhibiting other enzymes. However, the known inhibitions do not seem to be very serious and usually require high concentrations, whereas quite low concentrations of aminoacridines upset the synthesis and functioning of nucleic acids (see Chapter 22 in Albert, 1966).

Attempts have been made to link the antibacterial action of aminoacridines to their oxidation−reduction potentials (Breyer, Buchanan and Duewell, 1944). Thus E_0' of the first step in reduction at pH 7.3 lies between −0.47 and −0.92 V for the highly antibacterial substances 3-amino-, 9-amino-, and 3,6-diamino-acridine whereas the corresponding E_0' for the poorly antibacterial acridine and 1-, 2-, and 4-aminoacridine lies between −0.31 and −0.39 V.* Thus the small difference between a potential of −0.47 and −0.39 V would have to determine whether an acridine is antibacterial or not. That this is not the case is clearly shown in Fig. 10.3. Those acridines (e.g. 2-aminoacridine) which are feebly antibacterial at pH 7.3 (where they are poorly ionized) become as active as those that are highly antibacterial at pH 7.3 when the pH is lowered to a value (say 5.4) where both kinds of acridine are well ionized. Because this fall in pH does not lessen the difference in reduction potential between 2- and 3-amino-acridine, it must be concluded that reduction potentials do not affect anti-bacterial activity in the acridine series.

A hypothesis that antibacterial action of aminoacridines is due to free radicals produced during the first stage (1H) of reduction is also incorrect because *all* acridines produce these free radicals in the first stage (Kaye, 1950). Moreover, unreducible substances such as anthra-2-ylguanidine (Table 10.10) can produce a typical acridine-like antibacterial action. It is noteworthy that anthracenes show no reductive step even at −2.0 V (Zanker and Schnith, 1959). 1959).

(*c*) *Action on fungi.* Although, in common with most antibacterials, amino acridines have very little effect on fungi, a few of them have a strong trans-forming action on yeasts, in which they combine with the DNA occurring in the mitochondria, thus causing an inherited change (namely loss of all oxidative phosphorylation) without affecting the nucleus. The most active of these acridines is euflavine (3,6-diamino-10-methylacridinium chloride) (*6.5*) (Ephrussi and Hottinguer, 1950; Marcovich, 1953). However, the two amino-acridines found least toxic to human tissues and most useful in surgery (9-aminoacridine and 3,6-diaminoacridine) do not act in this way on yeast.

(*d*) *Use of acridines in today's medicine.* The antibacterial acridines, aminacrine and proflavine, find most employment as first-aid applications for wounds and

*These potentials refer to the normal hydrogen electrode. E_0 is defined in Section 11.4, and the above E_0' values are adjusted to pH 7.3.

burns, in ambulances and hospitals, and in medicated dressings for home use. Their popularity seems greatest in Britain, continental Europe, and Australia. Ethacridine (3,9-diamino-7-ethoxyacridine, 'Rivanol') is given orally for infections of the small and large intestines, mainly in Western Europe. Mepacrine ('Atebrin') is still much used in the treatment of malaria and lambliasis (giardiasis) in Asia and Africa (Section 10.3.5). Two acridine anti-cancer drugs, amsacrine and ledakrin, are making their way in the clinic (Section 10.3.4).

10.3.2 *Cationic antibacterials that have the aminoacridine type of action*

The relationship between cationic ionization and antibacterial activity, as demonstrated above in the acridine series, is valid also for at least seven other series, namely the three series of benzoacridines (*10.9*) to (*10.11*) of which 16 examples were studied, the phenanthridines (*10.12*), six examples, and the three series of benzoquinolines (*10.13*) to (*10.15*) of which 21 examples were studied (Albert, Rubbo and Burvill, 1949). Some of these examples are given in the first half of Table 10.10 (p. 407).

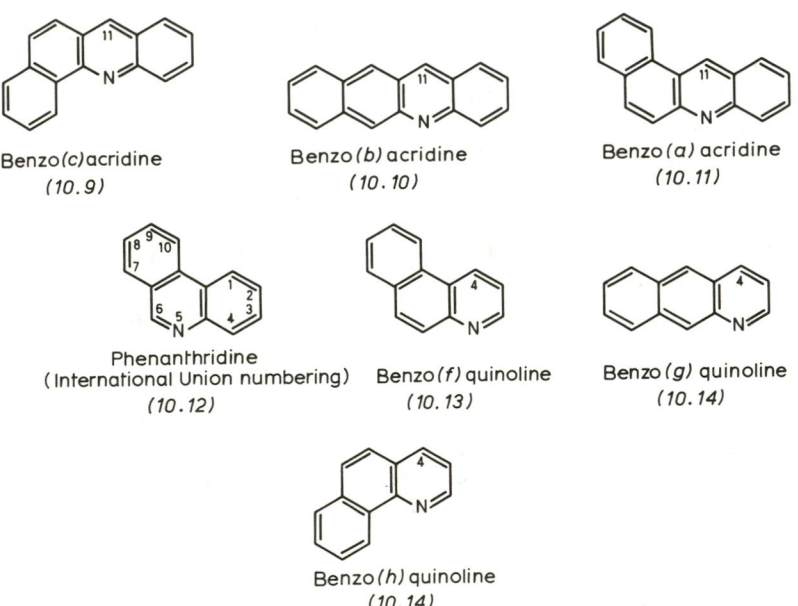

Benzo(c)acridine
(10.9)

Benzo(b)acridine
(10.10)

Benzo(a)acridine
(10.11)

Phenanthridine
(International Union numbering)
(10.12)

Benzo(f)quinoline
(10.13)

Benzo(g)quinoline
(10.14)

Benzo(h)quinoline
(10.14)

(*a*) *The requirement for flatness.* My colleagues and I found that removal of one or two rings from 9-aminoacridine (*10.16*) produced molecules which had no antibacterial action, even though they remained completely ionized. This may be seen in Table 10.10 by comparing 9-aminoacridine (*10.16*) with the structurally analogous 4-aminoquinoline (*10.17*) and 4-aminopyridine (*10.18*). The lack of antibacterial properties in the last two substances has been correlated with an insufficient area of flatness in these molecules (Albert *et al.*, 1949). If an envelope (as shown in Fig. 10.5) is drawn around the various nuclei, an area of only 28 Å² is obtained for quinoline and 17 Å² for pyridine, whereas acridine has an area of 38 Å².

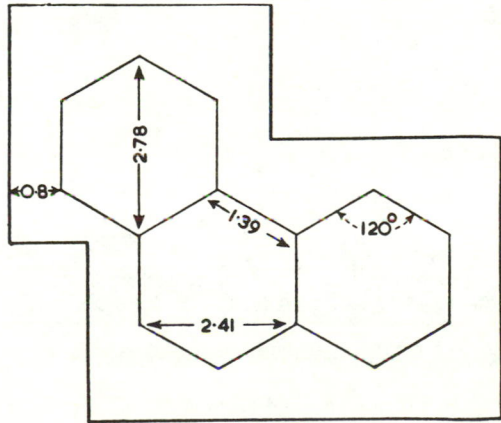

9-Aminoacridine
(1 : 160 000)
(10.16)

4-Aminoquinoline
(<1 : 5000)
(10.17)

4-Aminopyridine
(<1: 5000)
(10.18)

9-Aminotetrahydroacridine
(1: 5000)
(10.19)

4-Amino-2-styrylquinoline
(1: 80 000)
(10.20)

1: 160 000 (etc) = minimal bacteriostatic concentration, as in Table 10.10)

All the heterocyclic nuclei discussed so far are perfectly flat, because they are conjugated throughout (i.e. every second bond is a double bond). However, when 9-aminoacridine *(10.16)* is hydrogenated in one ring, to give 9-amino-tetrahydroacridine *(10.19)*, only 28 Å² of flat surface remain, because hydrogenated rings are always three dimensionally voluminous. Coincident with this change in flat area, the antibacterial activity almost completely vanishes.

This insight into the planarity required for activity suggested that it should be possible to create highly antibacterial quinolines and pyridines by the insertion of a group to increase the total area of flatness of the molecule. It is known, from

Fig. 10.5 Minimal rectangular envelope, drawn at a distance of 0.8 Å from the carbon and nitrogen atoms.

X-ray diffraction studies, that stilbene is a perfectly flat molecule*, hence it was thought that the addition of a styryl-group to 4-aminoquinoline [giving (*10.20*)] would produce a highly antibacterial compound. This indeed proved to be the case (see Table 10.10), and the addition of two styryl-groups to 4-amino-pyridine was similarly successful (Albert, *et al.*, 1949).

It remains to be pointed out that all the substances (*10.16*) to (*10.20*) are completely ionized at pH 7. Analogues of (*10.20*) that were not well ionized were not antibacterial. In large nuclei with a sufficient area of flatness (e.g. the acridine nucleus), cationic ionization is the only important limiting factor; but in the smaller nuclei (such as quinoline and pyridine), the minimal flat area of the molecule becomes another limiting factor.

Representations of the cations of 9-aminoacridine (*10.16*) and its tetrahydro-derivative (*10.19*) are shown in Fig. 10.6. These have been built from space-filling models which produce the correct interatomic and van der Waals distances. The non-coplanarity of the tetrahydro-derivative is clearly seen particularly in the side-view.

Why is the size of the flat area of these cations so important? Obviously, increasing the number of atoms in a molecule can increase its chances of

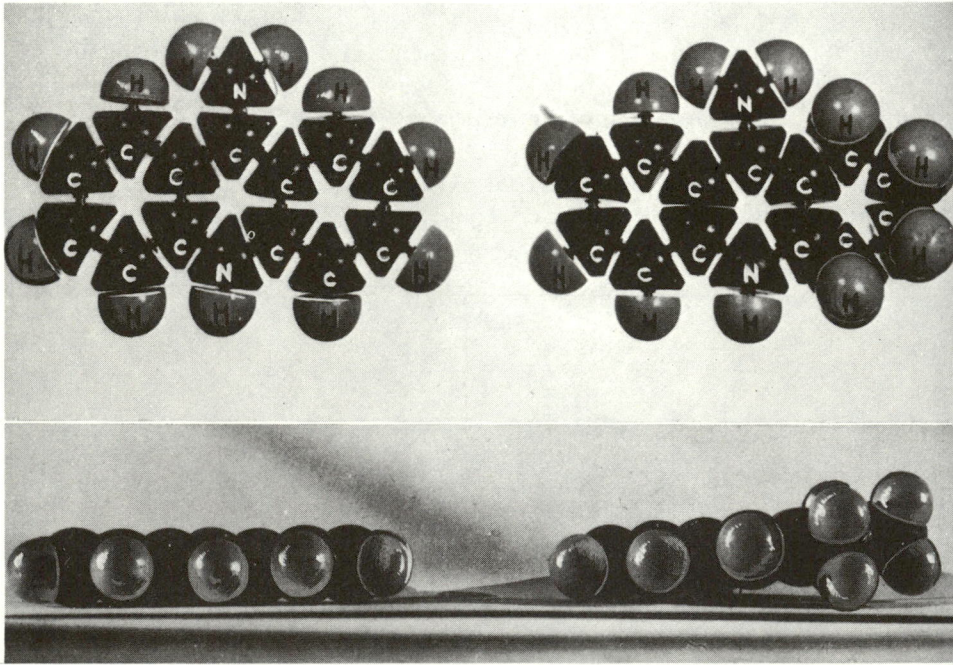

Fig. 10.6 Space models of the cations of 9-aminoacridine (the two examples on the left) and its 1,2,3,4-tetrahydro-derivative (examples on right) (see p. 403).

*Diphenyl, on the contrary, is not flat (Karle and Brockway, 1944).

adsorption. This is because the small van der Waals forces, each of which unites an atom of the agent to an atom of the receptor, become collectively a really strong force when a larger number of atoms are concerned on each side. This force is opposed by the molecule's kinetic energy of translation, which does not increase as the size of the molecule increases. The ion-pair, formed by the cation (the drug) and an anion on the receptor, would be very short-lived without supplementary bonds of this kind (Section 8.0). The special requirement of the present series, namely that the atoms in the drug should all lie *in one plane*, reveals that the atoms in the biological area with which they have to make an effective contact also lie in one plane.

It may be asked whether one may expect to obtain powerful antibacterials of the acridine type (i.e. acting at great dilution against both Gram-positive and -negative species, an action not decreased by the presence of protein) from *all* molecules having sufficient cationic ionization and area of flatness. A relevant substance is methylene blue, flat and highly ionized, but with a reduction potential so large that the cations are reduced by the metabolic products of most bacteria, such as *E. coli*; it is toxic to a related strain *Aerobacter aerogenes,* which cannot reduce it (the product of reduction is not ionized at pH 7). However, very few flat heteroaromatic nuclei yield highly ionizing amino-derivatives, because heteroatoma, in excess of one, tend drastically to reduce basic strength (Albert, Goldacre and Phillips, 1948; Albert, 1968).

We found that this difficulty can be overcome by building a short, highly basic side-chain on to a flat aromatic 38 Å^2 nucleus, so long as side-chain and nucleus are conjugated with one another. A guanidine or diguanide side-chain is suitable. In such a case, the nucleus need not be heteroaromatic: an aromatic hydrocarbon nucleus of the appropriate dimensions (e.g. anthracene) will do. Using this hypothesis, we soon found substances having a true 'acridine-type' antibacterial activity without any heterocyclic nucleus; the two anthracene derivatives in Table 10.10 are examples (Albert, Rubbo and Burvill, 1949).

It is surprising how closely the biochemical action of acridine analogues follows that of the acridines. Thus ethidium (*10.23*) inhibits the same two DNA-dependent polymerases as proflavine does, namely the one which synthesizes DNA and (although less strongly) the one that synthesizes RNA, by combining in each case with the DNA-primer (Elliott, 1963; Waring, 1965).

(*b*) *How acridines, and their analogues, are bound to nucleic acids.* It has been shown, *in vitro,* that proflavine (*10.7*) (3,6-diaminoacridine) is bound to DNA by two mechanisms, (a) a first-order reaction that reaches equilibrium at one proflavine per four or five nucleotides, and (b) a weaker, higher-order process that leads to the fixation of one proflavine molecule per nucleotide (Peacocke and Skerrett, 1956). The latter process apparently consists of an adsorption of further acridine molecules, very loosely, on to the outside of the DNA helix.

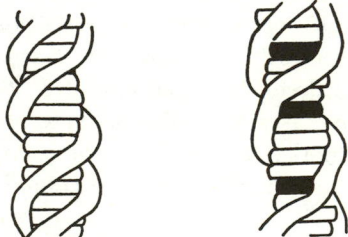

2-Anthrylguanidine
(10.21)

2-Anthryldiguanide
(10.22)

'Ethidium'
(10.23)

In 1961, Lerman suggested that the stronger attachment was caused by the *intercalation** of a 3,6-diaminoacridine molecule between two layers of base-pairs in such a way that the primary amino-groups were held in ionic linkage by two phosphoric acid residues of the Watson-Crick spiral, and the flat skeleton of the acridine ring rested on the purine and pyrimidine molecules to which it was held by van der Waals forces (Lerman, 1964a). An edgewise view of this arrangement is shown in Fig. 10.7. The need for a large flat area in the antibacterial molecules, as discussed above, was now explained, and so was the requirement for ionization.

It is evident from X-ray diffraction data that the base layers of DNA are, normally, making van der Waals contact with one another above and below, and that the distance between the centres of the atoms, in adjacent pairs on a strand, is 3.36 Å. This means that another 3.36 Å of space (6.72 Å in all) must be provided to admit the aminoacridine molecules, which have exactly the same thickness as those of the purine and pyrimidine bases. This space could be

Fig. 10.7 Sketches representing the secondary structure of normal DNA (left) and DNA containing intercalated proflavine molecules (right). The helix is drawn as viewed from a remote point, so that the base-pairs and the intercalated proflavine appear only in edgewise projection, and the phosphate deoxyribose backbone appears as a smooth coil. (Redrawn from Lerman, 1964b.)

*A word whose meaning, until then, had been confined to the insertion of February 29th into the leap-year calendar.

Table 10.10 Dependence of bacteriostasis on (a) ionization, and (b) flat area of molecule, in a series of cationic agents (For other examples and other bacteria, *see* Albert, Rubbo and Burvill, 1949).

Substance	*Min. bacteriostatic concentration for Strept. pyog. after 48 hours' incubation at 37°C. (Medium: 10% Serum broth; pH = 7.3)*	*Per cent ionized (pH 7.3; 37°C)*	*Flat area estimated as in Fig. 9.5 (sq. Å)*
Acridine	1 in 5000	1	38.5
9-Amino (*10.16*)	160 000	100	38.5
Benzo [*c*] acridine	< 5000	< 1	48.7
11-Amino	160 000	97	48.7
Benzo [*b*] acridine (*10.10*)	10 000	< 1	48.9
11-Amino	320 000	100	48.9
Benzo [*a*] acridine	< 5000	< 1	48.7
11-Amino	160 000	98	48.7
Phenanthridine (*10.12*)	10 000	< 1	38.3
6-Amino	40 000	50	38.3
3,6,8-Triamino	80 000	98	38.3
Benzo [*f*] quinoline (*10.13*)	5000	< 1	38.3
4-Amino	20 000	98	38.3
Benzo [*g*] quinoline	10 000	< 1	
4-Amino	40 000	100	38.3
Benzo [*h*] quinoline	< 5000	< 1	
4-Amino	80 000	96	
Quinoline	< 5000	< 1	27.9
4-Amino (*10.17*)	< 5000	98	27.9
4-Amino-2-styryl (*10.20*)	80 000	99	49.9
Pyridine	< 5000	< 1	17.4
4-Amino (*10.18*)	< 5000	98	17.4
4-Amino-2,6-distyryl	80 000	95	61.6
9-Amino-5,6,7,8-tetrahydro acridine (*10.19*)	5000	100	27.9
Phenylguanidine	< 5000	100	17
2-Anthrylguanidine (*10.21*)	40 000	100	38.5
Phenyl-N_1-diguanide	< 5000	100	17
2-Anthryl-N_1-diguanide (*10.22*)	40 000	100	38.5

(Albert *et al.*, 1949.)

provided, if the helix untwisted slightly,* and that turned out to be exactly what happened.

The steps by which Lerman reached this picture of intercalation will now be given. The interaction between proflavine and DNA gave a three-fold increase in *viscosity*. This change was attributed to the helix being not only extended, but also stiffened and straightened by the intercalated molecules. The DNA–proflavine complex was found to have a lower *sedimentation coefficient* than free DNA. This was attributed to loss of mass per unit length (a proflavine molecule has less than half the mass of the same volume of DNA). The above results were obtained in dilute aqueous solution. It was also found that fibres drawn from the complex gave much simpler *X-ray diffraction patterns* than those given by pure DNA. The meridianal spot, corresponding to the 3.4 Å separation between stacked layers, was retained but the new positions of the first equatorial reflections suggested that each DNA molecule was now more closely packed than pure DNA and hence had a smaller diameter (Lerman, 1961). In 1963, Lerman showed that the ratio of the fluorescence intensities of *flowing and stationary solutions* were compatible with perpendicularity of the acridine molecules to the helical axis. He next reported that the *rates of diazotization* of the primary amino-groups of proflavine were so much diminished in the presence of DNA as to suggest that it protected them from attack by the nitrous acid (Lerman, 1964). It was noted that an increased temperature was required to denature DNA after it had formed a complex with 9-aminoacridine (Lerman, 1964b) (cf. Chambron, below).

Meanwhile powerful support for the idea of intercalation began to arrive from other laboratories. *Radioautography* of DNA molecules (containing [³H]thymine), obtained from T2 coliphage, was performed before and after immersion in a dilute solution of proflavine. It was found that the aminoacridine had expanded the DNA strands from about 45 to 70 μm (Cairns, 1962). Equally convincing evidence came from the fact that the *melting-temperature* (T_m) of DNA was increased 20°C by proflavine, and that most of the bound proflavine was suddenly released when the complex 'melted' (i.e. the strands of the double-helix parted) (Chambron, Daune and Sadron, 1966). This phenomenon was confirmed by Kleinwächter, Bakarova and Bohacek (1969).

An X-ray *diffraction* study of the proflavine–DNA complex gave direct evidence of intercalation, by showing one molecule of the aminoacridine stacked parallel to the base-pairs in a 1:3 ratio (Neville and Davies, 1966). *Linear and circular dichroism* studies of flowing solutions of all five monoaminoacridines (complexed to DNA) confirm that the acridine cations lie in planes parallel to those of the base-pairs (Jackson and Mason, 1971). Finally, the *free-energy* of binding of aminoacridines to DNA was determined and found to be of the order

---

*This unwinding angle is somewhat variable, depending on the substance. For ethidium, it is 26°, as found from a high-precision buoyant-density gradient titration (Wang, 1974); for proflavine and the alkaloid ellipticine, it is rather less (Kohn *et al.*, 1975).

expected for intercalation, but far too high for any form of binding on the exterior of the DNA strands (Jordan, 1968).

In electrostatic terms, the space available for intercalation in DNA is a negatively charged field of large extent, one that receives contributions not only from the phosphate anions but also from all of the purine–pyrimidine bases except adenine, as well as fractional negative charges from the oxygen atoms of deoxyribose. It is into this negative field that the drug cations are drawn until they are in a position of near-electroneutrality. Then, van der Waals bonding with the bases ensures the final position as well as the relative permanence. In Fig. 10.8 can be seen details of the intercalation of proflavine, as learnt from X-ray diffraction analysis. The flat area of an average base-pair in a DNA or RNA spiral has been calculated to be about 50 $Å^2$, which is larger than the 38.5 $Å^2$ calculated for acridine by the method of Fig. 10.5. However, in a one-strand model, a little of each acridine molecule would be protruding. Lipophilic analogues, such as acridine orange (3,6-bisdimethylaminoacridine), not only intercalate but also cling tenaciously to the outside of the DNA fibre.

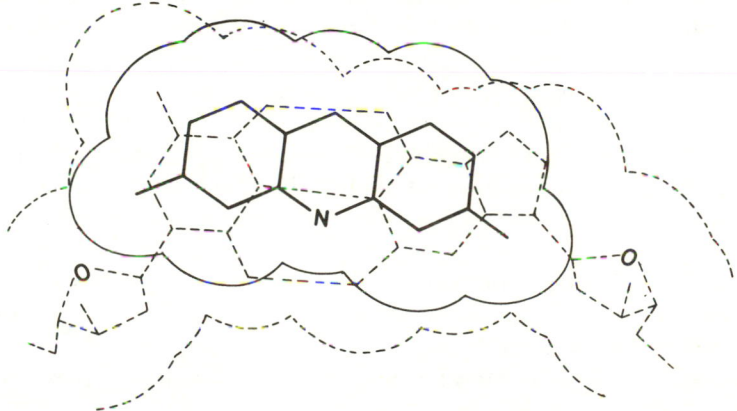

Fig. 10.8 Intercalation of 3,6-diaminoacridine into the base-pair layers of DNA. The left-hand ring of the acridine is almost superposed over cytosine, and the right-hand ring of the acridine is almost superposed over the pyrimidine ring of guanine. The van der Waals boundary of the acridine molecule is shown as an unbroken line. (Redrawn from Lerman, 1964b.)

(c) *Other substances which intercalate.* As often happens with new ideas, the hypothesis of intercalation was at first held to be unlikely. However (and this, too, is usual), no sooner had it become established than the action of all kinds of drugs began to be ascribed to it. To test these claims, a standard test for intercalation was introduced: observing if the suspected drug causes local uncoiling of the double-helix of a supercoil of circular DNA, such as that from coliphage (Waring, 1970). During the first additions of an aminoacridine, the number of right-handed supercoils steadily decreases; at a critical ratio of added molecules, the supercoils are all removed and the DNA molecule behaves as an untwisted open circle, of greater perimeter than before thanks to the intercalation. As further additions of the agent are made, the accumulated stretching of the DNA ring re-introduces supercoiling, but this time it is left-handed. Because supercoils are more compact than the open circles from which they are derived, they sediment more rapidly: hence the whole process is easily observed by changes in the sedimentation coefficient, which passes through a minimum (Waring, 1970).

Using this method, intercalation was demonstrated for some aminophenanthridines, including (*10.23*), for the anti-malarial drug chloroquine (*10.31*), for three carcinostatic antibiotics: nogalomycin, daunorubicin, and actinomycin D (*4.38*) (see Section 4.0). However, chlorpromazine and lysergic acid diethylamide (LSD), for which an action by intercalation had been suggested, gave no evidence of it. The following substances were found to interact strongly with DNA without any tendency to intercalate: spermine, streptomycin, diminazene, and mithramycin (Waring, 1970).

Other important examples of intercalation have been established. The red triphenylmethane dye parafuchsin (*10.5*) intercalates with DNA, but less tightly than proflavine does (Armstrong and Panzer, 1972). The antibiotic echinomycin has a polypeptide ring to which are attached two, widely spaced, quinoxaline rings which become intercalated simultaneously into two different sites of DNA. As a result, it causes the helix to unwind and to extend twice as much as happens with aminoacridines (Waring and Wakelin, 1974). Synthetic analogues have been made by inserting an acridine nucleus into each end of the molecules of putrescine (*11.6*) and spermine (*11.4*). The products inhibited DNA-directed RNA polymerase and elicited a higher T_m in DNA than singly intercalating aminoacridines did (Canellakis *et al.*, 1976). For anti-cancer properties of doubly intercalating substances, see Section 10.3.4.

For a review of the biological behaviour of intercalating drugs, see Schwartz (1979).

In closing this section, it is necessary to ask why aminoacridines, and cognate intercalating drugs, are so selective against bacteria and sparing of mammalian cells. It is established that intercalation, antibacterial properties, and percentage-ionization are three properties that run parallel in the acridine series (Jackson and Mason, 1971). The selectivity against bacteria seems to be due to two factors, the exposed situation of the solitary chromosome in bacteria (Section 5.3), and the preference shown by intercalating drugs for *circular* DNA (Waring, 1970). Apart from their selective action on bacteria, aminoacridines also act selectively on the following examples of circular DNA: they eliminate bacterial plasmids (Bouanchaud *et al.*, 1968); and they produce a heritable change in yeast by suppressing replication of DNA in mitochondria without affecting that of the nucleus (see end of Section 10.3.1, and also Hollenberg, Borst and van Bruggen, 1970).

Further, aminoacridines and aminophenanthridines can either suppress the kinetoplast in trypanosomes, or (in lower concentrations) distort its circular DNA (Riou and Delain, 1969). It has been calculated that, so long as these agents are removing superhelical turns from circular DNA, a negative free-energy change is occurring which increases the affinity for the agent, whereas the same DNA, once it is nicked and becomes a strand, has a decreased affinity (Bauer and Vinograd, 1970). It looks as though this is another example of selectivity by accumulation, the type discussed in Chapter 3.

10.3.3 *Cationic and anionic antibacterials with other types of action*

Many other antibacterials act in the form of cations but do not have the aminoacridine type of activity, the typical features of which are (a) bacteriostasis of both Gram-positive and -negative bacteria of many species, and at great dilution and (b) maintenance of antibacterial action in the presence of serum protein.

The *triphenylmethane dyes*, e.g. parafuchsin (*10.5*), crystal violet, brilliant green, and malachite green, are cationic antibacterials of a special chemical type (p. 391). The equilibrium between the cation and its base (actually a pseudo-base) is only slowly established. Thus there are both equilibrium and ionization constants to be taken into account. These have been determined by Goldacre and Phillips (1949). It can be seen from Table 2.6 that antibacterial action is linked with ionization in this series. However, unlike the aminoacridines, they have little activity against Gram-negative bacteria, and their activity against Gram-positive species is seriously depressed by serum protein (Goldacre and Phillips, 1949). The triphenylmethane pseudo-bases are liposoluble and penetrate cells easily. They have non-rigid structures whereas their cations are rigid and completely flat. The cations readily form covalent bonds with all kinds of nucleophilic reagents.

Aliphatic amines (including the *quaternary amines*) are also antibacterial, but are in no way acridine-like. They cause lysis of the cell membrane (see Section 14.3) and hence achieve a rapid bactericidal action, whereas the acridines, which are only slowly bactericidal, are used mainly for their bacteriostatic properties. Aliphatic amines with side-chains of C_{12} (or longer) have this lytic action well developed. They are non-rigid molecules which become easily adsorbed on the acidic groups of serum proteins: this wasteful effect is decreased if they are quaternized with at least one bulky substituent, such as a benzyl-group, as in Domagk's benzalkonium chloride ('Zephiran'). Cetyltrimethylammonium bromide (*14.16*) is another well-known example of this class of quaternary amine. In this series, also, the activity increases as the pH is raised, although the drug remains completely ionized throughout (Blubaugh, Botts and Gerwe, 1940; Gershenfeld and Milanick, 1941).

Salts of aliphatic sulfonic acids and ethereal sulfates are mildly antibacterial if the chain has 12 or more carbon atoms (e.g. sodium tetradecylsulfate). These function as the anion (Gershenfeld and Milanick, 1941; Putnam and Neurath, 1944).

Chlorhexidine (*10.24*), 1,6-bis-(*N-p*-chlorophenyldiguanido)hexane, is used as the gluconate for the disinfection of skin and mucous membranes. It is effective in high dilutions against Gram-positive and -negative organisms, and remains bactericidal in the presence of blood. See further Section 14.3.

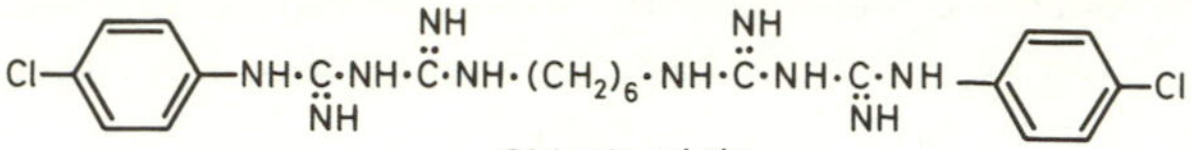

$$Cl-\langle\rangle-NH\cdot\overset{NH}{\underset{NH}{C}}\cdot NH\cdot\overset{\cdot\cdot}{C}\cdot NH\cdot(CH_2)_6\cdot NH\cdot\overset{\cdot\cdot}{C}\cdot NH\cdot\overset{NH}{\underset{NH}{C}}\cdot NH-\langle\rangle-Cl$$

Chlorhexidine
(10.24)

10.3.4 *Anti-cancer examples*

Aminacrine (9-aminoacridine) was found to act selectively, though quite weakly, against malignant cells. The effect was found intensified in 9-anilino-acridine without loss of cationic ionization at pH 7. From that starting point, the New Zealand team founded by Bruce Cain entered on a long process of fine tuning which led to the clinically useful drug amsacrine (*10.25*), 4-(9-acridinyl-amino)methanesulfon-*m*-anisidide. It is proving to be highly active against leukaemia and has a high therapeutic index in clinical trials (Legho *et al.*, 1979).

A wide variety of substituents can be used instead of the existing ones without losing antileukaemic action, so long as cationic ionization is conserved plus a small degree of lipophilicity as measured by partition coefficients (Denny, Atwell and Cain, 1979). The quantitative relationship between the strength of association with DNA and the effective dose (taken as 40% extension of life in mice with L1210 leukaemia) indicates that DNA is the site of action in this series (Baguley *et al.*, 1981). The association with DNA was measured by the displacement of ethidium (*10.23*) from calf thymus DNA, using fluorimetry.

The action of amsacrine on DNA soon leads to breaks in DNA strands which causes faulty chromosome segregation at mitosis, and death of the cell follows (Baguley *et al.*, 1981). The strong action on bone-marrow precludes the use of amsacrine in treating solid tumours. These workers have noted that 4,5-disubstitution in the acridine ring suppresses the bone-marrow effect without loss of cytotoxicity.

X-ray crystal analysis showed some double-bond character between C-9 and the anilino nitrogen atom, in agreement with earlier demonstration of electronic transmission between the acridine and the aniline ring systems. The acridine nucleus is seen to be intercalated between the layers of bases, and the anilino ring lies in the minor groove (Baguley *et al.*, 1981; Waring 1976).

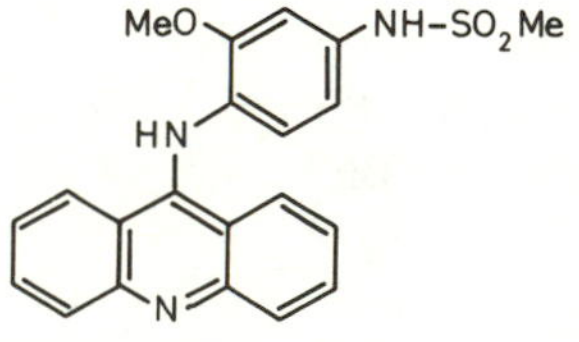

Amsacrine
(10.25)

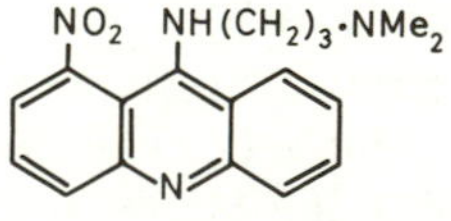

Nitracrin (Ledakrin)
(10.26)

Nitracrin (ledakrin) (*10.26*), discovered by Ledóchowski's team in Poland, has no action on bone-marrow and has shown clinically useful carcinostatic properties in colonic cancer, after surgery. A series of 30 patients showed no recurrence after 5 years (Bratkowska-Seniów *et al.*, 1976). This drug has a basic pK_a of 6.2 (nuclear) and 9.7 (side-chain) (cf. 7.9 and 10.5 for mepacrine). It is thought that the drug functions through reduction of the nitro-group to a hydroxylamino-group which cross-links DNA (Konopa, Koldej and Pawlak, 1976; Ledóchowski, 1976). Irritation at the site of injection is a minor disadvantage of this drug which seems to have a high therapeutic index.

For more on the use of acridines in treating cancer, see Denny *et al.* (1983).

Ellipticine, an alkaloid from a tropical evergreen tree, is 5,11-dimethyl-pyrido[4,3-*b*]carbazole (Woodward, Iacobucci and Hochstein, 1959). It has a planar ring structure and intercalates into DNA. Ellipticine is undergoing clinical trial in leukaemia. Echinomycin (quinomycine), chemically described in Section 10.3.2, is highly active against lymphocytic leukaemia and melano-carcinoma in the mouse without much action against any other form of cancer (Carter, Sakurai and Umezawa, 1981). Carzinophilin-A is an antibiotic whose structure is a complex bicyclic ring, formed by ether and ester linkages, with a basic nitrogen atom and two naphthalene rings. The latter intercalate into DNA and the rest of the structure occupies the minor groove. In addition, it has two aziridine rings that alkylate guanine groups in DNA after the other features of its molecule move this antibiotic into position. It has been used clinically against cancer in Japan, since the early 1960s, but its high toxicity to normal cells calls for caution (Lown and Hanstock, 1982).

Of the doubly intercalating aminoacridines (mentioned in Section 10.3.2), *NN'*-(9,9'-diacridinyl)-1,6-diaminohexane is under clinical investigation. Substances of this class intercalate so strongly that no free drug can be demonstrated (Kuhlmann and Mosher, 1981).

Doxorubicin (*4.27*) (adriamycin), and the related daunorubicin, are much used clinically in the treatment of solid tumours and leukaemias respectively. They are strong intercalators and act as cations thanks to the amino-group attached to the carbohydrate moiety. However, their mode of action is not fully understood and they have been placed in Section 4.0.4 with drugs that destroy DNA.

For a review of the behaviour of intercalating drugs in cancer, see Schwartz (1979).

10.3.5 *Anti-protozoal examples*

Many effective biocides, selective against protozoa, are the salts of strong organic bases, and one important trypanocide is the salt of a strong organic acid.

(*a*) *Flagellates*. Attention has already been drawn to the colossal consumption of food by trypanosomes, as necessitated by their constant movement (Section

4.9). This Achilles' heel is attacked by the arsenical trypanocides, especially melarsoprol (*13.3*), but their principal use is in a late stage of the disease. Another site for selective attack was indicated at the end of Section 10.3.2, namely the ability of amino-acridines and -phenanthridines, by combining with its circular DNA, to distort and eventually suppress the kinetoplast in these flagellates. The kinetoplast carries information to create mitochondria in the alternative form of the parasite which exists in the insect vector (a fly). Although drugs which injure the kinetoplast effectively prevent transmission to this secondary host, and hence can break an epidemic, they give no help to the infected patient. Rather, the curative effect of cationic drugs depends on their injury of nucleolus, ribosomes, and lysosomes, or their ability to suppress the parasites' excessive carbohydrate metabolism (Williamson, Macadam and Dixon, 1975).

What follows refers solely to African trypanosomiasis because *T. cruzi* is treated with nifurtimox (*6.20*), a non-ionized substance. The principal drugs used in preventing and treating African sleeping sickness are suramin (*6.8*), which is an anionic drug, and pentamidine (*10.27*) which is cationic*. Cattle, however, are treated with diminazene (*10.28*), ethidium (*10.23*), and quinapyramine (*10.29*), drugs suited also to the species that infect horses and camels. The biological nature of trypanosomiasis was described in Section 1.1 (p. 9). All known drugs have toxic side effects for the patient.

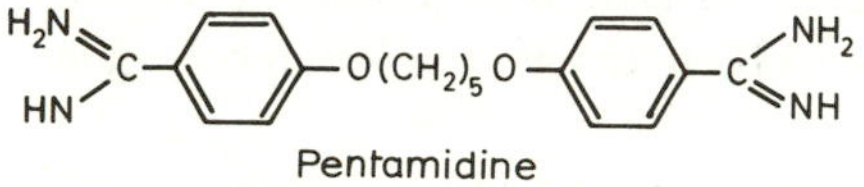

Pentamidine
(*10.27*)

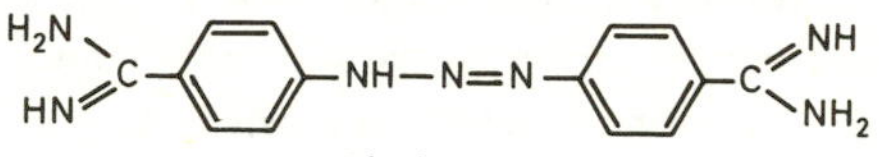

Diminazene
(*10.28*)

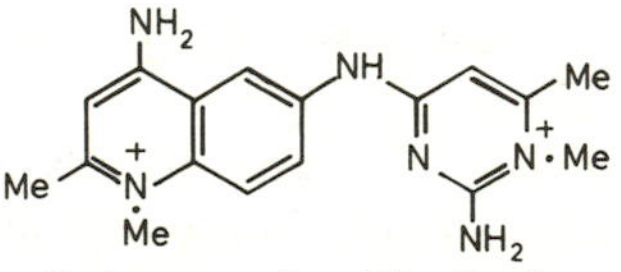

Quinapyramine (dication)
(*10.29*)

Because pathogenic trypanosomes are difficult to grow in culture, much work has been done on a related, easily grown, protozoon, *Crithidia oncopelti* (also known as *Strigomonas*). Recently, these difficulties have been mastered, and more relevant tests are being done with highly pathogenic trypanosomes (*Tr. rhodesiense*). This organism was examined *in vitro*, in the presence of drugs, for (a) loss of mobility as an index of injury to energy-yielding metabolism, and (b) infectivity for mice as an index of injury to cell-division. In a third test, an infection sustained in mice by this strain was treated with the same drugs. It was found that suramin (*6.8*) first injured the structure of ribosomes and then

---

*The aromatic amidines are strong bases (pK_a 11); quinapyramine and ethidium are quaternary (immonium) bases in equilibrium with their pseudobases (see p. 391).

suppressed cell division. Ethidium (*10.23*) and quinapyramine (*10.29*) ('Antrycide') injured the kinetoplast and suppressed cell-division. Pentamidine (*10.27*) primarily inhibited glycolysis, but also injured the kinetoplast (DNA) and nucleolus (RNA) (Williamson, Macadam and Dixon, 1975).

Pentamidine is rapidly absorbed by trypanosomes and at once exerts a trypanostatic effect on them, both *in vitro* and *in vivo* (Hawking, 1944). The parasites are finally killed by the body's defensive forces, some days later. [For the distribution of these and other amines in trypanosomes, see Williamson and Macadam (1965).] One indirect factor that helps a cure is the following. Those parasites that survive the direct effect of the drug remain infective, but they have a large number of drug-containing granules. When some of these parasites die, they liberate the drug, and also an antigen which causes the host to make an adequate amount of antibody. This indirect process is believed to explain the slow onset of the curative action of the diamidine drugs, and their prolonged prophylactic effect (Fulton and Grant, 1956). When the immune response of a rat was blocked by splenectomy, or by injecting copper, trypanosomiasis could no longer be cured by stilbamidine, suramin, or quinapyramine (Ormerod, 1961). Current medical practice is to protect residents in tropical Africa against trypanosomiasis by injecting pentamidine or suramin every 3 months.

Neither pentamidine nor diminazene intercalates (Waring, 1970), but each of them binds across the small groove of the DNA helix-rings in the kinetoplast, cross-linking by attachment of each amidine-group to a phosphate anion (Festy, Sturm and Daune, 1975). However, their selective damping down of carbohydrate metabolism by inhibiting α-glycerophosphate dehydrogenase, even at a dilution of 2 μM (Fairlamb and Bowman, 1975), plays a more important role in treatment, apparently by opposing the coenzymic effect of spermidine (*11.5*) (Bacchi, 1981).

Also for the action of diamidines on DNA, the aliphatic polyamines provide models. Spermine (*11.4*), so widely present in Nature (Cohen, 1971), has been shown by X-ray diffraction to combine with DNA by spanning the double-helix across the minor groove, and forming ionic linkages with the phosphate anions (Suwalsky *et al.*, 1969). The interval between phosphate groups is 10 Å across the minor groove, and 20 Å across the major. The exceptionally strong and durable nature of the ionic linkages given by amidines has already been indicated in (*8.4*).

Diamidines are also used to treat protozoal diseases other than trypanosomiasis in cattle, particularly the diseases grouped as 'piroplasmoses' caused by species of *Babesia* and *Theileria*. Apart from pentamidine and diminazene, good results are being obtained with amicarbalide, which is *NN'*-di-(3-amidinophenyl)urea, and the related drug, imidocarb. In humans, diamidines are successful in cases of leishmaniasis that have not responded to antimonials.

Reverting to trypanosomiasis, and the anionic drug, suramin (*6.8*), which is still much used in treating it, we note that the structure of this drug is remarkably specific in that the removal of the two methyl-groups abolishes activity completely. Nevertheless, quite unrelated polyanions, such as dextran sulfate have strong, if not quite medically useful, trypanocidal action. It has long been known that suramin abolishes the infectivity of trypanosomes without affecting their mobility (Roehl, 1926b). The electron microscope shows that this drug strikes hard and specifically at the parasite's ribosomes which lose their poly-somal character and form the 'cytoplasmic granules' described by the light microscopists. Fundamentally the attack seems to be on RNA polymerase; the nucleus, nucleolus, and kinetoplast remain unaffected and the synthesis of DNA continues (Macadam and Williamson, 1974).

Suramin is also used to kill *macro*filaria in the worm disease, onchocerciasis (p. 12) after the *micro*filaria have been subdued with diethylcarbamazine (*6.35b*).

(*b*) *Sporozoa*. Compared to flagellates, these protozoa lead a placid existence, but they have a more complicated life cycle. The genus *Plasmodium* causes the various types of malaria.

Acridines become strongly anti-malarial if substituted with a basic side-chain in the 9-position, but are inactive if quaternized. Mepacrine (*10.30*) (quin-acrine, 'Atebrin') was the first anti-malarial acridine, and the first synthetic anti-malarial to achieve widespread use. For further information, on con-nections between structure and action in the acridine anti-malarials, see Albert (1966).

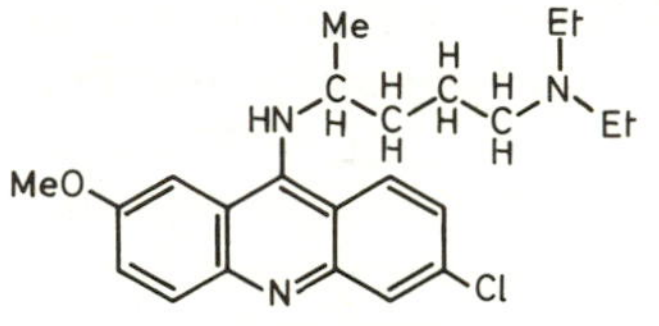

Mepacrine
(Quinacrine, 'Atebrin')
(*10.30*)

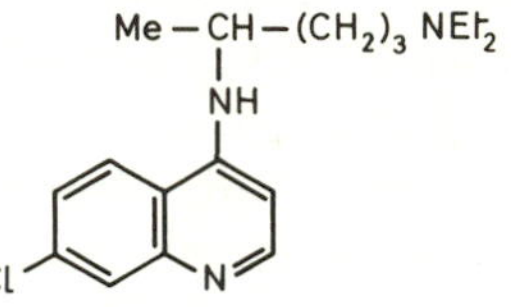

Chloroquine
(*10.31*)

Mepacrine was introduced in 1933, but the yellow colour that it imparted to the skin made it less liked by Caucasian people than the far less selective quinine. About 1940, the molecule was simplified in Germany to give the colourless analogue chloroquine (*10.31*) which, because of wartime secrecy, was independently rediscovered in the USA about 1942. Chloroquine (7-chloro-4-δ-diethylamino-α-methylbutylaminoquinoline) is now widely used in the treat-ment of malaria. A larger dose is required than for mepacrine, which is still manufactured. These anti-malarials are not trypanocidal, and the trypanocides are not anti-malarial.

Whereas the anti-malarials that interfere with folic acid metabolism (pyrimethamine, sulfadiazine, chloroguanide, see Section 9.3.3) injure the schizonts only at the time of their division, the nucleic acid combining drugs (especially mepacrine, chloroquine) and quinine injure the schizonts at all times (Josephson *et al.*, 1953) provided that they are in erythrocytes. Thanks to

their high selectivity, mepacrine and chloroquine are harmless to the mammalian and avian hosts in doses which are highly injurious to the intra-
erythrocytic schizonts (trophozoites). Unfortunately they are harmless to
exoerythrocytic schizonts, and also to the sporogenic (gamete, or sexual) stages.
The clue to this differentiation seems to lie in the fact that parasitized erythrocytes concentrate these drugs about one thousandfold. Thus a patient under
chloroquine treatment may have only 10^{-6} M of this drug in his bloodstream,
but as much as 10^{-3} M inside the erythrocytes (cf. Macomber, O'Brien and
Hahn, 1966).

There is general agreement that a major part of the anti-malarial action of
mepacrine and chloroquine is exerted by the drug's combining with DNA in the
parasite, the flat aromatic ring (in each example) undergoing intercalation, and
the basic group forming an ionic link with a phosphate group on the DNA
(O'Brien, Olenick and Hahn, 1966; Waring, 1970). Lerman (1963) had shown
that mepacrine readily became intercalated into DNA, just as the simpler
acridines described in Section 10.3.2. That chloroquine was similarly intercalated was shown by decreases in sedimentation rate, enhancement of
viscosity, and elevation of the melting profile. It reacted very little with single-
stranded DNA (Allison, O'Brien, and Hahn, 1965). Also chloroquine was
shown to inhibit bacterial DNA polymerase by combining with the DNA starter
(Cohen and Yielding, 1965). Naturally it would have been more interesting if
these tests had been done also with nucleic acid from *Plasmodia*. These three
anti-malarials inhibited incorporation of phosphate (^{32}P) into the DNA and
RNA of *Pl. gallinaceum* at 10^{-5} M in chicks and in parasitized whole blood;
Pl. berghei behaved similarly (Schellenberg and Coatney, 1961). It is noteworthy that the gametocidal 8-aminoquinolines do not inhibit phosphorus
incorporation.

The side chain of mepacrine and chloroquine, in which two nitrogen atoms
are separated by four carbon atoms, is reminiscent of putrescine (1,4-di-
aminobutane) which ornithine decarboxylase produces in the malarial parasite.

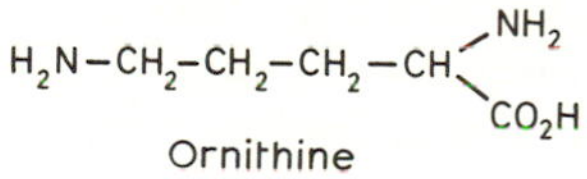

Ornithine
10.32

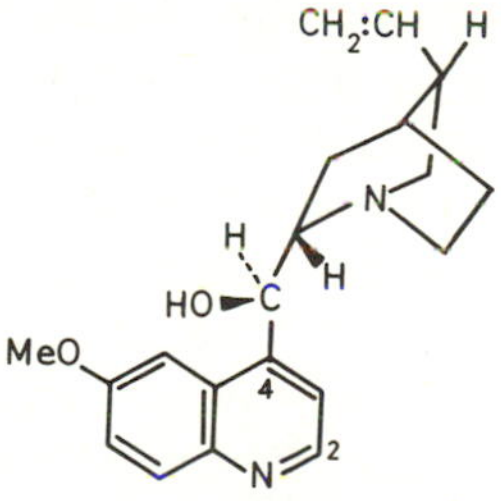

Quinine
(10.33)

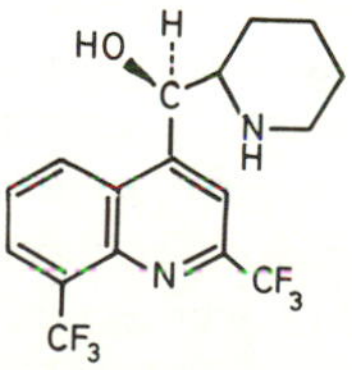

Mefloquine
(10.34)

Recently, it has been shown that chloroquine is a powerful inhibitor of the decarboxylation of ornithine (*10.32*) by this enzyme, a reaction that may play an important part in the therapeutic action (Königk and Putfarken, 1983). For the chemistry and biochemistry of polyamines, see Cohen (1971) and Taber and Taber (1983).

This side-chain, which is such an important feature of the molecules of mepacrine and chloroquine, is currently thought to bind two vertically adjacent phosphate anions on the *same* strand of DNA. Much simpler diamines, such as ethylenediamine, firmly bind to (and rigidify) single-stranded polynucleotides as well as helical DNA. The dication of ethylenediamine is just long enough to bind adjacent anions (7 Å apart), but it is too short to reach across the minor groove (10 Å) of the DNA double-helix and bind two strands together (Gabbay, 1968). The *N–N* distance in the side-chain of chloroquine, measured by X-ray crystallography, is 5.54 Å (Preston and Steward, 1970); that of mepacrine could differ little from this.

The ionization properties of the side-chain of mepacrine are also relevant to its activity. The pK_a of the nitrogen atom at the free end of the chain is 10.48, but only 7.92 at the fixed end (at 20°C, cf. 10.1 and 7.0 for ethylenediamine). Both ends of this side-chain are ionized at pH 7 (see Table 10.3), but the lower pK_a is greatly decreased by any shortening of the side-chain, with the result that both the ionization and the anti-malarial action fall steeply away (Albert, 1966). This drop is caused by the coulombic effect (repulsion between like charges varies inversely with the square of the distance).

Quinine (*10.33*) has been the mainstay in treating chloroquine-resistant cases of malaria. It intercalates into DNA very little and acts on the parasites in some quite different, but as yet unknown, way from chloroquine (Estensen, Krey and Hahn, 1969). Quinine is a wasteful drug because a high proportion of each dose is oxidized, in the 2-position, by the patient's liver and the product is inactive. This defect was overcome by the insertion of non-oxidizable substituents into that position. A particularly successful drug of this kind is mefloquine (*10.34*) which, in a single dose of 0.5 g, can produce clinical cure in patients whose life is threatened by chloroquine-resistant strains of *Plasmodium falciparum*. This drug does not react in any way with DNA (Davidson *et al.*, 1977).

The half-life of mefloquine in the human body averages 15 days. Unlike quinine, it has a very high therapeutic index. Intensive clinical studies have shown it to be safe and effective in curing both *falciparum* and *vivax* types of malaria, regardless of whether these are resistant to chloroquine. Fears that resistance to mefloquine would be initiated by indiscriminate use of this drug are limiting its *curative* use to areas where chloroquine-resistant strains of *Plasmodium* flourish. For *prophylactic* use in such areas, it is combined with the pyrimethamine–sulfadoxine pair to decrease likelihood of resistance developing, in accord with principles discussed in Section 6.5. For clinical details, see Schmidt *et al.* (1978); for the overall picture on mefloquine, see WHO (1983).

The pharmacologically active group, —O—C—C—N—, familiar to us from Section 7.2, has received considerable simplification in the transition from quinine (*10.33*) to mefloquine (*10.34*), but further simplification is possible along lines discussed in Section 7.3. The Walter Reed Army Institute of Research, in Washington D.C., which has fathered the mefloquine project, has also instituted researches to simplify both the side-chain and the nucleus, while retaining anti-malarial activity. The piperidine ring has been opened up to give a dialkylaminomethyl-group, while retaining excellent clinical activity against chloroquine-resistant strains, as in WR 30090, which is a quinoline analogue of (*10.36*).

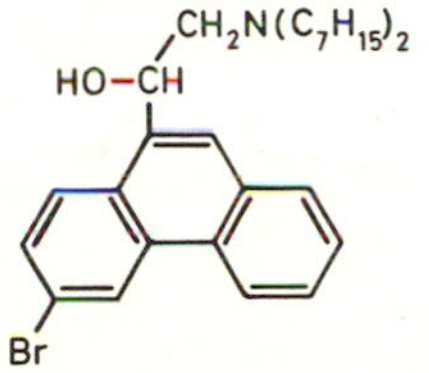

'Walter Reed 33063'
(*10.35*)

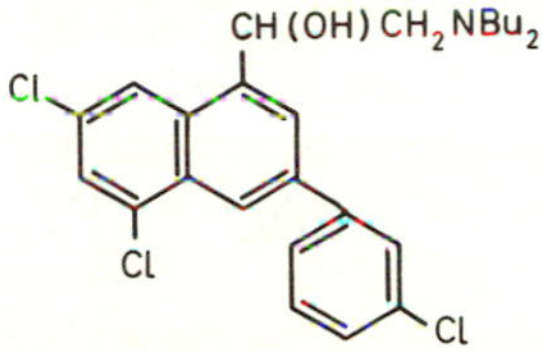

Naphthalene analogue of quinine
(*10.36*)

Similar side-chains can convert a non-heterocyclic nucleus into a good anti-malarial. Excellent results against *Pl. falciparum* infections were obtained, in laboratory and field, with the phenanthrene drugs WR 33063 (*10.35*) (Canfield and Rozman, 1974), and WR 122455 which is 3,6-bis(trifluoromethyl-α,2-piperidyl-9-phenanthrenemethanol (Rinehart, Arnold and Canfield, 1976). Naphthalene, too, can replace quinoline, because 1-[3-(*p*-chlorophenyl)-5,7-dichloronaphth-1-yl]-3-(dibutylamino)propanol (*10.36*) was highly active against *Pl. berghei* in mice, and was the exact therapeutic equivalent of the analogously substituted quinoline drug (Shamblee and Gillespie, 1979).

10.3.6 *Pharmacodynamic examples*

Acetylcholine (*7.4*), important transmitter of the nervous impulse between cells, is an aliphatic quaternary amine and hence completely ionized in all circumstances. The active form of nicotine (*7.26*), which mimics the action of acetylcholine at several sites, must be the cation because its methiodide (quaternized on the pyrrolidine nitrogen atom) is no less active (Barlow and Hamilton, 1962). The arecoline (*12.80*) cation, too, has potent acetylcholine-like properties (see Section 12.6). However, in experiments made (over the pH range 6.05 to 9.36) on the guinea-pig ileum, the neutral species of arecoline was found to have only 2% of the activity of the cation (Burgen, 1965).

The great majority of alkaloids, local anaesthetics, barbiturates, and neuroleptics have pK_a values between 6 and 8, so that both ionic and neutral species are present, in equilibrium, at physiological pH values. These substances will be considered in Sections 10.4 and 10.5. Among the alkaloids, atropine (pK_a 10) (*7.16*) and tubocurarine (*2.6*) are atypical in having a higher degree of ionization. Tubocurarine is a quaternary amine and hence completely ionized at all pH values. Other quaternary amines used in medicine are:

carbachol (*2.11*), neostigmine (*7.34*), decamethonium salts (*7.28*), suxa-methonium salts (*7.29*), and gallamine (*7.31*).

Anti-histamines and anti-depressants tend to have a pK_a about 9. For a study of the ionization of norepinephrine and its *N*-alkyl homologues, see Section 12.4.

Although many chemotherapeutic drugs function as anions, rather few pharmacodynamic drugs do so. However, the potent uricosuric activity of phenylbutazone (*10.37a*), a drug used in the treatment of gout, is due to the anion. As it has pK 4.4, and the pH of urine is 4.8 or higher, suboptimal concentrations of anion were found in the urinary system. This was overcome by the discovery of sulfinpyrazone (*10.37b*), whose lower and more satisfactory pK of 2.8, produced more anions and a better therapeutic effect (Burns *et al.*, 1960). (The improved pK comes about through hydrogen-bonding.) These drugs block the resorption of uric acid by the renal tubular cells.

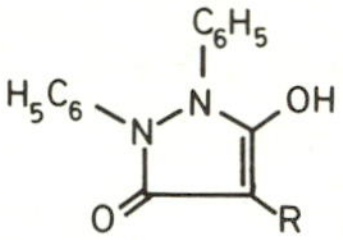

(a) Phenylbutazone (R = $-(CH_2)_3 \cdot Me$
(b) Sulfinpyrazone (R = $-(CH_2)_2 \cdot SO \cdot C_6H_5$
(*10.37*)

10.4 Substances that appear to be less active when ionized

There is no doubt that non-ionized substances can display striking physio-logical action, including substances such as ether, chloroform, and the powerful lachrymators bromoacetone and α-chloroacetophenone. Hence it did not seem unreasonable when Vermast (1921) noted that many weak acids exhibited their biological activity most fully at pH values where they were least ionized. An interesting example is the inhibition of cell-division in echinoderm eggs by salicylic acid (H. Smith, 1925). It can be seen from Fig. 10.9 that this acid is more active at pH 5 than at any other pH tested. At pH 5, a larger proportion of this acid ($pK_a = 3.0$) is in the form of molecules than at any higher pH, namely 0.99% (see Section 17.0). A rather simple explanation is that the molecules, but not the anions, of salicylic acid inhibit the division of these eggs. It would be expected that at still lower pH values salicylic acid would be a more powerful inhibitor, because it would contain a higher ratio of molecules to anions; one wonders why this region was not explored. However, complications often occur when such changes are attempted. Firstly, the organism may not continue to show indifference to pH. Secondly, the receptor for salicylic acid may undergo a change in ionization as the pH falls and would then hardly be likely to have the same affinity as before (see Section 10.6).

Similarly, it has been found (Clowes, Keltch and Krahl, 1940) that all members of a series of 30 barbiturates enter both eggs and larvae of the

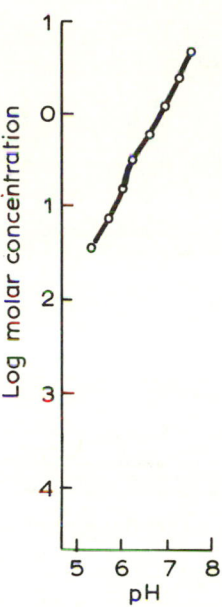

Fig. 10.9 The effect of pH on the concentrations of salicylic acid required to stop the cell division of *Echinarachnius parva*. Curve: total drug (= ions + neutral molecules). (Smith, 1925.)

sea-urchin *Arbacia* exclusively as molecules. Moreover, the resulting depressions of cell division and of·respiration were shown to be entirely due to molecules. Again, it has been shown for theophylline that the non-ionized form (as distinct from the anion) is the species causing stimulation of the turtle's heart (Hardman, 1962).

At present, there is anxiety that nitrous acid, which people consume in cigarette smoke and in preserved meats, may be uniting with secondary amines in the stomach, to form nitrosamine carcinogens (see Section 13.5 for carcinogenesis). In this connexion, it is interesting that the molecular form is the species of secondary amines that reacts fastest with nitrous acid, so that amines that are only feebly ionized at pH 2 would be the most susceptible to nitrosamine formation (Sander, Schweinsberg and Menz, 1968). These weak amines seem to be quite uncommon in the stomach contents.

When observing the effect of a weaker acid on biological material, it is usual to find that a constant amount of the substance is required, regardless of the pH of the medium, provided that the pH is at least one unit lower than the pK_a, thus ensuring that there is no ionization of the toxic agent. This is well illustrated in Fig. 10.10, which shows the effect of pH on the concentrations of phenol and acetic acid required to prevent the growth of various common moulds. Within the pH range of the experiment (2 to 6) it is seen that a constant amount of phenol is required, but a decreasing amount of acetic acid suffices as the pH

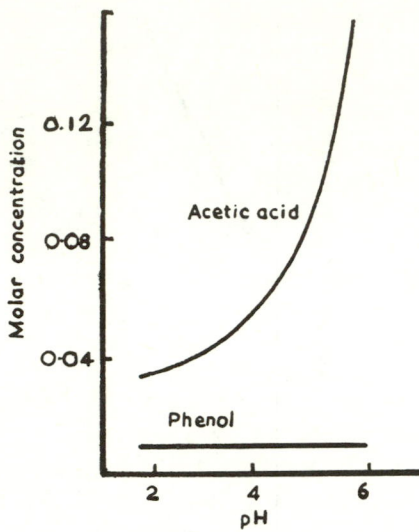

Fig. 10.10 The effect of pH on the concentration of phenol (pK_a = 9.9) and acetic acid (pK_a = 4.8) required to prevent the growth of common moulds. (Hoffman *et al.*, 1940, 1941.)

drops. This is so because the pK_a of phenol is 9.9, and hence it is non-ionized within the pH range of the experiments, whereas the pK_a of acetic acid is 4.8, and hence it is 90% ionized at pH 5.8, but only 9% ionized at pH 3.8, and so on.

All such experiments are unsatisfactory unless controls are set up to establish

Table 10.11 The connexion between ionization and the narcosis of *Arenicola* (Minimal anaesthetic doses, in g per 100 ml of sea-water, rendering this worm immobile after 5 min)

	pH 7.0	pH 8.0	pH 9.0
Non-electrolytes			
Isopropyl alcohol	2.5	2.5	2.5
Isopentanol	0.1	0.1	0.1
Chloroform	0.012	0.012	0.025
Chlorbutol	0.025	0.025	0.025
Weak bases (pK_a about 8.5)			
Cocaine	0.01	0.005	0.0025
Procaine	0.002	0.001	0.0005
Butacaine	0.001	0.0002	0.0002
Barbituric acids (pK_a about 8.0)			
Isoamyl, ethyl (amobarbital)	0.006	0.025	0.05
Propylmethylcarbinyl, ethyl			
(pentobarbital)	0.003	0.006	0.012
Diethylcarbinyl, ethyl	0.006	0.012	0.05
n-Amyl, ethyl	0.006	0.012	0.05

(Clowes and Keltch, 1931.)

if the change of pH has had an effect on the test-organism, as is often the case. This precaution was taken in experiments, summarized in Table 10.11, which deal with the narcotizing action of various substances on the worm *Arenicola* (Clowes and Keltch, 1931). It can be seen that the effect on non-electrolytes such as chloroform is independent of pH. These non-ionizing substances show that the changes of pH do not affect the worms. The weak bases, such as cocaine, behave differently. They become more effective as the pH is increased, i.e. in proportion as their ionization is suppressed. Similarly, the weak acids (four isomeric barbiturates) are more effective as the pH is decreased; again this corresponds to the suppression of their ionization. By washing the worms with sea-water, the narcotic action is readily reversed in each case.

The above studies of *Arenicola* were simplified by the fact of the pH change having no effect on the test-organism. Less fortunate were some (better un-named) workers who investigated the action of quinine derivatives on bacteria. They thought that they had shown that the neutral molecules were more active than the ions because these drugs became more effective as the alkalinity rose [*Z. Immunitäts.* (1922) **34**, 194]. Unfortunately they had overlooked the ionizing effect of alkali on the bacterial receptors (see Section 10.6).

10.5 Substances of which both ion and molecule play a part in the biological action

It is reasonably certain that many substances, particularly those whose pK_a lies in the range 6–8, penetrate into cells as molecules, even though they exert their biological action as ions (see Section 10.2). Fig. 10.11 illustrates this effect: it shows that the penetration of benzoic acid into yeast is inversely proportional to the percentage ionized.

Many agents are known in which (a) biological activity depends mainly on the proportion of neutral species present, and yet (b) the proportion which is ionized contributes also to the activity. Several examples of this behaviour will now be discussed.

When submitting a weak acid to biological test, it is usually found that a constant amount of the substance is required to produce a standard response at *all* pH values one unit or more below the pK_a. Under such conditions the ionization of the acid is slight (see Section 10.1), and hence the biological effect is due, in the first place, to the molecule. This effect is illustrated on the left-hand side of Fig. 10.12. However, if the pH is allowed to rise above the pK_a, an ever-increasing amount of the substance will be required to give the same response. When this response is analysed, one of two results is obtained, (a) a constant amount of the molecule is still required (cf. Section 10.4), or (b) an ever-decreasing amount of the molecule is required because the anion seemingly exhibits, although to a limited extent, the biological action of the molecule. Result (b) is illustrated by the action of benzoic acid on the mould *Mucor* (Fig. 10.12).

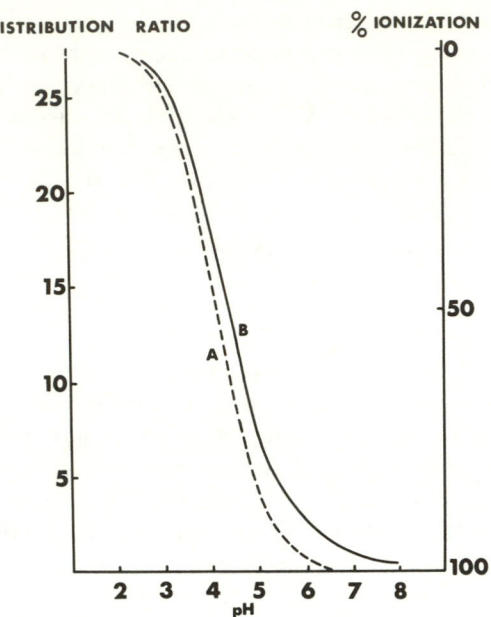

Fig. 10.11 The uptake of benzoic acid (pK 4.2) by baker's yeast at various pH values. The uptake is inversely proportional to the percentage ionized. (A) Ratio of distribution of total substance (cell/fluid). (B) Percentage ionization. (Bosund, 1960.)

This method of plotting ionic data was developed in Oxford by Simon (1950), who found that the majority of those substances that are most active when least ionized nevertheless have ions which exert some of the activity. For another example, the anti-fungal action of 2,4-dinitrophenol (pK_a 4.0), at various pH values, on the fungus *Trichoderma viride*, see Simon and Beevers (1952).

Many nitrogen heterocycles which have a ring-nitrogen that ionizes as an anion are potent inhibitors of the Hill reaction in photosynthesis (Section 4.6). It was found, though, that all activity was lost if the percentage ionized was too high, and this was traced to the inhibitors penetrating as the molecule, but acting as the anion. Suitable herbicides acting in this way were: imidazoles, benzimidazoles, purines, pyrazoles, indazoles, triazoles, and benzotriazoles (Büchel and Draber, 1969). Similarly dependent on pK are the 2-trifluoro-methylbenzimidazoles (Jones and Watson, 1965).

Similar results have been obtained with bases. Thus, pyrimethamine (*4.8*) which has pK_a 7.2, is best absorbed by cells from solutions alkaline enough to have a high proportion of neutral species. But the critical internal enzyme (dihydrofolate reductase) is inhibited only by the cation (Wood, Ferone and Hitchings, 1961).

Local anaesthetics are very dependent on a high proportion of molecules for transportation to the site of action, and this requirement confines clinically

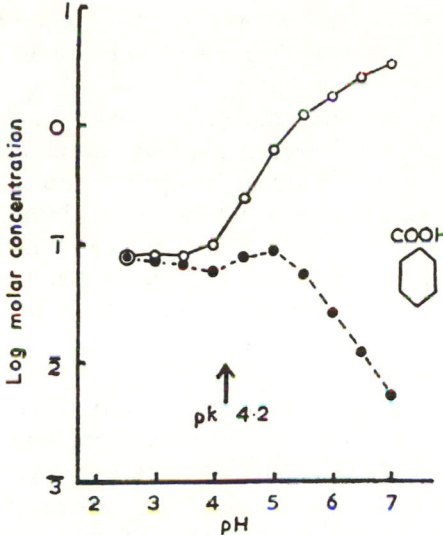

Fig. 10.12 The effect of pH on the concentrations of benzoic acid required to prevent the growth of the fungus *Mucor*. Upper curve: total substance (ions + neutral molecules). Lower curve: neutral molecules. (Cruess and Richert, 1929.)

useful examples to the weak bases. When it was found that the common local anaesthetics (which have pK_a values around 8.5) were more active when applied in alkaline solution, it was wrongly concluded that the neutral species was the active form (Trevan and Boock, 1927). This was disproved by soaking lengths of desheathed mammalian vagus nerve in cinchocaine (*7.15*). The preparation was then washed with pH 7 buffer. Cinchocaine, firmly bound to the inside surface of the nerve membrane, blocked electrical impulses. When the preparation was bathed in buffered saline at pH 9.5, conductivity was restored, but was lost again when pH 7 bathing was recommenced. Thus, the cation is the active species (Ritchie and Greengard, 1961, 1966; Narahashi and Frazier, 1971). Even quaternary amine analogues of the local anaesthetics were active if injected *inside* the nerve membrane, work done on the squid axonal membrane by Narahashi, Frazier and Moore (1972). Further, when the quaternization was effected with a benzyl group (to give increased lipophilicity), local anaesthetic activity could be demonstrated in these quaternized derivatives of cocaine, procaine, and cinchocaine, after external application (Nador *et al.*, 1953). It has to be borne in mind that these quaternary amines are completely ionized at all pH values. The slight anaesthetic action of benzocaine (*7.14*), which has pK_a 2.5, makes it clear that neutral species do possess some activity, although relatively little.

   Like the commonly used local anaesthetics, most alkaloids and anti-psychotic drugs have pK_a values about 8, so that they are about 16% non-ionized at pH 7.3. It is almost certain that these penetrate as molecules and act biologically as cations. The H_2-type histamine antagonists, like cimetidine (*9.58*), need to be largely ionized at pH 7, to exert their action, but if totally ionized at this pH, they cannot penetrate to their receptor (Black *et al.*, 1974).

The ionization as anions of many commonly used anti-inflammatory agents is positively correlated with their biological action provided that ionization is not complete and that they are sufficiently lipophilic to reach their prostaglandin-inhibiting site of action. Examples include indomethacin (pK_a 4.5) and the salicylates (Whitehouse and Dean, 1965). In the phenylindanedione (*9.39*) series, observation of the effect of varying the pK (on the anti-inflammatory action) pointed to a need for more accurate determinations of pK in the *ortho*-substituted aryl derivatives, which are by far the most biologically active (van der Berg *et al.*, 1975).

In the early days of research into varying the structure of the anti-bacterial sulfonamides, it was found that the activity almost disappeared if the sulfonamide group was methylated, an operation that prevented it from forming an anion. This was done, first, for sulfapyridine and sulfathiazole (Shepherd, Bratton and Blanchard, 1942). Consequently the relationship between ionization and anti-bacterial activity was explored by Bell and Roblin (1942) and their results are epitomized in Fig. 10.13. The optimal pK for the ionization, as an acid, of the sulfonamide group lay between 6 and 8. This suggested that penetration depends on the non-ionized species but biological action is ionic, as in other examples discussed above. More recent work, with a larger and more closely related set of sulfonamide antibacterials, confirms this optimum, although the curve is a little shallower (Seydel, 1981).

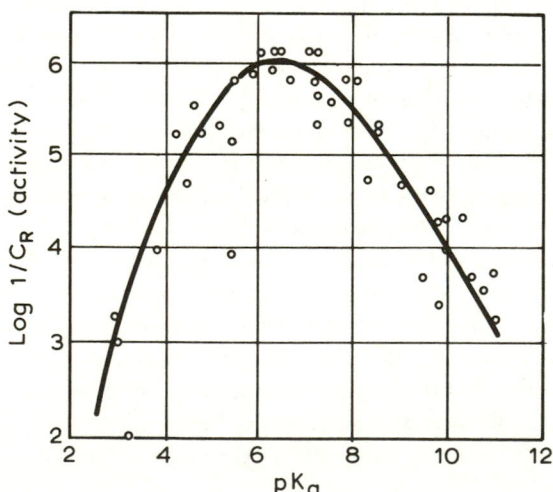

Fig. 10.13 Effect, on bacteriostatic action, of variation in the pK_a of a series of sulfonamides. Organism: *E. coli*. The substances on the left are the most highly ionized (as anions) at the pH of the test (pH 7, synthetic medium). (Bell and Roblin, 1942.)

That sulfonamides act as the anion but penetrate into the bacterial cell as the neutral species was later confirmed by comparing their behaviour in (a) a cell-free folate-synthesizing preparation of *E. coli* and (b) the intact cells of *E. coli*. The antibacterial action was directly proportional to percentage ionized in (a) but became dependent on lipophilicity, as well, in (b) (Miller, Doukos and Seydel, 1972).

10.6 The ionization of receptors

The pK_a values of receptors cannot be predicted in advance of experiment because their chemical nature is largely unknown. Obviously cationic drugs combine with anionic receptors which may have pK_a values of 2 or 7 (presence of phosphoric acid groups), 2 to 6 (carboxylic acid groups), or 10 (tyrosine-, pyrimidine-, cysteine-residues). Cationic receptors could have pK_a values of 4 (adenine), 7 (histidine), 10 (lysine), 13 (arginine).

(*a*) *Receptors outside cells.* By no means are all receptors within cells. For example, the frog's heart is affected (in opposite ways) by acetylcholine and methylene blue without penetration of the cells taking place (Clark, 1933). These and many similar observations are in harmony with current knowledge that the outside of the plasma membrane is rich in enzymes, and permeases. For example, yeast is known to have adenosine triphosphatase and several hydrolases on the outer surface.

The pK_a of a receptor on the outside of cells can often be studied by measuring the response to drugs over a range of pH values, provided that (a) the cell is known not to be injured by the pH changes and (b) that the ionization of the drug does not change within this range. For the application of this approach to learning something of the site of action of aminoacridine anti-bacterials on *E. coli*, see Section 10.3.1.

Not all workers have realized that the degree of ionization of a receptor can alter as the pH is changed. For example, the respiration of avian red blood cells (whole or haemolysed) was found to be inhibited 2.5 times as strongly by quinine at pH 10 as at pH 5. Because quinine has a pK_a of 8.4, it was concluded that the inhibition was caused by the molecule and not by the ion. The likelihood that an acidic receptor became more ionized at the higher pH (whereupon it could bind more quinine cations) was not mentioned [*Biochem. Zeitschr.* (1922) **128**, 169].

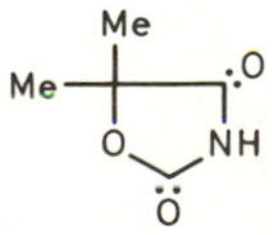

5,5-Dimethyloxazolidine-2,4-dione

(*10.38*)

(*b*) *Receptors inside cells.* A receptor may be supposed to be intracellular if (a) agents with lipophilic groups are more effective than examples without those groups, and (b) agents which yield only 70% of the active ionic species (at the pH of the test) are more active than those which are present completely as this species. Ionization studies then become more difficult, because the pH close to the receptor is the one of prime importance.

In the Waddell and Butler (1959) method for determining intracellular pH, three measurements are required: the external pH, and the extracellular and intracellular concentrations of an indicator *ion* to which the membrane is impermeable (it must be freely permeable to the neutral species). These authors used a colourless, fluorescent indicator 5,5-dimethyloxazolidine-2,4-dione, DMO (*10.38*), a weak acid of pK_a 6.13 at 37°C. By this method, the pH inside resting muscle (dog) was found to be 7.04, but inhalation of carbon dioxide brought it down to 6.6. Using a radioactive form of this indicator, Addanki, Cahill and Sotos (1967) found pH 7.74 inside resting mitochondria. The indicator is considered to be safe to use in patients.

Another method for measuring the pH inside cells is to observe the ^{31}P phosphate shift in n.m.r.; this procedure has an accuracy of ±0.02 unit of pH (Navon *et al.*, 1977). In another application of this method Roberts *et al.* (1980) found pH 7.1 for the cytoplasm of plant cells and 5.5 for the vacuoles. Their method was to note the chemical shift of the ^{31}P peak due to glucose 6-phosphate which traces a normal sigmoid titration curve (as in Fig. 10.1) when the pH is varied. Glucose 6-phosphate was chosen as the most abundant phosphorylated constituent of the plant cell.

In recent years, recessed-top microelectrodes have come into use for inserting into living cells and recording changes in pH during several hours (Thomas, 1974).

Work with indicators suggests that the pH of animal and plant cytoplasm is 6.8±0.2, and that these fluids are well buffered. Nucleoplasm was found to be 7.6. The external surfaces of organelles, if anionic (e.g. protein surfaces), can be richer in hydrogen ions than the cytoplasm because of a zeta potential effect. The cytoplasm of amoebae was found to be 6.7. The pH of plant vacuoles averaged 5.2, but was much lower in some species. The pH of cancerous tissues in mammals is often lower than that of the surrounding organ, a circumstance with a possible application in therapy (Section 4.2, p. 151). This low pH, which is connected with the high rate of anaerobic respiration, has led to more careful selection of pK_a values in designing carcinostatic drugs of the nitrogen-mustard type (Ross, 1961).

For infection with an animal virus, a pathway of low pH must temporarily exist to permit penetration into the host's intracellular vesicles. The lowered pH may be necessary to mediate a suitable conformational change in a protein (Helenius, Marsh and White, 1980).

That the pH within bacterial cells may vary with those of buffers in which they are placed is suggested by the following experiment. The tyrosine decarboxylase of *Strep. faecalis* has its maximal activity *in vitro* at pH 5.5. In the intact bacterium, it is only moderately active when the external environment is neutral or alkaline, but it becomes most active when placed in buffer at pH 5.0–5.5 (Gale, 1946). Again, in the turtle's heart, a linear relationship exists between the pH of extracellular water and intracellular water (ventricular tissue) in the range 6.5–9.5. Thus $pH_{int} = pH_{ext} + 2.98$ (Waddell and Hardman, 1960).

It is desirable in pharmacology to work with a buffer, even when the receptor is known to be an internal one. The external buffer brings about uniform conditions by (a) presenting the drug to the cell in a standard state of ionization, regardless of the form in which it was supplied, and (b) maintaining in a standard state the chemical groups responsible for the adsorption of the drug to the cell surface and its penetration into the cell.

The success of buffering depends on the proportion of cells with which the buffers come into contact. In bulky tissue preparations, this proportion may be very small, and the drugs can reach their receptors only after passage through other cells or intercellular fluids. The results from such experiments are not comparable with those obtained from experiments in which all the cells are in direct contact with the drug at a known pH (Simon and Beevers, 1951).

10.7 Conclusions

Consideration of the principles discussed in this chapter makes it clear that it is highly advantageous to know the ionization constants of all substances which are being examined biologically.

The next step is to discover which is the most effective, the ion or the molecule. This may be done in two ways, which are not alternative but complementary. In the first approach, the pK_a of the drug should be varied by appropriate substitution, taking full advantage of the chemist's ability to produce substances of any desired pK_a in almost any series (see Section 17.2). The advantage of this approach is that the ionization of the receptor is not interfered with, and the living cells are not removed from their optimal pH. The disadvantage is that the chemical change in the molecule may, of itself, be responsible for a changed biological response; this difficulty can be avoided by working with *two* analogues at each desired pK_a, and accepting the results only if they agree. In the second approach, the composition of the drug should be kept constant, but the pH of the medium varied. This has the advantage that the living cells are exposed to only one drug, but the effect of pH changes on the receptor and on the viability of the cell itself must be independently examined.

Because ionization, one of the commonest of variables, is brought under control by these methods, studies of this kind can greatly assist in the design of more meaningful experiments.

11

Metal-binding substances

Heavy metals, in traces, are essential for all forms of life. They are taken up by the living cell as cations, and their uptake is strictly regulated because most (or all) of them are toxic in excess. A remarkable specificity has been found: seldom can an excess of one essential metal prevent the damage caused by a deficiency of another. In fact, such an excess often increases the injurious effect of the deficiency.

Metal-binding substances, many of which function by chelation,* have furnished many useful drugs and other substances of value in selective toxicity. They are manufactured in huge quantities for this purpose, particularly those that are used in agriculture as fungicides. They function by upsetting the delicate balance that exists between the cations of heavy metals in nature. Some of them withdraw metals from living tissues, but many others reinforce the natural toxicity of a heavy metal.

Apart from their use as agricultural fungicides, metal-binding substances

---

*A substance which binds metals is called a *ligand*. When a metal is gripped, in a ligand, between any two of the elements, N, O, or S, a *chelate* ring is formed, and the metal is more tightly bound than when it is not a chelate, i.e. when it is not part of a ring. The term, coined by Morgan and Drew (1920) to describe this phenomenon, is derived from the Greek word *khele*, the crab's claw.

have found three types of use in veterinary and human medicine. Some which differentiate between vertebrates and their parasites make a valuable contribution to the fight against fungi, bacteria, and viruses (see Sections 11.7–11.9). Others, which may be described as antidotes, are used to distinguish between essential and poisonous metals (see Section 11.6). The third type favours normal over pathological processes, e.g. in rheumatic diseases, cardiovascular disease, and cancer (see Sections 11.9–11.11).

11.0 Metals in the living cell

Man, just like other vertebrates, requires cations of the following metals to facilitate a great many essential life processes. Moreover, many of these metals are essential for all other forms of life:

(a) The heavy metals: cobalt, copper, iron, manganese, molybdenum, and zinc, and in still smaller amounts: chromium, vanadium, nickel, and tin.

(b) The lighter, and usually more abundant, metals: calcium, magnesium, potassium, and sodium.

Of these, copper, iron, molybdenum, cobalt, and occasionally manganese, assist oxidation–reduction equilibria; zinc, magnesium and manganese are concerned with hydrolysis and with group-transfer; calcium plays its most important part in creating structures, flexible or rigid, but also it (and sometimes magnesium) can trigger a reaction, possibly by effecting a structural change. Sodium and potassium, because of their abundance, function as charge-carriers; they are only weakly bound and hence can exchange rapidly. Some examples of the biological functions of heavy metals will now be given [in the order of (a) and (b), above]. The position and valence of metals in enzymes is decided by electron-spin resonance and the new technique known as 'Extended X-ray Absorption for Fine Structure' (EXAFS), which is interpreted with the help of Fourier transforms and curve-fitting.

(*a*) *Cobalt* is an essential co-factor for several enzymes such as carboxytransphosphorylase (which transfers phosphate groups), glycol dehydrase (which effects a rearrangement), and the carboxypeptidase which destroys various kinins (inflammation-causing peptides that circulate in mammals) and plays a role in the digestion of polypeptides (Davies and Lowe, 1966). Cobalt is strongly bound in the large tetrapyrrole ring of the cobamide family of coenzymes. Two substances used for replacement therapy in pernicious anaemia, hydroxycobalamine and cyanocobalamine (vitamin B–12), are formed by the breakdown of these coenzymes during isolation from yeast or liver. These coenzymes catalyse the reduction of ribose nucleotides to deoxyribose nucleotides, the interconversion of β-methylaspartate and glutamate anions, the interconversion of methylmalonate and succinate anions, and the methylation of homocysteine to methionine.

In 1935, two groups of workers in Australia found that the vast, lush pastures that produced severe anaemia in sheep, in that country and in New Zealand, did so through lack of cobalt, an element that is required by plants in far smaller amounts. When the cobalt content of grass was less than 0.2 of a part per million, the sheep became weak and died. This debility was traced to a failure of microflora in the sheep's rumen to synthesize the cobalt-containing coenzymes (Marston, Allen and Smith, 1961). Addition of cobalt to the sheep's diet is now used to prevent this condition, and the productivity of the land has increased almost unbelievably. Fig. 11.1 shows the early stages of decline in a sheep with cobalt deficiency (Marston, 1949).

In the human, as little as one microgram of vitamin B−12 (intramuscularly) is sufficient daily therapy for pernicious anaemia. In a man weighing 70 kg, this represents only 0.000 02 p.p.m. of his weight.

(*b*) *Copper* is an essential constituent of many enzymes including phenolase (tyrosinase) which plays an important metabolic role in plants (including fungi) and in mammalian pigment cells, and also in providing the protein-tanning agent that hardens insect integument. Copper in the closely related laccase has

Fig. 11.1 Cobalt deficiency causes anaemia and wasting in sheep. The sheep on the left was reared on Australian pastures deficient in cobalt. The sheep on the right had adequate cobalt in pasture. (Marston, 1949.)

been shown by electron spin resonance spectroscopy to be bivalent in the resting state but to become univalent (Cu^+) while the enzyme is in use. Cytochrome oxidase contains two iron and two copper atoms per molecule; although the iron is porphyrin-bound, copper never is in Nature. Copper is also essential for the action of dopamine hydroxylase (a key enzyme in the synthesis of the catechol-amine hormones), monoamine oxidase, and ascorbic oxidase.

The green-coloured enzyme superoxide dismutase, present in every tissue of eukaryotes, contains both copper and iron linked to the same imidazole nucleus (His-61) in the apoenzyme. The copper undergoes a cycle of oxidation and reduction as it destroys toxic superoxide radicals ($\cdot O_2^-$) which arise from reactions using atmospheric oxygen.

In the superoxide dismutase of bovine erythrocytes, and possibly elsewhere, zinc and copper are both linked to a histidine moiety. Analogous enzymes containing manganese instead of copper occur in both mammalian and bac-terial organisms (Oberley and Buettner, 1979). For further reading on this highly unusual metalloenzyme, see Oberley (1982) and Rodgers and Powers (1981).

The effect of traces of copper on oat-seedlings, grown in a copper-deficient medium, is shown in Fig. 11.4. Plainly, too little copper is bad for growth, and so is too much. Until recognized as such, copper-deficiency was the cause of many a crop failure in the reclaimed areas of Holland and Denmark. Copper-deficiency in farm animals leads to anaemia, demyelination of the spinal cord, and loss of pigmentation. Excess of copper storage in the liver of sheep leads to haemolysis and death. Sheep, fed on a diet deficient in copper, lose the crimp in their wool. Because it is the crimp that makes fine wool saleable, this causes economic loss to farmers (see Fig. 11.2).

The well-established and much used *selective* action of copper against invert-ebrate animals, fungi, and algae depends on the superior arrangements that vertebrates have evolved for sequestering, transporting, and eliminating this very chemically active element. In a healthy mammal, more than 90% of all copper is bound to the protein, caeruloplasmin, which is synthesized in the liver and circulates in the bloodstream. Its physiological functions are to lower the concentration of cupric ions in the plasma to below the toxic level, to supply copper to the copper-containing enzymes, and to secure a measured release of iron from its stores. Any excess of the copper–caeruloplasmin complex is voided through the bile into the faeces. The non-caeruloplasmin copper in the serum is held, and made available, by serum albumin as described in Section 11.1 (p. 445).

The copper in caeruloplasmin is partly cuprous and partly cupric, as shown by electron-spin resonance. Absence of caeruloplasmin, which runs in families, is the basis of Wilson's disease which, if untreated, causes death in late child-hood through the destructive effect of cupric ions on the nervous system. For treatment with chelating drugs, see Section 11.6.

A bright blue copper-containing pigment, called plastocyanin, is involved in

Fig. 11.2 Copper deficiency in Australian pastures caused progressive loss of crimp in wool of a merino sheep, observed over 3 years. (Marston and Lee, 1948.)

photosynthetic electron transport of plants. Copper takes the place of iron in the blue respiratory pigment (haemocyanin) of certain molluscs and crustaceans. This element is essential for the production of haemoglobin in Man.

Copper is known to be concentrated several-fold (relative to other metals) in sympathetic nerve-endings, and in the synaptic vesicles of brain and other nerve tissue. There it seems to assist in storing norepinephrine in a ternary complex with ATP (Colburn and Maas, 1965), and 5-hydroxytryptamine is similarly stored (Roberts, 1966). Cupric ions, adenosine triphosphate, and norepinephrine form a tightly bound ternary complex (1:1:1) with a log K value of 12.15 [without the ATP, the (1:1) log K value is 10.25]. Epinephrine is bound a little more firmly, and dopamine rather less. The medicinally used analogues, such as amphetamine and phenylephrine, were relatively weakly bound. It is evident that both oxygen atoms of the catechol moiety play an important part in the binding. These adrenergic transmitters and drugs bind to the granules of synaptosomes in the ranking order of their stability constants (Rajan, Davis and Colburn, 1974).

For further reading on the role of copper in biology, see Sigel (1981).

(c) *Iron* is a vital constituent of the porphyrin enzymes: catalase, peroxidases, and the various cytochromes which are essential electron-transport agents for all living cells. Haemoglobin is the respiratory pigment in the red blood corpuscles not only of mammals but of birds, reptiles, amphibians, and fish. It is also found free in the plasma of some molluscs and annelid worms (some annelids have a green iron respiratory pigment, chlorocruorin). Myoglobin, the oxygen-transfer pigment of muscle, closely resembles haemoglobin.

Transferrin, a glycopeptide of mol. wt. 86000, binds ferric ions firmly between imidazole and tyrosine residues. It occurs in blood serum, where it masks ferric ions which, if free, would damage the heart. Transferrin is the only form of iron acceptable to reticulocytes for making haemoglobin. Like a similar β-globulin in hen's eggs (conalbumin), transferrin can so lower the concentration of ferric ions in a medium that bacteria can no longer grow in it. A similar polypeptide, lactoferrin, is found in sweat, tears, milk, and nasal and bronchial secretions. It is thought that the ability of these substances to prevent uptake of iron by bacteria can constitute an important defence against bacterial infections, especially as iron enhances the lethality of many bacteria, even as much as 100000-fold (Bullen and Rogers, 1969).

The principal iron-storage compounds of vertebrates are the proteins ferritin, found mainly in liver, and haemosiderin, found in spleen and muscles.

Iron is also an essential constituent of several non-porphyrin enzymes, e.g. aconitase, aldolase, and succinic dehydrogenase. Inhibition of the synthesis of glucose by tryptophan in animal cells depends on chelation. The tryptophan is metabolized to pyridine-2,3-dicarboxylic acid, which complexes the divalent iron necessary for the action of phosphoenolpyruvate carboxykinase (a key enzyme in the neogenesis of glucose) (Veneziale *et al.*, 1967).

Bacteria and fungi use the siderochromes, a remarkable series of phenols and hydroxamic acids, to forage for iron (more in Section 11.1).

Ferredoxins are proteins containing equal numbers of iron and sulfur atoms in each active centre. Some transfer electrons below the potential of the hydrogen electrode, e.g. ferredoxin FeS 1 at -305 mV, and ferredoxin FeS 2 at -20 mV. But another type (FeS s-3) works at the much higher potential of $+60$ mV (these figures are for E_0, versus the standard hydrogen electrode at pH 7.2) (Sweeney and Rabinowitz, 1980; Spiro, 1982). The 8Fe-8S ferredoxins are associated with the most primitive organisms (obligate anaerobic fermenters and photosynthesizers) where they are used for electron-transfer in the pyruvate phosphoroclastic system; the 4Fe-4S types probably came next in evolution and are found in sulfate- and nitrate-reducing bacteria. The later 2Fe-2S ferredoxins are found in plants and animals where they are essential for oxidative phosphorylation in mitochondria, for photosynthetic phosphorylation in chloroplasts, and for the synthesis of catecholamine hormones. The individual types are distinguished by e.s.r. and Mössbauer spectra.

Although iron is abundant in most soils, it is often unavailable to plants, particularly in basic soils. However, if the right bacteria are present, the iron becomes available because the micro-organisms solubilize it, with the aid of hydroxamate siderophores (Section 11.1). Such ferric organic complexes accumulate in the soil in sufficient concentration to supply the needs of plants (Powell *et al.*, 1980). In the absence of these helpful bacteria, ethylenediamine-tetra-acetic acid (EDTA), if sprayed on the soil, extracts iron by forming the EDTA–ferric complex, which is absorbed by the rootlets. Experiments with tomato plants, grown in an iron–EDTA medium labelled with ^{59}Fe, and with ^{14}C in the 2-position of the acetate-group, showed that the plant absorbs the intact complex, which is translocated. Later, the organic part is broken down by metabolism which leaves inorganic iron. When soil is poor in iron, ferric–EDTA is sprayed on the ground with the same result.

(*d*) *Manganese*, although a metal of variable valence, is rarely involved in biological oxidations. It is essential for the activity of many catabolic enzymes such as oxaloacetate decarboxylase, arginase, pyruvate oxidase, and prolidase, as well as other enzymes where large groups, such as sugar residues, are exchanged.

Manganese is essential for photosynthesis in plants, apparently facilitating that part of the Hill reaction which cleaves a hydroxyl anion to liberate molecular oxygen.

(*c*) *Molybdenum* has an essential function in the following enzymes: xanthine oxidase, aldehyde oxidase, nitrate reductase, and nitrogenase. In all of these it seems, on the basis of the electron-spin resonance spectrum, to be bound to sulfur and to undergo the valency change $Mo^{5+} \rightleftharpoons Mo^{6+}$; iron is usually present as well. Molybdenum is essential for nitrogen fixation, by the various species of *Rhizobium* and *Azotobacter* in plant root nodules (see Fig. 11.3).

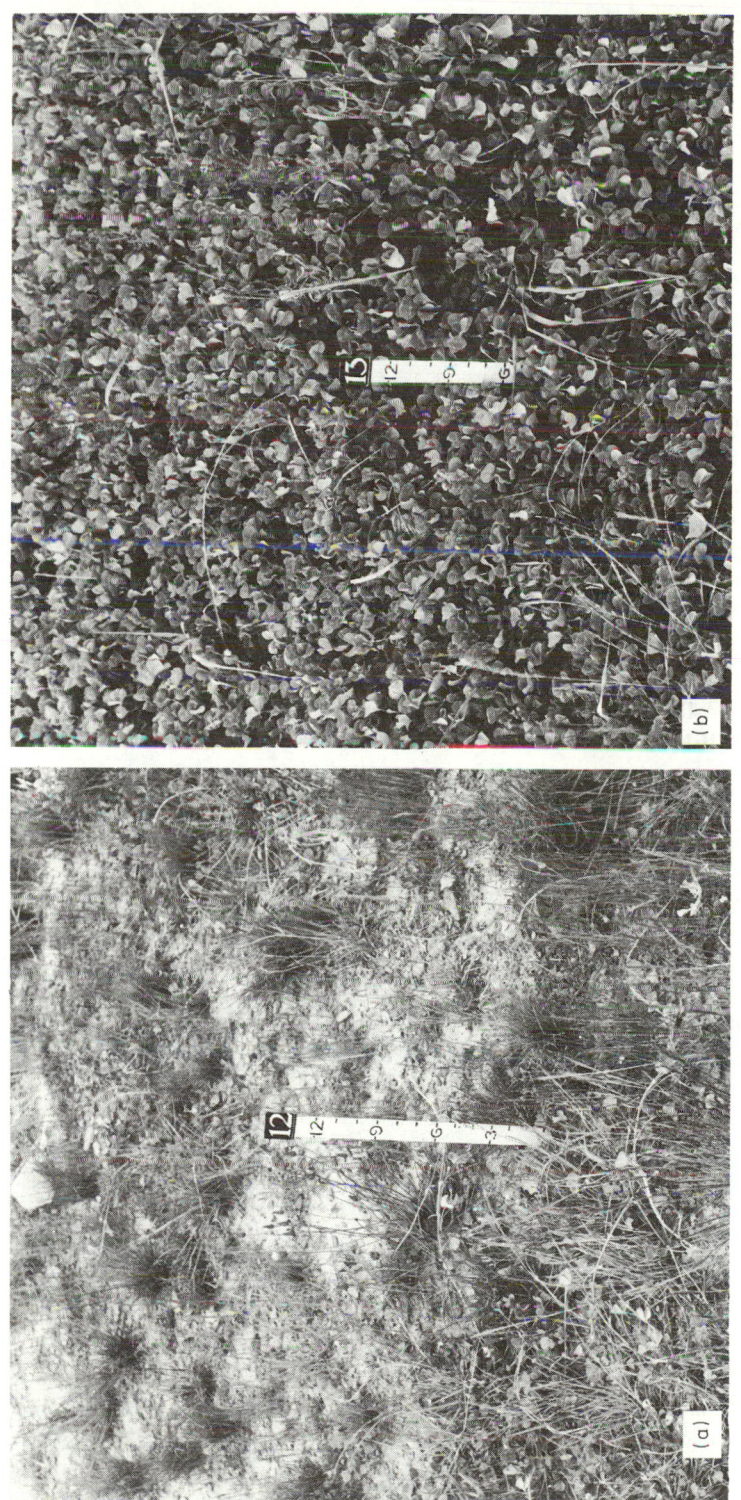

Fig. 11.3 Clover needs molybdenum for productive growth. (a) Subterranean clover growing on molybdenum-deficient soil. (b) Conditions as in (a), but 15 g of ammonium molybdate added per 1000 m² of soil.

Formate dehydrogenase, in the plasma membrane of *E. coli,* is a versatile enzyme which contains functional molybdenum, iron, and (tightly bound) selenium (Enoch and Lester, 1975).

(*f*) *Tungsten* is known to take the place of molybdenum in several related bacteria (Andreesen and Ljungdahl, 1974).

(*g*) *Zinc* is essential for the functioning of at least twenty different enzymes, and their functions are widely varied. They include the alcohol dehydrogenases of yeast and mammalian liver, glyceraldehyde phosphate dehydrogenase, phosphoglycomutase of yeast, alkaline phosphatase in bacteria, mammalian carboxypeptidase, carbonic anhydrase, AMP hydrolase, pyruvate carboxylase (yeast), and aldolase (yeast and bacteria). The alkaline phosphatase of *E. coli* has, in each molecule, four atoms of zinc: the two which maintain structure can be replaced by Mn^{2+}, Co^{2+}, or Cu^{2+}, whereas the other two atoms are essential for enzyme action (Trotman and Greenwood, 1971). Moreover, zinc is essential for the action of DNA and RNA polymerases in bacteria and phage, as found by dialysis against chelating agents (Coleman, 1974; Poiesz, Seal and Loeb, 1974).

Reverse transcriptase, isolated from several viruses, contains zinc in the proportion of one atom per molecule of enzyme (Vallee, 1975). One example, the RNA-dependent RNA polymerase of influenza B virus, is inactivated *in vitro* by *o*-phenanthrolines, selenocystamine, and other chelating agents (Oxford and Perrin, 1974).

The over-refined food of highly industrialized countries seems to produce zinc deficiency in the elderly, characterized by loss of acuity in taste and slow healing of cuts and abrasions. The mammalian kidney has a protein, metallothionein, which binds excess of zinc, cadmium, or mercury (Rupp, Voelter and Weser, 1974).

A deficiency of zinc causes serious disease in apple and citrus trees and grape vines. Lack of zinc in the soil causes poor yields of cereals. In the 1930s, the uneconomic sparse scrub in the Ninety Mile Plain in South Australia was converted to lush grasslands by regular aerial spraying with zinc salts; it now supports a large population of sheep.

(*h*) *Chromium* (Cr^{3+}), present in yeast as a complex with nicotinic acid and aliphatic aminoacids, is considered to be the 'glucose tolerance factor' essential for normal carbohydrate metabolism in man and acting by increasing the hypoglycaemic properties of insulin (Mertz, 1975). In highly industrialized countries where the food is plentiful but too refined, and again in developing countries, chromium deficiency is quite common (Gurson and Saner, 1971; Levine, Streeten and Doisy, 1968).

(*i*) *Vanadium*, essential for vertebrate nutrition (Lambert *et al.*, 1970), seems to be a natural inhibitor of Na^+ K^+−ATPase (Cantley *et al.*, 1978). In the urochordates (sea-squirts), animals which stand just below vertebrates in the phylogenetic tree, vanadium takes the place of iron in the oxygen-transport pigment of the blood.

(j) *Nickel* has been found essential for nutrition in pigs, also rats (Nielsen, 1975). The essential quantity is so much smaller than was required for the earlier-established essential metals, that searchers had to wait for the discovery of new analytical techniques, such as neutron activation. The enzyme urease (in beans) is nickel-dependent (Dixon *et al.*, 1975). Several bacteria have reductases that need nickel, such as *Methanobacterium* and *Desulphovibrio* (Thauer, 1980), and so do bacterial carbon monoxide dehydrogenases (Deake, Hu and Ward, 1980).

(k) *Tin* is thought to be essential for animal nutrition, but only in traces (Schwarz, Milne and Vinard, 1970).

(l) *Aluminium,* generally supposed to be biological inert, can, when solubilized in the acidic vacuoles of plants, show mild toxicity.

(m) *Calcium* is generously distributed in most living organisms (not much, however, in bacteria). The grosser accumulations of calcium include teeth, bones, shells, and the oxalate crystals of plants. Nature's use of calcium to rigidify load-bearing tissues is exemplified by tendons which are fragmented into fine collagen fibrils when their calcium is withdrawn, e.g. by a solution of EDTA, (*11.27*). Replacement of the lost calcium largely reverses this process. Cartilage is turned to bone by incorporating calcium and phosphate ions, a process which magnesium ions competitively reverse.

In mammals, both a thyroid and a parathyroid hormone exist to regulate the level of circulating calcium. Concentrations of calcium, inside and outside of cells are about 10^{-6} and 10^{-3} M respectively. Release of calcium, whether from outside the cell or from internal stores, can trigger a wide range of biological responses. These concentration levels are controlled by the plasma membrane, membranes of the endoplasmic reticulum (in muscle: the sarcoplasmic reticulum), and the inner membrane of mitochondria. One important function of calcium is to control the permeability of semi-permeable membranes which maintain their integrity in its presence and become porous and leaky in its absence. This control follows from the abundance of this ion, its divalent character, and the presence of phosphate anions in the membranes (Ames, Tsukada and Nesbett, 1967).

Another important role that calcium ions play is as a 'second messenger'. They can couple a chemical stimulus to a secretion, as in releasing neuro-transmitters after receipt of a nervous impulse, whether at synapses or at a neuromuscular junction (Miledi and Slater, 1966), or in the release of epine-phrine from the adrenal medulla at the instance of acetylcholine (Douglas,

1968). Likewise calcium is needed to couple the release of a neurotransmitter to a muscular contraction. Calcium seems also to initiate proliferation in many kinds of cells, such as muscle cells, lymphocytes, and fibroblasts. It also plays a role in phagocytosis. Those ionophores (Section 14.2) that transport calcium into cells can initiate many of these effects. Some think that, at the molecular level, calcium functions by giving ATPase access to its substrate (ATP), thus providing the energy for these various actions.

Calcium plays a vitally important part in possibly as many as three sequences of the contraction–relaxation cycle of vertebrate muscle (Taylor, Lymn and Moll, 1970). In the sarcoplasmic reticulum, calcium is stored in calsequestrin, a protein which holds, with high affinity, 43 atoms of calcium per molecule. The calcium is released in response to nervous impulses and triggers a sequence of reactions that lead to contraction of the muscle. After the contraction, the calcium is pumped back into this reticulum by the calcium/magnesium-dependent ATPase.

The essential role played by calcium in bringing about blood clotting has long been known. In some tissues, particularly in the lower animals, calcium can partly or completely replace sodium in carrying the inward current during nerve conduction.

The mitochondria of mammals (only) accumulate calcium and give it up again on demand, possibly controlling the calcification of tissues in this way. The heart's mitochondria accumulate calcium in exchange for sodium, but liver mitochondria exchange it for hydrogen ions, while keeping an independent channel for the slow exit of calcium. The contraction of heart muscle is triggered by the release of calcium from its mitochondria which, in turn, is brought about by sodium ions or cyclic adenosine monophosphate.

Calmodulin, an intracellular calcium-combining protein, is involved in many bodily processes such as secretion, activation of myosin kinase, and cyclic nucleotide metabolism. A similar protein, troponin-c, regulates conformational changes in skeletal muscle. The control of *skeletal* muscle contraction depends entirely on intracellular calcium. Hence those drugs such as nifedepine (Section 14.2) which block calcium channels, have no effect. On the other hand, *smooth* and *cardiac* muscles are much influenced by external calcium levels.

Calpain is a calcium-dependent cysteine protease which acts as an irreversible regulator in almost all mammalian cells (Murachi, 1983). The gla-proteins are important factors in blood clotting. They contain γ-carboxyglutamate residues which bind calcium (the γ-carboxy group is formed with the help of vitamin K). Other gla-proteins are present in the dentine of teeth and in bone matrix.

Biochemists make much use of aequorin to locate and measure traces of calcium in the various situations described above. This protein, responsible for the blue glow of the jellyfish *Aequorea*, fluoresces specifically in response to calcium (Shimomura and Johnson, 1969).

For the role of calcium in drug action, see Weiss (1978), and in biology see Anghilieri (1982) and Cheung (1980–2).

(n) *Magnesium* is the second commonest intracellular cation, just as calcium is the second commonest extracellularly. It is the irreplaceable link that keeps ribosomes intact, and attaches mRNA to ribosomes. It is a co-factor of all the enzymes that utilize ATP in phosphate transfer, and in many other enzymes concerned with group transfer or hydrolysis. It inhibits release of acetylcholine at the motor end-plate, and in many other ways acts as a reversible antagonist to calcium. In bacteria, magnesium is the most abundant bivalent metal. The essential central atom of the chlorophyll molecule is magnesium. The higher incidence of death from heart disease in soft-water areas seems to arise from lack of Mg^{2+}.

(o) *Potassium and sodium,* two ions apparently so similar physically, differ quantitatively in ability to penetrate cell membranes at rest and during excitation, in their affinities for active transport mechanisms, and in their ability to activate enzymes. For example, resting nerve membrane is about 30 times more permeable to K^+ than to Na^+ but, when active, it becomes 10 times more permeable to Na^+ than to K^+ (cf. Section 7.5.1). The size of the cations, hydrated or anhydrous, plays little part in this selectivity, which is governed by the free energy difference:

$$\Delta F_{a, \text{site}} - \Delta F_{b, \text{site}} - \Delta F_{a, \text{water}} + \Delta F_{b, \text{water}}$$

where ΔF_{site} is the free energy of attraction between the cation and the negatively charged site, and ΔF_{water} is the free energy of hydration of the cation. The principal variable governing Na^+/K^+ selectivity is the strength of the negative charges on the membrane as reflected in the pK_a values; Na^+ favours the stronger acids, and the crossover point occurs near pK_a 3 (Diamond and Wright, 1969).

Potassium is the essential intracellular ion of all living matter, and there is little sodium *inside* most cells except at the moment of stimulation. Terrestrial plants have little sodium anywhere.

Even univalent ions can be sequestered, and part of the Na^+ in living animal tissue is in an osmotically inactive form. It is held by polymers whose repeating units carry at least one acidic group (examples: nucleic acids, mucopolysaccharides). These sequestered sodium ions can instantly be liberated by ions of a higher valency, and the concentration measured with a sodium-responsive electrode (Palaty, 1966).

In recent years many experimentally useful agents have been found for altering permeability of membranes to these ions. Tetrodotoxin (Section 7.5.1) decreases permeability to sodium ions, and valinomycin (Section 14.2) promotes the uptake of potassium (Moore and Pressman, 1964). By forming a complex with the potassium ion, sodium tetraphenylborate ($NaBPh_4$) can dissociate tissues into single cells (Rappaport and Howze, 1966).

The healthy kidney secretes sodium while retaining potassium. Diuretics are the more valued the less they disturb this pattern (see Section 14.1).

(*p*) *Lithium* is not a normal cell constituent but, unlike any other known ion, it can replace sodium in facilitating nervous transmission although not efficiently. Also it is transported across isolated frog skin by the mechanism normally reserved for sodium (Schou, 1957). Its successful use in the treatment of chronic mania was initiated in Australia (Cade, 1949) and developed by Schou to the point where it has become the corner-stone of therapy for chronic mania. No lethargy accompanies its use. Unfortunately, the toxic dose is only about three times the therapeutic dose.

(*q*) *Cation antagonism.* Many diseases of plants and animals are traceable to maladjustments in the balance between metals. Soybean plants, treated with an excess of manganese, quickly develop signs of iron-deficiency which has to be corrected by dosing them with iron. If, on the contrary, they are grown in soil that is too rich in iron, they develop manganese-deficiency (Somers and Shive, 1942). In Britain, various pastures cause the economically serious disease in lambs called swayback. This is relieved by copper, and is associated with the presence of excessive amounts of zinc or lead in the grass. Other pastures, rich in molybdenum, cause signs of copper-deficiency in sheep (teart), relieved by the feeding of copper. Excess of zinc in the diet of rats causes an anaemia relieved by copper (Smith and Larson, 1946). Land in Holland which had been dressed with copper to prevent 'reclamation disease' (copper-deficiency) often produced crops with a marked deficiency in manganese, although the soil had the normal content of manganese. These antagonisms are reminiscent of the simple antagonism (calcium versus potassium) which is responsible for regulating the beat of the heart, as Ringer showed in 1883. A similar antagonism between magnesium and calcium governs the contractility of muscle.

Many of these antagonisms present a picture that is the inorganic equivalent of the metabolite analogues discussed in Chapter 9.

The poisoning effects of foreign metals are often traceable to cation antagonism. Thus lead, a well-known neurotoxin, displaces calcium from several parts of the nervous system, thus hindering release of neurotransmitters (Kober and Cooper, 1976). The widespread contamination by cadmium on the West Coast of Japan, coupled with a low calcium intake, caused the painful disease 'itai itai', a form of osteomalacia, around 1960. This turned out to be a straightforward case of cadmium–calcium antagonism (Friberg, 1974).

(*r*) *The biphasic response.* The response of organisms to heavy metals is known to be biphasic: given too little, the organism suffers severely; this is understandable from our knowledge of the large number of enzymes which cannot function without the appropriate trace-metal. But if the organism is given too much metal, a second phase of injury is seen, due to the toxic action of the excess.

This biphasic response is well illustrated by the action of copper on oats (Piper, 1942). Fig. 11.4 shows graphically that too much copper is as injurious as too little. These oat plants are seen growing in a series of vessels in which concentrations of copper vary from 0 to 3000 μg per litre: growth is seen to reach a maximum at 500 μg, and to fall away on either side of this figure. The growth of micro-organisms is often critically dependent on the concentration of one or more metal cations in the nutrient medium: too little and too much must be avoided to secure growth.

Fig. 11.4 Effect of copper on the height of oat-seedlings grown in nutrient copper-deficient medium. From left to right the quantities of copper present are nil, 3, 6, 10, 20, 100, 500, 2000, and 3000 μg per litre (see p. 442). (Piper, 1942.)

(s) *Metal-depleted media.* Newcomers to the trace-metal field are often surprised that it is necessary to deplete the medium before beginning experiments. Unfortunately 'chemically pure' and 'analytical grade' reagents are rich sources of the heavy metals. This is inevitably so, because a substance that is 99.99% pure would have 600 000 000 000 000 000 foreign molecules in each gram. This figure was calculated from the Avogadro number (6×10^{23} particles per mole of any substance) and assumes an average molecular weight of 100. Actually the makers of analytical reagents do not usually claim their products to be even as pure as 99.99%. Moreover, the metallic content of bacteriological media is normally high. A typical example of contamination was revealed by the algal flagellate *Euglena gracilis*. This cell requires only 4800 atoms (5×10^{-19} g) of cobalt for the formation of a new organism. In a preliminary experiment the 'analytically pure' iron salt used for making the culture medium was found to contain enough cobalt to furnish 33 times this amount (Hutner, 1949).

Metal-depleted medium, used in the work depicted in Fig. 11.4 was conveniently made as follows. The required nutrient salts were dissolved in water. This solution was shaken out with a solution of dithizone in chloroform. This operation was repeated many times, and finally the solution was shaken out with chloroform (to remove the dithizone) and finally aerated (to remove the chloroform) (Piper, 1942). Other substances with a high avidity for metals have been used similarly, e.g. 8-hydroxyquinoline for depleting bacteriological media (Rubbo, Albert and Gibson, 1950). Thirty-eight methods of metal depletion have been compared by Donald, Passey and Swaby (1952).

In experimental work, the need often arises to maintain a predetermined concentration of a given metal ion, a concentration that is not allowed to rise above this level nor fall below it. This situation is met by the use of metal ion buffers which maintain a steady pM just as hydrogen ion buffers maintain a steady pH. With their help, free metal ions are replenished (as they are removed by the reaction) from a reservoir of bound metal complex. The first complexing agents to be used for this purpose were citrate and tartrate ions, but much more application has been found for ethylenediaminetetra-acetic acid (EDTA) (cf. *11.27*), diethylenetriaminepenta-acetic acid (DTPA), and nitrilotriacetic acid (NTA). The necessary calculations will be found in Perrin and Dempsey (1974).

(*t*) *Non-metallic trace nutrients* include boron (for plants only), iodine and fluorine (for vertebrates), selenium, and silicon. Selenium forms an essential part of the important mammalian enzyme glutathione peroxidase which, in mammals, protects membranes from oxidation and is one of the microbiocidal enzymes in phagocytes (Stadtman, 1980). A selenium, iron, and molybdenum enzyme was described above under 'molybdenum'. Silicon is thought to contribute to the architecture and resiliance of connective tissues of vertebrates. In rats, silicon is essential for growth and development. It seems to be present as a silanolate, with Si-O-R bridges to such polysaccharides as heparin and hyaluronic acid (Schwarz, 1973).

For further reading on trace elements in nutrition, see Underwood (1977); and for the inorganic chemistry of biological processes, see Williams (1976).

11.1 Biochemical differences that can assist selectivity

Every biological system can be seen as an arena where a constant struggle is in progress for traces of heavy metal cations. Some of the strategies are ingenious. Vertebrate hosts, when invaded by micro-organisms, promptly decrease their level of plasma iron and zinc. Of all biologically essential heavy metals, iron is the most critical for bacterial welfare, as is zinc for yeasts and other fungi. Mammals decrease plasma iron by transferring it to the liver, at the same time decreasing intestinal absorption (Weinberg, 1974). In acute chronic infections, the level of plasma copper rises by 50 to 90% (Weinberg, 1972) and this hypercupraemia is often assumed to be anti-parasitic.

Hosts with tumours behave similarly. They have a decreased plasma level of iron (Konaka and Matsuoka, 1967) whereas that of copper is elevated (Mortazani et al., 1972). Some malignant tumours cause plasma zinc to fall (Davies, Musa and Dormandy, 1968), and this may be the host's doing, for many animal tumours are inhibited by a dietary deficiency of zinc (De Wys and Pories, 1972).

These observations lead to the question, 'What substances do living cells use for binding heavy metal cations?' There is quite a formidable array, of which the amino acids form a useful point of reference. Most of the amino acids which make up protein have almost exactly the same binding constants as glycine (11.1). Exceptionally, two common amino acids have outstandingly high binding affinity, namely histidine (11.2) through its imidazole ring and cysteine (11.3) through the thiol group (Albert, 1952). To these two examples must be added cystine, an intense, specific binding agent for copper (Hawkins and Perrin, 1963). It is currently understood that metals are bound to proteins mainly by histidine and cysteine residues in the protein. For example, the X-ray diffraction diagram of myoglobin shows a strong link between the iron and an imidazole nitrogen atom of histidine.

Computer analysis of 160 potentiometrically determined equilibria of Cu^{2+}, Zn^{2+}, and 17 amino acids (all simultaneously present) showed that 85% of the copper was present as a single complex (1:1:1 histidine/copper/cystine), and 67% of the zinc was bound by cysteine and histidine (Hallman, Perrin and Watt, 1971).

$H_2N \cdot CH_2 \cdot CO_2H$
Glycine
(11.1)

Histidine
(11.2)

$HS \cdot CH_2 \cdot CH(NH_2)CO_2H$
Cysteine
(11.3)

In the bloodstream, amino acids bind very little of the copper and zinc that are present. About 90% of the total copper is bound to the protein, caerulo-plasmin, giving a bright blue complex; it is so tightly bound that no exchange can be demonstrated. The concentration of exchangeable copper and zinc ions in blood plasma are 1.1 μM and 15 μM respectively (Agarwal and Perrin, 1976). Almost all of this exchangeable copper is bound to albumin through the NH_2-Asp-Thr-His- end (in bovine plasma) and by the NH_2-Asp-Ala-His- end (in human plasma), using the NH_2-group, two intervening peptide nitrogen atoms, and the imidazole nitrogen in the histidine residue. The conditional constant (expressed as log K^1) is 12.1 at pH 7.4 for the 1:1 copper–albumin complex, and 7.6 for the 1:1 zinc–albumin complex; however, the zinc is bound differently. Zinc, but not copper, is readily removed from plasma following injection of calcium–EDTA (cf. *11.27*).

About 12% of all known enzymes need a metal co-factor. How does the metal help? Metals seem often to act as a bridge between the substrate and the protein, which can activate the metal by withdrawing electrons from it. The excess positive charge, thus created on the metal, is well placed to withdraw electrons from the substrate. Thus the lowering of the free energy of activation is attributed to electronic deformation (in the substrate) mediated by the metal.

A somewhat similar picture has been obtained by electron-spin resonance studies of two manganese-containing enzymes: creatine kinase and muscle enolase. In the former, manganese was shown to form a bond to the coenzyme (adenosine diphosphate) and to the substrate, but not to the protein; in the latter, manganese seemed to act as a bridge between substrate and protein (Cohn and Leigh, 1962).

In some other cases, the sole role of the metal seems to be to ensure the correct tertiary folding of the protein, similar in effect to the S—S bond. This folding often brings together two or three amino acid residues that are situated far from one another in the extended polypeptide chain, but which can form the active site when assembled in this way. An example is zinc in the alkaline phosphatase of *E. coli*: when deprived of zinc, its circular-dichroism spectrum was radically changed by 6 M urea, as the native conformation gave way to a random coil (Trotman and Greenwood, 1971). For more about this enzyme see Section 11.0, under *Zinc*.

Although many enzymes (e.g. trypsin) function without the aid of a metal, those enzymes which require heavy metals usually hold them very tightly. In many cases, dialysis does not remove the metal; also, powerful chelating agents can enter the cells of micro-organisms without causing harm (see Section 11.7.1 below). This inaccessibility is more likely to be due to tertiary folding than to unusually high stability constants (see above).

Apart from amino acids, peptides and proteins, many other metal-binding substances play essential roles in all living cells. Firstly, there are the pteridines (including folic acid) and purines, whose stability constants have been determined (Albert, 1953; Albert and Serjeant, 1960). Riboflavine is most avid in the partly reduced state (Hemmerich, Veeger and Wood, 1965). Spermine (*11.4*), and the diamines spermidine (*11.5*) and putrescine (*11.6*) also compete.

$H_2N(CH_2)_3 NH(CH_2)_4 NH(CH_2)_3 NH_2$ $H_2N(CH_2)_3 NH(CH_2)_4 NH_2$ $H_2NCH_2 CH_2CH_2CH_2NH_2$

Spermine   Spermidine   Putrescine

(*11.4*)   (*11.5*)   (*11.6*)

All kinds of phosphates bind metals in living cells. Starting with inorganic phosphates, it is useful to note that the stability constants of orthophosphates (Ca, Mg, Zn, Cu^{2+}) have been redetermined recently, and a tendency was noted for 1:1 complexes, like $CaHPO_4$, to dimerize, e.g. to $Ca_2H_2 (PO_4)_2$ (Childs, 1970). For similar pyrophosphate data, see Wolhoff and Overbeck (1959). The stability constants of adenosine triphosphate complexes of Mg^{2+}, Ca^{2+}, Mn^{2+}, Co^{2+}, Ni^{2+}, Zn^{2+}, and Cu^{2+} have been determined (Perrin and Sharma, 1966), also values for AMP, ADP, UTP, GTP, CTP, and ITP (Phillips, 1966). Magnesium ions are tightly bound by the phosphate groups of ATP as shown by potentiometric titration and ^{31}P n.m.r. (Tuck and Baker, 1973). The stability constants of Cu^{2+}, Cd^{2+}, and Fe^{3+} with DNA have been found to be of the same order as those of AMP, which has far less avidity for cations than ATP (Bryan and Frieden, 1967). Some of the heavier metal cations are associated with nucleic acids in Nature, and it has been suggested that zinc ions are co-ordinated to purine bases when reversible winding and unwinding of DNA is taking place (Shin and Eichhorn 1968). Acids of the citric acid cycle also bind cations.

As far as is known, most of the above substances have stability constants similar in magnitude to those of the amino acids (Table 11.2, p. 457). Only the porphyrins exceed these figures: they hold iron so firmly that no exchange with radioactive iron can be detected (Hahn *et al.*, 1940).

It will be realized, from the above, that experiments involving trace-metals have no meaning if performed in the presence of phosphate or citrate buffers. However, one can use N-ethylmorpholine (b.p. 138–9°C) as a useful, non-chelating buffer for the pH range 7.0–8.2 (for other buffers, see Perrin and Dempsey, 1974).

Chelating agents characteristic of particular classes of organism. Studies in comparative biochemistry are constantly bringing to light new species differences in the binding and use of metals, some of which have been mentioned in Sections 4.2 and 11.0.

The bacterial cell, living aerobically, is faced with the problem of fulfilling its biological need for iron from an environment where most of the iron is in the form of highly insoluble ferric oxide (the solubility product is only 10^{-39} M). This plight led to the evolution of sequestering agents to dissolve and transport the iron, and make it biologically available. These siderochromes, as they are called, are confined to bacteria, yeasts and other fungi. The two commonest types are the phenolates, such as enterochelin (enterobactin) (*11.7*), and the hydroxamates, like deferoxamine B (*11.9*). Both types are weak acids with a pK_a about 9, and bind ferric iron very strongly (log K 45).

The most important member of the phenolate class of siderochromes, enterochelin, is a cyclic triester of 2,3-dihydroxyl-N-benzoyl-L-serine. It was isolated from *E. coli* by O'Brien and Gibson (1970), and its mode of action established by Gibson (in Australia) and Nielands (in the United States, 1974) using a series of mutants, each one of which reliably failed to carry out one of the biochemical steps. X-ray diffraction analysis showed that each ferric cation was bound to the six oxygen atoms supplied by the three catechol dianion moieties

(*11.8*), giving the purple complex (Anderson *et al.*, 1976). The bacterium excretes enterochelin which then forms the ferric complex from environmental iron; the complex is then absorbed by the bacterium, and hydrolysed with a specific enzyme to obtain the iron (O'Brien, Cox and Gibson, 1971).

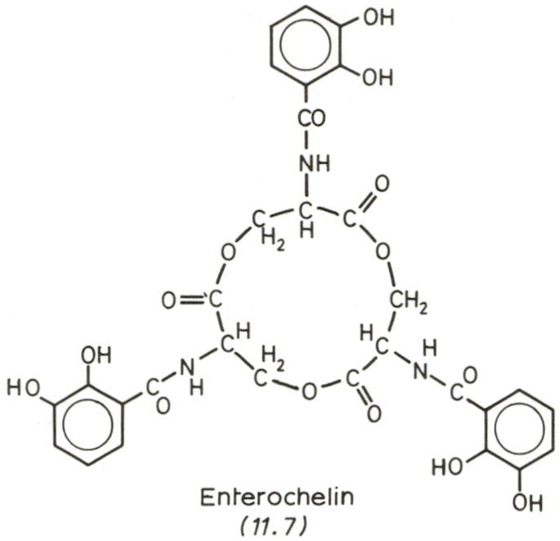

Enterochelin
(*11.7*)

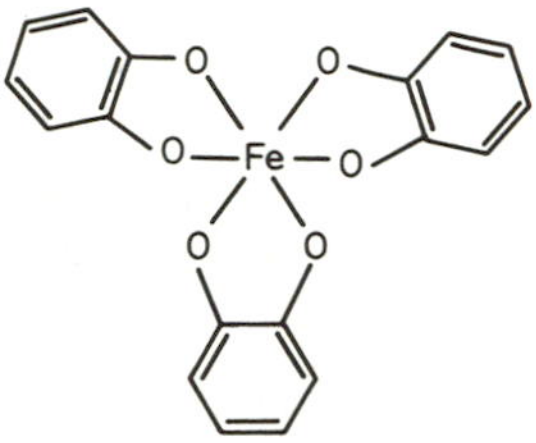

Detail from (*11.7*): enterochelin, after binding ferric ion
(*11.8*)

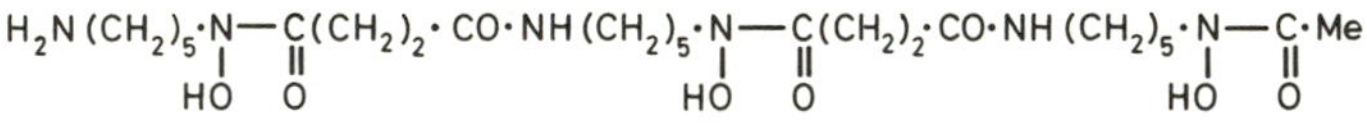

$$H_2N(CH_2)_5 \cdot N - \underset{\underset{O}{\parallel}}{\underset{HO}{|}}{C}(CH_2)_2 \cdot CO \cdot NH(CH_2)_5 \cdot N - \underset{\underset{O}{\parallel}}{\underset{HO}{|}}{C}(CH_2)_2 \cdot CO \cdot NH(CH_2)_5 \cdot N - \underset{\underset{O}{\parallel}}{\underset{HO}{|}}{C} \cdot Me$$

Deferoxamine
(*11.9*)

The deferoxamines, e.g. (*11.9*), are hexapeptides obtained from a *Streptomyces*. X-ray diffraction analysis shows a ferric ion held octahedrally by the six oxygen atoms of three hydroxamic acid groups. The log K, which is 31, greatly exceeds the 24 of EDTA (*11.27*), but is less than that of enterochelin (Schwarzenbach and Schwarzenbach, 1963).

Mycobactin T (*11.10*) is the siderochrome of the bacterium that causes human tuberculosis. Eight other mycobactins have been obtained from other species of *Mycobacteria*, a genus to which they are confined. They have only two

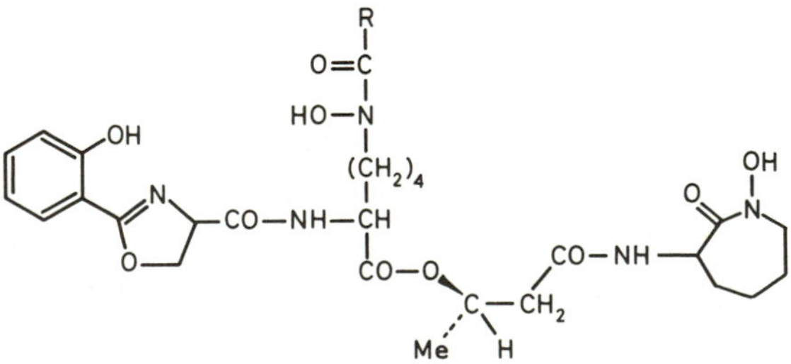

R = alkyl chain (C$_{19}$)
Mycobactin T
(*11.10*)

hydroxamate-groups, but a phenolic-group is present to aid the three-centre binding of ferric iron as strongly as in other siderochromes (Snow, 1970). These are the only natural iron scavengers that have good liposolubility.

Interesting chemotherapeutic possibilities are suggested by this knowledge that the iron-foraging of bacteria is very different from that of mammals. For information on the use of siderochromes as antidotes in iron poisoning, see Section 11.6.

Many fungi, such as *Aspergillus aerogenes*, excrete enterochelin whereas others (including *Aspergillus niger* and *Penicillium reticulosum*) angle for their iron with ferrichrome, a cyclic hexapeptide in which the iron is firmly held between three hydroxamate groups. Fungi contain many strongly chelating pyrones, such as kojic acid (*11.11*) which has the cumulative log stability constant (β_3) of 25 for ferric iron but their biological role is unknown, and this is also the case with the many polyhydroxyanthraquinones that fungi contain. All yeasts release the hydroxamic derivative rhodotorulic acid (*11.12*) to bring iron into the cell.

For a review of microbial iron metabolism, see Nielands (1974).

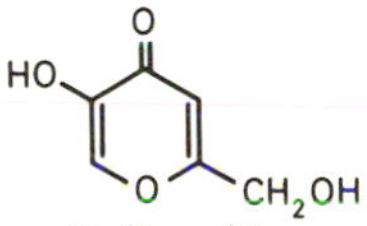

Kojic acid
(*11.11*)

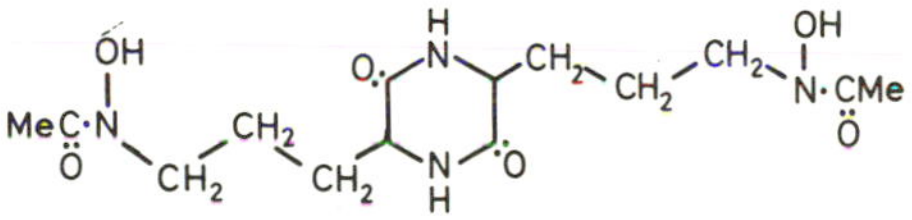

Rhodotorulic acid
(*11.12*)

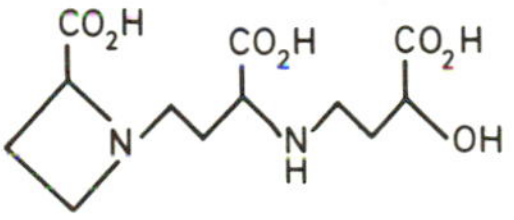

Mugineic acid
(*11.13*)

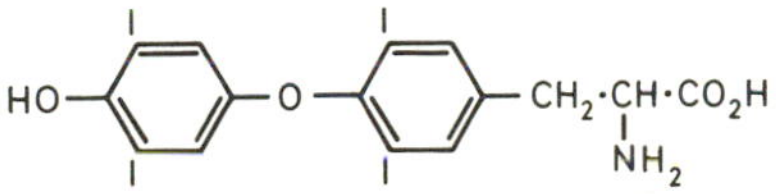

Thyroxine
(*11.14*)

Plants, too, have special iron-gathering chemicals. Many cereal grasses excrete mugineic acid (*11.13*) into the soil to gather ferric iron, the complex being absorbed and utilized. Mugineic acid is a derivative of the nitrogen ring-system, azetidine. Without a sufficient supply of iron, plants become chlorotic and go into decline (Mino *et al.*, 1981).

Several mammalian natural agonists are metal-binding, notably thyroxine (*11.14*), norepinephrine (*7.5*), and histamine (*7.6*). Only *mammalian* mitochondria accumulate calcium.

11.2 The chemistry of chelation

Biochemists who work with enzymes, many of which are specific for a particular metal, must regret that selectivity is quite rare among the Man-made reagents for metals. In fact, most chelating agents show approximately the same order of preference for metals, which goes as in Table 11.1.

Table 11.1 The ranking order of preference of most chelating agents

Fe^{3+}, Hg^{2+}	greatest avidity
Cu^{2+}, Al^{3+}	
Ni^{2+}, Pb^{2+}	
Co^{2+}, Zn^{2+}	
Fe^{2+}, Cd^{2+}	
Mn^{2+}	
Mg^{2+}	
Ca^{2+}	
Li^+	
Na^+	
K^+	least avidity

(Mellor and Maley, 1947.)

The first recognition of a general trend of this kind was made by Mellor and Maley (1947). Several of the bivalent metals, in the above scheme, follow one another sequentially in the periodic table, thus (atomic numbers in parentheses): (Mn (25), Fe (26), Co (27), Ni (28), Cu (29), Zn (30)). This is known as the first transition series and, in it, avidity steadily rises to a peak for Cu^{2+}. The increase in avidity follows the decline in ionic radius as the series is ascended (Irving and Williams, 1953). This relationship provides a rough clue to the above scheme. Because increasing the valence of a metal tends to diminish its radius, it is not surprising to find Fe^{3+} more avid than Fe^{2+}, or the tervalent ions at the top of the scheme and the univalent ions at the bottom (see Table 11.4 for ionic radii). However, many secondary factors interfere with the operation of so simple a rule. The important conclusion, though, is that most ligands cannot depart very far from the order of avidities shown in the above scheme.

Fortunately there are some exceptions and special attention will be given to those chelating agents, few in number, which depart from the above order. Before discussing such irregular examples, the principles of normal chelating behaviour will be reviewed. The chemical significance of heavy metals in biology arises from their ability to form bonds that are tighter than ordinary ionic bonds through being partly co-ordinate (Section 8.0).

A few ligands (known as *multi*dentate ligands, EDTA is an example) form more than one chelate ring with a metal ion when making a 1:1 complex. However, to simplify discussion, it is more convenient to deal first with the three principal kinds of *bi*dentate ligands, i.e. those forming a single ring in the 1:1 complex. These may have two electron-releasing groups (as in ethylenediamine and 2,2'-bipyridyl), in which case the charge on the metal cation is unaffected by chelation; or there can be one electron-releasing and one anionic group as in glycine, in which case the charge on the metal is diminished by one unit; or there can be two anionic groups as in oxalic acid, in which case the charge on the metal is decreased by two units. In general, chelation through oxygen or nitrogen takes place only when five- or six-membered rings can be formed, and of these, five-membered rings are much more stable. [However, chelation through sulfur enables stable four-membered rings to be formed (Peyronel, 1940; Deskin, 1958).] The three main types of chelation are illustrated in Fig. 11.5. The arrows in the ring imply that a normally unshared electron-pair is released from O, N, or S to the metal.

In the presence of excess ligand, 2:1 complexes can be formed. Ligands of the oxalic acid type use up their charge in forming 1:1 complexes, but these can unite further with the ethylenediamine type, forming a mixed complex (e.g. Watters and De Witt, 1960). A 1:1 complex of the glycine type can combine with another glycine-type ligand. An ethylenediamine type 1:1 complex can combine with any one of the three types of ligands.

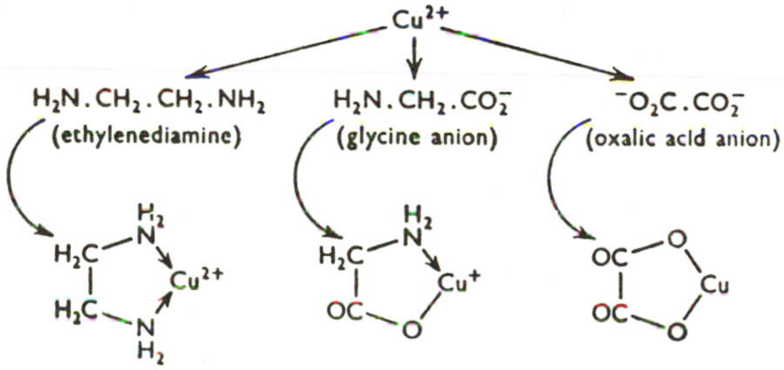

Fig. 11.5 Three main types of 1:1 complexes.

In general, bivalent copper is quadricovalent and is usually saturated when it has combined with two ligand molecules (whether the same or different). The same is usually true for bivalent calcium, magnesium, and manganese. But bivalent iron, cobalt, and zinc are sexacovalent towards ligands of the ethylenediamine type, and tervalent ions are sexacovalent towards the glycine types as well. The ligands are gathered around the metal in configurations which depend on the direction of the valence lines proceeding from the metal cation. Most of the above metals have usually been found to give tetrahedral complexes with the main types of ligand of Fig. 11.5, but Cu^{2+} prefers planar complexes, and Fe^{2+} and Fe^{3+} prefer octahedral complexes.

It should be carefully noted that the term 'ligand' does not apply to all of the organic material present, but only to that part of it which is in the *appropriate ionic form* for combining with the metal cation. For ethylenediamine, glycine, and oxalic acid, the ligands are the molecule, the monoanion, and the dianion, respectively. Hence, when stability constants are used to compare the relative avidity of ligands under physiological conditions, the pK_a values of the ligands must also be considered (see Section 11.3).

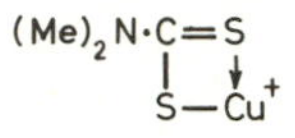

$(Me)_2 N \cdot C = S$
$S - Cu^+$

Copper dimethyldithiocarbamate complex
(11.15)

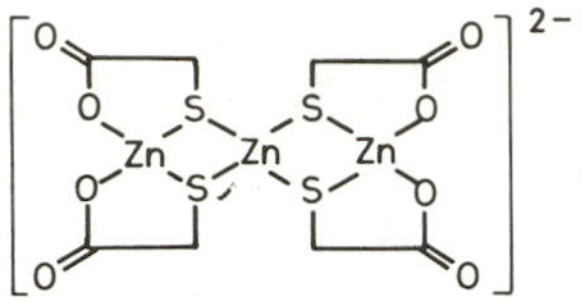

Zinc mercaptoacetic acid complex
(11.16)

Although sulfur can take part in simple four-membered rings such as in cupric dimethyldithiocarbamate (11.15), which is formed from cupric ions and dimethyldithiocarbamic acid (3.57), sulfur-containing ligands form an equilibrium mixture of many complexes. Thus zinc and mercaptoacetic acid (L^{2-}) form the series:

$$ZnL, ZnL_2^{2-}, ZnL_3^{4-}, Zn_2L_3^{2-}, Zn_3L_4^{2-}$$

The last of these $(Zn_3L_4^{2-})$ is one of the most stable and is shown as (11.16). Information about the above series was derived from computer analysis of titration data, a procedure which gave the following series for cysteine (L^{2-}) (Perrin and Sayce, 1968):

$$ZnHL_2^-, ZnH_2L_2, ZnL_2^{2-}, Zn_3L_4^{2-}$$

In more acidic solutions, protonated forms of the zinc–cysteine complex are present also (Shindo and Brown, 1965). It is evident that sulfur-containing ligands give a more complicated pattern of complexes than the simple patterns produced by oxygen- and nitrogen-containing ligands, which are more common and to which the next section will be devoted.

For detailed discussion of conformation and absolute configuration of chelated complexes, see Hawkins (1971). For books on principles and applications of metal chelation, see Bell (1977) and Perrin (1964); for a periodical series of books, see Sigel (1973 <).

11.3 Quantitative treatment of metal binding

In order to quantify the various degrees of tightness in binding (which vary between wide limits), stability constants are used. These are the constants governing the mass-action equilibrium between the ligand(s) and one ion of the metal. Thus for the 1:1 complex of glycine and cupric ion, the equilibrium is:

$$K_1 = \frac{[H_2N \cdot CH_2 \cdot CO_2 Cu^+]}{[Cu^{2+}][H_2N \cdot CH_2 \cdot CO_2^-]}$$

and for the 2:1 complex,

$$K_2 = \frac{[H_2N \cdot CH_2 \cdot CO_2CuO_2C \cdot CH_2 \cdot NH_2]}{[H_2N \cdot CH_2 \cdot CO_2Cu^+][H_2N \cdot CH_2 \cdot CO_2{}^-]}$$

where K_1 is the stability constant for the 1:1 complex, and K_2 is that for the 2:1 complex. In each case the product is in the top line, and the substances from which it is formed are in the lower line.

For many purposes the *cumulative stability constant* (β) is required; this is the product of the individual constants. If there are two individual constants, the product is designated β_2; if there are three constants, β_3. Stability constants are usually determined by potentiometric (glass electrode) titration of the ligand, in the presence and absence of the metal, and the results are processed by a set of rather complex calculations. For full details see Albert and Serjeant (1984).

In brief, the acidic group of the suspected chelating agent is titrated with alkali, and the pH is recorded after each tenth of an equivalent is added. Then a 1:1 mixture of the substance and a salt (e.g. copper perchlorate or nitrate) is titrated. If there is no complex formed, the new curve follows, in turn, the individual curves for the two components (see Fig. 11.6). But when chelation occurs, the hydrogen cations liberated by the chelation of the cupric cations

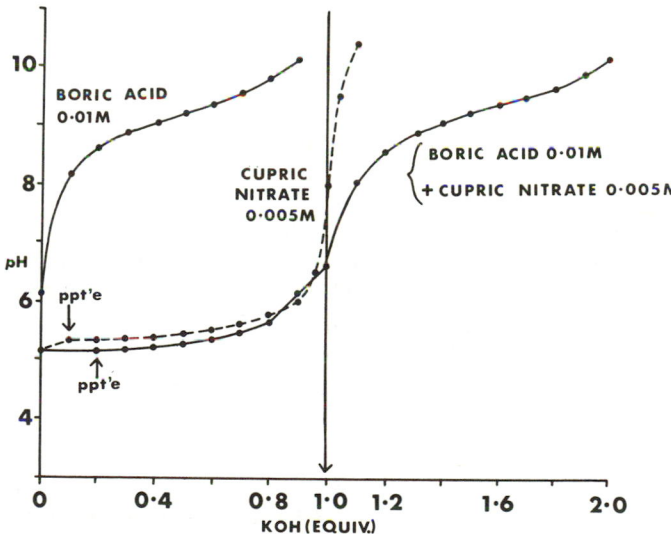

Fig. 11.6 Example of the use of potentiometry to see if a substance can chelate. The above curves show plainly that boric acid does not chelate. Contrast this with Fig. 11.7.

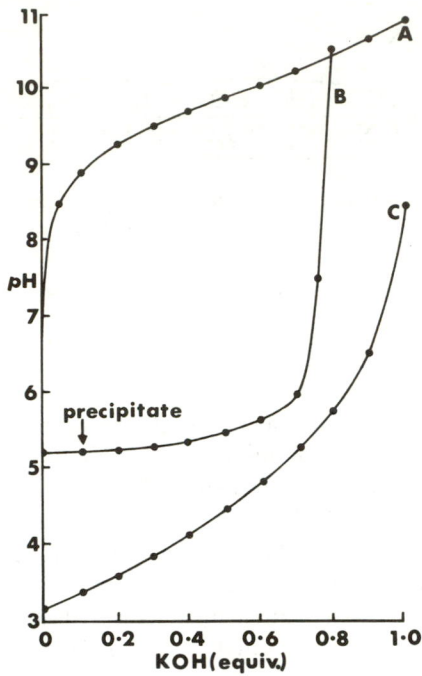

Fig. 11.7 The above curves show that glycine is a strong chelating agent. A, glycine (0.01 M); B, cupric nitrate (0.005 M); C, glycine (0.01 M) plus cupric nitrate (0.005 M).

displace the whole curve to lower pH values (see Fig. 11.7). Non-acidic substances, in the form of their salts with acids, may be similarly titrated with alkali.

Potentiometry can be used even for dilute solutions of sparingly soluble substances if unusual care is taken; e.g. 0.1 mM adenine, titrated in the presence of 8 μM cupric ions (in water), readily gave a precise constant (Albert and Serjeant, 1960). Solvents other than water should never be used if results of biological significance are required. Mixtures of water and an organic solvent give particularly misleading results (Albert and Serjeant, 1984).

The glass electrode, in potentiometric titration, can sometimes be advantageously replaced by a copper electrode when this is the metal of principal interest. For excessively insoluble complexes, potentiometry is replaced by exchange methods. Thus two ligands may be allowed to compete for one metal, or two metals for one ligand; the result is conveniently measured if one of the two components is isotopic (Schubert, 1956). Partition methods are useful when the constants are so high that the free [H$^+$] is large in comparison with total [H$^+$] (free and bound): in these circumstances, potentiometry is inaccurate. The method, in brief, is to shake a solution of the ligand (of known concentration in a water-immiscible solvent) with an aqueous solution of the metal cation (of known concentration in an aqueous buffer). The concentration of the ligand is then measured spectroscopically in the non-aqueous solution (McBryde, 1967). Of its own, ultraviolet spectrometry is not very useful. Other methods used in special circumstances are: potentiometry with cation-selective electrodes, infrared spectroscopy in deuterium oxide, magnetic resonance spectroscopy (both e.s.r and n.m.r.), optical rotatory dispersion, conductimetry, thermal relaxation (temperature jump), and polarography. For general purposes, straight potentiometry is superior to any of these.

It was not possible to calculate stability constants from titration data until Bjerrum (1941) showed that they are related to two variables ($\bar{n}$) and [L] by the following equation:

$$K_1 = \frac{\bar{n}}{(1-\bar{n})[L]} \qquad \text{(i)}$$

In this equation, [L] is the concentration of ligand, i.e. of the *chelating* species of the complex-forming agent. It is not calculable from [A] which is the known concentration of complex-forming agent at the start of the titration. Thus, if one were working with glycine, [A] refers to the concentration of glycine in all its three ionic forms plus its neutral form, in short, to the total material as taken from the bottle. But [L] refers only to the *anion* of glycine, because this is glycine's only chelating species. The definition of $\bar{n}$ (pronounced 'enbar') is the *fraction* of the ligand bound by one atom of metal. For a 1:1 complex, it ranges from 0 to 1.

The values of [L] and $\bar{n}$ change progressively during the titration, and are calculated as follows:

$$[L] = ([A]-[H^+]+[OH^-]+[HCl]-[KOH])/P$$
$$\bar{n} = ([A]-Q[L])/[M_0]$$

where [M_0] is the initial concentration of metal, [L] and [A] are defined above; [H^+] and [OH^-] are alternatives, depending on the pH and, similarly [HCl] and [KOH] are alternatives, depending on the titrant used. It remains to define P and Q:

$$\left. \begin{array}{l} P = [H^+]/K_a+2[H^+]^2/K_aK_a' \\ Q = [H^+]/K_a+[H^+]^2/K_aK_a'+1 \end{array} \right\} \begin{array}{l} \text{where } K_a \text{ is the first ionization} \\ \text{constant of the agent.} \end{array}$$

A simple view of what is happening during a titration is that the addition of alkali removes more and more hydrogen cations from the chelating agent, and these are replaced by metal cations. But, as the above equations indicate, the relationship is not a linear one.

The stability constant does not vary with pH, and the use of Equation (i) will give the same value for K_1, anywhere between $\bar{n} = 0$ and $\bar{n} = 1$, above which the following equation is substituted:

$$K_2 = \frac{\bar{n}-1}{(2-\bar{n})[L]}$$

Because β is the symbol for the product of individual stability constants, we write $\beta = K_1K_2$ for the combination of copper with glycine and for all other cases where a metal becomes saturated after combining with two portions of a ligand. For zinc and ethylenediamine, where the zinc becomes saturated after fixing three portions of the ligand, we write $\beta = K_1K_2K_3$, and, for the complex of a metal that is saturated when it has combined with one unit of ligand (for instance, zinc and EDTA), $\beta = K_1$.

For the large number of complexes in which the metal is saturated by two portions of ligand, the titration is followed by plotting values of $\bar{n}/(\bar{n}-1)$ [L] against the corresponding values of $(2-\bar{n})[L]/(\bar{n}-1)$, to yield a straight line of which the slope is β and the intercept is $-K_1$. This can even more conveniently be carried out by using a hand-held calculator equipped with a least-squares facility (Albert and Serjeant, 1984).

Because some combinations of metals and ligands take courses different from the regular type portrayed so far, it is important that each titration should be performed twice, so that two different ratios of metal to complexing agent are examined. If the same values of log K_1 and β are obtained, one can be sure that any competing equilibria are negligible.

Typical variations from the norm are (a) chelation by the zwitterionic species of an amino acid to give a positively charged saturated complex ('protonated complex'), and (b) complexes in which one affinity of the metal is satisfied by the hydroxyl ion ('hydrated species') (Perrin and Sharma, 1967). For such unusual cases, recourse should be made to the more detailed calculations afforded by a computer program (Leggett, 1983).

Many titrations performed on *one* metal and *two* different ligands, present together in the solution, can be dealt with quite simply (Perrin, Sayce and Sharma, 1967; Perrin and Sharma, 1968, 1969). When two ligands, A and B, are equally avid for a metal M, the mixed complex AMB is statistically favoured by a factor of 2 over simple complexes such as AMA. Similarly when three such ligands (A, B, and C) are present, the complex MABC is favoured by a factor of 6 (Watters and DeWitt, 1960).

The constants of a number of substances of interest to biological workers are given in Table 11.2. For other constants see the compilations edited by Sillén and Martell (1964, 1971) and Perrin (1979).

There is no certain way of predicting the stability constants of a substance in advance of synthesizing it, and making the measurements. The most useful guides in the prediction of approximate stability constants are as follows. In any one series, the more tightly a ligand combines with hydrogen ions (as measured by the pK values) the more tightly will it bind metals. [A pK value, if not already recorded in the literature, can be approximately calculated (Perrin, Dempsey and Serjeant, 1981).] However, this guide, which it must be emphasized holds only for a series of *related* substances, breaks down if bulky groups cause steric hindrance (see below).

When we have to compare members from *different* series, the most common

Table 11.2 Stability constants (logarithmic) of some complexing agents (in water at 20°C) (saturated constants reported as log β). The ratio is given as a pre-index: thus manganese forms a 2:1 complex with bipyridyl (log β = 6), reported as $^{2}6$; whereas it forms only a 1:1 complex with EDTA (log β = 13), reported as 13

Ligand	pK_a	Fe^{3+}	Cu^{2+}	Ni^{2+}	Zn^{2+}	Co^{2+}	Fe^{2+}	Mn^{2+}	Mg^{2+}
Glycine (11.1)	9.9, 2.4	10	8.5; $^{2}15$	6; $^{2}11$	5; $^{2}9$	5; $^{2}9$	4; $^{2}8$	3; $^{2}5.5$	2; $^{2}4$
Cysteine (11.3)	10.3, 8.4, 2.0	*	*	10; $^{2}19$	10; $^{2}18$	$^{3}16$	6	4	< 4
Histidine (11.2)	9.2, 6.0, 1.8	?	10.5; $^{2}19$	9; $^{2}16$	7; $^{2}12$	7; $^{2}13$	5; $^{2}9$	3.5	< 4
Histamine (7.6)	9.8, 6.0	4	10; $^{2}16$	7; $^{2}11$	5; $^{2}9$	5; $^{2}9$	4	3	?
Ethylenediamine (Fig. 11.5)	10.1, 7.0	?	11; $^{2}20$**	8; $^{2}18$	6; $^{2}12$	6; $^{3}14$	4; $^{3}9.5$	3; $^{2}5$	0.4
Ethylenediaminetetracetic acid (EDTA) (11.27)	11.0, 6.3, 2.7, 2.0	24	19	18	16	16	14	13	9
Pteroylglutamic acid (folic acid) (cf. 9.20)	8.3	?	$^{2}8$	$^{2}9$	$^{2}7.5$	$^{2}8$	$^{2}8$	$^{2}6$	?
Hypoxanthine (9.52)	8.9, 2.0	?	6	5	7	7	4	?	?
Guanosine	9.3	?	6	4	4.5	3	4	3	3
Adenine (4.3)	9.8, 4.3	?	$^{2}14$	4	$^{2}13$	4; $^{2}8$	?	3	3
8-Hydroxyquinoline (oxine) (11.30)	11.2, 3.8	12; $^{2}24$; $^{3}36$	12; $^{2}23$	10; $^{2}18$	8.5; $^{2}16$	9; $^{2}17$	8; $^{2}15$	7; $^{2}12$	4.5
o-Phenanthroline (11.18)	4.9	$^{3}14$	$^{2}20$	$^{3}24$	$^{3}17$	$^{3}20$	$^{3}21$	$^{3}10$	1
Bipyridyl (11.19)	4.3	?	$^{3}17$	$^{3}20$	$^{3}13$	$^{3}16$	$^{3}17.5$	$^{2}6$	0.5
Oxalic acid	4.2, 1.2	10	6	5.5	5	4.5	4.5	4	3***
Salicylic acid	13.4, 3.0	16; $^{2}28$	11; $^{2}19$	7; $^{2}12$	7	7; $^{2}11$	6; $^{2}11$	6; $^{2}10$	?
Tetracycline (11.36)	9.6, 7.8, 3.4	10; $^{3}25$	8; $^{2}13$	6; $^{2}11$	5; $^{2}9$	5; $^{2}10$	5; $^{2}9$	4; $^{2}8$	4
Isonicotinic hydrazide (isoniazid) (11.37)	10.8, 3.5, 1.9	?	8	5.5	5	5	?	?	?
Dimethyldithiocarbamic acid (3.57)	3.2	?	11; $^{2}22$	?	$^{2}9$	?	?	?	?

*Cysteine is oxidized by this cation.   **cf. $^{2}11$ for Cu^+.   ***Ca^{2+} is also ~ 3.

Sources: Tetracycline (Albert and Rees, 1956), isoniazid (Albert, 1956), salicylic acid (Perrin, 1958), bipyridyl and o-phenanthroline (Irving and Mellor, 1962), dimethyldithiocarbamic acid (Janssen, 1958), amino acids (Albert, 1950, 1952), other values (Sillén and Martell, 1964 and supplements).

disturbing factor, in the pK_a : log K_1 relationship, is resonance in a chelated ring. Here the most useful clue is to examine the double-bond composition of the ring : factors which lower double-bond character usually also decrease stability (Calvin and Wilson, 1945). Thus in the copper chelate of acetylacetone, the chelate ring contains two full double bonds and the metal is more tightly bound than in the salicylaldehyde chelate where, on account of the resonance of the benzene ring, one encounters bonds with only 50% double-bond character.

Help in the prediction of stability constants is expected to come from developments of ligand field theory. The core of this treatment is crystal field theory which postulates that the five d orbitals of heavy metals, although normally equal in energy, become differentiated when lying in the electrostatic field of the ligand. In particular, those d orbitals lying *in* the direction of the ligands are raised in energy, and those lying *away* from the ligands are lowered in energy. The donor electrons on the ligands repel the d electrons of the metal: this repulsion is reduced by the movement of d electrons into those d orbitals which are further from the ligands. For aromatic ligands and certain metals (notably iron, nickel, and cobalt) allowance must be made for π-bonding. With this modification, crystal field theory has been combined with molecular orbital theory to give what is called ligand field theory. For further readings, see Basolo and Pearson (1967).

Ranking order at a fixed pH. When members of a series of chelating agents have much the same ionization constants, the relative stability constants place the members in order of their ability to segregate the metal cations at any given pH value. However, when chelating agents of very different ionization constants have to be compared, stability constants do not give a good estimate of the relative competing value for cations. The reason is that, at the pH at which information is desired, these agents are likely to be ionized to different extents. The one which has the less affinity for metals (as shown by the lower stability constant) may, through a difference in pK_a, produce enormously more anion than the other agent. In this case, the substance with the less affinity for the metal may well be the one that gets most of it. This is so because chelation requires not only an affinity between ligand and metal, but also a ready supply

Table 11.3 Distribution of cations among some ligands in neutral solution. Figures in italics: proportional parts. Figures in brackets: log 1:1 stability constants (column to be read vertically)

Ligand	Cu^2	Fe^{2+}	Fe^{3+}
Glycine	*45* (8.5)	*30* (4.3)	*1* (10.0)
Salicylic acid	*1* (10.6)	*1* (6.6)	*400* (16.4)
Oxine	*200 000* (12.2)	*150 000* (8.0)	*200* (12.3)

(Perrin, 1958.)

of ligand anions from the agents (or of ligand molecules, if the agent is a base). This type of competition between stability constant and ionizing constant is illustrated in Table 11.3. In general, the chelating species is a dianion, if the agent is a diprotic acid such as oxalic acid; or it is a monoanion, if the agent is an ampholyte like glycine; or it is an uncharged molecule, if the agent is a diacidic base like ethylenediamine.

To calculate the percentage of a metal that is bound at a given pH, it is satisfactory to use the following calculation:

$$\text{Per cent metal bound as 1:1 complex} = \frac{100K_1WA}{1+K_1WA+K_1K_2W^2A^2}$$

$$\text{Per cent metal bound as 2:1 complex} = \frac{100K_1K_2W^2A^2}{1+K_1WA+K_1K_2W^2A^2}$$

Here A is the concentration of *un*complexed ligand. When the ligand is in great excess, the concentration of total ligand may be used in place of A; if otherwise, A may be calculated by successive approximations;

W is the factor: $1/(1+10^{(pK_{a1}-pH)}+10^{(pK_{a1}+pK_{a2}-2pH)})$.

For numerical values of stability constants, see Sillén and Martell (1964, 1971), Perrin (1979), Martell and Smith (1974–1982).

11.4 Chemical differences that can assist selectivity

Now that the regular features of affinity for metals have been reviewed, some interesting *irregular* features can be discussed. Selectivity for particular ions is dependent on such irregularities.

First, let us discuss the shapes of the chelated complexes. In most cases they are tetrahedral, but iron (both ferrous and ferric) forms octahedral complexes. However, the most remarkable contribution to selectivity is made by cupric ions which form planar complexes. This means that copper is uniquely sensitive to a type of steric effect imposed by bulky groups. Thus, in Table 11.2, the stability constants for the cupric complexes of bipyridyl (*11.19*), phenanthroline (*11.18*), and folic acid lie below those of the corresponding nickel complexes, which is contrary to the natural order discussed in Section 11.2.

Another type of steric effect arises when a bulky group is inserted near to the chelating groups. This prevents two ligands from approaching close enough to form a strong 2:1 complex, and, in the case of metallic cations of small diameter, the cation cannot be gripped. Thus, 2-methyloxine (*11.17*) binds all cations less strongly than oxine does, but the difference is greatest for aluminium (Irving and Rossotti, 1956), which has the smallest crystal radius (0.50 Å) of any common metal (see Table 11.4). Again, two molecules of 2,9-dimethyl-*o*-phenanthroline (*11.18*, R = CH_3) hinder one another too much to retain the ferrous ion. In this case the two ligand molecules push one another into planes where the nitrogen atoms can no longer form the octahedral complex required by Fe^{2+}; even the planar complex preferred by Cu^{2+} is formed only with difficulty. Yet the new planes are highly suitable for binding Cu^+ (which prefers tetrahedral complexes); the high stability of the cuprous complex forms the basis for the use of this substance in analysis.

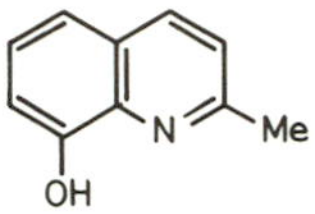

2-Methyloxine (11.17)	o-Phenanthroline (R = H) (11.18)	2,2'-Bipyridyl (11.19)	Two succinate dianions (11.20)

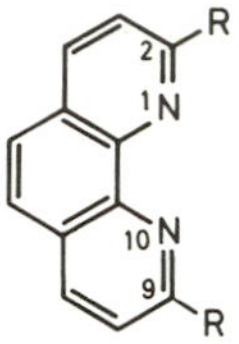

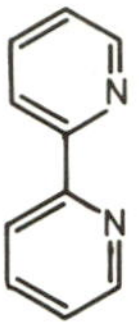

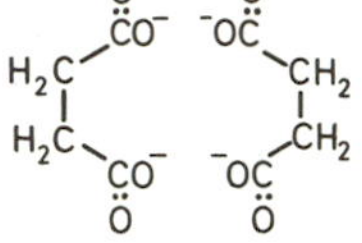

Let us return to those ligands that are too bulky to approach one another closely and hence cannot form a strongly held 2:1 complex with a cation of small diameter. The radius of the magnesium ions is 0.34 Å smaller than that of calcium (Table 11.4). Normally Mg^{2+} is more tightly held than Ca^{2+}, but this order is reversed by ligands that are too bulky for a pair of them to grip the metal. All dicarboxylic acids show this effect, e.g. succinic acid in (11.20) (note oxalic acid in Fig. 11.8). Tartaric acid is outstanding in its preference for Ca^{2+} over Mg^{2+} (Williams, 1952). Some relevant radii are given in Table 11.4. It should be noted that Fe^{2+}, Co^{2+}, Ni^{2+}, and Zn^{2+} have such similar diameters that they could not be separated through this steric effect.

The dimensions of the radii of anhydrous cations are constantly under revision (Ladd, 1968). All methods depend on inferences and assumptions. The data in Table 11.4 remain the most comprehensive and widely used. It will be

Table 11.4 Crystal radii of cations (Å)

Li^+	0.60	Be^{2+}	0.31	Al^{3+}	0.50
Na^+	0.95	Mg^{2+}	0.65	Fe^{3+}	0.64
K^+	1.33	Ca^{2+}	0.99	Ga^{3+}	0.62
Rb^+	1.49	Sr^{2+}	1.13	Tl^{3+}	0.95
Cs^+	1.69	Ba^{2+}	1.35	Co^{3+}	0.63
Cu^+	0.96	Mn^{2+}	0.80		
Tl^+	1.40	Fe^{2+}	0.76		
Ag^+	1.26	Co^{2+}	0.74		
NH_4^+	1.48	Ni^{2+}	0.72		
		Cu^{2+}	0.72		
		Zn^{2+}	0.74		
		Cd^{2+}	0.97		
		Hg^{2+}	1.10	U^{4+}	0.97
		Pb^{2+}	1.20	Pb^{4+}	0.84

(Pauling, 1960.)

appreciated that anhydrous radii are the relevant ones in chelation chemistry (hydrated radii are in Table 12.2).*

A selective enhancement of the bonding of iron is found in heteroaromatic ligands. This is caused by *back double-bonding*, i.e. a π-bonding in which unused electrons in a suitably placed $3d$ orbital are contributed by the iron to empty molecular orbitals in the ligand. The result is that electrons flow in the direction shown in (*11.22*), instead of the normal direction, as in (*11.21*). In consequence, the ligand–metal bonds are reinforced and the complex gains extra stability. Ferrous iron usually gives complexes which are paramagnetic and almost colourless; but with heteroaromatic ligands, containing doubly bound nitrogen, Fe^{2+} forms deeply coloured complexes (usually red) with loss of paramagnetism. The effect is seen (in Table 11.2) for bipyridyl, folic acid, and *o*-phenanthroline, where the constants for ferrous iron greatly exceed those for zinc. Nickel and cobalt are also capable of back double-bonding, but to a smaller extent.

$$R \cdot NH_2$$
$$\downarrow$$
$$Fe^{2+}$$

Normal electron flow
(*11.21*)

Back double-bonding
(*11.22*)

Examination of the log K values for *o*-phenanthroline and 2,2'-bipyridyl (in Table 11.2) reveals the action of *two* principles which have been described above. The preference of Cu^{2+} for a planar structure does not readily admit a third molecule of ligand. Hence Cu^{2+} takes an unusually low place in the table, below Ni^{2+}. Remarkably high β values are seen for Fe^{2+}. These arise from the outstandingly high third† partial constant (K_3), where a spin-paired complex admits of back double-bonding. The net result is that Fe^{2+} takes an unusually high place above Zn^{2+} and Co^{2+}. The extraordinary result of the combination of these two principles is that Fe^{2+} is bound as strongly as Cu^{2+} (Irving and Mellor, 1962).

Another kind of differentiation is illustrated in the plot of stability constants against atomic number in the series: calcium, magnesium, manganese, iron, cobalt, nickel, copper, and zinc (Fig. 11.8). This curve rises most steeply for ligands that use two nitrogen atoms for chelation (e.g. ethylenediamine), less steeply for those that chelate with one nitrogen and one oxygen atom (e.g. glycine), and least steeply for those which use two oxygen atoms (e.g. oxalic acid) (Irving and Williams, 1953). The curves for substances of these different classes cross one another between Mn^{2+} and Fe^{2+} in Fig. 11.8. Not only Mg^{2+} and Ca^{2+}, but also Fe^{3+}, Tl^{3+}, Mo^{5+}, and V^{5+} show a preference for combining with oxygen rather than nitrogen.

*The reactions which have been discussed up to this point are *equilibrium* steric effects. For *kinetic* steric effects, see Wilkins (1962).
†Binding between two nitrogen atoms (as in ethylenediamine) tends to produce 3:1 complexes.

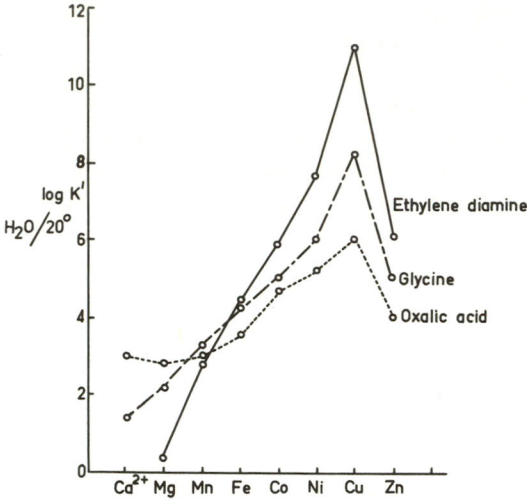

Fig. 11.8 The relative affinities of bivalent cations for oxygen and nitrogen types of chelating agent. The logarithms of the first stability constants are plotted against rising atomic numbers.

Given the choice of combining with an oxygen ligand or a sulfur ligand, most metals prefer the former. However, Cu^+, Ag^+, Au^+, Hg^{2+}, As^{3+}, and Sb^{3+} prefer sulfur to oxygen, thus explaining the successful use of dimercaprol (*11.23*) as an antidote for poisoning by several of these cations. Three other cations, Cu^{2+}, Ni^{2+}, and Co^{2+}, slightly prefer sulfur to oxygen if the sulfur is non-ionic, as in an organic sulfide.

Another factor, which can cause the relative affinities of a series of metals to change, is the reduction–oxidation potential* of a metal, which is always altered by chelation in a metal of variable valency (e.g. Cu, Fe, Co, Mn, Mo, V). It can even result in the metal quickly changing, after chelation, to a valency higher or lower than the one originally chosen. The scope can be gleaned from Table 11.5. (It should be borne in mind that the scale of potentials ranges from about +2 V for the most powerful oxidizing agents to about −2 V for the most powerful reducing agents.) Cobalt provides an interesting example of this effect. Normally cobaltous salts are stable in aqueous solution whereas a cobaltic salt is instantly decomposed by water with the evolution of oxygen. However, upon chelation with ethylenediamine the potential falls so drastically that the

---

*E_0 is the oxidation–reduction potential, a constant calculated at pH 0 by the following expression:

$$E_0 = E_H - \frac{RT}{nF} \ln \frac{[ox]}{[red]}$$

where E_H is the measured potential, R is the gas constant, T the absolute temperature, F the faraday, n the number of electrons concerned, and ln is the natural logarithm. At any other pH than 0, the symbol E′ is used, and the pH specified. Potentials obtained by polarography are often quoted 'versus the saturated calomel electrode', and need the addition of 0.246 V to put them on the above (normal hydrogen electrode) scale.

cobaltous complex becomes readily oxidized to the more stable cobaltic state.

The potential of a complex is determined by (a) the ionic charge on the ligand, (b) back-bonding of metal to ligand, and (c) the crystal field effect (Perrin, 1959). For example, concerning (a) it is known that when the ligand is anionic the higher valence state of the metal is favoured; concerning (b) it is known that back-bonding from iron favours the lower valences; and concerning (c) it is known that the crystal field effect favours Cu^{2+} and Fe^{2+} to other valences of these metals. Unfortunately, these factors cannot yet be integrated to predict the potential of a complex in advance of measurement.

Table 11.5 The effect of ligands on reduction–oxidation potentials. (Mainly at 20°C.)

Simple ion	E_0 (potential in volts)	Complexes* with:	E_0
$Fe(H_2O)_6{}^{2+} = Fe(H_2O)_6{}^{3+}$	+0.77	[1]EDTA	+0.14
		[1,2]Salicylic acid	+0.20 −0.22
		[6]Cyanide ion	+0.36
		[1]Glycine	+0.38
		[1,2]8-Hydroxyquinoline	+0.52 +0.27
		[1,2]Oxytetracycline	+0.57 +0.43
		[3]Bipyridyl	+1.06
		[3]o-Phenanthroline	+1.06
$Co(H_2O)_6{}^{2+} = Co(H_2O)_6{}^{3+}$	+1.84	[6]Cyanide ion	−0.83
		[3]Ethylenediamine	−0.22
		[6]Ammonia	+0.14
$Cu(H_2O)_x{}^{+} = Cu(H_2O)_4{}^{2+}$	+0.71	[2]Ethylenediamine	−0.38
		[2]Glycine	−0.16
		[2]Bipyridyl	+0.12
		[1,2]Methylthioethylamine	+0.19 +0.24
		[1,2]Pyridine	+0.20 +0.27
		[2]Ammonia	+0.34
		[1,2]Imidazole	+0.26 +0.35

*The pre-indexes have the same meaning as in Table 11.2 (e.g. [3]Bipyridyl means the 3:1 complex of bipyridyl). Data taken principally from Perrin (1959), and Hawkins and Perrin (1962).

Table 11.5 shows that some complexes are more powerful oxidizing agents than the free metal cations. Such complexes could cause the destruction of vital metabolites, e.g. the copper diethyldithiocarbamate chelate destroys thioctic acid (2.28) (Sijpesteijn and Janssen, 1959).

A more subtle aspect of ligand–metal interaction is that the bound metal may change the selectivity of the organic ligand by: (a) changing the electron distribution in the ligand, (b) making a chemically active centre of the ligand, (c) forcing the ligand molecules into a particular conformational form, (d) providing a conducting pathway for electron-adding or -removal, (e) increasing the liposolubility of the ligand, and hence helping it to penetrate into a living cell.

See Section 14.2 for ionophores which, without any chelation, transport univalent ions (e.g. Na^+) through natural membranes.

11.5 The various modes of biological action of chelating agents (an introduction)

The biphasic response of organisms to metals, so neatly illustrated by Piper's oats (Fig. 11.4), suggests two distinct mechanisms for the action of a chelating agent in biology: (i) the removal of metals from the cell, or the masking of metals within the cell, and (ii) the imposition on the cell of metals in greater quantity (or at a higher oxidation potential) than normal. Further sub-divisions can be made, depending on whether the metals in question are essential or toxic.

(a) *Mechanism (i)*. Most of the chelating agents that have achieved a biological use by this mechanism are antidotes, designed to mask or remove toxic metals that have accidentally been ingested by higher mammals. They are dealt with further in Section 11.6.

Few cases are known where a metal-binding agent is, of itself, highly injurious to an organism. Thus oxine (*11.30*) can enter the cells of bacteria and fungi without apparent injury (see Section 11.7.1). Such freedom from damage arises from the normal steric and affinity factors which allowed the active site to accumulate and retain the metal.

The most thoroughly investigated example of injury by masking is that caused by hydrogen cyanide. This agent binds the free valencies of the iron in cytochrome oxidase without detaching it from its four-bond contact with the porphyrin nucleus. Thus, this enzyme is prevented from uniting with its substrate, and respiration comes to a standstill. In many species this causes immediate death of the organism.

(b) *Mechanism (ii)*. How this mechanism has been used to supply iron to trees has been recounted in Section 11.0. Another example is the injection into lime-deficient patients of the calcium gluconic acid complex, which slowly breaks down and liberates ionic calcium. The best investigated examples of mechanism (ii) are found among the chelating bactericides and fungicides (see Section 11.7). Data necessary for the understanding of mechanism (ii) will now be discussed under two headings, 'the co-operative effect' and 'the partition effect'.

(i) *The co-operative effect*. That a metal can become much more chemically active after chelation is evident enough in the oxygen-binding properties of haemoglobin and the oxidizing powers of the haem enzymes. Inorganic salts of iron have catalase- and peroxidase-like properties, but these are enormously increased upon incorporation in the porphyrin nucleus attached to a specific protein. Similarly, cuprous ions catalyse the aerial oxidation of ascorbic acid, but this effect is immensely magnified in the enzyme ascorbic oxidase (Table 10.4).

Proteins are not always necessary to elicit this co-operative effect. In fact it often happens that, by adding metal-binding substances to sequester a metal, the very metal-catalysed effect that it is desired to suppress becomes aggravated. The offending metals are usually those which can undergo a change of valency, particularly copper and iron. The co-operative effect is most likely to be encountered if insufficient metal-binding agent is added, so that the complex is unsaturated. Some examples will now be given.

In the presence of hydrogen peroxide, the 2:1 complex of o-phenanthroline (*11.18*) with cuprous ion instantly cleaves double-stranded DNA. The reaction is sustained by the recycling of copper through its two valencies. This is shown by the lack of fission by 2,9-dimethyl-o-phenanthroline which, for steric reasons, binds cupr*ous* copper only (Pope *et al.*, 1982). The oxidative blackening of dihydroxyphenylalanine in the presence of cupric sulfate is greatly accelerated if o-phenanthroline is added (Isaka, 1957). The auto-oxidation of glutathione in an extract of ocular lens tissue (catalysed by trace of iron in the extract) is accelerated by EDTA (Pirie and Van Heyningen, 1954). Both o-phenanthroline and bipyridyl increase the rate of iron-catalysed decomposition of hydrogen peroxide by a hundredfold: these are, in fact, models for haemoprotein enzymes.

The copper-catalysed hydrolysis of di-isopropyl phosphorofluoridate is much accelerated by amino acids, ethylenediamine, o-phenanthroline, and bipyridyl. The best proportions are those which give a 1:1 complex (Wagner-Jauregg *et al.*, 1955). EDTA does not assist this hydrolysis, indeed EDTA is not a co-operator with copper, and can prevent oxine from acting as one (Byrde and Woodcock, 1957).

The mode of action of many selectively toxic agents is due to this co-operative effect, as will be demonstrated in Section 11.7. Metals usually have more specificity in the chelated state than when free as inorganic salts.

(*ii*) *The partition effect.* It is known that cell membranes carefully regulate the intake of heavy-metal cations; even cations essential in traces for nutrition are toxic in excess.* But uncharged complexes are liposoluble and hence should readily penetrate cell membranes in a way which the cell cannot regulate. Fig. 11.5 shows that, for a divalent metal, the complexes of the oxalic acid type, and 2:1 complexes of the glycine type, are uncharged and hence should penetrate cells readily. [Complexes of polydentate substances like EDTA, (*11.27*), often have hydrophilic groups in excess of those with which the metal can combine, and such complexes are not expected to penetrate ordinary cell membranes.] The liposoluble complexes enable the chelating agent to transport the metal into the cells in quantities larger than would normally occur. Some metal chelates act on the *outside* of micro-organisms, whose characteristic negative charge causes them to attract, and accumulate, positively charged complexes

---

*Iron is more toxic than is commonly believed; large oral doses of ferrous sulfate cause hepatic necrosis in Man within 48 hours (Luongo and Bjornson, 1954). The toxic action of encephalomyelitis virus, which is rich in iron, has been ascribed to viral transport of this metal across the blood—brain barrier which is normally impervious to it (Racker and Krimsky, 1947).

such as (a) the glycine type of chelating agent furnishes when unsaturated (e.g. 1:1 complexes of divalent metals) and (b) the ethylenediamine type furnishes at all degrees of saturation (see Fig. 11.5).

Chelation has been proposed to suppress a host-toxic effect, in a potentially useful drug, by masking the offending group with a metal (Gosálvez *et al.*, 1978). For more on the use of chelation to obtain masking effects, see Perrin (1970).

11.6 Diminution, by chelation, of the toxic effect of a metal. Antidotes

Several chelating agents are regularly used in hospitals as antidotes for occupational poisoning by metals, for chronic metal intoxication arising from therapy or household contamination, or to hasten the excretion of radioactive elements. They have been used in this way only since 1945. These antidotes circulate in the bloodstream without causing much depletion of the body's essential heavy metals; this is, of course, a matter of dosage. To ensure that little of the antidote enters cells, and that it is rapidly excreted, the molecules of antidotes are provided with polar (preferably ionizable) groups such as —OH, —CO_2H, —SH, —NH_2. These are present in such excess that at least one group remains free after the agent is saturated with metals. In addition to these considerations, an antidote is usually so designed that its chelated complexes cannot penetrate from the bloodstream into cells and are easily excreted by the kidneys.

The first antidote, dimercaprol (*11.23*), was invented in 1940 to counter poisoning by an arsenical war gas (Peters, Stocken and Thompson, 1945), and hence was at first called British antilewisite (or BAL). Today it is much used to treat poisoning by compounds of gold, mercury (in inorganic or organic combination), antimony, and arsenic. Binding occurs as shown in (*11.24*). This antidote is given intramuscularly every four hours during the first day, and then in accordance with the patient's needs. Not only is the toxic nature of these elements masked by the dimercaprol, but they are excreted still bound to it.

CH_2—SH	CH_2—S	CO_2H	Me
|	| Hg	|	|
CH—SH	CH—S	HC—SH	Me — C —SH
|	|	|	|
CH_2—OH	CH_2—OH	HC—SH	CH—NH_2
		|	|
		CO_2H	CO_2H
Dimercaprol	Capture of a mercuric	Dimercaptosuccinic	Penicillamine
(*11.23*)	ion by dimercaprol	acid	(*11.26*)
	(*11.24*)	(*11.25*)	

More recently, 2,3-dimercaptosuccinic acid (*11.25*) has been introduced because it has the following advantages over dimercaprol: it is active orally, penetrates the blood–brain barrier, and actually removes methylmercury ion from the brain (Aaseth and Friedheim, 1978). This ion is often consumed, by whole communities, from dressed seed or from fish exposed to mercury-containing industrial waste. Another orally active analogue, undergoing trials, is 2,3-dimercaptopropane-1-sulfonic acid.

D-Penicillamine (*11.26*), similarly used in copper poisoning, is non-toxic. It is employed in Wilson's disease, an inherited error of metabolism in which dietary copper cannot be eliminated, but accumulates in brain, kidneys, and liver in such quantities that death occurs in the early years of life. Oral administration of penicillamine prevents this and can even reverse the pathological lesions produced by the copper ions. The copper–penicillamine complex is readily voided in the urine (Walshe, 1968). Penicillamine, being effective orally, is convenient to use in the later stages of treating chronic lead poisoning with intravenous EDTA.

Calcium EDTA (*11.28*), injected as its sodium salt, sodium calcium edetate, is a most effective remedy for lead poisoning (Bessman, Rubin and Leikin, 1954). Sodium edetate (the same substance minus the calcium) was used in an attempt to decalcify stenosed valves in the heart; even though it also caused a fall in blood calcium, the patients were none the worse after long courses of injections, e.g. 3 g (as a 0.5% solution) every 24 hours for 5 days a week, and 3 weeks per month (Seven, 1960). Sodium edetate has been used successfully to overcome digitalis-induced cardiac arrhythmia by restoring K^+/Ca^{2+} balance (Szekely and Wynne, 1963).

$$\underset{\substack{\text{EDTA}\\ \text{(Ethylenediaminetetra-acetic acid)}\\ (11.27)}}{HO_2C \cdot H_2C \cdot N \overset{\overset{\displaystyle CO_2H \quad CO_2H}{\underset{\displaystyle |}{CH_2}\quad \underset{\displaystyle |}{CH_2}}}{\underset{\underset{\displaystyle H_2 \quad H_2}{C-C}}{}} N \cdot CH_2 \cdot CO_2H}$$

$$\underset{\substack{\text{Calcium edetate}\\ (11.28)}}{HO_2C \cdot CH_2N \overset{\overset{\displaystyle O_2C \qquad CO_2}{\underset{\displaystyle H_2C}{}\quad Ca\quad \underset{\displaystyle CH_2}{}}}{\underset{\underset{\displaystyle H_2C-CH_2}{}}{}} NCH_2 \cdot CO_2H}$$

$$\underset{\substack{\text{Cysteamine}\\ (11.29)}}{\overset{CH_2-SH}{\underset{CH_2-NH_2}{|}}}$$

Contamination of atomic-energy workers by plutonium must be treated promptly, because the radioactivity can destroy the capillaries that are distributing the antidote. Agents that remove iron, usually remove plutonium, and deferoxamine (*11.9*) is recommended. A useful supplement is diethylene-triaminepenta-acetic acid (DTPA) (Smith, Chapman and Marlow, 1969). Both substances are given parenterally.

Parenteral deferoxamine is the standard treatment for iron poisoning (see p. 448), and in chronic cases may have to be continued for years. Iron poisoning tends to set in after the repeated blood transfusions required in two genetically determined diseases, thalassaemia (Cooley's anaemia) and sickle-cell anaemia. In thalassaemia, the beta chain of the haemoglobin molecule is produced in an inadequate amount, so that the blood transports oxygen inefficiently. Patients do not survive childhood without repeated transfusions, yet frequently die in mid-teens or early twenties from the excess of inorganic iron that accumulates in kidneys, liver, and heart from these transfusions. This disease is most common among populations bordering the Mediterranean Sea, and in Southeast Asia. In some parts of Nigeria, as many as 30% of the population carries the gene for sickle-cell anaemia, a disease in which haemoglobin is not deficient but much of it is chemically different. About 0.15% of black children in the United States have the disease and, although the childhood mortality used to be high, some now live to have children of their own. In this disease, blood transfusions are needed only during the painful hypoxic crises.

Other iron-chelating agents undergoing trial are rhodotorulic acid (*11.12*), cheaply made from yeasts but still requiring injection, 2,3-dihydroxybenzoic acid which is part of the enterochelin molecule (*11.7*) and is active orally, and ethylenediamine-bis(2-hydroxyphenylacetic acid) which is related to EDTA.

Beryllium poisoning is treated with large doses of sodium salicylate, or smaller doses of aurine tricarboxylic acid. The latter, which has three salicylate moieties linked to a central carbon atom, does not assist the excretion of this metal, as salicylate does, but keeps it, in the tissues, as a very insoluble complex.

Sodium diethyldithiocarbamate, cf. (*3.57*), is a useful remedy in thallium poisoning (Sunderman, 1967).

2-Mercaptoethylamine (*11.29*) (cysteamine) effectively prevents radiation injury. It has been shown that, in a mouse irradiated with 750 rad (whole-body single dose), significant increases of iron and copper occur in the bone-marrow, adrenals, spleen, liver, lungs, and thymus, and of iron in muscle and kidney (Yendell, Tupper and Wills, 1967). One school of thought assigns the radio-protective properties of cysteamine to its ability to sequester the heavy-metal cations released from cells by irradiation (Jones, 1960). Other hypotheses are summarized by P. Brown (1967).

High-energy radiation produces, from water, the hydroxyl free-radical anion ($\cdot OH^-$) which attacks DNA. It has been suggested that this process continues by producing hydroperoxides of the pyrimidine bases of DNA, e.g. 5,6-dihydro-6-hydroperoxy-6-hydroxythymine. The hypothesis supposes that these hydro-peroxides react with transition metal ions to form free radicals which chemically change neighbouring bases, and this leads to mispairing during replication (Thomas *et al.*, 1976).

It is clear that effective masking of iron and copper ions would break any destructive chain reaction (oxidative) which they may catalyse. Other sub-stances, all modelled on cysteamine, are available for preventing radiation sickness. An example is 2-mercaptoethylguanidine, generated in the body from *S*-(2-aminoethyl)isothiuronium salts (Doherty, Shapira and Burnett, 1957).

Radioactive cations can, in some cases, be removed from the body by exchange dilution: the same cation (*not* radioactive) is injected as a complex of moderate stability. Thus zirconium citrate has been used to remove radioactive zirconium (Schubert, 1957).

Hydroxyurea exerts its valued anti-cancer action by diminishing a biological (*not* a toxic) effect of a metal in the uneconomic species. There it sequesters the iron required by the hydrogenase that converts ribosides to deoxyribosides (see Section 4.0, p. 125).

For a book on the use of chelating agents in heavy-metal poisoning, see Levine (1978).

11.7 Augmentation, by chelation, of the toxic effect of a metal

8-Hydroxyquinoline (*11.30*) has a similar antibacterial spectrum to penicillin. Although by no means so selective, it and its derivatives have been in regular use, since about 1895, for topical application in wounds. It has the advantage over many other antibacterials of acting quickly and of being fungicidal as well; but it is not very active against Gram-negative bacteria. Ointments containing 8-hydroxyquinoline are used in dermatology, e.g. in rashes due to resistant staphylococci, in secondarily infected eczemas, and also as a fungicide. The complex which 8-hydroxyquinoline forms with copper is much used for proofing structural materials, including tents, against fungal attack.

When it was found in 1944 that 8-hydroxyquinoline exerts its destructive effect on micro-organisms through chelation (Albert, 1944; Albert *et al.*, 1947), a new prospect was opened up that many other antibacterial and anti-fungal substances might be found to act in this way. It was soon shown that *oxine* (as 8-hydroxyquinoline had long been called by analytical chemists) had little activity in media depleted of heavy metals (Rubbo, Albert and Gibson, 1950; Albert, Gibson and Rubbo, 1953). Thus it acts, not by removing a metal essential for life, but by forming a lethal complex with a metal accidentally present (see Section 11.7.1). Several substances chemically related to oxine (see 11.7.2), and others chemically unrelated (see 11.7.3), were subsequently shown to act against micro-organisms by this mechanism, which is the 'co-operative effect' discussed in Section 11.5. It seems that only metals of variable valence can activate these agents.

11.7.1 *Mode of action of 8-hydroxyquinoline (oxine)*

That the antibacterial action of oxine (*11.30*) is due to chelation was shown as follows (Albert *et al.*, 1947). Oxine has long been used in chemical analysis as a chelating agent, and it has outstanding chelating properties (Table 11.2). The other six hydroxyquinolines, all isomers of oxine, did not chelate at all when tested, and they were without antibacterial action. Yet even two parts per million (M/100 000) of oxine prevented the growth of staphylococci and streptococci. Further, the *O*- and *N*-methyl-derivatives (*11.31*) and (*11.32*) of oxine were found to be without chelating action (as expected, because —CH$_3$, unlike H, cannot exchange with a metal), and neither of them was anti-bacterial. Thus the connexion between chelation and antibacterial action was established.

It remained to be shown whether the toxic action of oxine was due to the withdrawal of essential metals, as had been suggested (Zentmyer, 1944), or whether it actually increased the toxic action of metals normally present in the medium. The latter proved to be the case for both the bacteriostatic and the bactericidal actions (Rubbo *et al.*, 1950; Albert *et al.*, 1953), as was indicated in the outline of this work presented in Section 2.3 (p. 37). The first clue came from concentration quenching.

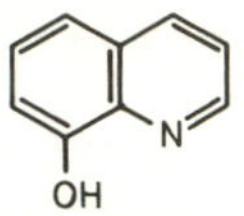

8-Hydroxyquinoline (oxine)   8-Methoxyquinoline   Oxine methochloride
        (11.30)   (11.31)   (11.32)

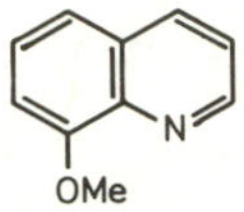

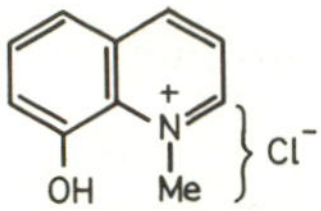

(a) *Concentration quenching.* It is highly unusual for the effect of a biologically active substance to decrease as the concentration is increased. However, oxine shows this phenomenon to an unprecedented degree. As will be seen from Table 11.6, staphylococci which are killed in an hour by M/100000 oxine are not killed (even in 3 hours) by M/1600 oxine (in fact, even a saturated solution, which is M/200, will not kill them). There was, however, a degree of toxicity after 24 hours (Albert *et al.*, 1953). Streptococci behaved similarly. The meaning of this concentration quenching became evident when it was found that it occurred in broth, but not in distilled water.

Table 11.6 The effect of increasing concentration on the bactericidal action of oxine in broth ('concentration quenching'). *Staph. aureus* in meat broth at pH 7.0–7.3 (20°C)

Concentration of oxine $1/M$	Growth after exposure			
	0	1 h	3 h	24 h
800	+++	+++	+++	+
1600	+++	+++	+++	+
3200	+++	+++	+	+
6400	+++	+++	+	−
12 000	+++	+	+	−
25 000	+++	+	−	−
50 000	+++	+	−	−
100 000	+++	−	−	−
200 000	+++	+++	+++	+++

The bactericidal test in this, and the following tables, is based on that of Miles and Misra (1938). At the end of the given time, samples were withdrawn, diluted, and inoculated on a dried blood–agar plate. The plates were read after 48 hours at 37°C. Symbols: −, no growth; +, up to 50 colonies; ++, 50–150 colonies; +++, uncountable. (Albert, Gibson and Rubbo, 1953.)

(b) *Experiments with oxine in distilled water.* The viability of staphylococci for at least 24 hours in distilled water permits some decisive tests to be made. It will be seen, from the distilled water experiments in Table 2.3 (p. 37) that oxine (0.01 mM) is biologically inert, but becomes bactericidal in the presence of a similar quantity of iron. Clearly the toxic agent is neither oxine nor iron, but the oxine–iron complex. When broth replaced water, no added iron was necessary because it was present in the medium. When the concentration of oxine was increased to 1.25 mM, the bactericidal action disappeared. This was attributed to formation of a non-antibacterial 2:1 oxine–iron complex (*11.33*) because, when sufficient extra iron was added to the broth so that the 1:1 complex (*2.25*) was re-formed, full bactericidal properties were restored (Albert *et al.*, 1953).

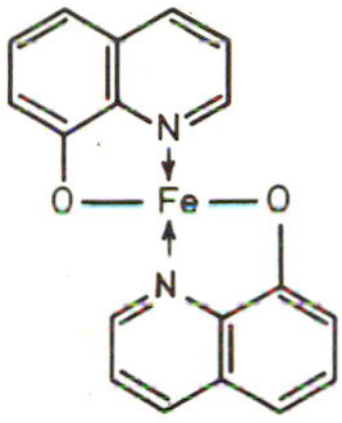

The 2:1 oxine ferrous
complex (saturated)
(*11.33*)

(*c*) *The metals co-toxic with oxine.* In the absence of added heavy metals, oxine enters the bacterial cell (*Staph. aureus*) without harming it (Beckett, Vahora and Robinson, 1958). Similarly, it enters the fungal cell (*Aspergillus niger*) without causing any harm (Greathouse *et al.*, 1954). (The latter investigation was done with two varieties of radioactive oxine, one from [^{14}C]aniline, and the other from [^{14}C]glycerol, labelled in the benzene and the pyridine ring, respectively.) Yet when a suitable metal was made available at the same time, these organisms were severely injured.

Damage to Gram-positive bacteria has been observed only when one of the following cations was present in the medium: Cu^{2+}, Fe^{2+}, or Fe^{3+}, and of these, iron seemed the more important. In aerobic cultures ferrous oxine enters into rapid equilibrium with the ferric form. Oxine has only a feeble action on most Gram-negative bacteria and shows no clear requirement for a particular metal (Rubbo *et al.*, 1950). Oxine damages mycelial fungi only when cupric ions are present in the medium, and iron cannot replace copper for this purpose (Anderson and Swaby, 1951; Block, 1956); this is true also for yeasts (Nordbring-Hertz, 1955).

(*d*) *Antagonism by cobalt of the toxic action of oxine–iron.* It is evident that the addition of a large excess (say, 200 equivalents) of an inert metal could prevent the toxic action of oxine–iron, if the stability constant of the new complex was greater than, or not much less than, that of oxine–iron; thus, by the law of mass-action, the oxine should almost entirely combine with the inert metal. As expected, experiments showed that cadmium, cobalt, zinc, and nickel were protective under these conditions, and manganese, magnesium, and calcium were ineffective (see Table 11.2 for stability constants).

Cobalt, however, has a unique position. Not only is it protective in large amounts, but in traces also. As little as M/25 000 cobaltous sulfate completely prevents the *bacteriostatic* effect of M/100 000 oxine (Rubbo *et al.*, 1950): this antagonism is illustrated in Fig. 11.9. Table 11.7 shows how effective cobalt is in preventing the bactericidal action of oxine–iron. It is only a little less effective against oxine–copper.

Cobalt also protects yeasts against oxine–copper (Nordbring-Hertz, 1955), but mycelial fungi are not protected, apparently because of their slow rate of growth (Anderson and Swaby, 1951). However, see the protective effects of cobalt on DMDC (dimethyldithiocarbamic acid) (p. 478). Cobalt uniquely protects trypanosomes against the lethal action of oxine, as shown in Table 11.8 (Williamson, 1959).

What is the explanation of this protective action of cobalt? At first it might seem that the cobalt combined with the oxine and thus denied it to the iron. But if this were so, nickel would be still more effective because the stability constant of nickel–oxine is much higher than that of cobalt–oxine (Table 11.2; also Albert, 1953). Yet nickel has no protective action at low concentrations.

A better clue comes from knowledge that several vital cell constituents, particularly mercapto-compounds [such as dihydrolipoic acid (*2.28*), see Section 11.7.3] are easily oxidized by atmospheric oxygen if traces of iron or copper are present. These oxidations lead to the formation of hydrogen peroxide and superoxide radicals ($^{\cdot}O_2^-$) which, in the presence of the metal cations, produce a fulminating chain reaction so that a very small amount of metal can catalyse widespread destruction. In some model reactions of this kind, traces of cobalt have been found to act as a chainbreaker which greatly moderates the destruction (see Fig. 11.10)*.

Table 11.7 Protective action of cobalt against the bactericidal action of oxine–iron and oxine–copper. *Staph. aureus* in metal-depleted broth at pH 7.3 (20°C) M/25 000 oxine present in every tube

	Conc. of metal added, 1/M			*Growth after exposure*			
Tube	*FeSO₄*	*CuSO₄*	*CoSO₄*	0	2 h	4 h	24 h
1	nil	nil	nil	+++	+++	+++	+++
2	50 000	nil	nil	+++	–	–	–
3	50 000	nil	50 000	+++	++	++	+++
4	nil	50 000	nil	+++	–	–	–
5	nil	50 000	50 000	+++	–	–	–
6	nil	50 000	10 000	+++	+++	+++	+++

(Albert *et al.*, 1953.)

*For information on chain reactions, see Dainton (1966).

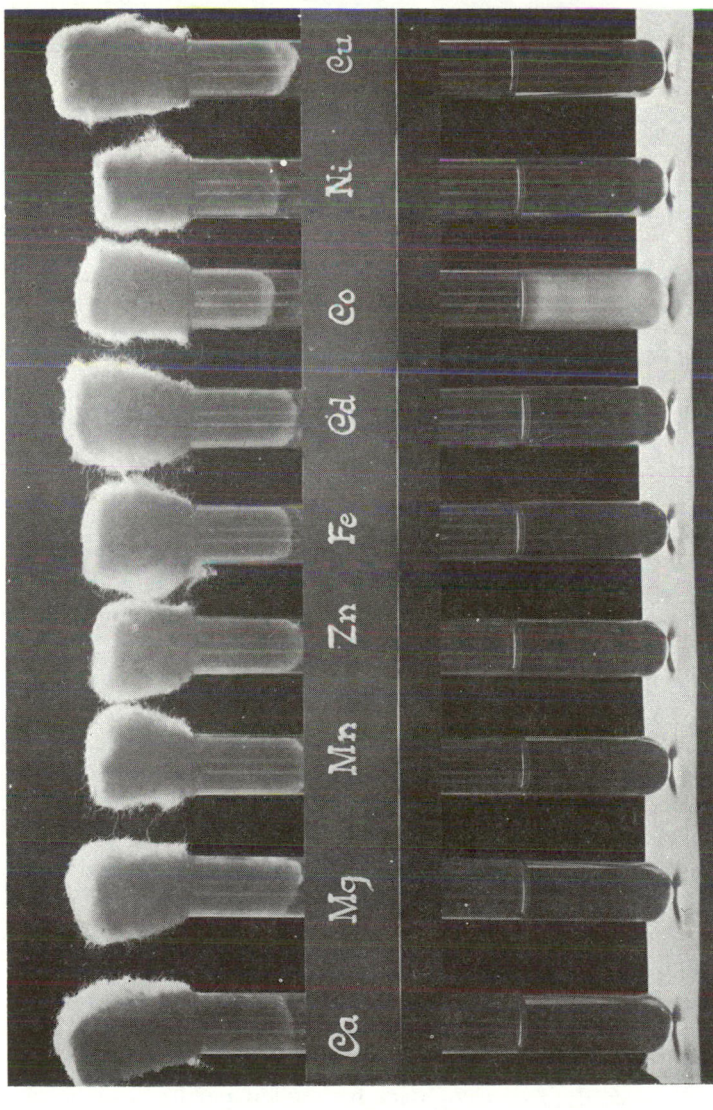

Fig. 11.9 The antagonism between oxine and traces of cobalt. *Staphylococcus aureus* in nutrient broth at pH 7.2 (see p. 472).

Table 11.8 Protective effect of cobalt against the trypanocidal action of oxine. *T. rhodesiense* in horse-serum glucose saline at 37°C, incubated for 4 hours

Tube	Concentration $1/M$			Trypanosomes surviving (per cent of control)
	Oxine	Co^{2+}	Any one of the following $Cu^{2+}, Ni^{2+}, Zn^{2+}$ $Fe^{2+}, Mn^{2+}, Mg^{2+}$	
1	nil	nil	nil	100
2	800 000	nil	nil	< 1
3	800 000	400 000	nil	117
4–9	800 000	nil	400 000	< 1

(Williamson, 1959.)

Oxine is a co-operative chelating agent (as defined in Section 11.5). Thus a mixture of inorganic iron and oxine catalyses the aerial oxidation of the —SH groups in nucleoproteins from rat liver and from fish eggs, whereas inorganic iron is ineffective on its own (Bernheim and Bernheim, 1939). The superior catalytic powers of oxine–iron probably spring from the rearrangement of the orbitals of the ferric cation caused by chelation (the unusual colours, red for ferrous–oxine and green for ferric–oxine, are evidence of rearrangement).

It is reasonable to assume that the toxic species of oxine is the 1:1 ferrous complex because it is unsaturated; in other words it has the unused combining power that is necessary in a catalyst. The 2:1 complex (*11.33*), on the other hand, is saturated and unlikely to be catalytic. (In aerobic systems, both ferr*ous* and ferr*ic* species are present in equilibrium; it follows that the 1:1 and 2:1 ferr*ic* complexes are unsaturated and presumably catalytically active, whereas the 3:1 complex is saturated and likely to be inert.) This concept reinforces a similar conclusion made on the basis of concentration quenching (p. 470).

The site of action of oxine–iron in bacteria is unknown, but a strong clue is afforded by the site in fungi, namely the oxidative destruction of dihydrolipoic acid (see p. 478). More general information on the site is afforded by data on liposolubility. Derivatives of oxine having a low oil/water partition coefficient are not antibacterial. Thus although oxine-5-sulfonic acid has the same stability constants as oxine (Albert, 1953), unlike oxine, it has no tendency to pass from water into lipids; also, it has no antibacterial properties whatsoever. To find if the high partition coefficient of oxine was essential for its antibacterial action, a series of uncharged derivatives of oxine having low partition coefficients were synthesized (Albert and Hampton, 1952, 1954) and tested (Albert *et al.*, 1954). These aza-oxines, which are dealt with in Section 11.7.2, showed that antibacterial action fell and rose, as partition coefficients fell and rose in response to small changes in molecular structure. Thus a high partition coefficient plays a very important part in determining the action of oxine and related substances.

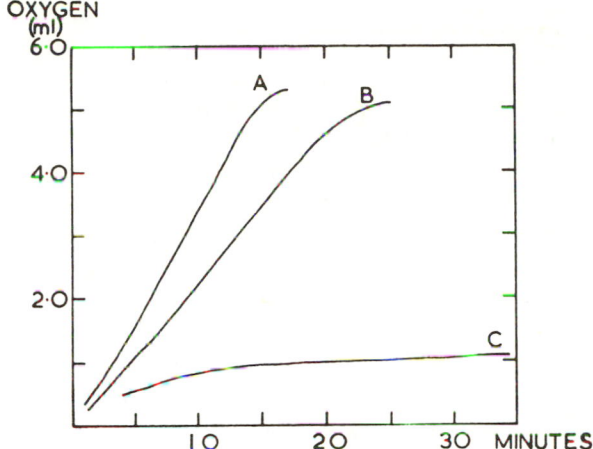

Fig. 11.10 The protective action of cobalt upon the copper-catalysed oxidation (by O_2) of cysteine ($M/40$) ($20°C$). A, $M/10\,000$ cupric sulfate; B, $M/100\,000$ cupric sulfate; C, as B, but with $M/500$ cobaltous sulfate. (Baur and Preis, 1936.)

This is evidence that the action of oxine takes place *inside* the cell, or at least within the cytoplasmic membrane.

(*e*) *Derivatives of oxine*. Halogenation of oxine improves the performance against Gram-negative bacteria, and many such derivatives are in use. Halquinol (a mixture of 5-chloro- and 5,7-dichloro-oxine) and chlorquinaldol (5,7-dichloro-2-methyloxine), which are also issued as 'Quinolor' and 'Steroxin' ointments, respectively, are used in dermatology. 5,7-Di-iodo-oxine (di-iodohydroxyquin, 'Diodoquin'), found in many pharmacopoeias, and 5-chloro-7-iodo-oxine (clioquinol, chinoform, iodochlorhydroxyquin, 'Vioform') are used for bacterial dysentery. Between 1955 and 1970, some 10000 Japanese people suffered from subacute myelo optic neuropathy (SMON), a painful, disabling, and sometimes fatal disease, thought to have been brought on by clioquinol. The outbreak led to suspension of sales in Japan, and protracted litigation. However, Meade (1975) found the data correlated poorly. In July 1977, the British Government's Committee on Safety of Medicines reviewed the evidence and concluded that the hazard of SMON was insignificant at the dosage approved for travellers' diarrhoea. For further discussion on potential hazards of halogenated oxines, see Gilman, Goodman and Gilman (1980, p. 1064) who advise caution.

5-Methyloxine ('Intetrix') was successfully used in France for outbreaks of cholera (Zribi and Ben-Rachid, 1973). This substance is highly active *in vitro* against *Plasmodium falciparum*, even against strains resistant to chloroquine (Scheibel and Adler, 1982).

11.7.2 *Substances chemically related to oxine*

It was found that a pyridine or benzene ring can be annelated to the oxine molecule without loss of activity; however, annelation to the 2,3-face introduced steric hindrance to chelation (Albert *et al.*, 1947).

Increasing the partition coefficient of oxine fourfold, by inserting lipophilic substituents, did not appreciably improve the *in vitro* action: this physical property must already be maximal in oxine (Albert, Rees and Tomlinson, 1956). It is therefore more interesting to know what happens when the partition coefficient is lowered. In oxine, this is suitably managed by synthesizing analogues where one or more =CH-group is replaced by the far more hydrophilic =N-atom (Albert and Hampton, 1952, 1954). Although these substances also have trivial names, such as 8-hydroxy-cinnoline, -quinazoline, and -quinoxaline, they can properly (and for the present purposes most conveniently) be termed 2-aza-oxine, 3-aza-oxine (*11.34*), and so forth.

Nitrogen and oxygen atoms lower (and halogens and alkyl-groups raise) partition coefficients. It can be seen from Table 11.9 that (as expected) the substitution of =N— for =CH— lowered the partition coefficient. With this lowering, the antibacterial action fell, often to vanishing point. Next, without removing the extra nitrogen atom, the partition coefficient was raised (by inserting a small alkyl-group, never exceeding three carbon atoms in length) until it reached, and even exceeded, that of oxine. When this was done, the antibacterial action also rose until it reached a maximum, as shown in Table 11.9 (Albert *et al.*, 1954).

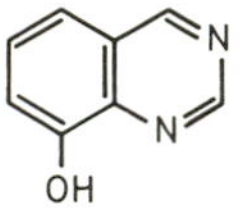

3-Aza-oxine
(11. 34)

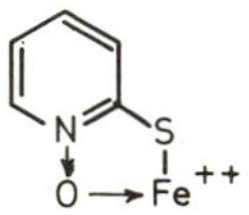

Pyrithione complex
(11.35)

Because of the inductive effect of the second nitrogen atom, neither the high stability constant nor the high antibacterial action of oxine is quite reached (the latter falls short by only one twofold serial dilution). However, the last seven substances in Table 11.9 all have stability constants of the same order. For these seven substances, the antibacterial action runs parallel to the partition coefficient, which establishes the correlation.

11.7.3 *Substances acting like oxine but not chemically related*

The *N*-oxides of pyridine, quinoline, and benzoquinoline are antibacterial provided that an ionizable group is present in the 2-position to make chelation possible. A mercapto-group has been found best for this purpose. Thus 2-mercaptopyridine-*N*-oxide (1-hydroxypyridine-2-thione) (*2.26*) (pyrithione, 'Omadine') is as intensely antibacterial as oxine. Although the chelated

Table 11.9 The fall and rise of bacteriostatic action with fall and rise of partition coefficient. *Strept. pyogenes* in meat broth at pH 7.3 (37°C)

Substance	Partition coefficient oleyl alcohol/water	Lowest inhibitory dilution 1/M	log first stability constant (Ni²⁺)
Oxine	67	200 000	9.8
5-Aza-oxine	<0.02	<800	5.8
7-Aza-oxine	0.1	<800	6.7
6-Aza-oxine	1	<800	5.9
3-Aza-oxine	5	13 000	7.6
2-Aza-oxine	6	13 000	7.8
4-Aza-oxine	8	6400	7.6
4-Methyl-2-aza-oxine	16	25 000	8.1
4-Methyl-3-aza-oxine	17	50 000	7.9
4-Propyl-3-aza-oxine	135	100 000	7.9
7-Allyl-3-aza-oxine	310	100 000	7.9

(Albert *et al.*, 1954.)

complex, e.g. (*11.35*), has an entirely different structure from that of oxine, e.g. (*2.24*), the mode of action is judged to be the same by the three following criteria: both substances are bactericidal only in the presence of iron, this action is prevented by cobalt, and also by an excess of the substance itself (Albert *et al.*, 1956). Its innocuousness in the absence of iron is shown in Table 11.10. It is much used as a fungicide in ointments and soaps. In a shampoo base, it has proved very effective against *Pityrosporum ovale*, one of the commonest causes of dandruff. It is also much used as an industrial fungicide (textiles, cooling systems, cutting oils).

Dimethyldithiocarbamic acid, or DMDC (*3.57*), is a potent fungicide much used in agriculture as the sodium ('NaDDC'), iron ('Fermate' or ferbam), and zinc ('Zerlate' or ziram) salts. Formula (*11.15*) shows the 1:1 copper complex.

Table 11.10 The innocuousness of 2-mercaptopyridine-*N*-oxide in the absence of iron. (Bactericidal test, in glass-distilled water). *Staph. aureus* pH 6–7 (20°C)

2-Mercaptopyridine-N-oxide (pyrithione) (2.26) 1/M	Fe³	Growth (plated out after 1 hour)
80 000	nil	+++
nil	80 000	+++
80 000	80 000	−

+++ means prolific growth with uncountable colonies.
−means no detectable growth.
(Albert *et al.*, 1956.)

The Australian work on oxine, described above, led to a parallel study on DMDC in Denmark in which this substance was shown to act anti-fungally only as its copper complex (Goksøyr, 1955). This work was taken up and extended in Holland (Sijpesteijn, Janssen and van der Kerk, 1957). Table 11.11 shows some typical results from the latter workers.

Table 11.11, which contains evidence of concentration quenching reveals the necessity of a metal ion (copper) for the action of DMDC on *Aspergillus niger*. Iron is ineffective. Several other moulds behave similarly (Sijpesteijn and Janssen, 1959). It will be evident from Table 11.11 that the effective proportions of metal and chelating agent are critical, as in oxine (Table 2.3, and Albert *et al.*, 1956). The Dutch workers attribute the zone of no inhibition (which lies between 10 and 50 p.p.m. of DMDC) to the extraordinary insolubility of the 2 : 1 copper complex. (The 2 : 1 copper complex of DMDC is soluble in water only 0.01 p.p.m., and this restricts the supply of agent to the cell.) The third zone, a toxic one, appearing in concentrations higher than 50 p.p.m., has been attributed to an intrinsic toxic effect of DMDC, unconnected with metal-binding (Sijpesteijn *et al.*, 1957).

Table 11.11 Effect of copper on growth inhibition in *Aspergillus niger* by sodium dimethyldithiocarbamate (3 days at 24°C). *Medium*: Glucose, salts, and vitamins in glass-distilled water at pH 7

Cu^{2+} 1/M	Sodium DMDC							
		1/M						
	nil	700 000	280 000	140 000	70 000	28 000	14 000	7000
nil	+	+	+	+	+	+	+	+
83 000	+	+	+	+	+	+	+	+
25 000	+	+	−	−	−	±	+	+

(Sijpesteijn and Janssen, 1959.)

Cobalt antagonizes the action of sodium dimethyldithiocarbamate on *Aspergillus niger* just as it does that of oxine (copper-catalysed) on yeast (p. 472) (Dr A. Kaars Sijpesteijn, personal communication, 1969). [For the copper-activated bactericidal properties of DMDC, see Liebermeister (1950).]

The accumulation of pyruvic acid in treated *Aspergillus* points to the molecular site of action of DMDC, oxine, and pyrithione, namely catalysis of the oxidative destruction of dihydrolipoic acid [thioctic acid (*2.28*)] (Sijpesteijn and Janssen, 1959). This is the essential coenzyme for oxidative decarboxylation of pyruvic acid by dihydrolipoylacetyltransferase, a component of the multi-enzyme complex known as pyruvate dehydrogenase.

For further reading on the dimethyldithiocarbamates, see Thorn and Ludwig (1962). Some agriculturalists prefer to use the zinc and iron complexes of

DMDC, principally because of the excellent adhesion to plants which enables the complexes to resist long periods of rain. Others prefer tetramethylthiuram disulfide, which is the disulfide obtained by oxidizing DMDC, and which is slowly reduced to DMDC under field conditions.

The following oxine-type chelating agents, tested *in vitro* on *Plasmodium falciparum*, were at least as highly inhibitory as quinine (used as control): oxine, 5-methyloxine, 2-mercaptopyridine-*N*-oxide, 2-mercaptoquinoline-*N*-oxide, and sodium diethyldithiocarbamate (Scheibel and Adler, 1982). It is not known whether 2,2'-bipyridyl has the oxine-type of biological activity, but derivatives of it show concentration quenching, including derivatives active against *Mycoplasma gallisepticum* in culture (Linschoten *et al.*, 1984). Mycoplasmas, the smallest of the self-replicating prokaryotes, lack cell walls and require cholesterol for growth (two differences from bacteria). *M. gallisepticum* is a frequently encountered cause of respiratory illness among chickens reared in a confined space. This organism is also inhibited by *o*-phenanthroline (*11.18*), and particularly strongly when there are methyl-groups placed *ortho* to the ring-nitrogen atoms. Stoichiometric quantities of copper had to be present for biological activity to take place. It was concluded that the active agent is the cuprous complex (Antic *et al.*, 1977).

1,10-Phenanthroline and its *C*-methyl-derivatives have general antibacterial properties. The metal complexes are also antibacterial, but the *speed* of action becomes greater with metals of decreased binding power (see Table 11.2). Bactericidal action was greatly enhanced by the addition of six methyl-groups (Shulman and Dwyer, 1964; Butler *et al.*, 1969).

11.8 The tetracyclines

Tetracycline (*11.36*) and its derivatives are octahydro-naphthacenes. The tetracyclines are among the most used of all drugs in treating systemic bacterial infections. They cannot act like oxine because their action on bacteria is slow and is not promoted by iron (Albert and Rees, 1956). They have very little action on fungi. [See Section 4.1 (p. 144) for background material on tetracyclines.]

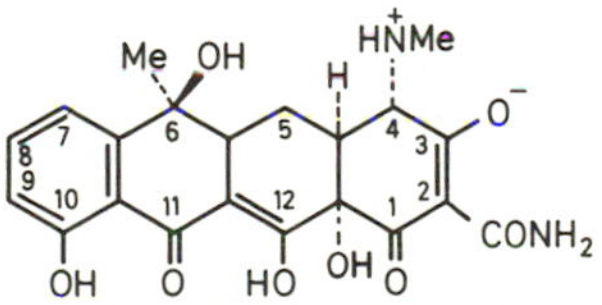

Tetracycline
(*11.36*)

That the tetracyclines were effective chelating agents was first demonstrated by Albert and Rees, 1956. Their stability constants were found to be similar to those for glycine (see Table 11.2).

Tetracycline is a strong acid in which the hydroxy-group shown at position 3 has pK_a 3.3 (it is part of a tricarbonylmethane system formed from the oxygen atoms in positions 1, 2, and 3). The remaining pK_a values (7.8 and 9.6) represent simultaneous ionizations from both (a) the dimethylamino-group, and (b) the phenolic β-diketone system which embraces positions 10, 11, and 12 (Leeson, Krueger and Nash, 1963; Rigler *et al.*, 1965). The *simultaneous* dissociation of two groups, whose pK_a values are separated by less than two pK units, is well known from work on cysteine (Edsall *et al.*, 1958).

Studies of the structure–activity relationships of tetracyclines show that many changes can be made at positions 5, 6, and 7 without harming the antibacterial action (McCormick *et al.*, 1960; Blackwood, 1970). The following details are more critical. The tricarbonylmethane area, as defined above, plays a prominent part in metal binding and is essential for antibacterial activity. ^{1}H n.m.r. spectra show line-broadening in this area when Ca^{2+}, Cu^{2+}, Mn^{2+}, and Co^{2+}, are added to a solution of tetracycline; but Mg^{2+} gave such an anomalous signal as to suggest that it, uniquely, caused a conformational change before binding (Williamson and Everett, 1975). It is known that the phenol–diketone area (based on positions 10, 11, 12) must be present and free for Mg^{2+} and Ca^{2+} to be bound (Mitscher, Bonacci and Sokoloski, 1968; Doluizio and Martin, 1963). It is also known that the stereochemistry of the 4a–12a fold [defined as the junction between the third and fourth rings, reading (*11.36*) from left to right] is critical. (These last three correlations, taken together, define the locus of magnesium cations in the chelated complex.) Finally, the dimethylamino-group is essential for *in vivo* (but not for *in vitro*) antibacterial properties.
not for *in vitro*) antibacterial properties.

As was pointed out in the introduction to Chapter 3, the selectivity of tetracyclines rests on their being accumulated far more by bacteria than by mammalian cells. There is evidence that these drugs form a liposoluble complex with magnesium in the bacterial plasma membrane, and that this leads to their availability to the cytoplasm (Franklin, 1971; Dockter and Magnuson, 1973; Franklin and Snow, 1981). The decisive site of action in bacteria is on the ribosomes, which are also rich in magnesium, and whose protein-synthesizing activity is brought to a halt by these antibiotics. Although nickel, cobalt, and zinc ions (which bind tetracyclines much more tightly than magnesium) do not reverse the action of these drugs, magnesium does so (Weinberg, 1954), and fluorescence studies have shown that chelation of Mg^{2+} in bacterial ribosomes precedes the inhibition of protein synthesis by tetracyclines (White and Cantor, 1971).

Because of their chelating action, tetracycline drugs are inactivated in the patient's bowel by any dietary calcium or magnesium ions, whether from milk or from antacid medication. Through such mishaps, many patients have lost the potential benefit of these antibiotics. Tetracyclines are usually given orally. Tetracycline, itself, is still much prescribed, but there are also lower-dose forms available: demeclocycline and methacycline, and a sub-class of these which require less frequent dosing: doxycycline and minocycline.

11.9 Substances whose biological action is at least partly due to chelation

Many other chelating agents have found widespread employment in medicine. Although, because of their constitution, they could not fail to combine with heavy metals in the tissues, enough is not always known of their mode of action to decide whether they act by chelation alone. Yet, no matter how much or little their chelating properties may govern the mode of action, these properties deserve study because of their role in absorption and the causation of side effects.

(a) *Isoniazid* (*11.37*). The discovery of this substance, the sheet-anchor of tuberculosis therapy, was outlined in Section 6.3.1 (p. 224). It was shown in 1956 that this drug, the hydrazide of isonicotinic acid, is an effective chelating agent with stability constants of the same order as glycine (Albert, 1956). The complexes (*11.39*) of isoniazid are formed through the anion (*11.38*). No anion can be formed by the methyl-derivative *N*-methyl-*N*-isonicotinoylhydrazine (*11.40*), and this substance has virtually no anti-tubercular activity (Cymerman-Craig and Willis, 1955; Cymerman-Craig *et al.*, 1955). Thus isoniazid inhibited *Mycobacterium tuberculosis* H_{37} Rv. (*in vitro*, in the presence of 10% serum) at M/5 000 000, but substance (*11.40*) was inactive.

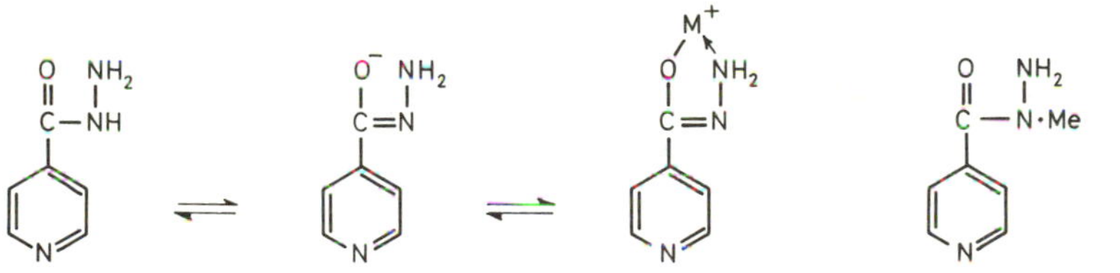

| Isoniazid | Anion | Metal-complex | Methyl derivative |
| (11.37) | (11.38) | (11.39) | (11.40) |

In the presence of cupric ions, isoniazid is taken up much more rapidly by *M. tuberculosis* H_{37} Rv., in culture (Youatt, 1962). Blood serum copper (see p. 433) is maintained at a low level, but, bearing in mind the small number of bacteria that sustain an infection and the length of the treatment, quite sufficient copper is available to convert a therapeutic concentration of the drug into its chelate. However, the action of isoniazid cannot be due *entirely* to chelation, because neither of its two isomers has any notable action against *M. tuberculosis*, even *in vitro*, although their stability constants are as high, or even higher (Albert, 1956). This is shown in Table 11.12. It is known that isoniazid enters the cells of non-resistant tubercle bacteria, and does not penetrate the cells of resistant strains; also the common bacterial species such as *Staph. aureus* and *E. coli*, which are never affected by isoniazid, do not absorb it (Youatt, 1958).

It seems fairly certain that chelation with trace copper makes the drug liposoluble enough to penetrate the most lipophilic of all bacterial integuments, but the question remains: what happens next? The answer was first outlined by Krüger-Thiemer (1957), and has since become firmly supported (Seydel *et al.*,

1976a). In this explanation, isoniazid is seen as the transport form (pro-drug) of the true drug isonicotinamide to which it is oxidatively degraded. This amide is then built into an analogue of nicotinamide adenine dinucleotide (NAD). The replacement of NAD by this analogue is held to disturb lipid metabolism. Undoubtedly, isoniazid at a therapeutically effective concentration, causes a decrease in mycobacterial NAD leading to a breakdown in the synthesis of mycolic acid [the typical waxy material that forms a protective coating around the H_{37} strain of the bacterium (Winder and Collins, 1970)].

Table 11.12 Physical and anti-tubercular properties of hydrazides

Hydrazide	pK$_a$ hydrazide group	log stability constant* 1:1				Relative activity on M. tuberculosis H$_{37}$ Rv.	
		Cu^{2+}	Ni^{2+}	Co^{2+}	Zn^{2+}	In vitro	In vivo
Isonicotinic**	10.77	8.0	5.5	4.8	5.4	1	1
Nicotinic**	11.47	8.7	6.0	5.4	?	0.001	—
Picolinic**	12.27	12.4	10.7	9.6	8.4	0.017	—
Benzoic	12.45	9.0	6.3	?	?	0.002	—
Cyanoacetic	11.17	8.5	6.0	5.3	?	0.008	0.2

*For some 2:1 stability constants, *see* Albert (1956).
**These three substances are isomers. They have the hydrazide group in the 4-, 3-, and 2-position respectively.
(Albert, 1956.)

Seydel and his colleagues (1976b) synthesized the isonicotinic analogue and found it identical with material isolated from the H_{37} strain of *M. tuberculosis*. They presented the following additional evidence: only those strains of *M. tuberculosis* which contain peroxidase are susceptible to injury by isoniazid (the highly pathogenic strains, if not drug-resistant, contain abundant peroxidase). Peroxidase turns isoniazid to isonicotinic acid *in vitro* (cell-free experiment) and, during isoniazid therapy, isonicotinic acid accumulates in the bacterial cells. Nicotinic acid and nicotinic hydrazide can antagonize the bactericidal action of isoniazid *in vitro*.

It was once thought that isoniazid acted on *Mycobacteria* by combining with pyridoxal. This was soon disproved, but some side effects on the patient have been traced to this reaction.

(*b*) *Other chelating anti-tuberculous drugs.* Thioacetazone (*p*-acetamidobenzaldehyde thiosemicarbazone) (*11.41*), also known as amithiozone, is specific for the tuberculosis bacterium but much weaker in action than isoniazid. There is no cross-resistance beteen the two drugs. The thiosemicarbazone group chelates heavy metals, and the anti-tubercular action of thioacetazone is enhanced by copper, which it reduces to the cuprous state (it binds zinc more strongly) (Stünzi, 1982).

The anti-tubercular properties of 2-acetylpyridine thiosemicarbazones are enhanced by co-ordination with iron or copper (Scovill, Klayman and Franchino, 1982).

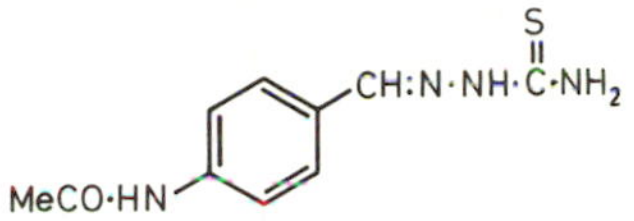

Thioacetazone
(11.41)

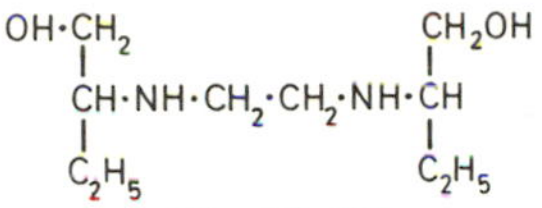

Ethambutol
(11.42)

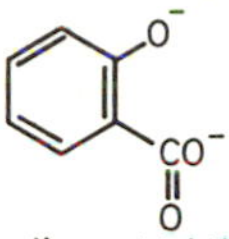

Salicylic acid (dianion)
(11.43)

To avoid resistance setting in, tuberculosis is preferably treated with a combination of isoniazid and rifampicin (4.37) (Fox, 1977). Because of the expense of the latter, ethambutol (11.42) is often substituted and is very effective although it is almost inactive against other species of bacteria. Ethambutol is simply a hydroxyalkyl-derivative of ethylenediamine (Table 11.2) and chelates metals in the same way. Its selectivity is attributed to steric hindrance arising from the α-branching of the side-chains (Shepherd and Wilkinson, 1962). This drug is metabolized to the corresponding dicarboxylic acid, which is a far stronger binder of copper and zinc (Cole, May and Williams, 1981).

p-Aminosalicylic acid and streptomycin, former adjuvants used with isoniazid, have almost been abandoned. Salicylic acid is a strong chelating agent (Table 11.2) provided that it is exposed to conditions alkaline enough to form the ligand which is the dianion (11.43) (Perrin, 1958). It has a mildly anti-bacterial and anti-fungal action; its widespread use in loosening calloused skin is attributed to its acting as a milder form of phenol. Derivatives such as salicylanilide, salicylamide, and salicylic aldehyde, which cannot produce a carboxy-anion, are much weaker chelating agents (Perrin, 1958). Aspirin, where the phenolic group (only) is blocked, has no chelating power, but is fairly rapidly hydrolysed in the body to salicylic acid which Sih and Takeguchi (1973) think may chelate the copper in prostaglandin synthetase in the therapy of rheumatic diseases. The antibiotic antimycin has a 3-formamidosalicylamide structure, and strongly binds ferric iron (Farley, Strong and Bydalek, 1965).

Nalidixic acid (4.33) binds heavy metals between the carboxylic- and the oxo-groups (Crumplin, Midgley and Smith, 1980). It is not known if this plays a part in its biological action, which is inhibition of the synthesis of DNA (Section 4.0.5, p. 136). The antibacterial action of kojic acid (11.11), a pyrone extracted from certain fungi, is enhanced by metallic cations (Weinberg, 1957). Bacitracin, a polypeptide antibiotic (Section 13.2), loses its antibacterial action against *Staph. aureus* in the presence of EDTA; the action is restored by bivalent cations (Adler and Snoke, 1962).

(c) *Anti-viral drugs.* Phosphonoacetic acid (p. 131) forms a six-membered ring with zinc and other divalent metals (Stünzi and Perrin, 1979). It is thought that methisazone (6.19) and its analogues owe their anti-viral action (see Section 6.3.2) to chelation in which the metal is held between the oxygen atom and the second nitrogen of the side-chain (O'Sullivan and Sadler, 1961). This drug reduces copper to the cuprous condition; zinc is the metal that it binds most strongly (Stünzi, 1982). Methisazone inactivates many kinds of tumour-causing viruses of the RNA type, in tissue culture (Levinson, Woodson and Jackson, 1971). Thiosemicarbazones of 2-acetylpyridine strongly inhibit the *in vivo* replication of herpes simplex virus, types 1 and 2, without greatly affecting cellular DNA or protein synthesis (Shipman *et al.*, 1981).

(*d*) *Anti-fungal examples.* Apart from the powerful anti-fungals mentioned in Section 11.7, and salicylic acid (above), special value is attached to simple derivatives of tropolone (11.44) which occur in the heartwood of several species of conifers. These substances combine with copper, often accidentally present, and provide several decades of protection against fungal attack, justifying the erection of unpainted buildings. Monoisopropyl derivatives of (11.44), the thujaplicins, are common examples, and the site of action is thought to be inhibition of the formation of acetyl-coenzyme A from acetate ions (Raa and Goksøyr, 1966).

(*e*) *Anti-cancer examples.* Thiosemicarbazones have been mentioned above in connexion with their anti-viral and anti-tubercular properties. They also have anti-cancer activity, as already mentioned for 5-hydroxypicolinicaldehyde thio-semicarbazone (4.12), in Section 4.0.1. Such substances powerfully inhibit ribonucleoside diphosphate reductase, the enzyme which converts ribo- to deoxyribo-nucleotides, leading to inhibition of DNA synthesis. Interaction of these drugs with the enzymically-required iron is thought to underlie their action (Agrawal *et al.*, 1972). Hydroxyurea (4.10), which is used in the clinics for treating melanoma, acts in the same way (Section 4.0.1).

Because 5-hydroxypicolinicaldehyde thiosemicarbazone, like thioacetazone (11.41), is only bidentate (i.e. makes only two bonds to each atom of metal), better bonding has been sought with tridentate derivatives of this aldehyde (Ankel and Petering, 1980), and with quadridentate bisthiosemicarbazones of 1,2-dicarbonyl compounds such as 3-ethoxy-2-oxobutyraldehyde (Petering, 1972). It has been demonstrated that cupric ions can transport this substance into the living cell (Sartorelli and Creasey, 1969).

Bleomycin (4.28), the most clinically used of all metal-binding anti-cancer drugs, uses its bound iron to activate atmospheric oxygen and, finally, degrade DNA (Section 4.0.4).

D-Penicillamine (11.26) has a favourable chemotherapeutic effect on adeno-carcinoma implanted into rats. This is attributed to inactivation of the super-oxide dismutase that is controlled by copper and zinc, a loss for which normal cells are compensated by the manganese-controlled analogous enzyme in which cancer cells are deficient (Okuyama and Hitoshi, 1981; Oberley, 1982).

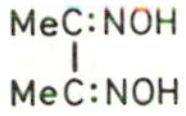

Dimethylglyoxime (*11.45*), which showed no activity against Ehrlich's carcinoma and Crocker sarcoma mouse test-systems, did so when a minute, non-toxic amount of Cu^{2+} ions was added to the system (Takamiya, 1960).

For the clinically useful platinum complexes see Section 11.10.

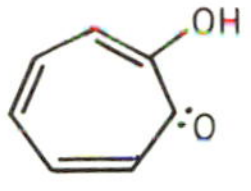

Tropolone	Dimethylglyoxime
(*11.44*)	(*11.45*)

Concanavalin A, a protein (isolated from jack-beans) which binds carbo-hydrates and has the general properties of a 'lectin', agglutinates and inhibits growth of malignant cells (Sharon and Lis, 1972), but it also agglutinates erythrocytes. Each monomeric unit of concanavalin A has one site that binds calcium ions and another that binds Zn^{2+}, Co^{2+}, or Mn^{2+}, and *both* sites must be occupied by the appropriate metal for biological activity to occur.

(*f*) *Anti-protozoal examples.* Bialamicol (*11.46*) was introduced in 1950 as the first synthetic drug *directly* to attack the parasite in amoebiasis. It has chelating properties of the glycine type (Dill *et al.*, 1957). 2-Acetylpyridine thiosemi-carbazones cure *Pl. berghei* infections in mice at the low dose of 20 mg/kg (Klayman *et al.*, 1979).

The following chelating agents cured all mice infected with lethal inocula of *Trypanosoma brucei* when given intravenously: caffeic acid (3,4-dihydroxy-cinnamic acid), neocupreine (2,9-dimethyl-*o*-phenanthroline) and the 2-pyridylhydrazone of pyridin-2-al (Shapiro *et al.*, 1982). These agents were thought to deprive the parasites of the essential iron in their glycerophosphate oxidase which, uniquely in these organisms, effects terminal oxidation in the metabolic pathway.

(*g*) *Anthelmintic examples.* Even quite simple *NN'*-chelating agents have anthel-mintic properties, notably 2,2'-bipyridyl (*11.19*) and *o*-phenanthroline (*11.18*) (Baldwin, 1948b; Dwyer *et al.*, 1952). The much used vermicide, thiabendazole (*6.36*), strongly chelates transition-metal cations between the two nitrogen atoms (Kowala *et al.*, 1971).

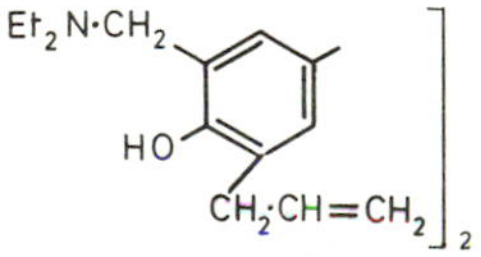

Bialamicol	Hydralazine
(*11.46*)	(*11.47*)

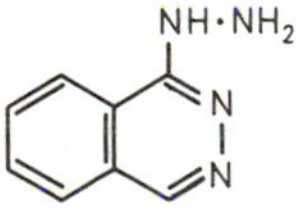

(*h*) *Pharmacodynamic examples.* The metal-binding properties of epinephrine, thyroxine, and histamine were discussed in Section 11.1. Those polypeptide hormones that have a bisulfide (—S·S—) group are potentiated by divalent metal cations. Thus the contractile effect of oxytocin on the uterus is potentiated by Co^{2+}, Mn^{2+}, and Ni^{2+}, but less by Ca^{2+} and Cu^{2+}. It is thought that the metal forms a ternary complex with the hormone and its receptor (Schild, 1969).

Cortisone is sometimes credited with chelating power; but, from inspection of the formula, this seems unlikely, and no proof has been produced. The suggestion that indolyl-3-acetic acid exerts plant-growth-regulating effects through chelation has been discounted by showing that it lacks chelating properties (Perrin, 1961).

Many metal-binding agents are able to reduce raised blood-pressure in Man apparently by direct action on muscles controlling small blood-vessels (Schroeder, Perry and Menhard, 1955). Examples are: hydralazine (*11.47*), dimercaprol (*11.23*), and the sodium thiocyanate liberated when sodium nitroprusside is injected intravenously to resolve a crisis of hyperpiesia. Given orally, hydralazine is one of the most frequently prescribed remedies for moderately severe hyperpiesia. It chelates transition-metal cations between the two nitrogen atoms, as in ethylenediamine. Analogous hydrazine-derivatives of other nitrogenous heterocycles have similar therapeutic properties.

Amrinone (5-amino-3,4'-bipyridyl-6-one) is beginning to be used for oral treatment of congestive heart failure (Benotti *et al.*, 1978). It can bind metal cations between the 5- and 6-substituents. See Section 14.2 for ionophoric cardioactive agents.

The anti-diabetic drug phenformin (β-phenylethylbiguanide) forms chelates with cupric ions with avidity similar to that of glycine. Among related compounds, the ability to chelate copper runs parallel to the anti-diabetic activity which is effected by inhibition of hepatic gluconeogenesis (Foye, O'Laughlin and Duvall, 1961).

11.10 The special case of robust complexes

In aqueous solution, the complexes of most metal cations exist in dynamic equilibrium with their components. If we disturb this equilibrium, another one is instantly formed. It is quite otherwise with *robust complexes* which persist for hours (or even days) under conditions favourable to their decomposition; any biological properties that they may have are strikingly different from those of their components. Robust complexes are formed where metal ions have 3,4 (low spin), 5, or 6 *d* electrons provided that formation of the complex involves large values of ligand-field stabilization energy. Metals most prone to form robust complexes are the transition metals: platinum, iridium, osmium, palladium, rhodium, ruthenium, also (but not so frequently) nickel, cobalt, and iron. The halide and, particularly, the cyanide anions most readily form robust complexes with these transition metals, then come other nitrogen anions, whereas oxygen-presenting types show the effect least.

The curariform action of some robust chelates was discovered by Beccari

(1938) who found that a doubly charged co-ordination complex of ferrous iron and three molecules of 2,2'-bipyridyl caused mild paralysis and respiratory failure in rabbits. Later he showed that rabbits, as well as frogs, excreted this complex unchanged (Beccari, 1941). Extending this work, Dwyer, Gyarfas and O'Dwyer (1951) examined complexes of iron, nickel, cobalt, ruthenium, and osmium and found that their most active example was the 2,2',2''-terpyridyl complex of ruthenium which had one tenth the activity of tubocurarine. Only positively charged complexes were active. Working with the isolated rat diaphragm, or the whole mouse, they found that the paralysed muscles, responded to direct electrical stimulation and the paralysis was reversed by physostigmine, just as with tubocurarine. In these complexes there are no chemically active groups exposed, and the redox potentials are out of the biological range. It has to be concluded that the geometry and the charge of these robust complexes is responsible for their biological effects, and that the role of the metal is no more than that of holding the remaining atoms together in a configuration complementary to that of the receptor (Shulman and Dwyer, 1964). Examples more active than tubocurarine have since been made by arranging for *two* positive charges to be present at the same distance apart as in tubocurarine (Taylor, Callahan and Shaikh, 1975).

In 1970, it was shown that cis-dichlorodiammine platinum (II) (*11.48*) could effect regression of large, solid sarcoma-180 tumours (Rosenberg and Van Camp, 1970). This substance, now known as cisplatin, has become established as the best treatment for testicular and ovarian tumours, often curative. It is also useful in treating cancer of the bladder, head, and neck (Rosencweig *et al.*, 1977; Einhorn and Williams, 1979). The major toxic effects seen during cisplatin therapy are vomiting, myelosuppression, ototoxicity, and (worst of all) kidney damage. The drug is given intravenously and the nephrotoxicity is countered by ensuring a copious flow of urine.

Cisplatin is a pro-drug. Biological action begins only after both chlorine atoms have been replaced by hydroxy groups. The corresponding *trans* isomer is inactive. The mode of action is thought to be a binding of two strands of DNA, as well as increased phosphorylation of chromatin in the nucleus. Analogues with better chemotherapeutic index have been found by replacing the two ammonia molecules by two molecules of cyclohexylamine (Connors *et al.*, 1972). At the present time, clinical trials are proceeding with carboplatin (*11.49*) which is giving as good a remission rate, for ovarian cancer, as cisplatin without the latter's nephrotoxicity or ototoxicity, and it causes less nausea (Calvert *et al.*, 1982).

For more on platinum-containing anti-cancer agents, see Roberts and Pera (1983).

Spirogermanium (*11.50*), a newly introduced robust complex, seems useful in treating tumours of the colon and ovary. It is one of the few carcinolytic agents that do not attack the bone-marrow (myelosuppression). Large doses show toxic effects on the central nervous system (Slavik *et al.*, 1982).

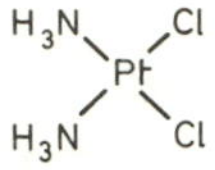

Cisplatin
(11.48)

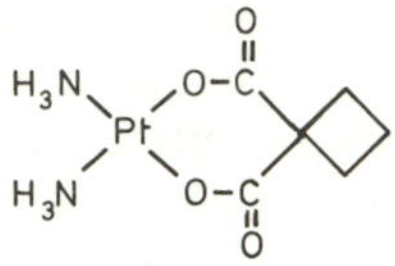

Carboplatin
(11.49)

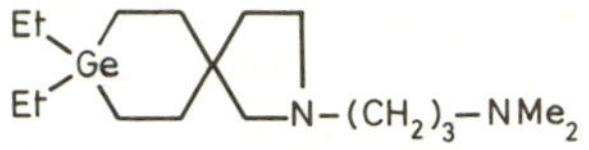

Spirogermanium
(11.50)

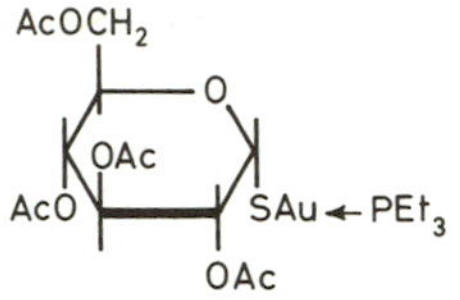

Auranofin
(11.51)

The gold complexes, reserved for treating the severest manifestations of arthritis, all contain the monovalent state of this metal, linked to sulfur. This arrangement can be classed as robust only in so far as it is slow to form, although quick to exchange. Until recently, all such preparations had to be injected, but auranofin (11.51) is now available for oral therapy (Berglöf, 1977). It is, chemically, 1-thio-β-D-glucopyranosato(triethylphosphine)gold-2,3,4,6-tetra-acetate.

11.11   Fundamental considerations in designing new chelating agents. Promising avenues of application

The starting-point for these considerations must be the hard fact that the majority of chelating agents have no biological action. For example, very few of the metal-binding agents commonly used in analytical work are antibacterial (Albert, et al., 1947; Schraüfstätter, 1950).

No chelating agent can be expected to be active, in a biological environment, unless its stability constants are at least as high as those of the common amino acids, e.g. glycine in Table 11.2. However, no good purpose may be served by flying to the other extreme and producing agents with very high stability constants: such substances are liable to become lost by saturation before reaching the desired site of action. For some applications, at least, moderate ease of cation exchange is necessary (Schubert, 1957).

At some stage in the design it must be decided whether penetration into cells is to be desired or to be avoided. If desirable, lipophilic groups should be added to the molecule. For this purpose carbon, halogens, hydrogen, and sulfur are considered as lipophilic, and nitrogen and oxygen as hydrophilic; the effect can be followed by determining oil/water partition coefficients. The effect of adding first hydrophilic, and then lipophilic, groups to oxine is illustrated in Table 11.9. Complexes of the ethylenediamine type (see Fig. 11.5), which preserve the original charge on the metal, need to be supplied with lipophilic groups to acquire ease of penetration; in fact they need more of this help than would be necessary for an uncharged complex.

Very small changes in a molecule can produce large changes in partition coefficients. Thus, the insertion of an extra ring-nitrogen into oxine to give 3-aza-oxine (8-hydroxyquinazoline) makes the molecule much more hydrophilic and it lowers the oil/water partition coefficient from 67 to 5 (see Table 11.9). The addition of a side-chain of only three carbon atoms (the propylgroup) restores the partition coefficient.

Apart from this use of partition coefficients, uptake of an agent by cells can be forced in another way: by using ligands which resemble natural substrates such as amino acids, carbohydrates, choline, purines, and pyrimidines.

Steric and other chemical differences which can assist selectivity were discussed in Section 11.4.

Many biologically active chelating agents are unsuited for internal therapy in Man simply because they are not selective enough. One must always proceed cautiously with metal-binding substances because some of them, such as sodium dimethyldithiocarbamate (Kadota and Midorikawa, 1951), injure the islets of Langerhans and thus cause diabetes in experimental animals. The 5-amino-, also the 2-methyl-derivatives of oxine act similarly (Kadota and Abe, 1954).

Some instances are recorded where chelating agents have been able to overcome drug resistance (Section 6.5, p. 263).

Current topics in chelation research embrace all the subjects mentioned in Section 11.9, plus the following: new methods for detoxification and decontamination of tissues injured by chemically toxic or radioactive metals, attempts to combat the element of chelation in dental caries, and to prevent the loss of calcium from bones that occurs in old age, the cause of so much precocious senility.

12

Steric factors

The size and shape of a molecule can play a very important part in its biological action. For some types of action a flat molecule is required, whereas other types of action require a three-dimensionally bulky molecule. The benzene ring (*12.1*) is flat, and the six bonds leading from the ring to all attached atoms or groups also lie in the plane of the paper. All other conjugated systems (i.e. those where every second bond is a double bond) are similarly flat.

On the other hand, aliphatic and alicyclic structures are non-planar. Thus cyclohexane (*12.2*), which differs from benzene in having six more hydrogen atoms, is non-planar and the bonds leave the ring at different angles (only four of the six carbon atoms lie in the plane of the paper). Atropine (*7.16*) is an example of a drug which has a non-planar ring, whereas 9-aminoacridine (*10.8*) is typical of the drugs which have flat molecules. Both types of architecture are common in drugs, and sometimes both types occur in the one molecule, as in mepacrine (*10.30*) and nicotine (*7.26*).

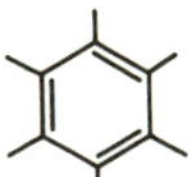

Benzene
(*12.1*)

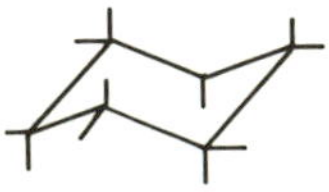

Cyclohexane
(*12.2*)

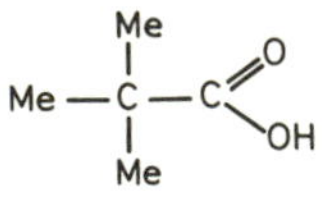

$$Me-\underset{\underset{Me}{|}}{\overset{\overset{Me}{|}}{C}}-C\underset{OH}{\overset{O}{\diagup}}$$

Trimethylacetic acid
(*12.3*)

Bretylium
(*12.4*)

12.0 Some fundamental considerations

A biological requirement for a minimal area of flatness (in the molecule of a chemotherapeutic drug) was first observed in members of the aminoacridine series, which require about 38 $Å^2$ of planar surface in order to exhibit anti-bacterial properties (Albert, Rubbo and Burvill, 1949). How this requirement was traced to the need for these molecules to lie on the flat purine and pyrimidine rings of DNA was described in Section 10.3.2.

Each benzene or pyridine ring in a molecule permits 2–3 kcal/mol (8–12 kJ/mol) of adhesion, through van der Waals bonding, provided that it can rest on a flat area in the neighbourhood of the receptor. If it is desired to replace such a ring, in an agent, by a more hydrogenated ring, the following relationships become relevant. The cyclopentyl ring, because of the crowding caused by substituents (even by hydrogen atoms) is nearly flat; this also applies to the five-membered ring of ribose. The cyclopentenyl ring is flat, but the cyclo-hexenyl and cyclobutyl rings are puckered, and the cyclohexyl ring (12.2) is very much puckered. Even if the cyclobutyl ring were flat, it could have only two-thirds of the van der Waals attraction of a benzene ring. Another source of loss of van der Waals energy arises from the out-of-plane situation of the first atom of each substituent in a non-benzenoid ring.

Replacement of phenyl-groups has sometimes been carried out as a test for the necessity for a flat structure in a drug molecule. When the phenyl ring in amphetamine, and related phenylethylamine and phenylethanolamine drugs, was replaced by cyclo-butyl, -pentenyl, -hexenyl, and -hexyl rings, a decrease in the intensity of pharmacodynamic action was found in each case (Burger *et al.*, 1961, 1963).

Steric hindrance to effecting a simple chemical reaction is well known in preparative work. For example, it is much more difficult to esterify trimethyl-acetic acid (12.3) than acetic acid. Steric hindrance to the hydration of a double bond was discussed in Section 2.5 (p. 46). Steric influences on solubility and several other properties were discussed in Section 2.5.1. Steric interference between two groups, one on either side of a double bond, can destroy the planarity of this linkage thus introducing new physical and chemical properties into the molecule. In Section 13.1 we shall see how steric hindrance to enzymic hydrolysis has been built into the molecules of several of the semi-synthetic penicillins. Again, lidocaine (7.13), which was designed to hydrolyse less readily in the bloodstream than existing local anaesthetics, was given the two ortho-placed methyl groups, and durable anaesthesia resulted.

An example of a steric hindrance necessary to achieve the desired pharmaco-logical response is afforded by bretylium (12.4), used as a blood-pressure-lowering drug. The bromine in this substance can be replaced by chlorine, iodine, a methyl- or a nitro-group without losing the biological effect, but the latter disappears when the bromine is replaced by hydrogen. This suggests that an *ortho*-substituent is necessary to force the basic side-chain out of the plane of the benzene ring (Boura, Copp and Green, 1959). The estrogenic properties of stilbestrol depend on the steric influence of the two ethyl groups, as explained in Section 12.2.

The solution to any problem with a stereochemical aspect requires access to molecular models. Of these, there are two main kinds. The first is the *skeletal* or *framework* model such as those devised by Dreiding or Kendrew. These indicate the centres of bonds that join atoms, and are useful to find conformations suitable for interaction between two molecules. The other type of model, *space-filling* (e.g. CPK, or Courtauld), shows both the shape of the molecule and the volume that it occupies. This kind is very useful for showing the overall shape, surface and volume of a molecule. With practice, a chemist can learn to see a conformational drawing as a three-dimensional skeletal shape, and eventually as a space-filling molecule. There are also the CCS models, which are fundamentally skeletal models that can be quickly converted to space-filling types and back again (Clarke, 1977).

For a helpful book on model building, see Walton (1978).

Larger structures may be built with the 'Molecular Design' kits offered by Academic Press and used in either the skeletal or the space-filling mode. It is claimed that two turns of DNA can be built with such a kit in 6 hours, and a molecule of tRNA in 3 days. Firm locking of otherwise freely rotating bonds is a feature of these models.

Instead of building solid models, their images can be created on a cathode-ray tube by the use of computer-controlled modelling with a graphics system such as the Evans and Sutherland Multi-user Picture System. This is operated by a high-resolution calligraphic (line-drawing) mono- or poly-chromatic computer-fed terminal. Chemical structure input is effected with a light pen or digitizing tablet. With its aid, the molecular outlines of various drugs and their receptors can be created, and each independently fitted to the other. The resolution is better than in domestic television receivers.

In actual use, one projects an image of a receptor in one colour and of an agonist in another colour, with all van der Waals bonds in correct proportions, and rotates the receptor and drug images with respect to one another. When 'best fit' has been achieved, one can rotate the complex to study various aspects of it. Coloured stereoptical pairs of photographs are now appearing in the literature, e.g. dihydrofolate reductase and its inhibitors (Hansch *et al.*, 1982). Thyroxine (*11.14*), whose reaction with its first receptor, prealbumin, was described in Section 2.4, can be seen, thus combined, in coloured stereoptics (Blaney *et al.*, 1982).

In such computer graphics, the three-dimensional effect is arranged by making fainter those parts of the image that should be distant from the viewer. Alternatively a *pair* of images can be commanded, for viewing with stereo-spectacles.

In a new, and highly sophisticated, extension of this facility, molecules are represented not as nuclei but in terms of electron densities. The shape of the molecule is delicately suggested as a net covering the exterior of these charge-points. Alternatively, the molecules can be shown skeletally, as in Dreiding models, with the electron density superimposed in a second colour. The electrostatic fields generated around the molecule can be indicated by joining negative potentials in blue and the positive ones in white (Richards and Sackwild, 1982).

In this work, details of the receptor, particularly if it forms part of an enzyme, may be obtainable from the Cambridge or the Brookhaven crystal-structure databanks (see Section 17.4).

The next three Sections, 12.1 to 12.3, will discuss some of the biological consequences of optical isomerism, geometrical isomerism, and conformational behaviour.

12.1 Optical isomerism

Carbon, far more than any other atom, is responsible for the principal forms of stereoisomerism dealt with in this and the following two sections. Most of the examples of *optical* isomerism arise from the presence in the molecule of a carbon atom with four single bonds, each of which is connected to a different kind of atom or group. Such a carbon atom, i.e. one with four different kinds of substituents, is called *asymmetric*.

The atom of carbon, although spherical, has its four bonds evenly (and rigidly) disposed, as if at the corners of a tetrahedron, namely a solid with four faces, all meeting at 109°. It may be tilted in all directions without changing the order of the four substituents. Thus in (*12.5*), BCD may be thought of as a triangular plane resting on a table, and A as an apex much nearer the eye. In other presentations of the same tetrahedron, B, C, or D in turn can be made the apex, leaving the other three groups in the plane. The presentation (*12.5*) can be tilted forward to (*12.6*) so that CD forms a line on the table, and AB is another line, parallel to the table, but above it. Presentation (*12.7*), an intermediate state between (*12.5*) and (*12.6*), has CD on the table, B a little nearer the eye, and A much nearer the eye. These are all presentations of the *same* molecule, but if two substituents are interchanged, as in (*12.8*), a different substance (namely, the other member of a chiral pair) is produced. The consequences of this can conveniently be examined with the help of lactic acid (*12.9*).

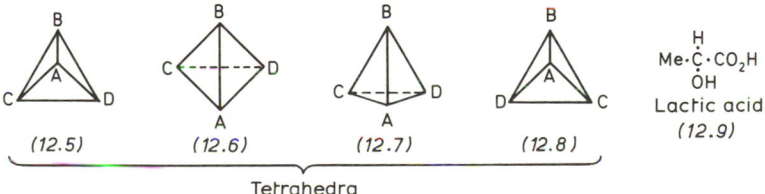

(12.5)   (12.6)   (12.7)   (12.8)

Tetrahedra

Lactic acid
(12.9)

When four different atoms, or groups, are attached to a central carbon atom, as in lactic acid (*12.9*), two different spatial arrangements, (*12.10*) and (*12.11*), are possible. One arrangement is the non-superposable mirror-image of the other, and hence they are related as one's right hand is to one's left. From the Greek word *kheir* (a hand) has been fashioned the word 'chiral', and such pairs composed of a left-handed and a right-handed member are referred to as *chiral pairs* or enantiomers. Older names were: enantiomorphs, and optical antipodes.

The two members of a chiral pair have identical chemical properties. They also have all physical properties identical, except this one: each rotates the plane of polarized light in the opposite direction to the other, but to an equal degree. Thus for a molecule to be chiral, it must have this potentiality for optical isomerism. The biological interest springs from the fact that optically active substances are absorbed selectively on optically active surfaces, such as protein, and hence although the two optical isomers of an agent often show the same biological effect, they usually show it to a very different degree.

$$
\begin{array}{cc}
\overset{\displaystyle B}{\underset{\displaystyle A}{C\cdots\vert\cdots D}} & \overset{\displaystyle B}{\underset{\displaystyle A}{D\cdots\vert\cdots C}} \\[2mm]
(12.10) & (12.11)
\end{array}
$$

For simpler representation on paper, it is more covenient to think of the four bonds, which unite the four different substituents to the carbon atom, as proceeding from the *centre* of the atom but to the same corners as in (*12.5*) to (*12.7*). Then (*12.10*) becomes an acceptable representation of one of the enantiomers of lactic acid, and (*12.11*) of the other. The wedges imply that the atoms at their broad edges are above the plane of the paper and those at the end of the lines of the dashes are below [(*12.10*) is presented in the same aspect as (*12.6*)]. In a crowded molecule, or for speed of handwriting, the wedges may be replaced by thick lines as in (*7.11*); but the dashes should not be replaced by three dots, the symbol for a hydrogen bond.

Often the use of heat, or of acid or alkali, converts one enantiomer into a mixture of equal parts of both enantiomers, which causes optical activity (as measured in a polarimeter) to disappear. The substance is then said to have been racemized. Both D- and L-lactic acid occur in nature, unracemized*.

A recent (1980) survey showed that, of 2000 most commonly used drugs, about half were chiral. Of the synthetic examples, only 88 were marketed, optically resolved, as single enantiomers. Of 550 agrochemicals, studied at the same time, only 103 were chiral.

If a molecule has *two* asymmetric carbon atoms, provided that these are non-identical, there can be *four* possible stereoisomeric forms, and hence two completely different racemates. The relationship between an enantiomer from

---

*Resolution, the reverse of racemization, is effected with an optically active ion of opposite charge, or with an enzyme which destroys one enantiomer (Wilen, 1971; Wilen, Collet and Jacques, 1977). Sometimes, chromatography on an optically active support, such as a sheet of paper, suffices (Albert and Serjeant, 1964).

one racemate and one from the other cannot be a mirror-image, and members of such a pair are called diastereoisomers. The four ephedrines (*12.12*) to (*12.15*) present such a set of diastereoisomers (Witkop and Foltz, 1957). It will be seen that more than one change has to be made in passing from (*12.12*) to (*12.13*) to conserve the mirror-image relationship. A more quickly written notation (*12.42*) will be mentioned at the end of Section 12.3.

                OH   NHMe   H   H
                 | | | |
        Ph—C — C—Me   Ph— C — C—Me
                 | | | |
                H   H   OH   NHMe

   (–)-Ephedrine (anti-asthmatic drug)   (+)-Ephedrine (not used in medicine)
                (3R, 2S)   (12.13)
                (12.12)

                H   NHMe   OH   H
                 | | | |
        Ph—C — C—Me   Ph — C — C—Me
                 | | | |
                OH   H   H   NHMe

        (+)-Pseudoephedrine   (–)-Pseudoephedrine
        (local vasoconstrictor)   (not used in medicine)
                (3S, 2S)   (12.15)
                (12.14)

            The four diastereoisomeric forms of ephedrine

12.1.1 *The allocation of chirality*

From the time of Biot and Pasteur, any substance which rotated the plane of polarized light to the left was designated (−), and its enantiomer similarly (+). This classification, although merely descriptive, is still used when nothing more fundamental is known about a substance (the equivalent small letters, *l* and *d*, are no longer in use). However, by the end of the nineteenth century, it was realized that this designation does not get to the heart of the matter, which is as follows. An asymmetric molecule must have either a right-handed (clockwise) or a left-handed twist, but the polarimeter can only reveal that a twist is present and is quite unreliable for specifying its direction. This is because the direction of rotation is influenced by other groups present, and by the solvent, the temperature, and the wavelength of the polarized light. Because of these variables, many substances can yield both (−) and (+) rotations. Even when rotations were limited to water at 25°C, and the D-line of a sodium lamp, many anomalies were found. The nineteenth century closed with strong efforts to establish a *standard* by which all substances with a right-handed twist could be grouped together regardless of the polarimeter sign!

In 1906, Martin Rosanoff, a young instructor at New York University, suggested (+)-glyceraldehyde as a standard of 'absolute configuration' whose twist was to be designated by a small capital D (Hudson, 1948). He chose glyceraldehyde as being the simplest possible substance with the properties of a sugar. At that time, it had not been isolated, but was synthesized and resolved by Wohl and Momber (1917). Because this substance [HO.CH$_2$CH(OH) CHO] is also formed by the action of the body's enzymes on glucose and fructose, it has become the sheet-anchor for carbohydrate chemists. All substances that can be related to this standard by *genetic change*, namely degradation or synthesis by gentle reactions that avoid groups attached to the chiral atom, have been admitted to the D-series.

Examples of current usage of the 'absolute' notation based on glyceraldehyde are: D-(+)-glucose for blood-sugar, D-(−)-fructose for fruit sugar and D-(−)-deoxyribose for the sugar in DNA.

Because there was only a 50% chance that the choice of D for glyceraldehyde could correspond to a natural right-handed twist, efforts were made to find new physical techniques that could yield the desired information. No help was obtainable from X-ray-crystallographic diffraction. Although that technique faithully records all bond lengths and bond angles, these measurements do not differ between members of a chiral pair! Luckily, in 1951, the Dutch chemist, Bijvoet found a new technique that led to assignment of the first natural configuration. This was the first time that X-rays had been adapted to distinguish between a model and its inversion.

What Bijvoet did was to introduce a phase-lag by using a narrow waveband to excite the fluorescence of *one* atom near the chiral centre. Using sodium rubidium tartrate, it was shown that dextrorotatory tartaric acid has a right-handed twist. Because this enantiomer had already been related to D-(+)-glyceraldehyde by genetic change, the 'absolute configuration' designated as D turned out to be *absolute* after all (Bijvoet, Peerdemar and Van Bommel, 1951).

The Bijvoet technique is too burdensome for *routine* use in establishing configurations. About 200 examples of its use will be found in the *Atlas of Stereochemistry* by Klyne and Buckingham (1978), and the listing is continued in the periodical, *Molecular Structure and Dimensions*, issued by the Crystallographic Data Centre, Cambridge, England, and the International Union of Crystallography. Ordinary X-ray-diffraction analysis becomes helpful when more than one chiral centre is present provided that the configuration of the first centre is known.

Protein chemists had long felt the need for an absolute standard for amino acids. In 1950, D-(+)-glyceraldehyde was converted, by genetic change in five steps, to D-(+)-serine. This showed that the (−)-serine, isolated from the hydrolysis of proteins, was L-(−)-serine. This result helped to fix all the amino acids of protein as members of the L-family, although some (such as alanine) show the (+) rotation (Brewster *et al.*, 1950).

Contrary to what one sometimes hears or reads, Nature does not favour one particular twist. For example, (+)- and (−)-quartz are equally common and so are (+)- and (−)-lactic acids. Bacteria are as mortally injured by deprivation of D- as of L-alanine (Section 5.3) and, although a mixture of D- and L-amino acids cannot be fitted into the structure of a protein, they are often found side-by-side in polypeptides which they tend to cyclize (Section 14.3). The *Cinchona* tree produces in its bark the two enantiomers known as (+)-quinidine and (−)-quinine (*10.33*) whose secondary alcoholic group has an asymmetric carbon atom. Both isomers are used in medicine; quinidine in cardiac arrhythmia and quinine in malaria.

Less reliable than genetic change and far less reliable than a Bijvoet analysis is the attempt to establish the direction of molecular twist by optical rotatory dispersion, where the optical rotations of the compound are plotted against the various wavelengths of light at which they were determined (Djerassi, 1960). This method can be misleading except for quite simple molecules.

The *sequence rule* was introduced by Cahn, Ingold and Prelog (1956) as an aid to nomenclature of optically active molecules. The authors stated that use of the symbols (*R*) and (*S*) would remedy ambiguities that arose when the D and L system was applied outside the carbohydrate and amino-acid series. They also claimed that it would systematize the storage of stereochemical information and assist retrieval, e.g. for building a molecular model. The utility of the system is greatest when two or three chiral atoms are present in the same molecule. A simplified version of the sequence rule is presented by Cahn (1964).

To use this rule, one begins with the chiral centre of lowest number in the chemical formula, numbered according to IUPAC rules. Its chirality may be taken from the *Atlas of Stereochemistry* or else determined by procedures discussed above. Next, the four atoms joined to the asymmetric carbon atom are named a,b,c,d in order of decreasing atomic weight*. If two of them are carbon atoms, priority is determined by the nature of the atom to which each is joined. One then has to imagine that the substituent of lowest weight (d) lies in the axis of a steering wheel. The three other groups are then disposed around the rim of the wheel in the order a → b → c, and in the direction that their known chirality requires. The wheel is then rotated in the direction a → b → c. If this rotation is to the right, the configuration is catalogued as (*R*) from the Latin word *rectus* (right); but if the rotation is to the left, the configuration is (*S*) or *sinister*. Examples of these designations are seen in ephedrine (*12.12*) and pseudo-ephedrine (*12.14*). Natural threonine is (2*S*,3*R*).

The sequence rule has been well received and widely applied. It appears at its most unhelpful when, in a related series of compounds, formulae that are clearly analogous receive different sequence designations (Cahn and Dermer, 1979). For example, most of the amino acids derived from the hydrolysis of protein are (*S*) including serine, but cysteine (which differs from serine only in having S for O), is graded (*R*). This anomaly of the sequence rule arises from the precedence that it gives to sulfur, weightwise; similar anomalies arise from its preference of oxygen over nitrogen on the same grounds. With this background in mind, Cahn and Dermer recommend that the sequence rule be not used for carbo-

---

*The three authors arbitrarily chose *decreasing* weight to make their (*R*) correspond to D (in D-(+)-glyceraldehyde.

hydrates, amino acids, steroids, cyclitols, and lipids all of which families have satisfactory rules of their own.

Turning to structural formulae, three different kinds are in use for depicting carbohydrates. Of these, the Fischer projection for glucose (12.16) is the oldest. Here the carbon chain is written vertically with the aldehyde (hemiacetal) group near the top. The groups projecting to left and right of the carbon chain are considered to be in the plane of the paper. The enantiomer with the highest-numbered asymmetric carbon atom *on the right* is regarded as having the D configuration.

Later, Haworth introduced the more realistic representations that bear his name, e.g. (12.17). In these, the acetal group is kept on the right. The edge of the ring that is nearest the reader is represented by thickened lines. This makes it clear which of the substituents are above and which below the plane of the ring; for example the —CH_2OH group is *above*, and the hydrogen atom that is attached to the same carbon is *below* the plane of the ring (the carbon atoms of the ring are not shown). However, it is known from X-ray diffraction analysis that the rings of hexoses are not flat, but puckered. This knowledge leads to conformational representations, such as (12.18). Each type of representation has its uses, depending on the circumstances.

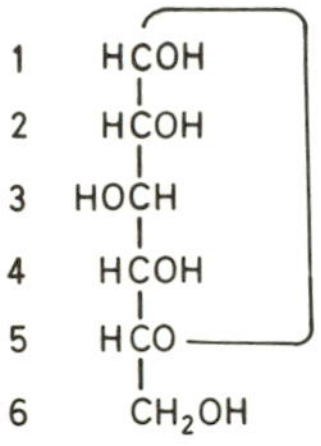

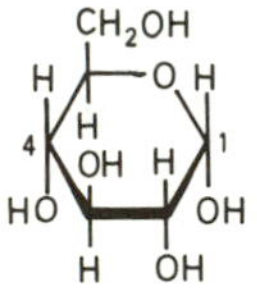

Haworth's representation
of D-glucose
(12.17)

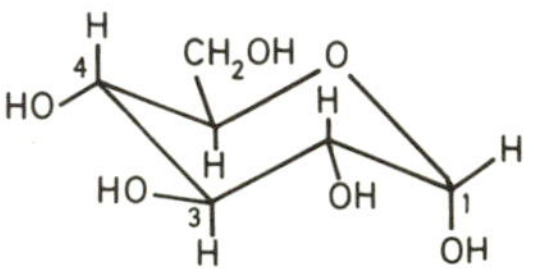

Conformational representation
of D-glucose
(12.18)

D-Glucose
(12.16)

The *Atlas of Stereochemistry* lists the absolute configurations of about 3000 organic substances, with derivations. Apart from optical isomerism, it lists many examples whose chirality is owed to isotopic substitution, to chiral axes or planes, or to chiral centres other than carbon (Klyne and Buckingham, 1978).

12.1.2 *Pharmacological aspects of chirality*

Cushny (1926) was the first to realize that differences in the biological activity of enantiomers were caused by one chiral form fitting the receptor much better than the other one did. When there is no difference in biological activity, we must assume that the asymmetric atom is not involved in contact with the receptor or else makes only a two-point contact with it. Examples of enantiomers with identical pharmacological action are the (+)- and (−)-forms of barbiturates (2.34) which, when made optically active by incorporating an asymmetric carbon atom in a 5-alkyl-group, act equally strongly (Kleiderer and Shonle, 1934). Both (+)- and (−)-cocaine (7.11) are equally powerful local

anaesthetics (Gottlieb, 1923). Many such examples are known among synthetic
drugs. Thus (+)- and (−)-chloroquine (*10.31*) have equal anti-malarial action
(Riegel and Sherwood, 1949). In such cases, it is assumed that the agent and the
receptor make only a two-point contact at the asymmetric carbon atom, or else
that this atom is not involved in the contact.

When one optical enantiomer is more biologically active than the other,
antagonism between them is rarely found. This is because the space-relation-
ship required for adsorption on the receptor is the very one altered by passing
from D- to L-forms, or vice versa. For this reason, a mixture of two optical
antipodes (or the racemized substance) usually has the averaged potency of
both constituents, and there is no antagonism. Among the few exceptions to this
generalization are D-histidine which inhibits the hydrolysis of L-histidine by
histidase (Edlbacher, Bauer and Becker, 1940), and 2,4,5-trichlorophenoxy-
propionic acid (*12.19*), the plant hormone effect of whose D-form is greatly
decreased by the L-enantiomer (Smith, Wain and Wightman, 1952). The latter
example seems to be due to the fact that such auxins require *two* consecutive
receptors, of which the first may not be stereospecific (Section 2.4, p. 41). A
tremendous number of examples are known where there is a very large difference in pharmacological action between enantiomers. Thus D(−)-isopropyl-
norepinephrine (isoprenaline) has 800 times the bronchodilator effect of its
L(+)-isomer (Luduena *et al.*, 1957). Similarly, the natural D(−)-isomer of
epinephrine has 12–20 times the activity of its enantiomer on various test-
objects (Tainter, 1930; Blaschko, 1950). Again, the L(+)-form of acetyl-β-
methylcholine is about 200 times more active on gut than the D(−)-form.
Nicotine presents a very unusual feature: the difference in activity between the
natural L(−)-form and the stereoisomeric D-form varies with the test-site, so
that equipment ratios for the two enantiomers vary from about 1:1 to 1:40.
However, where, as is usual, one form is more active, that is always the L-form
(Barlow and Hamilton, 1965).

Many adrenergic amines have been examined for the effect of optical
isomerism on pressor activity. It is thought that D-epinephrine makes a three-
point contact with its receptor through the following three groups: (a) the
amino-group, (b) the benzene ring with its two phenolic hydroxyl-groups, and
(c) the alcoholic hydroxyl-group in the side-chain. The biologically weak
optical isomer, L-epinephrine, can make contact by only two groups (see Fig.
12.1, in which the shapes and sizes of the symbols have been chosen arbitrarily).
Deoxyepinephrine (epinine) should therefore have much the same activity as
L-epinephrine, as is indeed the case (Easson and Stedman, 1933). This hypothesis has been confirmed by further examples (Badger, 1947; Stedman, 1947).

Many agents, which are biologically highly active but lack the molecular
structure necessary for optical isomerism, display differences in activity
between isomers once the possibility of isomerism is introduced. Thus the auxin
types of plant-growth regulators usually lack an asymmetric carbon atom and
hence are incapable of furnishing pairs of optical isomers. However, when an

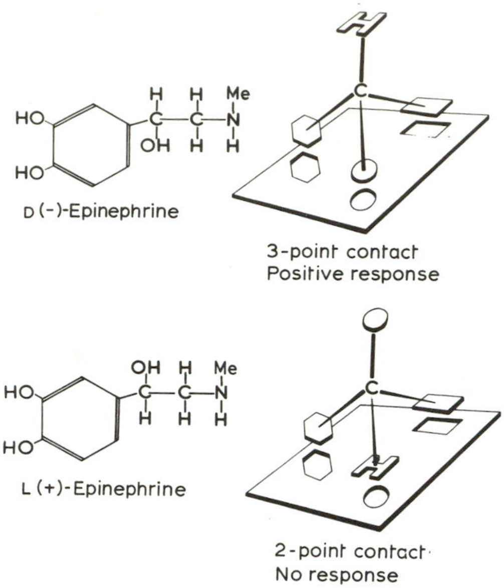

Fig. 12.1. Diagram illustrative of contact between optical isomerides and a surface complementary to one of them.

asymmetric carbon atom was deliberately introduced, as in α:2,4,5-trichlorophenoxypropionic acid (*12.19*), it was found that the biological activity was confined mainly to the D(+)-form. This led to the formulation of the hypothesis that auxin-like action requires a three-point contact as depicted for epinephrine in Fig. 12.1.

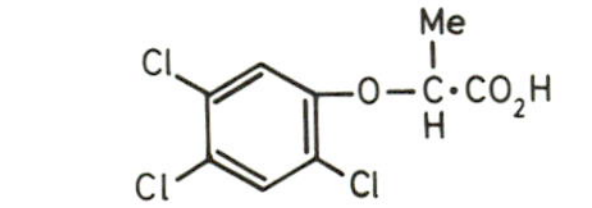

Trichlorophenoxypropionic acid
(*12.19*)

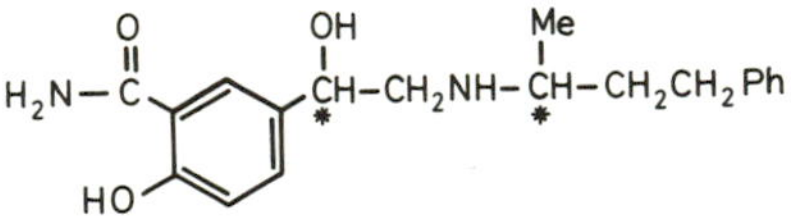

Labetalol
(*12.20*)

Propranolol (*12.56*), which was the first clinically useful β-adrenergic-blocking agent (Section 12.4) used in the treatment of hypertension, has most of this activity in the laevorotatory (*S*) isomer, but both enantiomers are equal as membrane stabilizers and local anaesthetics (Barrett and Cullum, 1968).

Labetalol (*12.20*), a recent analogue, has four stereoisomers because it has two centres of asymmetry (marked with stars). The (*R,S*) and the (*S,S*) isomers are virtually inactive, the (*S,R*) isomer has most of the α-adrenergic blocking activity, and the (*R,R*) has virtually all of the β-adrenergic blocking activity and is the only isomer with anti-hypertensive activity comparable to propranolol (Gold *et al.*, 1982).

Optical isomerism can occur even in the absence of an asymmetric carbon atom when inhibition of rotation creates a centre of asymmetry in the molecule. The best-known examples are those biphenyls which are separable into (+) and (−) isomers when each of the two benzene rings has a bulky *ortho*-substituent. These *ortho*-groups sterically interfere with free rotation and hence introduce the necessary element of asymmetry. Both *ortho*-groups may be the same. The simplest compound to be resolved is biphenyl-2,2′-disulfonic acid (*12.21*).

In allene, where the two double bonds are perpendicular to one another, derivatives such as (*12.22*) can be optically resolved. In one example, X was the benzene and Y the naphthalene ring.

For further reading on stereochemistry, see Mislow (1965); for molecular asymmetry (its presence and its absence) in biology, see Bentley (1969).

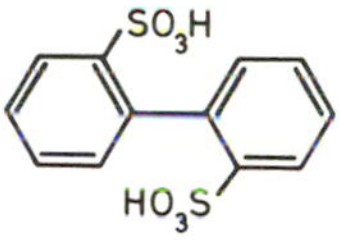

| A chiral biphenyl | A chiral allene | (*12.23*) | (*12.24*) |
| (*12.21*) | (*12.22*) | (cis) | (trans) |

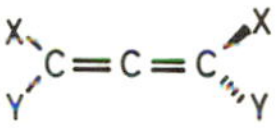

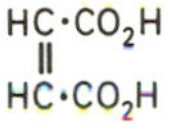

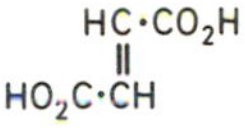

12.2 Geometrical isomerism

Geometrical isomers are found when rotation of atoms in a molecule is restricted by a double bond. Such a pair of geometrical isomers is exemplified by maleic (*12.23*) and fumaric (*12.24*) acids (respectively *cis-* and *trans-*). Geometrical isomers, in spite of a similar fundamental structure, are not related to one another as an object is to its mirror-image and do not rotate the plane of polarized light. In general, *cis-* and *trans-* forms have very different physical properties. For example, maleic acid (*12.23*) has mp. 130°C, and pK_a 1.9, and it is very soluble in cold water (79 g/100 ml), whereas its geometrical isomer, fumaric acid (*12.24*) has the constants, respectively, of 287°C, 3.0, and 0.7 g/100 ml. Not surprisingly, geometrical isomers have very different biological properties (Butler, 1944), and hence it is important, when inspecting a new chemical formula, to recognize any possibility for this kind of isomerism.

The separation of *cis-* and *trans-* isomers is easily effected by crystallization or chromatography. There is no universal method for interconverting members of such a pair, but often heat produces the more stable and light the less stable isomer. Human vision depends on the conversion by light of the 11-*cis*-isomer of retinal to the 11-*trans*-form. As soon as the excitatory beam is shut off, this carotenoid pigment reverts to the *cis*-form, thus terminating the impulses relayed to the brain (Gilardi *et al.*, 1971).

The ring of cyclopentane is almost flat and gives rise to *cis-* and *trans-* isomers, as though it were a big double bond. The ring of cyclohexane, although not flat, is flat enough to give this result. Thus both *cis-* (*12.25*) and *trans-* (*12.26*) forms of diaminocyclohexane exist, and each is available commercially. Both optical and geometrical isomerisms can occur in the one molecule, and the *trans*- isomer has been resolved into (*S,S*) (*12.27.*) and (*R,R*) (*12.28*) chiral isomers. However the *cis*- geometrical isomer cannot be optically resolved because it has a plane of symmetry. The benzene ring, having only one substituent in each position, cannot give rise to geometrical isomerism.

| *cis*- | *trans*- | *trans*-(*S,S*)- | *trans*-(*R,R*)- |
| (*12.25*) | (*12.26*) | (*12.27*) | (*12.28*) |

Stereoisomeric forms of 1,2-diaminocyclohexane

Sometimes it has been difficult to decide which two substituents out of the four at the ends of a double bond should be selected for determining whether the configuration is *cis* or *trans*. The sequence rule prescribes that the two heaviest atoms should be selected and writes Z (from the German *zusammen*) for *cis* and E (from the German *entgegen*) for *trans*. In formulae with many possibilities for geometrical isomerism, many authors write *r* after the numeral of the lowest-numbered substituent, and then *c*- and *t*- before the numeral for each position known to be *cis* or *trans*, respectively.

The enlargement of plant cells by indol-3-ylacetic acid (*4.82*) is imitated by many of those carboxylic acids in which the carboxy group makes an angle with the aromatic ring. Geometric isomerism provides one way for meeting this requirement, and it has been found that *cis*-cinnamic acid is active, whereas the *trans*-isomer is inactive (Haagen-Smit and Went, 1935). Again, 2-phenylcyclo-propane-1-carboxylic acid and 1,2,3,4-tetrahydro-naphthalidene-1-acetic acid are active only in the *cis*-form (Veldstra and Van der Westeringh, 1951). In these examples, it is easily seen from molecular models that the ring and carboxylic acid groups are planar in the *trans*-isomer (inactive) but non-planar in the *cis*-isomer (active). Veldstra was the first to point out this connexion between non-planarity and plant-growth activity. Another way to introduce non-planarity is by steric hindrance. Thus, although benzoic acid is kept flat by resonance, and is biologically inactive, 2,6-dichlorobenzoic acid and 8-methyl-1-naphthoic acid are non-planar, and biologically active (Veldstra, 1963).

In all work on auxin analogues, the carboxy group can be replaced by other electron-attracting groups ($-CN$, $-NO_2$, $-SO_3H$) with only moderate loss of biological action. For a summary of connexions between structure and action in this series, see Koepfli, Thimann and Went, 1938; Veldstra, 1963). The action of auxins requires sequential receptors (p. 41).

The geometrical isomerism of steroids has attracted much attention. This family of naturally occurring, and largely saturated, substances has the general structure (*12.29*). The formula shows the numbering of the carbon atoms and the lettering of the four rings. In all naturally occurring steroids, the junction

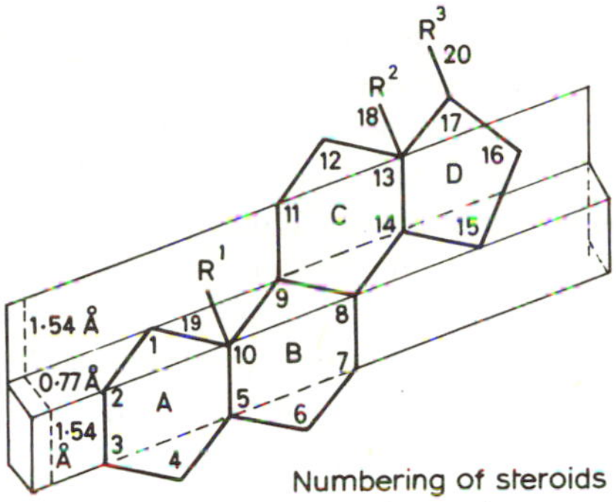

Numbering of steroids
(12.29)

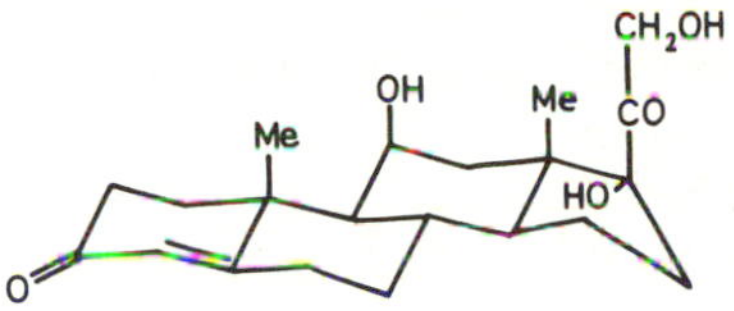

Hydrocortisone
(as seen from front edge)
(12.30)

between rings B and C is *trans* and they are both locked in a chair conformation. In the cardiac glycosides the junction between rings C and D is *cis*, but in the animal hormones, sterols and bile acids this is *trans*. Most biologically active steroids have a *trans* junction between rings A and B, and are therefore said to belong to the '5α' (formerly 'allo') series. That each ring is puckered can best be seen from the edge view, such as (*12.30*).

The meaning of '5α' is that the hydrogen atom in position 5 lies below the general plane of the rings. All substituents which lie below this plane in other positions are designated α, and those that lie above this plane are called β. The α-substituents are represented by dotted lines, and the β-substituents by thick lines. These symbols, α- and β-, are used, with similar meaning, for other polycyclic structures, also, such as triterpenes and alkaloids. The complexity of such structures makes the application of (*R*) and (*S*) nomenclature too difficult.

In general, a high degree of any kind of mammalian biological activity, in the steroid series, is correlated with a lack of α-substituents along the edge of the molecule which runs from 1 to 17, and a lack of β-substituents from the bottom edge (from position 4 to 15). The edge view of the hydrocortisone molecule (*12.30*) illustrates the above generalization (Sarett, Patchett and Steelman, 1963). It is thought that interaction of steroid hormones with proteins takes place through the uncluttered underside (or α-surface) of the molecule. See Section 2.4 (p. 42) for specific protein transportation, as the first step in the biological action action of steroid hormones.

The various steroids differ from one another mainly by variations in the nature of R^1, R^2, and R^3 (*12.29*), but extranuclear substituents are often present, and sometimes a degree of unsaturation is found. Cyclohexenone structure in ring A is usually necessary for progesterone, androsterone, and corticosteroid activity. Cortisone-type action requires, generally, oxygen atoms at positions 3, 11, and 17, as well as the characteristic —CO.CH$_2$OH group in position 17. Androgenic and corticoid activity are highly dependent on these details of structure, but progestational activity persists when the 17-acetyl group is changed to the unnatural α-configuration, and replacement of the methyl-group in position 18 by an ethyl-group actually furnishes increased pro-gestational activity, as in norgestrel, the 'minipill' oral contraceptive.

Of all steroid hormones, the estrogens are the least dependent on structure. Provided that ring A is aromatic, and carries the acidic hydroxy-group in the 3-position, the rest of the steroid structure assumes only secondary importance. The simple and very effective (although not quite benign) benzene analogues, particularly diethylstilbestrol (*12.31*) do not share so much of the shape of the estrogenic steroid molecule as was thought when the first member was introduced in 1938. X-ray diffraction studies reveal a *trans*-structure, distorted by steric hindrance from the methylene fragments of the two ethyl-groups. As a result, the two benzene rings make a dihedral angle of 63° with the central ethylene framework, and the total picture is not at all like a steroid. However, the distance between the two oxygen atoms is roughly similar, namely 12.1 Å in diethylstilbestrol, and 10.7 to 11.1 Å in the steroidal estrogens: but all these molecules are too rigid to accommodate to a fixed distance on the receptor. As it is known that hydrogen-bonding to the receptor by two oxygen atoms is essential for estrogenic activity, the required flexibility must exist *in the receptor* (Weeks, Cooper and Norton, 1970). That the action of diethylstilbestrol is also dependent on its molecular thickness was first suggested by Oki and Urushibara in 1952. It has since been measured as 4.5 Å, identical to that of steroidal estrogens (Weeks *et al.*, 1970).

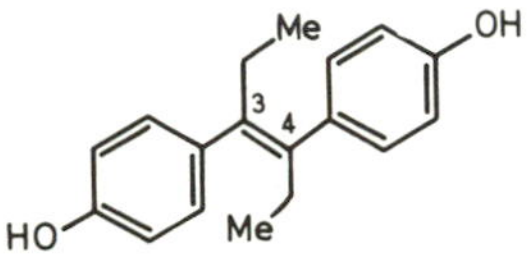

Diethylstilbestrol
(*12.31*)

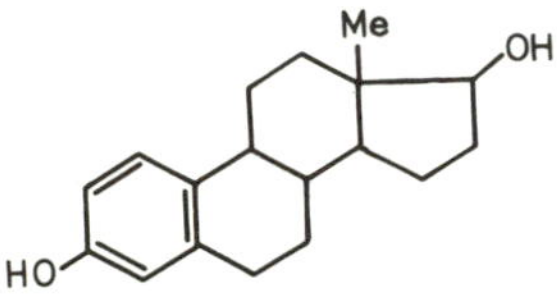

Estradiol
(*12.32*)

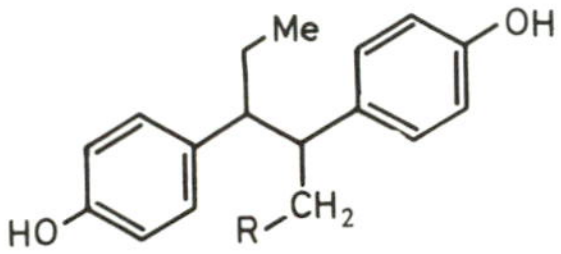

(a) Hexestrol (R = Me)
(b) Norhexestrol (R = H)
(*12.33*)

Diethylstilbestrol (3,4-*bis*-4-hydroxyphenylhexa-3-ene) (*12.31*) was introduced in 1938 as a powerful and inexpensive substitute for the principal female hormone estradiol (*12.32*) (Dodds *et al.*, 1938). Highly active orally, which the natural hormone is not, and longer-acting, this synthetic drug was hailed as a milestone in endocrine therapy, Since then it has often received a bad press, sometimes because of its widespread use in increasing the weight of farmers' livestock, but also because it has on occasion produced cancer in the daughters of women treated with it during pregnancy. In spite of this, diethylstilbestrol remains a safe drug if not given in the first trimester of pregnancy (and this applies equally to the natural hormone), and is still widely prescribed. For the use of its phosphate in treating prostatic cancer, see Section 4.2, p. 151. The estrogenic properties of diethylstilbestrol are fully retained in its dihydro-derivative, hexestrol (*12.33a*) which has the (3R,4S) configuration. It combines more tightly with the estrogen-binding protein (Section 2.4) than diethylstilbestrol does, but its lower homologue, norhexestrol (*12.33b*) binds more strongly still (Landvatter and Katzenellenbogen, 1982).

The cardioactive glycosides are discussed in Section 14.1. For further reading on the chemistry and stereochemistry of steroids, see Shoppee (1964); for the biochemistry and pharmacology, see Briggs and Christie (1977).

The geometrical isomerism of 4-aminocrotonic acid, which helped to define the active conformation of the neurotransmitter, γ-aminobutyric acid, is described in Section 12.7.

12.3 Conformational behaviour

Even when a bond is perfectly free to rotate, as every *single* bond is, an infrared spectrum often reveals that the atoms of the molecule assume various preferred positions. The rules governing conformational analysis were outlined by Hassel in 1943 and consolidated by Barton in 1950. Other methods which have greatly helped in conformational analysis include X-ray (and electron) diffraction, microwave spectra, dipole moments, chemical reactivity (see below), optical rotation (optical rotatory dispersion, circular dichroism), and n.m.r., often with application of the Karplus equation (see Section 17.3). The idea of conformation stands in contrast to the idea of configuration. Stereoisomers that are interconvertible by rotation of atoms or groups around a single bond are called 'conformational isomers' or (for short) 'conformers' or even 'rotamers'. Those that would require bond-breaking for interconversion are called 'configurational isomers', and we have seen many examples in Sections 12.1 and 12.2. Although, during a physical measurement, we see evidence for the existence of a pair of conformers, each with its own physical characteristic, we cannot hope to isolate them as separate entities, because of the speed of interconversion. However, as we shall see later, we can use a combination of conformational isomerism with optical isomerism, in a complex enough molecule, to separate a pair of isomers which do not have identical physical properties.

12.3.1 *Conformation related to a pair of substituents*

When two substituents, heavier than hydrogen, take up positions as far from one another as possible, their relative conformation is called '*staggered*' (formerly, 'trans', which is confusing because of its use in geometrical isomerism). When these two groups move opposite one another, their relative conformation is called '*eclipsed*' ('*gauche*' is intermediate). These conformations are illustrated here by 1-chloropropane in (*12.34*) and (*12.35*).

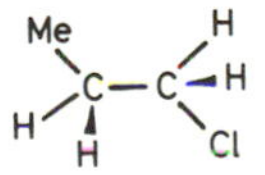

Chloropropane
(staggered)
(*12.34*)

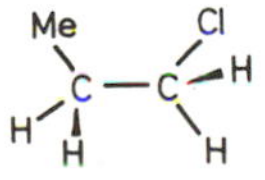

Chloropropane
(eclipsed)
(*12.35*)

This conformational analysis can be extended to alicyclic rings, both with respect to substituents and ring-shape. As regards the latter, cyclohexane (*12.2*) can exist in three conformations: chair (*12.36*), boat or tub (*12.37*), and twist or half-chair (*12.38*).

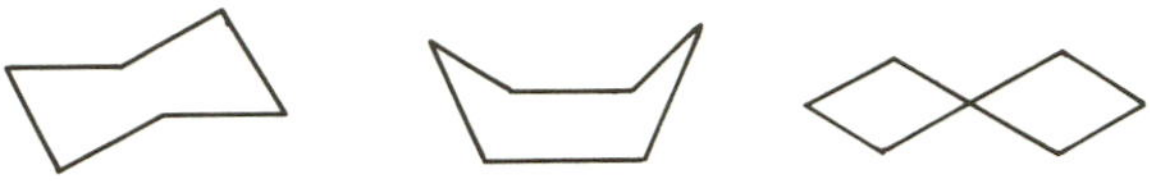

Chair   Boat   Twist
(12.36)   (12.37)   (12.38)
The three conformers of cyclohexane

The chair form is less strained and hence highly preferred (in the chair form, each axial hydrogen atoms is 2.5 Å from the other two axial hydrogens on the same side of the ring). The twist form, intermediate between boat and chair, and even the strained boat form, can be stabilized with two or more fused rings if appropriately substituted. Thus there are two forms of decahydronaphthalene, not spontaneously interconvertible, whose structure has been determined by electron-diffraction analysis. The *trans* form has m.p. $-30°C$, and b.p. $_{100}$ 117°C, and consists of two *trans*-fused chair rings. The *cis* isomer has m.p. $-43°C$, and b.p. $_{100}$ 124°C, and consists of two *cis*-fused chair rings. At elevated temperatures and in the presence of a catalyst, the *cis*-form is converted to the *trans*-form. This is a case of geometrical isomerism with respect to the bridgehead carbon atoms, but each ring remains conformationally mobile.

Much interest is attached to the CO—NH bond because it is fundamental to peptide structure. In proteins this is always *eclipsed* (*trans*), although peptides that contain proline or are otherwise tertiary amides can contain, in equilibrium, a high proportion of the *staggered* (*cis*) form, even 40% if the CO is part of an ionized carboxyl group at the end of the peptide (Rabenstein and Anvarhusein, 1982). In secondary formamides, the peptide bond exists as an equilibrium mixture of *eclipsed* and *staggered* conformers in the proportion of 92 to 8 for the *N*-methyl example (*12.39*) (*12.40*). Their co-existence shows two separate ^{1}H n.m.r. signals. When the substituent on the nitrogen atom is made more bulky, the proportion of the *staggered* conformer rises, becoming 18% when the methyl is replaced by a *t*-butyl group (La Planche and Rogers, 1964).

(eclipsed)   (staggered)
(12.39)   (12.40)
Two conformers of *N*-methylformamide

12.3.2 *Conformation of a single substituent*

The two principal conformations into which substituents fall are classified as *equatorial* (i.e. in the general plane of the ring) and *axial* (i.e. perpendicular to this plane). In a mono-substituted cyclohexane, the equatorial isomer predominates because this form has least interference from hydrogen atoms. The two conformers have the *chair* form and, although they give separate signals in physical measurements, they are in such rapid equilibrium as to be physically inseparable. In multi-substituted rings, some groups will necessarily be in the axial positions. Because axial groups are subject to greater steric hindrance, equatorial hydroxyl- and carboxyl-groups are the more readily esterified and the product the more readily hydrolysed. Often a conformational change is superimposed on geometrical isomerism, as in the pair: cocaine (*7.11*) and ψ-cocaine, the former having an axial, and the latter an equatorial methoxycarbonyl group.

Many other formulae embodying conformational information have been used in earlier chapters of this book, e.g. atropine (*7.16*), nicotine (*7.26*), morphine (*7.35*), permethrin (*6.54*), and penicillin (*13.4*), and other formulae of this kind will be used later in this chapter.

In thyroxine (*11.14*) and the more biologically active tri-iodothyronine, the iodine atoms in the 3- and 4-positions force the two rings into a conformation in which they are perpendicular to one another. Further work, with analogues, indicates that this arrangement is essential for thyroid function (Dietrich *et al.*, 1977). See Section 2.4 for other stereochemical data on how thyroid hormones fit their receptor.

Emetine and cycloheximide are two molecules that look quite dissimilar at first glance, but have conformationally similar areas which cause both of them to inhibit protein synthesis in the ribosomes of most living cells (Section 4.1, p. 147).

Acetylcholine has the same conformation in aqueous solution (as determined by n.m.r. in D_2O) as in the solid state (as determined by X-ray crystallography) (see Section 12.6). How widely does this correlation extend? Byrn, Graber and Midland (1976) have reviewed the literature on this and, as X-ray diffraction analysis is exceedingly slow, they have devised a simple test: the infrared spectrum in the solid state is compared with that in chloroform. Although this test requires further examination, it is worth recording that they found very similar infrared spectra under both conditions for choline chloride and for the anti-histaminic drug, methapyrilene, whereas they found very different spectra for histamine and for the anti-histaminic drug, diphenhydramine.

12.3.3 *Mutual adaptation of drug and receptor*

Even though we may find out, as in the previous paragraph, what conformation a drug has in solution and what it has in one very special example of the solid state, namely when it is adsorbed on itself, we still do not know what its conformation is on the receptor. Some receptors are enzymes, and cases are noted where a substrate has undergone a change of conformation on contact with its enzyme. For example strong distortion on binding to the enzyme lysozyme was shown by a tetrasaccharide fragment of murein (Section 5.3) when observed by Fourier-transform n.m.r. (Sykes, Patt and Dolphin, 1971). On the other hand, no difference in conformation was found for N-acetyl-L-tryptophan when it combined with chymotrypsin, compared by both ^{13}C n.m.r. and X-ray crystal data (Rodgers and Roberts, 1973).

A related question is: Can some drugs induce a conformational change in their receptors? This aspect is discussed under the heading 'Allosteric hypothesis' in Section 7.5.2. Certainly some substrates cause conformational changes in enzymes, clearly seen in X-ray-diffraction studies (see, for example, the account of carboxypeptidase in Section 9.0). Another conformational change, studied by the same technique, is caused by the addition of oxygen to haemoglobin (to give oxyhaemoglobin). This reaction involves a movement of the proximal histidine residue towards the plane of the porphyrin ring by about 0.85 Å (Fig. 12.2). This movement sets other tertiary-structural changes in motion because the iron atom is rigidly linked to a histidine residue. What happens is that Helix F of the globin is moved towards Helix H in the centre of the molecule and consequently expels the tyrosine moiety (140) from its pocket between helices F and H. The expelled tyrosine drags arginine (141) with it, thus breaking the latter's salt linkage. This conformational change facilitates the approach of a second oxygen atom by loosening the tight structure of the four haem units in the haemoglobin molecule.

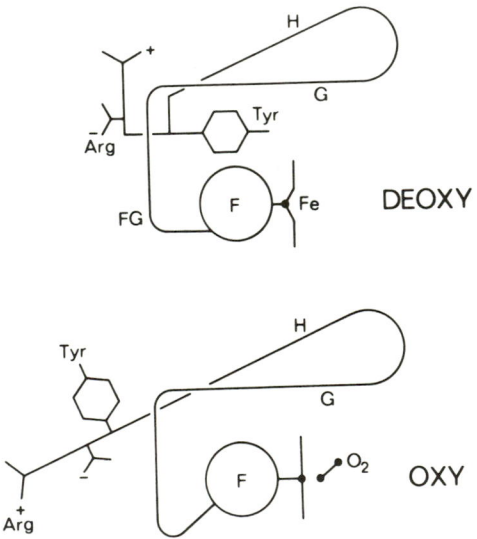

Fig. 12.2 Diagram of conformational change in tertiary structure of the haemoglobin sub-units on reaction with oxygen. (Perutz, 1970, modified in correspondence.)

An example nearer to drug action: the equilibrium conformation of bovine pancreatic RNAase is changed to different extents by the inhibitors cytidine 2'-,

3'-, and 5'-monophosphate, as seen in an n.m.r. study, and the change is proportional to the inhibition (Meadows, Roberts and Jardetzky, 1969).

Conformational specificity has been demonstrated by the coenzyme NADH (nicotinamide adenine dinucleotide) (*12.41*). Some apoenzymes that use this coenzyme remove only the axial (and others the equatorial) hydrogen atoms in the 4-position, as can readily be shown by replacing each of these atoms, in turn, by a deuterium atom. The absolute conformation of the labile hydrogen atoms in NADH were determined by Cornforth *et al.* (1962).

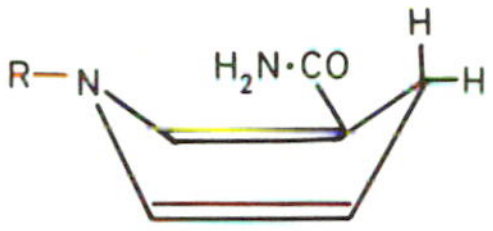

NADH
(R is ribosyldiphosphoadenosine)
(*12.41*)

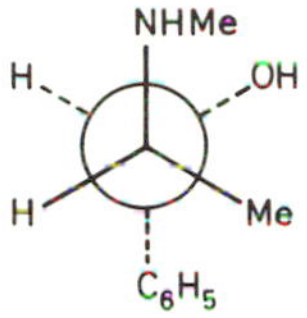

Ephedrine
(Newman projection)
(*12.42*)

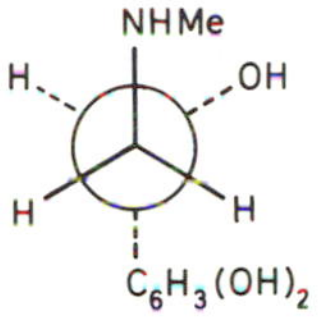

Epinephrine
(adrenaline)
(*12.43*)

It is important to keep in mind that a change in conformation can alter all physical properties and hence biological activity. This becomes important when a pair of conformers can be isolated and tested separately. This rare, but useful, event occurs most often when a chiral centre is also present in the molecule, and is in such a position that optical resolution gives one enantiomer that presses upon and distorts the framework of the molecule. But this goal can be achieved by other conditions of steric congestion that can be engineered in saturated bicyclic molecules (see Section 12.8, p. 539 for example). As a result, one isomer can be more lipophilic than the other, or a stronger base. Hence biological activity can depend, in the last analysis, not on the change in shape of the molecule but on the new physical properties conferred by this change. If a drug-designer finds these new physical properties advantageous, he should be able to provide them in other ways, independent of conformation.

A shorthand notation that displays a good deal of information about optical isomerism, or conformation, or both together, is the Newman projection, exemplified in (*12.42*) which may be compared with the usual representation (*12.12*). These projections are most useful when both of two consecutive framework carbon atoms are chiral. To use this notation, these two carbon atoms are superposed, so that six groups apparently radiate from one point. For purposes of comparison, it can be used also for related substances that have only *one* chiral centre, as with epinephrine (*12.43*).

Discussions of conformation often require knowledge of the torsion angle. This, in a group of atoms XABY is the angle that the plane XAB makes with the plane ABY. The four torsion angles of acetylcholine are discussed in Section 12.6, p. 525.

For further reading on conformation, see Barton and Cookson (1956) and Eliel *et al.* (1965). For the conformation of macromolecules, see Hopfinger (1973), and of heterocylces, see Riddell (1980).

For further reading on general aspects of stereochemistry, see Eliel and Basolo (1968), also Bentley (1969) who emphasizes the biological applications. For drugs with heterocyclic nuclei, see Armarego (1977).

The remainder of the present chapter summarizes what is known of the nature of several *receptors*, important in pharmacodynamics, much of the information being derived from stereochemistry. Receptors are the ultimate goal of much drug-designing.

12.4 Catecholamine receptors

The pioneer work on the connexion between constitution and activity in the phenylethylamine series of sympathomimetic drugs was carried out by Barger and Dale (1910). It has since become clear that *direct* action of the hormone and neurotransmitter type, is strongest in examples with hydroxy-groups in the 3- and 4-positions of the benzene ring, i.e. the 'catecholamines' such as nor-epinephrine, epinephrine and dopamine. The *indirect* action of examples without these embellishments has been outlined in Sections 9.4.3 (p. 358) and 7.6.3 (p. 300). As recounted in Section 12.1 the D-catecholamines (Fig. 12.1) have much more biological activity than their L-enantiomers.

The existence of two different adrenergic receptors was first suggested by Ahlquist (1948) Those responses evoked most readily by noradrenaline (norepinephrine), less by adrenaline (*7.43*), still less by isoprenaline (*12.44*) and least by *N-t*-butylnorepinephrine are credited to α-receptors; those for which this sequence runs in the reverse direction are credited to β-receptors. Typical α-responses are: constriction of blood-vessels, stimulation of the uterus, relaxation of the intestine. Typical β-responses are: dilatation of blood-vessels, relaxation of the uterus, stimulation of muscle glycogenolysis, production of tachycardia (Levy and Ahlquist, 1961).

The ideal way to characterize catecholamine receptors is to isolate them. Although this has not yet been accomplished in more than traces, considerable progress has been made with specific labelling. [^{3}H]Dihydroergocryptine, an

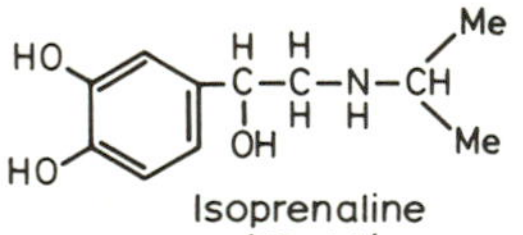

Isoprenaline
(*12.44*)

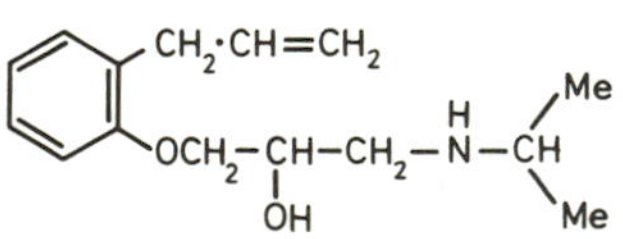

Alprenolol
(*12.45*)

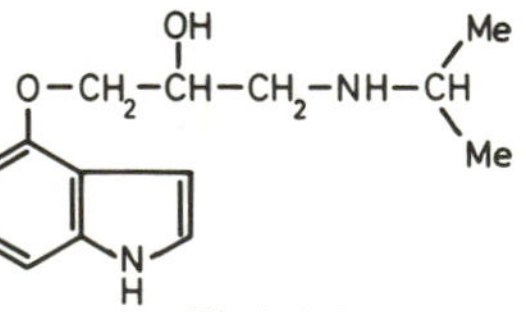

Pindolol
(*12.46*)

ergot alkaloid which is a potent α-adrenergic antagonist, was used to label (reversibly) the receptors on uterine smooth muscle. With its help, the relative specificity and potency of α-adrenergic agonists and antagonists at the receptor sites were found to correspond closely with evaluations made on the whole organ (Williams, Mullikin and Lefkowitz, 1976).

A similar exploration of the β-receptor has been made with $(-)[^3H]$dihydro-alprenolol, a labelled derivative of the potent β-adrenergic antagonistic drug alprenolol (12.45). It selectively and reversibly bound β-adrenergic receptors in rat adipocytes. With this marker, it could be seen that β-adrenergic agonists and antagonists competed for the binding sites stereospecifically and in the same ratios as is found in therapy (Williams, Jarett and Lefkowitz, 1976). Another potent β-adrenergic antagonistic drug, pindolol (12.46), has supplied a useful (reversible) marker, namely its C-$[^{125}I]$iodohydroxybenzyl-derivative. It was allowed to react with turkey erythrocyte membranes, and the labelled receptors were isolated, although only on a micro-scale (E. Brown et al., 1976). The same marker was applied to human fibroblasts and to rat glioma cells. Its displacement by β-adrenergic agonists and antagonists was found to be stereospecific and to occur in the same relative potencies as have been established in laboratory animals (Maguire et al., 1976). The binding of these labelled antagonists was saturable, also it could be prevented by prior application of the same substance unlabelled. Measurement of dissociation constants showed that the number of β-adrenergic receptor sites on an avian erythrocyte is about 1000.

Epinephrine (adrenaline) is (R)-$(-)$-1-(3,4-dihydroxyphenyl)-2-methyl-aminoethanol. The preferred conformation is the fully extended, staggered form in the solid state (Carlstrom and Bergin, 1967) and also in solution (Ison, Partington and Roberts, 1973). The ionization of three catecholamines and three of their agonistic analogues will now be discussed. Although norepin-ephrine and its N-alkyl homologues exist mainly in the form of cation between pH 6 and 8, the presence of acidic and basic groups in the one molecule, groups whose ionization constants lie within 2 pK units of one another and are of opposite charge, results in a complex pattern of minor ionic species. By methods similar to those used for tyrosine (Edsall, Martin and Hollingworth, 1958), four microscopic constants were obtained from the two apparent constants. The existence of *four* constants derives from the fact that the amino group has two constants depending on whether a phenolic group is simultaneously ionized or not. A similar circumstance provides the first-to-ionize phenolic group with two constants also. (Ionization of the first phenolic group represses ionization of the other phenolic group by Coulomb's principle.)

Here are the results for norepinephrine. The macroscopic constants were 8.63 (phenolic) and 9.73 (basic). With the help of some ultraviolet spectrometry, the following microscopic constants were calculated: pK_{1Z} 8.78, pK_{2Z} 9.58, pK_{1N} 9.16, and pK_{2N} 9.20, where Z and N refer to the zwitterion and the neutral species respectively. From these four constants, the top line of Table 12.1 was calculated. It can be seen that the cation is the major species present in this, and five other adrenergic agonists (Ijzerman *et al.*, 1984). Earlier work by Sinistri and Villa (1962) had already pointed in this direction. Thus it is most likely that the cation is the bearer of the biological action. Moreover there is no difference in ionization pattern between α- and β-agonists. (As a matter of less interest, the second-to-ionize phenolic group had pK_a 12.9.)

Table 12.1 Ionization of natural catecholamines and related agonists. (Percentage, in water at pH 7.4 and 25°C)

Agonist	Cation	Zwitterion	Molecule	Anion
Norepinephrine	94.4	3.9	1.6	0.03
Epinephrine	95.5	4.5		0.01
Isoprenaline	94.8	5.2		0.01
Salbutamol	97.9	1.5	0.6	
Terbutaline	95.6	4.4		0.01
Orciprenaline	95.5	4.5		0.01

(Ijzerman *et al.*, 1984.)

Until a workable laboratory preparation of adrenergic receptors (something corresponding to the acetylcholine micro-sacs of p. 28) is available, further efforts to understand receptor function must concentrate on the structure–activity relationships of catecholamine agonists and their antagonists in a variety of biological systems. The principal mode of attachment of the catecholamines to all their receptors is thought to be by an ionic bond. The union between catecholamine and receptor brought about by this bond is likely to be reinforced further by a hydrogen bond from the β-hydroxy-group, the absence of which strongly diminishes the action, as was discovered by Barger and Dale (1910) (see discussion to Fig. 12.1).

(*a*) *The α-receptors.* In smooth muscle, the α-receptors act by increasing the permeability of the cell membranes to inorganic ions (Bülbring and Tomita, 1969). In intestinal muscle this increase favours K^+, so that hyperpolarization followed by muscular relaxation sets in. In most other kinds of smooth muscle, the permeability increase extends to Na^+ and Ca^{2+}, so that the membrane potential falls, and excitation is followed by contraction.

In recent years, it has become possible to distinguish two types of α-adrenoreceptor: an α_1-receptor which is postsynaptic, and a α_2-receptor which is usually presynaptic but can also be found postsynaptically. The former receptors mediate vasoconstriction and the latter bring about resorption of previously released norepinephrine. Drugs which are α-*agonists* are used to raise blood-pressure in hypotensive states, and as general vasoconstrictors in dental anaesthesia, nasal congestion, and conjunctivitis of the eye. Norepinephrine is the natural agonist, but phenylephrine (*12.47*) has some advantages. This substance, which does not occur naturally, has the constitution of epinephrine minus one phenolic group. First investigated by Barger and Dale (1910), phenylephrine is now known to have an almost purely α_1-agonistic action. It is a much-prescribed drug. Compared to norepinephrine, it is a little weaker but the effect lasts longer, it is active by mouth, and does not stimulate the central nervous system. Phenylephrine acts directly on the postsynaptic adrenergic receptor as a pure agonist; it is not a false transmitter (see p. 302 for definition).

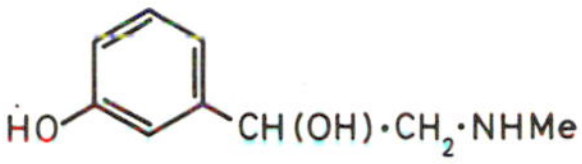

HO—⟨ring⟩—CH(OH)·CH$_2$·NHMe

Phenylephrine
(12.47)

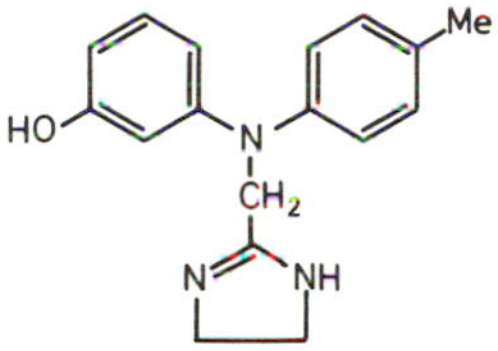

HO—⟨ring⟩—N—⟨ring⟩—Me
|
CH$_2$
|
⟨imidazoline N NH⟩

Phentolamine
(12.48)

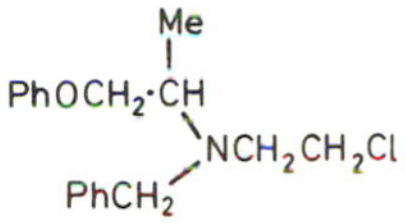

Me
|
PhOCH$_2$·CH
\
NCH$_2$CH$_2$Cl
/
PhCH$_2$

Phenoxybenzamine
(12.49)

The original *α-antagonists* did not distinguish between α_1- and α_2-types and found only a few therapeutic applications, but they gave some help in frostbite and other, more long-term, cases of impaired circulation. Typical examples were the imidazolines such as phentolamine (*12.48*) and the chlorethylamines such as phenoxybenzamine (*12.49*). The latter became activated by self-quaternization (defined on p. 577) and caused irreversible alkylation of α_1-receptors.

More recently, pure α_1-antagonists have been introduced to control hypertension, notably prazosin (*7.57*) and the chemically similar indoramin (Docherty, MacDonald and McGrath, 1979). These drugs are valued for treating hypertension in the many asthmatics who regularly take a β-agonist, or for those patients who do not respond well to the currently favoured combination of a diuretic and a β-antagonist. Because they cause some postural hypotension, α_1-antagonists are not prescribed for uncomplicated cases. A very interesting new drug is labetalol (*12.20*) which displays both α_1- and β_2-blocking properties. Its potential can be seen from the following considerations.

α_1-blockers lower blood-pressure by inhibiting the stimulant effect of norepinephrine at vascular α-adrenoceptors. Hence they dilate peripheral arterioles and so decrease peripheral resistance, an action which effectively lowers blood-pressure in hyperpiestic patients. However, there are two unwelcome side effects: postural hypotension where the patient faints upon rising to his feet, and reflex tachycardia. As for the β-blockers, their anti-hypertensive effect is accompanied by reduction in cardiac output which reflexly adds a little to peripheral resistance. However, drugs that can block both α_1- and β-receptors show an additive beneficial effect on the patient, while cancelling the side effects of either drug given separately (Richards and Prichard, 1978).

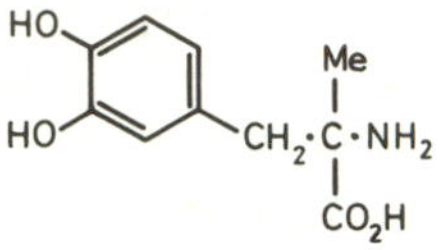

HO, HO—⟨⟩—CH₂·C·NH₂ with Me, CO₂H
Methyldopa
(12.50)

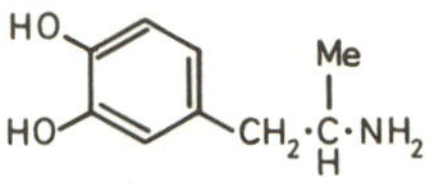

HO, HO—⟨⟩—CH₂·C·NH₂ with Me, H
Methyldopamine
(12.51)

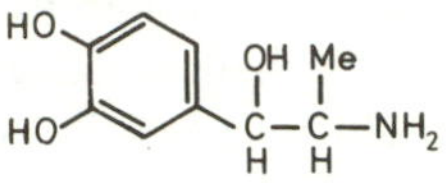

HO, HO—⟨⟩—C–C—NH₂ with OH Me, H H
α-Methylnorepinephrine
(12.52)

Two α_2-agonists, clonidine and methyldopa (α-methyldopa), occupy a favoured place in the treatment of the severer forms of hypertension. Both drugs have their site of action in the brain where they are selectively accumulated.

Methyldopa (12.50) is 2-methyl-3′,4′-dihydroxyphenylalanine, and is used orally (Gillespie et al., 1962). In the central nervous system, it is decarboxylated to methyldopamine (12.51) and this is hydroxylated (much as in Fig. 9.2) to α-methylnorepinephrine (12.52). Both of these metabolic products are thought to reduce high blood-pressure by central stimulation of α_2-adrenergic receptors (Iversen, 1967; Freed, Quintero and Murphy, 1978). These receptors appear to be in an inhibitory neuron whose activation depresses the peripheral sympathetic system.

Clonidine (12.53) acts at a similar site (Kobinger, 1978; Timmermans and Van Zwieten, 1977). It is 2-(2,6-dichloroanilino)-2-imidazoline; the tautomer depicted here is the one favoured in the crystal as revealed by X-ray-diffraction analysis. All significant departures from the structure of clonidine lead to considerable loss of potency. For example, replacement of the nitrogen bridge by —CH₂— lowers activity 25-fold. The two chlorine atoms make the molecule sufficiently lipophilic to penetrate the blood–brain barrier and, being in the 2,6-positions, they force the molecule into a non-planar conformation that has a torsion angle of 75° between the two rings. This twist causes a lowering of pK_a by interfering with the resonance of the guanidine moiety. This effect ensures that some neutral species is always present in the ionic equilibrium mixture, a condition that assists penetration (Stähle, 1982).

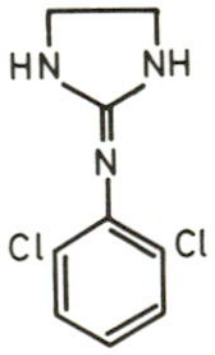

HN⌐NH, N, Cl⟨⟩Cl
Clonidine
(12.53)

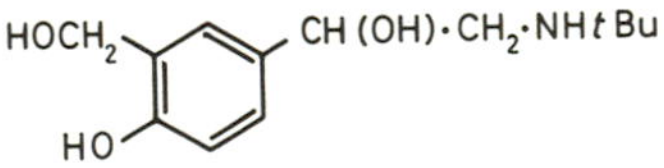

HOCH₂—⟨⟩—CH(OH)·CH₂·NH*t*Bu, HO
Salbutamol (Albuterol)
(12.54)

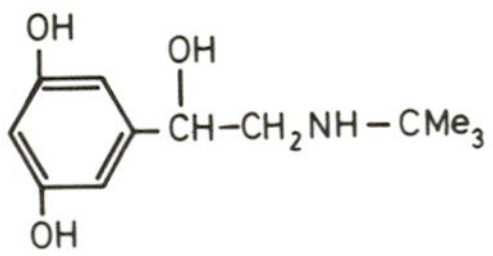

OH, OH, CH–CH₂NH—CMe₃, OH
Terbutaline
(12.55)

Little clinical use has yet been made of α_2-antagonists. They are being looked into as possible anti-depressants which might allow norepinephrine to accumulate in the brain (Caroon et al., 1982).

(b) *The β-receptors.* Two kinds of β-receptors have been recognized (Lands et al.,

1967). The β_1-receptors increase the force and rate of contraction of heart muscle, dilate coronary blood-vessels, and relax smooth muscle in the gastro-intestinal tract, whereas β_2-receptors relax smooth muscle in the bronchi, uterus, and the arteries that supply skeletal muscle. The β_1-receptor exists in a less lipophilic environment than the β_2-receptor (Basil *et al.*, 1976).

Excellent *agonists* have been discovered in recent years for use as broncho-dilators by asthmatic patients who inhale them as aerosols. The design of these agonists was based on isoprenaline (*12.44*) whose structure was altered to restrict the action to β_2-receptors. This achievement avoided the β_1-activated tachycardia which was a side effect of earlier drugs. Care was taken to reject the catechol-like arrangement of hydroxy-groups which provides a site for degrad-ation by the enzyme, COMT. The β_2-type activity requires only one phenolic group. Typical examples are salbutamol (*12.54*) (albuterol) and terbutaline (*12.55*). In general, activity increases with the size of the *N*-alkyl group, up to *t*-butyl.

Striated muscle fibres (as in the diaphragm) and uterine muscle lack sym-pathetic innervation and have receptors of the β_2-type. They respond strongly to the hormone (epinephrine) and not to the neurotransmitter (norepi-nephrine). Sympathetically innervated tissue, on the other hand (e.g. heart and duodenum), has an excess of β_1-over β_2-receptors.

Many adrenergic *inhibitors* are used for the treatment of moderately elevated blood-pressure, alone or in combination with a diuretic. They have come to be known as the β-blockers and have made a tremendous contribution to the health of the community. Early attempts to find β-blocking drugs produced only partial agonists. It became clear that a greater departure must be made from the structure of norepinephrine. Success came when J.W. Black and his colleagues discovered the side-chain —OCH$_2$—CH(OH)—CH$_2$NHR (where R is isopropyl or *t*-butyl) that confers purely antagonistic β-type activity on an aromatic hydrocarbon (Black *et al.*, 1964; Le Count, 1982).

The most prescribed drug in this category is propranolol (*12.56*), 1-iso-propylamino-3-(1-naphthyl)-2-propanol, whose unprecedented clinical prop-erties were announced by Prichard and Gillam (1964). It appears to act in three

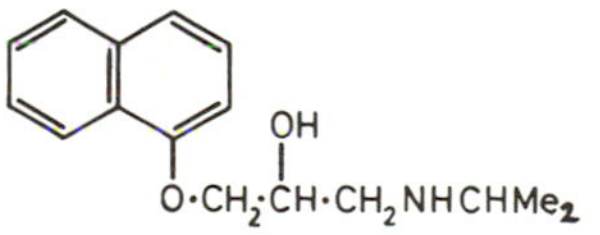

O·CH$_2$CH·CH$_2$NHCHMe$_2$

Propranolol
(12.56)

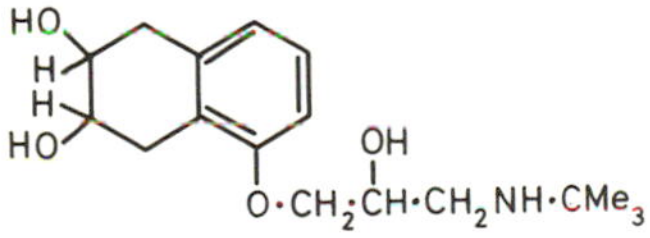

Nadolol
(12.57)

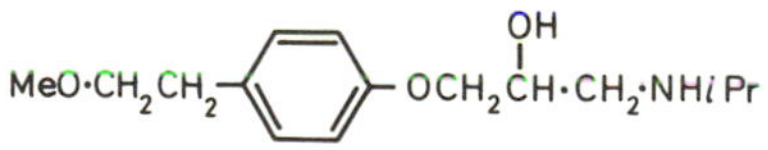

O·CH$_2$CH·CH$_2$NH·CMe$_3$

MeO·CH$_2$CH$_2$— —OCH$_2$CH·CH$_2$NH*i*Pr

Metoprolol
(12.58)

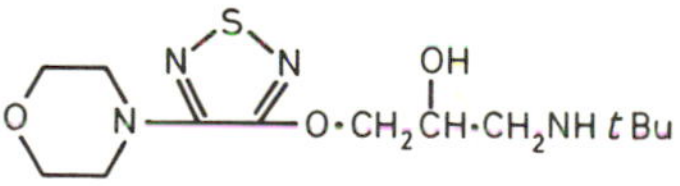

Timolol
(12.59)

O·CH$_2$CH·CH$_2$NH *t*Bu

ways, simultaneously. Thanks to its lipophilicity [P (octanol/water) is 3.6], it enters the brain and has a small central effect. A larger effect is exerted on the β_1-receptors of the heart whose rate is decreased and also on the β_2-receptors of the vascular muscles which the drug relaxes (Lewis and Haeusler, 1975). Another established use of propranolol is to support the ischaemic heart in angina. For a historic account of the discovery of the β-blockers, see Le Count (1982).

It is impossible here to mention all the marketed drugs whose molecules are modifications of that of propranolol. Nadolol (*12.57*) is pharmacologically similar but longer acting. Atenolol and metoprolol (*12.58*) act somewhat more selectively on the cardiac β_1-receptors, and some physicians prefer these for patients with bronchial difficulty*. Pindolol (*12.46*) and oxprenolol are classed as pseudo-agonists (term defined on p. 30) and show β_1-stimulatory effect on the heart. Alprenolol (*12.45*) and sotalol rather resemble propranolol in action. Timolol (*12.59*) also resembles propranolol, but is much more powerful. It is also used as eye drops for reducing intraocular tension in glaucoma. In a 3-year study in Norway, timolol was shown to decrease deaths after heart attacks when taken regularly (Pedersen, 1981).

(*c*) *The nature of sympathomimetic receptors.* The β_1- and β_2-receptors, when appropriately stimulated, are known to trigger adenylate cyclase so that this enzyme produces adenosine $3',5'$-cyclic monophosphate (cAMP) (*12.60*). It has been found that the β-receptors and the enzyme are situated close together, the receptor being in the outside layer of the membrane, near to the adenylate cyclase which is on the inner side. As soon as the receptor binds the neurotransmitter, the cyclase (with the help of guanosine triphosphate) produces cAMP (Perkins, 1973; Levitzki, 1977; Citri and Schramm, 1980). When adenylate cyclase is blocked with guanyl-5'-yl imidodiphosphate, there is a conformational change that inhibits the β-receptor from taking up agonists, but its uptake of antagonists is unaffected (Lefkovitz, Mullikin and Caron, 1976). It is noteworthy that these antagonists [including propranolol (*12.56*)] block the β-site without blocking adenylate cyclase; also the enzyme can be inactivated by heat without affecting uptake on the β-site (Schramm *et al.*, 1977). Injection of cAMP can bring about physiological responses similar to those evoked by the catecholamines.

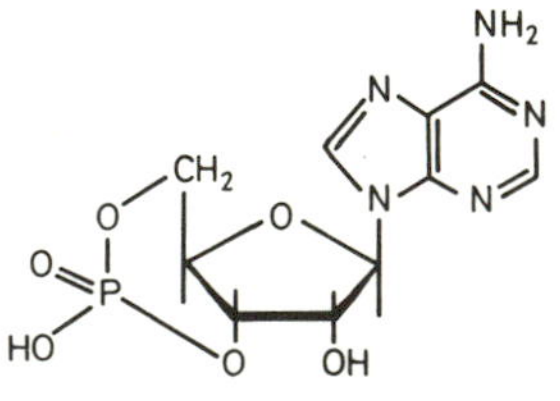

Cyclic AMP
(*12.60*)

---

*A small change in (*12.58*) can confine action to β_1 receptor (Machin *et al.*, 1983).

The α_2-receptor, when stimulated, inhibits adenylate cyclase (in a reaction that needs GTP) and the level of cAMP falls. The α_1-receptor, when stimulated, propagates the nervous impulse by liberating calcium ions, and does not affect adenylate cyclase.

(*d*) *Cyclic adenosine monophosphate*. The physiological task of cAMP, when released indirectly by catecholamines, is to energize the enzyme protein kinase which, it is thought, may phosphorylate a specific postsynaptic membrane protein, thus creating a pore for the passage of inorganic cations, and this pore might be abolished by the rapid enzymatic hydrolysis of the phosphoester-group (Greengard, 1976). Clearly, much work remains to be done to establish the molecular basis of the postsynaptic potential. The methylated xanthines, such as caffeine (*7.52a*) and theophylline, specifically hinder the destruction of cAMP by phosphodiesterase, but only weakly.

The stimulant effect of *caffeine* on the central nervous system seems to follow from a strong antagonistic action on the adenosine receptors (Daly, Brons and Snyder, 1981).

Apart from mediating the action of catecholamines, cAMP and its congener cGMP are common secondary-transmitters of hormone action. For example, they play the main part in liberating insulin from the pancreas *in vivo*. *Tolbutamide* (*12.61*), and similar sulfonylureas (such as chlorpropamide, tolazamide, and glibenclamide), which are used to alleviate the *diabetes* of elderly people, somewhat selectively inhibit pancreatic phosphodiesterase and thus preserve cAMP from destruction (Goldfine, Perlman and Roth, 1971). Activity within a series of sulfonylureas is positively correlated with lipophilicity and also with binding by serum albumin (Seydel, Ahrens and Losert, 1975). These cyclic phosphates also mediate the diuretic effect of vasopressin, and the action of ACTH and many other hormones (Sutherland, Øye and Butcher, 1965).

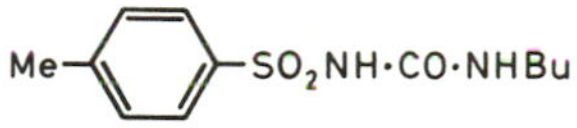

$$\text{Me}\!-\!\!\left\langle\!\!\!\bigcirc\!\!\!\right\rangle\!\!-\!\text{SO}_2\text{NH}\cdot\text{CO}\cdot\text{NHBu}$$

Tolbutamide
(*12.61*)

(*e*) *Dopamine*. Though less is known about this neurotransmitter (*7.48*) than is known about norepinephrine, the nature of its receptors is currently the subject of vigorous research. Human diseases are associated with both deficiency and excess. In Parkinson's disease, the nigro-striatal region of the brain lacks dopamine, and medication with levodopa is designed to restore it. Schizophrenia is still considered to be a disturbance of dopamine metabolism, characterized by excess (Haracz, 1982). The neuroleptic drugs such as chlorpromazine, used for controlling this condition, block the dopamine receptor in the corpus striatum. Preliminary work has been done on isolation of dopamine receptors.

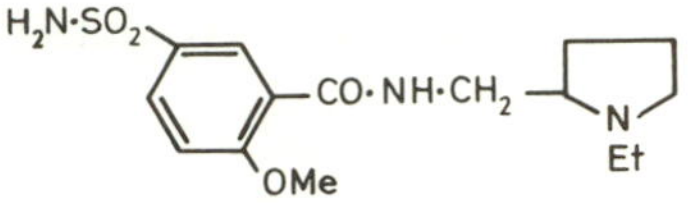

H₂N·SO₂ ... ―CO·NH·CH₂― [pyrrolidine with N-Et]

Sulpiride
(12.62)

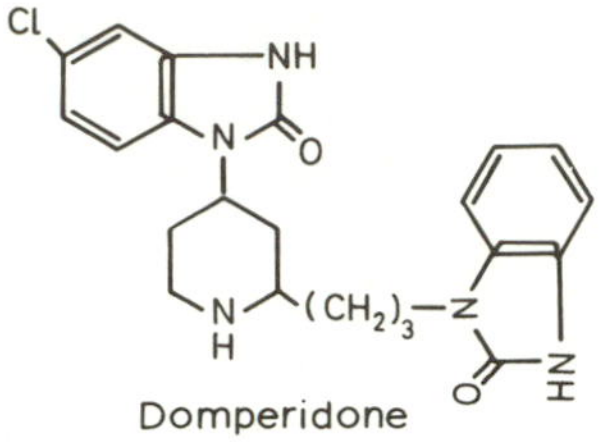

Domperidone
(12.63)

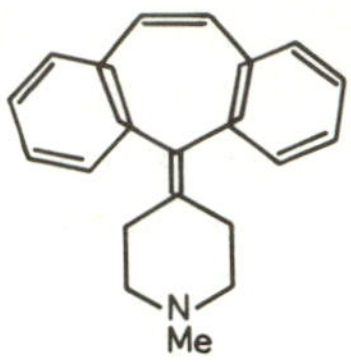

Cyproheptadine
(12.64)

A universally agreed classification of dopamine receptors has yet to be achieved. A widely quoted attempt is that of Kebabian and Calne (1979) which depends on whether the receptor activates (D-1), or does not activate (D-2), adenylate cyclase. Possible ways to overcome difficulties inherent in this approach have been suggested by Leff and Creese (1983). In an alternative classification of peripheral dopamine receptors, Goldberg and Kohli, (1983) distinguish DA_1 and DA_2 receptors. The former effect vascular dilatation by muscular relaxation. For this receptor, apomorphine is only weakly antagonistic, whereas (R)-sulpiride (12.62) is a strong antagonist and domperidone (12.63) is inactive. The latter receptor (DA_2), when excited, inhibits the action of any norepinephrine that has been released from sympathetic nerves. For this receptor, apomorphine is a strong agonist and (S)-sulpiride and domperidone are potent antagonists. In the central nervous system, the dopamine receptor appears to be DA_2 in type.

The side-chain of dopamine is held in a rigid cyclic conformation in 2-amino-6,7-dihydroxyl-1,2,3,4-tetrahydronaphthalene. This substance, injected into the brain of a conscious rat, is taken up by brain synaptosomes where it mimics the actions of dopamine (Elkhawad and Woodruff, 1975). A hypothetical, three-dimensional model of a D-2 receptor was worked out by Olsen et al. (1981).

For medication of Parkinson's disease with a dopamine precursor, see Section 3.6.1; for the use of dopamine antagonists in treating schizophrenia, see Section 12.9.

(f) 5-Hydroxytryptamine (serotonin) (7.49) is a phenolic amine, and hence related to the catecholamines, if only distantly. Little is known about its receptors, of which there appear to be two types. Of these, 5-HT₁ is defined as the site that binds 5-hydroxytryptamine, and other agonists, leading to contraction of peripheral blood-vessels. The 5-HT₂ receptor is recognized principally for its binding of specific inhibitors such as methysergide (a complex derivative of ergot) and cyproheptadine (12.64) which prevent uptake of isotopically labelled 5-hydroxytryptamine (Fozard, 1983).

Neither 5-hydroxytryptamine nor related agonists have found a clinical use. The tricyclic anti-depressants (Section 12.9) seem to act by inhibiting uptake of both 5-hydroxytryptamine and the catecholamines. Two antagonists of 5-

hydroxytryptamine are regularly used in the prophylaxis of migraine and alleviating the dumping syndrome that follows gastrectomy: cyproheptadine (*12.64*) and methysergide (1-methyl-lysergic acid butanolamide), both given orally.

For further reading on adrenoceptors and their catecholamine agonists and antagonists, see Kunos (1981–3), Cooper and Bloom (1982) and Yamamura, Enna and Kuhar (1978).

12.5 Acetylcholinesterase

The enzyme acetylcholinesterase, although available in a pure state, has so high a molecular weight (320 000) that X-ray diffraction has elucidated only a little of its structure. Amino acid analysis of the *C*-terminal groups reveals that two different kinds of chain are present (each of the same molecular weight) and that each kind is present twice in the molecule (Leuzinger, 1971). This enzyme is obtainable, in a fairly pure state, by extracting the electric organ of the eel *Electrophorus electricus*. It is a general ester-hydrolysing enzyme, but differs from other esterases in the far greater efficiency of its action on acetylcholine.

Acetylcholine (*12.65*) is a permanent cation, whose degree of ionization cannot be changed by variations in pH. Investigation of the molecular mechanism of its hydrolysis began with Wilson and Bergmann (1950) who found the dissociation constant of the acetylcholine–enzyme complex to be 2.6×10^{-4}. By observing changes in the efficacy of the enzyme with changing pH,* they concluded that the enzyme had a basic group of pK_a 7.2 (believed to be the imidazole ring of a histidine residue) and an acidic group of pK_a 9.3 (believed to be a tyrosine residue). The enzyme–substrate complex, they suggested, was formed by an ionic linkage between the quaternary ammonium-group of acetylcholine and the anion of the pK_a 9.3 group on the enzyme protein, and by a simultaneous dipole–dipole bond between the double bound nitrogen of the imidazole ring (on the enzyme protein) and the fractional positive charge of the carbon atom in the C═O group of the ester. Thus the enzyme was said to have two binding sites, an *anionic site* which bound the cation of the substrate, and an *esteratic site* which first bound and then hydrolysed the ester-group ('esteratic' was coined as an adjective of the noun 'esterase', Wilson and Bergmann, 1950).

Fig. 12.3 The active patch on acetylcholinesterase (Wilson, 1962; after Wilson and Bergmann, 1950). (G stood for 'glyoxaline', an old synonym for imidazole.)

*This method of divining the pK_a values of enzyme sites is not without pitfalls (Kosower, 1962). A change in ionization of a group on the enzyme protein can precipitate a conformational change.

Eventually, the active patch on AChase came to be represented by the sort of diagram that Haldane introduced in 1930 for enzymes generally, as in Fig. 12.3 (Wilson, 1962). By the use of some fairly rigid inhibitors, such as the *cis*- and *trans*-isomers of 2-trimethylammoniocyclopentanol iodide, it was concluded that the two binding sites on the enzyme were about 2.5 Å apart (Friess and Baldridge, 1956).

In a later modification of this hypothesis, it was postulated that the carbonyl oxygen atom of acetylcholine formed a covalent bond with the hydroxyl-group of a serine residue. This suggestion followed from the discovery that organic phosphates, such as isofluorophate (*13.26*), phosphorylated a serine-group of the enzyme and that the sequence of amino acids at this site is Glu-Ser-Ala (Schaffer, May and Summerson, 1954; see Section 9.0). It was concluded that the acetyl-group of acetylcholine is attracted to, and acetylates, the hydroxy-group of a serine residue in the enzyme. After that, the imidazole-group, located on another fold of the enzyme protein, assists hydrolysis of the acetylated group (Wilson and Cabib, 1956).

$$Me_3\overset{+}{N}-CH_2-CH_2-O-\overset{\overset{O}{\parallel}}{C}Me \qquad Me_3C-CH_2-CH_2-O-\overset{\overset{O}{\parallel}}{C}-Me$$

Acetylcholine (cation)   3,3-Dimethylbutyl acetate
(*12.65*)   (3,3,3-Trimethylpropyl acetate)
(*12.66*)

When it was found that 3,3-dimethylbutyl acetate (*12.66*) and acetylcholine are equally easily hydrolysed by acetylcholinesterase, it was realized that the van der Waals forces of the methyl-groups (surrounding the cation of acetylcholine) could be more important than a positive charge in binding the natural transmitter to the enzyme (Whittaker, 1951). In fact, little credence is now given to the 'anionic site' of Fig. 12.3 (O'Brien, 1971), which is currently looked on as a lipophilic site which attracts the highly lipophilic end of acetylcholine, quite regardless of its being cationic.

In the absence of enzymes, acetylcholine, in cold aqueous solution, is stable to acid, but unstable to alkali above pH 10. The velocity constants for its hydrolysis show that the charge on the nitrogen atom attracts the attacking hydroxyl ions to the neighbourhood of the ester-group; hence acetylcholine is much more susceptible to alkaline hydrolysis than ordinary esters such as ethyl acetate (Butterworth, Eley and Stone, 1953).

There is enough acetylcholinesterase in the end-plate of the muscle to split a thousand million molecules of acetylcholine in a millisecond, that is a thousand times as many molecules as are needed to depolarize the end-plate (Nachmansohn, 1940).

Apart from inhibition by organic phosphates and other acylating agents (Section 13.3), acetylcholinesterase can also be reversibly blocked by simple quaternary ammonium compounds. As the size of the alkyl-groups in these was increased, the increased affinity (measured as $-\Delta F$) contributed by each methylene-group (CH_2) was found to be about 300 cal/mol for bisquaternary compounds. For a given chain length, the *mono-* was bound more strongly than the *bis*-quaternary inhibitor (Bergmann and Segal, 1954). For a study of the hydrophobic areas on cholinesterases, see Kabachnik *et al* (1970).

A different method for terminating the action of acetylcholine occurs in the hearts of bivalent molluscs which have no cholinesterases. The release of ACh from the nerve-ending causes the muscle to excrete an ATP-like substance which lowers the sensitivity of the ACh receptors, apparently by an allosteric change. This mechanism persists in the hearts of higher animals, but is overshadowed by the more efficient AChase (Turpaev and Sakharov, 1973).

12.6 Acetylcholine receptors

Acetylcholine is the natural transmitter at several different kinds of synaptic site, e.g. (a) the somatic nerve → voluntary muscle junctions, (b) the ganglionic synapses which are nerve → nerve junctions in the autonomic system, (c) such postganglionic nerve-endings as are parasympathetic. In addition it has some transmitting duties in the central nervous system, e.g. between spinal cord root fibres and Renshaw cells.

Nicotine mimics acetylcholine at sites (a) and (b), whereas muscarine does so at site (c) (see Table 7.1). Hence it is usual to divide acetylcholine receptors into 'nicotinic' and 'muscarinic' receptors, a useful classification first made by Dale (1914). The work of muscarinic and nicotinic receptors is carried out on very different time scales. A single nervous stimulus affects muscarinic receptors for at least 500 ms, a long duration that is preceded by a latency of about 100 ms. In contrast to this, nicotinic receptors at voluntary neuromuscular junctions are stimulated for only 0.2 ms, and even the ganglionic synapses, which are slower, average only 60 ms. Cyclic GMP is thought to be a necessary mediator of muscarinic reponses. One consequence of interest is that smooth muscle reacts far more slowly than voluntary muscle. Heart muscle is distinguished from both by the fact that acetylcholine increases its polarization (and hence slows the heart), whereas it decreases polarization in other types of muscle.

Because the ACh receptor does not hydrolyse acetylcholine, the esteratic site of acetylcholinesterase (see Fig. 12.3) must be absent. Other fundamental differences in the two sites are indicated by the following: (a) dimethylbutyl acetate (*12.66*) is a good substrate for the enzyme, but barely activates the receptor which requires a basic group for marked activity; (b) muscarine is not a substrate for the enzyme and yet it is a powerful agonist for the muscarinic receptor; (c) di-isopropyl phosphorofluoridate (isoflurophate) (*13.26*) binds to the active site of AChase but not to an ACh receptor; (d) acetyl-bungarotoxin specifically binds to the ACh nicotinic receptor but not to the enzyme.

The next few pages will be devoted to the muscarinic receptor followed by several pages on the nicotinic receptor. Which is the easier to study? The structural requirements of the nicotinic receptor seem to be less demanding, for an agonist, than those needed to satisfy the muscarinic receptor. Also, the relative abundance of the purified nicotinic receptor, as extracted from fish, gives it a certain advantage. On the other hand, the fact that the nicotinic receptor has more sites (operating at different rates) increases the complexity of its investigation.

12.6.1 *The muscarinic receptor for acetylcholine*

The muscarinic receptor is remarkable not only for the long duration of its response to ACh but also for the ready onset of this response (Greengard, 1976). This receptor can be extracted from bovine brain by using only the caudate nucleus in which most of the brain's acetylcholine is found. This nucleus forms part of the corpus striatum, a structure that nestles in the main cleft of the cerebrum. The nucleus was extracted with aqueous sodium cholate and the extract, after ultracentrifugation, submitted to electrophoresis. A reagent that binds strongly and specifically to the muscarinic receptor was then added. This reagent is the [^{3}H]benzilic ester of quinuclidin-3-ol (*12.67*), which evolved from choline benzilate, a product of the search for specifically muscarinic esters of choline. The structure of choline is contained in that of quinuclidinol. After purification of this complex, the reagent was slowly dialysed away, leaving the muscarinic receptor. Muscarinic antagonists (atropine, hyoscine, isopropamide) prevent the receptor from binding the reagent, whereas nicotinic antagonists (tubocurarine, hexamethonium) and agonists (nicotine) do not interfere (Carson, 1982; Yamamura and Snyder, 1974).

The muscarinic receptor has also been extracted from heads of the fruitfly, *Drosophila* (Dudai and Ben-Barak, 1977), whereas the housefly has, in its head, a quite different type of ACh receptor (p. 536).

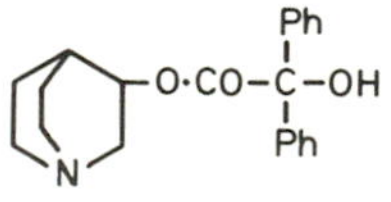

Quinuclidin-3-yl benzilate
(12.67)

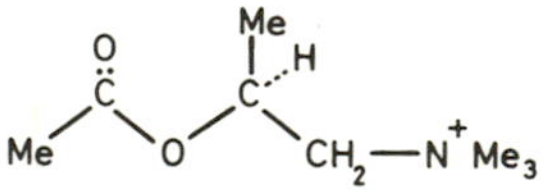

L(+)-Acetyl-β-methylcholine
Methacholine
(12.68)

Bethanechol (cation)
(12.69)

Most of what we know of the structure of the muscarinic receptor has been gleaned from studies of structure–activity relationships among agonists that mimic acetylcholine at the postganglionic synapses. These are the receptors traditionally termed muscarinic, but it must be kept in mind that this is a name of historical origin. Muscarine has no use in human medicine. Not much attention can be given to information derived from antagonists because these are, from their very nature, only partly dependent on structure. However, the antagonist pirenzepine (a benzodiazepinone) has revealed that there are two sub-types of this receptor.

The most typical muscarinic agonist is methacholine (*12.68*), L-acetyl-β-methylcholine. This simple molecule, which differs from that of acetylcholine only in possessing one extra methyl-group, has the full muscarinic action of that neurotransmitter, but none of the latter's nicotinic action. Interestingly, it is more than 200 times as active as its D-enantiomer (Ellenbroek and Van Rossum, 1960). In human medicine, methacholine is mainly used to overcome postoperative retention of the bowels. Bethanechol (*12.69*), constructed as a hybrid of methacholine and carbachol (*2.11*), has an almost purely muscarinic action and is similarly used in the clinic. The great advantage of these drugs over acetylcholine is that they are not quickly hydrolysed by acetylcholinesterase.

The equilibrium distance of the quaternary nitrogen atom of acetylcholine from the negatively charged group of the receptor has been calculated as 3.29 Å from the difference in free energy of the receptor-interaction of (a) acetylcholine (*12.65*) and (b) dimethylbutyl acetate (*12.66*). The latter has a non-basic head-group isosteric with the basic head-group of acetylcholine. This distance is practically identical with the distance of closest approach found by molecular models (Burgen, 1965).

In recent years, the conformations of many of the more rigid ACh agonists have been determined by X-ray crystallography in order to predict (a) which of the many conformations, assumable by the loosely jointed molecule of acetylcholine, is the one active at a given receptor, and hence (b) the stereochemistry of the receptor itself. Because of the known deformability of receptors (see Section 12.3), this approach is of only limited value but must be discussed.

A two-dimensional projection of the X-ray-diffraction diagram of acetylcholine (bromide) is shown in (*12.70*) (Canepa, Pauling and Sörum, 1966). The interatomic distance from one N-methyl carbon atom to the ethereal oxygen (i.e. the oxygen that links two carbons) is 3.02 Å in the original, and from the nitrogen to the ethereal oxygen is 3.29 Å (both these distances are shorter than usual). The two oxygen atoms are co-planar with the three carbon atoms that are nearest the right-hand side of (*12.70*). Thus the molecules of acetylcholine and muscarine (*12.72*) (X-ray data by Jellinek, 1957) have remarkably similar shapes in the solid state. However, the conformation of the acetylcholine molecule could be very different in aqueous solution, where it would be unconstrained by neighbouring molecules of the same substance. Hence its conformation was examined in deuterium oxide solution by analysing the vicinal coupling constants derivable from the proton magnetic resonance spectrum (Culvenor and Ham, 1966). The results (*12.71*) support the X-ray analysis except that the ester group has a more normal ester conformation. The *gauche* arrangement of the $^{+}$NCCO sequence, which is the prominent feature of (*12.70*), (*12.71*), and (*12.72*), is that preferred by very many 1,2-disubstituted ethanes in solution. It must be concluded that the preferred conformation of acetylcholine is remarkably normal: yet contact with the receptor could change the conformation completely.

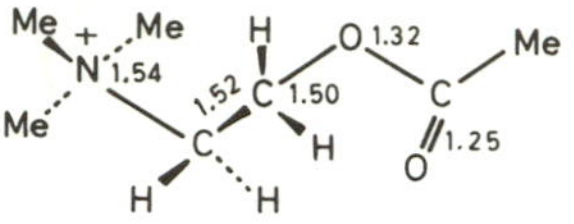

Acetylcholine
(X-ray diffraction data; bond lengths in Å)
(12.70)

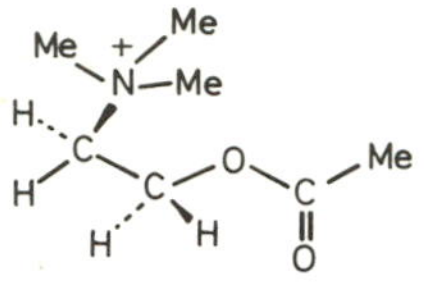

Acetylcholine
(p.m.r. data)
(12.71)

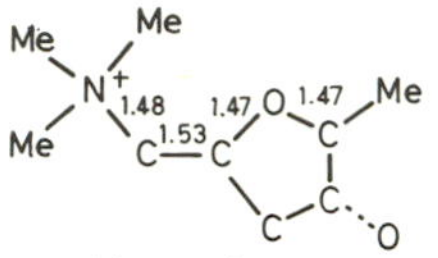

Muscarine
(X-ray diffraction data)
(12.72)

The rigid nature of the ring-system in muscarine makes more definite information available about the dimensions of the muscarinic receptor of acetylcholine than can be obtained from the latter directly (Waser, 1961). The seven stereoisomers of muscarine have been examined, but only natural L(+)-muscarine has strong acetylcholine-like activity. The nitrogen atom must be quaternary, and the ring-oxygen atom must not be replaced by sulfur, otherwise all activity is lost. These facts led Waser to postulate that muscarine is bound to the receptor by the nitrogen and the ring-oxygen atoms. Although the conformational freedom of muscarine is much less than that of acetylcholine, the trimethylammonium side-chain is free to move over almost the whole area of the rest of the molecule (Waser, 1961).

Values for the equipotent ratios of L-muscarine (relative to acetylcholine), at various postganglionic cholinergic receptors, vary from about 0.1 to 5.4. Marked activity (among isomers and analogues of muscarine) is restricted to those substances in which the arrangement of the methyl, the hydroxyl, and the onium side-chain groups is the same as in muscarine.

It may now be asked what further structural modifications of the acetylcholine and muscarine molecules are compatible with the possession of strong acetylcholine-like activity. The data are only semi-quantiative, pending laborious experimental work. This is because the 'activity' of each substance is made up of two factors, the *efficacy* and the *affinity* (see p. 293). Thus the cation of dimethylaminoethyl acetate (*12.73*), which is the unquaternized analogue of acetylcholine (*12.65*), has been described as having almost no muscarinic

Me$_2$N·CH$_2$·CH$_2$·O·Ċ·Me
Dimethylaminoethyl acetate
(12.73)

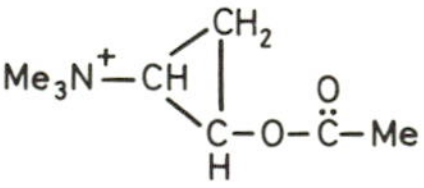

2-Acetoxycyclopropyl trimethylammonium (cation)
(12.74)

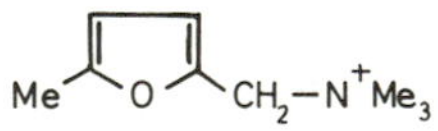

5-Methylfurmethide
(12.75)

activity. Actually it has a higher intrinsic activity than acetylcholine, and its poor performance is due to its feeble affinity, which is a thousand times less than that of acetylcholine (Gloge, Lüllmann and Mutschler, 1966).

2-Acetoxycyclopropyl trimethylammonium iodide (*12.74*) is of special interest as a substance in which the N—C—C—O structure of acetylcholine is held rigidly by covalent bonds. It was obtained as a mixture of four isomers, whose configurations were determined by X-ray diffraction (Chothia and Pauling, 1970). The (+) *trans*-isomer had the full muscarinic activity of acetylcholine, but the other isomers had very little activity. (The active isomer had only 1% of the activity of acetylcholine at nicotinic sites, and the other isomers were even less active.) It was concluded that, at least for muscarinic activity, acetylcholine must adopt the conformation (see p. 505) possessed by the active (1*S*, 2*S*) isomer of (*12.74*). 5-Methyl-2-trimethylammoniomethyl-furan (*12.75*) (5-methylfurmethide, or 5-methylfurtrethonium) has about the same amount of muscarinic activity as acetylcholine and almost no nicotinic activity (Armitage and Ing, 1954). As with muscarine, the ring is flat and rigid, and (thanks to the double bonds) the —CH_2 portion of the $CH_2N^+Me_3$ group is held firmly in the plane of the paper and hence the shape of the molecule is largely defined. It should be noted that the hydroxy-group of muscarine is *not* represented in this molecule, and no stereoisomers are possible. The X-ray crystallography of this substance has been reported (Baker *et al.*, 1971) and the details in part resemble those reported for muscarine and acetylcholine, and yet so strongly differ in other ways as not to advance knowledge of the muscarinic conformation. Loss of the 5-methyl-group from (*12.75*) greatly lessens the muscarinic activity. As this loss must cause depletion of electrons from the ethereal oxygen atom, but no change in conformation, one may suspect that the electron-distribution exerts more control on muscarinic activity than conformation does. Also furmethide and 5-methylfurmethide exemplify the five-atom side-chain rule (see below).

Great difficulty has been encountered in trying to prepare completely rigid agonists. A closer look at the great conformational latitude of the acetylcholine molecule (*12.76*) may be appropriate here, to help follow the literature. The conformations of acetylcholine depend on the four torsion angles: C5-C4-N-C3, O1-C5-C4-N, C6-O1-C5-C4 and O2-C6-O1-C5. The first of these (from analogy with all similar structures) is probably invariable, an anti-planar extended chain with torsion angle of 180°. This invariance is caused by the steric hindrance exerted by the *N*-methyl-groups, and the restraint on closer packing imposed by van der Waals radii (see Section 8.0, p. 314). The last of these parameters (O2-C6-O1-C5) is also fixed because all ester-groups are planar, due to the C6-O1 bond. X-ray-diffraction studies have shown that the other two torsion angles can vary greatly in the crystals of various acetylcholine agonists.

```
         C1                    C7
         |                    /
  C3 — N — C4 — C5 — O1 — C6
         |                    \
         C2                     O2
  Acetylcholine numbering
         (12.76)
```

R — N^+Me_3

Quaternary amines
(12.77)

Such studies have suggested to one group of workers that muscarinic activity requires only that C1 and C7 of acetylcholine must be on the same side of the molecule (Baker *et al.*, 1971). Others, from similar evidence (coupled with that obtained from antagonists, which is necessarily of less significance), have postulated that acetylcholine reacts with muscarinic receptors through the ester oxygen atom as well as the ammonium cation (Beers and Reich, 1970).

(a) The cationic head of acetylcholine. It is useful to pause here and compare acetylcholine cations with inorganic cations likely to be present at the receptor surface. Whereas potassium has a stimulant action on all of the muscle, the action of acetylcholine is normally confined to the small end-plate region. The shielding effect of the alkyl-groups (on the nitrogen atom in a quaternary amine) ensures that the ion is virtually anhydrous in aqueous solution (Robinson and Stokes, 1959). Thus the effective ionic radius in solution may be taken as the same as the radius obtained from X-ray crystallography. It is seen from Table 12.2 that the tetramethylammonium ion has a radius of 2.41 Å, and this must also be the radius of the cationic head of acetylcholine.

Table 12.2 The size of some cations

Cation	Anhydrous radius (from crystal data)[a] Å	Presumed hydrated radius (from mobility in water at 25°C)[b] Å
Li^+	0.60	3.7
Na^+	0.95	3.3
K^+	1.33	
NH_4^+	1.48	—[d]
NMe_4^+	2.41[c]	
Mg^{2+}	0.65	4.4
Ca^{2+}	0.99	4.2

[a] Pauling, 1960; [b] according to the Law of G.G. Stokes, with corrections by R. Robinson and R. Stokes, 1959; [c] Johnson, 1960; [d] imprecise value (1.3 Å): outer layer of water is present but only feebly held, as in Cs^+ and Rb^+.

Whereas the radius of every inorganic ion, in the anhydrous state, is well established, those of low mass are strongly hydrated in solution so that their effective radius is much greater. Just how much greater cannot be said because an exact method for measurement is lacking. However, the figures for Li^+ and Na^+ in the last column of Table 12.2 are indicative. The cationic head is the widest part of the acetylcholine molecule. What happens if it is made wider? First, let us note the acetylcholine-like effect of simple aliphatic quaternary amines. These have only a feeble action on muscarinic sites, although their action on nicotinic sites is considerable (see below). Tetramethylammonium salts, for example, have only about one-thousandth of the activity of acetylcholine on the gut and heart (Clark and Raventós, 1937). Tetraethylammonium salts, and still higher homologues, are only antagonists. However, the series of alkyltrimethylammonium salts (*12.77*) is full of interest. Affinity for the receptor increases as the alkyl chain is lengthened, but efficacy falls from tetramethyl-ammonium to ethyltrimethylammonium, and then climbs to a maximum at *n*-pentyltrimethylammonium [tested on mammalian gut: Stephenson (1956); Van Rossum and Ariëns (1959); on frog's heart: Raventós (1937); on dog blood-pressure: Alles and Knoefel (1939)]. This, and related discoveries, led to the formulation of the 'five-atom rule' (see below). However, even the most potent members have no more than 1% of the activity of acetylcholine.

(*b*) *Alterations to the cationic head of acetylcholine.* Successive replacement of methyl-groups, in acetylcholine (*12.65*) by either hydrogen or ethyl causes a steep decline in parasympathomimetic activity of all kinds (Ing, 1949). It was pointed out above that replacement by hydrogen lowers affinity but increases efficacy. Ethyl, on the other hand, seems to increase affinity but lower efficacy (Barlow, Scott and Stephenson, 1963). Tertiary amines, like (*12.73*), are at least 99% ionized at pH 7.3. Hence the loss in affinity is not due to loss of basic strength. However, the acetylcholine receptors seem to require strong van der Waals linkages with the head-group of acetylcholine (Holton and Ing, 1949). All evidence suggests that a basin-like depression exists in the receptor and is shaped for the maximal van der Waals contacts from four carbon atoms tetrahydrally disposed on the quaternary nitrogen atoms (Belleau and Puranen, 1963).

The adverse effect of replacing three methyl- by three ethyl-groups has been observed in many other series of substances with acetylcholine-like properties (Barlow, 1968).

$$R-\overset{\overset{O}{\|}}{C}\cdot Y \longleftrightarrow R-\overset{\overset{O^-}{\|}}{C}:Y^+$$

Resonance hybrids
(*12.78*)

Oxotremorine
(*12.79*)

Although *simple* tertiary amines do not show marked muscarinic properties, this deficiency can be remedied if the remainder of the molecule binds power-fully to the receptor. Thus increased binding power can be given to the carbonyl-group if the lone pair of electrons on the oxygen atom is more strongly delocalized than it is in esters ($Y = OR$) in the resonance hybrid shown in (*12.78*). Amides ($Y = NH_2$ or NR_2) support this resonance much better, because nitrogen is better equipped than oxygen to carry a positive charge. The infrared absorption frequency of the carbonyl-group is considered to be a measure of this effect. As it decreases from the value typical of esters (1735 cm^{-1}) to that of amides (1690 for free, and 1650 for associated, amides), the muscarinic activity of tertiary bases rises steadily until, as in (*12.79*), it almost reaches that of quaternary bases (Bebbington, Brimblecombe and Shakeshaft, 1966). [Oxotremorine is a muscarinic agonist, used in experimental pharmacology; the amide function is contained in the pyrrolidone ring (Cho, Haslett and Jenden, 1962)]. Two other muscarinic agonists that consist of a tertiary amine associated with an ester group will now be mentioned. Arecoline (*12.80*), in the form of betel nut, has been chewed for its euphoric effect by natives of the East Indies from early times. Its strong muscarinic action becomes comprehensible when viewed as a 'reversed acetylcholine' of which (*12.85*) is the prototype. It also has some nicotinic-type action, but is not used in human medicine. Pilocarpine (*12.81*), a weaker muscarinic agonist, has its ester group in the cyclic (or lactone) form. It is used mainly to decrease intraocular tension in glaucoma.

Arecoline
(*12.80*)

Pilocarpine
(*12.81*)

Analogues of acetylcholine in which the nitrogen atom is replaced by phosphorus or arsenic have only from 1 to 10% of the activity of acetylcholine (at a variety of sites). This replacement of nitrogen does not alter the bond angles, but increases the mean distance between the methyl-groups by 27 to 35%, because of the increased length of the P—C and As—C bonds (Holton and Ing, 1949).

(*c*) *The five-atom side-chain rule.* The need for a chain of *five* atoms in order to achieve efficacy in acetylcholine-like substances was first conceived by Alles and Knoefel (1939), and consolidated by Ing (1949). In acetylcholine, and its analogues of the general type $R-^+NMe_3$, the most active member of any series of homologues is usually the one where R is a *five*-atom chain (excluding hydrogen atoms). This was discussed above for the case where R is an alkyl-group, but it is equally true if some of the carbon atoms are replaced, e.g. by oxygen. Thus acetylcholine (*12.65*) is much more active than formyl- or propionyl-choline; and butyryl- and valeryl-choline have scarcely any muscarinic activity. Both the nitrous and the nitric esters of choline, which have such five-atom chains, have considerable muscarinic activity (Dale, 1914). The acetic esters of $HO \cdot CH_2 \cdot ^+NMe_3$ and of $HO \cdot (CH_2)_3 \cdot ^+NMe_3$ are both less active than that of choline ($HO \cdot (CH_2)_2 \cdot ^+NMe_3$ (Hunt and Taveau, 1911).

Similarly ethoxycholine [the ethyl ether of choline (*12.82*)] is more active than either the methyl or propyl ether (Dale, 1914); also, the *n*-propyl ether is the most active in the $HO \cdot CH_2 \cdot ^+NMe_3$ series. In the dioxolane (or 'acetal') series of substances with muscarinic action, e.g. (*12.83*), the compound in which R is Me is far more active than those in which R is H or Et (Fourneau *et al.*, 1944). The isomer illustrated [L(+)-*cis*] is six times as potent as acetylcholine (Belleau and Lacasse, 1964). For the factors underlying the five-atom rule, see under *'Antagonists of acetylcholine'*, below. (None of the substances mentioned in this sub-section has found a place in medicine.)

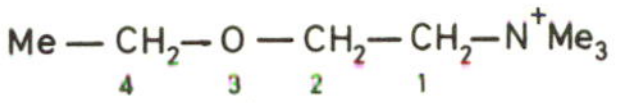

$$Me - CH_2 - O - CH_2 - CH_2 - \overset{+}{N}Me_3$$
$$\qquad\quad 4\qquad\quad 3\quad\ 2\qquad 1$$

Ethoxycholine (cation)
(12.82)

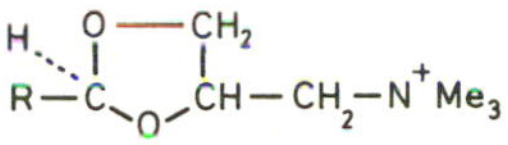

Dioxolane agonists
(12.83)

(*d*) *The optimal position for an oxygen atom in the side-chain.* The ether (*12.82*) has approximately 1 to 10% of the activity of acetylcholine, and is more active at muscarinic than at nicotinic sites. This activity is diminished if the oxygen atom is moved from the 3- to the 2- or 4-position. The ketone (*12.84*) has little muscarinic (but from 0.2 to 100% of the nicotinic) action of acetylcholine on various test preparations. If the carbonyl-group is moved from the 4- to the 3- or 2-position, the action is diminished (Ing, Kordik and Williams, 1952). Thus both the ethereal and carbonyl oxygen atoms have maximal activity in the positions where they occur in acetylcholine (*12.65*).

This accords with measurements of models made by Beers and Reich (1970), who concluded that muscarinic and nicotinic receptors bind to the nitrogen atom and to an oxygen atom, ethereal in the first case, ketonic in the second, in acetylcholine. Also, that these receptors span corresponding structures in all muscarinic and nicotinic agents and antagonists. Their distances, 4.4 Å to ethereal oxygen and 5.9 Å to ketonic oxygen, are a little longer than X-ray measurements suggest, possibly because the models have not allowed for partial double-bond character in the ester region. However, the relativity of their two distances seems established. They visualized that the receptors used hydrogen bonds for binding to the oxygen atoms.

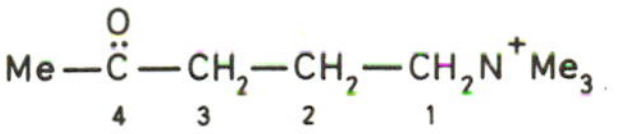

Ketone analogue of acetylcholine
(12.84)

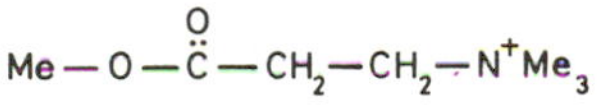

'Reversed acetylcholine'
(12.85)

The methyl ester of β-trimethylammoniopropionate (*12.85*) may be regarded as acetylcholine in which the order of the ethereal and carbonyl oxygen atoms has been reversed. It is a very poor substrate for acetylcholinesterase, and does not inhibit this enzyme. However, it has strong muscarinic and moderate nicotinic properties (Bass *et al.*, 1950).

(*e*) *Antagonists of muscarinic receptors.* These can be expected to shed less light on the nature of receptors than agonists have done, because it is the nature of an antagonist to cover a receptor without fitting it exactly. The most studied antagonist is atropine (*7.16*), and its response to structural modification was discussed on p. 274.

The well known principle that a stimulant may be turned into an antagonist by increasing the molecular weight is seen in the alkyltrimethylammonium homologues (*12.77*) (Table 7.2). Kinetic studies showed that all members of this series go on to the muscarinic receptor at the same rate, but the rate of dissociation from the receptor falls as the number of methylene-groups increases (see Section 10.3.2, p. 405 for explanation). Hence, the lower members are pure muscarinic stimulants, but when $R = C_6H_{13}$ some residual atropine-like blocking is seen as well, and when $R = C_{12}H_{25}$ the action is purely atropinic (Paton, 1961). The intrusion of antagonistic action when R is > 5 may explain why a five-membered side-chain gives maximal muscarinic action in so many series (see above). The fall-off in activity that occurs when the chain is shortened by removing the terminal methyl-group may be due to the presence of a corresponding lipophilic area on the receptor, as suggested by Chothia (1970).

Muscarone, obtained by oxidizing the secondary alcoholic-group in muscarine (*12.72*) to a carbonyl-group, not only has increased muscarinic properties, but considerable nicotinic properties also. It has both ethereal and carbonyl oxygen atoms, but more widely separated than in acetylcholine.

For therapeutic uses of muscarinic antagonists, see under atropine in Section 7.3.

12.6.2 *The nicotinic receptors for acetylcholine*

Compared to its muscarinic analogue, the nicotinic receptor responds to stimulation with only a short reaction. The winning of large quantities of the nicotinic receptor from electric fish was described in Section 2.1 (see also Fig. 2.1). This glycoprotein can be recognized with the help of a Formosan snake toxin called α-bungarotoxin, used as its [³H]acetyl derivative. This reagent binds specifically to the acetylcholine-binding site of the receptor. The complex with this toxin is also used for purifying the receptor, as it is reversed by long dialysis. In 1980, this receptor was re-sited in an artificial planar lipid bilayer to demonstrate its permease nature. It was found, by X-ray diffraction, to span this membrane, protruding slightly from both surfaces. It had a diameter of about 8.5 nm, and it included a 0.65 nm ion-translocating pore. This reconstituted permease responded to ACh by facilitating passage of cations, an effect blocked by tubocurarine (Nelson *et al.*, 1980).

Each permease rosette (see Fig. 2.1) turns out to be formed of five parallel strands which surround the ion pore. Two of the strands, named the α-subunit, are considered to contain the whole of the receptor. The three remaining strands seem to be purely structural.

The mRNA that codes for the α-subunit (obtained from *Torpedo californica*) was made to generate its complementary DNA, and this was cloned. In this way, the mRNA could be prepared in quantity, and the sequence of the nucleotides mapped. This operation spelled out the order and nature of the amino acids in the α-subunit.

This subunit, which contains 437 amino acid residues, has a mol. wt. of 50 116. The disulfide bridge that is known to be near the ACh-binding site is thought to be formed of the two cysteine residues, 128 and 142. For the ACh-binding site itself, the authors favour the aspartic acid residue 138 (and their second choice is glutamic acid-129) in conjunction with histidine-134. Glucose is most likely attached to the nitrogen atom of asparagine residue 141 (Noda et al., 1982).

In Nature, no less than two molecules of ACh must combine with the α-protein to evoke a response. One microsecond pulse of ACh produces, in skeletal muscle, an apparently instant response that lasts for 10 μs. These figures suggest that the receptor *directly* controls the micropore, and that no phosphorylation step is required. Thus the nicotinic response is much simpler and more direct than that postulated by Greengard for the muscarinic response (p. 522).

The above account is of the nicotinic receptor obtained from the electric organ of fish, but very similar material has been isolated from mammalian skeletal muscle (Dolly and Barnard, 1977; Froehner, Reiness and Hall, 1977). A similar protein, of mol. wt. about 86 000, was obtained from the cerebral cortex of the guinea pig. The binding of [³H]bungarotoxin to this receptor was inhibited by tubocurarine and gallamine (Bosman, 1972).

For more on the nicotinic receptors, see reviews by Karlin (1980) and Conti-Tronconi and Raferty (1982).

Long before nicotinic receptors had been isolated, many studies were made of structure−activity relationships among agonists that act on acetylcholine activated sites traditionally termed nicotinic. This name is of historical origin but nicotine has no use in human medicine. Nicotine (*7.26*) has long been known to mimic acetylcholine at three sites in vertebrates: (a) at the motor end-plate of voluntary muscle in the neuromuscular junction; (b) at the ganglia of both sympathetic and parasympathetic nerves; and (c) in the cerebral cortex. Site (a) is the most accessible to it, as is evident when a mammal is poisoned by nicotine. However, in insects, which have no acetylcholine at the neuro-muscular junction, nicotine exerts its principal effect on the ganglia of the central nervous system (Yaeger and Munson, 1945). Nicotine is now much less used as an insecticide. Nicotine-like insecticides have been synthesized (Kamiura et al., 1963): for high activity, an intact pyridine ring is required with a basic substituent (in the 3-position) which may be aliphatic, as in (*7.27*), but must not be quaternized because this prevents penetration. For an effect in mammals, quaternization of the pyrrolidine ring-nitrogen is permissible, but brings no advantage. The fact that nicotine, in excess, has an anti-acetylcholine (blocking) action on ganglia adds a difficulty to discovering the nature of the nicotinic receptor by using this alkaloid and its analogues (Barlow and Hamilton, 1962). The euphoric effect, to which so many cigarette smokers are addicted, seems to involve site (c) (see above).

At this stage, it seems important to name the *simplest* substance that has a strong nicotinic action, free from muscarinic activity. This, the fundamental nicotinic agonist, is tetramethylammonium chloride (*12.77*, R = Me). It occupies the same position, as a simple standard, in the nicotinic series that methacholine (*12.68*) does in the muscarinic series.

Tetramethylammonium salts and acetylcholine salts are equipotent at autonomic ganglia sites (Burns and Dale, 1915), but the former have only one-hundredth of this potency at the skeletal neuromuscular junction which is also a nicotinic site, and only one-thousandth at smooth muscle junctions (muscarinic). Only transient demands are made on the transmitter at autonomic ganglia, whereas a nerve–muscle junction requires a lingering action greatly facilitated by molecules that permit hydrogen-bonding as acetylcholine does through the ketonic oxygen atom. Apparently the tetramethylammonium cation is the only necessary effector in acetylcholine: the rest of the molecule has evolved to supply extra bonding for those actions that do not take place quickly.

In the alkyltrimethylammonium cation series (of which tetramethyl-ammonium is the foundation member) maximal activity is reached at the *n*-pentyltrimethylammonium salts which are about eight times as active as acetylcholine (and about as active as nicotine) at nicotinic receptors (Willey, 1955). D-Lactoylcholine has strong nicotinic, but little muscarinic, potency (Sastry, Lasslo and Pfeiffer, 1960). Phenolic ethers of choline, such as (*12.86*), have a strong nicotinic (but little muscarinic) activity (Hey, 1952). However, this ratio is reversed by inserting methyl-groups into the two *ortho*-positions of the benzene ring, a reminder that nicotinic action is easily repressed by steric hindrance to which muscarinic action is indifferent (note the effect of the *C*-methyl-group in methacholine, p. 523).

Choline phenyl ether
(*12.86*)

Acetylthiocholine
(*12.87*)

Perhaps it has been too easily assumed that nicotinic agonists require *gauche* conformation (defined in Section 12.3). Admittedly the great majority of these agonists have been found to carry a $-^{+}N \cdot C \cdot C \cdot O-$ group in the *gauche* (synclinal) conformation (Baker *et al.*, 1971; Culvenor and Ham, 1966; Canepa *et al.*, 1966), whether studied in the solid state or in solution. One rare exception, carbachol (*2.11*) is *staggered* (anti-planar) in the crystal but *gauche* in solution (Barrnas and Clastre, 1970; Baker *et al.*, 1971). However, the *gauche* hypothesis is challenged by agonists containing the $-^{+}N \cdot C \cdot C \cdot S$ and $-^{+}N \cdot C \cdot C \cdot Se-$ groups which prefer the *staggered* (anti-planar) conformation, both in the crystal (Shefter and Mautner, 1969) and (on p.m.r. evidence) in solution (Cushley and Mautner, 1970). For example, cholinethiol and *S*-methylthiocholine (both *staggered*) are potent agonists whereas their oxygen analogues choline and *O*-methylcholine (both *gauche*) are not (Mautner, Bartels and Webb, 1966). Moreover, replacement of the carbonyl oxygen of ACh by sulfur leaves the conformation *gauche* (Mautner, Dexter and Low, 1972), and both acetylcholine and acetylthiocholine (*12.87*) are highly active in various biological tests (Scott and Mautner, 1967; Mautner *et al.*, 1966; Mautner, 1969). Thus it looks as if the resting conformation is not so important an ingredient of agonist activity as the electron distribution. It has been calculated that a receptor site need expend no more than 1 kcal (4.18 kJ) of energy to change the preferred *gauche* form of acetylcholine into the *staggered* form (Liquori, Damiani and de Coen, 1968).

In an immunochemical approach to learning more about the receptor site, an antibody was prepared against choline phenyl ether (*12.86*) in the rabbit. Unfortunately, when this antibody was used as a model for nicotinic receptors, it was found unable to distinguish between muscarinic and nicotinic agents, nor between agonists and antagonists (Marlow, Metcalf and Burgen, 1969).

(*a*) *Antagonists of nicotinic receptors*. No purely nicotinic agonist is used in medicine. Carbachol (*2.11*) has both nicotinic and muscarinic properties, but the latter preponderate. Anti-cholinesterase drugs permit medication of the patient with his own acetylcholine, which means simultaneous muscarinic and nicotinic therapy. Such treatment is given with physostigmine (*2.8*) for short-lived (emergency) treatment in glaucoma, with organophosphates (Section 13.3) for more durable treatment of this permanent affliction of old age, and neostigmine (*2.10*) for overcoming the neuromuscular block in myasthenia gravis. The last-named drug has some direct nicotinic action on voluntary muscle.

Although the stimulation of nicotinic receptors plays such a small part in medication, there is widespread use of nicotinic antagonists in surgery where they provide deep muscular relaxation which enables a lower dose of general anaesthetic to be used. Tubocurarine (*2.6*) introduced for this purpose in 1942, has been supplemented by other quaternary ammonium cations, and two main types are known. Members of the first class (e.g. tubocurarine) prevent acetylcholine from depolarizing the post synaptic membrane but do not themselves possess any acetylcholine-like activity at this site. They are competitive with acetylcholine. The synthesis, liberation, and breakdown of acetylcholine are not interfered with, but this transmitter is excluded from its receptor, and hence profound muscular relaxation occurs. Because the cationic heads of tubocurarine (2.6) (which has only *one* quaternized amino-group) and gallamine (*7.31*) seem to be too sterically hindered to fit into an acetylcholine receptor, Waser (1960) assumed that these molecules merely cover a pore in which the acetylcholine receptor lies. The second class of antagonist at the neuromuscular junction, e.g. decamethonium (*7.28*) and suxamethonium (*7.29*), prevent acetylcholine from depolarizing the postsynaptic membrane but possess enough acetylcholine-like activity to cause depolarization before neuromuscular blockade sets in. These substances have unhindered trimethylammonium-groups; hence at least one end should be able to combine with the anionic site on the acetylcholine receptor, while the whole molecule is bulky enough to block the pore.

There is remarkable specificity in the affinities of these bis-onium salts towards the three main kinds of acetylcholine receptor. All of them are practically inactive at the preganglionic terminal; but activity at ganglia and at the neuromuscular junction is dramatic if, for each of these sites, the right homologue has been chosen. One hundred times that dose of decamethonium which completely blocks the neuromuscular junction does not affect the ganglia. Likewise, 100 times the dose of hexamethonium which blocks ganglia does not affect the neuromuscular junction (Paton and Zaimis, 1949, 1952). This specificity is illustrated in Table 12.3.

Table 12.3 The specificity of bis-trimethylammonium cations, of different chain-lengths, for two important acetylcholine receptors.* (Potencies are relative; 1.0 is maximal for each site)

Type of cholinergic block	Number of methylene groups								
	4	5	6	7	8	9	10	11	12
Nerve–nerve junction	0.01	0.8	1.0	0.1	0.02				
Nerve–muscle junction	—	—	—	—	0.07	0.7	1.0	0.55	0.2

*Anti-nicotinic action. (Paton and Zaimis, see text.)

Ganglionic-blocking nicotinic drugs, such as hexamethonium (7.30), were introduced about 1950 for the treatment of hypertension. Their action has been found to be too general, with multiple side effects, and they have been replaced by the β-adrenergic blockers and by drugs which directly relax muscle in small blood-vessels.

A drug which can combine with two sites at the one time should have a very high affinity because (a), when one end of the molecule dissociates and hence leaves the surface temporarily, the attached part must hold the liberated end within striking distance for recombination, and (b), van der Waals forces should cause the molecule to be strongly held on the biological surface that lies between the two binding sites. The introduction of ester groups into the molecules of bis-onium neuromuscular blockers strongly increases their ability to produce depolarizing inhibition. The structures in question (12.88) are esters of choline with dicarboxylic acids. Of these, succinylcholine (7.29) is much used in surgery. It is valued for the ease with which it is slowly hydrolysed by serum cholinesterase, thus helping the patient to recover. (See p. 96 for a new drug, atracurium, which is hydrolysed by water alone.)

The enhanced action introduced by the ester groups of (12.88) indicates the presence of hydrogen-bonding groups on the receptor. In this series, the second trimethylammonium-group can be replaced by hydrogen without much reduction in potency (Danilov et al., 1974). One is led to expect that one trimethylammonium-group in decamethonium could be replaced by another hydrophilic group (such as methoxy) without harm to the action, and that only *one* acidic group on the receptor is required for binding each molecule of nicotinic antagonist.

$$Me_3\overset{+}{N}-CH_2CH_2-O-\overset{O}{\overset{\|}{C}}-(CH_2)_n-\overset{O}{\overset{\|}{C}}-O-CH_2CH_2\cdot\overset{+}{N}Me_3$$

Esters of choline with dicarboxylic acids
(12.88)

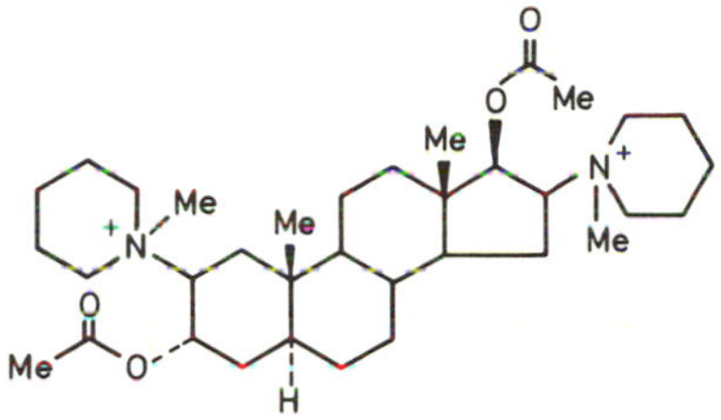

Pancuronium (dication)
(12.89)

$$Me_3\overset{+}{N}\cdot CH_2\cdot CH_2\cdot O\cdot\overset{O}{\overset{\|}{C}}\cdot NH\cdot(CH_2)_6\cdot NH\cdot\overset{O}{\overset{\|}{C}}\cdot O\cdot CH_2\cdot CH_2\overset{+}{N}Me_3$$

Carbolonium (dication)
(12.90)

Evidence of the distance that separates the two binding groups of the ACh receptor is provided by pancuronium bromide (12.89) ('Pavulon'), a potent non-depolarizing neuromuscular blocking agent that has proved clinically useful (Baird and Reid, 1967). The evident rigidity of this molecule has been confirmed by n.m.r. in solution. X-ray crystallography reveals that the distance between the two nitrogen atoms is 11.08 Å, which should be compared with 10.7 Å, the N^+-N^+ distance in the crystals of N, O, O'-trimethyl-d-tubocurarinium di-iodide, and presumably this distance is the same in tubocurarine. It must be pointed out that pancuronium has much more of the acetylcholine structure than is found in the bis-onium series, in fact it has the sequence $Me \cdot CO(:O) \cdot CHR \cdot CHR \cdot N^+R_3$ twice over (Savage et al., 1970).

In the crystal, where it lies fully extended, decamethonium (7.28) has an inter-nitrogen separation of 13.7 Å (Lonsdale, Milledge and Pant, 1965), but it seems to be more puckered in solution. Succinylcholine, flexible like all depolarizing agents, has given values from 7.8 to 11.9 Å for the inter-nitrogen distance in the crystal, depending on which anion forms the salt!

A cholinoreceptor sensitive to N-16-N compounds appeared early in the course of evolution, and is more widely spread in the Animal Kingdom. By 'N-16-N' is meant those bis-onium compounds where the two nitrogen atoms are separated by a chain of 16 atoms, not necessarily all carbon; examples: suberyldicholine and sebacinyldicholine; also carbolonium bromide (12.90) ('Imbretil') which is used a little in human surgery. The neuromuscular junctions of molluscs, echinoderms, and protochordates (Table 1.1) are inhibited by '16', but not by '10' compounds; some annelid worms respond to '10' inhibitors but all are inhibited by the '16' types. Amphibians respond more strongly to '16' than to '10' inhibitors; higher vertebrates are roughly equally sensitive to both types. The receptors for both '16' and '10' compounds, when both are present, must share some atoms, because it is not possible to block one type of receptor without blocking the other (Khromov-Borisov and Michelson, 1966; Michelson and Zeimal, 1973). These results have led to a model in which each of four active sites on the muscle receptor contribute one anionic group to each corner of a square; in this hypothesis, the sides of the square anchor the '10' structures such as suxamethonium whereas the '16' structures sit on the diagonals (Khromov-Borisov and Michelson, 1966).

For more on the evolution and uses of nicotinic antagonists, see Section 7.3.

(*b*) *A third type of cholinergic receptor.* Extraction of the heads of houseflies gave two ACh-receptor proteins which had a hybrid nicotinic–muscarinic character. They had mol. wt. 94 000 and 64 000 respectively and both responded to several typical nicotinic and muscarinic agonists and antagonists. Yet neither was blocked by either α-bungarotoxin or quinuclidinyl benzilate, considered to be the typical blocking agents for nicotinic and muscarinic receptors, respectively (Tripathi, Tripathi and O'Brien, 1979).

For more on the binding of cholinergic neurotransmitters (and their antagonists) to receptors, see Triggle and Triggle (1976), Cooper and Bloom (1982), and Yamamura, Enna and Kuhar (1978).

12.7 The GABA receptor and the benzodiazepines

Of all the neurotransmitters in the central nervous system, γ-aminobutyric acid (GABA) (*12.91*) has lent itself most readily to investigation and hence is the best understood. GABA mediates transmission of impulses in local inhibitory neurons in parts of the brain that lie near to the spinal cord, and it also mediates presynaptic inhibition within the spinal cord. At least, half a dozen other centres in the brain, including the major output cells of the cerebellar cortex, appear to depend on GABA as a neurotransmitter of inhibitory impulses. GABA mediation is usually detected with the convulsant alkaloids picrotoxin and bicuculline, neither of which is used in medicine (Curtis *et al.*, 1971).

GABA transmission has been further explored with simpler molecules such as muscimol (which mimics GABA), hydroxylaminoacetic acid (which inhibits its synthesis), and 2,4-diaminobutyric acid which inhibits presynaptic recapture of used GABA (Johnston, 1978; Iversen, 1978). Nipecotic acid (*12.92*) (piperidine-3-carboxylic acid), too, inhibits the presynaptic uptake of GABA without any effect on its postsynaptic receptor. However, the isomeric piperidine-4-carboxylic acid (isonipecotic acid) is a specific and very effective agonist for the postsynaptic GABA receptor. This may mean that GABA has different conformers combining with these two targets (Krogsgaard-Larsen, Scheel-Krüger and Köfod, 1979).

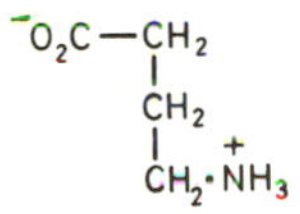

GABA
(12.91)

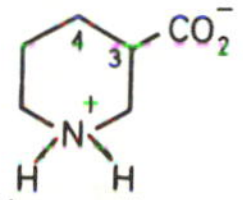

Nipecotic acid
(12.92)

Cis-4-Aminocrotonic acid, an analogue of GABA, has the folded conformation (*12.94*) and is biologically inactive, whereas the *trans*-isomer (*12.93*) acts
as efficiently as GABA in the mammalian central nervous system. This shows
that the extended, rather than the folded, conformation of GABA is the important one in neurotransmission (Johnston *et al.*, 1975).

The normal physiological effect of GABA is exerted on a *compound receptor*, a
structure more complex than the acetylcholine nicotinic receptor. Within the
compound structure there have been located a receptor for GABA, one for the
benzodiazepines, and a third which can bind picrotoxin: all three are situated
around the mouth of the chloride ion (Cl^-) channel (Olsen, 1982). Originally, it
was thought that there was more than one type of GABA receptor (Hill and
Bowery, 1981) and benzodiazepine receptor. However, careful kinetic studies
showed that benzodiazepine receptors exist as *single* populations that have two
conformations. Isomerization between these two states is induced by benzodiazepine agonists such as diazepam (*12.95*), and also by the benzodiazepine
antagonist known as RO 15-1788 (*12.96*) (Chiu, Dryden and Rosenberg, 1982).
Activation of the high-affinity conformation of the benzodiazepine receptor
enhances GABA transmission and increases chloride conductance. The benzodiazepine antagonist (*12.96*) produces neither of these effects, presumably
through its having an unfavourable steric configuration. The convulsant,
methyl 6,7-dimethoxy-4-ethyl-β-carboline-3-carboxylate, see (*12.99*), binds to
the benzodiazepine channel in a way that may lock the chloride channel shut
(Chiu and Rosenberg, 1983).

Trans-4-Aminocrotonic acid
(12.93)

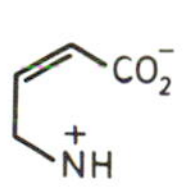

Cis-4-Aminocrotonic acid
(12.94)

Diazepam
(12.95)

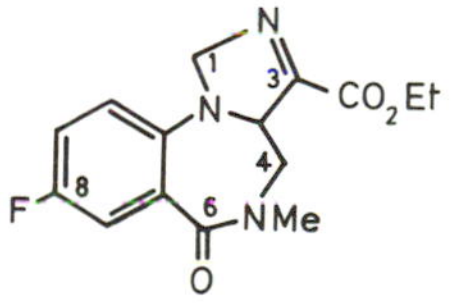

RO 15-1788
(12.96)

The only GABA analogue that has yet found a place in human medicine is baclofen, which is the β-(p-chlorophenyl) derivative of GABA. It is used to relieve spastic states following CNS injury or multiple sclerosis (Davidoff, 1978).

Benzodiazepines. The tale of these widely prescribed drugs began when Sternbach discovered chlordiazepoxide (*12.97*) ('Librium') in 1959. Its first clinical use was to alleviate anxiety (Tobin and Lewis, 1960). The more potent analogue, diazepam (*12.95*) ('Valium') soon followed and was tested clinically (Randall, 1961). Many other examples were marketed. As a class, the benzo-diazepines are sedative, hypnotic, anti-convulsant, anxiety-suppressing, muscle-relaxing, and suitable for pre-operative medication. In different examples, one or other of these properties comes more to the fore. Those that are marketed as hypnotics, and hence are given in higher doses, have been selected as not having too much duration of action: in this class one notes nitrazepam (*12.98*) flunitrazepam, flurazepam, and temazepam. Clonazepam is preferred for treating epilepsy. In Section 3.6, it was noted that chlordiazepoxide, diazepam, and temazepam all break down in the body to nordazepam (*3.47*) which has a half-life in the plasma of 90 hours and seems to be responsible for a good deal of the activity of the three pro-drugs.

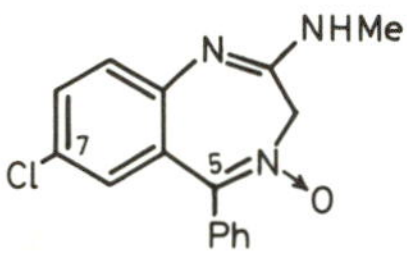

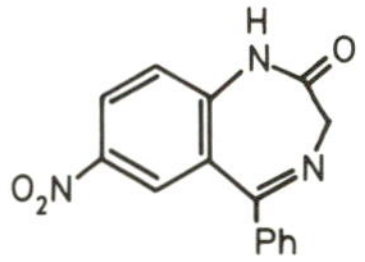

Chlordiazepoxide	Nitrazepam	Ethyl carboline-3-carboxylate
(*12.97*)	(*12.98*)	(*12.99*)

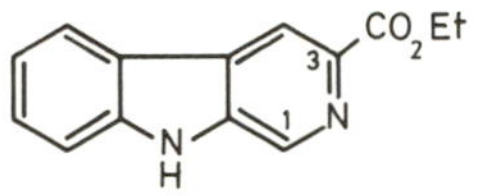

In moderate doses, the action of the benzodiazepines is remarkably like that of the barbiturates, but they are more selective because larger doses do not introduce the toxicity to the patient seen in barbiturate medication. Intra-venous diazepam can be used in place of thiopental for the induction of general anaesthesia. It has been found that barbiturates act on the picrotoxin receptor in the GABA–receptor complex (Olsen, 1982) and, in this way, synergize the action of GABA. This effect is stereospecific for in pentobarbital it is confined to the (S)-isomer (Huang and Barker, 1980). It would seem that this type of activity plays only a minor part in medication with barbiturates which behave on the whole as structure-independent lipophiles (see p. 622).

The benzodiazepines were originally marketed as 'tranquillizers', then as 'minor tranquillizers', but are best classified among the sedatives. They are of no use in treating psychoses.

We have seen how benzodiazepines bind near to, but not on, the GABA-binding site. From this position, they can increase the permeability of brain membranes to chloride ions, and thus enhance GABA's effect in stabilizing the electric potential of the cell, a factor that reduces its excitability (Simmonds, 1980). This raises the question: What normally occupies the benzodiazepine site in the GABA receptor? Some workers think that it is a protein called modulin (Costa, Di Chiara and Gessa, 1981) or nepenthin (Woolf and Nixon, 1981), but others nominate the ethyl carboline-3-carboxylic ester (*12.99*), isolated from human urine, because it has affinity for the benzodiazepine receptor (Braestrup, Nielsen and Olsen, 1980).

Ivermectin, a macrocyclic lactone which is being increasingly used as a highly selective insecticide and vermifuge, is an agonist of GABA in the peripheral nervous systems of insects and nematodes. It does not act on flukes and tapeworms, which have alternative neurotransmitters. Mammals are protected by the fact that their GABA-using system is confined to the central nervous system, protected from ivermectin by the blood–brain barrier. So far, its use has been confined to farm animals. As it is lipophilic, a single dose, oral or sub-cutaneous, lasts for several weeks.

For further reading on benzodiazepines, see Garattini, Mussini and Randall (1973) and Usdin *et al.,* (1983).

12.8 Morphine and the opioid receptors

The term opioid (synonym, *narcotic analgesic*) includes both natural and syn-thetic agents with morphine-type pain-relieving properties. Their action is exerted on the central nervous system and on smooth muscle. Morphine, the principal analgesic alkaloid of opium, is the standard by which other opioids are measured. It is still used to subdue pain, cough, and nervous diarrhoea, and to bring comfort to patients recovering from an operation. Section 7.3 dealt with the shape of the morphine molecule, also the search for opioids with less complex structure and those with the analgesic, but not the habituating (euphoric) properties of morphine. The first of the morphine antagonists, nalorphine (*7.35*), was also discussed in that section. Although it was originally thought that the aromatic ring in all synthetic analgesics must have an axial conformation (Beckett, 1959; Beckett and Casy, 1962), it is now clear that the steric relationship of the phenyl ring is often unimportant. It must be realized that earlier representations of the receptor site were made, in ignorance of this fact, before the ready availability of X-ray diffraction, nuclear magnetic resonance, and optical rotatory dispersion. Particularly when applied to the more rigid molecules, these new methods have thrown strong light on the problem.

The most convincing demonstration that analgesic action is independent of the conformation of the aromatic ring was furnished by the synthesis of 1-methyl-4-phenyl-*trans*-decahydro-4-propionoxyquinoline, where the two con-formers [(*12.100*) and (*12.101*) where the phenyl group is equatorial and axial, respectively] *are separable* (on alumina). Because of close crowding, they cannot be interconverted. Neither conformer was optically active (but each should be optically resolvable into a pair of enantiomers). Although possessing only one-tenth of the analgesic action of morphine (mouse), there was no difference

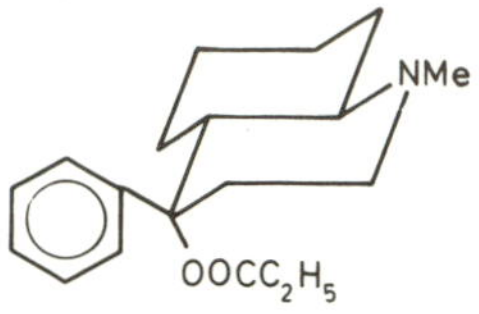

(12.100)

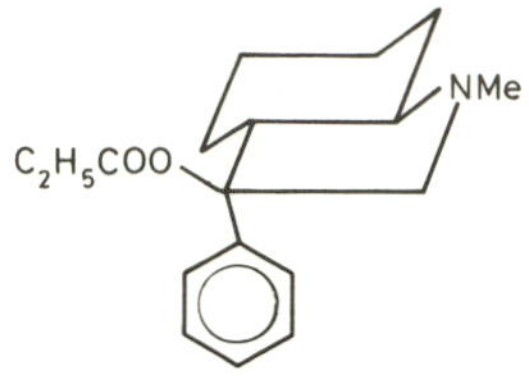

(12.101)

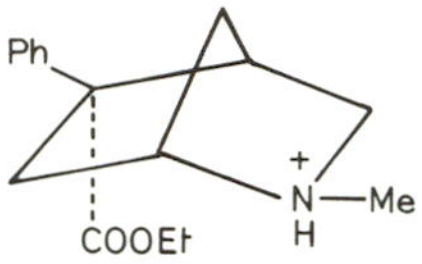

(12.102)

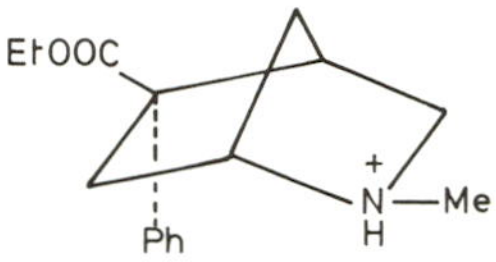

(12.103)

Two pairs of stereoisomeric opioids

in potency between the two conformers (Smissman and Steinman, 1966).

Another pair of rigid, non-interconvertible isomers have provided further insight into the problem. These are the two epimeric 2-methyl-5-phenyl-5-carboethoxy-2-azabicyclo [2.2.1] heptanes [(12.102) exophenyl and (12.103) endophenyl], which may be regarded as rigid analogues of pethidine (7.37). The endophenyl-epimer, twice as potent an analgesic as pethidine, was four times as potent as the exo-epimer (mouse). Analysis of the brain showed that it was penetrated twice as readily by the endo-epimer; in addition to a difference in partition coefficients between the two epimers, a difference of 0.16 unit was found in the pK_a values (Portoghese, Mikhail and Kupferberg, 1968). This work provides another example of a change in conformation of the aromatic ring having no drastic effect on analgesic activity. More importantly, it gives a timely reminder that physical properties change when conformations change, and that an altered physiological response may spring from these new properties rather than the mere change in shape of the molecule. On the evidence of this pair of substances, the conformational requirement for analgesia is low: the distance from the centre of the benzene ring to the nitrogen atom is about 6 Å for (12.102) but only 4 Å for (12.103), a truly enormous difference.

Because the molecule of morphine is rigid and undeformable, it should be capable of yielding information about the dimensions of groups in the morphine receptor. The first step is to nominate, in the morphine molecule, those features which observations on simpler analogues have indicated to be essential for analgesic action. There seem to be three such features: the benzene ring, the nitrogen atom, and possibly the α,β-unsaturated secondary alcohol. Using existing X-ray-diffraction data, French workers have laboriously calculated many of the distances separating these centres in morphine and codeine, and corresponding ones in alphaprodine and methadone (Jung and Lami, 1970; Jung, Koffel and Lami, 1971). Collectively, these measurements indicate a receptor of such elastic dimensions, that there must be more than one kind of analgesic receptor.

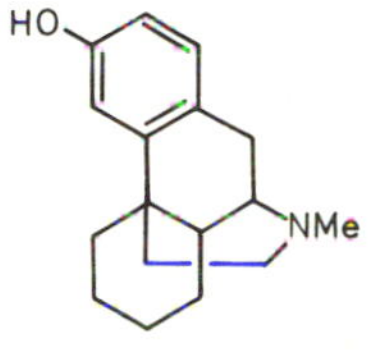

Levorphanol
(12.104)

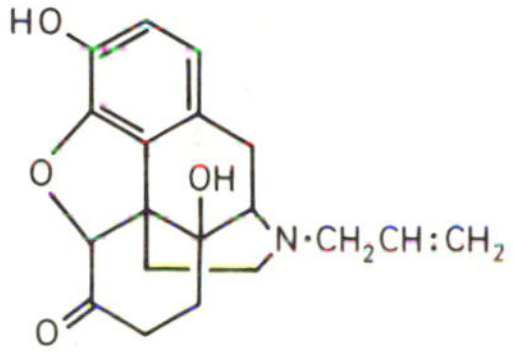

Naloxone
(12.105)

(a) *The opioid receptors.* Avram Goldstein first demonstrated the existence of an opioid receptor in the mouse brain by studying the stereospecific interaction of levorphanol (a morphine congener) (*12.104*) and the antagonist naloxone (*12.105*), on subcellular fractions (Goldstein, Lowney and Pal, 1971). Partial purification furnished the receptor as a lipoprotein of estimated mass 60 000 daltons. It was suggested that combination of receptor with agonist (but not with antagonist) produces a conformational change that initiates the analgesic response. The existence of more than one kind of opioid receptor was suspected at this stage (Lowney *et al.*, 1974). Pert and Snyder (1973) confirmed the presence of opiate receptors in nervous tissue and went on to demonstrate a differential effect exerted by sodium ions on the binding of opioid agonists and antagonists by the opioid receptor. Sodium ions, by increasing the number of sites available for antagonists, were thought to be reversing a conformational change favoured by agonists (Pert and Snyder, 1974).

The next question concerned the normal, physiological function of the opioid receptor, insofar as morphine is foreign to the animal organism. Hughes (1975) extracted from brain a pentapeptide (H-*Tyr-Gly-Gly-Phe-Met*-OH), called *Met*-enkephalin, which was accompanied by a small amount of a less active analogue in which the methionine was replaced by leucine. Both enkephalins mimic the ability of morphine to block muscular contractions (electrically evoked) of the ileum (guinea pig) and vas deferens (mouse). They also inhibit the stereo-specific receptor binding of an opiate antagonist, [^{3}H]naloxone (*12.105*), in brain homogenates. When put into the brain of a living mouse (with a cannula, because the enkephalins do not cross membranes, and are hydrolysed in 10 minutes by brain enzymes), the more powerful of the two, *Met*-enkephalin, was found only one-twentieth as potent as morphine as an analgesic. Repeated applications produced tolerance; also, it was found to be cross-tolerant with morphine (Hughes *et al.*, 1975).

By serial replacement, it was found that the three active parts of the enkephalins are the —OH and —NH of the tyrosine fragment, and the lipo-philic area in the phenylalanine residue (successfully replaceable by other lipophilic groups) (Gorin *et al.*, 1980). Swiss workers who made stepwise changes to the molecule of *Met*-enkephalin, produced 'FK 33–824' which is 1000 times as potent as morphine by intra-cerebroventricular injection in laboratory animals (Roemer *et al.*, 1977), but has anaphylactoid side effects in Man. A French group then discovered thiorphan, an (artificial) dipeptide that inhibits the enzyme, enkephalinase*, thus causing the body to receive an overdose of its own enkephalins. Chemically, this analgesic (*12.106*) is DL-3-mercapto-2-benzylpropanoylglycine (Roques *et al.*, 1980). This raises the question, Do some analgesics act, not on the enkephalin receptors, but on enkephalinase? It may be relevant that naloxone blocks the action of thiorphan.

Another series of naturally occurring analgesic polypeptides was uncovered by the hydrolysis of brain lipotropin (Guillemin, 1978). The most active analgesic, called *β-endorphin*, was constituted of the amino acids of lipotropin from 61 to 91. It produced an analgesia in rats that was long-lasting and twice as intense as that produced by morphine. Naloxone (*12.105*) instantly abolished this analgesia. This endorphin contains the amino acid sequence of *Met*-enkephalin on residues 61–65, but it is now known that the enkephalin does not arise from the endorphin. Still more recently, *dynorphin*, a still more powerful analgesic, 200 times as potent as morphine, has been isolated from the posterior pituitary gland, and shown to be a heptadecapeptide (Goldstein *et al.*, 1981). For a review on the natural roles played by dynorphin, the endorphins, and the enkephalins in the modulation of pain, see Terenius (1982).

Although no clear picture has yet emerged, there is a current tendency to think of enkephalins as neurotransmitters, and of β-endorphin as a hormone that suppresses output of other neurotransmitters. It is puzzling that the human body liberates an analgesic under some painful conditions but not others.

Different kinds of opioid receptors have been distinguished in mammals. This classification, which is based on pharmacological behaviour, may be accepted (but with reserve) until the receptor structures are better known. There may, for example, be grouping, or conformational equilibria of a smaller number of receptors (Lord *et al.*, 1977; Iwamoto and Martin, 1981).

The most important receptor is called mu (μ) because morphine binds to it and thereby initiates analgesia. The natural enkephalins bind only weakly to the mu, but strongly to the delta, site. However, a highly selective ligand for the mu site, namely D-Ala^2,MePhe^4, Gly-ol^5-enkephalin, has been made by altering a natural enkephalin in three positions. Previously the mu receptor was identified by labelling with (tritiated) dihydromorphine. Naloxone acts as an antagonist of this receptor, and also of the sigma and kappa receptors. Pentazocine, which is used clinically as an opioid analgesic, is an antagonist of the mu receptor, but an agonist of the kappa and sigma receptors. Morphine also binds to the kappa (κ) receptor which produces spinal, but not general, analgesia and sedation.

The sigma (σ) receptor, activation of which causes hallucination and respiratory stimulation, is activated by nalorphine and pentazocine. It is uniquely stimulated by *N*-allylnormetazocine, but no unique antagonist is known.

The delta (δ) receptor appears to be physically attached to the mu receptor (Smith, Lee and Loh, 1983). β-Endorphin binds to both the mu and the delta receptor. Enkephalins bind mainly to the delta receptor (Chang *et al.*, 1981;

*This enzyme hydrolyses the bond between *Gly* and *Phe*.

Snyder 1980). The delta receptor modulates relaxation of the gut more power-fully than it produces analgesia. Etorphine is a typical agonist. This receptor can be specifically labelled with the agonist D-Pen^2,DL-Pen^5-enkephalin, where 'Pen' stands for dimethylcysteine. A specific antagonist is NN-diallyl-Tyr-Gly-Gly-(CH$_2$S)-Phe-Leu(OH) (Shaw *et al.*, 1982). The mouse vas deferens possesses strong delta but weak mu activity, whereas the guinea-pig ileum is rich in mu but poor in delta activity; both are much used as differentiating test-material (Chang *et al.*, 1981).

Hydrolysis of the casein in milk liberates a tetrapeptide, H$_2$N-Tyr-Pro-Phe-Pro(OH), whose moderate analgesic activity is exerted on the mu, but not on the delta, receptor (Chang *et al.*, 1981). The (apparent) hormone, dynorphin acts on the kappa (κ) receptor.

The partial agonist, nalorphine (*7.35b*) was dealt with on p. 279.

(*b*) *Opioids in food.* We have mentioned an opioid tetrapeptide arising in the digestion of casein. Delicate, modern analytical methods have demonstrated, and confirmed beyond doubt, that traces of morphine are constantly ingested by Man from such sources as lettuce, mother's milk, cow's milk (its origin has been traced to hay), in quantities such as 500 ng per kg (wet weight) (Hazum *et al.*, 1981). Coffee has been shown to be rich in a non-peptide substance (mol. wt. under 3500) which strongly binds to the mu receptor as a partial agonist or antagonist (Boublik *et al.*, 1983).

(*c*) *Tolerance and physical dependence.* Investigation of these phenomena has been narrowed by the observation that there is no increased breakdown of opioids in tolerant animals (Klee and Streaty, 1974). Likewise, the number and binding affinity for opioid receptors in morphine-dependent rats are the same as in controls. A study of the effect of opiates on cultured nerve-cells suggests that morphine causes inhibition of adenylate cyclase and that this enzyme becomes overactive when the drug is withdrawn. Hence cyclic adenosine monophos-phate (cAMP) accumulates, as can be seen in the brain of rats with symptoms of morphine abstinence. Moreover, injection of cAMP into the cerebroventricle intensified the abstinence syndrome. Again, purines, which inhibit phospho-diesterase and hence allow cAMP to build up (checked on homogenates of rat brain), produce symptoms of morphine withdrawal in rats which have never been exposed to morphine (Sharma, Klee and Nirenberg, 1975; Collier and Francis, 1975).

The physical and the subjective symptoms that follow withdrawal of opioids from the dependent subject are rapidly reversed by clonidine (*12.53*) without causing euphoria. This result seems to be connected with a finding that chronic morphine administration in the rat increases the number of α-adrenergic receptors in the brain, and that clonidine is an efficient antagonist for these receptors (Redmond and Kleber, 1980). Some clinics are using clonidine as part of their withdrawal programme.

$HS-CH_2-CH(CH_2Ph)CO-NH-CH_2CO_2H$

Inhibitor of enkephalinase
(12.106)

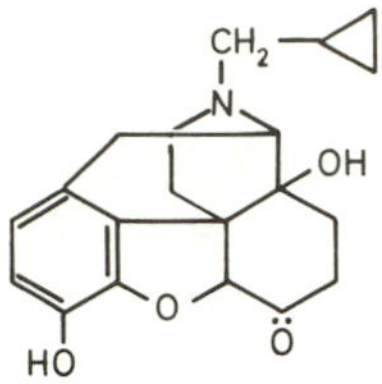

Naltrexone
(12.107)

The United States pioneered the treatment of heroin addiction by methadone (7.38), given orally (Dole, Nyswander and Kreek, 1966). This treatment avoids the emotional swings induced by heroin and is more hygienic. Treatment is preferably residential, and the dose (1 mg of methadone for each 2 mg of heroin) is reduced at the rate of about 20% a day. A minority of patients cannot adjust to this treatment, whereas others demand a faster cure and are given the antagonist naloxone (12.105), followed by the longer-acting naltrexone (12.107). Many patients relapse when faced with withdrawal symptoms which may continue for several weeks, such as insomnia, malaise, and hyperactivity of the bowels. They can return to a methadone clinic and attempt a slower withdrawal regimen.

Clinical trials are continuing with methadyl acetate which has an apparent half-life of 72 hours, as against methadone's 20 hours (Ling, Klett and Gillis, 1978).

Although the search is always in progress for strong analgesics without euphoric (and hence potentially addictive) properties, the management of the postoperative patient in hospital wards requires the euphoric type, such as morphine, a switch to a purely analgesic substance being made a little later. Some powerful analgesics are being discovered which do not, apparently, act through the opioid receptors. For example, there is meptazinol (12.108) which is strongly analgesic, causes no euphoria, and evinces less respiratory depression than morphine or pentazocine (Pearce and Robson, 1980). Nantradol (12.109) is powerfully analgesic but not antagonized by naloxone. This phenanthridine drug underwent clinical trial as an anti-emetic in cancer chemotherapy (Johnson and Milne, 1980) and was withdrawn.

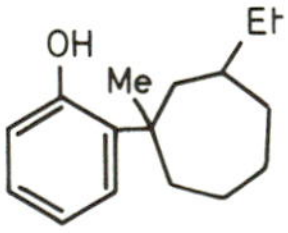

Meptazinol
(12.108)

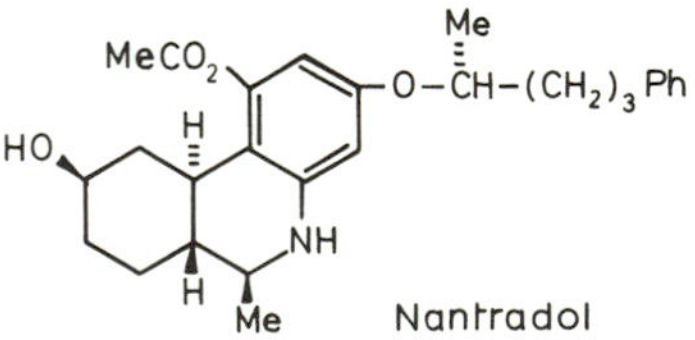

Nantradol
(12.109)

(d) *Miscellany.* The state of analgesia in mice, brought about by acupuncture and thought to be caused by release of endorphin, is lost when naloxone is injected (Mayer *et al.*, 1977; Pomeranz and Chiu, 1976). Patients receiving intracerebral electric stimulation, for relief of chronic pain, produce raised levels of circulating opioids (Terenius and Wahlstrom, 1978).

Although the induction of sleep is not an important action of the opioids, it is appropriate to mention the sleep-inducing nonapeptide *'dsip'* isolated from the rabbit brain. It is *Trp-Ala-Gly-Gly-Asp-Ala-Ser-Gly-Glu* and has been tested on human volunteers in five different kinds of sleep studies. All patients awoke refreshed (Schoenenberger and Schneider-Helment, 1983).

For further reading on structure–activity relationships in opioids, see Szekely (1982) and Kosterlitz, Collier and Villareal (1972).

12.9 Psychotherapeutic agents

By the mid-1950s, it was realized that even the severest of psychiatric illnesses were far more amenable to medication than to psychotherapy. Thanks to these agents, most patients can soon be returned to their home, and quite a high proportion of them become capable of earning their living. This demonstration that mental illness is biochemical in nature does not overlook the precipitating effect of adverse factors, at home or at work (Baldessarini, 1977).

The major psychoses ('madness' or 'insanity') are divided into two types: schizophrenias and manic-depressive disorders. The schizophrenias are characterized by disordered thinking, emotional withdrawal and, often, audio or visual delusions. Delirium, when present, can lead to violence. In the manic-depressive disorders, memory and orientation are retained but emotion, thought, and behaviour are highly disordered. This illness can be unipolar (e.g. depressive only) or bipolar (alternating between both modes).

12.9.1 *Neuroleptic (anti-psychotic) drugs*

Although valuable as first-aid in an outburst of mania (manic-depressive disorders), these drugs find their principal use in the treatment of schizophrenia, formerly held to be the most intractable form of insanity. Before 1952, nearly two-thirds of schizophrenic patients admitted to hospital were still there 2 years later, whereas now less than 10 % remain away from home. The medication that changed this situation so dramatically was the phenothiazine derivative, chlorpromazine (*12.110*) introduced by Delay, Deniker and Harl (1952). This drug controlled the delusions and hallucinations and ended the need for physical restraint. Asylums quickly lost their air of uncontrollable disorder, their stenches and screaming, as patients were discharged to take up light work, often as a gardener's assistant. Mental hospitals in so many cities were repainted and handed over for other civic purposes. Yet these were not *complete* cures because the patient had to continue on medication indefinitely and often was incapable of taking much interest in the world outside himself.

Although the action of chlorpromazine is so powerful and calming, it does not cloud consciousness, a distinction from the general biological depressants of

Chapter 15. The molecule of chlorpromazine was subjected to many changes, one of which was to provide a long-acting form for discharged patients. Thus an injection of 25 mg of fluphenazine enanthate, dissolved in oil, provides maintenance therapy for 2 to 3 weeks. Fluphenazine is 10-{3-[4-(2-hydroxyethyl)-piperazin-1-yl]propyl}-2-trifluoromethylphenothiazine, obviously derived from chlorpromazine which is 10-(3-dimethylaminopropyl)-2-chlorophenothiazine (*12.110*).

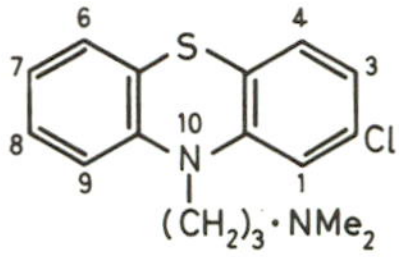

Chlorpromazine
(*12.110*)

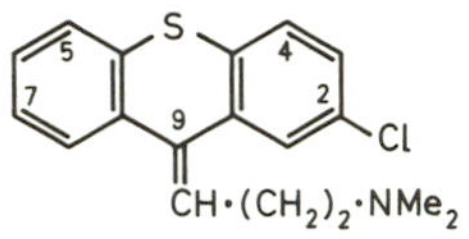

Chlorprothixene
(*12.111*)

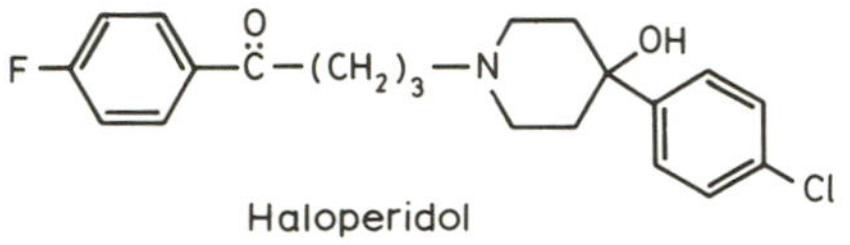

Haloperidol
(*12.112*)

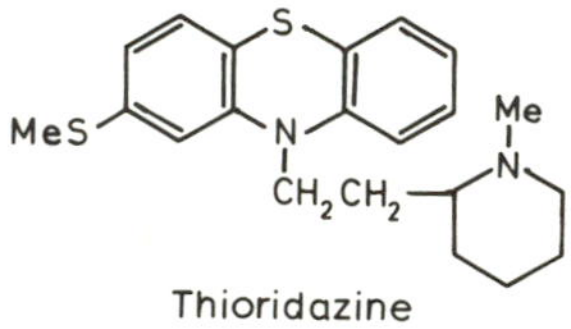

Thioridazine
(*12.113*)

The neuroleptic (at first called 'tranquillizing') action of these phenothiazines was later found in other chemical series, notably among the thioxanthenes [e.g. chlorprothixene (*12.111*)]. and butyrophenones [e.g. haloperidol (*12.112*)]. At first, the dose of the butyrophenones was much lower than was needed for the other two categories, but now these too have low-dose examples such as perphenazine, trifluoperazine, fluphenazine, and thiothixene.*

The principal action of neuroleptic drugs appears to be the blockade of the dopamine receptor in brain, demonstrated on the mouse brain (Carlsson and Lundqvist, 1963; Seeman, 1977) and in the schizophrenic patient (Horn and Snyder, 1971). Confirmatory evidence is supplied by the neuroleptic drug, flupentixol, a close relative of chlorprothixene (*12.111*). The double bond, arising from the 9-position, permits geometrical isomerism. The clinical benefits of this drug are confined to the *cis* isomer, and this is the only one that blocks the dopamine receptor *in vitro* (Crow and Johnstone, 1977). Parkinson's disease is characterized by dopamine deficiency in the brain, and is treated by administering dopamine (p. 103). But schizophrenia is characterized by a deleterious accumulation of dopamine in the forebrain, and possibly elsewhere.

*Dose reduction followed replacement of the metabolizable dimethylamino-group by piperazine.

Hence it is not surprising that side effects of the treatment of schizophrenia with neuroleptic agents resemble symptoms of mild parkinsonism. These extra-pyramidal symptoms, which include tremor, and rigidity, are relieved by reducing the dose, also by giving an anti-cholinergic drug such as trihexy-phenedyl (*7.24*). After long treatment with neuroleptic drugs, about 20% of the patients exhibit *tardive dyskinesia* which usually takes the form of repeated, compulsory facial movements. In the elderly, the condition seems to be per-manent. It occurs less often in treatment with low dose (as opposed to high dose, see above) drugs, one of the best being thioridazine (*12.113*) in the pheno-thiazine series, also (though it is still under trial) timiperone in the buryro-phenone series. Of similar promise, although still under trial, is sulpiride which is 5-(aminosulfonyl)-*N*-[(1-ethyl-2-pyrrolidinyl)methyl]-2-methoxybenzamide (*12.62*), so chemically different from earlier remedies (Bratfos and Haug, 1979).

A role for norepinephrine systems in schizophrenic pathology has been suggested (Mason, 1983) and may lead to a different therapy. The red free-radicals to which phenothiazines are readily oxidized (Borg and Cotzias, 1962) do not appear to play any part in their neuroleptic action.

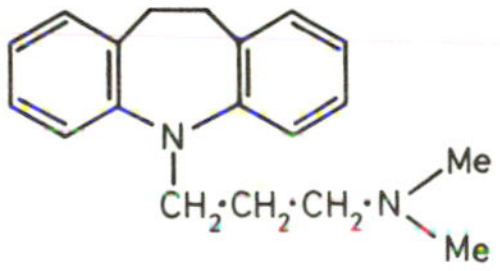

Imipramine
(12.114)

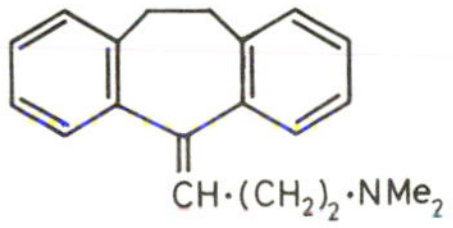

Amitriptyline
(12.115)

12.9.2 *Drugs for controlling mood*

In 1949, Cade, in Australia, showed that lithium salts would control the manic phase of manic-depressive psychosis. This was the first effective psycho-therapeutic agent of any kind, ante-dating even chlorpromazine. Possibly on account of the remoteness of his country, Cade's discovery was slow in gaining acceptance but is, today, the standard treatment (see further, Section 11.0).

The depressive phase of this psychosis was first brought under control by Kuhn (1958) who found that the tricyclic compound, imipramine (*12.114*) restored a normal mood to severely depressed mental patients. Since then, it and the related drug, amitriptyline (*12.115*) have become the standard treat-ment for this condition. Results are obtained only after 1 or 2 weeks' admin-istration, and treatment usually continues for 6 months. High doses must be avoided in patients with heart disability which these drugs can accentuate.

Because good anti-depressive and good neuroleptic activity is found in sub-stances with a nucleus of three fused rings, but the two properties never occur in the one compound, an exploration was made of structure–action relations in the two systems (Wilhelm and Kuhn, 1970). The topology of the tricyclic skeleton determines the type of activity, but this is modified by the side-chain.

The shape of the skeleton (ring-system) is determined by four steric parameters: the bending angle α, the annellation angle β, the torsion angle γ, and the interatomic distance δ. Of these variables, *the bending angle* is most responsible for giving the drug its characteristic pharmacological action. A relatively flat molecule, such as a phenothiazine or thioxanthene ($\alpha = 25°$), has mainly neuroleptic (schizophrenia controlling) properties, whereas a decidedly bent skeleton, such as a dibenzocycloheptadiene (55°), imidodibenzyl (55°), or dibenzodiazepin (65°), produces thymoleptic (anti-depressive) drugs. It will be realized that imipramine is an imidodibenzyl and amitriptyline a dibenzocycloheptadiene. The bending angle α is defined as the angle which the planes of the two lateral aromatic rings would make if projected towards one another. The *annelation angle* β is defined as the rotation of the two lateral aromatic rings away from a straight line. It is 0° for dihydroanthracene, 60° for dihydrophenanthrene, and 40° for imidodibenzyl. The *torsion angle* γ measures any twisting of the planes of the two lateral aromatic rings relative to one another. Torsion is introduced by the presence of a seven-membered central ring; it varies only between 0° and 20° in the drugs under discussion.

For good neuroleptic properties, the side-chain should be unbranched and exactly three carbon atoms in length. It will be appreciated that, as in thioridazine (*12.113*), one of these carbon atoms can be incorporated in a piperidine ring. Some anti-histaminic drugs, notably promethazine, have been built on a phenothiazine nucleus, but they require a side-chain of exactly two carbon atoms. Anti-depressant drugs can tolerate a side-chain of either two or three carbon atoms. X-ray-diffraction analysis has shown that this chain needs to be free to take up a position where its terminal basic group can be near (for psycholeptic action) or actually overlap (for anti-depressant action) a benzene ring. These structures seem to indicate small, and yet smaller, receptor cavities for these two kinds of biological activity.

Patients who cannot shake off the initial anti-muscarinic side effects (such as dryness of the mouth) of imipramine and amitriptyline are placed on the corresponding secondary amines, desipramine and nortriptyline respectively, which have a methylamino terminus to the side-chain.

The valuable properties of the tricyclic anti-depressant drugs seem to stem from an ability to increase the availability of norepinephrine and 5-hydroxytryptamine. Closer study has revealed both a pro- and anti-5-hydroxytryptamine effect, as well as a pro- and an anti-norepinephrine effect, but at different times in the treatment or at different loci in the brain (Koe, 1983; Fillion and Fillion, 1981). Many scientists are working in this field and we must await the results of their efforts.

The tricyclic anti-depressants usually have a pK_a of 9.5 to 10 and a partition coefficient of 4 in the system octanol/water. Improvement in their therapeutic performance has followed two directions, both leading to products without effect on the heart. First came the tetracyclic anti-depressants such as maprotiline (*12.116*) and mianserin (*12.117*). Then there are anti-depressant drugs with much simpler nuclei such as nomifensine (*12.118*). These newer products

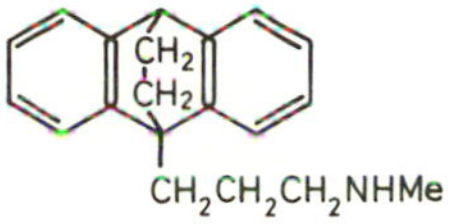

Maprotiline   Mianserin   Nomifensine
(12.116)   .(12.117)   (12.118)

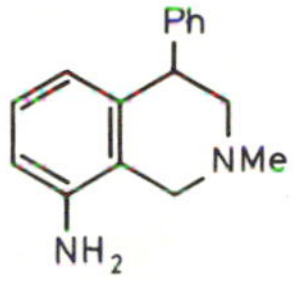

may not necessarily act by the same biochemical mechanisms as imipramine and amitriptyline and have yet, except perhaps for cardiac patients, to demonstrate superiority.

Patients who do not respond to the tricyclic anti-depressants have two alternative therapies available, both of which increase the availability of neuro-transmitter amines in the brain. The monoamine oxidase inhibitors (Section 9.4, p. 359) are highly effective but impose dietary restraints and have often caused dangerous surges in blood-pressure. Electroconvulsive therapy, which may sound brutal, but is actually painless to the patient, can achieve lasting improvement (Green, Heale and Grahame-Smith, 1977).

12.10 Conclusion

The study of steric factors has contributed much to the understanding of structure–action relationships in biologically active agents, and is also beginning to provide information about the nature of the various receptors, some of which have been discussed in this chapter. Yet other receptors have been investigated by similar competitive methods, among them those for insulin and thyroxine. It may safely be said that the present interest in this subject will continue to grow as its ability to throw light on the action of agonists is appreciated.

Investigations into the nature of receptors often begin with binding studies, using labelled molecules related to the appropriate drug. In this approach, false results have often arisen from the ever-present sites of loss, on which the ligand can be tenaciously adsorbed. Mere isolation of a fraction which has bound the ligand provides no evidence that a receptor is on hand, unless it is shown to perform the tasks characteristic of this receptor. It must be shown that the appropriate agonists and antagonists are bound and exchanged exactly as in Nature, as a preliminary to demonstrations of ionic control, or whatever is kinetically characteristic of that receptor. In short, care needs to be exerted not to make binding studies say more than they logically can.

Meanwhile, as if in expectation of early, definite knowledge of receptors (perhaps even of a ready supply of purified receptors), the old, conjectural diagrams of receptor sites (like Fig. 12.3) are vanishing from the literature (but for a collection of them, see Ehrenpreis, Fleisch and Mittag, 1969).

For further reading on steric factors, see Bentley (1969).

13

The covalent bond in selective toxicity

Although the majority of biologically active substances combine only loosely with receptors and are easily released by washing, a few agents combine by covalent bonds which are of a more durable character (see Section 8.0). Covalent bonds involving carbon can be broken by great heat and also by powerful chemicals, but usually not by mild reagents at temperatures compatible with life.

Marginally easier to break are some covalent bonds that link atoms other than carbon, although a swamping excess of the reagent is usually required. Two examples: the restoration by mercaptan antidotes of cells poisoned by arsenicals (exchange of one As—S bond for another), and rescue of a victim poisoned by an acetylcholinesterase inhibitor, using an oxime antidote (exchange of one P—O bond for another), as described in Sections 12.0 and 12.3 respectively.

One aspect of covalent bond formation has already been adequately treated, namely the degradation of a pro-drug to the true active agent (see Section 3.6). Also, many covalent bonds are made and broken by enzymes in the normal metabolic reactions of the cell, and also in disposing of foreign materials (see Section 3.5). However, the present chapter is concerned solely with the formation of a covalent bond between an agent and the complementary atoms of its receptor.

13.0 Arsenicals, antimonials, and mercurials

The arsenicals were the first drugs to be recognized as acting through formation of covalent bonds. Ehrlich (1909) suggested that some arsenicals combine with essential mercapto-groups (also called sulphydryl-groups, thiol-groups, or SH groups) in the parasite. This suggestion arose from the fact that arsenoxides very readily gave the reaction shown in (13.1), even in the presence of a great excess of water. Moreover this reaction is the only one available to arsenoxides under biological conditions. The three levels of oxidation of arsenic are exemplified in (13.2), (6.4) and (6.3), which are respectively the pentavalent and the two trivalent (arsenoxide and arsenobenzene) states. It is noteworthy that the arsenobenzene state is at a lower level of oxidation than the arsenoxide state, but the arsenic atom is trivalent in both*.

Although Ehrlich had found that pentavalent arsenicals did not act in the body until reduced to arsenoxides, it had not occurred to him that arseno-benzenes acted only after oxidation to arsenoxides. This was established between 1920 and 1925 by Swiss-born Carl Voegtlin and his co-workers in the United States Public Health Service (for a review, see Voegtlin, 1925).

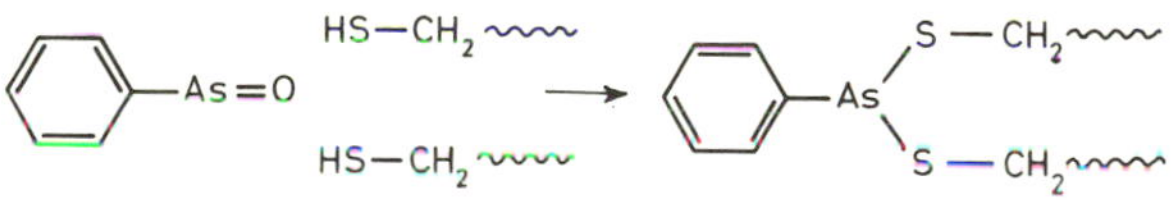

How arsenical drugs combine with mercaptans

(13.1)

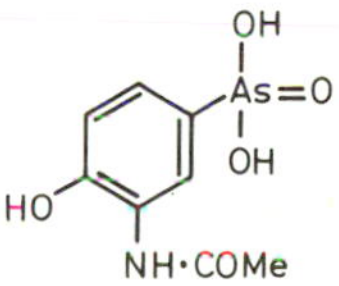

Acetarsol

(13.2)

Voegtlin showed that the arsenoxide level is the only one at which organic arsenicals are therapeutically active. In Fig. 13.1 it can be seen that oxo-phenarsine (6.4) brings about a dramatic lowering of the trypanosome population in the rat's bloodstream within half an hour of administration, but arsphenamine (6.3), which is the corresponding arsenobenzene, requires more than 5 hours to accomplish the same result. This indicates that arsphenamine has to be changed to a more active substance before it can act. Voegtlin found that, when arsphenamine was shaken with air for a few minutes, it was con-verted to its arsenoxide (oxophenarsine) which then exerted the typical rapid action. The same result was obtained when arsphenamine was incubated with oxidizing tissues, such as fresh liver.

Fig. 13.1 also illustrates how the action of arsenoxides is antagonized by substances containing thiol-groups (such as glutathione, cysteine, or sodium thioglycollate). This effect is not durable if the amount of thiol is small, because the complex can hydrolyse to its original components. With larger amounts of

*Arsenobenzenes have been shown by X-ray analysis to be trimeric in the solid state (Hedberg, Hughes and Waser, 1961), and ebullioscopy indicates that benzenearsenoxides are associated fourfold (Blicke and Smith, 1930). The loose associations do not affect this discussion.

thiols, the effect is prolonged because of the mass-action law effect. Voegtlin used several tests that demonstrated an abundance of mercapto-groups in trypanosomes.

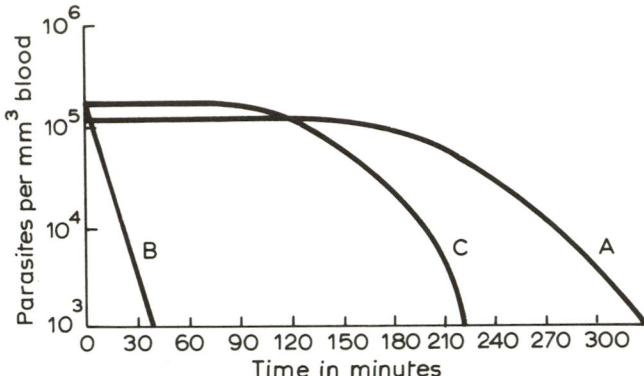

Fig. 13.1 The parasiticidal effect in trypanosome-infected rats of (A) arsphenamine ('Salvarsan'), (B) oxophenarsine ('Mapharside'), and (C) oxophenarsine plus reduced glutathione. (Redrawn from Voegtlin, 1925.)

The reaction of arsenoxides with thiol-groups is reversible, and parasites that have been treated with several lethal doses of arsenic can be saved if subsequently treated with many equivalents of a thiol-compound.

It is now accepted that death of the parasite, because it occurs so quickly, often in less than 30 minutes, must be derived from the drug's interference with respiration and glycolysis, and not from any attack on nucleic acids or protein synthesis (Macadam and Williamson, 1974). This leads to the question, to what are the vital thiol-groups of the parasite attached? At first Voegtlin suspected glutathione, later, the thiol-groups of proteins (Rosenthal and Voegtlin, 1930). Current opinion favours enzymic thiol-groups, many of which have been shown, in pure isolated enzymes, to be strongly inhibited by arsenoxides and protected by thiols.

The most arsenic-sensitive of known enzymes are (a) lipoic acid dehydrogenase, and (b) the α-oxidases (e.g. pyruvate oxidase) which use dihydrolipoic acid (*2.28*) (thioctic acid) as a coenzyme. Lipoic acid dehydrogenase has two cysteine residues on different chains of protein which are kept adjacent by tertiary folding (Massey, Hofmann and Palmer, 1962). Arsenicals can bind these cysteine thiol-groups as in (*13.1*). In (b), the lipoic acid (6,8-dimercaptooctanoic acid) is easily bridged by arsenicals as in (*13.1*) (Peters, 1963).

How do organic arsenicals injure the parasite without harm to the host? It was pointed out in the introduction to Chapter 3 that the parasites have a relatively higher abundance of free SH-groups than the host's cells. Additional selectivity is derived from analogous enzymes and from favourable distribution, as follows. It has been observed that several enzymes in trypanosomes, e.g.

phosphopyruvate kinase, are highly sensitive to arsenicals, whereas the corresponding mammalian enzyme is much less affected (Grant and Sargent, 1960). In addition, much selectivity is conferred by the aromatic-ring structure of the drug. Inorganic and aliphatic arsenicals are no less toxic to trypanosomes than are the aromatic arsenicals; but only the aromatic arsenicals are selective, and they are most selective if substituted as in (6.4). This conclusion is supported by drug-resistance studies (see Section 6.5) which point to selective adsorption.

(a) *How arsenicals act in syphilis.* Voegtlin's studies, extended by his colleagues Rosenthal (1932) and Eagle (1939), have shown that arsenical drugs act on spirochaetes by the same mechanism as on trypanosomes. No matter at what level of oxidation the drug may be, it has first to be converted to the arsenoxide level and then it combines with essential thiol-groups in the parasite. For a review of the connexion between constitution and selective action in arsenicals designed for treating syphilis, see Eagle (1951).

Because Ehrlich had thought that arsenoxides were too toxic for human use, syphilis was treated for a quarter of a century with unnecessarily large doses of arsenic, in the form of arsphenamine ('Salvarsan') and neoarsphenamine ('Neo-salvarsan'). A small portion of those massive doses was converted to the potent arsenoxides in the body; the greater part engaged in side reactions, dangerous for the patient. Eventually, Tatum and Cooper (1934) showed that *oxophenarsine*, the arsenoxide corresponding to arsphenamine, was a safer drug than 'Salvarsan' because, although it was more toxic to the host, it was even more toxic to the parasite and hence smaller doses were effective. This substance (*m*-amino-*p*-hydroxyphenylarsenoxide) (6.4) ('Mapharsen') made it possible to cure syphilis with small, safe doses of arsenic. Penicillin has now taken the place of arsenic in treating this disease.

(b) *Chemistry of the mercapto-group's antidotal action.* Why is the arsenic–sulfur bond so readily split by mercaptans? The arsenic atom, quite unlike those of nitrogen, carbon, and oxygen, has a vacant *d* orbital, into which can go a lone pair of electrons from the sulfur atom of cysteine. This leads to a higher-valency transition-complex having three* sulfur atoms united to one arsenic atom. Such a complex is unstable, and each As—S bond has a roughly equal chance of breaking. Thus there is a one in three chance that the newly formed bond will survive. If an excess of cysteine is present and this process is allowed to continue, there would soon be very little of the arsenical drug bonded to the receptor (a simple mass-action effect). A well-known analogue of this situation is the hydrolysis of silicon tetrachloride, which is rapidly decomposed by water because silicon has a vacant *d* orbital. By contrast, carbon tetrachloride is stable because carbon has no vacant *d* orbital.

*Two sulfur atoms are supplied by the receptor, as in (13.1), and the third comes from the attacking thiol (e.g. cysteine).

Some types of arsenic–sulfur bonds are stronger than those which we have been considering. Lewisite (ClCH:CH·AsCl$_2$), the most vesicant of the arsenical war gases, produces lesions in Man that are not reversed or even averted by any monothiol. The work done by Peters and his colleagues at Oxford from 1923 onwards led to the discovery that lewisite injures the skin and exerts widespread toxicity by blocking —SH— groups of pyruvate oxidase (Peters, 1963). Injuries caused by lewisite can be completely reversed by those dithiols which have two —SH— groups close together (Peters, Stocken and Thompson, 1945). Thus an unusually firm combination with —SH— groups can be reversed by an unusually potent antidote. The great virtue of having two mercapto-groups in the one molecule is that one bond will always hold when the other is temporarily broken by thermal agitation. The best of these dithiols (2,3-dimercaptopropanol (*11.23*); dimercaprol) has since proved to be an excellent hospital antidote for poisoning (see Section 11.6).

(*c*) *Arsenicals as antibacterials.* Until 1944, it was generally supposed that arsenicals had no effect on common pathogenic bacteria. This misconception arose from failure to test the substances at the correct level of oxidation. When this was done, it was found that arsenicals are capable of strong antibacterial

Table 13.1 The effect of organic arsenicals on bacteria. Greatest dilutions completely inhibiting visible growth in 48 hours at 37°C. (*Medium*: Peptone broth, containing 10% serum; pH 7.2).

	Organism				
Substance	Cl. welchii	Strept. haem. A	Staph. aureus	E. coli	Proteus
Acetarsol (*13.2*) (*m*-acetamido-*p*-hydroxyphenyl arsonic acid)	*	*	*	*	*
Oxophenarsine (*6.4*) (*m*-amino-*p*-hydroxyphenyl arsenoxide)	1:160000	1:80000	1:160000	1:10000	1:10000
Neoarsphenamine (*m*-amino-*p*-hydroxy-arsenobenzene-*N*-methylenesulfoxylate)	1:10000	1:10000	1:10000	*	*
Oxophenarsine in 0.1% thioglycollate broth	1:5000	1:10000	*	*	*
Mercuric chloride (for comparison)	1:40000	1:160000	1:40000	1:80000	1:80000

*Signifies not inhibitory, even at a concentration of 1:5000.
(Albert, Falk and Rubbo, 1944.)

activity (Albert, Falk and Rubbo, 1944). The results, in Table 13.1, show that the active level for bacteriostasis is the arsenoxide level, and that the conditions can convert a little material to this from the arsenobenzene level, but none from the pentavalent level. The antibacterial action of the arsenoxide is seen to be neutralized by a mercaptan (sodium thioglycollate).

(*d*) *The place of arsenicals in medicine.* When Ehrlich launched arsphenamine in 1910, the use of arsenical drugs in medicine quickly reached its zenith. This great discovery not only provided a remedy for syphilis, a hitherto incurable disease, but was actually the first drug that could eliminate an infection caused by a bacterium, a triumph that kept flagging hopes alive until Domagk's discovery of the sulfonamides in 1935. But more significant than all of this was Ehrlich's being able to point to the chemical nature of the union of a drug with its receptor, an insight that provided the only solid example of a concept that is now completely accepted but which, for seemingly interminable decades, had little else than the As—S bond to hold it together.

A decline in the use of arsenic followed the introduction in 1943 of penicillin for the treatment of syphilis. Today, the clinical use of arsenic is almost confined to the treatment of those advanced cases of African sleeping sickness where the brain has become infected. The preferred drug is a masked arsenoxide, melarsoprol (*13.3*), which easily penetrates into the central nervous system although the usual trypanocidal drugs cannot do this (Friedheim, 1951). This drug is the condensation product of an arsenoxide with 1,2-dimercaptoethane: after it has passed the blood—brain barrier, the arsenoxide is liberated. Melarsoprol usually effects a rapid cure and is without the dangerous optic toxicity of the formerly used remedy, tryparsamide (*6.7*)

Melarsoprol

(13.3)

(*e*) *Antimonial drugs.* It is generally assumed that these act on important mercapto-groups after reduction, where necessary, to the trivalent condition, but this hypothesis has not been adequately tested. Certainly, the union between antimonial drugs (trivalent) and schistosomal phosphofructokinase is not reversed by mercaptoethanol (Mansour and Bueding, 1954), but the more powerful effect of dimercaprol does not seem to have been tested. For further reading on antimonials, see their effect on carbohydrate metabolism in Table 4.6 (Section 4.4) and their therapeutic use in leishmaniasis and schistosomiasis (Sections 6.3.3 and 6.3.5, respectively).

(*f*) *Mercurials.* Mercurials, such as mercuric chloride and phenyl mercuric nitrate, exert their antiseptic action on bacteria by combining with essential SH groups (Fildes, 1940b). The bacteria appear to be dead, but are easily revived by treatment with a thiol such as thioglycollic acid or even hydrogen sulfide (Chick, 1908).

For a great many years, mercurial compounds were the most powerful of all known diuretics, but the introduction of chlorthiazide in 1958, followed by other types of purely *organic* diuretics, made the mercurial diuretics seem toxic and inefficient by comparison and they are no longer used. Mercurial diuresis depended on the liberation of mercuric ions by the acidity of the kidney fluids and the blockade by these ions of a key mercapto-group in the kidney enzyme responsible for resorption of sodium chloride.

13.1 The penicillins

Bacteria stand in sharp contrast to vertebrate cells by having a cell wall, and this wall is of a unique chemical character (see Section 5.3). The most vulnerable component of this wall is acetylmuramic acid (*5.9*) linked to polypeptides such as (*5.10*). Penicillin (*13.4b*) acts on bacteria by blocking the incorporation of these acetylmuramic peptides into new cell wall (Rogers and Mandelstam, 1962). It was found that the concentration of drug required to inhibit cell wall synthesis and that required to inhibit growth in the same organism agree within a two- to three-fold range for both Gram-positive and -negative bacteria.

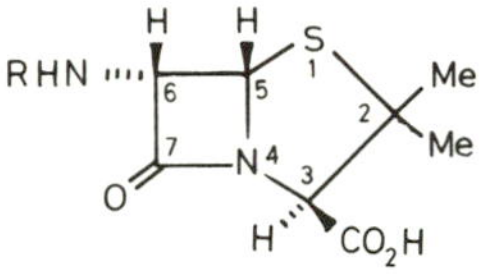

(a) 6-Aminopenicillanic acid (R = H)
(b) Benzylpenicillin (R = Ph CH$_2$CO−)
(*13.4*)

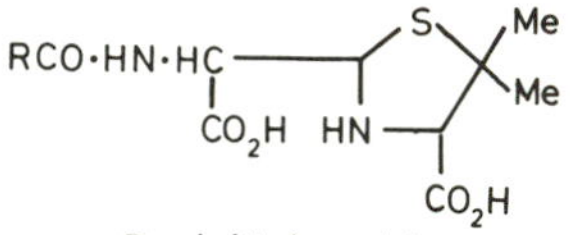

Penicilloic acids
(*13.5*)

Bacteria are somewhat unusual in having a high internal pressure. Penicillin kills *growing* bacteria by causing them to burst because of this high pressure which a defective wall can no longer sustain. However, they do not burst if placed in a non-absorbable medium of high osmotic pressure (e.g. 0.3 M-sucrose solution). Lederberg (1957) worked out these details (of the lysis of bacteria by penicillin) using *E. coli*, and they have been confirmed in other species. The bursting of bacterial cells by penicillin is apparently assisted by an enzyme known as mucopeptidase which is present in almost all bacteria. When the biosynthesis of this enzyme is halted by chloramphenicol, penicillin is much less effective (Rogers, 1967).

Other antibiotics that interfere with cell-wall synthesis are cycloserine (*5.11*) (see Section 9.4.3), vancomycin, bacitricin, and many substances chemically closer to penicillin, all discussed in Section 13.2.

Many penicillins have been isolated from the liquor in which active strains of

Penicillium notatum are growing (see Section 6.3.1). They have the general structure (*13.4a*), differing only in the nature of the side-chain. The variety known as benzylpenicillin soon became the standard, and is always to be understood when 'penicillin' is specified (*5.13; 13.4b*)*.

The early workers with penicillin were inconvenienced by its tendency to hydrolyse to the corresponding penicilloic acid (*13.5*), by alkali, cupric ions, or β-lactamase (usually called penicillinase), and to penillic acid, and penicillamine (*11.26*). In natural biosynthesis it is formed by the condensation of 6-penicillamine, D-valine, and phenylacetic acid. The presence of the strained four-membered lactam ring in penicillin makes it a powerful, but specific, acylating agent; this ring readily opens between C-7 and the nitrogen atom (Johnson, Woodward and Robinson, 1949). X-ray-diffraction analysis shows that the two rings of penicillin are almost at right-angles to one another. Nevertheless, the strain, and loss of resonance-stabilization, of penicillin and cephalosporin have in the past been overestimated (for a review that sets out the facts, see Proctor, Gensmantel and Page, 1982).

Studies on the inactivation of benzylpenicillin have shown that this substance has a strong preference for acylating β-aminomercaptans, much less affinity for other amines, and none for other mercaptans. The product [e.g. (*13.6*) from β-mercaptoethylamine] is acylated on the amino-group. The function of the mercapto-groups seems to be to form a hydrogen bond with the oxygen atom of the lactam and hence increase the polarization of the C=O bond in the penicillin.

Experiments with ^{35}S penicillin show that resistant strains of staphylococci, even those that produce no penicillinase, take up no penicillin from solution, but susceptible strains of various species combine with from 200 to 750 molecules per cell. This amount is held tightly, cannot be washed away, and does not exchange with non-radioactive penicillin (Rowley *et al.*, 1950; Maass and Johnson, 1949). The first molecules of penicillin are bound by inert material but, when this is saturated, the penicillin combines covalently with a receptor playing a key role in the biosynthesis of cell wall. Mammalian cells do not take up penicillin, but the bacterial cell binds it in less than 2 minutes. The penicillin-binding component is on the exterior, in the cytoplasmic membrane where the cell wall is synthesized. This component occurs in a phospholipid-containing fraction of staphylococci (Cooper, 1956).

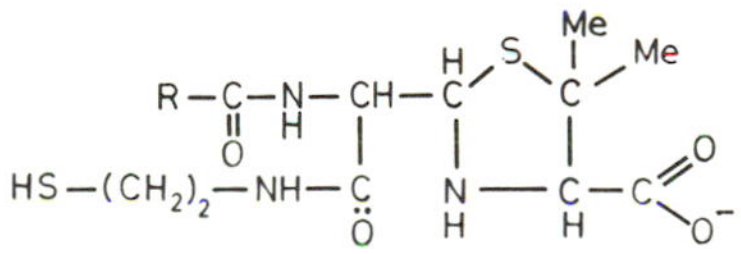

Product from β-mercaptoethylamine and a penicillin

(*13.6*)

*Penicillin is (3*S*), (5*R*), and (6*R*) (Pitt, 1952; McGregor, 1974).

The main site of action of penicillin is a membrane-bound transpeptidase in the bacterial cytoplasmic membrane. It performs two tasks: initially, it attracts the D-alanyl-D-alanine end of the monomer which is waiting to be polymerized to form new cell-wall. This monomer is a peptidoglycan, that is to say it consists of muropeptide units (Section 5.3) bound together through the condensation of their carbohydrate components. D-Alanine is the recognition signal for this enzyme. The next step is an acylation of the enzyme by the peptidoglycan, with loss of one molecule of D-Alanine:

$$\text{Enzyme} + \text{R-D-Ala-D-Ala} \rightarrow \text{R-D-Ala-enzyme} + \text{D-Ala}$$

There is some evidence that the site of this acylation is the —OH-group of a serine residue.

The transpeptidase then becomes regenerated by linking the penultimate D-alanyl residue of the peptidoglycan molecule (to which it is attached) to the terminal amino-group on the chain of four glycine residues (attached to lysine) on a second peptidoglycan molecule (slight variants of this linkage occur in different bacterial species). The reaction may be summarized:

$$\text{R-D-Ala-enzyme} + \text{Gly} \rightarrow \text{R-D-Ala-Gly} + \text{enzyme}$$

(Franklin and Snow, 1981; Gale *et al.*, 1981).

This transpeptidase, of which a cell-free specimen was obtained from *E. coli*, is irreversibly inhibited by penicillin at a concentration that is lethal to the intact bacterial cell (Izaki, Matsuhashi and Strominger, 1966). These authors suggested that penicillin is taken up by the transpeptidase site (of the enzyme) which it irreversibly acylates by using the β-lactam ring as the acylating agent (Izaki, Matsuhashi and Strominger, 1968). There seems to be general agreement on this concept. Support for this thinking comes from X-ray analysis of a D-alanyl-carboxytranspeptidase from *Streptomyces*-R 61, in the absence and presence of *o*-iodophenyl-penicillin, and of a slow-reacting model substrate: α-t-butoxycarbonyl-ϵ-trifluoroacetyl-L-lysyl-D-glutamyl-D-alanine. Both substrates occupied the same site and had their carboxy group applied to the terminal amino-groups of two helices of the enzyme (Kelly *et al.*, 1982).

Tipper and Strominger (1965), by comparing molecular models of penicillin and D-alanyl-D-alanine claimed that the former could effectively function as a metabolite analogue of the latter. Thus the N' to N'' distances in these molecules, shown as projections of molecular models in (*13.7*) and (*13.8*), are identical, namely 3.3 Å. Also the N'' to C' distances are similar in both models, namely 2.5 Å. From N' to C' is 5.4 Å in the penicillins and 5.7 Å in alanyl-alanine, distances which are not identical but still quite similar. The D-configuration of the penicillin is maintained by the sulfur atom in conjunction with the carbon atom that is common to both rings. [Epimerization at the 6-position, see (*13.4*), causes loss of all antibacterial activity.] The most reactive bond in the penicillin molecule (the β-lactam bond) corresponds in position to the peptide bond joining the two D-alanine residues. However, there are discrepancies: the

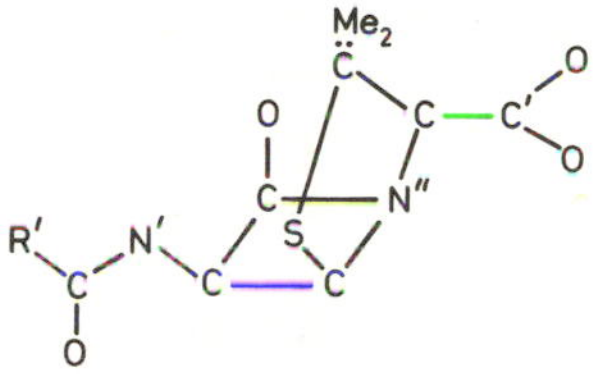

The penicillins
(projection from molecular models)
(13.7)

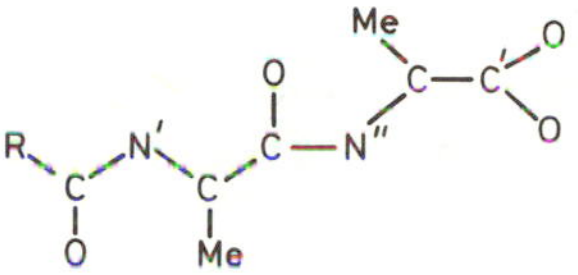

(Acyl)-D-alanyl-D-alanine
(13.8)

bond linking the two alanine residues is about 25% longer than the lactam bond. Again, the C-CO-N angle in D-Ala-D-Ala is 117°, but only 91° in penicillin. Moreover, the β-lactam ring has a dihedral angle around the lactam bond close to 135°. This is far from the angle of the peptide bond (180°) and also far from that of the tetrahedral transitional state (see Section 9.5) which D-alanyl-D-alanine would be expected to assume on the enzyme (Lee, 1971). Rather more serious evidence against this structural analogue hypothesis is that 6-methyl benzylpenicillin (cf. *13.4*), although an even better structural analogue of D-Ala-D-Ala, is completely inactive against bacteria (Boehme *et al.*, 1971). This seems at variance with the Strominger hypothesis of penicillin action, but no better hypothesis has emerged.

(*a*) *Modified penicillins.* Benzylpenicillin (penicillin G) is still the most generally useful of all antibiotics employed in daily medical practice, and is still the most used form of injectable penicillin, just as penicillin V, i.e. phenoxymethylpenicillin (*13.4*, R = $C_6H_5OCH_2CO$—), is the most prescribed form of oral penicillin. In 1957, 6-aminopenicillanic acid (6-APA) was isolated from the fermentation liquors of *Penicillium chrysogenum* (Batchelor *et al.*, 1961) and proved to be an ideal starting material for the production of many semi-synthetic analogues of penicillin. In many of these, 6-APA is acylated with a group similar to that of the phenylacetyl-group in benzylpenicillin (*13.4b*), but carrying a sterically inhibiting substituent in the methylene (CH_2)-group. These analogues, less susceptible to acid hydrolysis in the stomach than benzylpenicillin, were the forerunners of the contemporary specialized penicillins. Two other goals sought in modifying the traditional penicillins were: resistance to hydrolysis by β-lactamases (present in resistant bacteria), and an extended range that would embrace Gram-negative bacteria in addition to the Gram-positive types against which penicillin is most active. A further achievement was to include any two (but, so far, not the whole three) of these desirable properties in a single molecule. In the following, we shall review the properties of some of the most prescribed of the modified penicillins.

Amoxycillin (*13.9a*) and ampicillin (*13.9b*) resemble penicillin V in being well absorbed orally (amoxycillin, best of all), principally because they are resistant to gastric acidity, and (like penicillin V) they are inactivated by β-lactamase.

Where they show superiority to penicillins G and V (the two traditional penicillins) is in their broad spectrum, which embraces many Gram-negative types. However, it is pointless to prescribe them unless a Gram-negative organism is concerned, because of the discomfort that can be caused by interfering with bowel flora. Hetacillin has similar properties to ampicillin, but shows no advantage. The inclusion of a basic group seems to be what gives the amoxycillin family its broad-spectrum properties.

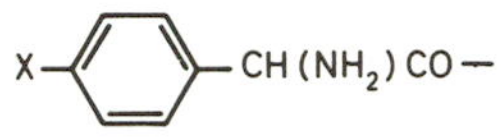

X—⟨benzene⟩—CH(NH₂)CO—

(a) Amoxycillin (X = HO−)
(b) Ampicillin (X = H−)
Insert (as R−) into *(13.4a)*
     (13.9)

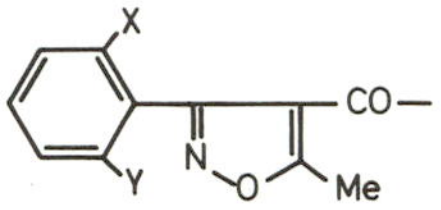

⟨oxazole structure with X, Y, CO—, N, O, Me⟩

(a) Oxacillin (X = Y = H)
(b) Cloxacillin (X = H ; Y = Cl)
(c) Dicloxacillin (X = Y = Cl)
Insert (as R−) into *(13.4a)*
     (13.10)

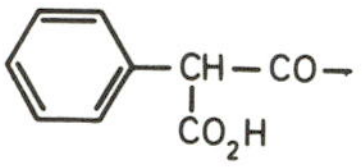

⟨benzene⟩—CH—CO—
         |
         CO₂H

Carbenicillin
(13.11)
(Insert as before)

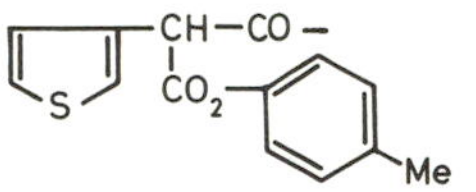

⟨thiophene⟩—CH—CO—
              |
              CO₂—⟨benzene⟩—Me

Ticarcillin
(13.12)
(Insert as before)

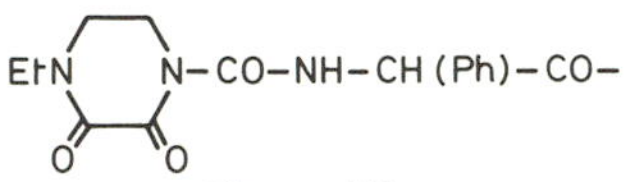

EtN⟨piperazinedione⟩N—CO—NH—CH(Ph)—CO—

Piperacillin
(13.13)
(Insert as before)

It may seem a far cry from these broad-spectrum penicillins to the methacillin family which are specialized to deal only with *Staphylococcus aureus*, and (of this species) only members that are resistant because of the β-lactamase that they carry. The need for these specialized penicillins arose when this resistant organism had established itself in most of the larger hospitals as a highly contagious source of nosocomial* infection. Methicillin [6-(2,6-dimethoxy-benzamido)penicillanic acid], the foundation member of this family, was susceptible to gastric acid hydrolysis and was soon replaced by the oxacillins *(13.10)* all of which are active orally. Of these, dicloxacillin *(13.10c)* is considered the most powerful and is certainly very effective; yet even this drug is not the equal of the original penicillin G for non-resistant staphylococci. Nafcillin, another member of the methicillin family, is now little used because of poor oral availability.

Sad to say, there are also strains of *Staphylococcus aureus* which owe their resistance to another factor, possibly a change in cell-wall composition. Such strains, which are becoming increasingly common in hospitals, are also resistant to cephalosporins, erythromycin, the aminoglycosides, and the tetracyclines. For them, the only effective antibiotic is a rather unselective one, vancomycin (p. 565), but the synthetic anti-bacterials, such as nitrofurantoin and trimethoprim are usually effective.

To discover what enables the methicillin family to withstand degradation by

*Meaning 'contracted in hospital' from the Greek, nosos (disease), and nosokomeion (hospital).

β-lactamase, a study was made using X-ray analysis supplemented by n.m.r. and infrared spectroscopy (Blanpain *et al.*, 1980). The essential feature was shown to be rigidity of the side-chain derived from van der Waals interactions supplemented (in the oxacillins) by conjugation between the amide, isoxazole, and phenyl substituents. This rigidity apparently prevents the enzyme's active site from conforming to the lactam group in the drug.

Next, we come to the carbenicillin family which is distinguished for its attack on Gram-negative organisms, including types not susceptible to members of the amoxycillin family. These extra organisms include *Pseudomonas*, *Enterobacter*, *Proteus*, and *Klebsiella*, opportunistic organisms that are widely and harmlessly distributed, but which severely attack hospital patients whose immune systems have been partly suspended for therapeutic reasons (e.g. in performing a graft). Members of the carbenicillin family have many disadvantages: they are inactive orally, easily destroyed by β-lactamase, and without effect on some common Gram-positive organisms, such as *Staphylococcus aureus*. However, they are often life-saving for the infected patient. Carbenicillin (*13.11*) carries a carboxylic acid-group which, when esterified, gives orally active products. Ticarcillin (*13.12*) and piperacillin (*13.13*) are more recent, and more powerful, members of this family.

From the above account, it is evident that the therapeutic range of the original penicillin G has been considerably extended through the new semi-synthetic products. These are, however, more costly, and should not be prescribed without good reason. An economy can be effected, when giving a member of the amoxycillin family orally, by prescribing oral probenecid (*3.16*) as well, because this decreases renal elimination of penicillins.

All penicillins that have achieved clinical success have an aromatic ring near to the acylating group (7—CO). This requirement, present also in the cephalosporins, suggests that a corresponding flat area is present in the receptor.

A new development in prescribing penicillins is to give clavulanic acid (*13.14*) at the same time. This is a product, isolated from *Streptomyces clavuligerus*, which has a β-lactam structure, but only a little anti-bacterial action. On the other hand, it is a strong inhibitor of β-lactamases, and hence is synergistic with ampicillin and amoxycillin, with which it is most often administered.

Pro-drugs which break down to penicillins after administration were discussed in Section 3.6 (p. 101).

For further reading on the β-lactam antibiotics, see the book by Morin and Gorman (1982).

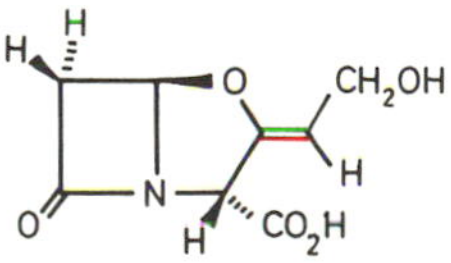

Clavulanic acid
(*13.14*)

13.2 Cephalosporins. Other β-lactam inhibitors of the formation of new cell walls.

Some time after Brotzu (1948) had demonstrated antibiotic properties in the fungus *Cephalosporium acremonium* (from a sewage outfall in Sardinia), Newton and Abraham (1956), in Oxford, isolated the active principle, cephalosporin C, and showed that it had the constitution (*13.15*) in which X was aminoadipic acid and Y was the acetoxy-group (—O·COMe). The nucleus, which consists of a β-lactam ring fused to a dihydrothiazine ring, seemed strangely reminiscent of the penicillin nucleus. However, the antibacterial action of cephalosporin C was too slight for medicinal use. Taking a clue from penicillin, the activity was greatly improved when a side-chain bearing an aromatic ring replaced the aliphatic side chain in the 7-position. To achieve this, cephalosporin C is regularly hydrolysed to 7-aminocephalosporanic acid [(*13.15*) in which X = H and Y = —O·COMe], which sufficiently resembles 6-aminopenicillanic acid (*13.4a*) to suggest what should be done to provide semi-synthetic cephalosporins for the clinic.

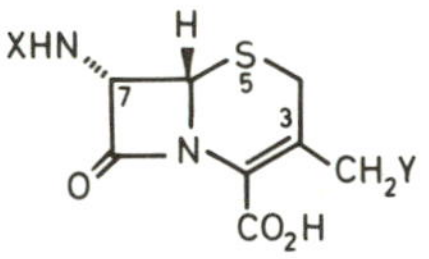

The cephalosporin skeleton
(*13.15*)

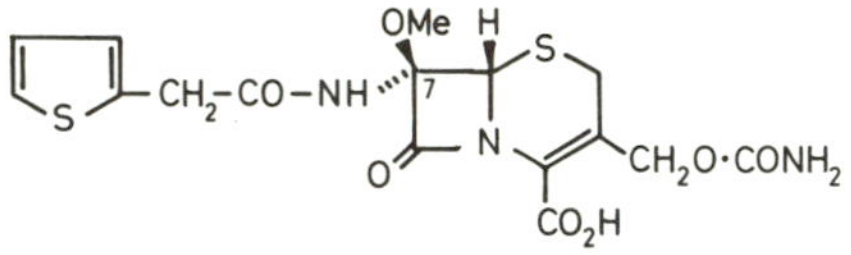

Cefoxitin
(*13.16*)

Like the penicillins, the cephalosporins act by blocking the D-alanine-recognizing polymerase responsible for the synthesis of new bacterial cell wall (Abraham, 1962). Because the cephalosporins are more expensive than the penicillins, and generally less active, they are not considered a first choice in treating Gram-positive infections. Yet they are useful for the rare patient who is sensitive to penicillin although sensitivity to cephalosporins is also sometimes found. More often, they are used against penicillin G-insensitive strains of *Staphylococcus aureus* which carry the type of β-lactamase which attacks only penicillins (the so-called penicillinases). However, other types that are cephalosporinases are often encountered in this organism, and a prior test must be made in the pathology laboratory. The later development of cephamandole and cefoxitin (see below) provided alternatives to the carbenicillin family of penicillins for fighting Gram-negative infections.

The principal types of cephalosporins will now be described. Table 13.2 gives the chemical structures, and International Names (note that many names now beginning 'cef' were formerly 'ceph').

Cefalotin (formerly cephalothin), the first of the semi-synthetic cephalosporins, is the least affected by staphylococcal β-lactamase, and it is still one of the most used. Cefalotin is not well absorbed when given orally and, as the

Table 13.2 Cephalosporin antibiotics

Name	Insert into (13.15) 7-position	3-position
Cefalotin*	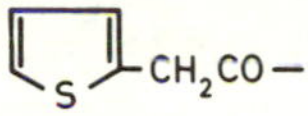 $-CH_2CO-$	$-O \cdot COMe$
Cefaloridine	(as above)	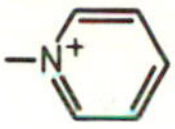
Cefalexin†	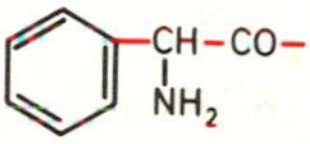	$-H$
Cefamandole	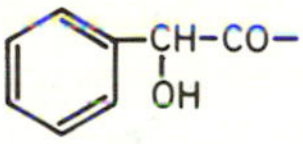	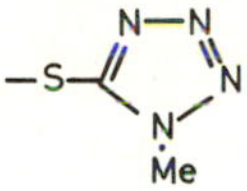
Cefotaxime	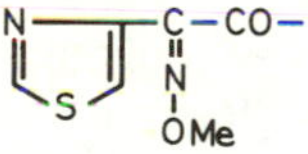	$-OMe$
Cefuroxime	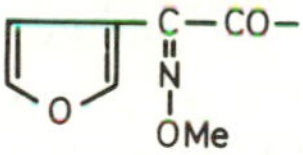	$-O \cdot CONH_2$

*Cefapirin acts similarly.
†Similar action, after oral dosage, from cefaclor, cefadroxil, and cefradine.

intramuscular injections are painful, it has to be given intravenously. *Cefalori-dine*, another early example, resembles cefalotin in action, but the molecule has been modified to allow painless intramuscular injections. Unfortunately it is somewhat nephrotoxic. *Cefalexin*, the first of the orally active cephalosporins, owes this advantage to the glycine side-chain, as in ampicillin (*13.9*). It is not so potent as cefalotin, but otherwise resembles it in being most active against Gram-positive organisms. Probenecid (*3.16*) is often given at the same time as these antibiotics to retard their elimination into the urine.

Cefamandole was the first cephalosporin to kill Gram-negative organisms, an action that seems to be intensified in the newer *cefotaxime* and *cefuroxime*. These drugs, all of which must be injected, are useful in nosocomial infections and hence provide alternatives to the carbenicillin family (p. 561). They also have value as urinary antiseptics.

Cefoxitin (*13.16*) is, surprisingly, obtained not from the fungus but from bacteria (*Streptomyces* sp.) which provide a cefomycin with the same aminoadipic acid side-chain as in cephalosporin C. Replacement of this side-chain gave cefoxitin, which is highly active against Gram-negative (much less against Gram-positive) species. It has to be given intramuscularly. Against the anaerobe, *Bacteroides fragilis*, a contemporary cause of many troublesome bowel infections, it is more active than any cephalosporin. Cefoxitin is stable to β-lactamases secreted by Gram-negative organisms, a property ascribed to the steric hindrance provided by the double substitution in the 7-position (the 7-methoxy-group is present in the original bacterial product).

For further reading on cephalosporins, see the review by Neu (1982) and the book by Morin and Gorman (1982).

The mode of action of β-lactam antibiotics. The double bond in the six-membered ring of cephalosporins seems to activate the β-lactam group in the same way that the torsional effect of the five-membered ring in penicillins does (Sweet and Dahl, 1970). Moving this bond to the 2,3-position gives a biologically inactive molecule. It is thought that potentiation of the acylating power of cephalosporins by the 3,4-double bond is effected by electron delocalization through enamine resonance outside the lactam ring (Sweet and Dahl, 1970). When other types of β-lactam antibacterials became available, however, indications were seen that this was not the total specification for combination between drug and cell-wall polymerase. In the first place, it was surprising that as simple a nucleus as azetidin-2-one could confer a penicillin-like action, as in aztreonam (*13.17*) (Kamiya *et al.*, 1981). This semi-synthetic antibiotic is under clinical investigation (Cohen, 1983). Fine-tuning of the specifications gained pace with investigations of the penems and carbopenems by the Woodward school in Harvard.

2-Pentylpenem (*13.18b*) is a totally synthetic molecule whose nucleus differs from that of penicillin by inclusion of a double bond, placed where it occurs in the cephalosporins. It is as active as penicillin against bacteria and is moderately stable. In fact, only the (5*R*) enantiomer is anti-bacterial, as is the case with penicillin (Lang *et al.*, 1979). The structure of this compound is remarkably simple for a β-lactam anti-bacterial; moreover, much of the anti-bacterial power remains when the pentyl group is removed, as in (*13.18a*). The Woodward group examined, by X-ray analysis, the length of the 3,4- and the 4,7- bonds, and they concluded that the nitrogen was truly pyramidal (as in penicillin), that is to say, it was completely released from the flat configuration that resonance imposes on a normal amide group. The nitrogen atom in the two isomeric carbapenems (*13.19*) and (*13.20*) also turned out to be pyramidal. All three nuclei had a moderate stability to hydrolysis, much as in penicillin. When the anti-bacterial properties were examined, (*13.18a*) and (*13.9*) proved to be highly active whereas (*13.20*) lacked antibacterial properties (Pfaendler *et al.*, 1981). The Woodward school concluded that neither moderate reactivity of the β-lactam group, nor pyramidal geometry for the nitrogen atom are sufficient to explain or predict antibacterial activity in penicillin-like compounds. Cohen (1983) suggests that the missing factor is the conformation of the carboxylic acid group, which he supposes must be equatorial to generate activity, as in (*13.18*) and (*13.19*), and it must not be axial, as in (*13.20*). In penicillin, the conformation of this group, in solution, is not fixed.

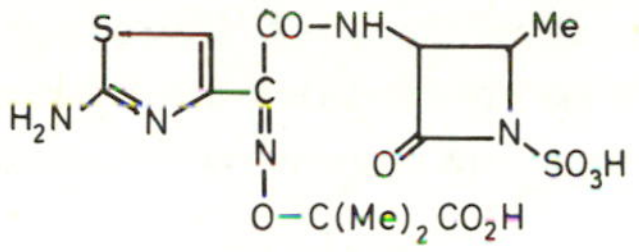

Aztreonam
(13.17)

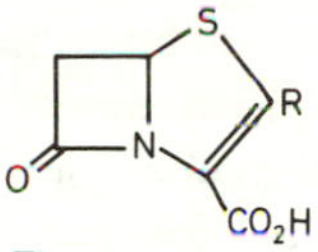

The penems
(a) R = H
(b) R = C_5H_{11}
(13.18)

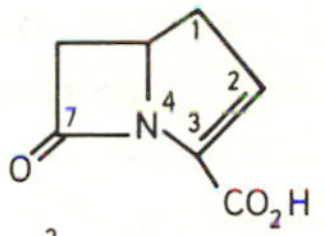

Δ^2- Carbapenem
(13.19)

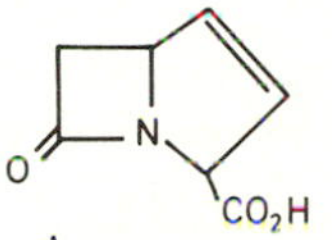

Δ^1- Carbapenem
(13.20)

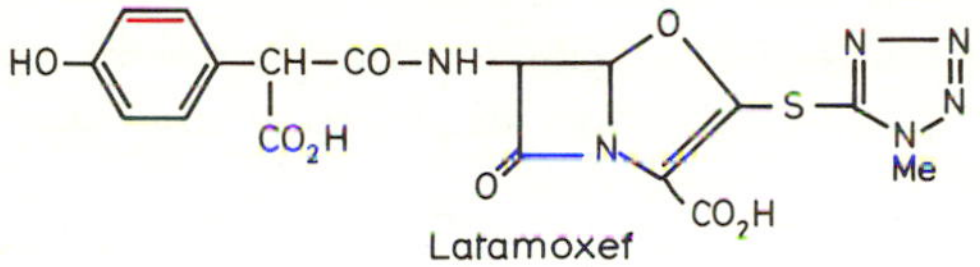

Latamoxef
(13.21)

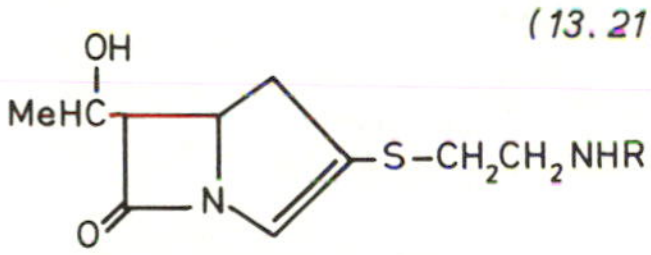

(a) Thienamycin (R = H)
(b) Amidine derivative (R = —CH:NH)
(13.22)

Some of the newer β-lactams have entered clinical trial (see also clavulanic acid, p. 561). Latamoxef (Moxalactam) (*13.21*) is a semi-synthetic analogue of the cephalosporins in which sulfur has been replaced by oxygen (compare structure with that of cefamandole, Table 13.2). It penetrates into the cerebrospinal fluid better than any cephalosporin, and is being used in meningitis (Schaad, 1981; Neu, 1982). Thienamycin (*13.22a*) is a carbapenem from *Streptomyces cattleya*. It has a broad spectrum and is highly active against *Pseudomonas*. Unfortunately, because it acylates its own amino-group, it is unstable in concentrated solution. This defect has been overcome by issuing it as the amidine analogue (*13.22b*). This combines two useful properties, it inhibits β-lactamases, and it inhibits synthesis of bacterial cell wall (Tally, Jacobus and Gorbach, 1978).

For further reading on the newer β-lactam antibacterials, see Neu (1982).

Vancomycin (mol. wt. about 1500) is a chlorinated polyphenyl ether antibiotic, having sugar and amino acid side-chains, and obtained from *Streptomyces orientalis*. For the molecular structure, see Sheldrick *et al.* (1978). It combines with the D-alanyl-D-alanine portion of the monomer awaiting polymerization to give new bacterial cell wall, and hence growing cells die by rupture of the plasma membrane (Anderson *et al.*, 1965). This drug must be given intravenously and never in conjunction with other drugs that share its tendency to nephrotoxicity and ototoxicity. Vancomycin is largely reserved for patients infected by resistant staphylococci whose resistance is not due to β-lactamase.

Bacitracin, extracted from *B. subtilis*, is a decapeptide containing four D-amino acid residues and a thiazoline nucleus. Its point of attack is the formation of a complex with the sugar-transporting undecaprenyl pyrophosphate in the plasma membrane, an early stage in cell-wall synthesis briefly referred to in Section 5.3 (Stone and Strominger, 1971). Too nephrotoxic for internal use, bacitracin is a valuable constituent of antiseptic ointments. It requires a divalent metal for its attack (Section 11.9).

Fosfonomycin (*13.23*), 1,3-epoxypropylphosphonic acid, was isolated from a *Streptomyces* strain. By mimicking phosphoenolpyruvate, it combines covalently with a cysteine residue in UDP-acetylglucosamine-pyruvyl transferase. In this way, it holds up an early stage in the biosynthesis of peptidoglycan, and hence prevents new cell wall being formed. In Italy, it is used in paediatrics, e.g. for meningitis caused by Gram-negative bacteria (Kahan *et al.*, 1974).

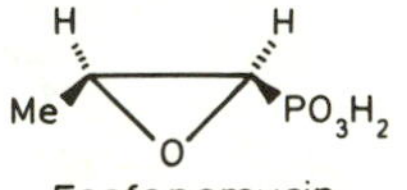

Fosfonomycin
(13.23)

13.3 Organic phosphates and carbamates

The substances discussed in this section have become important because of their property of acylating esterases, particularly acetylcholinesterase (AChase). The death of insects correlates well with the inhibition of their AChase, provided that thoracic (but not brain) isoenzyme is tested (Bigley, 1966; Farnham, Gregory and Sawicki, 1966). Similarly the toxicity of organophosphates to mammals is directly proportional to the inhibition of their AChase (Aldridge and Barnes, 1952).

In 1936, Gerhard Schrader began, as a young man, his search for new insecticides in the Bayer Company, Germany, and his studies soon led him to the organophosphates. By 1944, he had produced the (not very selective) tetraethyl pyrophosphate which replaced imported insecticides in beleaguered Germany. He went on to discover (in 1945) parathion (*13.24*), which the victorious Allies were quick to market (Schrader, 1963). Parathion had an improved selectivity, although still far below what today's standards require.

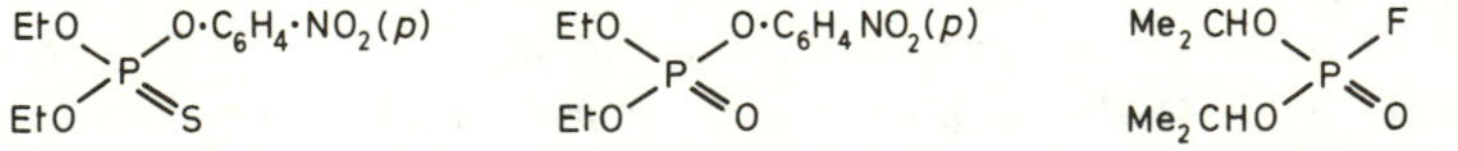

Parathion
(13.24)

Paraoxon
(13.25)

Isoflurophate
(13.26)

(a) *The organophosphates: chemistry of action.* As early as 1932, Lange and von Kreuger, who discovered dialkyl phosphorofluoridates such as (*13.26*), complained of breathlessness, sensitivity to light, and temporary loss of consciousness. Schrader (1963) recalled that his work with organophosphates in 1936 had caused some (reversible) loss of vision. His biological colleague H. Kükenthal could only have attributed these symptoms to inhibition of AChase*. To investigate this reaction, the phosphorylation of the enzyme, use was made of the strong β-emission of ^{32}P which has a half-life of 14 days. Enzyme studies showed that isoflurophate (*13.26*) ('Dyflos') effectively di-isopropylphosphorylated the hydroxy group of a serine residue in the enzyme, which is also the catalytic site for acetylcholine. Hydrogen fluoride was split off in this reaction (Schaffer, May and Summerson, 1954). When the acylated enzyme was hydrolysed, *O*-di-isopropylphosphorylserine was isolated. The accepted chemical name for isoflurophate (*13.26*) is di-isopropyl phosphorofluoridate. It was formerly called di-isopropyl fluorophosphonate, but the nomenclature of organophosphates has been revised to make it more self-consistent. Isoflurophate has been used in treating glaucoma by reduction of intraocular tension, but there are newer and safer drugs.

The action of organophosphates on AChase was investigated also with paraoxon (*13.25*) which Gage (1953) showed to be the active form of parathion (*13.24*). Similarly to the results with isoflurophate, it was found that paraoxon diethylphosphorylates the hydroxy group in serine, at the same time ejecting *p*-nitrophenol (Hartley and Kilby, 1952). O'Brien (1971), summing all available evidence, concluded that AChase has only one *catalytic* site (the serine hydroxy group) but that its substrates require a secondary site as well, occupation of which causes a conformational change in the enzyme, favourable to operation of the mechanics of the catalytic site. He listed three kinds of these secondary sites, one used by acetylcholine and also by tetraethylammonium salts, another used by the organophosphates, and a third by the carbamates. O'Brien made it clear that the secondary site postulated in the Wilson and Bergmann model of this enzyme (Fig. 12.3) is, in fact, not anionic but lipophilic.

The kinetic studies of Aldridge (1950) and Main (1964) showed that the first result, when the enzyme and an organophosphate are brought together, is the formation of a reversible complex which they measured by an affinity-for-enzyme constant, K. This complex breaks down by phosphorylating the enzyme and shedding the leaving group as it does so, a reaction that shows only a small tendency for reversal. This phosphorylation is measured by kp, the phosphorylation (or inactivation) constant. Values of K and kp have been determined for many organophosphorus compounds and for several sources of AChase.

*During the Second World War, several organophosphates were developed in Germany as chemical warfare agents ('nerve gases') but were never used. Of these (soman, tabun, and sarin), the last-named was considered the most potent (Schrader, 1963; Windholz, 1976), an unmeritorious triumph of counter-selectivity.

These kinetic studies have formed a basis for systematizing structure–action relationships. For example, in a series of analogues of paraoxon (*13.25*), in which the alkoxy groups were varied from methyl to butyl, and the electron-attracting substituent in the benzene ring was varied, neither the nature of the alkoxy nor the leaving groups affected *kp*. This was equally true for variations made in the four alkyl groups of the malaoxon molecule (*13.27*). However, the *K* values for the paraoxon series were 2 to 10 times greater than those for the malaoxon series. It seems that these chemical changes could alter the affinity for the enzyme but did not change the phosphorylating power (Chiu, Main and Dauterman, 1969). Up to a point, anti-AChase activity is greatly increased by electron-attracting properties built into the leaving group; but excess creates a product so easily hydrolysed by water as to be impractical (Aldridge and Barnes, 1952; Aldridge and Davison, 1952). Below the maximum for electron attraction, these authors found a linear relationship between (a) *log* hydrolysis constant and (b) the *log* bimolecular rate-constant for the reaction between agent and enzyme.

Effective insecticides have been found among the phosphinic, phosphoric, and phosphonic esters. In a phosphinic ester, both alkyl groups are *directly* attached to the phosphorus atom; in a phosphoric ester, both are attached through oxygen; and a phosphonic ester is intermediate in this respect.

The essentials of enzyme phosphorylation and its reversal are shown in Scheme 13.1.

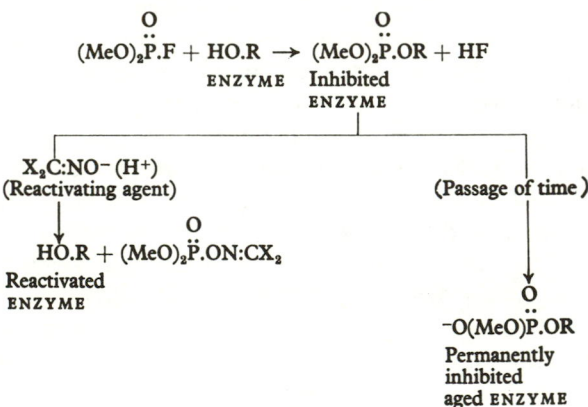

Scheme 13.1 Inactivation of an enzyme by phosphorylation, and its reactivation if not 'aged'.

If the inhibited enzyme has not 'aged', it can be re-activated by the anionic group of a hydroxylamine (see below). The phenomenon of ageing was discovered by Hobbiger (1955), who noted that esters of secondary alcohols evoke this phenomenon faster than those of primary alcohols. The correct interpretation, namely that ageing is proportional to the hydrolysis of an alkoxy-group to an anion, came later (Berends *et al.*, 1959). Coulombic repulsion prevents the response of the aged, inactivated enzyme to the antidote.

In insects poisoned by organic phosphorus compounds, assays for acetylcholinesterase show that this enzyme becomes increasingly inhibited during the first hour and the levels of free acetylcholine rise sharply (Smallman and Fiske, 1958). For further information on the effect of these agents on the nervous system of insects, see Section 6.4.1 (p. 243).

Acetylcholinesterases isolated from various insect species differ greatly in their susceptibility to phosphorus insecticides (R. O'Brien, 1967). Access to this enzyme is more difficult in insects than in mammals; hence it is understandable that ionized analogues have proved not to be insecticidal because foreign ions do not readily penetrate membranes (Section 3.2).

(*b*) *The organophosphates: improving their selectivity.* The insecticide industry realized that, to achieve selectivity, new kinds of organic phosphorus compounds must be found, ones that would be differently metabolized in insects and vertebrates, i.e. they must either be inert substances which only insects can activate, or they must be active substances which only mammals can detoxify. Schrader's parathion (*13.24*) (*O,O*-diethyl *O-p*-nitrophenyl phosphorothioate) had a wide, though insufficient, margin of safety which depended upon the presence in the insect of an enzyme that changed P:S to P:O, thus producing paraoxon (*13.25*) which is the true toxicant. This change, which occurred to a smaller degree in vertebrates, provided a margin of selectivity. It was later found that the margin of safety could be much improved by replacing the two ethoxy-groups in parathion by two methoxy-groups, giving 'parathion-methyl' which was introduced in 1954.

A surprising but welcome increase in selectivity followed Schrader's discovery of fenitrothion. This differs from parathion-methyl only by having a methyl-group *ortho* to the nitro-group and, chemically, is *O,O*-dimethyl *O*-3-methyl-4-nitrophenyl phosphorothioate. Fenitrothion has an LD_{50} of 500 mg/kg compared to only 8 mg for parathion and 20 mg for parathion-methyl. Its oxygen analogue, fenitrooxon, strongly inhibited the AChase of houseflies at low concentrations which had little effect on the analogous enzymes in bovine erythrocytes and mouse brain (Hollingworth, Fukoto and Metcalf, 1967). There have been improvements, too, in selectivity between economic and uneconomic insects. For example, the *O,O*-di-isopropyl analogue of paraoxon (*13.25*) inhibited AChase of flies but not that of honeybees, an advantage attributed to a steric effect (Camp, Fukoto and Metcalf, 1969).

While these developments of the parathion molecule were slowly taking place, success had crowned a more ambitious project, namely to introduce *two* safety mechanisms into the one molecule. This was achieved by the discovery of malathion (*13.27b*) which is *O,O*-dimethyl *S*-1,2-bisethoxycarbonylethyl phosphorodithioate, introduced in 1950 by the American Cyanamid Company. The actual toxicant is malaoxon in which P:S is replaced by P:O. Malathion is rapidly de-esterified in mammals by carboxyesterases and the resultant acid is quickly eliminated; whereas the insects perform the P:S to P:O change, but carry out very little de-esterification. Thus two metabolic changes which lead to favourable selectivity take place in the one substrate (Krueger and O'Brien, 1959). The P:S to P:O change in insects is effected by microsomes in the gut, nerve cord, and fat body. Insect microsomes can be isolated and used in experiments (Nakatsugawa and Dahm, 1965). Although carboxyesterases do occur in insects, they are relatively inactive. It is important to keep in mind that an insect receives a much higher dose (w/w) than any vertebrate is likely to.

$$\begin{array}{c} MeO \\ \diagdown \\ P \\ \diagup \diagdown \\ MeO R \end{array} \begin{array}{l} S-CH \cdot CO_2Et \\ CH_2 \cdot CO_2Et \end{array}$$

(a) Malaoxon (R=O)
(b) Malathion (R=S)
(13.27)

As can be seen from Fig. 13.2, the mouse converts only a small proportion of a dose of malathion into the toxic malaoxon, whereas the cockroach not only converts much more but retains the toxin longer. This graph needs to be read in conjunction with Table 13.3 which shows that the mouse can withstand a huge dose of malathion because it does not readily convert it to the toxic derivative malaoxon, whereas the cockroach, and particularly the housefly, are excellent converters. Similarly mammalian carboxyamidases are capable of removing amide-groups which insects are almost unable to hydrolyse (see dimethoate, below).

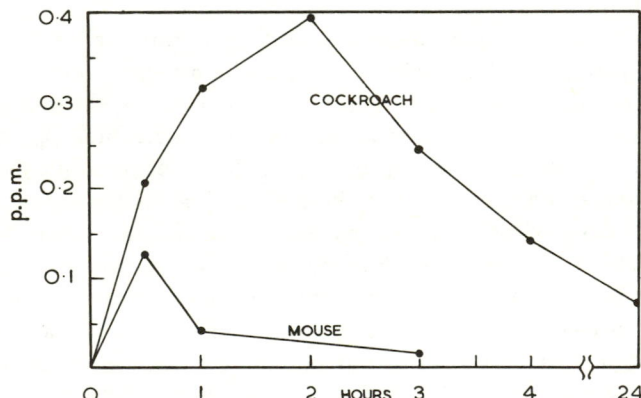

Fig. 13.2 Malaoxon level after injecting 30 p.p.m. of [^{32}P-]Malathion. (Krueger and O'Brien, 1959.)

Malathion has a fine safety record and enjoys widespread use. Another insecticide with two inbuilt safeguards is the Swiss product, diazinon, which is *O,O*-diethyl *O*-(2-isopropyl-6-methyl-4-pyrimidinyl) phosphorothioate (*13.28*),

Table 13.3 Toxicity of phosphorus insecticides to mammals and insects. (LD_{50} in p.p.m. w/w)

Species	Malathion	Malaoxon
Mouse (i/p)	1590	75
Housefly	30	15
Cockroach	120	15

(Krueger and O'Brien, 1959.)

discovered in 1952. Mammals quickly *C*-hydroxylate the isopropyl-group, leading to rapid elimination in the urine, whereas insects cannot do this readily but convert P:S to P:O, resulting in excellent selectivity (Miyazaki *et al.*, 1970; Krueger, O'Brien and Dauterman, 1960). Diazinon has proved extraordinarily successful, when sprayed on the hindquarters of sheep, in ridding them of the maggots of blowflies which cause intense irritation and subsequent loss of condition.

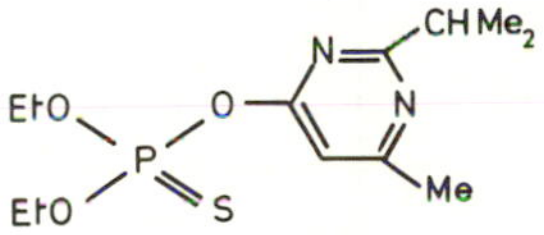

Diazinon
(13.28)

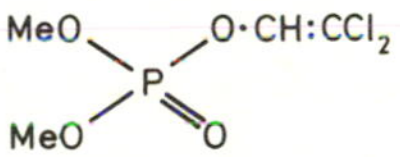

Dichlorvos
(13.29)

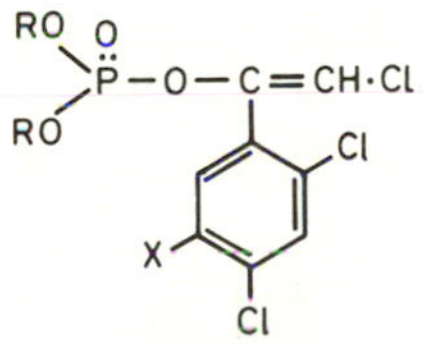

(a) Chlorfenvinphos (R=Et, X=H)
(b) Tetrachlorvinphos (R=Me, X=Cl)
(13.30)

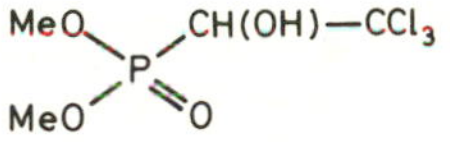

Trichlorofon ; Metrifonate
(13.31)

The 1950s witnessed a search for detoxicating mechanisms other than those requiring a *S* to *O* change. It was found that an aliphatic phosphate ester sufficed, provided that it bore a short side-chain with two or more electron-attracting features such as a double bond and a halogen atom. An early, and still much used example is dichlorvos (*13.29*) introduced by the Shell Company in 1955, and used as a domestic fly-killer. Chemically, it is dimethyl 2,2-dichlorovinyl phosphate. When inhaled by mammals, or absorbed through the skin, both plasma and the liver hydrolyse it rapidly to dimethyl phosphate and monomethyl 2,2-dichlorovinylphosphate; the latter is then quickly hydrolysed, first to dichlorovinylphosphoric acid, then to inorganic phosphate (Casida, McBryde and Niedermeir, 1962). To insects, however, its toxicity is cumulative (Hodgson and Casida, 1962). The American Conference of Government Industrial Hygienists set a human threshold limit value, for dichlorvos in air, of 1 μg per litre (Ashe, 1964). 'Vapona' slow-release strips, when used as directed (one strip per 1000 ft³), give off this concentration initially, but for the major

Table 13.4 Comparative toxicity of organophosphorus insecticides. (LD_{50} is the dose lethal to 50% of the animals.)

Substance	LD$_{50}$ for rat (mg/kg) Oral (acute)	Dermal	LD$_{50}$ for insects Megoura viciae (vetch aphid) in p.p.m.	Tetranychus telarius (mite) in p.p.m. Adults	Eggs	Pieris brassicae (larvae) (cabbage butterfly) in lb/acre	Musca domestica (adult housefly) (female) in mg/ft^2
Tetraethyl pyrophosphate	0.5	20	2	75	—	—	—
Parathion	3	50	2	65	—	—	—
Schradan	5	50	—	—	—	—	—
Dichlorvos	25	75	30	35	550	< 0.016	< 0.1
Dimethoate	200	750	3	2	60	0.7	1.8
Diazinon	300	> 1200	12	14	175	0.04	34
Trichlorophon	650	> 2800	—	—	—	—	—
Menazon	1200	> 500	4	90	50	> 10	—
Malathion	1400	> 4000	18	125	175	0.44	8.9
Haloxon	(900–2000)	?	—	—	—	—	—

(Partly from Edson *et al.*, 1964, 1966, also personal communication from Dr E.F. Edson.)

part of the life of the strip, the concentration is about 0.1 μg per litre. Although dichlorvos pest-strips went through a period of public suspicion, current opinion is that they are safe to use, even in households and kitchens.

A variation on the dichlorvos molecule produced chlorfenvinphos (*13.30a*) [2-chloro-1-(2,4-dichlorophenyl)vinyl diethyl phosphate] used to kill a wide range of insect pests on major crops; one application lasts about 3 weeks on foliage, or 2 months on the soil. Under the name 'Supona', it is used as a dip for sheep and cattle to give long-lasting freedom from ecto-parasites (flies, lice, keds, ticks, and mites). Much less toxic to mammals, but not so long-lasting, is tetrachlorvinphos (*13.30b*), used in agriculture and for ectoparasites.

Some mammalian and insect toxicities of organic phosphates are compared in Table 13.4. It can be seen that the lowering of mammalian toxicity, achieved in recent years, has not been at the expense of insect toxicity. For further data, see Tolkmith (1966).

The 1982 WHO recommendation for spraying against anopheline mosquitoes, in the tropical anti-malarial campaign, is fenitrothion (see p. 569). Against the blackfly that conveys onchocerciasis-causing worms in West Africa, they counsel the larvacidal sprays temephos (*O,O,O,O*-tetramethyl *O,O*-4,4'-phenylenethiophenylene diphosphorothionate) and chlorophoxim (*O,O*-diethyl 2-chloro-α-cyanobenzylideneamino phosphorothionate). The phosphorus insecticides recommended by the Food and Agricultural Organization for the Integrated Pest Management Programme are trichlorofon (*13.31*), malathion (*13.27b*), diazinon (*13.28*), chlorpyrifos (*13.32*), phosphamidon (*13.33*) and dimethoate (*13.35*).

Some organic phosphates have been made so selective that they may be taken by mouth. Thus, trichlorofon (*13.31*) is given orally (as a single dose) to cattle to protect them against the hide-piercing warble fly. This substance, known as metrifonate when used medicinally, is dimethyl 2,2,2-trichloro-1-hydroxy-ethylphosphonate. It is a pro-drug, for it is dehydrated and isomerized to dichlorvos (*13.29*), in the body.

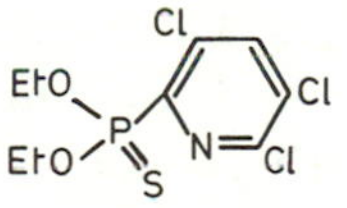

Chlorpyrifos
(13.32)

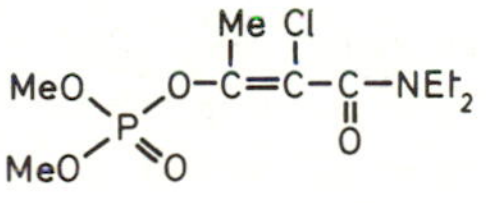

Phosphamidon
(13.33)

(*c*) *Systemic organophosphate insecticides.* About 1950, Schrader discovered the first insecticide which, when applied to the soil, was taken up by the roots and translocated to all parts of the plant. This gave protection from aphids and other sucking insects for up to to 6 weeks whereas the plant, lacking any nervous system, remained completely unaffected. Another selective advantage of this type of toxicity is that no insect can be killed unless it bites the plant, thus sparing the lives of beneficial insects. Schrader's original preparation, demeton

('Systox'), was a mixture of two isomers: *0,0*-diethyl *0*-ethylmercaptoethyl phosphorothioate (*13.34a*) and the corresponding thiolate (*13.34b*). The Bayer company eventually marketed these two components separately, as well as methyl-demeton in which both ethoxy groups had been replaced by methoxy groups. By then, the American systemic agent, dimethoate (*13.35*) was well established, and a Swiss analogue, phosphamidon (*13.33*) followed about 1960. These latter two enjoy the current FAO recommendation.

Systemic insecticides are valued over unselective contact insecticides which kill *all* insects, both good and bad, and hence interfere with pollination and contribute to the starvation of field birds.

(a) $(EtO)_2 P(:S)-OCH_2CH_2 SEt$

and

(b) $(EtO)_2 P(:O)-SCH_2CH_2 SEt$

Demeton
(*13.34*)

$$MeO \diagdown \atop MeO \diagup P \diagdown_{S}^{S \cdot CH_2 \cdot \overset{O}{\overset{\|}{C}} \cdot NHMe}$$

Dimethoate
(*13.35*)

Dimethoate also has an intrinsic selectivity, for it is far more toxic to insects than to mammals. This favourable effect was found, surprisingly, to depend little on differences in $S \rightarrow O$ conversion (as in malathion and diazinon), but to rely mainly on preferential operation of mammalian amidase (Krueger, O'Brien and Dauterman, 1960). This discovery introduced an expanded feeling of latitude in designing selective organophosphate insecticides.

(*d*) *Organophosphate anthelmintics.* In 1951 there appeared, from Schrader's laboratory, an organophosphate that was highly selective as a vermifuge when given orally to cattle. This agent was coumaphos (*13.36a*) ('Asuntol'). The coumarin component was suggested by the metabolic ease with which cattle handle the high coumarin content of clover. Two changes made to the molecular structure, in England, gave haloxon (*13.36b*) which is remarkably non-toxic to mammals; the LD_{50} in the rat is 900 mg/kg, and lambs easily tolerate five times the anthelmintic dose (N. Brown *et al.*, 1962), a selective effect for which the $H \rightarrow Cl$ change was found to be more responsible than the $O \rightarrow S$ change. Haloxon is used, orally, to kill nematode worms in cattle and sheep, oxyurid and ascarid worms in horses, and several varieties of worms in pigs and poultry. The high selectivity of haloxon is traced to the principle of analogous enzymes (Section 4.2). This agent irreversibly inhibits the AChase of the worm *Haemonchus contortus*, whereas the corresponding enzyme in sheep erythrocytes is only temporarily blocked and recovery is rapid (Lee and Hodsden, 1963); calf AChase recovered halfway in 30 minutes (Pickering and Malone, 1967).

Dichlorvos (*13.29*) is given orally as a broad-spectrum anthelmintic for swine, horses, and dogs. For the use of metrifonate in human schistosomiasis, see p. 232.

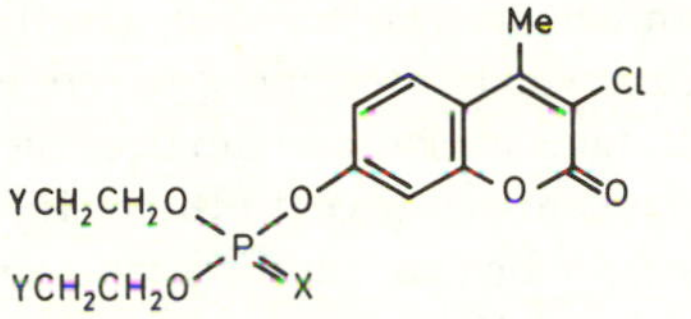

(a) Coumaphos (X = O, Y = H)
(b) Haloxon (X = S, Y = Cl)
(13.36)

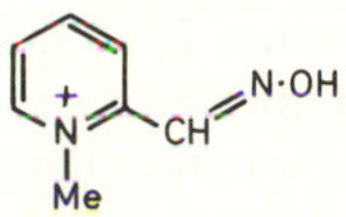

Pralidoxime
(13.37)

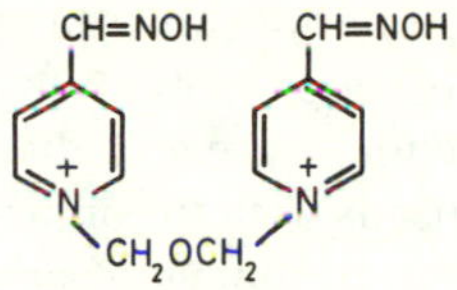

Obidoxime
(13.38)

(e) *Antidotes.* Because the acute toxicity to Man of many of the phosphorus insecticides is high, first-aid remedies are kept on hand. The most useful of these is an injection of atropine (which acts primarily on muscarinic sites) followed by an oxime specific for nicotinic sites. These oximes re-activate phosphorylated acetylcholinesterase in patients (Holmes and Robins, 1955) just as with the isolated enzyme (Wilson and Meislich, 1953; Childs *et al.*, 1955). A nucleophilic attack by the reagent upon the phosphorus atom causes the bond between this atom and the oxygen atom of the serine moiety to break and simultaneously a new covalent bond is formed with the hydroxylamine, as in Scheme 13.1. Much experience has been obtained with the antidote, pralidoxime (*13.37*) (2-PAM), which is the *anti*-form of pyridine-2-aldoxime methochloride. It is effective orally, but much more effective parenterally. Obidoxime (*13.38*) is a newer and more powerful antidote, usually administered intramuscularly (Hobbiger and Vojvodić, 1966). Carbamate poisoning is not benefited by oximes.

A severe type of toxicity which does not involve AChase is shown by the triaryl phosphates such as tricresyl phosphate, formerly much used as a plasticizer, and also by some fluorine-containing organophosphates. These cause delayed and persistent (up to 2 years) neurotoxicity characterized by swollen neurons. For this condition, oximes are no remedy.

For further reading on the chemistry of organophosphate insecticides, see Fest and Schmidt (1982).

(f) *Carbamates (urethanes).* From 1947 onwards, carbamate insecticides have been brought very much to the fore on the grounds that they are less toxic to humans and more selective among insect species. These substances act by acylating acetylcholinesterase. The $R_2N \cdot CO$-group is presumably transferred to the hydroxyl-group of serine, thus forming a carbamyl-derivative of the enzyme (Hobbiger, 1954; Wilson, Harrison and Ginsburg, 1961). Carbamylated AChase recovers activity spontaneously and much faster than when phosphorylated, provided that the *N*-alkyl group is kept small (Davies, Campbell and Kearns, 1970). On the average, the human recovery rate is faster from carbamate than from organophosphate poisoning. The half-life of most *N*-methylcarbamate insecticides is about 15 minutes compared with, say, 8 hours for many of the most used phosphorus equivalents. Incorporation of any other alkyl group has been found disadvantageous. To have an electron-attracting substituent in the molecule, a feature which intensifies the action of organophosphate insecticides, is of no advantage in the carbamates. Steric factors have an enhanced importance in the carbamate series (O'Brien, 1971). Just as with the organophosphates, the carbamates form an addition compound with the enzyme before carbamoylating it (O'Brien, 1968).

Carbamate insecticides tend to be, on average, more hydrophilic than phosphorus insecticides and have proved excellent for systemic use, because avoidance of (solubility) extremes favours a good uptake by plant tissues as well as insect nerves. Ionic carbamates, such as those used in human medicine, do not penetrate into insect nerves.

The first carbamate insecticide to make a strong impact on agricultural and veterinary practice was carbaryl (*13.39*) ('Sevin'), introduced by Union Carbide Corporation in 1959. Chemically, it is 1-naphthyl *N*-methylcarbamate. The mammalian toxicity is low (LD_{50} 540 mg/kg, oral, rat). Carbaryl is destroyed in the mammal by ring-hydroxylation followed by conjugation of the phenolic group (Fukuto, 1972). It is used systemically, e.g. in plants to kill leaf-eating caterpillars. Propoxur (*13.40*) ('Baygon') followed in 1963. It is a simpler molecule, 2-isopropoxyphenyl *N*-methylcarbamate. Methomyl (*13.41*), an aliphatic example, was introduced in 1966. Carbamate insecticides are much used in conjunction with the synergists that boost the action of pyrethrins (Section 3.5).

The FAO recommend the following carbamates for Integrated Pest Management: carbaryl and methomyl. Carbaryl and propoxur are much used for ectoparasites in domestic pets. Monocrotophos (*13.42*) combines a vinylogous carbamate group with a phosphate group in the one molecule. It is used as an agricultural insecticide, applied by aerial spray, over vast areas, without apparent harm to Man.

For further reading on carbamate insecticides, see O'Brien (1971) and Kuhr (1976). For carbamate fungicides, see Section 6.4 (p. 248).

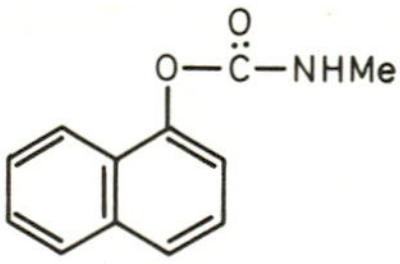

Carbaryl
(13.39)

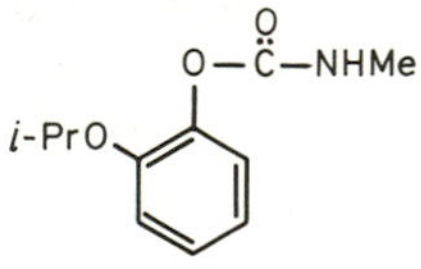

Propoxur
(13.40)

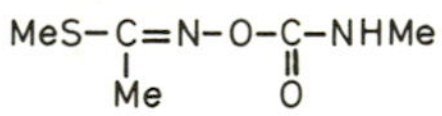

MeS–C=N–O–C–NHMe
Methomyl
(13.41)

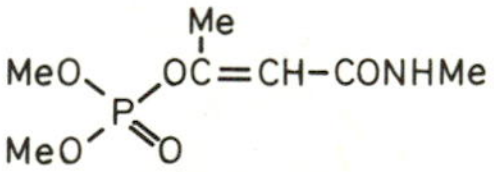

Monocrotophos
(13.42)

In human medicine, carbachol (carbamoylcholine chloride) (*2.11*), which is isosteric with acetylcholine (*2.7*), is used to block acetylcholinesterase and is slowly hydrolysed by it. It exhibits both the muscarinic and nicotinic actions of acetylcholine in a more prolonged and intense form. Other examples used in human medicine (Section 7.6.3) include physostigmine (*2.8*) and neostigmine (*2.10*) for myasthenia and glaucoma. Carbaryl does excellently in pediculosis.

13.4 Alkylating agents

All alkylating agents are toxic to mammals; methyl iodide and diazomethane provide examples. However, there are substances with *two* alkylating groups which are selectively toxic to mammalian tumours. In fact, these 'nitrogen-mustards' as they came to be called, were the first drugs of all anti-cancer drugs. How they were discovered, under wartime conditions about 1942, is told in Section 6.3.6.

It is the chemical nature of alkylating agents to esterify the phosphoric and carboxylic acid anions present in the living cell, and to etherify the anions of phenols and mercaptans. The more powerful of these agents can also alkylate neutral amino groups (primary, secondary, and even tertiary). The reaction usually takes place by a S_N1 (first-order nucleophilic substitution) reaction (Ross, 1962). We shall explore this reaction for chlormethine (mustine, mechlorethamine), the first nitrogen-mustard to be used clinically, and one that is still valued for treating Hodgkin's disease and other lymphomas. It is used intravenously, either alone or (more often) in the MOPP regimen (i.e. with vincristine, procarbazine, and prednisone). On contact with water, mechlorethamine (*4.25*) slowly cyclizes to an aziridinium cation (*13.43*), which is the true drug. Although much of the dose of mustine becomes hydrolysed or fixed on sites of loss, enough reaches the cancerous cells to produce a powerful anti-proliferative effect. The toxic effect of the nitrogen-mustards shows a small but definitely favourable selectivity, largely confined to the bone-marrow, blood-stream, and lymphatics (Ross, 1962). For *solid* tumours, which these drugs do not massively influence, their use must be preceded by surgical excision of the core. This feeble selectivity is enhanced by physical localization, such as injection directly into the tumour, or the use of tourniquets, or temperature gradients.

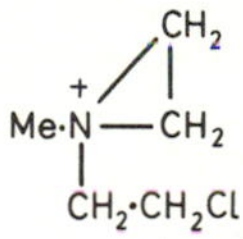

Intermediate in action
of chlormethine
(*13.43*)

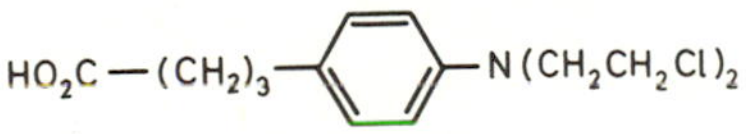

Chlorambucil
(*13.44*)

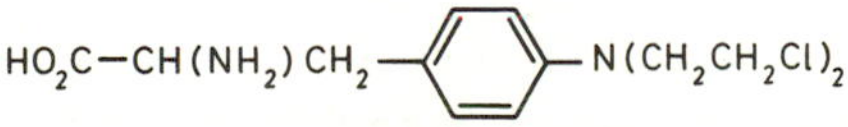

Sarcolysine (melphalan)
(*13.45*)

In the S_N1 type of reaction characteristic of the nitrogen-mustards, the cation of the drug, e.g. (*13.43*) which is formed relatively slowly, attacks the anion of the biopolymer rapidly, and the product is neutral; alternatively this cation attacks the neutral amine (again rapidly) and expels a hydrogen ion. The rate of reaction depends on the concentration of the site to be alkylated, but not on that of the nitrogen-mustard.

Chlorambucil (*13.44*) [*p*-bis-2-(chloroethyl)aminophenylbutyric acid], discovered in 1953 (Everett, Roberts and Ross), exemplifies the 'aromatic nitrogen-mustards'. These, because of the low basic strength of the aniline-type nitrogen atom, can form aziridinium salts only very slowly. The result is that they reach their target sites before becoming dissipated by side reactions. Actually, chlorambucil is an extreme case: it is the slowest acting and least toxic of the nitrogen-mustards in clinical use. It has proved highly suppressive in chronic lymphocytic leukaemia, and useful for continuation at home, by oral therapy, of treatments begun in hospital with cyclophosphamide or chlormethine.

A related drug, L-sarcolysin (*13.45*) (melphalan) was designed in the hope that its amino acid structure (it is a derivative of phenylalanine) might lead to activated passage through membranes (see Section 3.2, p. 70). This apparently occurs, because the D-stereoisomer is inactive (Bergel and Stock, 1954). It is intermediate in activity between mustine and chlorambucil. Mannomustine [1,6-bis-(2-chloroethyl)amino-1,6-dideoxymannitol] represents a similar attempt to secure activated uptake, in this case with reference to glucose (Vargha *et al.*, 1957). Although effective, it seems to offer no special advantage.

A clinically valuable alkylating agent of quite another kind is busulphan (*13.46*) (1,4-dimethanesulfonyloxobutane), discovered by G. Timmis. It is highly effective in, and almost specific for, chronic myelocytic leukaemia, and has few side effects (Haddow and Timmis, 1953). Its mode of action is apparently the alkylation of, and eventual stripping of sulfur from, cysteine residues of proteins and peptides (Roberts and Warwick, 1961). Busulphan acts, without prior ionization, by an S_N2 reaction, i.e. one whose rate depends on the concentrations of both the alkylating agent and the nucleophilic site on the biopolymer.

$$MeSO_2-O-(CH_2)_4-O-SO_2Me$$
Busulphan
(13.46)

If the nitrogen-mustards are, in general, to be regarded as pro-drugs, then cyclophosphamide (*3.38*) is the pro-drug of a pro-drug because it cannot form an aziridinium cation, like (*13.43*), until the phosphonolactone ring is opened, as happens nearer to the site of desired action (see Section 3.6, p. 101). Following this reaction with ^{31}P revealed that the alkylating action of this aziridinium ion is accelerated as the pH falls from 7 to 5. In this way, selectivity is exercised against the more acidic malignant cell (Eagle, Zon and Egan, 1979).

Cyclophosphamide can be given orally and produces good clinical results in several malignant conditions including Hodgkin's disease, lymphosarcoma, Burkitt's lymphoma, acute lymphoblastic leukaemia of childhood, and after surgery for carcinoma of the breast in premenopausal women (Fairley and Simister, 1965). It is the most used of all alkylating agents in the treatment of cancer.

At therapeutic dose levels, chlormethine and cyclophosphamide inhibit the synthesis of DNA (Wheeler and Alexander, 1969). To explain the need for *two* alkylating groups, it was suggested that these agents act by cross-linking two strands of DNA, thus preventing them from replicating. That the join was between the 7-nitrogen atom of two guanine residues was made particularly likely when it was demonstrated that di(2-chloroethyl) sulfide attacks DNA in such a way that the hydrolysate contains much di(guanin-7-ylethyl) sulfide. RNA was much less affected (Lawley and Brookes, 1967). What is thought to happen between chlormethine and DNA is shown in Scheme 13.2. It had long been recognized that, to have anti-cancer action, a drug requires two alkylating functions (e.g. —Cl), separated by about five atoms, as in substances (*4.25*) and (*13.45*) (Haddow *et al.*, 1948). These alkylating functions must not be so reactive that they can be hydrolysed by water, and yet they must not be so firmly attached that they fail to alkylate the biological receptors. The idea that bifunctional molecules of this kind act by forming cross-linkages in the chromosomes is consistent with this critical distance of separation of the functions and the frequent mitotic irregularities seen under the microscope (Goldacre, Loveless and Ross, 1949). This may not seem a mode of action that could lead to high selectivity, but the selectivity of nitrogen-mustards is only moderate.

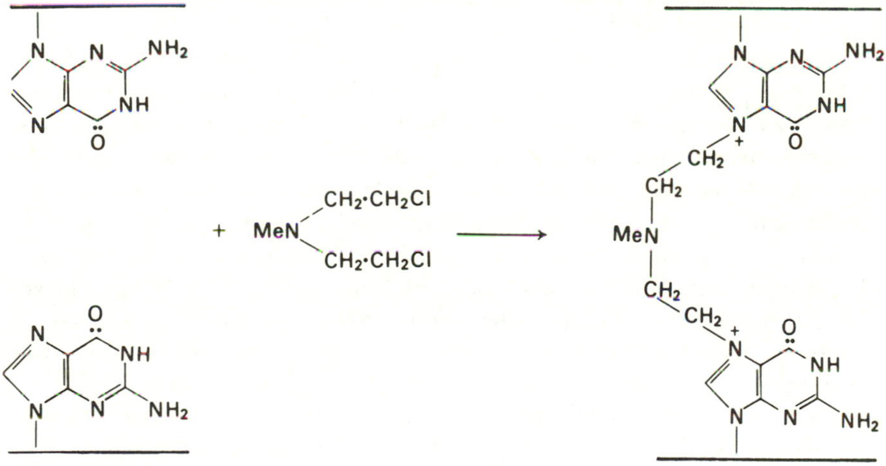

Scheme 13.2

Mitomycin C (*4.26*) is a violet-coloured antibiotic which mammalian metabolism converts to a bifunctional alkylating agent that irreversibly unites two strands of DNA (Iyer and Szybalski, 1964). It has found limited application in cancer of the stomach and bowel but has a high toxicity for the patient.

Bifunctional alkylating agents resemble ionizing radiations in not being specific for any particular stage of the cell cycle, and working better in large, well-spaced doses than in smaller, frequent ones.

In the constant search for new types of alkylating agents, the nitrosoureas were introduced (Johnston, McCaleb and Montgomery, 1963) and have proved active against a range of tumours. Their liposolubility enables them to penetrate the blood–brain barrier, and good results with some types of brain tumours have been reported. Unfortunately, in all cases, the dose must be kept small because nitrosoureas have a strong tendency to cause myelosuppression. The most used examples are the orally administered lomustine (*13.47*), introduced as CCNU, which is 1-(2-chloroethyl)-3-cyclohexyl-1-nitrosourea, and its methylated analogue, semustine. A nitrosourea antibiotic, streptozocin, is sometimes used in pancreatic tumours.

The action of the nitrosoureas seems to be mainly on DNA which is alkylated with a non-enzymic degradation product, the 2-chloroethylcarbonium ion ($ClCH_2CH_2^+$) (Cheng *et al.*, 1972; Connors and Hare, 1974; Carter, Sakurai and Umezawa, 1981). A second product of this spontaneous degradation is cyclohexyl isocyanate ($C_6H_{11}N:C:O$) which has been found to carbamoylate (i.e. form ureas with) proteins, but therapeutically plays a minor role.

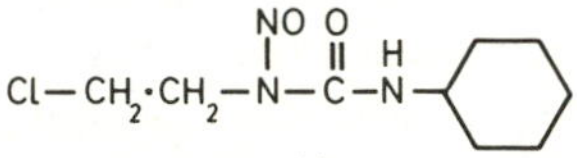

Lomustine
(*13.47*)

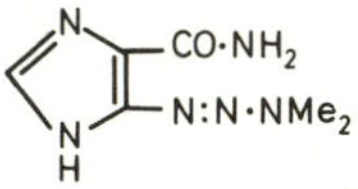

Dacarbazine
(*13.48*)

Another pro-drug that releases an alkylating agent (in this case by the action of liver *e.r.*) is dacarbazine (*13.48*), which is 5-(3,3-dimethyl-1-triazeno)imidazole-4-carboxamide (Shealy and Krauth, 1966). Given intravenously, it is twice demethylated giving 5-aminoimidazole-4-carboxamide (a normal body constituent) and the methyl radical, detected by mass spectroscopy. This radical is the true drug which methylates the guanine of RNA in the 7-position, also attacks protein, but affects DNA very little (Bono, 1976). It has achieved about 20% success in obtaining regression of malignant melanoma, a disease for which, perhaps, no other remedy is even as effective as this. Myelosuppression prevents physicians from increasing the dose (Carter and Friedman, 1972).

Some fungicides that alkylate mercapto-groups in essential enzymes were described in Section 6.4.2, e.g. captan (*6.58*) and chlorothalonil (*6.59*).

For further information on the alkylating anti-cancer drugs, see Wilman and Connors, 1983. The carcinogenic action of alkylating agents is dealt with in Section 13.5.

13.5 Lethal synthesis

The name 'lethal synthesis' was devised by Peters to describe the biochemical alteration of fluoroacetic acid, which is not toxic in itself but is built up by mammalian metabolism to a toxic substance, fluorocitric acid. This change, which occurs in the tricarboxylic acid cycle, depends on fluoroacetic acid being sterically so similar to acetic acid that it is accepted by a series of enzymes and undergoes similar enzymic changes until it ends as fluorocitric acid (*13.49*) (Peters *et al.*, 1953; Liébecq and Peters, 1949). The toxic effect is due to this acid blocking the enzyme which dehydrates citric acid to cis-aconitic acid. Hence fluorocitric acid acts as a metabolite analogue of citric acid.

What distinguishes lethal synthesis from the cases of 'degradation before action' discussed in Section 3.6 is simply that it *is* synthesis and not degradation: a substance with only two carbon atoms has been made into one with six. Moreover, the raw material for this synthesis passes through at least three enzyme systems, and is changed a little by each, before it becomes toxic.

Fluoroacetic acid occurs in a South African plant *Dichapetalum*, and has killed many cattle in that country. The toxic action of fluoroacetic acid does not depend on any chemical reactivity on the part of fluorine, but on the small *size* of the fluorine atom (Table 9.1) (Bartlett and Barron, 1947), which deceives the earlier, poorly discriminating enzymes into treating it as a hydrogen atom whereas the later enzymes have more inbuilt precision. Fluoroacetic acid, which is used to exterminate rabbits in their burrows, is not very selective and presents a hazard to human beings.

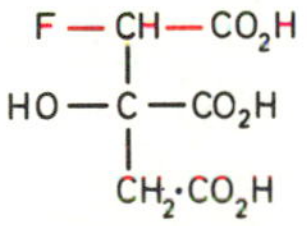

Fluorocitric acid (*13.49*)

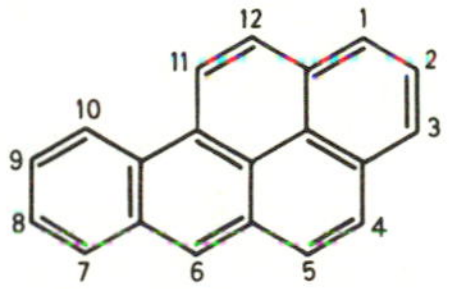

Benzo [*a*] pyrene (*13.50*)

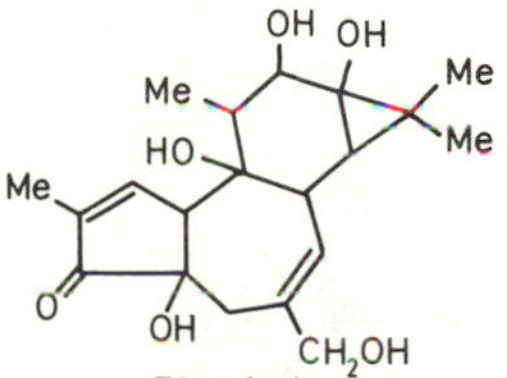

Phorbol (*13.51*)

Carcinogens are counter-selective chemicals, often harmless in themselves, which the host's metabolism can elaborate into the true carcinogenic agents. The first link between cancer and the chemical environment occurred in 1775, when Potts traced the frequent scrotal cancers of young chimney-sweeps to contact with tar and soot (the lads were commonly stripped of their clothes and impelled up the flues). A century and a half passed before the causative chemical was isolated from coal tar and found to be benzo[*a*]pyrene (*13.50*) (Cook, Hewett and Hieger, 1933).

Until that time, it had been assumed that cancers were the inevitable accompaniment of ageing. Today, because of the results of comparative studies of populations, the triggering of cancer is seen to be largely environmental. This follows from the great variation in the amount and types of cancer found in different communities in different parts of the world. The manifestations of cancer in migrants, too, changes in proportion as they adapt their life-style to the country of adoption. The International Agency for Research on Cancer (1976) reported that 80% or more of all human cases are activated by exposure to sunlight, tobacco, alcohol, or polluted air (cf. Cairns, 1975). Genetic pre-disposition is recognized as being important, too. A new example of chemical carcinogenesis is identified when a cluster of cases of a particular kind of cancer has been reported for a trade or pastime.

An important advance in understanding the carcinogenic process was made when the Israeli scientist, Berenblum (1941) showed that it was a two-stage process. He found that, after *promotion* by a seemingly inert substance, an otherwise insignificantly small dose of a carcinogenic hydrocarbon caused *initiation* of the growth of a tumour. Many have followed up this discovery, and the most used promotor has been croton oil, or its active principles which are esters of phorbol (*13.51*), a tetracyclic poly-alcohol. One of the most used of these promotors is TPA, 12-*O*-tetradecanoylphorbol 13-acetate. It has been suggested that these promotors work by activating protein kinase-C in plasma (Castagna *et al.*, 1982). Berenblum noted that no cancer resulted if the initiator was applied *before* the promotor (Berenblum, 1974).

Today, many agents are known which make cells grow faster and with less self-cohesion (Section 5.1). This 'transformation', as it is called, can be brought about with certain viruses, vegetable lectins, and a few laboratory chemicals. Transformation is reversible, but if not reversed it is heritable, and hence has the nature of a mutation (a genetic change transmitted to progeny). Trans-formed cells do not spontaneously become malignant, but require the inter-vention of a carcinogen.

Quite recently, a different two-stage carcinogenetic process was discovered. Known carcinogens like dimethyl sulfate or benzo[*a*]pyrene (*13.50*) are used, in this process, to change healthy cells into a strain that proliferates indefinitely. Yet, other than having acquired an unlimited life-span for their progeny, these cells retain all the properties of normal cells and, unlike the transformed cells discussed above, they have normal growth characteristics. Yet these cells can be converted to malignant cells in a single step by incubating them with an oncogene (such as the one known as EJ). In this way, the cells become overtly cancerous, ignoring all restraining signals from healthy neighbour cells (Newbold and Overell, 1983). (The carcinogenic chemical in this process can be replaced by a different oncogene.) It is now thought likely that some human cancers are caused by the mutation of normal genes to oncogenes (Fox, 1983).

In some cases, the difference in chemical composition between an oncogene and its equivalent normal gene is minimal. Thus bladder cancer can follow upon the change in a single nucleotide that specifies a single amino acid of difference, namely the substitution of valine for glycine as the twelfth residue from the amino-terminal of the protein (Tabin *et al.*, 1982). Although it seems unlikely that most carcinogenic activation of genes will turn out to be as simple as this, it may be recalled that a single amino acid change in another protein (valine for glutamate in haemoglobin) is the cause of sickle-cell anaemia (Pauling *et al.*, 1949).

Perhaps the most astonishing thing about carcinogens is how very few substances are known *for certain* to cause cancer in Man. There are two factors at work here: one is the great difficulty of getting the relevant information and the other is what seems to be Man's superior repair mechanisms so that there is (seemingly) less information to obtain. The International Agency for Research on Cancer (IARC), an agency of the World Health Organization, is steadily engaged in sifting the evidence. Up to 1982, they had examined 585 substances claimed to be likely causers of cancer in Man. Of these, only 147 were found to be carcinogenic in experimental animals, and for only 44 was evidence found that they could cause cancer in humans. Most of this evidence was obtained by analysing the health records of factory workers. An example: men engaged in manufacturing benzidine (4,4'-diaminobiphenyl, an intermediate for dyes) had 25% and 75% excess incidence of bladder cancer after (respectively) 15 and 20 years' exposure. The manufacture of benzidine is now illegal in most industrialized countries. However, there are still some countries that make it, and convert it to technically very desirable dyes, such as Direct Black 38, which break down in human metabolism and liberate the benzidine.

Another example: IARC found that workers exposed, in a working lifetime, to 100 p.p.m. of benzene in the factory air sustained 140–170 per 1000 excess deaths from leukaemia. The use of benzene as a solvent is now widely banned, but as benzene is an essential starting material for countless drugs and other useful substances, its manufacture and use are tolerated, but only under carefully monitored conditions.

Other chemicals found to be carcinogenic in Man were (a) some *inorganic materials*: arsenic, asbestos (principally blue asbestos), cadmium, chromates, haematite (one kind of iron ore), nickel (refining only); (b) *some organic intermediates* for dyes and plastics: 2-naphthylamine (now banned), 4-aminobiphenyl, vinyl chloride (through its metabolite:chloroethylene oxide); (c) *drugs*: azathioprene, chlorambucil, cyclophosphamide, diethylstilbestrol, melphelan, MOPP (defined on p. 577). Cancer-prone *occupations*, where the risk seems to be chemical are: manufacturing and repairing shoes, manufacturing rubber, making furniture, the manufacture of benzaldehyde by chlorinating toluene, any trade that brings contact with soot, tar, or crude oils.

The general public is similarly at risk if smoking cigarettes or eating food contaminated with the fungal product, aflatoxin. They may also be at some risks from the drugs listed under (c), much less so since timing and dosage are so carefully controlled. No evidence was found to place the following items on the list although each one has been subjected to adverse publicity: caffeine, artificial sweeteners, formaldehyde, and chlorinated hydrocarbons (International Agency for Research on Cancer, 1982).

For a carcinogen to produce overt cancer in Man may take 10 to 20 years, even though he be constantly exposed to it in the home or workplace. In the mouse, with a life-expectation of about 2 years, many months must pass before a suspected carcinogen can be confirmed or denied. What is happening in this interval? Apart from the uncertainty that surrounds the promotor phase in Berenblum's two-stage initiation of malignancy (p. 582), lies the fact that many carcinogens need to undergo one or more covalent chemical reactions so that the pro-carcinogen (as administered) can become the ultimate (also called proxi-mate) carcinogen, which actually effects the malignancy. The process involves an increase in molecular weight leading to a positively charged product (Miller, 1970). Such carcinogens are electrophilic reagents (like the nitrogen-mustards) ready to form covalent links with the bases of DNA. Unlike the nitrogen-mustards, they cannot cross-link two DNA strands, and this lack may make them more destructive than useful (Brookes, 1966).

Because laboratory rats and mice are so much more prone to cancer than are human beings, no unambiguous criterion exists to extrapolate to Man the results of experiments on rodents, although a suitable formula is constantly sought (International Agency for Research in Cancer, 1982). There are strong indications that Man's DNA-repair process is more efficient than that of rodents. In fact, evidence from radiation exposure suggests that the repair of injured genetic information in Man is quite remarkable. Schul, Otake and Neel (1981), reporting for the Radiation Effects Research Foundation, Hiroshima and USA, describe a study of the populations of Nagasaki and Hiroshima. Forty years after the intense radiation of these cities, careful investigations con-tinuously performed over the last 34 years found *no heritable defects*. In short, parental exposure conferred no statistically significant mutations on the off-spring of whom the following characteristics were studied: untoward outcome of their pregnancies, deaths in the children, and frequency of children with chromosomal aberrations, all compared to a matched population from a non-irradiated area.

Let us now look at the chemical changes that carcinogenic aromatic amines undergo, in mammals, as they are metabolized from pro- to ultimate-carcino-gens. In general the amino group is first hydroxylated (giving a hydroxyl-amine), but the hydroxy-group is transferred to a neighbouring (*ortho*) carbon atom where it is converted to the sulfate ester, which then hydrolyses to the carbenium ion, such as (*13.52*) from 2-acetamidofluorene (Cramer, Miller and Miller, 1960). These carbenium ions seem to be the ultimate carcinogens. Injection of inorganic sulfates into rats receiving *N*-hydroxy-2-acetamido-fluorene greatly increases the toxicity of this intermediate (De Baun *et al.*, 1970). Parallel changes have been demonstrated in the carcinogens benzidine and 4-acetamidobiphenyl (Miller, Miller and Hartmann, 1961), 2-aminonaph-thalene (Boyland, Dukes and Grover, 1963), and 4-hydroxyaminoquinoline-1-oxide (from the bioreduction of 4-nitroquinoline-1-oxide) (Endo, Ono and Sugimura, 1970). 4-Dimethylaminoazobenzene (called butter yellow, formerly used for colouring margarine) is metabolically changed to the monomethyl

Example of ultimate carcinogen
(cation)
(13.52)

Aniline mustard
(13.53)

analogue, which undergoes similar changes to those described for 2-acetamido-fluorene (Poirier *et al.*, 1967).

The carcinogenicity of the *polycyclic hydrocarbons* depends on their becoming bound, after metabolism, to nucleic acid, as first shown by Brookes and Lawley (1964). The DNA of mitochondria seems more affected than that of the nucleus (Nudd and Wilkie, 1983). Benzo[*a*]pyrene (*13.50*), that universal contaminant of industrialized cities, is metabolized by mammalian cells to give the corresponding 7,8,9,10-tetrahydro-7,8-dihydroxy 9,10-epoxide, which binds strongly to DNA (Sims *et al.*, 1974), a result that has been widely confirmed. The epoxy ring undergoes nucleophilic opening to form a link between the 10-position of the hydrocarbon and the exocyclic amino-group of both guanine and adenine in the DNA (Straub *et al.*, 1977).

Boyland (1950) was the first to suggest that metabolic conversion to epoxides may be the essential step in the carcinogenicity of the polycyclic hydrocarbons. At that time, a rival hypothesis held sway: that union with the target took place through a 'K region' whose contribution to the carcinogenicity rose as its electron density rose (Robinson, 1946; Pullman and Pullman, 1955). This hypothesis was based on knowledge that several polycyclic hydrocarbons, such as 1,2-benzanthracene, were not carcinogenic, but became so when a methyl group was inserted in particular positions. Examples of the K region are the 9,10-double bond in phenanthrene, and the 4,5-bond in (*13.50*). Today the K-region hypothesis is out of favour, and the epoxidation approach is being found applicable to other carcinogenic polycyclic hydrocarbons.

The re-entrant part of a hydrocarbon molecule is called the 'bay region', an example of which can be seen between positions 10 and 11 in (*13.50*). Epoxidation is most likely to occur in the bay region, but the bulk of neighbouring substituents can prevent it. The anthracenes that correspond to the carcinogenic phenanthrenes necessarily lack a bay region and are not carcinogenic.

The mutagenic effect of the following drugs, many of them anthelmintics, is thought to be effected by a chemical change caused by intestinal bacteria: hycanthone, amoscanate, metrifonate, oxamniquine, mebendazole, metronidazole, and diazepam. It can be prevented by treatment of the experimental animal with anti-oxidants of the kind sanctioned for prepared foods, e.g. 2-*t*-butyl-4-hydroxyanisole (BHA) and 6-ethoxy-2,2,4-trimethyl-1,2-dihydroquinoline (ethoxyquin). Even carcinogens such as benz[*a*]pyrene, nitrosamines, and urethane have also been countered in this way. It is thought that these anti-oxidants act by activating glutathione-*S*-transferase and epoxide hydratase (Bueding *et al.*, 1979).

Concerning *alkylating agents*, the medicinally useful ones (see previous section) usually alkylate the 7-position of guanine. By way of contrast, the highly carcinogenic examples, using a S_N2 mechanism, alkylate on oxygen. Thus ethylnitrosourea and ethyl-*N'*-nitro-*N*-nitrosoguanidine attacks *O*-2 of cytosine and uracil in RNA, and *O*-2 of thymine and *O*-6 of guanine in double-stranded DNA (Singer, 1976).

For general reading on carcinogenesis, see Freudenthal and Jones (1976) and Arcos and Argus (1968 <); for carcinogenesis arising through the action of radicals on lipids, and consequent peroxidation, see McBrien and Slater (1982); for legal regulatory aspects concerning drugs and food additives, also the USA Delaney Clause, see Coulston (1979).

(*a*) *Testing for carcinogenicity.* Tests using whole animals, of which the rat is favoured, take from 2 to 3 years and are costly particularly because very large colonies must be used to get statistically meaningful results. Hence, it has been attractive to turn to short-term tests, even though they may only distantly reflect the situation in Man. There is, for example, a current interest in tests for mutagenesis based on this thought: carcinogenesis resembles mutation because different chemicals cause different types of cancer, and each kind then reproduces without further variation. One ingenious, and very popular test of this kind utilizes histidine-requiring mutants of the bacterium, *Salmonella typhimurium*, which die when any attempt is made to grow them on a histidine-free medium. However, an appropriate mutagen, if included in the culture, can cause a frame-shift back-mutation which allows the bacteria to produce histidine once again (Ames, Sims and Grover, 1972; Ames *et al.*, 1973).

Newcomers to this field might be excused for thinking that there would be little chance of a foreign chemical effecting the back-mutation. Nevertheless, it has proved to be quite commonplace and is, for example, given by many substances that have a rigid nucleus. In practice, the Ames Test (as it has come to be called) uses mutants chosen from capsule-less variants of the bacterium, because the carbohydrate coating may otherwise prevent free diffusion of the drug. It is usual to run a duplicate test including (or on material subjected to) the S9 microsomal fraction from rat liver, to pick up a possibly carcinogenic metabolite formed by contact with mixed-function oxidase (see p. 89).

Some disconcerting false positives and false negatives have been furnished by the Ames test. It is most unfortunate that two constituents of animal cells, cysteine and glutathione, give a positive test, because these substances are known to play a part in *preventing* carcinogenesis (Glatt, Protić-Sabljić and Oesch, 1983). Again, dioxin (p. 42), whose often violent activity seems to be of a mutagenic character at least in susceptible mice, is not classed as a mutagen in the Ames test. These irregularities serve to remind us that the Ames test is only an initial sorting, and that there has been a high degree of agreement between mutagenicity discovered in this test and carcinogenicity in rodents.

There is a growing trend to couple the Ames test with a second short-term test, such as an assay for chromosome damage in Man, using lymphocytes cultured from blood. Both tests, run side-by-side, can provide clear answers in 6 weeks. A much-used alternative is the Styles cell-transformation test (Purchase *et al.*, 1976). This depends on the biological peculiarity that most mammalian tissue-cells fail to grow in semi-solid agar, a medium that coats them and prevents them from joining together; under these circumstances they simply die. However if a substance that can cause cell transformation (p. 582) is supplied, growth (as single cells) can freely take place. For his test, Sykes used human liver cells and hamster kidney cells. This test has, on occasion, proved complementary to the Ames. For example, the Ames test pronounced 2-naph-thylamine positive but butter yellow negative, whereas the Styles test reports the exact opposite (both substances are known human carcinogens).

It is only too easy to suppose, and this thought is much nurtured by the mass-media, that every mutagen is a potential carcinogen. Not to be led too rapidly down this seductive path, we should bear in mind that the kind of mutagenicity which the Ames test is monitoring (namely, ability to effect a frameshift) is not the kind of 'mutagenicity' that cancer cells exhibit (inde-pendence of growth and acceleration of the cell cycle). One cannot fail to be impressed by series in which, as the carcinogenicity goes up (among the different members), the mutagenicity comes down (and vice versa). Such a series is constituted by 15 derivatives of aniline-mustard (*13.53*) (Leo *et al.*, 1981). These substances were simultaneously examined in a collaborative programme between the Mutagenesis Screening Section of the Frederick Cancer Research Center, Maryland, and two Californian Schools. They were graded for *mutagenicity* by the Ames test (with and without prior incubation in the S9 microsomal fraction of liver), and for *carcinogenicity* by a count of the increase in lung tumours in a strain of mice prone to this malignancy (a 6-months' test). It was found that the 4-phenoxy derivative ranked as the highest for mutagenicity and among the lowest for carcinogenicity, whereas the reverse was found for the 3,5-diureido ($-NH \cdot CO \cdot NH_2)_2$ analogue.

That such possibilities exist for *selective mutagenicity* is excellent news for all who are concerned by the steady increase in agent-resistance among un-economic species.

(*b*) *Other examples of lethal synthesis.* Several analogues of the purine and pyrimidine bases of nucleic acids have been dealt with in Section 4.0 (pp. 125–140) where their clinical uses are outlined. The most successful are those that undergo lethal synthesis in the cell, by being built up into analogues of deoxy-ribosides and -ribotides which then compete with the normal nucleosides and nucleotides. In this category are to be found the anti-cancer drugs 5-fluorouracil (*3.3*), cytarabine (*4.13*), and 6-mercaptopurine (*3.14*), the anti-viral drugs viadarabine (*4.16*) and acyclovir (*4.19*), the immuno-suppressant azathioprine (*3.41*), and the oral fungicide flucytosine (*4.23*). Viewed with a mixture of gratitude and caution is 5-iododeoxyuridine (*4.14*), gratitude because it was the first successful anti-viral drug, and caution because it is built into the body's DNA and hence, as a potential mutagen, not to be used internally. Also dealt with in Section 4.0 are analogues which affect RNA synthesis: 8-azaguanine (*4.40*) and 6-azauracil (*4.41*), the former an anti-cancer drug, the latter an agricultural fungicide.

The (fortunately rare) sensitization of patients to penicillin occurs apparently through acylation of the ϵ-amino-group of lysine, in γ-globulin, by the β-lactam ring of the drug (Hamilton-Miller and Abraham, 1971).

13.6 Miscellaneous examples

The pharmacological effects, on the nervous system, of reserpine, and of the hydrazines that inhibit monoamine oxidase (p. 360), are irreversible. These facts suggest that covalency is involved.

The hormone that synthesizes thyroxine seems to be a protein bearing a sulfenyl iodide group (—CSI). It is thought that sulfur-bearing heterocycles such as 2-thiouracil (*13.54*) and its 6-propyl- and 6-methyl-derivatives, that are clinically used to control thyrotoxicosis, do so by covalently binding to this group, forming a S-S bond and liberating HI (Cunningham, 1964; Jirousek and Pritchard, 1971).

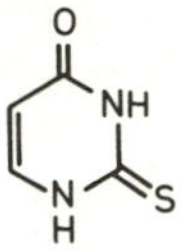

2-Thiouracil
(*13.54*)

Carbon tetrachloride, formerly used against the intestinal parasite, hookworm, and still in use as a solvent and fire-extinguisher and as a vermifuge in cattle, sheep, and poultry, causes extensive liver damage in humans. Apparently a flavoprotein enzyme converts this agent to the free radical, ˙CCl_3 (Slater, 1966). Free-radical scavengers, such as propyl gallate, promethazine, or *NN'*-diphenyl-*p*-phenylenediamine, were found to suppress this reaction strongly (Slater and Sawyer, 1971). Lipid peroxidation is known to require free radicals both for initiation and propagation. Even low concentrations of carbon tetrachloride can stimulate peroxidation in washed suspensions of liver at 37°C. The principal damage caused to liver by carbon tetrachloride is peroxidation of the double bond of the β-aliphatic chains in the phospholipids of endoplasmic reticulum membrane, leading to the irreparable breakdown of this structure (May and McCay, 1968). For futher reading, see McBrien and Slater (1982).

Chlorine in aqueous solution can exist as an *element* only below pH 2. In less acid solutions it is converted to hypochlorous acid:

$$Cl_2 + H_2O = HOCl + H^+ + Cl^-$$

The pK of hypochlorous acid is 7.2, and the disinfectant activity is proportional to the concentration of neutral molecule. The marked efficiency of chlorine, which is active at 0.2 p.p.m. in water, indicates that its action is rather specific. It has been suggested that, after conversion to hypochlorous acid, which penetrates the cell as the neutral molecule, the sulfhydryl (—SH)-groups of essential enzymes are attacked. The aldolase of *E. coli* is one of the essential enzymes of glycolysis sufficiently sensitive to hypochlorous acid to explain the bactericidal action of this substance (Knox *et al.*, 1948). Some organisms, presumed to have been killed by hypochlorous acid, have been revived by sodium thiosulfate (Mudge and Smith, 1935).

Hypochlorous acid reacts with ammonia to give the less active substance, chloramine (NH_2Cl), which is often used where a depot effect is required, e.g. in swimming pools. Similarly, hypochlorous acid forms substituted chloramines with amino acids and proteins, both outside and inside the cell. These can also act as depots, because they are in equilibrium with hypochlorous acid. Finally an artificial chloramine, *p*-toluenesulfonylchloramide (Chloramine-T) is used by campers as a source of hypochlorous acid.

Iodine, in neutral aqueous solution, exists mainly as such and the sporicidal action is due to elementary iodine. At pH 8.5, the hypoiodous acid : iodine ratio rises to 1 : 1, with a consequent drop in sporicidal activity. Hypoiodous acid (p*K* 10) is a weak acid. Iodine does not form a depot substance with ammonia, but its complex (KI_3) with potassium iodide has depot properties. A great excess of potassium iodide is required to hold iodine as this tri-iodide in aqueous solution (Wyss and Strandskov, 1945).

Formaldehyde is believed to injure bacteria by combining with the amino-groups of proteins, which are thereby changed in nature and function. Potassium permanganate and hydrogen peroxide can break down all kinds of organic matter by their violent oxidizing action. Hydrogen peroxide can initiate a self-propagating chain reaction in the presence of an oxidizable substrate (e.g. a thiol, or ascorbic acid) and a trace of a heavy-metal cation (e.g. iron).

Difluoromethylornithine (*9.88*) and other irreversible enzyme blockers (both ASDIIs and IMBIs) are dealt with in Section 9.7.

14

Surface chemistry.
The modifiction of membranes
by surface-active agents

The intense cytological studies of the last two decades have shown that cells, as well as most of the organelles within them, are covered with lipoprotein membranes (see Section 5.4). Life is possible only because of the presence of these biphasic membranes (lipophilic within, and hydrophilic on both outer surfaces), which provide for the separation of reactants. They also impose on the sequence of reactions an order that would otherwise be almost impossible to arrange.

Often in, but sometimes apart from, these membranes, enzymes offer surfaces of the very greatest biological selectivity. Macromolecules, including enzymes and other large protein molecules, whether particulate or dissolved in the bulk phase, present interfaces for reactions. Human blood serum, which has 100 m² of protein surface in each cubic centimetre, exemplifies the large interfaces available in protein 'solutions'.

14.0 Surface phenomena *in vitro*

The short-range (van der Waals) forces, described in Section 8.0, exert an attraction between all molecules which are in contact with one another. The presence of these forces in liquids becomes particularly evident at surfaces. Molecules in the bulk of the liquid are subject to these forces equally and in all directions. However, the molecules located at an air–water interface experience almost no force from the gas phase, so that the attractive forces of the liquid phase are unopposed. Hence the molecules of the surface layer tend to be pulled into the bulk phase, and the surface is forced to assume the least area possible (this explains the spherical shape of gas bubbles and drops of liquid). The situation is shown in Fig. 14.1(a). Constant exchange is taking place between those solvent molecules that are in the surface and those in the bulk phase.

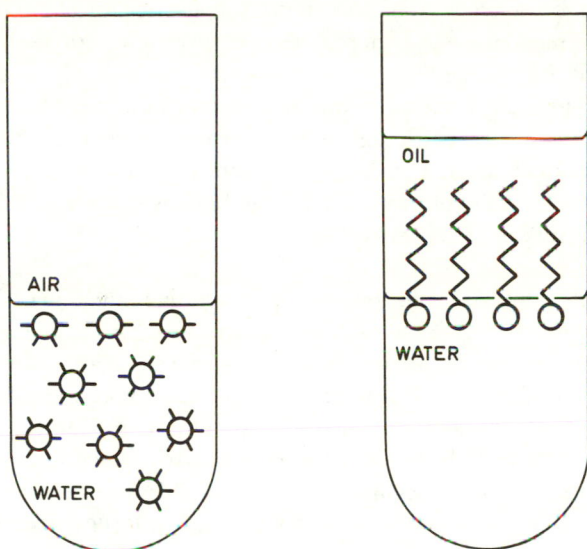

Fig. 14.1 (a) Diagram showing the unequal attractions experienced by molecules at the air–water interface. (b) Orientation of an amphiphilic substance at an oil–water interface.

A liquid–liquid interface (i.e. the surface between two immiscible liquids) behaves something like an air–water surface, but the contrast in pull exerted on the interfacial molecules is naturally much less. The surface tension at such an interface is often near to the difference of the individual tensions of the two liquids at the air–liquid interface.

Amphiphilic substances tend to become concentrated at interfaces. The molecules of these substances often consist of a long hydrocarbon chain attached to a short head which is polar. In most cases this head owes its polarity to the presence of oxygen or nitrogen atoms, and the lone pairs of electrons on these atoms form hydrogen bonds with water molecules. The hydrocarbon tail, on the other hand, can only be accommodated in the aqueous phase by breaking hydrogen bonds between water molecules, which energetically oppose this fission. Hence amphiphilic molecules find their position of least energy at the oil–water interface where the hydrophilic head groups can remain in the water, and the lipophilic chains can enter the oil where they associate freely with the similar chains of the solvent (see Fig. 14.1(b)). It will be evident that such a concentration of the amphiphile must stop as soon as a monomolecular layer of it is present at the interface. However, such a monolayer is always the site of turbulent exchange with other molecules of the amphiphile which are seeking places at the interface. Such an interface has a reduced surface tension and can easily become deformed. Both soluble and insoluble substances can be accumulated at interfaces.

It has long been recognized that factors operating at surfaces, especially orientation of molecules and increased concentration, can lead to chemical behaviour differing markedly from that observed in a bulk phase. For example, the molecules in a surface film of the *trans*-isomer of an unsaturated aliphatic acid can be pressed together so tightly that permanganate ion in the underlying aqueous phase does not reach the double bonds. Hence only the *cis*-isomer is oxidizable under these conditions, whereas both isomers are oxidized at the same rate in a bulk phase (Rideal, 1945).

Micelles. Dilute aqueous solutions of surface-active substances have normal physical properties, but at a higher concentration (characteristic for each substance) there occurs an abrupt change in surface tension, osmotic pressure, and electrical conductivity. These changes are due to the formation of a new, dispersed phase which takes the form of aggregates named micelles. These are often roughly spherical: the hydrocarbon chains are in the interior of the sphere and the hydrophilic groups occupy the outside of the micelles in contact with the solvent water. The lowest concentration at which micelle formation occurs is called the *critical micelle concentration,* an abrupt transition point.

For further reading on colloidal dispersion and micelles, see the review by Fisher and Oakenfull (1977), and the book by Tanford (1980).

14.1 Surface phenomena and drug action. Diuretics. Cardiac glycosides

Because most drugs are acting at surfaces, it is often said that the physical properties of these agents would have more biological significance if measured after adsorption on monomolecular films. Interesting results have been obtained in this way, but too few to form a coherent picture. This slow development is due not only to the scarcity of workers with both physical and biological backgrounds, but it also arises because some of the fundamental laws of surface chemistry remain incompletely known.

14.1.1 *Micellar equilibria*

A section of surface chemistry that offers few difficulties, and can greatly assist the study of selective toxicity, is the investigation of equilibria between monomers and micelles in aqueous solution. For example, the ions of agents with molecular weight over 150 usually polymerize reversibly to form micelles in concentrated solution. Examination of the effect of substituents on the *critical*

micelle concentrations (c.m.c.), and relating this to biological activity, is a potentially rewarding field. Convenient methods for the investigation of c.m.c.
include high-precision optical interferometry, deviations from Beer's Law (proportionality between concentration and optical density) in the ultraviolet,
search for maxima in specific conductivity (a micelle is a better conductor than
the ions of which it is composed), and detection of a sudden, relative decrease in
gegen-ion (e.g. Cl^- or Na^+) concentration as the concentration of the drug is
increased.

The use of soaps to solubilize phenols in water, for use as disinfectants,
depends on the formation of *mixed micelles* of the soap and the phenol. It has been
found that variation in the proportion of soap to phenol leads to several zones, as
shown in Fig. 14.2 (Berry, Cook and Wills, 1956). The first of these exists below
0.03 M potassium laurate and is poorly bactericidal; the maximal bactericidal
effect is obtained when the soap reaches 0.03 M, a figure which is identical with
the critical micelle concentration for this soap. It was concluded that this
bactericidal action is a combined attack of the phenol (mainly) and the soap on
the protoplasmic membrane (see Section 14.3 for membrane damage). The next
zone, up to 0.045 M soap, is one of greatly diminished bactericidal effect. The
interpretation is that the phenol has entered the micelles, many more of which
must have formed, and hence little of it is available for disinfection. When still
more soap is introduced, a final zone (vigorous disinfection) appears, due to the
toxicity of the soap itself. All the phenols commonly used as disinfectants,
including *p*-chloro-*m*-xylenol, form zones like these.

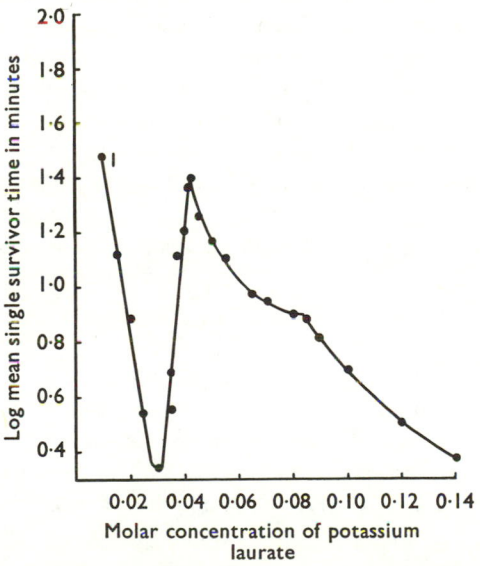

Fig. 14.2 Bactericidal action at 20°C of 4-benzylphenol (0.0016 M) in aqueous potassium
laurate (concentration varied, as shown). Organism: *E. coli*, 2×10^9 organisms per ml.
(Berry, Cook and Wills, 1956.)

The anthelmintic action of phenols is similarly potentiated by soaps and again it is important to avoid excess of the soap, because micelles can retain most of the phenol and deny it to the worms (Alexander and Trim, 1946). Below the critical micelle concentration of the soap, an adjuvant effect of the soap on the phenol is observable. No soap penetrates into the worms.

The formation of mixed micelles is well known in physical chemistry. An example: potassium myristate (C_{14}) develops micelles at one-quarter the concentration at which potassium laurate (C_{12}) does. Because of mixed micelle formation, as little as 15% of potassium myristate halves the critical micelle concentration of potassium laurate (Klevens, 1948).

A related phenomenon is the accumulation by serum albumin of drugs of sufficiently high oil/water partition coefficients (Brodie and Hogben, 1957) as discussed in Section 3.4 (p. 84). The following recounts a further variant. Several species of bacteria require oleic acid for growth, but are inhibited by more than faint traces. However, in the presence of serum albumin, they will tolerate several hundred times the usual bacteriostatic concentration. This is because each molecule of albumin can bind nine molecules of fatty acid so tightly that the acid is no longer even haemolytic. Yet the mixed micelles remain in equilibrium with enough free oleic acid to nourish the bacteria (Davis and Dubos, 1947).

The destructive effect of phenol on the human skin was ameliorated by the introduction of lipophilic (alkyl- or chloro-) groups into the molecule, a change initiated by Bechhold and Ehrlich (1906). Because these are lipophilic groups, those phenols which are least harmful to the skin are the most strongly antagonized by serum, because, having higher oil/water partition coefficients, they enter the albumin core. Solutions of *p*-chloro-*m*-xylenol, a typical chlorinated phenol, are widely used for rapidly disinfecting intact skin and mucous membranes, but are harmful if used out of this context.

Micelle formation of phospholipids and bile salts plays an important part in the transport, by solubilization, of fatty material in the bloodstream and small bowel respectively.

For further reading on the biological importance of micelles, see Elworthy, Florence and Macfarlane, 1968.

14.1.2 *Plasma membranes*

Lamellar, and partly micellar, mosaics of lipids and proteins form a thin, semi-permeable membrane around the exterior of every cell, and also around each organelle in the cell (see Section 5.4). The permeability of natural membranes was discussed in Section 3.2. The interaction of these membranes with diuretics, cardiac glycosides, and other ionophoric effects will now be considered.

For more reading on membranes, see Robertson (1983).

14.1.3 *Diuretics*

Changing the permeability of the membrane that lines the kidney tubules is fundamental to the action of most diuretics. An activated process normally pumps sodium ions from the urine into the renal tubules. Until 1957, a strong diuretic effect could be obtained only by injecting an organic mercurial, such as mersalyl. In the kidney, these agents liberated mercuric ions, which combined with a mercapto (—SH) group in a permease active in cation transport (Clarkson, Rothstein and Sutherland, 1965). Currently used diuretics contain no mercury. The most-used one is hydrochlorothiazide (*14.1a*) (de Stevens *et al.*, 1958), a more powerful diuretic than chlorothiazide (*14.1b*) which was discovered a year earlier (Novello and Sprague, 1957). These benzothiadiazide diuretics (familiarly called 'the thiazides') are based on 1,2,4-benzothiadiazine-1,1-dioxide. Their effect is exerted high in the rising distal segment of the kidney tubules, where they prevent the membrane from resorbing sodium ions (and hence, passively, chloride ions). Although they are primary sulfonamides, very little of their action is exerted on carbonic anhydrase, unlike that of acetazolamide (*9.65*). The thiazides increase the excretion of potassium ions also, a side effect causing tiredness in the patient. To prevent this, triamterene (see below) is given simultaneously. The principal use of the thiazides, given orally, is for reducing elevated blood-pressure by sheer loss of water, often the only treatment needed in mild cases of hyperpiesia.

A few years later, the 'high ceiling' diuretics were discovered. The most-used of these, furosemide (frusemide) (*14.2*), elicits a peak diuresis far greater than can be obtained with any thiazide. It is used to reduce oedema (dropsy) (Muschaweck and Hajdú, 1964). It acts by inhibiting resorption of chloride ions (and hence sodium ions, passively) in the ascending renal tubule at a site below where the thiazides act. There is no action on carbonic anhydrase. Ethacrynic acid (a phenoxyacetic acid) acts similarly, but is harder to control.

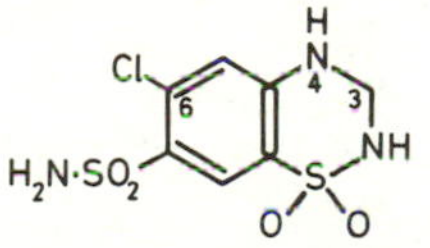

(a) Hydrochlorothiazide (as drawn)
(b) Chlorothiazide (has 3,4-double bond)
(*14.1*)

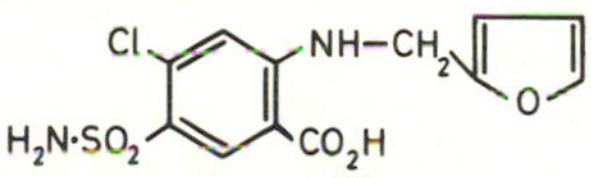

Furosemide
(*14.2*)

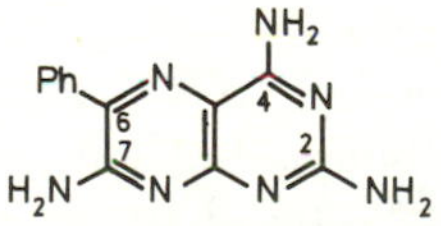

Triamterene
(*14.3*)

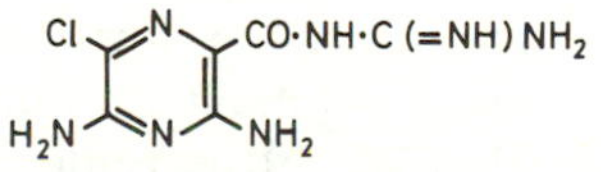

Amiloride
(*14.4*)

Triamterene (*14.3*) is a potassium-retaining diuretic which inhibits both the excretion of potassium and the resorption of sodium, high in the ascending tubule. Contrary to what was thought earlier, its action is not connected with that of aldosterone. Triamterene is a satisfactory diuretic on its own, but is often administered as a corrective alongside the thiazides to counter their potassium-wasting effect. Although a pteridine, it has no anti-folate activity (Wiebelhaus *et al.*, 1965). Another potassium-sparing diuretic is amiloride (*14.4*) which has a pyrazine ring and a basic side-chain.

For further reading on diuretics, see Cragoe (1983).

14.1.4 *Cardiac glycosides*

These highly valued natural products are used to slow the heart while strengthening its beat, properties not available elsewhere in cardiac therapy. The most used sources of these glycosides are two species of *Digitalis* (foxglove) and various species of *Strophanthus* (for ouabain), whereas lily-of-the-valley and squill are no longer used. All the cardiac glycosides have a 19-carbon steroid nucleus (see Section 12.2 for chemistry). The β-hydroxy-group at C-14 ensures the unusual *cis* junction of rings C and D which some think is important for the positive inotropic (muscular-contracting) action which these drugs reinforce in patients with failing hearts. A much-used example, digitoxin, is the glycoside of a steroid aglycone called digitoxigenin (*14.5*), and has one molecule of glucose and three of digitoxose bound in the 3-position. The sugar-free genins, although taken up and distributed faster than the glycosides, are not used therapeutically because they are metabolized too fast, by forming sulfates and glucuronides on the 3-hydroxy group (Thomas, 1981). Apart from this protective effect, the sugar moiety that is near to the steroid nucleus plays a part in binding the drug to the receptor. Activity is increased if this sugar is replaced by an aminoalkyl side chain (Thomas, 1981).

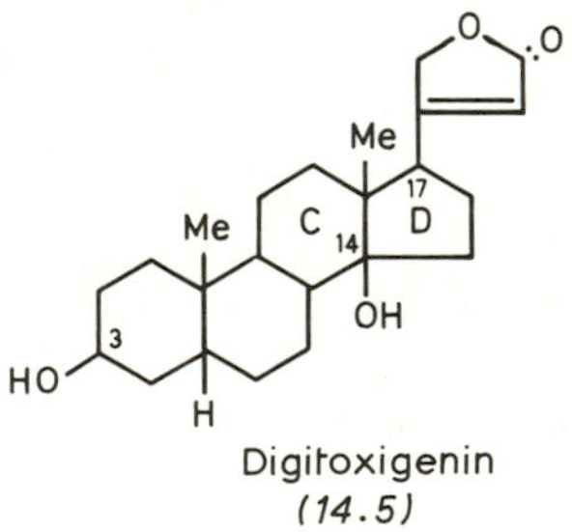

Digitoxigenin
(*14.5*)

After oral administration, some molecules of the glycoside reach the heart where they strongly inhibit the enzyme sodium–potassium adenosine triphosphatase (Bonting, 1970). Either this enzyme is the glycoside receptor, or it undergoes a conformational change by being next to the receptor. When fractions containing this enzyme were extracted from the hearts of various mammals, dosed with tritiated glycosides, it was found that those species (such as Man and dog) which are highly sensitive to cardiac glycosides, showed a high degree of association between the glycoside and the enzyme. Other species (such as rat) with only a low sensitivity to the glycosides, showed little affinity for the analogous enzyme (Schwartz, Matsui and Laughter, 1968). The same authors showed that chemically related but inotropically inactive glycosides did not become bound to the human enzymes.

It is commonly supposed that when this enzyme becomes inhibited, the concentration of sodium ions builds up in the cell, causing extra calcium ions to be liberated (Klaus and Lee, 1969). (The nerve impulse which initiates each heart beat does so by releasing about 50 micromoles of calcium ions per kilogram of heart. This Ca^{2+}, by inhibiting troponin, leads to a contraction of the heart muscle.)

The prominence of the lactone ring in cardiac glycosides prompted the examination of simpler lactones for inotropic activity. Following preliminary work by Chen in 1942, Giarmann (1949) examined 28 simple, non-steroidal lactones and found one with about 5% of the activity of the best cardiac glycosides. However, later work showed that the lactone ring in the glycosides can be substituted by unsaturated ester or nitrile groups without loss of biological activity (Thomas, Boutagy and Gelbart, 1974).

If the lactone ring is to be replaced, an essential requirement is a co-planar arrangement of atoms, as in —CH : CH—CR=A, where A is a heteroatom (O or N), and R may be H, but must not be larger than the methoxy group. Activity is lost if the ethylenic bond is reduced, or if the conjugated system is extended. It is suggested that attachment to the receptor begins with attraction between C-20 (the atom attached to C-17) and the positively micro-charged heteroatom (Thomas, 1981).

Other substances with a digitalis-like action include prednisolone bis-guanylhydrazone and cassaine, one of the alkaloids of *Erythrophleum* (Thorp and Cobbin, 1967). These structures may seem to be chemically unrelated, but their side-chains fit the above receptor model. In each case, the large nucleus may play an important part by 'freezing', in the receptor, a conformational change associated with the cation-pump cycle. In the steroids, this action is associated with the lipophilic α-underface of the ring (Thomas, 1981).

An important question remains: why do cardiac glycosides selectively affect the heart although they are distributed from the bloodstream to the kidneys and liver, organs rich in ATPase?

The margin of safety with all cardiac glycosides is very small, and it is most unlikely that, if they were newly discovered drugs, they would receive a government licence for clinical use. Nevertheless, patients, many of them elderly and frail, have had their lives extended for many years by the intelligent use of these agents, and fatalities are rare. For all that, discovery of new, more-selective types, preferably with a non-steroid nucleus, would be welcomed.

14.2 Ionophores

The lipid bilayer of natural membranes presents a complete barrier to the free diffusion of inorganic ions which, being strongly hydrated, lack lipophilicity. However, ionophores are usually on hand to facilitate this transport. There are two kinds of ionophore, the mobile and the stationary. The latter ('ion pumps') consist of a water-filled channel spanning the bilayer. Much of what we know of ionophores has been learnt from the mobile type, derived from microbes (e.g. valinomycin, gramicidin). Such foreign ionophores, if effective, are toxic to mammalian cells except in low doses.

Although diuretics and cardiac glycosides are concerned with the transfer of ions, they do not appear to be ionophores. Even the sodium–potassium-activated adenosine triphosphatase seems only to be an adjunct. Thus the sodium pump uses this membrane-bound enzyme to remove sodium from the cell and introduce potassium. For this process, the energy comes from hydrolysis of ATP. The enzyme binds and releases the cations at different stages of the pump's cycle. Conformational changes during the cycle cause cations to be released on a side of the membrane *opposite* to the one through which they entered. Part of the mechanism is as follows: the hydrolysis of ATP requires a sodium-dependent phosphorylation of an aspartyl group on the enzyme followed by potassium-dependent transfer, by a phosphatase, of this phosphoryl group from aspartate to water. Some workers think that vanadate anion, which competes with ATP for binding sites, may regulate this enzyme physiologically, at least in some species. A related enzyme, calcium–magnesium-activated adenosine triphosphatase, similarly pumps magnesium into the cell, and calcium out. The nature of mammalian ionophores, of the mobile kind, is little known.

In contrast to the process of chelation (Chapter 11) in which a hydrogen ion on the ligand is replaced by an inorganic cation, transportation by ionophores requires only a stepwise replacement of the water molecules (attached to the co-ordination sphere of the cations) by the ionophore. The energy of desolvation must be compensated by the ionophore's binding energy. First the solvated cation must strike, through thermal agitation, the polar ligand groups of the ionophore, which then enfolds the cation, so that the complex presents a wholly lipophilic exterior to the plasma membrane which then accepts it. The conformational change of enfolding the cation is fast, but has been measured (for valinomycin) by such relaxation techniques as sound-wave absorption, or fast temperature jumps (Grell, Eggers and Funck, 1972).

Several fungal metabolites have been found which enhance the passive uptake of potassium. Some, like nonactin and monactin (*14.6*), are macro-tetrolides; others, like valinomycin, are depsipeptides. Valinomycin (*14.7*) (from *Streptomyces fulvissimus*) is a macrocycle composed of three residues of each of L-valine, D-valine, L-lactic acid, and D-α-hydroxy-isovaleric acid, linked alternatively by ester and amide bonds to form a 36-membered ring (Shemyakin *et al.*, 1963). It is soluble in lipids but insoluble in water. The perimeter of this flat molecule is lipophilic whereas the interior has a large hole surrounded by uncharged hydrophilic groups. The uptake of K$^+$ produces a conformational change which yields a bracelet-shaped molecule.

Valinomycin assists potassium ions to penetrate erythrocytes, mitochondria (Moore and Pressman, 1964), and bacterial plasma membranes (Harold and Baarda, 1967) but has little effect on the permeability of sodium, lithium, or hydrogen ions. Nonactin acts similarly, but monactin also slightly assists the permeability of sodium (Henderson, McGivan and Chappell, 1969). Valinomycin is active against cancer in mice (Carter, Sakurai and Umezawa, 1981).

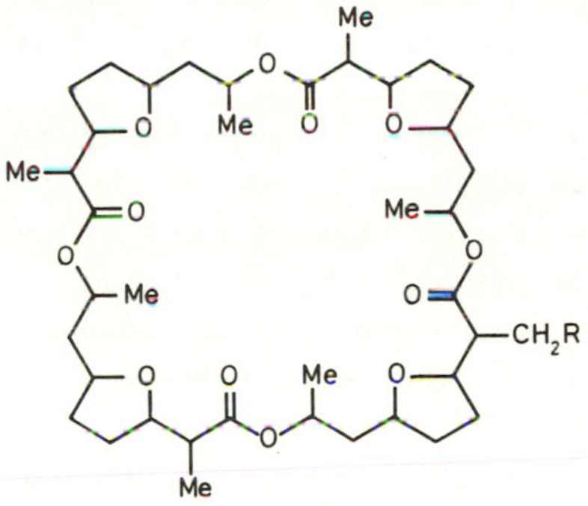

(a) Nonactin (R = H)
(b) Monactin (R = Me)
(14.6)

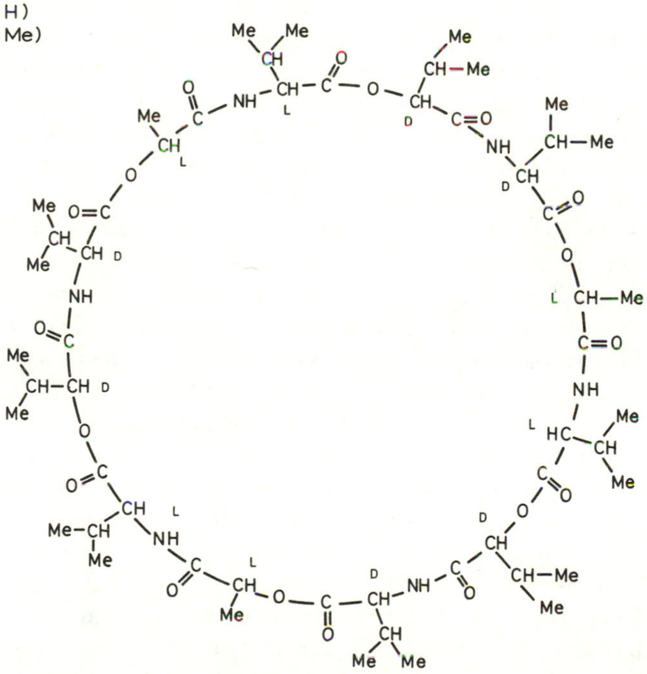

Valinomycin
(14.7)

X-ray-diffraction studies have shown that nonactin co-ordinates a potassium ion by four ether and four carbonyl oxygen atoms (Dobler, Dunitz and Kilbourn, 1969), whereas valinomycin uses six carbonyl oxygens. The previously hydrated potassium ion is thus provided with a totally lipoid exterior which enables it to pass more easily into the cell membrane. In the collisions of this complex with the distal membrane–water boundary, a small percentage of the molecules pass through, and some of the potassium ions manage to exchange the nonactin molecule for water (this is the slow stage of the penetration). It will be realized that an anion (say, Cl^-) must accompany K^+ on these travels, but at a little distance, and constantly undergoing exchange. The preference of nonactin for K^+ (over Na^+) depends on the lower energy of dehydration of the former ion rather than on the dimensions of the central cavity (Prestegard and Chan, 1970).

Of greater interest, because it may indicate how potassium pores are formed and controlled in mammalian physiology, is alamethicin, a linear peptide from the fungus *Trichoderma viride*. It has 20 peptide-linked components which, except for the terminal L-phenylalaninol, are all amino acids: eight of them are α-aminoisobutyric acid residues, and the others are normal protein constituents. X-ray crystal structure analysis shows that the residues facing in one direction are hydrophilic and those of opposite situation are lipophilic.

Model-building indicates that this substance can form aggregates of eight molecules, oriented to form a hydrophilic *pore* of about 4.5 Å in diameter which can be blocked by the glutamine residue (Gln-7) in response to a change in dipole moment. What is known for sure is that alamethacin inserts itself spontaneously into lipid membranes and that its pore, opened by voltage gating, admits the passage of inorganic cations (Fox and Richards, 1982).

Simple and inexpensive synthetic chemicals, known as 'Crowns', show great specificity in transporting alkali (and alkaline earth) metal cations from aqueous to lipophilic solvent layers. One of the most active (*14.8*), is essentially two molecules of 1,2-dihydroxycyclohexane fused with two molecules of di(ethylene glycol). The complexing power of the Crowns is relatively low, even though the cation is gripped by six oxygen atoms (Izatt *et al.*, 1971). The Crown illustrated here prefers potassium to sodium, but one with a slightly smaller ring prefers sodium, and a still smaller one lithium. Conformational changes often accompany their uptake of salts (Pedersen, 1970).

After Pedersen's discovery of the Crowns, in the USA, the Cryptands were developed in France. These are mixed ether–amine macrocycles that do not merely encircle inorganic cations, as Crowns do, but surround them in all directions, leading to tighter binding (Lehn, Sauvage and Dietrich, 1970). Example (*14.9*) is a very strong potassium binder. Because the Crowns and Cryptands tend to be host-toxic, they have found little application so far in biology. Example (*14.8*), for instance, is a skin and eye irritant; higher homologues affect the central nervous system (Pedersen, 1972). In spite of this disadvantage, they can be viewed as models for the larger and costlier naturally occurring ionophores. The more lipophilic the Crown or Cryptand, the more it prefers a univalent over a bivalent cation; thus, sodium is preferred over calcium although both anydrous cations have the same diameter.

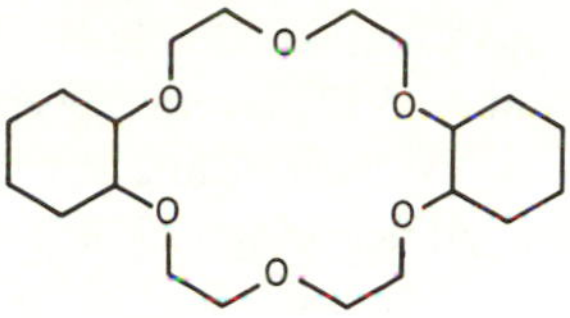

Dicyclohexyl-18-crown-6
(14.8)

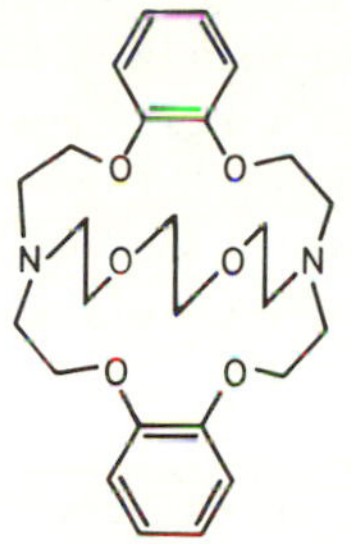

CH₂CH₂OEt

Me CH₂OCH₂CONC₇H₁₅
 C
Me CH₂OCH₂CONC₇H₁₅

CH₂CH₂OEt
Lithium ionophore
(14.10)

Cryptand-2.2.2
(14.9)

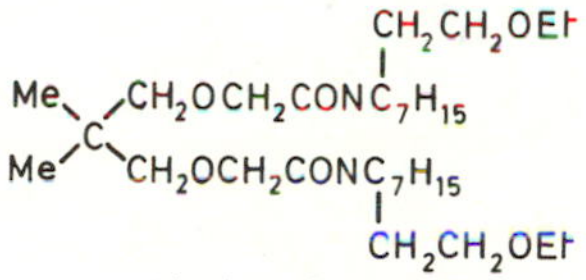

The Crowns have been used as models for the transport of anions across membranes, against a concentration gradient (cf. Type 2 transport, p. 68). The anions transported were N-benzoylated amino acids and small peptides; the membrane was a stirred layer of chloroform bounded on each face by water. Because the Crowns had little solubility in water, they remained mainly in the chloroform. At the first water/chloroform interface, potassium ions were introduced. This led to extraction of the ions into the chloroform layer to form a ternary complex (anion, K^+, ionophore) which released both ions into the post-membrane aqueous phase. The depleted ionophore continued the cycle by extracting more anion and cation from the first aqueous phase, thus starting another round of the cycle (Tsukube, 1982). For anion-complexing Cryptands, see Dietrich et al. (1978).

An acyclic ionophore (14.10) for lithium cations forms a hexadentate lipophilic envelope around the ion in an octahedral arrangement. The selectivity factor for Li^+ over Na^+ was 40, a figure exceeding anything previously attained. Given intravenously, it increased uptake of lithium by the rat brain (Shanzer, Samuel and Korenstein, 1983).

(a) Anionic ionophores. The neutral ionophores, when provided with an inorganic cation, carry a positive charge and hence (like the cationic anti-bacterials in Section 14.3) are likely to injure natural membranes. However, about one hundred natural ionophores have been isolated that have a carboxylic acid group, and some of them will now be described.

Avatec (*14.11*), which is a tetrahydropyranyl tetrahydrofurylheptyl deriva-tive of salicylic acid was isolated from a *Streptomyces*. X-ray diffraction showed that, apart from the salicylic hydroxy-group, every oxygen atom was bound to the cation (barium, in this case). Many derivatives have been made and their partition coefficients and ionization constants recorded (Westley *et al.*, 1973). This substance readily releases calcium ions (and also catecholamines) in biological membranes (Pressman and Guzman, 1974). Its inotropic effect on the heart (Osborne, Wenger and Zanko, 1977) is reminiscent of that evoked by digitalis, but the selectivity seems to be borderline. As 'Lasalocid', it is in regular use as a remedy for the protozoal disease coccidiosis in fowls.

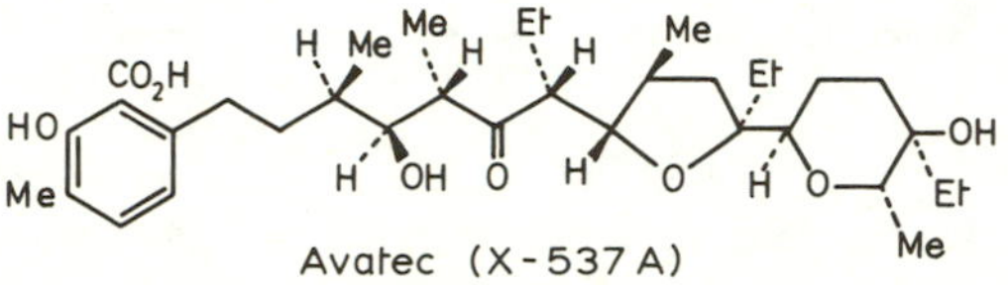

Avatec (X-537A)
(14.11)

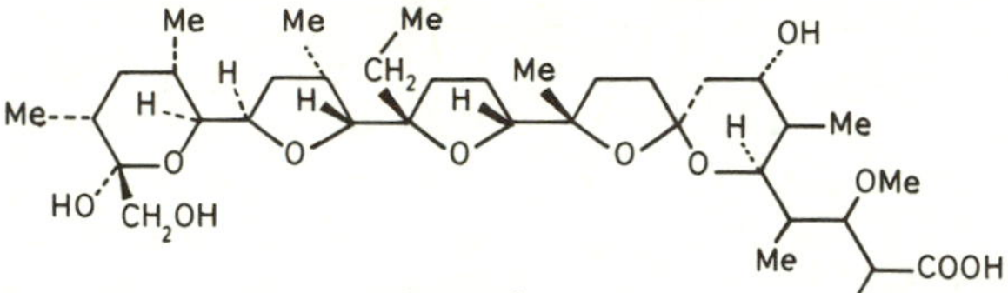

Monensin
(14.12)

Monensin (*14.12*), isolated from *Streptomyces cinnamonensis*, is a fully hydro-genated pyranyl-furyl-furyl-dioxaspirodecyl derivative of butyric acid, mol. wt. 670 (Lutz, Winkler and Dunitz, 1971). The sodium salt is more soluble in hydrocarbons than in water. It is used to promote growth of cattle, acting by changing active transport in anaerobic rumen bacteria, which favours survival of strains which produce nutritive aliphatic acids. It is also used to treat coccidiosis in chicks. For another example, see gramicidin D (p. 605).

Although the microbial ionophores, discussed in this section, are useful tools in the study of ion transport, the highest selectivity in biological processes may be associated with protein types, such as *porin*, which has been extracted from *E. coli* and purified. It forms channels of about 9 Å diameter in the lipid bilayer of these organisms (Chen *et al.*, 1982).

(*b*) *Calcium-channel blockers.* In 1964, Fleckenstein found that verapamil (*14.13*) mimics the effect of calcium-withdrawal in muscle (Fleckenstein, 1977). In the intervening period many other substances, of diversified structure, but all of them antagonizing the functioning of calcium channels, were discovered. Today we can distinguish two series. Type A examples [e.g. verapamil, nifedipine (*14.14*) and other 1,4-dihydropyridines, and diltiazem] are highly selective against Ca^{2+}, whereas Type B examples [e.g. prexylamine, fendiline, and terodiline (which are all diphenylpropylamines)] also block the Na^+ current. In a subdivision of this system, verapamil and diltiazem are equiactive in cardiac and vascular smooth muscle, whereas nifedipine (and the other dihydro-pyridines) are selective for vascular smooth muscle (Towart and Schramm, 1984).

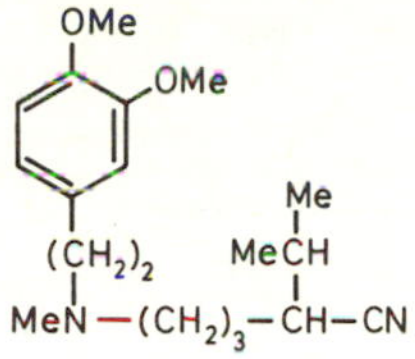

OMe
OMe
Me
(CH₂)₂   MeCH
MeN—(CH₂)₃—CH—CN

Verapamil
(14.13)

NO₂
MeCO₂   CO₂Me
Me   N   Me
H

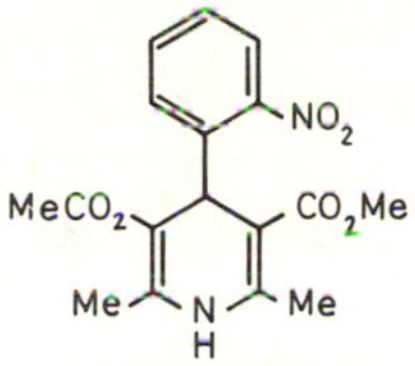

Nifedipine
(14.14)

Nifedipine is used, often with propranolol (p. 515) to lower high blood-pressure. Verapamil, which has more patient-toxicity, is used in treating angina and cardiac arhythmia. Inhibitors of sodium and potassium uptake have been described on p. 595. For further reading on the calcium channel blockers, see Janis and Triggle (1983) (a review) and Rahwan and Witiak (1982) (book). Nifedipine loses its action if there is no *ortho*-substituent in the benzene ring, suggesting that it is important for the pyridine and the benzene rings to be non-coplanar, giving a bulky, stopper-like molecule. What is certain is that no chelation takes place. Replacement, in nifedipine, of the nitro- by a trifluoro-methyl-group converts this blocker to the agonist known as BAY-K 8644 (Towart and Schramm, 1984). For further reading on calcium-channel blockers, see Flaim and Zelis (1982) and Fleckenstein (1983).

For further reading on transport in biomembranes, see Antoloni, Gliozzi and Gorio (1982) and Martonosi (1982).

14.3 The injury of membranes by biologically active agents

Cholesterol, in various proportions, is a natural constituent of mammalian plasma membranes but, if the proportion is allowed to increase, membrane function is usually diminished. Thus the membrane that surrounds the sarco-plasmic reticulum vesicles in muscle progressively loses the calcium-transport-ing function of its ATPase when cholesterol starts replacing the phospholipids. Similarly, ox-heart mitochondria, when exposed to cholesterol, progressively lose the activity of ATPase, succinate dehydrogenase, and β-hydroxybutyrate dehydrogenase (Warren *et al.*, 1975).

The alkylphenols [e.g. (14.15)], the long-chain quaternary amines [e.g. (14.16)], and the polypeptide antibiotics all have the property of cracking bacterial cell walls (Gale and Taylor, 1947). Experiments conducted on proto-plasts in hypertonic media indicate that all three classes of substances dis-integrate cytoplasmic membranes and that the effect on cell walls is secondary

Hexylresorcinol
(14.15)

Water-repelling tail

Cation   Anion

Cetyltrimethylammonium bromide (CTAB)
(14.16)

to this. The disruptive effect of these substances has also been studied by measuring the time taken for the cell contents to be extruded. This is conveniently done by plotting absorbance at 260 nm against time, as in Fig. 14.3, which shows the effect of hexylresorcinol *(14.15)* on *E. coli*. The action is almost complete in 2 minutes. Similar results have been recorded for quaternary amines such as CTAB *(14.16)* (Salton, 1951). The nature of these disintegrations of the membrane seems to be the creation of pores, which are large enough to permit cytochrome *c* to diffuse out.

One of the mildest of these plasma-membrane-injuring drugs is polymyxin B *(14.17)*, which is injected intramuscularly into patients with infections caused

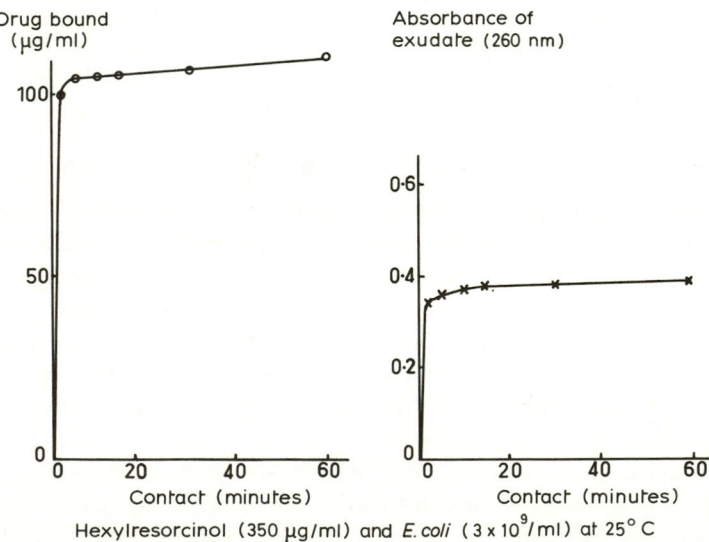

Hexylresorcinol (350 μg/ml) and *E. coli* (3×10^9/ml) at 25°C

Fig. 14.3 The quick rupturing effect of hexylresorcinol on *E. coli* (3×10^9 organisms per ml). (Beckett, Patki and Robinson, 1959.)

by Gram-negative bacteria, particularly *Pseudomonas*, a bacterium unaffected by most other antibacterial agents. It is a cyclic heptapeptide with a tripeptide side-chain *N*-acylated by a lipophilic acid (Suzuki *et al.*, 1964), and is obtained from a strain of *Bacillus polymyxa*. Other polymixins, with two D-amino acids, are particularly kidney-damaging (Wilkinson and Lowe, 1966). Because of the presence of six residues of 2,4-diaminobutyric acid in each molecule, the poly-myxins are quite basic. This property, and the long hydrocarbon chain of the acylating group, give to these antibiotics a structure and amphiphilic properties akin to those of CTAB. The principal use of alkylphenols, quaternary amines, and polypeptide antibiotics is as *topical* antibacterials.

$$
\begin{array}{c}
\text{Dab} \longrightarrow \text{D-Phe} \longrightarrow \text{Thr}\\
\diagup \qquad\qquad\qquad |\\
\text{Moa} \longrightarrow \text{Dab} \longrightarrow \text{Thr} \longrightarrow \text{Dab} \longrightarrow \text{Dab}\\
\diagdown \qquad\qquad\qquad\\
\text{Thr} \longrightarrow \text{Dab} \longrightarrow \text{Dab}
\end{array}
$$

Polymixin B₁

(Moa = 6-methyloctanoic acid, Dab = diaminobutyric acid,
Thr = threonine, D-Phe = D-phenylalanine)

(14.17)

$$
\begin{array}{c}
\text{L-Orn} \longrightarrow \text{L-Leu} \longrightarrow \text{D-Phe}\\
\nearrow \qquad\qquad\qquad \searrow\\
\text{L-Val} \qquad\qquad\qquad \text{L-Pro}\\
\uparrow \qquad\qquad\qquad \downarrow\\
\text{L-Pro} \qquad\qquad\qquad \text{L-Val}\\
\nwarrow \qquad\qquad\qquad \swarrow\\
\text{D-Phe} \longrightarrow \text{L-Leu} \longrightarrow \text{L-Orn}
\end{array}
$$

Gramicidin S
(14.18)

Some other antibiotics derived from bacillary bacteria will now be discussed. A cyclic decapeptide antibiotic, gramicidin S (*14.18*), has two residues of D-phenylalanine as its sole unusual amino acid. X-ray diffraction, combined with n.m.r., has shown that the structure is a ladder of which four N—H ... O bonds form the rungs (Conti, 1969). Gramicidin S is membrane-damaging but does not transport cations. Tyrocidine B has a rather similar structure. Gramicidin D, Dubos's original gramicidin, is a linear polypeptide of 15 amino acids, many of which are D. It injures bacteria by the removal of potassium (as valinomycin does, but more strongly), without injury to the plasma membrane. Gramicidin A, the principal constituent of commercial 'gramicidin' is similar (Haydon and Hladky, 1972). It is used in antibacterial ointments. Pairs of gramicidin D molecules form a helical ion channel that is 5 Å in diameter and 32 Å long.

Although the alkylphenols, the quaternary amines, and the polypeptide antibiotics are surface-active, this property alone cannot cause biological action (Luduena *et al.*, 1955). Ordinary surface-active substances of the three charge-types (cationic, anionic, and neutral) are highly antibacterial, slightly antibacterial, and inert, respectively, when examples of equal surface-activity are compared. Thus *high surface-activity does not of itself cause damage,* but only helps a substance to become concentrated on the bacterial surface. Any damage to bacteria arises from structure-specific loosening of the components of the cell membrane.

The membrane-injury being discussed in this section is essentially the collapse of a mixed film, which has been much studied in model systems. If the adsorbed surface-active substance merely forms a new monolayer *over* the film, nothing more unusual may occur than that separate areas of the film sometimes adhere to one another (the phenomenon of agglutination). To collapse a film, the adsorbed agent must neutralize the film's characteristic charge (e.g. by an adsorbed cation if the film is anionic), or the agent must enter *into* the film, as happens when the agent is more surface-active than the components of the film, or can form stronger bonds with one of the components of a mixed film and hence displace the other component. In the latter way, red blood cells are haemolysed by saponins or phospholipids (Schulman and Rideal, 1937). It is possible for a film to be made stronger by interpenetration, thus avoiding collapse, but normal biological function may no longer be maintained. Further examples of membrane-injuring agents will now be reviewed.

Two much-used agricultural fungicides are cationic surface-active agents: dodine (*n*-dodecylguanidine) and glyodin (2-heptadecyl-2-imidazoline). These rupture the cytoplasmic membranes of some fungi but kill others by merely increasing the permeability of this membrane, after which they enter the cell and destroy *intra*cellular membrane structure (Somers and Pring, 1966).

Chlorohexidine (*10.24*), a much-used local antiseptic, was found to release the cell contents of *E. coli* when these bacteria were placed in a 1:10000 aqueous solution for 5 minutes at 20°C, and most of them died (Rye and Wiseman, 1966). Both tetrachlorosalicylanilide and hexachlorophane released 260 nm-absorbing material (a sign of damage to the cytoplasmic membrane) from bacteria which then died. Both substances are liposoluble skin disinfectants (Woodroffe and Wilkinson, 1966; Joswick, 1961).

Streptomycin, gentamicin, and the other aminoglycoside antibiotics exert their bactericidal action by suppressing protein synthesis in ribosomes. However, because of their highly polar composition, it would not be expected that they could penetrate cell membranes by passive diffusion. Fortuitously, normal electron transport associated with oxidative phosphorylation effectively transports these glycosides into the bacterial cell. After a short time, the rate greatly increases presumably through injury of the plasma membrane by the drug (Anand, Davis and Armitage, 1960; Bryan and Van Den Elzen, 1977).

The discovery of the polyene antibiotics, isolated from species of *Streptomyces*, found in soils about 1950, ushered in a treatment for systemic fungal infections that was biochemically unique (Dutcher, Boyak and Fox, 1954).

These polyene anti-fungal agents are macrolides containing a highly hydroxylated portion in each molecule and from four to seven conjugated double bonds in a lactone ring. According to the number of these double bonds, they are called tetraenes, heptaenes, etc. The conjugated portion is unsubstituted and all-*trans* in configuration. It can be seen from the formula of nystatin (*14.19*) that the ring has a relatively planar lipophilic section and a less rigid hydrophilic section. Various substituents occur in different members of the family: e.g. amino-sugars, carboxyl-groups, epoxides, aliphatic and aromatic side-chains. Polyene antibiotics are lethal to the majority of fungi, protozoa, and algae. Bacteria are quite unaffected, and the higher plants and animals are intermediate in susceptibility. The clinical selectivity of these fungicides rests on the dissimilarity between cholesterol (in all animal membranes) and the plant sterols.

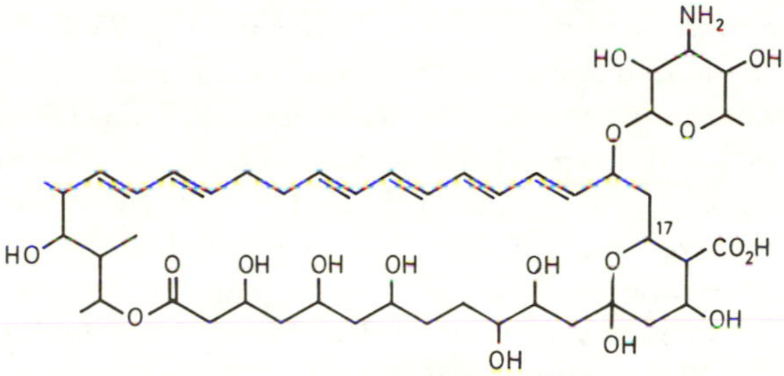

Nystatin
(*14.19*)

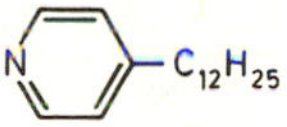

4-Dodecylpyridine
(*14.20*)

These polyenes injure fungi by binding to the ergosterol and sitosterol of the plasma membranes (Hamilton-Miller, 1973). *Mycoplasma laidlawii* was unaffected by filipin when grown in the absence of the sterol, but lysed when grown in its presence (Weber and Kinsky, 1965). When polyene antibiotics were injected under mixed lipids, present as a monolayer in a surface-trough, it was found that the polyene interpenetrated the film and increased its area. Reorientation of the sterol component, caused by interaction with the polyene, seemed to make the film leaky (Demel, Van Deenen and Kinsky, 1965). Mitochondrial and nuclear membranes are not affected by the polyenes, nor is the cell wall (Kinsky, 1962).

A whole spectrum of damage to the cytoplasmic membrane is shown by the various polyenes, from filipin, which produces a severe damage and bursting, to N-acetylcandidin, which produces only a leakage of potassium ions and sorbose, and N-succinylperimycin, which released only potassium (Borowski and Cybulska, 1967). In general, members of lower molecular weight cause the most damage (Cirillo, Harsch and Lampen, 1964). Both nystatin and amphotericin B (*5.14*) produce ion-permeable, water-filled pores of radius about 4 Å in fungal membranes (Dennis, Stead and Andreoli, 1970).

The two most clinically useful members are (a) nystatin* (*14.19*) whose

*Named for New York State Public Health Research Institute.

composition was worked out by Birch *et al.* (1964) and Chong and Rickards (1970), and (b) amphotericin *(5.14)*. The latter's structure, determined by X-ray diffraction (Ganis *et al.*, 1971), is based on a 35-membered heptaene lactone ring, a long thin molecule, with hydrophilic and hydrophobic halves well separated, and ten hydroxy-groups capable of forming hydrogen bonds.

Nystatin is given orally (often as a suspension for it is only feebly water-soluble) for treating candidiasis of the throat, vagina, and rectum. It is not absorbed from the intestine. Amphotericin B is administered intravenously as a colloidal suspension to treat deep-lying, systemic fungal infections of the lung, bone-marrow, and meninges. Its selectivity is not high but, so far, there is no effective substitute.

The miconazole family of anti-fungal agents, which interfere with formation of the fungal plasma membrane, were dealt with in Section 6.3.4 (p. 230).

Attempts are being made to find agents that will rupture the membranes of cancer cells selectively. For example amines, such as 4-dodecylpyridine *(14.20)*, with a pK_a value between 5 and 9, are selectively accumulated by lysosomes because these have an acidic sap, as have cancer cells (p. 151). The amines enter the cells freely, being non-ionized and lipophilic, but are trapped as cations by the acidic sap and cannot escape. Because the cations (but not the neutral species) are detergents, the lysosomal membranes become ruptured, and it is hoped to adapt this phenomenon to anti-cancer therapy. 4-Dodecylpyridine is non-toxic to mice after intraperitoneal injection of 0.5 g/kg (Firestone, Pisano and Bonney, 1979).

Much of the pain and inflammation experienced by sufferers from gout is caused by the contents of lysosomes which rupture when crystals of sodium urate, phagocytosed from the bloodstream by polymorphonuclear leucocytes, weaken the plasma membrane of these cells.

14.4 The preservation of membranes by biologically active agents

Biological membranes are under constant threat of rupture by noxious substances present in the same, or neighbouring, cells. These materials (normally restrained by another membrane) can become liberated as the result of a pathological process, such as an infection, allergy, or auto-immune phenomenon. Of these substances, histamine produces a violent but short-lived reaction, but there are long-acting analogues.

Before discussing what is being done to protect vital membranes, it is useful to note some details of the mechanism by which one cell can damage another. Chief of the offenders are the mast* cell and its circulating analogue: the basophil leucocyte. Each mast cell has up to 5 million receptors for the antibody known as IgE. The mere binding of IgE to the cell does not trigger a release of

---

* *Mast* (from the German for well-fed) *cells* occur near the exterior of tissues; their function is little understood.

noxious substances; what is needed is that a circulating antigen should first join on to the IgE. This union activates dormant channels which pump Ca^{2+} into the cell, a process which releases (*inter alia*) bound histamine. The formation of these channels is facilitated by a lipid, phosphatidylcholine, present in the cell membrane.

The bronchospasm of human asthma is caused by much longer-acting substances than histamine. Formerly one spoke of SRS (slow-reacting substance), but the injury is now known to arise from a family of four leukotrienes, prostaglandin-related substances liberated from the human lung. Most typical is leukotriene D (*14.21*) (Morris *et al.*, 1980).

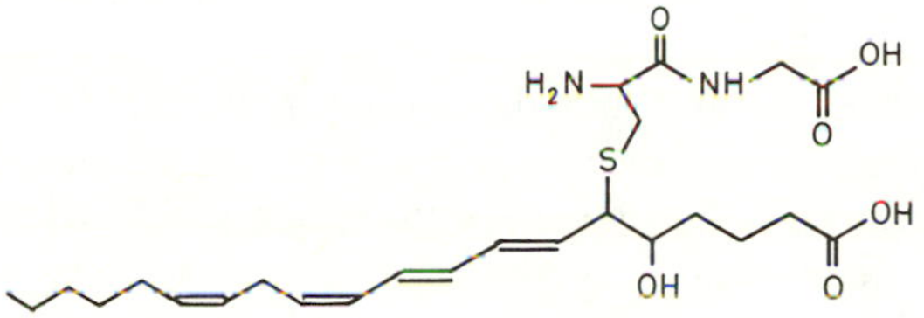

Leukotriene D
(14.21)

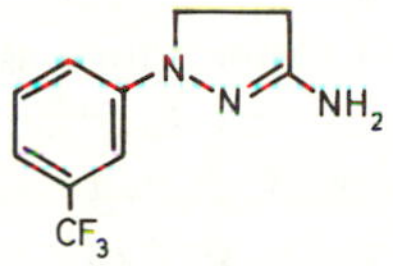

3-Amino-1-(*m*-trifluoromethyl-phenyl)-2-pyrazoline
(14.22)

A potent and selective inhibitor of biosynthesis of the leukotrienes is the 5,6-methano derivative of leukotriene-A_4 ('methano' implies a three-membered carbocyclic ring formed by the bivalent substituent: CH_2) (Nicolaou, Petasis and Seitz, 1981). Another approach is afforded by the pyrazoline drug (*14.22*) known as BW 755C which went into clinical trial in 1980. It inhibits an enzyme which synthesizes the leukotrienes (Higgs, Flower and Vane, 1979).

In the absence of a clinically acceptable anti-leukotriene drug, most cases of asthma have been successfully controlled by daily insufflation of sodium cromoglycate (*14.23*) (cromolyn sodium). This drug, released in England in 1968, prevents rupture of pulmonary mast cells, possibly by interfering with calcium transport (Johnson, 1978). It was discovered, during other work in the chromone series, when one of the investigators, Roger Altounyan, himself an asthmatic subject, experienced prophylaxis after accidentally inhaling the fine dust of a new chromone carboxylic acid (Cox, 1967; Howell and Altounyan, 1967). While this drug inhibits attacks, it cannot reverse them.

The action of cromolyn sodium is highly specialized, for it does not protect the mast cells of *skin*. It is not absorbed when taken by mouth. It has been used with some success in preventing hay fever and as a prophylactic (given orally) against food allergens in the gut.

Many potent analogues of (*14.23*) have since been found with molecules only half as large. They all seem to require an acidic group, a planar structure, and a ring-nitrogen or -oxygen placed diametrically opposite an oxo-group (C:O). Apart from stabilizing mast cells, some also keep the plasma membranes of lymphocytes intact. In most cases, the results of clinical trials are awaited. Potent examples have been found in the 8-azapurine series, such as 2-*o*-prop-

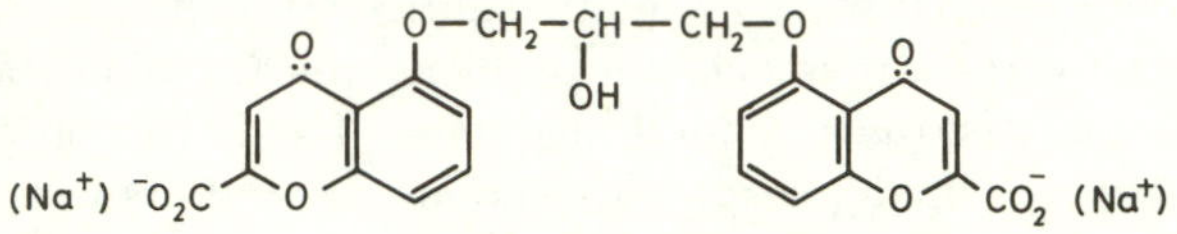

Sodium cromoglycate
(cromolyn sodium)
(14.23)

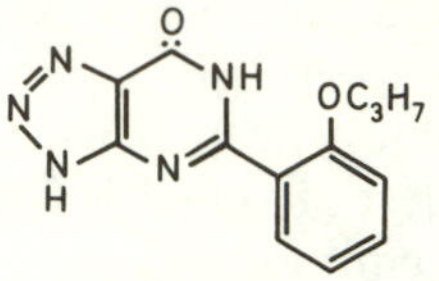

2-o-Propoxyphenyl-8-
azapurin-6-one
(14.24)

oxyphenyl-8-azapurin-6-one (*14.24*), which is 40 times as potent as sodium cromoglycate (Broughton *et al.*, 1975). In this molecule, the propoxy-group exerts sufficient steric hindrance to force the phenyl-group into the plane of the azapurine nucleus, thus extending the flat area.

Many other injurious agents are liberated in pathological conditions, including bradykinin (and other kinins), some of the prostaglandins (although others have a healing function), and even potassium ions in certain situations. Calcium plays an important part in maintaining the integrity of membranes. In dermatology, zinc and aluminium preparations are much used to repair damaged tissues and accelerate healing. The stabilization of membranes by corticoid drugs in dermatology is well established (Weissman and Dingle, 1961).

The anti-epileptic action of phenobarbitone and phenytoin (*15.4*) is usually thought to be a stabilization of a membrane in the central nervous system. The inhibition by sodium valproate of the aggregation of blood-platelets and the anti-arrhythmic action of phenytoin on the heart are seen as a wider manifestation of their membrane-stabilizing powers. To what extent sedation and general anaesthesia in the human are examples of membrane stabilization may be judged by reading Chapter 15.

Progress in treating arthritis and other inflammatory diseases by membrane stabilization seems to be at hand.

15

Biological activity
unrelated to structure

As has been abundantly shown in the foregoing chapters, most biologically active substances are highly sensitive to small changes in structure (see, for example, the introduction to Chapter 2). However, no such strictures exist for one class of substance: the biological depressants. These are substances which depress many cellular functions in most forms of life. In Man they are used as hypnotics and general anaesthetics. In higher concentrations, if volatile, they make useful insecticides, particularly for the fumigation of stored grain. Anti-mitotic effects (mentioned in Section 15.1) are evident at low concentrations and seem to be independent of structure.

15.0 General biological depressants (hypnotics, general anaesthetics, and volatile insecticides)

General biological depressants are substantially non-ionized substances. They may be hydrocarbons (aliphatic or aromatic), chlorinated hydrocarbons, alcohols, ethers, ketones, sulfones, weak acids, weak bases, or aliphatic nitro-compounds. Aldehydes, esters, strong acids, and strong bases usually act differently. Many hypotheses have been put forward to explain the action of depressants, and these will be briefly reviewed here.

The first coherent hypothesis, developed independently by Ernest Overton (1901) and Hans Meyer (1899), had these three principles;

(a) all chemically indifferent substances which are soluble in lipids have depressant properties;

(b) the depressant effect is quickest and strongest in cells which are particularly rich in lipids;

(c) the depressant effect, even in a series of unrelated substances, increases with rising partition coefficient between a lipid and water. The more lipophilic substances have the higher coefficients and the greater depressant action.*

This hypothesis still forms the basis of current thinking, but it is now recognized that the correlation of paragraph (c) is not linear but parabolic (Hansch *et al.*, 1968, Hansch, 1971, pp. 297, 300). This follows from the fact that any substance that is so lipophilic that it has virtually no solubility in water cannot have depressant properties, because it will accumulate in the first oily 'site of loss' which it encounters and be unable to leave it.

The most important use of depressants is in human beings, as hypnotics and (*15.1b*), is widely used. After intravenous injection into the forearm, the patient sensation, in still larger doses muscular relaxation. The introduction of specific muscle relaxants, like suxamethonium (*7.29*), which is short-acting, and tubo-curarine (*2.6*), which is long-acting, made it unnecessary to push the dose of a general anaesthetic so high as was done before 1940. In industrialized countries the most used anaesthetic is *halothane* (introduced, in Britain, about 1955) which is 2-bromo-2-chloro-1,1,1-trifluorethane. Almost as much used, though more recently introduced, is *enflurane* which is 2-chloro-1,1,2-trifluoroethyl difluoromethyl ether. Less used, because of suspected renal toxicity, is *methoxy-flurane*, another halogenated ether. These anaesthetics are more hypnotic than analgesic, and hence are usually administered with about 20% of nitrous oxide, which is far more analgesic than hypnotic. The following anaesthetics, each thought to be outstanding in its day, are no longer used in industrialized countries, yet they retain some advantages for emergency use in the field, or in less-developed countries where the newer (complicated and expensive) anaes-thetic apparatus is not available, nor are the skilled specialists to utilize it: ether (diethyl ether), divinyl ether, ethyl chloride, cyclopropane, ethylene, trichloro-ethylene, and chloroform. Thus, thanks to the muscle-relaxing drugs, anaes-thetics that would have been considered too light, 25 years ago, now dominate the scene. The patient has gained enormously from this advance in selectivity, and the faster recovery time that it permits.

For induction to anaesthesia with halothane, the barbiturate, thiopental (*15.1b*), is widely used. After intravenous injection into the forearm, the patient usually falls asleep within 15 seconds.

Analyses of nerve membranes have brought to light the unusually high ratio of cholesterol to phospholipid, namely about 1 to 3 (Chacko *et al.* 1976). Because the spinal cord and brain are clad in lipid-rich membranes, a lower dose of a depressant is effective for the central nervous system as compared with the musculature. Thus the partition coefficient of halothane is 6.80 for brain (grey matter), but only 2.92 for muscle (tissue/gas, 37°C) (Lowe, 1968). In detail, the site of action of a general anaesthetic in the central nervous system is not exactly known, some workers favouring the polysynaptic sites, others the cytoplasmic membrane. The physical and molecular nature of this site will be discussed later in this chapter.

*'Narkose', in the original German.

Regardless of structure, anaesthetics and hypnotics have a lipophilicity that approximates to $\log P = 2.0$, the property being lost when this figure is raised or lowered (Hansch, 1971, p. 300). Table 15.1 presents a cross-section of relevant data. In addition, many anaesthetic gases have had partition coefficients determined by gas chromatography (Hansch *et al.*, 1975, Leo *et al.*, 1975); agreement was excellent. For general information on partition coefficients, see Sections 3.3 and 17.1.

Table 15.1 Partition coefficients of anaesthetics and hypnotics used in human medicine (See Table 17.3 for coefficients of other substances)

Drug	$\log P$ (octanol/water)
Halothane	1.81
Methoxyflurane	2.21
Chloroform	1.97
Trichloroethylene	2.29
Diethyl ether*	0.77
Dipropyl ether	2.03
Ethchlorvynol	2.00
Barbiturates	
Ethyl, isoamyl (amobarbital)	1.95
Ethyl, 1-methylbutyl (pentobarbital, 'Nembutal')	1.95
Allyl, 1-methylbutyl (secobarbital)	2.15
Ethyl, phenyl (phenobarbital)	1.42
Ethyl, 1-methylbutyl-2-thio (thiopental)†	3.00
Chlordiazepoxide ('Librium')‡	2.44
Diazepam‡	2.82

*Five times weaker an anaesthetic than dipropyl ether.
†Thought to have an additional source of activity.
‡For details of site of action, see p. 537

(Hansch *et al.*, 1968; Glave and Hansch, 1972.)

Kurt Meyer, a son of Hans Meyer, refined and extended his father's work (Meyer and Hemmi, 1935). Table 15.2 shows a selection of these authors' correlations of partition coefficients with biological depression. Because the test-object is a non-mammalian vertebrate, and most of the tested substances are not used in human medicine, these results help to show how widely the principle extends. In Table 15.2, substances of widely different partition coefficients were studied, so that the biologically effective concentration has to be considered. To estimate the concentration of the agent in the cell lipids, results in the first two columns of figures have been multiplied. The products, in the last column, are uniform over a twofold span. Yet earlier results, in which olive oil was used as the lipid, were more variable but led to the same conclusion (see

Table 2.1; of historical interest). The complete Meyer and Hemmi series of 17 substances, from which those in Table 15.2 were selected, have a correlation coefficient (r) of 0.99, a figure that signifies virtually complete agreement (Hansch, 1971). In Section 3.3, it was explained how results obtained with the help of one non-aqueous solvent can be converted for comparison when a different solvent was employed: the same ranking order is usually preserved.

Table 15.2 Correlation of minimal hypnotic dose with partition coefficient

Substance	Partition coefficient (oleyl alcohol/water)	Concentration immobilizing tadpoles, mole/litre (water)	Calculated depressant concentration, mole/litre (cell lipid)
Ethanol	0.10	0.33	0.033
n-Butanol	0.65	0.03	0.020
Valeramide	0.30	0.07	0.021
Benzamide	2.50	0.013	0.033
Salicylamide	5.90	0.0033	0.021
Phenobarbitone	5.90	0.008	0.048
o-Nitroaniline	14.0	0.0025	0.035
Thymol	950.0	4.7×10^{-5}	0.045

(Meyer and Hemmi, 1935.)

(a) *Ferguson's contribution.* In 1939, the correlation of depressant action with lipophilicity was incorporated into a broader (thermodynamic) generalization by Ferguson, who formulated the problem in terms of ΔG, the change that occurs in free energy when a molecule passes from an existing state to an equilibrium. The derivative of G used in this correlation was the partial molar free energy known as the *chemical potential* (μ). Lewis and Randall (1923) had shown that

$$\mu = \mu_0 + RT\ln A$$

where μ_0 is the chemical potential of the standard state and A is the (thermodynamic) *activity*, as opposed to the *concentration*. It follows that the change in μ is proportional to the thermodynamic activity. This led Ferguson (1939) to point out that the chemical potentials of isohypnotic solutions must be the same for all hypnotic substances.

Because few relevant thermodynamic activity values were known, and they are usually difficult to determine, Ferguson employed the following relationship (suitable for volatile compounds only):

Let the thermodynamic activity of a pure substance (i.e. one existing under its own vapour pressure) be taken as unity, then the thermodynamic activity of any concentration of its vapour *in air* is equal to the relative saturation of the vapour. Thus,

$$A = p_t/p_s \qquad \qquad \text{(i)}$$

where p_t is the partial pressure of the vapour in air, and p_s is the saturated vapour pressure of the substance at any given temperature.

Simply stated, any component that is distributed between two phases at equilibrium, must have the same partial pressure in each phase.

A possible weakness in this approach is the assumption of a biological equilibrium. Yet it is remarkable, when a general anaesthetic is being administered to a patient, how rapidly equilibration seems to occur. If the flow of anaesthetic gas is turned down, that patient at once achieves a measure of consciousness, and when the former flow-rate is resumed, the patient is very quickly back on the lower plane of anaesthesia. Actually, because of the complexity of the systems concerned, these are only 'steady states' but, to a first approximation, Ferguson felt justified in treating them as equilibria, in order to make use of simple thermodynamic principles.

Given this assumption of equilibrium, it follows that the thermodynamic potential of the depressant is the same in all phases. This means that one has only to measure the thermodynamic activity in the external phase to learn the thermodynamic activity in the biophase, for both activities must be the same. This is true even though the location and chemical nature of the biophase are unknown (by *biophase* is meant the nerve membrane, or other site, where the depressant exerts its characteristic physiological action). The nature of the biophase is discussed on p. 619.

Equation (i) gives A for volatile substances, applied as *vapour*. The thermodynamic activity *in aqueous solution* is roughly equal to the relative saturation of the solution, provided that the substance is not exceedingly soluble.

Hence the important figure for the correlation of depressants is the *relative saturation* of these substances in the vehicle (air or water) in which they are administered, and not the *concentration*. This principle has been confirmed on a variety of biological test-objects (Ferguson, 1939, 1951; Ferguson and Pirie, 1948; Burtt, 1945).

It has been found that many structurally non-specific agents produce equal biological effects at roughly the same relative saturation. Thus, a substance soluble in water to the extent of 1% will exhibit, in a 0.2% solution, the same degree of depressant activity as a substance soluble to the extent of only 0.1% does in a 0.02% solution. It is evident that all changes in structure that lower water solubility will tend to intensify depressant action in the sense that the new substance will act at a lower concentration than its parent. In ascending a series of homologues, aqueous solubilities decrease from member to member, and the concentrations of these required to produce a specified biological effect rapidly decrease. However, in such a series, the thermodynamic activities necessary to produce a given degree of depressant action are found to increase gradually until a member is reached that is effective only in saturated solution (i.e. a solution that is in equilibrium with undissolved material and hence is at a thermodynamic activity of 1). The next higher homologue inevitably has little or no biological action (Ferguson, 1939, 1951).

Ferguson's principle has found greatest application in understanding the non-structure-specific insecticides. Table 15.3 shows the effect of various hydrocarbons, halogenated paraffins, nitro-hydrocarbons, and weak bases on the wireworm (larva of *Agriotes*). These substances were applied as vapours diluted with air. It is seen from this Table that when equimolar concentrations are compared in column two, a 4000-fold variation in activity appears. However, a much more uniform result is obtained when *proportional saturations* are compared by calculating the least fraction of the saturated vapour pressure required to kill the test-animal. Here (within the much narrower limits of a ninefold variation) a reasonably constant figure (around 0.5) is obtained. This means that the air becomes lethal when it is about half saturated. On the other hand, non-Fergusonian substances such as hydrocyanic acid, carbon bisulfide, ammonia, and strong organic bases give low and non-concordant figures (e.g. ammonia, 0.00008).

Table 15.3 Agreement between the relative saturations of the toxic concentrations of various substances, tested on wireworms

Substance	Toxic concentration, mmoles/litre lethal in 1000 min. at 15°C	p_s (vapour pressure at 15°C, mm)	p_t/p_s (relative saturation of toxic concentration)
Monomethylaniline	3.7	0.22	0.3
Dimethylaniline	6.6	0.28	0.4
Pyridine	76	10.4	0.1
Bromoform	94	3.2	0.5
Tetrachloroethane	141	4.2	0.6
Chlorobenzene	200	6.8	0.5
Toluene	420	17.0	0.4
Nitromethane	710	23	0.6
Benzene	775	58	0.2
Heptane	800	27	0.5
Chloroform	1040	128	0.2
Trichlorethylene	1200	52	0.4
Carbon tetrachloride	1600	73	0.4
Hexane	3000	96	0.6
Dichlorethylene	3100	230	0.2
Pentane	16000	320	0.9

(Ferguson, 1939.)

Were the figure 0.5 a constant for the action of structurally non-specific substances on every kind of cell, it would be idle to expect them to show any *selectivity* in their toxicity. However, this 'relative saturation' varies, for different cells, between 0.01 and 1. For example, it is approximately 0.03 for the production of narcosis in mice and tadpoles and for the inhibition of the development of sea-urchin eggs. A higher value, 0.5, has been found for the minimal concentrations of aniline, phenols, aliphatic alcohols, and ketones required to kill *B. typhosus* in aqueous solution, i.e. approximately half-saturated solutions are required (Ferguson, 1939).

Table 15.4 may be viewed as analogous to Table 15.2, but expressed in Fergusonian terms. In it, six depressants which vary in potency over a 400-fold range can be contained within a twofold range of thermodynamic activities. It dramatically illustrates the difference between concentration and activity as a basis for calculation.

Table 15.4. Suppression of motility of tadpoles. (Liquid phase work)

Substance	Minimal depressant concentration, moles/litre	Minimal depressant thermodynamic activity, $\times 10^2$
Ethanol	0.41	2.8
Acetone	0.28	3.0
Diethyl ether	0.03	4.0
Butanol	0.02	2.1
Chloroform	0.001	1.9
Octanol	0.0001	2.8
	400-fold	2-fold

(Brink and Posternak, 1948.)

Ferguson's principle enables those working with toxic agents to discover if the biological action is structurally specific or not. This is done by finding the 'relative saturation' value of the candidate agent, and comparing it with the values of five or six substances known not to be structurally specific (such as any in Tables 15.1–15.3). If there is no agreement, in the biological test selected, the candidate agent is almost certainly acting by a specific mechanism. In that case, a long programme of work will be required to elicit the mechanism. However, if the test discloses a non-structure-specific mode of action, a much shorter programme should suffice to perfect the discovery.

In the four decades of discussion that followed its introduction, Ferguson's principle has not been significantly enlarged, or even further clarified. Its daring assumption that partial pressures can be substituted for thermodynamic activities has not been falsified; on the other hand, it did not predict the *parabolic* biological response to increasing partition coefficients (Hansch, 1971). At the present time, the incorporation of partition coefficients into a quantitative structure–action programme, statistically tested (Hansch and Fujita, 1964; Hansch, 1971), is more widely used in seeking perfection of a discovery (see Sections 3.3, and Chapter 16). However, these two methods, both of them based on thermodynamics, are not mutually exclusive.

(*b*) *Anaesthetic action of inert gases.* One of the most exciting and challenging aspects of the study of anaesthetics was the early realization that chemically inert gases can be as profound biological depressants as any substance in the preceding Tables (provided the comparison is made in thermodynamic activities). This is well illustrated by the anaesthetic properties of nitrogen. Deep-sea divers, more than a hundred years ago, experienced dizziness and unco-ordination, symptoms which were eventually traced to breathing, under pressure, the nitrogen component of their compressed air (Unsworth, 1861). When several experienced divers were subjected, in a laboratory, to mixtures of oxygen (20%) with either nitrogen, or argon, or both, these mixtures had no effect at atmospheric pressure, but a pronounced hypnotic effect at pressures corresponding to depths of 90–300 feet (Behnke and Yarbrough, 1939). These two chemically inert gases are lipophiles, and the oil/water partition coefficient of argon is even higher than that of ether.

Helium, the lightest of the inert gases, is free from narcotizing properties which increase with molecular weight. Argon (mol. wt. 40) is more narcotizing than nitrogen (mol. wt. 28). When a diver is 200 feet below the surface of the sea, he is working under seven times the pressure to which he is accustomed in ordinary life. If air is used for breathing, the extra nitrogen, more lipophilic than oxygen, decreases the ability to think clearly and to take sensible decisions. To prevent this 'nitrogen narcosis', a helium–oxygen (4:1) mixture is now used and this has been found compatible with clear decision-taking, even at a depth of 1000 feet.

Xenon (mol. wt. 131), at atmospheric pressure, causes temporary loss of reflexes in mice (Lawrence *et al.*, 1946). In this experiment, the gas was effective at a thermodynamic activity of only 0.01 and thus was highly potent. But would it produce anaesthesia in Man? To find out, xenon was administered to patients undergoing operations in the University of Iowa's Hospital (Cullen and Gross, 1951). It was diluted with 20% oxygen, and administered without prior medi-cation (no barbiturates or muscle relaxants). Loss of consciousness was prompt, relaxation of muscle good, and recovery rapid. The degree of anaesthesia was as deep as with ethylene but not quite so deep as with ether. Surgeons performed orchidectomy, and ligation of the Fallopian tubes, and considered xenon a satisfactory light anaesthetic. This result leads to the important generalization that anaesthesia by inert molecules, of which xenon is outstanding for its high potency and low chemical reactivity, takes place *as soon as a constant fraction of some phase of the nerve cell is occupied by foreign molecules.*

(*c*) *Exactly where do hypnotics act?* Structurally non-specific agents seem to act by becoming accumulated in some vitally important part of a cell, thus disorganizing a chain of metabolic processes. In short, they act *simply as foreign bodies*. Obviously they owe the property of being accumulated by cells to a favourable physical property conferred upon them by their chemical constitution. The required physical property depends only on the balance between two common variables in molecular structure, namely hydrophilic atoms and lipophilic atoms. Suitable proportions of these atoms can always be reached by ascending innumerable homologous series of many different chemical types.

What is the nature of this biophase, so sensitive to anaesthetics? The amoeba, a monocellular protozoon, is often taken as the simplest possible model for a nerve cell. Hiller (1927) showed that amoebae were anaesthetized when placed in a dilute solution of a depressant, but not when the depressant was injected into the animal. When the depressant was allowed to diffuse away from the amoebae, their sensitivity was restored to normal.

In mammals, the most sensitive phase is certainly a similar lipoprotein membrane, one that is localized in the central nervous system and which may be a mitochondrial, synaptosomal, or plasma membrane. Accumulation of a depressant in such a membrane could cause swelling. The simple physics of swelling is that lipophilic substances become intercalated between the hydrocarbon side-chains of proteins and possibly lipids in a membrane (see Section 5.4.1). Although some expansion of lipoprotein membranes by inert gases (Clements and Wilson, 1962) and of human erythrocyte membranes by halothane and ether (Seeman and Roth, 1972) have been demonstrated, these amount to only a few parts per thousand (using clinically achieved concentrations of the anaesthetics), a dimensional change that Franks and Lieb (1981) consider too small to form a basis for explaining general anaesthesia. The following evidence points to the depressant's becoming simply adsorbed to the sensitive areas, thus incidentally occluding the openings of ion channels (Franks and Lieb, 1982).

(*d*) *Effects of temperature and pressure.* The potency of an anaesthetic in the gas phase decreases with increasing temperature because heat drives the agent off the target site into the gas phase. This points to simple adsorption as the most likely mode of action of depressants and it is incompatible with (a) the critical volume hypothesis (Miller *et al.*, 1973), (b) the phase-transition hypothesis (Trudell and Hubbell, 1976), and (c) the lipid bilayer fluidity hypothesis (Metcalfe, Seeman and Burgen, 1968), each of which would imply that a rise in temperature should increase anaesthetic potency. The critical volume hypothesis held that anaesthesia occurred when the depressant caused the target site to expand beyond a certain volume, which would lead to the ion pores being squeezed shut. The lipid bilayer hypothesis required an increase in fluidity, but this was not demonstrable in clinically effective concentrations of the depressant. The phase-transition hypothesis failed when it was shown that two equally potent anaesthetics shifted the solid/fluid phase equilibrium of lipid bilayers in *opposite* directions (Pringle, Brown and Miller, 1981).

Although pressure can introduce anaesthesia, *excess* pressure can abolish it. Thus, newts, anaesthetized with nitrogen under 34 atmospheres, were restored to consciousness by increasing the pressure to 140 atmospheres. Similarly newts anaesthetized (without pressure) by ether, halothane, or sodium pentobarbitone, were deanaesthetized by applying hydrostatic pressure outside their bodies (Lever *et al.*, 1971). This effect of excess pressure was at first interpreted as a decrease in critical volume (Miller *et al.*, 1973), but a simpler explanation is preferred: the anaesthetic molecule, lying adsorbed to the target site, is squeezed away from it (Franks and Lieb, 1982).

(*e*) *Disproven hypotheses.* The problem of how depressants act on the target area has, at one time or another, occupied some of the best minds of this century. Traube (1904) held that the action was proportional to the lowering of surface tension at the air/water interface. Warburg (1921) suggested that a hypnotic covers and inhibits oxidizing enzymes, just as a thin film of grease can protect iron from rusting. These hypotheses were effectively refuted by Meyer and Hemmi (1935). Other plausible hypotheses, any of which may still yield some facet of the truth, have just been referred to, and others will now be summarized.

Contrary to what had earlier been supposed, there is no parallel between the anaesthetic doses of a series of depressants and their inhibition of the various enzymes isolated from the respiratory chain (see review by Butler, 1950). The concentrations of depressants required to inhibit respiration in tissues, *in vitro*, has always proved to be much higher than those used to produce surgical anaesthesia in humans. Thus the observed drop in cellular respiration caused by depressants is more likely to be a result of reduced nervous activity rather than a cause of it (McIlwain, 1962).

Linus Pauling and Stanley Miller independently proposed that each molecule of an anaesthetic acts as a stabilizing centre around which water molecules join to form a more compact state of water which produces anaesthesia. It was thought that these denser water molecules may form a cover over nerve cell-membranes, thus hindering the passage of ions which could, in turn, impede depolarization. In support of these hypotheses, the authors claimed that a close correlation existed between the partial pressures required for anaesthesia by various gases and the partial pressures of their *hydrates*, in fact that the former divided by the latter pressure equalled a constant (Pauling, 1961; Miller, 1961). [An exposition of the structure of water was given in Section 3.1 (p. 64).]

Concerning hydrates, although hydrocarbons and the rare gases form true hydrates, these are stable only under several atmospheres of pressure (cf. Stackelberg, 1954). In the Pauling hypothesis, it is assumed that anaesthetics form ice clathrates like propane does. Because they could not be stable at 37°C and atmospheric pressure, a further assumption had to be made: that the clathrates were stabilized by the charged side-chains of protein molecules, leading to a decrease in neural conduction. It will be noted that these hypotheses resemble the others described above in one important way: they connect depressant action with a process of reversible disorganization. However, the water-organization hypotheses have certain weaknesses, which will now be discussed.

To find out for certain if a special structure is induced in water by non-polar groups, Grunwald and Ralph (1967) measured the rate-constant for breaking the hydrogen bond that unites a tertiary amine to water, as in R_3N-HOH. Here, R was a hydrocarbon group whose size was systematically increased. The results showed no evidence of the induction of ice-like structure by the hydrophobic groups. Another difficulty preventing acceptance of the hydrate hypotheses is that the cavity within ice clusters can accommodate only *spherical* molecules and hence, when a hydrocarbon series is ascended, the ability to form hydrates is lost whereas anaesthetic activity increases. Diethyl ether, most typical of the anaesthetics, is a rather flat, butterfly-shaped molecule that forms no crystalline hydrate.

It is important to make sure that any proposed hypothesis applies equally well to the fluorocarbons which are the most used of all general anaesthetics. Actually, fluorinated substances provide an excellent test of the hydration hypothesis because they are noted for the weakness of their intermolecular associations. The tendency of the various anaesthetics to form hydrates was appropriately compared by plotting the reciprocal of the dissociation pressure at 0°C against anaesthetic potency. It was then seen that fluorinated compounds were much less hydrated than other anaesthetics when compared in equi-anaesthetic concentrations (Miller, Paton and Smith, 1965).

In another investigation of the comparative value of lipid and hydration hypotheses (Eger *et al.*, 1969), the minimal anaesthetic concentrations of the following agents were plotted against the relevant physical properties: carbon tetrafluoride, sulfur hexafluoride, nitrous oxide, xenon, cyclopropane, fluorexene (trifluoroethyl vinyl ether), diethyl ether, enflurane (see above), halothane, chloroform, and methoxyflurane (see above). (These anaesthetics have been arranged here in order of increasing lipid/water solubility.) The results of this study showed an excellent correlation between anaesthetic potency and lipophilicity, and little correlation with hydrate formation. Thus, although the m.a.c. (minimal anaesthetic concentration) varied from 0.0023 to 26 atmospheres (a range of 11 000-fold), the lowest product of m.a.c. and the partition coefficient was 1.44 and the highest 2.86, a difference of only two-fold. On the other hand, the ratio of m.a.c. to hydrate dissociation pressures varied from 0.12 to 6.4 (a range of 53-fold), apart from the fact that the following anaesthetics apparently formed *no* hydrate: halothane, ether, methoxyflurane, fluorexene, and enflurane.

The value of reviewing rival hypotheses of the mode of action of general anaesthetics is that even the best of currently accepted explanations, e.g. that of Franks and Lieb (1982), while adequate for hypnotics may be an over-simplification for general anaesthetics. The latter are required to have both hypnotic and analgesic properties, such as was manifested by diethyl ether in its day.

However, current practice is to use halothane (which has a largely hypnotic effect) mixed with nitrous oxide (which has a largely analgesic effect).

(*f*) *Barbiturates*. The barbiturates were introduced in 1903 as hypnotics to replace simple aliphatic amides and sulfones. In the course of time, the therapeutic index of barbiturates has been much improved by making one side-chain of such a nature that it would be degraded in a few hours by the body's detoxifying mechanisms (see Section 3.5). The first barbiturate hypnotic, known as barbitone ('Veronal'), was not susceptible to metabolic oxidation. It was excreted only slowly and this led to over-medication of the patient. Typical of the later, readily-oxidizable types is pentobarbital (*15.1a*) ('Nembutal') which owes its comparative safety to the branch in the side-chain; this assists degradation (a double bond in the side-chain is equally effective).

To secure rapidity of onset, the partition coefficient of barbiturates has been raised by one of the three following devices: (a) increasing the number of saturated carbon atoms in the side-chain, (b) substituting a sulfur for an oxygen atom, as in the intravenous anaesthetic, thiopental (*15.1b*), or (c) by a single *N*-alkylation. Convulsant properties arise if lipophilicity is pushed too high, e.g. if a side-chain is lengthened beyond six carbon atoms or if *both* nitrogen atoms are methylated. Molecules which are more lipophilic still, such as dibenzyl-barbituric acid, are simply inactive.

A decline in prescribing barbiturates as hypnotics stems from tardy recognition that they are insidiously habituating, and that even 12 hours after a single dose, error in judgment, loss of attention, and lapses in driving performance are common. To a large extent, their place has been taken by the benzodiazepines (see Section 12.7), which are subject to similar disadvantages, including dependence, but have a higher LD_{50}. Chronic insomnia benefits more from psychological treatment than from medication with hypnotics.

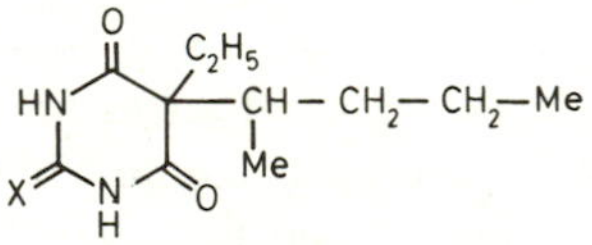

(a) Pentobarbital ($X = O$)
(b) Thiopental ($X = S$)
(*15.1*)

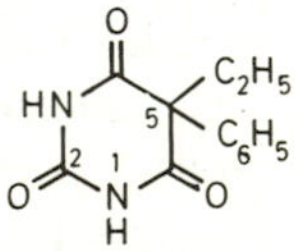

Phenobarbital
(*15.2*)

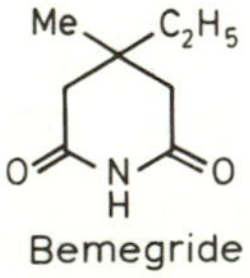

Bemegride
(*15.3*)

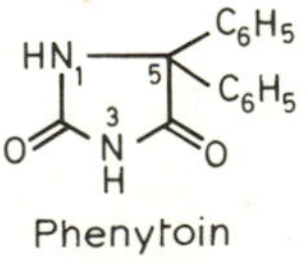

Phenytoin
(*15.4*)

Do barbiturates have a mode of action different from other hypnotics? Although they become adsorbed on the GABA receptor (Section 12.7), their action there seems to be qualitatively different from that of benzodiazepines, and seemingly weaker. In fact, the action of barbiturates corresponds so closely to that of other depressants with non-specific structures (see Table 15.1) that, at clinically effective doses, they may simply act by a Franks and Lieb (1982) mechanism (p. 619). It is evident, though, that dependence, and withdrawal symptoms, are connected with the GABA system (Ho and Harris, 1981).

The small amount of ionization of which the hypnotic barbiturates are capable does not contribute to their action (Mautner and Clemson, 1970; Hansch and Anderson, 1967).

When the free prescribing of barbiturates began to be denounced, manufacturers were quick to market similarly substituted piperidines (i.e. rings with one nitrogen atom instead of two) such as methyprylone ('Noludar') and glutethimide ('Doriden'), without any notable gain in selectivity (Mautner and Clemson, 1970; Hansch *et al.*, 1968).

Bemegride ('Megimide'; β,β-methylethylglutarimide) (*15.3*), a strong antagonist of barbiturates, has proved useful in cases of poisoning. The structure is somewhat reminiscent of that of the barbiturates, but its analeptic action is not specifically directed against barbiturates. In Section 7.5.2, it was pointed out that increasing the size of the alkyl-group converts these stimulants into hypnotics (Laycock and Shulman, 1967), whereas the contrary is true for barbiturates. Conformational patterns responsible for convulsive effects of barbiturates have been demonstrated by computer-graphic pattern analysis (Andrews *et al.*, 1983).

(*g*) *Anti-epileptic drugs*. Phenobarbitone (*15.2*), in addition to its hypnotic properties, is a powerful anti-epileptic drug. The phenyl-group is essential for this action, which is available without the hypnotic effect in phenytoin (*15.4*) (5,5-diphenylhydantoin), and mephenytoin (*3.32*). Anti-epileptic drugs act by specifically curbing the high-voltage, high-frequency brain distribution of grand mal (Greengard and McIlwain, 1955). How they begin to do so is indicated by the specific inhibitory action of phenytoin on the resting sodium permeability in the central nervous system (Perry, McKinney and De Weer, 1978). Sodium valproate (dipropylacetate) acts differently, apparently inhibiting an enzyme that degrades GABA (Bruni and Wilder, 1979).

(*h*) *Cannabis indica*. Δ^9-Tetrahydrocannabinol (THC) (*15.5*), the principal active constituent of marihuana, has both likenesses and disparities when compared to general biological depressants. Its high partition-coefficient (about 8000 in the octanol/water system) seems responsible for the slowness of its kinetics and its persistence in the body. It keeps subjects in a mildly narcotized state for hours without being able to effect a higher degree of anaesthesia. This result is reminiscent of the cut-off effect found when the molecular weight of aliphatic alcohols is steadily increased (Section 2.5.1 and Fig. 2.3) (Paton, 1975).

The 3'-hydroxy- and 3',11-dihydroxy-derivatives of THC, which arise during metabolism, conserve many of the pharmacological properties of the parent substance (Handrick et al., 1982). Because alternative numbering systems exist, it should be noted that, here, 3' refers to the side-chain and 11 to the methyl-group of (15.5).

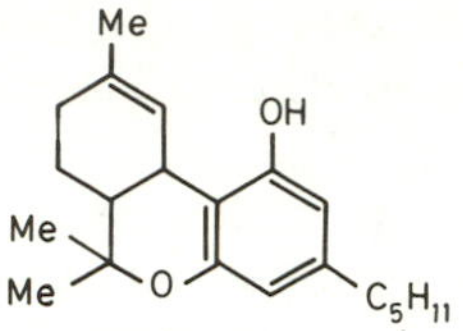

Tetrahydrocannabinol
(15.5)

In low dosage, the sedative effects of smoking marihuana are similar to those elicited by barbiturates, but a more prominent euphoria coupled to uncontrollable flow of thoughts and disturbed perceptions of time and space are evident. The effects of one cigarette last from 2 to 3 hours, during which time the subject is too poorly co-ordinated to be able to drive a car safely (judgment impaired and reaction-time slowed). Ethanol enhances this impairment of performance. In higher dosage, marihuana produces confusion of past with present, anxiety, and hallucinations (Thacore and Shulka, 1976).

15.1 Mitotic disorganizers

Many simple biological depressants have an adverse effect on mitosis. They allow the chromosomes to divide, but the daughter chromosomes remain side by side instead of moving to two opposite spindle-poles. Thus the spindle is disorganized and the chromosomes do not proceed beyond metaphase. Among the large number of inert substances that act in this way are: hexane and other aliphatic hydrocarbons, alcohols, ether, chloroform, acetone, paraldehyde, acetamide, urethane, acetophenone, benzophenone, acetanilide, benzene, chlorobenzene, and sulphonal (Östergren, 1951); also nitrogen and argon under pressure (Ferguson, Hawkins and Doxey, 1950).

This effect, which appears to be a straightforward case of biological depression, is readily reversed when the concentration of the agent falls. Colchicine, a mitotic disorganizer of another kind (Section 5.4.7) gives results at a very low thermodynamic activity and is sensitive to small changes in molecular structure. Again, substances which cause drastic and irreversible changes such as chromosome breakage (e.g. phenols, coumarins, and mustard gas) have physical properties quite different from those of the structurally indifferent depressants.

16

The perfection of a discovery

Let us suppose that a substance has been discovered with a rare and much desired biological property. It may have been found by application of the principles of selectivity, or from a chance laboratory observation, or by a tedious mechanical screening programme. Usually it is neither potent nor selective enough to be thought of as a candidate drug, heading towards clinical trial, nevertheless it can be a very promising lead.* What steps can be taken to improve it, by finding the most selective example among its many chemical relatives? Moreover, how can this goal be reached by making the fewest possible examples?

Some useful preliminaries can be extracted from earlier chapters. For example, tests can determine whether the ion or the molecule is the biologically active species (Section 10.7). Whether it can chelate with a metal cation may be suggested by a glance at the structure (Section 11.2) followed by a quick pH test (Section 11.3). Whether the molecule is structurally non-specific can be found by comparing the minimal hypnotic activity with that of an established example such as chloroform, or by similar comparison of partition coefficients (Section 15.0).

From this point onwards, the paths diverge. One pathway, as used in this book so far, is to pursue the scientific study of the cytological, biochemical, and distributive clues that provided the lead in the first place. The other pathway is the statistical correlation of several physical properties of the lead molecule with its biological actions. It must be emphasized that statistics lies remote from experimental science and can, at best, indicate only a probability. However, these statistical methods are much used, particularly in industry, and it is proposed to give a brief account of correlation analysis (quantitative structure–activity relationships, or QSAR) which embodies them.

*In this use, 'lead' rhymes with 'seed' and not with 'said' (unlike Pb).

16.0 Multiple regression analysis

The most popular of these approaches is *multiple regression analysis*, introduced by
Hansch in 1968. In this method, the most favoured variables are (i) partition
coefficients (P), from the system octanol/water; (ii) the sigma (σ) and rho (ρ)
values derived from Hammett's Linear Free Energy Equation (Section 17.2);
and (iii) Taft's steric factors (E_s) which are used to determine what space for
substituents exists between a nominated small region of the lead molecule and
its receptor (Section 16.2). These variables are correlated in the following
equation:

$$\log \mathrm{I}/C = -k(\log P)^2 + k_2(\log P) + k_3\sigma\rho + k_4 E_s + k_5$$

In essence this is essentially the regression equation given in Section 3.3 (p. 80),
but with two extra terms: $k_3\sigma\rho$ and $k_4 E_s$. We should now proceed to discuss, in
turn, the three principal descriptors of the equation, namely P, $\sigma\rho$, and E_s.

C (the lowest biologically effective concentration) and P (the partition coefficient) were discussed in Section 3.3 which may usefully be re-read at this point.
The required value of P may already have been experimentally determined and
the review by Leo *et al.* (1971), which contains several thousand such values,
should be consulted. If the value is not yet known, it should be determined
experimentally as outlined in Section 17.1. If a hasty answer is required, or if
experimental resources are lacking, a value of P can be constructed from the
fragmental constants listed in Table 17.2, but this can only yield an approximation. Before having recourse to such calculations, the reader is advised to
become familiar with the circumstances in which they are not applicable, or
need much modification (Rekker, 1977).

Hammett's sigma and rho values (also Taft's sigma values for aliphatic
compounds) are discussed in Section 17.2. The most frequently needed values,
experimentally determined, are in Table 17.4 (p. 647) and a great many others
will be found in Appendix A.1 of Perrin, Dempsey and Serjeant's book (1981),
in which will also be found Appendix A.6, a useful list of sigma constants for
heteroatoms in heterocyclic rings (note also the help available on p. 87 for
obtaining a rho constant when the nucleus is not benzene). Sigma and rho
constants are also listed by Hansch and Leo (1979).

When applying sigma and rho values to regression equations, one is at liberty
to insert the sigma value of a single substituent, or of any combination of
substituents (by addition of sigma values), until the computer yields a high r
(correlation coefficient) for one of the many values of k_3 that it is programmed to
generate.

The principal defect of sigma values is that they do not differ outstandingly
from one another. This is because they are located along a scale that is only 1.8
logarithmic units from one end to the other. Moreover, little reliance can be put
on the second place of decimals, because sigma values are obtained by sub-

tracting two logarithms of which one or both of the two numbers is likely to be insecure in its second decimal place (Albert and Serjeant, 1984). In spite of these shortcomings, sigma values present the most often-used method for introducing the electronic properties of the molecule into the regression equation, which would otherwise be unrepresentative of a very important property of medicinal substances.

An alternative approach, apparently not yet attempted, is to insert the ionization constant directly into the regression equation. Apart from avoiding the error that can arise in the Hammett process of subtracting two similar quantities, this modification is more representative of electron distribution in the whole molecule, although proceeding from only a portion of it. Better still, in examples likely to be partly ionized at the biological pH value, the percentage ionized could usefully be entered as a more exact descriptor. For non-ionizing substances, it may be possible to use the dipole moment, particularly as a set of local dipoles which may be calculated from electron-density diagrams as obtained, say, from Pariser–Parr calculations (Kier, 1971).

E_s has been, so far, the most used of the steric descriptors. These are intended to introduce, into the regression equation, those properties of (a) the gross bulk and (b) the regional details of shape, which give a drug its access to, and (for important portions of a molecule) its affinity for, the receptor. E_s is actually Taft's steric parameter which he extracted from experimental work of an entirely different character. This was his study of the reaction equilibria of molecules that carried a bulky group adjacent to a chemically (but non-biologically) reactive centre. Hansch introduced this descriptor into his equation and has used it often; as an alternative, when it did not work, he has employed molecular refractivity. Both of these descriptors represent bulk as a single numeral: the Taft factor applying to a single substituent (Taft, 1956), whereas molar refractivity can be applied either to a substituent or (by summation) to the whole molecule (Vogel, 1948; Hansch et al., 1973).

Occasionally one or other of these descriptors has given a satisfactory correlation but, in general, the situation is too complex for either to suffice. In fact, it takes little imagination to see that not one solitary numeral but a whole series of them may be necessary to define those properties which make a molecule acceptable to a receptor. Chemists have tended to ignore this final (steric) term in the Hansch equation by the simple expedient (conscious or otherwise) of keeping it constant, namely by working on a set of compounds whose steric properties are allowed to vary only minimally. Yet, one has to realize, if only from the huge difference in biological effect that R and S stereoisomers exert, that steric terms must often be the most important of all. Yet they remain the most elusive! At the present time, many attempts are being made to discover better approaches to quantifying the steric component, and these are discussed in Section 16.2.

The degree of hydrogen-bonding has been proposed as an extra term in the regression equation (Fujita, Nishioka and Nakajima, 1977), and other candidate terms are listed by Hansch and Leo (1979). However none of these is in regular use.

(*a*) *Planning for validity.* Because correlation analysis is so expensive of computer time, it is essential that the raw data should be sufficient, both in kind and in quantity, to produce results that can yield valid conclusions. The first consideration is to have a statistically significant number of compounds in both the *exploring* and the *confirming* sets. Each set needs at least five compounds for each descriptor included in the regression equation. Thus, if one has decided to keep the steric features of the molecules fairly constant, ten compounds would be the minimum for each set, namely five for the partitioning descriptor and five for the Hammett constants. Altogether, then, one needs twenty distinct, although chemically and structurally related, examples, and they should be equipped with substituents that cover a wide range of values for the two selected descriptors. Next the *P* and Hammett symbols are replaced by the appropriate numerals, and the data for 10 examples inserted into a least-square computer program, to yield values for the coefficients (*k*).

The results are then expressed in terms of *n*, *s* and *r*. Here *n* is 10, *s* is the standard deviation for the regression, and *r* is the correlation coefficient, which should ideally be 1.00, but is acceptable if 0.95 or over. The second set of 10 compounds is now provided with the coefficients (*k*) obtained by this exercise (no computer-usage enters into this half). If a similarly low *s* and high *n* are obtained, the task is concluded, because one knows which descriptors are significant in the structure–action relationships of this set of compounds, and in which proportions they exert this significance (Topliss and Costello, 1972; Hansch, Unger and Forsythe, 1973). One can now proceed to a bolder version of the project in which much wider variations (including steric ones) are allowed between members of a set. However, if the confirming set does not furnish *s* and *r* values of as high quality as did the exploring set, it is a case of 'back to the drawing board' for replanning the project.

In choosing the first compounds to submit to multiple regression analysis, tremendous help can be derived from Craig's sigma-versus-pi scatter diagram (Fig. 16.1). Representative compounds must be drawn from *each* of the four quadrants to discover the σ/π region of maximal activity.

In deciding which descriptors to include in a regression equation, some preliminary inspection is practised. If there are two examples in the set that have widely differing sigma values but approximately the same degree of biological activity ($1/C$), it would be tempting to exclude sigma values and hence be free to work with a smaller set of examples. Such reasoning may be deceptive, as in the following case. A series of phenols with pK_a values ranging from 11 (weak acids) to 5 (stronger acids) was being tested for toxicity to fish in an environmental study of the pollution of lakes. It was found that toxicity was roughly proportional to log *P*, and incorporating sigma values did not smooth the results to give good *s* and *r* values. Reverting to benchwork, careful study

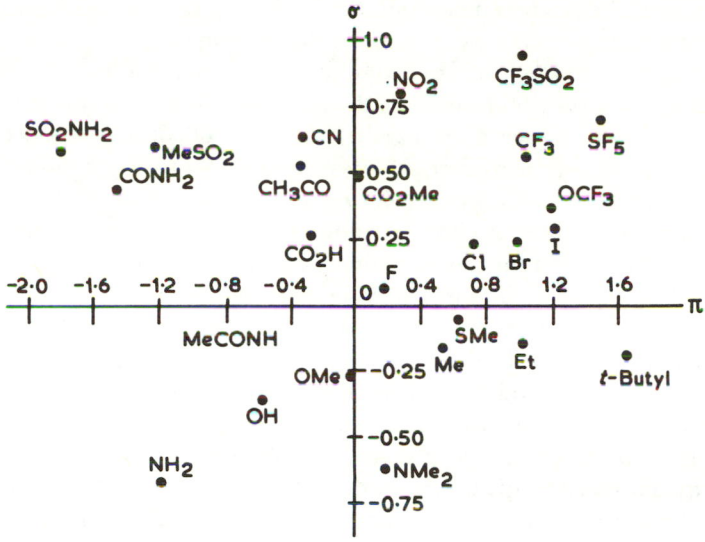

Fig. 16.1 A cluster assembly, made by combining Hammett σ and Hansch π values of some commonly used para-substituents. (Craig, 1971; redrawn by Redl, Cramer and Berkoff, 1974.)

showed that, as the pK_a fell, and the phenol consequently became more ionized at the biological pH, so it became more toxic in any test where penetration of a biological membrane was not required. However, the ion had little tendency to pass into a non-aqueous phase, so that in studies with intact fish, the two effects of ionization *nearly* cancelled (Saarikoski and Viluksela, 1982).

(*b*) *Processing the results.* To obtain the standard deviation (*s*), each difference (between a result and the arithmetic mean of all results) is squared. The products are added, and the sum is divided by the number of examples taken. The square root of this quotient is the standard deviation. In a physical–chemical experiment, ±0.02 would be a common range for a standard deviation, small because the nature of the calculation cancels outlying readings. However, because it is the nature of biological results to vary rather widely, much larger deviations are tolerated in their regressions (±0.15 is common). The interest of the standard deviation is that it shows the limits within which the results vary about the mean. It is also used in the calculation of *r*, the correlation coefficient. If *r* is 0.9, it is implied that 81% of the variation in *x* is attributable to changes in *y*; similarly when *r* is 0.95, the attribution rises to 90%. For more on statistics of this kind, see Crossland (1981).

(*c*) *Working without a computer.* Those who prefer to work without computers or statistics may derive great help from the 'Decision Tree' of Topliss (1972). He begins with the case where the lead-molecule is dominated by a benzene ring. In this case, one should first make the p-chloro-derivative and find if it is less or more active than the parent lead. If the potency is increased, the gain can be attributed to $+\pi$, $+\sigma$, or both together (for signs of substituents, see Tables 17.2 and 17.3). If the chloro-derivative is equipotent with the parent, this should be interpreted as a small favourable π-effect (almost all drugs show a small biologically-favourable π-effect) counterbalanced by an unfavourable σ-relationship. To test this interim conclusion, the 4-methyl-derivative should be made. This ought to show increased potency because it is $+\pi$ and $-\sigma$. If the 4-Me-derivative is less potent than the 4-chloro analogue, an unfavourable steric effect may be assumed for *para*-substitution. Hence analogues should be made with chlorine, and methyl, in the 3-position. Finally if the 4-chloro-derivative is significantly less potent than the parent, the most likely cause is control by $-\sigma$, because $-\pi$ is rare. This should be tested by making the 4-dimethylamino-derivative, where the effects of $-\sigma$ would become more evident. Topliss goes on to suggest how a side-chain could be similarly manipulated, starting with steady increases in π, while keeping steric effects low by preferring, e.g. a cyclobutyl substituent to open-chain groups. For a further development along these lines, see Topliss (1977).

(*d*) *Some thoughts on regression analysis.* The Baconian revolution of 300 years ago has taught us that an appeal to experiment yields far more useful knowledge than the pursuit of truth by argument (Bacon, 1620). Statistics is a branch not of science but of mathematics and, while its role in servicing medical science is much valued, it is at best a poor substitute for benchwork. Multiple regression analysis is at its best when it is not practised in isolation but is used to persuade research into new, not clearly foreseen, directions. Hence its most valuable achievement may turn out to be the exposure of areas of neglect. In one example, a good lead turned up, a quinoline substituted in the heterocyclic ring. Chemists were asked to submit, say, 20 examples with alternative substituents, and when they had done this they felt satisfied that they had submitted a representative cross-section. In this, they were self-deceived. Inspection of the σ values, long before any regression analysis was begun, showed that one end of the Hammett σ-scale was unrepresented. This happened because any standard synthetic reaction, that seemingly can make an endless variety of products, is always partial to producing examples up one horn of the Hammett scale, and (for good Hammett-related reasons) neglecting the other. The chemists therefore had to find a quite different reaction to complete their project, and as a favourable reaction did not exist, they were obliged to discover one. Appreciation of this concept has caused no small revolution in the thinking of the chemical staff of drug-seeking teams.

It is both healthy and reassuring that multiple regression analysts view one another's work with a highly critical eye (e.g. Cammarata *et al.*, 1970; Martin,

1970). The following quotations from the Pomona School are worth keeping in mind. 'Space limitation has not allowed a discussion of many instances where no combination of constants has yielded even moderate correlations' (Hansch, 1971). After cautioning against overreliance on statistical criteria to the neglect of common sense, Unger and Hansch (1973) add: 'Without such a qualitative perspective, one is apt to generate statistical unicorns, beasts that exist on paper, but not in reality. For example, it has recently become all too clear that one can correlate a set of dependent variables using random numbers as independent variables'. In this connexion, see also Topliss and Costello (1972).

For further reading on the free energy relationships discussed above, see Chapman and Shorter (1972), and also the reading list at the end of this chapter.

16.1 Alternative methods

(a) *The Free and Wilson additive method.* This is a refinement of an old hypothesis that biological activity is an additive property to which every substituent in the molecule makes a mathematically predetermined contribution (Free and Wilson, 1964). As Hansch (1971) remarked, this approach is completely unencumbered by any attempt to apply thermodynamics or physical chemistry. Instead, a computer is used to determine substituent constants by employing the method of least-squares as a data-fitting device to find 'best numbers' for each substituent in a drug when furnished with biological data from a standard laboratory test. From these constants, estimates are made of the degree of biological activity of *combinations* of substituents. The claimed advantage of the Free−Wilson method is that it can be used when no physical data are available. However, in like circumstances, the terms in a multiple regression analysis equation can be taken from tables. A disadvantage of the Free−Wilson method is that it requires many examples bearing a given substituent in order to obtain a valid solution to equations with very many unknowns. For that reason, the method is usually confined to development of those drugs or pesticides where large numbers of closely related analogues are available. For further reading on this method, see James (1974) and Purcell, Bass and Clayton, 1973. Little mathematical involvement is required.

(b) *Molecular orbital calculations.* Since the early 1950s, when the Pullmans introduced quantum mechanics for the study of carcinogenicity in polycyclic hydrocarbons, several attempts have been made to derive QSAR results by calculations made along similar lines. The principal obstacles to advancement are (a) the high cost of computer time and (b) the abstract nature of the resulting descriptors which practising chemists and biologists have difficulty in picturing. Moreover, the large number of parameters generated for each compound necessitate availability of a large number of analogues to yield valid correlations. So far, the most satisfactory use of this method seems to be the correlation of *conformation* with biological activity as, for example, in developing new agonists and antagonists related to the natural neurotransmitters (Weintraub and Hopfinger, 1973). The information yielded by molecular orbitals is mainly electronic, but some attempts to direct it into the area of steric properties will be referred to in the next Section (16.2). There can be no doubt that quantum chemistry will play an important part in explaining and predicting drug action, but at present there is a confusing array of methods in use which makes difficult any comparison between results from various workers.

For further reading, see Richards (1983), Kier (1971) and Purcell, Bass and Clayton (1973).

(*c*) *Molecular connectivity.* Petroleum chemists have long used calculations about branching in molecules to predict such physical properties as boiling-point, viscosity, and heat of formation, qualities which vary greatly between isomers but which were found to depend on the degree of compactness of the molecule. In the course of time, this approach was quantified in the form of matrices (Randić, 1974), and Kier applied it to biology (Kier and Hall, 1976). The method has the advantage that its calculations can be made quickly, on a hand-held calculator, and that no physical property has to be measured. It can be used alone for lipophiles, which furnish much the same information as Hansch's log P, as shown in studies of general anaesthetics and the narcosis of tadpoles. For amphiphiles, it may be used as a rough measure of the bulkiness of a molecule, in which case it can be combined with log P (experimentally determined) and with Hammett coefficients, in a multiple regression equation. However, as a steric term it has the grave disadvantage of not being able to reflect fine stereochemical alternatives. Thus it does not register any difference between *cis* and *trans* isomers, between R and S optical antipodes, or between conformers, all properties on which specific drug actions often strongly depend. It also has the disadvantage of requiring many compounds in a set, because of the large number of indicator variables* which it introduces, one of them for each atom (other than hydrogen) in the molecule.

A molecular connectivity calculation is begun by drawing the structural formula as bonds only, omitting hydrogen. The atom at the end of each bond, or at the junction of bonds, is called a vertex, and it is assigned a value equal to the number of non-hydrogen atoms to which it is connected. Hence, for carbon atoms, the possible numbers are 1,2,3, or 4, and they are called 'vertex valences'. Next the valences situated at the two ends of each bond are multiplied to give what is termed the 'vertex product' or 'edge-term', denoted as v. The square root of each edge-term is then converted to its reciprocal. The sum of the reciprocals gives chi (χ), the *index of connectivity* for that compound. That of the heptane called trimethylbutane is $\chi 2.943$, made up of $2 \times 0.577 + 3 \times 0.5 + 0.289$. Normal heptane has $\chi 3.414$. The difference between these two values is a

---

*Unlike a descriptor, which may have a whole range of values, an *indicator variable* is a quality that is either present or absent.

measure of the 'piled up' character of the more branched molecule.

To apply their method to most kinds of drugs, Kier and Hall have had to devise vertex valences for the heteroatoms nitrogen, oxygen, and sulfur, not considered in earlier work. These values were found sometimes empirically, sometimes with the help of a regression program. Because the bond angles of heteroatoms are wider than those of tetrahedral carbon, the vertex valences are larger. Difficulties were encountered in heteroaromatic molecules containing more than one heteroatom.

16.2 Steric considerations

To express, numerically, the volume and shape of a molecule (either the whole compound or its individual regions) is a goal still far from achievement, but eagerly sought by correlation analysts. With molecular models, busily hand-manipulated, it is easy to visualize the fit of an agonist on to what is known of the architecture of its receptor, and the stereochemical picture is instantly compre-hensible. At a much more sophisticated level, this quick understanding is conveyed by the Evans and Sutherland colour picture system (Section 12.0, p. 492). By such visual methods it is possible to model, plausibly, the touching of two complementary regions, one on the drug molecule and one on the receptor, regions that are complementary for lipophilicity, for electric charge, and/or for exact steric fit. In this rough way, visual control is exercised over the folding and fitting, which may involve many atoms, each one to a different degree. Likely conformational changes, too, can be explored in both drug and receptor.

These insights, incomplete as they must be, are more difficult to attain mathematically, and often hard to visualize when expressed numerically as the coefficients of unfamiliar descriptors. A remedy which the bench chemist can take is to synthesize analogues of the most active drug, each analogue equipped with a bulky substituent inserted in a different place. The biological tests then give information about the space available around each position in the mol-ecule. Many have attempted this approach, but it is tedious. Hence the frequent endeavours to find short cuts to the goal by calculation. A selection of these methods will now be reviewed.

Some of the methods are concerned with the overall molecular shape, which seems to be all that is required for some types of biological results. For instance, Holan found that analogues of both DDT and pyrethrin were effective insecti-cides so long as they preserved the *wedge shape* (and size) of the original molecules (wedges, like minute crowbars, to prize open the insects' sodium channels) (Section 7.6.5, p. 305). It seems, though, that most agents depend more on the shape of selected region(s) of the molecule. For them, methods have been devised that extract data for molecular *portions*, and try to determine which region is specific (and *why* it is significant) for the biological activity. Both approaches are fuelled by a strong desire to optimize the activity of known agents.

16.2.1 *Approaches directed mainly to overall dimensions*

The chlorinated cyclodiene insecticides act in a biologically different way from the DDT—pyrethrin group of insecticides and have different steric requirements. Soloway (1965) examined 106 cyclodienes and concluded that, for high activity, they should be nearly spherical. Proceeding to detail, he projected a model of aldrin on to a plane and obtained the outline of a highly modulated circle. Molecules which could just traverse through an aperture of these dimensions, see (*6.47*), scored well as insecticides. In addition, the presence of two electron-attracting substituents (usually chlorine atoms) was necessary.

More useful than these general ideas of shape would be the calculation of *van der Waals volumes* of a molecule, namely the volume it occupies up to the ends of its van der Waals radii (Section 8.0.4). This is usually attempted indirectly through molar refractivity (see below). Attempts have been made to find these volumes by covering Corey—Pauling—Koltun models with adhesive cellulose tape, sinking them in water, and reading the displacement (Leo, Hansch and Jow, 1976), but several pitfalls have been pointed out (Pearlman, 1980). The computation of van der Waals volumes of molecules, from geometrical data, is beset with difficulties (such as occlusion of parts of the atoms, depicted as spheres), even when using the recommended method of summed lunes*. Because molecular volumes are linearly related to molecular surface areas, which can be somewhat more easily computed, Pearlman (1980) recommends that the latter be used in correlation analysis.

An undeservedly neglected approach to molecular volume is embodied in the *'similarity coefficients'* of Amoore (1964), which were designed to correlate odour with molecular structure. Briefly, he photographed molecular models from three directions, all at right-angles to one another. On these photos, he inscribed radii at 10° intervals and then integrated the 108 measurements. This gave reasonably good correlations in spite of the inbuilt imprecision of Amoore's biological test which was an odour-judging panel of human beings. Because of its simplicity, Amoore's method deserves some trials on pharmacological problems, with more objective testing.

Balaban's group in Romania devised three methods for relating a drug to its receptor from consideration of volume relationships. The methods are known briefly as MSD, MTD, and MCD, arranged in order of increasing complexity.

To use MSD (*minimal steric difference*), one must select the most active drug for the biological effect under study. The structural formula is then drawn, as bonds only, neglecting all hydrogen atoms. All candidate drugs of the same chemical type are drawn similarly, on transparent material, and superimposed on the original drawing. Next, the non-superposable atoms are marked with an asterisk, disregarding any difference of flatness or elevation (e.g. no distinction

---

*Lune, the segment of a sphere corresponding roughly to one of the 'quarters' of an orange. Pearlman computed high numbers of these, per atom, reducing the volume where a neighbouring atom interfered (van der Waals repulsions are partly overcome *within* molecules).

is made between benzene and cyclohexane). Each asterisked atom is awarded a volume penalty (or 'misfit number') of 1, if C, N or O; but larger atoms forfeit more, e.g. 1.5 for S and Cl, and 2 for I. Although differences in bond lengths and bond angles are to be neglected, one is cautioned not to superimpose a bonded pair of atoms upon a non-bonded pair, because covalent bonds (usually 1 to 2 Å) are so much shorter than van der Waals contact distances (3–4 Å).

Balaban's thesis is that the ·affinity of a drug for its receptor is a *linearly decreasing* function of the sum of non-overlapping volumes of a candidate drug and the receptor cavity. The most active drug is considered to reflect the general shape of the receptor cavity. However, if a substituent in the most active drug is not repeatedly covered by a substituent in the candidate drugs, it should be removed because (Balaban calculates) 0.85 kcal of energy has to be spent to deform the conformation of the receptor to accommodate it. Such an operation should produce a more active drug. This (and the other two methods) are explained more fully in the book (Balaban *et al.*, 1980). MSD is most suitable for those large molecules of which only a small part is to be changed. The decapeptide known as luteinizing-hormone-releasing hormone became 2.48 times as active when the terminal glycinamide was replaced by methylamine. The methylamine attracted no MSD penalty, but glycinamide was awarded three, and this suggested other ruses to improve the biological effect of the natural hormone (Fujino *et al.*, 1973, for synthesis and biological assay; Balaban *et al.*, 1980, for MSD analysis). For smaller molecules, MSD does not perform as well as molecular refraction or Verloop's method (see below). For more on MSD, see Simon *et al.*, 1984.

Proceeding to Balaban's second method, MTD (*minimal topological difference*), which succeeds best with cyclic and other rigid molecules, the most active of known drugs is again selected and the structure drawn as bonds, omitting all hydrogen atoms. However, in this method, bond angles and steric features are utilized. Formulae of the candidate drugs are then superimposed, and the heights of atoms above and below a median plane are recorded. The aim is to depict a complex 'hypermolecule' which is intended to reveal the three-dimensional shape of the receptor. In these superpositions, an endeavour is made to place a polar group over a polar group, and a hydrogen-bondable atom over a similar one. Scoring is by awarding 1 to those atoms judged to touch the receptor's cavity wall; those lying within the receptor cavity score 0, and those outside the cavity score -1. These allocations are made by inspection, or by a computerized program printed in Balaban's book. The sum of $+1$ values correlates well with biological activity, and has been applied to estrogens (Popoviciu, 1978). Correlation was excellent ($r = 0.971$) for a series of 25 steroids bearing oxygen in both the 3 and 17 positions, but this sank to 0.741 when 25 other steroids were tested, and non-steroid estrogens fared only moderately. The best features of MTD are that it can point to substituents or regions that lie outside the presumed receptor cavity, and which may profitably be eliminated, and that it can report unused receptor space and hence suggest new shapes for candidate drugs which might make better use of it. MTD works particularly badly with flexible (aliphatic) molecules, where MSD is preferable.

Balaban's third method MCD (*Monte-Carlo Design*) is designed to calculate the van der Waals volume of a molecule in cubic Å. This begins by programming a computer to furnish a large set of random numbers, which are used, in conjunction with a notated formula of a drug molecule, to compute a very large number of its radii and, from them, the volume. Difficult to use though this method may be, it has furnished sufficient results to support the use of molar refractivities as measures of van der Waals volumes (see below).

16.2.2 *Approaches directed mainly to specific regions*

When studying equilibria in the hydrolysis of esters, and the steric hindrance of a neighbouring group on this process, Taft (1953) extracted his well-known *steric parameters* (E_s). Although he put them to no biological use, Hansch, about 1965, began to insert them into multiple regression equations as a measure of the steric hindrance that a substituent might exert on the adsorption of a drug's most active group on to the receptor. For these values, see Taft (1953) and Hansch and Leo (1979). E_s values may range from 0 for methyl, through -0.9 for phenyl and -1.5 for *t*-butyl, to -2.1 for trichloromethyl, and still larger values are possible.

Several Taft values can be used in one regression equation, as was done by Kutter and Hansch (1969) who studied the preventive effect of substituted benzoic acids on the combination of ovalbumin antigen with its antibody, each benzoic acid being incorporated as a hapten in the antigen. They obtained this multiple regression equation:

$$\log K = 0.863 E_s^o + 0.081 E_s^m + 0.446 E_s^p - 0.695$$
$$n = 22 \qquad s = 0.177 \qquad r = 0.974$$

In this equation, there is a term for each of the three positions of substitution, and it is evident that the *ortho* and *para* positions are the most significant. Little benefit was obtained by adding a log P or a Hammett term, yet the index of correlation (r) is excellent.

Molar refractivity, in spite of its name, has to do with single substituents, although their values can be used together to create an image of the whole molecule. The preferred method for calculating 'MR', as it is abbreviated, is the use of the Lorenz and Lorentz equation in a form which yields cubic centimetres per mole:

$$MR = MW(n^2 - 1)/d(n^2 + 2)$$

where d is the specific gravity of the liquid phase and n is the refractive index. Many values, variously derived, are listed by Vogel (1948), and more are in Hansch and Leo's book (1979). Hansch introduced MR in 1973, as an alternative to E_s, for the steric term in multiple regression analysis (he sounded a note of warning that MS also has an electronic component, although the steric component predominates) (Hansch *et al.*, 1973).

In *Verloop's 'sterimol' method*, the structural formula of each group in the drug molecule (hydrogen included) is, in turn, converted to single-line notation which is fed into a computer, atom by atom. The data on bond length, bond angles, and van der Waals radii are added. This enables computation of the van der Waals envelope around the submitted group. The program is then commanded to provide a numerical value (L) for the *length* of the group, starting from point of attachment, and measured at right angles to the long axis, values for the *minimum breadth* (B_1) and the *maximum breadth* (B_4). The Dutch authors also provided tables of (L) and (B) values for all common groups and for many sequences of groups. For example, the —CH_2CN group has the sterimol values: 3.99 (L), 1.52, 1.90, 1.90, and 4.12 (B_{1-4}) (Verloop, Hoogenstraaten and Tipker, 1976).

No conformational information is incorporated in this method except that the authors always programmed the lowest-energy (most extended) form of the molecule, leaving its further manipulation to the receptor. Returning to Kutter and Hansch's analysis of the effect of benzoic acids in preventing antigen–antibody combination (see above), Verloop and colleagues (1976) replaced this with the following analysis:

$$\log K = -1.46B_1^o + 0.50B_4^p + 0.84$$
$$n = 36 \qquad s = 0.302 \qquad r = 0.955$$

It is noteworthy that (L) was found irrelevant here and that they were able to include many substances which Kutter and Hansch had had to discard as 'outliers'.

Turning to a series of larvicides, the Dutch workers obtained the best correlation by combining partition and Hammett-sigma values with two 'sterimol' terms: $(L)^2$ and (B_4). In subsequent work, the Verloop method was subtly refined to increase predictive powers; a new table of substituent values was issued, and the arithmetical correspondence between (B_1) and E_s was noted (Verloop, 1981, 1983).

In using the Verloop method, it is desirable to synthesize a series of analogues where the maximal (also minimal) width occurs at various points along the long axis of the molecule.

A new approach to finding steric terms for regression equations is provided by *Hopfinger's molecular shape analysis*. His principal descriptor is the *overlapping steric volume* (V_o), which is the fraction of molecular volume shared by a candidate drug and its most biologically active conformer (Hopfinger, 1980). Values of V_o are obtained by elaborate calculations in which an approach by molecular mechanics (using both rotational and translational components) is preferred over molecular orbital methods.

When, as is usual, the molecules have flexibility, the CHEMLAB programme (available through CIS, see Section 17.4) is used to vary the torsional rotational angle(s), at 30° intervals, until one or more conformation of minimal energy is located. These energies (expressed in kcal/mol) are plotted, as a function of the angle, to locate major potential wells. The pattern of this plot for the most biologically active members gives a standard for ranking the other members, by energy differences. Where a member has more than one stable conformer, each is treated as an individual entry in the pairwise comparisons. The information for each member is combined, in matrices, with bond-lengths (van der Waals) and bond-angles to obtain a print-out of the shared molecular volumes. Although this method reports on whole molecules, it uses regional configuration data.

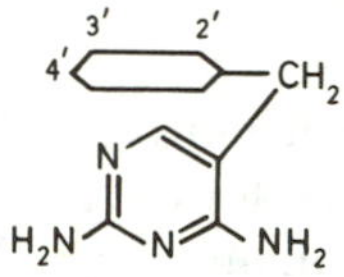

2,4-Diamino-5-benzylpyrimidine
(16.1)

Hopfinger's method is illustrated by his treatment of a series of 2,4-diamino-5-benzylpyrimidines (16.1), of which the most outstanding member is the antibacterial drug, trimethoprim (9.32), discussed on p. 352. The test, inhibition of bovine liver DHFR (dihydrofolate reductase)*, furnished the following equation:

$$\log 1/C = -21.3V_0 + 2.39(V_0)^2 + 0.44\pi_{3,4} + 52.2$$
$$n = 23 \qquad s = 0.1 \qquad r = 0.93$$

In those members which are highly active against E. coli DHFR, the benzene ring of (16.1) makes an angle of 30° with the methylene bridge which makes an angle of 90° with the pyrimidine ring (as decided by the above conformational analysis). Hence the benzene ring, in these compounds, is roughly perpendicular to the pyrimidine ring which it almost covers. There was no preferred conformation detected for members which were principally active against the mammalian enzyme.

In the above equation, π is related to log P (see p. 81), and the subscripts 3 and 4 indicate that the π values for the 3- and 4-positions were summed. Attempted inclusion of Hammett's sigma gave no help. Hopfinger's results almost exactly duplicate those of Blaney et al. (1979) who derived the equation:

$$\log 1/C = 0.62\pi_3 + 0.32\sigma_{3,4,5} + 4.99$$
$$n = 23 \qquad s = 0.15 \qquad r = 0.93$$

*This is a test for toxicity to patient, because trimethoprim has little effect on *mammalian* DHFR (see Table 4.2).

The state of the art of regression analysis is illuminated by these two regression equations, which give correlations of equal excellence: Hopfinger's by incorporating all possible steric information and the Californians' by the careful exclusion of this property.

Hopfinger also examined a set of derivatives of 2,4-diamino-6-dimethyl-1-phenyl-1,3,5-triazine, of which the 4-chloro-derivative, cycloguanil (*3.35*), is an outstanding anti-malarial drug. Here, C refers to 50% inhibition of DHF reductase from a mammalian tumour. The 35 examples covered a 10 000-fold spread in inhibition. Better correlations were obtained with S_o than with V_o, the former being defined as $V_o^{2/3}$ which is a term with the same dimensions as, but quite different from, surface area (the author warns us) and hence difficult to visualize (Hopfinger, 1980). Nevertheless, it gave good correlation in the following equation:

$$\log 1/C = -1.47S_o + 0.02(S_o)^2 + 0.38\pi_{3,4} - 17.1$$
$$n = 25 \qquad s = 0.40 \qquad r = 0.96$$

This should be compared with the Californians' results for a larger series incorporating all the above compounds:

$$\log 1/C = 0.89\pi_3 - 0.13(\pi_3)^2 + 0.15\text{MR}_4 + 6.62$$
$$n = 65 \qquad s = 0.33 \qquad r = 0.91$$

Both of these interpretations are confined to the steric and partitioning descriptors.

Another series of DHFR inhibitors was investigated by molecular shape analysis: the 2,4-diaminoquinazolines (Battershell, Malhotra and Hopfinger, 1981). This gave the QSAR equation:

$$\log 1/C = 0.35S_o - 0.002(S_o)^2 + 0.49\pi_{5,6} - 0.09\theta - 6.95$$
$$n = 35 \qquad s = 0.36 \qquad r = 0.97$$

Here, θ is the torsional angle formed by a 5 or 6 phenyl group, if present.

Hopfinger has, more recently, published a new method of molecular-shape analysis based on *molecular-potential energy fields* (Hopfinger, 1983). These are computed as the sum of van der Waals attractions *and* repulsions, and the electrostatic interactions between electric charges at fixed positions. As before, no account is taken of the receptor except that the most active drug is taken roughly to mirror the receptor in its structure. The descriptor is obtained by subtracting the fields of pairs of superposed drug molecules. This method was applied to the above-described set of 2,4-diamino-5-benzylpyrimidines, with similar results provided that a $\pi_{3,4}$ term was included.

16.2.3 *Concluding remarks on steric factors*

In this important area, more work is in progress than can be described here, but at least the highlights have been displayed. Much remains to be done. Meanwhile, as purified receptors become available in sufficient quantity, more of the effort may be diverted to experimental work, particularly X-ray analysis, as is happening with the receptors for acetylcholine (p. 530) and the drugs that inhibit dihydrofolate reductase (p. 350).

For further reading on the strategy and tactics of perfecting a lead by quantitative structure–activity relationships (correlation analysis) of drugs, see Topliss (1983), Hahn (1975), and Martin (1978, 1979). For more information on the physicochemical properties of drugs and how to estimate them, see Hansch and Leo (1979), Yalkowsky, Sinkula and Valvoni (1980), and Lyman, Rechl and Rosenblatt (1982).

17

Some numerical assistance

17.0 Table: Calculation of percentage ionized, given pK$_a$ and pH

The first table ever to list the values for percentage ionized, as calculated from pH and pK_a, appeared in the first edition of *Selective Toxicity* (1951), and is reproduced below. For acids it is calculated as follows:

$$\% \text{ Ionized} = \frac{100}{1 + 10^{(pK_a - pH)}}$$

whereas for bases, the following equation is used:

$$\% \text{ Ionized} = \frac{100}{1 + 10^{(pH - pK_a)}}$$

17.1 Table and discussion: Fragmental constants and partition coefficients

This section supplements the discussion on partition coefficients in Section 3.3.

Partition coefficients of solids and liquids are conveniently measured in a 250 ml glass-stoppered centrifuge tube. This is charged with the weighed solute and such proportions of the two solvents as are suggested by previous runs or analogous solutes. The tube is inverted 100 times in 5 minutes, although equilibrium will probably occur within the first 2 minutes. The mixture is then centrifuged, allowed to settle, and inspected for traces of emulsification (if this has occurred, a fresh start must be made). Either phase can now be analysed for its content of solute (Leo, Hansch and Elkins, 1971).

Partition coefficients of gases are measured by bubbling the gas through a mixture of the two solvents, then withdrawing a sample of the organic layer and

Table 17.1 Calculations of percentage ionized, given pK_a and pH

pK_a − pH	*If anion*	*If cation*
−6.0	99.99990	0.0000999
−5.0	99.99900	0.0009999
−4.0	99.9900	0.0009990
−3.5	99.968	0.0316
−3.4	99.960	0.0398
−3.3	99.950	0.0501
−3.2	99.937	0.0630
−3.1	99.921	0.0794
−3.0	99.90	0.09991
−2.9	99.87	0.1257
−2.8	99.84	0.1582
−2.7	99.80	0.1991
−2.6	99.75	0.2505
−2.5	99.68	0.3152
−2.4	99.60	0.3966
−2.3	99.50	0.4987
−2.2	99.37	0.6270
−2.1	99.21	0.7879
−2.0	99.01	0.990
−1.9	98.76	1.243
−1.8	98.44	1.560
−1.7	98.04	1.956
−1.6	97.55	2.450
−1.5	96.93	3.07
−1.4	96.17	3.83
−1.3	95.23	4.77
−1.2	94.07	5.93
−1.1	92.64	7.36
−1.0	90.91	9.09
−0.9	88.81	11.19
−0.8	86.30	13.70
−0.7	83.37	16.63
−0.6	79.93	20.07
−0.5	75.97	24.03
−0.4	71.53	28.47
−0.3	66.61	33.39
−0.2	61.32	38.68
−0.1	55.73	44.27

analysing it by gas−liquid partition chromatography (Leo *et al.*, 1975; Hansch *et al.*, 1975).

For a table of 5800 partition coefficients, see Leo, Hansch and Elkins (1971). See Table 17.3 for a representative selection of these values.

Table 17.2 presents a useful selection of fragmental constants (**f**) derived from octanol/water partition coefficients (*P*). Other values can be constructed from the table. These **f** values, which replace the former π values, tend to be

$pK_a - pH$	If anion	If cation
0	50.00	50.00
+0.1	44.27	55.73
+0.2	36.68	61.32
+0.3	33.39	66.61
+0.4	28.47	71.53
+0.5	24.03	75.97
+0.6	20.07	79.93
+0.7	16.63	83.37
+0.8	13.70	86.30
+0.9	11.19	88.81
+1.0	9.09	90.91
+1.1	7.36	92.64
+1.2	5.93	94.07
+1.3	4.77	95.23
+1.4	3.83	96.17
+1.5	3.07	96.93
+1.6	2.450	97.55
+1.7	1.956	98.04
+1.8	1.560	94.44
+1.9	1.243	98.76
+2.0	0.990	99.01
+2.1	0.7879	99.21
+2.2	0.6270	99.37
+2.3	0.4987	99.50
+2.4	0.3966	99.60
+2.5	0.3152	99.68
+2.6	0.2505	99.75
+2.7	0.1991	99.80
+2.8	0.1582	99.84
+2.9	0.1257	99.87
+3.0	0.09991	99.90
+3.1	0.0794	99.921
+3.2	0.0630	99.937
+3.3	0.0501	99.950
+3.4	0.0398	99.960
+3.5	0.0316	99.968
+4.0	0.0099990	99.9900
+5.0	0.0009999	99.99900
+6.0	0.0000999	99.99990

about 0.2 higher but not consistently so. They are selected from 100 values based on 1000 log P measurements.

The lipophilic effect of hydrocarbon fragments is evident from the table. When a carbon–carbon double bond is present, it makes a definite (-0.55) contribution to hydrophilicity whereas a C—C triple bond makes a much larger one (-1.42). The aliphatic equivalent to phenyl, as a lipophilic substituent, lies between ethyl and propyl, whereas naphthyl approximates to

Table 17.2 Some common fragmental constants

A. No carbon (hydrogen absent)	Aliphatic	Aromatic		C. With carbon (hydrogen absent)	Aliphatic	Aromatic
—I	0.56	1.43		—C—	0.18	
—Br	0.26	1.21		—CN	−1.05	0.19
—Cl	0.06	0.93		—CON <	−2.88	−2.02
—F	−0.49	0.38		—CO—	−1.67	−0.81
—N <	−2.07	−0.91		—CO₂—	−1.27	−0.41
—NO₂	−0.93	−0.07		—CO₂(ion)	−5.00	−4.14
—O—	−1.59	−0.43		—SCN	−0.34	
—SO₂N		−2.39		—CCl₃	2.09	
—SO—	−2.75	−2.05		—CF₃	0.75	1.33
—SO₂—		−1.87		—O·C(:O)—	−0.41	−0.99
—S—	−0.51	0.11				

B. No carbon (hydrogen present)				D. With carbon (hydrogen present)		
—H	0.17	0.17		—CH₃*	0.69	
—NH—	−1.82	−0.95		—CH₂—	0.53	
—OH	−1.47	−0.31		—CH <	0.33	
—NH₂	−1.42	−0.85		—CH=CH₂	0.91	
—SH	0.00	0.62		—C₆H₅		1.84
—SO₂NH₂	−1.48	−1.94		—C₆H₄		1.66
				—CONH	−2.43	−1.56
E. Heterocyclic				—O·CONH—	−1.92	−1.34
Imidazolyl-	−0.10			—CO₂H	−0.94	−0.08
Pyrrolyl-	0.60			—CONH₂	−1.97	−1.10
Pyridyl-	0.53			—NHSO₂CF₃		1.22
Quinolinyl-	1.82					
Acridinyl-	3.11					
Indolyl-	1.89					
Benzimi-						
dazolyl-	1.19					
Uracilyl-	−1.31					
Barbituryl-	−1.55					

*Hydrocarbon substituents on aromatic rings are given aliphatic values.
(Rekker and de Kort, 1979.)

pentyl. The hydrophilic properties of lone pairs of electrons are clearly reflected in the **f** values of nitrogen- and oxygen-containing groups.

17.2 Table and discussion: Electronic effects in molecules (Hammett and Taft sigma values)

A knowledge of the effects of various substituents on any ionizable group allows each substituent to be classed as electron-attracting or electron-releasing, and

Table 17.3 Partition coefficients of drugs between octanol and water. (See Table 15.1 for coefficients of hypnotics and general anaesthetics which average $\log P = 2.0$)

Substance	Structural formula	log P
Tetracycline	11.36	−1.47
Adenosine	9.72	−1.10
Sulfadiazine	9.9	−0.13
Caffeine	7.52	−0.07
Sulfamethizole	9.13	0.54
Morphine	7.35	0.76
Ephedrine	12.12	0.93
Prednisolone		1.42
Atropine	7.16	1.79
Benzylpenicillin	5.13	1.83*
Procaine	7.8	1.87
Diphenylhydantoin	15.4	2.47
Erythromycin	4.47	2.48
Diazepam	12.95	2.82
Naphthalene		3.37
Progesterone		3.87
Cinchocaine (dibucaine)	7.15	4.40
Imipramine	12.114	4.62
Chlorpromazine	12.110	5.16

*Corrected for ionization, but as this is a strong acid (pK_a 2.8), the *effective* log *P* is in the *minus* region.
(Leo, Hansch and Elkins, 1971.)

placed in ranking order. This operation was quantified, about 1940, by the introduction of Hammett's *Linear Free Energy Equation*:

$$\log K - \log K_o = \rho\sigma$$

where K_o is the ionization constant of benzoic acid, which is the standard, K is the ionization constant of a benzoic acid bearing the substituent under investigation, ρ (rho) is a constant pertaining to the nucleus under investigation, and σ (sigma) is the constant pertaining to the substituent (ρ was given the value of 1.00 for the ionization of benzoic acid). The subject has been lucidly explained by Hammett (1970). A list of σ values was compiled by Exner (1972) and critically discussed. A very convenient list is presented by Perrin, Dempsey and Serjeant, 1981.

(*a*) *Hammett's sigma constants for aromatic substituents.* A longer list of these has been put together specially for multiple regression analysis (Hansch and Leo, 1979). The selection given in Table 17.4 should suffice to illustrate how substituents can be ranked according to their electronic effects. The *meta* constants in the first column of figures depend mainly on the inductive effect. They have greater reproducibility than the *para* values (in the second column) which, while weaker in inductive influence, have an additional mesomeric influence which can cause a value to vary in proportion to any 'through-resonance' that a pair of suitable substituents may permit. Through-resonance, discussed by Clark and Perrin (1964), is illustrated, for *p*-nitrophenol, in (*17.1*).

Through-resonance in *p*-nitrophenol
(17.1)

In Table 17.4, the values marked with a minus sign are electron-releasing; all others are electron-attracting. It will be seen at once that the whole span of values is less than two units; hence quite small changes in σ values may indicate fairly large electronic effects. Much attention should not be given to the second place of decimals, because σ values are the difference between two pK_a values each of which may have a small uncertainty in the second place. Again, if σ values are recalculated (using an average ρ) from sources other than the ionization of benzoic acid, they show differences from those of Table 17.4 although the ranking order remains unchanged. When zwitterion formation is possible (as in *m*-aminobenzoic acid where there is much, although *p*-amino-benzoic acid has only a little), the σ-values for *ions*, given at the botton of Table 17.4, should be substituted. In that case, Bjerrum's electrostatic distance term should be incorporated in the form of *r*, the distance in Å between the charges (Hoefnagel, Hoefnagel and Wepster, 1978), thus:

$$\text{Substituent effect} = \rho\sigma \times 3.1/r$$

This effect can assume special importance in aliphatic compounds, where the distance between charges is often larger.

Through-resonance achieves its largest increases in σ_p values when a -nitro-, -cyano-, or -aldehydo-group is placed *para* to a -hydroxy- or -amino-group. Thus the usual value of 0.78 for a nitro-group jumps to a 1.24 in these circumstances, and the 0.44 usual for —CHO goes to 1.03. These and other examples are listed by Perrin, Dempsey and Serjeant (1981) in their Appendix 4.

It was mentioned above that σ_m values are not free from a small mesomeric (resonance) effect. Taft and Lewis (1959) took the first steps to separate the two components of σ_m by comparison with the σ values which Taft had obtained for aliphatic acids (see below) and which were necessarily resonance-free. This separation was further refined by Swain and Lupton (1968) (who were the first to show that the field effects of *m*- and *p*-positioned groups were quite different) into $\mathfrak{F}$ (field) and $\mathfrak{R}$ (resonance) effects*. A systematic arithmetical error in the

*For deficiencies in Swain-Lupton approach, see Reynolds and Topson, 1984.

Table 17.4 Sigma substituent constants

	From benzoic acid		From acetic acid
	Meta substituents σ_m	Para substituents σ_p	σ^*
—NHMe	−0.30	−0.84	−0.81
—CH_2SiMe_3	−0.17	−0.27	−0.31
—NMe_2	−0.15	−0.83	0.32
—t-Bu	−0.09	−0.15	−0.30
—Me	−0.06	−0.14	0.00
—NH·OH	−0.04	−0.34	0.30
—NH·CO·NH_2	−0.03	−0.24	1.31
—NH·NH_2	−0.02	−0.55	0.40
—H	0.00	0.00	
—NH_2	0.00	−0.57	0.62
—C_6H_5	0.05	−0.01	0.75
—OMe	0.11	−0.28	1.81
—NH·COMe	0.12	−0.09	1.40
—OH	0.13	−0.38	1.34
—SMe	0.14	0.00	1.56
—SH	0.25	0.15	1.68
—N_3	0.27	0.15	2.62
—CO·NH_2	0.28	0.31	1.68
—CO·OMe	0.32	0.39	2.00
—CO·OH	0.35	0.44	2.08
—CHO	0.36	0.44	2.15
—COMe	0.36	0.47	1.81
—OCF_3	0.36	0.33	
—SCF_3	0.38	0.50	2.75
—F	0.34	0.06	3.21
—Cl	0.37	0.24	2.96
—Br	0.39	0.22	2.84
—I	0.35	0.21	2.46
—O·COMe	0.39	0.31	2.56
—CCl_3	0.40	0.46	2.65
—SCN	0.41	0.52	3.43
—CF_3	0.46	0.53	2.61
—SO_2NH_2	0.46	0.57	2.61
—CN	0.62	0.70	3.30
—SO_2Me	0.64	0.73	3.68
—NO_2	0.74	0.78	4.25
—SO_2CF_3	0.76	0.95	4.50
Ions			
—S^-	−0.36		
—CO_2^-	0.09	−0.05	−1.06
—NH_3^+	0.67	0.53	3.76
—NMe_3^+	0.99	0.96	4.55

(Perrin, Dempsey and Serjeant, 1981.)

latter work has been ironed out by Hansch *et al.* (1973) who present a list of nearly 200 such values. The ranking order of *resonance* (mesomeric) effects is: (most negative) —NH$_2$, —OMe, F, —NHAc, Cl, Me, tBu, Ph, H, (neutral), —NMe$_3^+$ (neutral), —CO$_2^-$, —CO$_2$H, —NO$_2$ (most positive). Swain and Lupton also gave the following ranking order of *field* (inductive) effects in aromatic substituents (necessarily the same as in aliphatic substituents): (most negative) —CO$_2^-$, tBu, Me, H (neutral), —NH$_2$, Ph, —OMe, —NHAc, —CO$_2$H, Cl, F, —NO$_2$, —NMe$_3^+$ (most positive). For more on the separation of field and resonance contributions, see Taft and Grob (1974).

The sigma values of *ortho*-substituents present problems similar to those of *para*-substituents plus, very often, those introduced by steric distortion and by hydrogen-bonding. The most determined attack on this multifactorial problem was made by Fujita and Nishioka (1975) who set σ_o as equal to σ_p, plus a fraction of the Swain and Lupton field constant (to deal with the greater proximity), plus a fraction of Taft's steric factor (E_s, see p. 627). When hydrogen-bonding is present, as in o-nitrophenol, a further correction must be made. A table of σ_o values has been published for 35 substituents that are *ortho* to an amino-, a hydroxy-, or a pyridino substituent (Perrin, Dempsey and Serjeant, 1981, Appendix 5).

(*b*) *Taft's sigma constants for aliphatic substituents.* Following Hammett's success with aromatic compounds, Taft sought to obtain a reasonable measure of the inductive effect of substituents in aliphatic compounds. Of his various alternative suggestions, the sigma-star (σ^*) values (Taft, 1956) have proved most useful both in predicting a pK_a and in multiple regression analysis. Originally, Taft obtained them by comparing the rates of hydrolysis of α-substituted methyl acetates by acid and then by base. An approximate, but more convenient derivation is:

$$\sigma^* = (-\Delta pK_a - 0.06)/0.63$$

where ΔpK_a is obtained by subtracting, from the pK_a of acetic acid, the pK_a of the acetic acid which bears the substituent whose sigma-star value is required. The coefficients have been designed to yield 0 for a methyl substituent.

When a substituent is moved out, down an aliphatic chain, its effect is attenuated. Only 40% of it is transmitted by one —CH$_2$-group, and hence only 16 and 6% (respectively) by two or three such groups. In such calculations, —CH=CH— has the same attenuating power as —CH$_2$—.

Numerically, the sigma-star values are much larger than σ_m values whose effect is attenuated by the greater distance. Table 17.4 has a selection of sigma-star values in the last column. More comprehensive tables are in Perrin, Dempsey and Serjeant (1981) and in Hansch and Leo (1979). In dealing with heterocyclic substituents, it should be recalled that the doubly bound nitrogen (as in pyridine) is strongly electron-attracting, whereas the singly bound nitrogen (as in pyrrole) is strongly electron-releasing.

(c) *Hammett's rho constants.* Hammett's original use of ρ (rho) values was to convert substituent constants, obtained in reactions where a covalent bond was broken, to σ values comparable to those obtained from the pK_as of substituted benzoic acids, where rho is uniformly taken as 1.00. Later, rho was used to convert values obtained by subtracting, from the pK_a of phenol (9.99), the pK_a of a substituted phenol, using $\rho = 2.23$. In reverse, the pK_a of 3-nitrophenol can be calculated as 9.99 minus 2.23×0.74 (0.74 is from Table 17.4), which gives 8.34 (the experimentally determined value is 8.36). The rho value for aniline is 2.81. A table of sigma values derived from phenol and aniline is available (Hansch and Leo, 1979).

Another use of rho is to convert sigma values, derived from benzoic acid, for use in other nuclei. This was first achieved for naphthalene; the results were then extended to quinoline, and then to other heterocycles [Perrin, 1965c; Perrin, Dempsey and Serjeant, 1981 (their Section 7.2)]. Although rho continues to appear in *formal* statements of Hansch's multiple regression equation, it is usually only indirectly present, namely as a *nucleus-modified* sigma value.

17.3 Table and discussion: Nuclear magnetic resonance

(a) *Introduction.* The magnetic moments possessed by atomic nuclei with odd values of either mass or atomic number cause them to *precess* when placed in a powerful magnetic field. The frequency of this precession, which is directly proportional to the field strength, is characteristic of the type of nucleus and falls within the radio-frequency range. Of the elements most commonly encountered in organic chemistry, ^1H, ^{13}C, ^{14}N, ^{19}F, and ^{31}P give this effect, whereas ^{12}C and ^{16}O do not. Because the most readily accessible apparatus is specialized for the detection and measurement of ^1H, the nuclear magnetic resonance (n.m.r.) discussion will begin with that of the hydrogen nucleus.

(b) *Proton magnetic resonance.* A solution of the compound to be studied is irradiated in a strong magnetic field with an electromagnetic beam of radiofrequency 2.35 T (T = tesla)* or more. The transitions to a higher energy level which this irradiation causes in each hydrogen nucleus are detected by measuring the absorption of energy from the transmitting circuit. This absorption unbalances a radio-frequency bridge, and the out-of-balance signal is then amplified, detected, and recorded.

The field experienced in each hydrogen nucleus is less than that applied by the instrument because of the diamagnetic shielding effect of the extranuclear electron cloud. In the original spectrometers, the magnetic field was varied while a constant radiofrequency was maintained. In modern spectrometers, this frequency is applied, at constant magnetic field, in the form of a high-powered pulse that excites all nuclei simultaneously. The data are collected in the form of decay of induction with time. They are then automatically averaged and

---

*2.35 T equals 100 MHz or 23 500 cycles per second.

transformed, by the *Fourier* procedure, from a *time* domain to a *frequency* domain. This process yields an absorption spectrum very similar to that given by earlier instruments, but much better resolved. The difference between the resonance frequencies of various hydrogen nuclei is called the 'chemical shift' and is usually measured in parts per million. The value 0.00 p.p.m. has been arbitrarily assigned to the resonance of the hydrogen atoms in tetramethylsilane, enabling a table of delta (δ) values to be drawn up, e.g. Table 17.5. A high δ value indicates a decrease in the applied field necessitated by much shielding in the molecule. (In an earlier scale, tetramethylsilane was given the value of 10, and the results are recorded as tau values ($\tau = 10 - \delta$).

The commonest use of n.m.r. in chemistry is to decide whether certain groups are present or absent. Hence it is much used in finding the constitution of newly synthesized or isolated substances. It is also valuable in investigating potentially tautomeric molecules. When the tautomeric equilibrium is reached slowly (as usually occurs when the mobile proton is attached to a *carbon* atom in one tautomeric form), signals from both tautomeric forms are recorded independently, and a kinetic study may be made. The equilibrium is reached almost instantaneously when the mobile proton is attached to a *nitrogen* or *oxygen* atom.

Table 17.5 Approximate δ values for hydrogen nuclei in different molecular environments

Environment	δ(p.p.m.)	Environment	δ(p.p.m.)
CH_3—C	0.9	NH_2—Ar	3.5*
C—CH_2—C $\big\}$	1.2–1.5	CH_3—O—CO	3.8
C—CH(R)—C		CH_3—O—Ar	3.8
H_2N—CR_2	1.6*	CH_2 : C	4.7
CH_3C : C	1.8	HR(R) : C	5.3
CH_3—C : O	2.0	HO—C	5.3*
CH_3—S	2.0	H_2N—CO	7.0*
CH_3—N	2.2	H—Ar	7–8
CH_3—Ar	2.4	HO—Ar	7.7*
CH_3—N(R)—CO	2.8	R·CHO	9.8
CH_3—O	3.3	R·COOH	10.8

*Hydrogen-bonding can change these values (see text).

Because *hydrogen-bonding* alters the environment of a hydrogen nucleus, it can have a large effect on the δ value. For example, concentrated solutions of alcohols in carbon tetrachloride usually give δ 5.3 for the —OH proton, but when they are diluted (or heated) sufficiently, the signal moves downfield in the direction of 0.5 as intermolecular bonding is lost. However, substances with intramolecular hydrogen bonds retain their high δ values on dilution, and this is diagnostic. By running the spectrum at various temperatures, inter- and intramolecular hydrogen-bonding can be even better differentiated, even in complex molecules such as polypeptides (Kessler, 1982).

Apart from the shielding effects mentioned above, the external field of the instrument can induce electron currents which strongly shield acetylenic hydrogen atoms (δ 2.3) whereas aromatic protons are strongly deshielded (δ 7 to 8).

The peaks of a proton spectrum are often split into doublets, triplets or quartets because of spin–spin interaction between hydrogen nuclei of different delta values. The splitting is recorded as a frequency, in the form of J, which is the spin–spin coupling constant. Coupling of hydrogen nuclei through σ-bonds is strong but rapidly attenuated by distance, whereas couplings transmitted through π-electron systems in aromatic molecules are relatively small although not greatly diminished by increased distance.

The most useful *solvents* in proton magnetic resonance studies are those that are free from hydrogen atoms (e.g. carbon tetrachloride), those that give proton signals at very low field (e.g. trifluoroacetic acid), and those that are completely deuterated (e.g. deuterated dimethyl sulfoxide, acetone, chloroform). In work of biological application, water is the preferred solvent, but this injects a swamping hydrogen signal in mid-field. This signal, fortunately, can be removed by a pulse method. When deuterium oxide is used as the solvent, it contributes no signals, but it removes useful signals from rapidly exchanging hydrogen atoms attached to nitrogen, oxygen, or sulfur.

For a collection of analysed spectra, see Simons and Zanger (1972). For an introduction to ^{1}H n.m.r., see Jackman and Sternhell (1969). The ^{1}H n.m.r. of nitrogen heterocycles is reviewed by Batterham (1973).

(*c*) *Non-proton magnetic resonance*. Second only to ^{1}H n.m.r. in frequency of use, ^{13}C n.m.r. gives information that is complementary to that furnished by the proton spectrum. Because the natural abundance of ^{13}C is only 1.1%, a larger specimen tube is used. The naturally weak signals are enhanced by rapid time-averaging prior to the Fourier transform. The application of appropriate jamming frequencies provides single signals free from the usual C—H coupling (C—C coupling presents no problem). Compared to ^{1}H n.m.r., ^{13}C spectra show fewer local effects, and ring currents and anisotropy are proportionately smaller. With these aids, ^{13}C spectroscopy presents signals that are well-spaced, clear, and easily interpreted.

The range usually observed is 0–220 p.p.m., assigning 0 to tetramethylsilane (TMS) as internal standard. At high yield, methyl groups appear (10–30 p.p.m.), followed by CH and NCH_2 at 50 p.p.m., then aromatic and heteroaromatic carbon signals at about 125 p.p.m., and finally carbon signals from carboxylic acids and amides at about 170 p.p.m. Thus the signal for carbon (in any particular C and H substituent) tends to occur at the same fraction of the total scan as would this substituent's hydrogen signal on the usual ^{1}H scan of 0–12 p.p.m. For biological work, water is often the ideal solvent, as it permits the use of buffers. However, carbon-containing solvents can be used provided that they give no signal in the region being studied.

For a book on interpreting ^{13}C spectra, see Wehrli and Wirthlin (1976).

Natural abundance of ^{15}N is only 0.37%, but the Fourier transform adjunct has made its n.m.r. spectroscopy practicable. Using the ammonium ion as a standard, the range usually scanned is from 0 to 800 p.p.m. A large specimen tube is used. Alkylamines provide signals near 0, amides near 100, purines and pyrimidines near 200, nitro-groups at about 350, and nitroso-groups about 750 p.p.m. On the whole, the distribution of signals is uneven, most of them being found at high field.

The unshared electron pairs of nitrogen endow it with properties denied to carbon. Protonation changes the resonance pattern, both in magnitude and direction, in ways that are characteristic of each type of nitrogen-containing group. Also the resonance positions are usually much more sensitive to solvent changes than are those of a similarly placed carbon atom, and this has diagnostic uses. The examinations are often run in water, dimethyl sulfoxide, or 2-trifluoroethanol. Alternative internal standards are nitromethane or the nitrate anion. The geometries of alkaloids and synthetic drugs have been clarified by ^{15}N work, and it was found that *cis* isomers have a more shielded nitrogen atom than *trans* isomers. N . . . H hydrogen-bonding plays a critical role in many drug-receptor unions that are under investigation.

For a book on interpreting ^{15}N spectra, see Levy and Lichter (1979).

With its natural abundance of 100%, ^{31}P spectroscopy is much used in biological studies, even in the living human subject. The signals are somewhat weaker than those given by ^{1}H and ^{19}F, but quite clear. Phosphoric acid is generally used as an external standard and the scale ranges from -5 to $+25$ p.p.m., with phosphates and phospholipids giving signals near 0, nucleoside phosphates from 5 to 222, and pyrophosphate diesters quite sharply at about 12: in sum, a rather bunched distribution. Spin–spin coupling to hydrogen atoms tends to be large and variable. Water is a practicable solvent. The many uses of this technique include investigations of phosphorus insecticides, ATP, and the nucleic acids.

A natural abundance of 100% is also helpful in ^{19}F spectroscopy which is conducted in small-diameter spinning tubes. The scale ranges from 0 to 150 in one direction and 0 to -150 in the other with trifluoroacetic acid taken as 0. Resolution is excellent, but the signals are subject to long-range environmental effects and can be hard to interpret. The uses are for fluorine-containing drugs, such as fluothane, or as a label (say, as a trifluoroacetyl-group) for a fluorine-free drug.

(*d*) *Relaxation methods.* These can provide steric data about a key atom in a molecule by temporarily suspending the influence of a neighbouring atom. One useful technique of this kind provides T_1, the spin-lattice relaxation time. Sequences of radiofrequency pulses are used to produce a non-equilibrium magnetization, usually on a CH group undergoing ^{13}C n.m.r. examination. The time taken for return to equilibrium conditions can provide a measure of the dynamic properties of a drug in solution, such as rotational motion of the molecular backbone, or internal rotations in a side-chain. For an application of this technique, see the work by Allerhand and Komorowski (1973) on gramicidin S.

Another much-used technique is the nuclear Overhauser enhancement (NOE) which, again, is usually studied on a particular CH group under ^{13}C n.m.r. conditions. NOE is the increase in signal strength usually shown by one atom while a neighbouring atom is being irradiated at its own resonance frequency. The conclusions are surer if *two* neighbouring atoms can be treated in this way. This technique has provided useful information on molecular motion, on conformation of large molecules such as peptides, and on internuclear distances (approximate).

(*e*) *Applications of n.m.r. in biological and medicinal chemistry.* There are two main uses. The first of these is to establish configurations, or even conformations, of small biologically active molecules in solution. Many such problems have been solved by use of the Karplus equation which correlates the coupling constant between vicinal protons with the torsional angles (defined in Section 12.3) between different carbon–hydrogen planes (Roberts, 1968). In this way, each of the four isomers of cocaine (*7.11*) was assigned a conformation (Sinnema *et al.*, 1968). Another example was the determination, by p.m.r., of the conformation of acetylcholine in aqueous solution (see p. 523). A study of the antihypertensive drug, captopril (*9.50*), by ^{1}H and ^{13}C in water, at a range of pH values, showed that this substance existed as an equilibrium mixture of two isomers with respect to the conformation across the amide bond. The *trans* conformation was the most abundant, the actual proportion of this form depending on the state of ionization both of the carboxy- and the mercapto-groups. The carboxylic-group of the *trans* conformer was less acidic than that of the *cis* conformer by 0.66 pK unit. The results were interpreted to indicate intramolecular hydrogen-bonding between the amide carbonyl-group and the carboxylic proton in the *trans* conformer (Rabenstein and Anvarhusein, 1982). For other conformational applications, see Casy (1971).

The second use of n.m.r. in drug studies, and it is one of the most powerful physical techniques for this purpose, is to investigate the reaction between a small molecule and a biopolymer. Because many of the resonances in the spectrum of the biopolymer will overlap, special techniques are used, such as perturbation of the resonances by introducing paramagnetic ions such as the lanthanides (e.g. lanthanum, europium), or spin-labels which are stable organic free radicals (e.g. the nitroxides). Solvent relaxation enhancement (e.g. with T_1, as described on p. 652) is similarly useful. Alternatively, the drug can be derivatized with a group containing ^{13}C or ^{19}F. For more on these methods, see Jardetzky and Roberts (1981).

The use of spin–spin relaxation will be discussed next. Part of the n.m.r. spectrum of an agent undergoes broadening after adsorption has occurred on to a macromolecule. Briefly, the technique is based on the fact that the relaxation rate $(1/T_2)$ – and hence the width of the n.m.r. signals – depends on the motion of the molecule which gives rise to the line. If the motion of the molecule is restricted, the relaxation time of the nuclei is shortened and the width of its n.m.r. signals is increased. Thus when a small molecule is bound to a macro-molecule the lines of its n.m.r. spectrum broaden by a factor of 100–1000, very roughly in proportion to the increase in effective molecular weight. If only a part of the small molecule becomes attached, its signal is broadened selectively while signals of the unattached parts of the molecule remain narrow by comparison. The part of the molecule firmly held on the macromolecule is therefore readily identifiable (Jardetzky and Jardetzky, 1962). Thus, when atropine (*7.16*) was bound by acetylcholinesterase, the signals for both phenyl- and *N*-methyl-groups were broadened, and equally so, showing that both ends of the molecule were firmly bound. Physostigmine (*2.8*) was similarly shown to be firmly bound by both of its *N*-methyl-groups (Kato, Yung and Ihnat, 1970; Kato, 1975).

Two general types of complexes have been distinguished, (a) van der Waals complexes where the binding (also the n.m.r. relaxation rate for the bound molecule) increases with increasing polarity of the solvent, and (b) electrostatic complexes where the reverse is true. Examples of the former are the binding of penicillin side-chains and the sulfanilamide moiety to serum albumin (Fischer and Jardetzky, 1965); examples of the latter are the binding of choline deriva-tives to an anti-choline antibody. Such an antibody was prepared against phenoxycholine to provide an immunochemical model for cholinergic recep-tors. Unhappily the specificity of this antibody turned out to be too low to permit differentiation between agonists and antagonists, or between muscarinic and nicotinic cholinomimetics (Marlow, Metcalf and Burgen, 1969).

Particularly convenient for study by this technique is the common type of complex for which the binding constants K_s falls in the range 10^3–10^5. Such complexes exist in rapid equilibrium with the uncomplexed species, the rate of complex formation k_R in these systems being of the order of 10^8–10^{10} litres mol^{-1} s^{-1}, and the rate of dissociation k_D, by virtue of the relationship $K_s = k_R/k_D$, of the order 10^3–10^5 s^{-1}.

Actinomycin (*4.38*), when examined by ^{1}H, ^{13}C and ^{31}P n.m.r. in the presence of deoxyoligonucleotides, revealed intercalation by the chemical shifts, caused by the ring-current fields of the phenoxazone nucleus, in the purine–pyrimidine resonance signals of the nucleotide. This led to detailed understanding of the solution structure of the intercalated components, and established that inter-calation occurred at the site of the central GC base-pairs (Patel, 1974). This work was extended, using ^{15}N-enriched actinomycin (Brown, Shafer and Mirau, 1982).

For a review of n.m.r. studies of the binding of methotrexate (*9.29*) and of 2,4-diaminopyrimidine, both of them inhibitors, to the dihydrofolate reductase of *L. casei*, see Roberts *et al.* (1977).

The ^{31}P n.m.r. spectra of buffered aqueous solutions of phosphoramide mustard (*3.39*), which is the active metabolic product from cyclophosphamide

(*3.38*), afforded first-order rate constants for the conversion to aziridinium cation product (see p. 101). Variation of rate with pH showed a tenfold faster loss of (*3.39*) between pH 7.4 and 6.0 and this suggested that the increased acidity of cancer cells may confer selectivity (Engle, Zon and Egan, 1979). Cyclophosphamide is currently given as the racemate, but the asymmetric phosphorus centre has been resolved into (+) and (−) enantiomers, easily distinguishable by either ^{1}H or ^{31}P n.m.r. in the presence of europium, a method with advantages over the traditional optical rotation measurements (Ludeman, Bartlett and Zon, 1979).

To conclude on a cautionary note, in rapidly exchanging systems, n.m.r. data alone are not suffiecient to distinguish between different possible structures of the formed complex. However, the data can suggest which associations are the more likely ones, and then some adjunct such as X-ray analysis may be used to decide between them (for a review, see Williams, 1977).

For further reading, see Cohen (1980), Campbell and Dobson (1979), Jardetzky and Roberts (1981), and Dweck (1977).

17.4 Searching the literature

Students often ask: 'If I have the name of a drug, how can I find out more about it?' To accomplish this, it is better to start with the generic name, such as 'acyclovir' rather than the trade name which, in this case, is 'Zovirax'. However, both names are indexed in the 28th edition of *Martindale's Extra Pharmacopoeia* (Reynolds, 1982) which also gives the full chemical name, but not the structural formula. It gives a thorough coverage of uses and side effects (with references to recent literature), doses, preparations, and synonyms (including trade names). For structural formulae, see *The Merck Index*, 10th edition, 1983. This work, not strictly alphabetical, is best approached through its index. Further chemical information can be obtained from *Heilbron's Dictionary of Organic Compounds*, 5th edition, 1982 which, in seven volumes, gives the chemical structure (including absolute configuration), also chemical and spectroscopic properties of an immense number of organic compounds. Chemical Abstracts Service Registry Numbers (CAS) are also included. For generic and trade names of agrochemicals, see Worthing (1983).

Goodman and Gilman's Pharmacological Basis of Therapeutics (the 6th edition is by Gilman, Goodman and Gilman, 1980) gives a full account of each drug in current use with emphasis on biology. There are good physiological (and sometimes biochemical) details of the mode of action and many structural formulae are included. The comparative worth of various drugs for the same illness is carefully evaluated. *Burger's Medicinal Chemistry*, 4th editon in three volumes, emphasizes chemistry rather than biology. The effect on the action caused by exchanging substituents is kept well to the fore (Wolff, 1981).

For very recently introduced drugs, see *Drugs in Current Use and New Drugs*, a soft-covered annual which reflects the American scene. Because of official

caution, the latter can lag behind the world scene which can be glimpsed in
Drugs printed in Basel, Switzerland (original work in 1971, but with annual
supplements; the entries are very short).

More quantitative data can be obtained either from reference books or from
computerized databanks. Let us deal first with the books. A selection of *ionization constants* is given in Albert and Serjeant (1984). Exhaustive compilations of
ionization constants are listed on p. 383. Recently determined values may be
located through *Chemical Abstracts* through the entry: 'Ionization in Liquids' in
the General Subject Index, or under the name of the particular substance in the
Chemical Substance Index. For the prediction of ionization constants from the
chemical structure, see Perrin, Dempsey and Serjeant (1981). The standard
compilations of the *stability constants* of chelated metals are given at the end of
Section 11.3 (p. 459).

A useful list of 5800 *partition coefficients* is given by Hansch, Leo and Elkins
(1971). They are presented as log *P* values in the system octanol/water or, if
determined in another system, they are transposed to octanol/water values. The
book *Substituent Constants for Correlation Analysis* (Hansch and Leo, 1979) has two
very useful tables. Table VI-1, headed 'Well-characterized Aromatic Substituents' contains the following information for 166 groups:

1. Hydrophobic substituent constants (π) (the possibly more useful hydrophobic fragmental constants are in Appendix III),
2. A statement of whether the substituent is hydrogen-bonding, hydrogen-
   bondable, or neither,
3. Molar refractivity,
4. Swain and Lupton's Field and Resonance values, and
5. Hammett's σ_m and σ_p values.

Table VI-2 provides the following values for 110 'Well-characterized Aliphatic Substituents':

1. Hydrophobic fragmental constants,
2. and 3. (As 2 and 3, above), and
4. Swain and Lupton's Field Values. The expected sigma-star values are in
   Appendix I.

Values of Taft's E_s, for both aliphatic and aromatic substituents are in
Appendix I.

The *Registry of Toxic Effects of Chemical Substances* (RTECS) is published
annually by the US National Institute for Occupational Safety and Health
(Ohio). The computerized version, which contains data on over 40 000 compounds, is available via the TOXLINE system of the US National Library of
Medicine and the Chemical Information Service (CIS), sponsored jointly by
the National Institutes of Health and the Environmental Protection Agency,
Washington. Within CIS there exists also a powerful structural search system
known as SANSS (Structure and Nomenclature Search System) which provides

rapid on-line access to about 200 000 substances. With its help, one can select a cohort (sub-population) of substances that have one (or even more) structural feature in common with a candidate substance. The search, at about $75 an hour, is started by supplying the structure, sub-structure, Chemical Abstracts Registry Number (CAS), *or* the name, which may be full, partial, left-truncated, right-truncated, a synonym, a trade name or a trivial name.

A more exhaustive search for cognate substructures can be carried out with the CAS Online System, where more than 600 000 substances are searchable through *Chemical Abstracts*; in France the DARC system is used similarly. In this approach, some biological data will be delivered directly, but more can often be obtained by inserting the CAS number of the substance into the US National Institute of Health's MEDLARS system (available without charge in many countries through local Departments of Health).

Reverting to the NIH/EPA Chemical Information System (CIS), this has recently added a new database called CESARS (Chemical Evaluation Search and Retrieval System) which includes physical and chemical properties, toxicity, carcinogenicity, and environmental fate (degradation).

Perhaps it is unnecessary to warn readers that the toxicity data obtainable from these systems is *un*selective, even counter-selective in nature.

Databanks for nucleic acid sequences are available: see Walgate (1982) for Europe, and Lewin (1982) for the USA. A protein crystal-structure databank is maintained by the Brookhaven National Laboratory, Long Island, NY 11973, USA.

An X-ray single crystal search system (XTAL) is maintained jointly by NIH/EPA. This draws on tables published by the US National Bureau of Standards and the Cambridge (England) Crystallographic Data File (CCDF). Researchers in England will know of the Cambridge Crystallographic Centre, University Chemical Laboratory, Lensfield Road. This Centre has 30 000 entries resulting from a complete literature search, polymers excluded. They are most easily searched through CSSR (Crystal Structure Search Retrieval) c/o Daresbury Laboratory of the Science and Engineering Research Council, Warrington WA4 4AD, UK, an approach which permits display as stick models.

Many of these complex and powerful retrieval systems, although originating in a few geographical locations, are available via terminals located in many countries throughout the world. Used intelligently, these facilities can save time and increase productivity. Their great disadvantage is that they tend to favour search through similarity of *structure* whereas the greater dependency of biological properties on *physical* properties is now acknowledged. Identical physical properties can, of course, be supplied by very different structures. A forerunner of a different approach is offered by another component of the Chemical Information Service (see above) known as LOGP. This supplies the logarithms of the partition coefficients of a great many substances, which can either be searched by name to get the value, or by a desired value to get a set of names. It also provides references to the literature.

References

Aaseth, J. and Friedheim, E. (1978) *Acta Pharmacol. Toxicol.*, **42**, 248.
Abeles, R. and Maycock, A. (1976) *Acc. Chem. Res.*, **9**, 313.
Abelson, H. and Penman, S. (1972) *Nature New Biol.*, **237**, 144.
Abraham, E. (1962) *Pharmacol. Rev.*, **14**, 473.
Adams, E. and Macleod, I. (1977) *Medicine, Baltimore*, **56**, 315.
Adams, M., McPherson, A., Rossmann, M., Schevitz, R. and Wonscott, A. (1970) *J. Molec. Biol.*, **51**, 31.
Adamson, R., Bridges, J. and Williams, R. (1966) *Biochem J.*, **101**, 71P.
Adcock, E. (1940) *J. Exper. Biol.*, **17**, 449.
Addanki, S., Cahill, F. and Sotos, J. (1967) *Nature, Lond.*, **214**, 400.
Adler, R. and Snoke, J. (1962) *J. Bact.*, **83**, 1315.
Adler, S. and Tchernomoretz, I. (1942) *Ann. Trop. Med. Parasit.*, **36**, 11.
Aeschlimann, J. and Reinert, M. (1931) *J. Pharmacol.*, **43**, 413.
Africk, J. and Fulton, J. (1971) *Brit. J. Dermatol.*, **84**, 151.
Agarwal, M. (ed.) (1979) *Antihormones*, Amsterdam: Elsevier.
Agarwal, R. and Perrin, D. (1976) *Agents and Actions*, **6**, 667.
Agarwal, R., Spector, T. and Parks, R. (1977) *Biochem. Pharmacol.*, **24**, 693.
Agrawal, K., Booth, B., Moore, E. and Sartorelli, A. (1972) *J. Med. Chem.*, **15**, 1184.
Agrawal, K. and Sartorelli, A. (1975) *Antineoplastic and Immunosuppressive Agents* (Sartorelli, A. and Johns, C., eds), Berlin: Springer.
Ahlquist, R. (1948) *Amer. J. Physiol.*, **153**, 586.
Aigami, K., Inamoto, Y., Takaishi, N. and Fujikura, Y. (1976) *J. Med. Chem.*, **19**, 536.
Aigami, K., Inamoto, Y., Takaishi, N., Hattori, K., Takatsuki, A. and Tamura, G. (1975) *J. Med. Chem.*, **18**, 713.
Aito, A. (ed.) (1978) *Conjugation Reactions in Drug Biotransformations*, Amsterdam: Elsevier.
Albert, A. (1944) *Med. J. Austral.*, **i**, 245.
Albert, A. (1950) *Biochem. J.*, **47**, 531.
Albert, A. (1952) *Biochem. J.*, **50**, 690.
Albert, A. (1953) *Biochem. J.*, **54**, 646.
Albert, A. (1956) *Nature, Lond.*, **177**, 525.
Albert, A. (1966) *The Acridines, their Preparation, Properties and Uses*, 2nd edn, London: Edward Arnold.
Albert, A. (1967) *Angew. Chem., Internat. edn*, **6**, 919 (Review).
Albert, A. (1968) *Heterocyclic Chemistry, an Introduction*, 2nd edn, London: Athlone Press.
Albert, A. (1976) *Adv. Heterocycl. Chem.*, **20**, 117 (Review).
Albert, A. (1980) *J. Chem. Soc., Perk. 1*, p. 2918.
Albert, A., Armarego, W. and Spinner, E. (1961) *J. Chem. Soc.*, pp. 2689, 5267.
Albert, A. and Brown, D. (1954) *J. Chem. Soc.*, p. 2060.

Albert, A., Brown, D. and Cheeseman, G. (1952) *J. Chem. Soc.*, (a) p. 1620; (b) p. 4219.

Albert, A., Falk, J. and Rubbo, S. (1944) *Nature, Lond.*, **153,** 712.

Albert, A., Gibson, M. and Rubbo, S. (1953) *Brit. J. Exper. Path.*, **34,** 119.

Albert, A. and Goldacre, R. (1946) *J. Chem. Soc.*, p. 706.

Albert, A. and Goldacre, R. (1948) *Nature, Lond.*, **161,** 95.

Albert, A., Goldacre, R. and Phillips, J. (1948) *J. Chem. Soc.*, p. 2240.

Albert, A. and Hampton, A. (1952) *J. Chem. Soc.*, p. 4985.

Albert, A., and Hampton, A. (1954) *J. Chem. Soc.*, p. 505.

Albert, A., Hampton, A., Selbie, F. and Simon, R. (1954) *Brit. J. Exper. Path.*, **35,** 75.

Albert, A. and Howell, C. (1962) *J. Chem. Soc.*, p. 1591.

Albert, A., Howell, C. and Spinner, E. (1962) *J. Chem. Soc.*, p. 2595.

Albert, A. and Rees, C. (1956) *Nature, Lond.*, **177,** 433.

Albert, A., Rees, C. and Tomlinson, A. (1956) *Brit. J. Exper. Path.*, **37,** 500.

Albert, A. and Reich, F. (1961) *J. Chem. Soc.*, p. 127.

Albert, A., Rubbo, S. and Burvill, M. (1949) *Brit. J. Exper. Path.* **30,** 159.

Albert, A., Rubbo, S. and Goldacre, R. (1941) *Nature, Lond.*, **147,** 332.

Albert, A., Rubbo, S., Goldacre, R. and Balfour, B. (1947) *Brit. J. Exper. Path.*, **28,** 69.

Albert, A., Rubbo, S., Goldacre, R., Davey, M. and Stone, J. (1945) *Brit. J. Exper. Path.*, **26,** 160.

Albert, A. and Serjeant, E. (1960) *Biochem. J.*, **76,** 621.

Albert, A. and Serjeant, E. (1964) *J. Chem. Soc.*, p. 3357.

Albert, A. and Serjeant, E. (1984) *The Determination of Ionization Constants* (3rd edn), London and New York: Chapman and Hall.

Aldridge, W. (1950) *Biochem. J.*, **46,** 451.

Aldridge, W. (1958) *Biochem. J.*, **69,** 367.

Aldridge, W. and Barnes, J. (1952) *Nature, Lond.*, **169,** 345.

Aldridge, W. and Davison, A. (1952) *Biochem. J.*, **52,** 663.

Alexander, A. and Trim, T. (1946) *Proc. Roy. Soc. B.*, **113,** 220.

Allen, J., Atherton, F., Hall, M., Hassall, C., Holmes, S., Lambert, R., Nisbet, L. and Ringrose, P. (1978) *Nature, Lond.*, **272,** 56.

Allen, J. and Lees, L. (1980) *Antimicrob. Agents Chemother.*, **17,** 1973.

Allen, L., Huffman, J., Cook, P., Meyer, R., Robins, R. and Sidwell, R. (1977) *Antimicrob. Agents Chemother.*, **12,** 114.

Allen, N. (1981) *Chem. Eng. News*, **59,** (Aug. 31), p. 9.

Allen, T., Brown, D., Cowden, W., Grigg, G., Hart, N., Lamberton, J. and Lane, A. (1984), *J. Antibiot.*, **37,** 376.

Allerhand, A. and Komorowski, R. (1973) *J. Amer. Chem. Soc.*, **95,** 8228.

Alles, G. and Knoefel, P. (1939) *Univ. Calif. Publ. Pharmacol.*, **1,** 187.

Allison, J., O'Brien, R. and Hahn, F. (1965) *Science*, **149,** 1111; *Antimicrob. Agents. Chemother.*, p. 310.

Allison, R. and Hahn, F. (1977) *Antimicrob. Agents Chemother.*, **11,** 251.

Alt, F., Kellems, R., Bertino, J. and Schimke, R. (1978) *J. Biol. Chem.*, **253,** 1357.

Alving, C., Steck, E., Chapman, W., Waits, V., Hendricks, L., Swartz, G. and Hanson, W. (1978) *Proc. Natl. Acad. Sci., U.S.A.*, **75,** 2959.

American Cancer Society (1983) *Cancer Facts and Figures*, New York: Amer. Cancer. Soc.

American Thoracic Society (1980) *Amer. Rev. Respir. Dis.*, **121,** 611.

Ames, A., Tsukada, T. and Nesbett, F. (1967) *J. Neurochem.*, **14,** 145.

Ames, B. and Dubin, D. (1960) *J. Biol. Chem.*, **235,** 769.

Ames, B., Durston, W., Yamasaki, E. and Lee, F. (1973) *Proc. Natl. Acad. Sci., U.S.A.*, **70,** 2281.

Ames, B., Sims, P. and Grover, P. (1972) *Science*, **176,** 47.

Amoore, J. (1964) *Ann. N.Y. Acad. Sci.*, **116,** 457.

Anand, N., Davis, B. and Armitage, A. (1960) *Nature, Lond.*, **185,** 23.

Anderson, B., Buckingham, D., Robertson, G., Webb, J., Murray, K. and Clark, P. (1976) *Nature, Lond.*, **262,** 722.

Anderson, B. and Swaby, R. (1951) *Austral. J. Sci. Res. B.*, **4,** 275.

Anderson, J., Matsubashi, M., Hoskin, M. and Strominger, J. (1965) *Proc. Natl. Acad. Sci., U.S.A.*, **53,** 881.

Anderson, K. and Liao, S. (1968) *Nature, Lond.*, **219,** 277.

Andrea, T., Cavalieri, R., Goldfine, I. and Jorgensen, E. (1980) *Biochemistry*, **19,** 55.

Andreesen, J. and Ljungdahl, L. (1974) *J. Bact.*, **120,** 6.

Andrews, P., Mark, L., Winkler, D. and Jones, G. (1983) *J. Med. Chem.*, **26,** 1223.

Anghilieri, L. (ed.) (1982) *Role of Calcium in Biological Systems* (3 vols), Boca Raton, Florida: CRC Press.

Ankel, E. and Petering, D. (1980) *Biochem. Pharmacol.*, **29,** 1833.

Anon. (1974) *WHO Chronicle*, p. 386.

Anon. (1983) *Nature, Lond.*, **302,** 280.

Antic, B., Van der Goot, H., Nauta, W., Balt, S., de Bolster, M., Stouthamer, A., Verheul, H. and Vis, R. (1977) *Eur. J. Med. Chem.*, **12,** 573.

Antoloni, R., Gliozzi, A. and Gorio, A. (eds) (1982) *Transport in Biomembranes: Model Systems and Reconstitution*, New York: Raven Press.

Anton, A. (1960) *J. Pharmacol.*, **129,** 282.

Anton, A. and Solomon, H. (eds) (1973) *Ann. N.Y. Acad. Sci.*, **226,** 1.

Anzai, K. and Suzuki, S. (1961) *J. Antibiot., Japan A*, **14,** pp. 253, 340. (*Chem. Abs.* 1962, 56, pp. 8849, 10677).

Arcamone, F., Cassinelli, G., Fantini, G., Grein, A., Orezzi, P., Poli, C. and Spalla, C. (1969) *Biotechnol. Bioeng.*, **11,** 1101.

Arcos, J. and Argus, M. (1968–) *Chemical Induction of Cancer*, New York: Academic Press.

Ariëns, E. (1954) *Arch. Internat. Pharmacodyn. Thér.*, **99,** 32.

Ariëns, E. (1960) in *Adrenergic Mechanisms* (Vane, J., Wolstenholme, G. and O'Connor, M., eds) London: Churchill.

Ariëns, E. (1964) *Molecular Pharmacology*, New York: Academic Press.

Ariëns, E., van Rossum, J. and Simonis, A. (1957) *Pharmacol. Rev.*, **9,** 218.

Armarego, W. (1977) *Stereochemistry of Heterocyclic Compounds* (2 vols), New York: Wiley-Interscience.

Armitage, A. and Ing, H. (1954) *Brit. J. Pharmacol.*, **9,** 376.

Armstrong, D. (1966) *Proc. Natl. Acad. Sci., U.S.A.*, **56,** 64.

Armstrong, G., Bradbury, F. and Standen, H. (1951) *Ann. Appl. Biol.*, **38,** 555.

Armstrong, R. and Panzer, N. (1972) *J. Amer. Chem. Soc.*, **94,** 7650.

Arndt, R., Schulz-Harder, B. and Schulz-Harder, J. (1982) *Biochem. Pharmacol.*, **31,** 3120.

Arunlakshana, O. and Schild, H. (1959) *Brit. J. Pharmacol.*, **14,** 48.

Ashe, H. (1964) *Arch. Environ. Health*, **9,** 545.

Ashton, F. and Crafts, A. (1981) *Mode of Action of Herbicides*, 2nd edn, New York: Wiley.

Audus, L. (ed.) (1976) *Herbicides*, 2nd edn (2 vols), New York: Academic Press.

Avery, O., MacLeod, C. and McCarty, M. (1944) *J. Exper. Med.*, **79,** 137.

Aviado, D., Brugler, B. and Bellet, J. (1968) *Exper. Path.*, **23,** 294.

Awad, S., Downie, J. and Kiruluta, H. (1979) *Canad. J. Surg.*, **22,** 515.

Axelrod, J. (1955) *J. Biol. Chem.*, **214,** 753.

Aziz, S. and Knowles, C. (1973) *Nature, Lond.*, **242,** 417.

Bacchi, C. (1981) *J. Protozool.*, **28,** 20.

Bacon, F. (1620) *Novum Organum*, Lugd. Batavorum (Leyden).

Bacon, J., Milne, B., Taylor, I. and Webley, D. (1965) *Biochem. J.*, **95,** 28C.

Baddiley, J., Hancock, I. and Sherwood, P. (1973) *Nature, Lond.*, **243,** 43.

Badger, G. (1947) *Nature, Lond.*, **159,** 194.

Baguley, B., Denny, W., Atwell, G. and Cain, B. (1981) *J. Med. Chem.*, **24,** 520.

Bailey, A. and Mansfield, J. (eds) (1982) *Phytoalexins*, London: Blackie; New York: Halstead.

Baird, W. and Reid, A. (1967) *Brit. J. Anaesth.*, **39,** 755.

Baker, B. (1967) *Design of Active-Site Directed Irreversible Enzyme Inhibitors*, New York: Wiley.

Baker, B., Lee, W., Martinez, A., Ross, L. and Goodman, L. (1962) *J. Org. Chem.*, **27,** 3283.

Baker, B. and Patel, R. (1964) *J. Pharm. Sci.*, **53,** 717.

Baker, B. and Shapiro, H. (1966) *J. Pharm. Sci.*, **55,** 308.

Baker, D., Beddell, C., Champness, J., Goodford, P., Norrington, F., Smith, D. and Stammers, D. (1981) *FEBS Lett.*, **126,** 49.

Baker, P. (1968) *Brit. Med. Bull.*, **24,** 179.

Baker, R., Chotia, C., Pauling, P. and Petcher, T. (1971) *Nature, Lond.*, **230,** 439.

Balaban, A., Chiriac, A., Motoc, J. and Simon, Z. (1980) *Steric Fit in Quantitative Structure—Activity Relations*, Berlin: Springer.

Baldessarini, R. (1977) *Chemotherapy in Psychiatry*, Cambridge, Mass.: Harvard University Press.

Baldoni, J., Villafranca, J. and Mallettee, M. (1980) *Fed. Proc.*, **39,** 2568.

Baldwin, B., Clarke, C. and Wilson, I. (1968) *Biochim. Biophys. Acta*, **162,** 614.

Baldwin, E. (1948a) *An Introduction to Comparative Biochemistry*, 3rd edn, Cambridge: University Press.

Baldwin, E. (1948b) *Brit. J. Pharmacol. Chemother.*, **3,** 91.

Ball, W. and French, O. (1935) *Bull. Univ. Calif. Agric. Exper. Station*, No. 596.

Ballard, B. and Nelson, E. (1962) *J. Pharm. Sci.*, **51,** 915.

Banerjee, S., Yalkowsky, S. and Valvoni, S. (1980) *Environ. Sci. Tech.*, **14,** 1227.

Bard, R. and Gunsalus, L. (1950) *J. Bact.*, **59,** 387.

Barger, G. and Dale, H. (1910) *J. Physiol.*, **41,** 19.

Barlin, G. (1982) *The Pyrazines*, New York: Wiley-Interscience, p. 5.

Barlow, R. (1968) *Chemical Pharmacology*, 2nd edn, London: Methuen.

Barlow, R. (1980) *Quantitative Aspects of Chemical Pharmacology*, London: Croom Helm.

Barlow, R. and Hamilton, J. (1962) *Brit. J. Pharmacol.*, **18,** pp. 510, 543.

Barlow, R. and Hamilton, J. (1965) *Brit. J. Pharmacol.*, **25,** 206.

Barlow, R. and Ing, H. (1948) *Brit. J. Pharmacol.*, **3,** 298.

Barlow, R., Scott, K. and Stephenson, R. (1963) *Brit. J. Pharmacol.*, **21,** 509.

Barnes, J., Magee, P., Boyland, E., Haddow, A., Passey, R., Bullough, W., Cruickshank, C., Salaman, M. and Williams, R. (1957) *Nature, Lond.*, **180,** 62.

Barnett, J., Ralph, A. and Munday, K. (1970) *Biochem. J.*, **116,** 537.

Barr, P., Jones, A., Verhelst, G. and Walker, R. (1981) *J. Chem. Soc., Perk. 1*, p. 1665.

Barrett, A. and Cullum, V. (1968) *Brit. J. Pharmacol.*, **34,** 43.

Barrett, P. (1974) *Proc. Brit. Weed Control Conf.*, **12,** 229.

Barrett, W., Rutledge, R., Plummer, A. and Yonkman, F. (1953) *J. Pharmacol.*, **108,** 305.

Barrnas, Y. and Clastre, J. (1970) *Compt. rend. Acad. Sci., Paris*, **270C,** 306.

Bartlett, G. and Barron, E. (1947) *J. Biol. Chem.*, **170,** 67.

Barton, D. and Cookson, R. (1956) *Quart. Rev. Chem. Soc.*, **10,** 44.

Bartz, Q. (1948) *J. Biol. Chem.*, **172,** 445.

Barza, M. and Scheife, R. (1977) *J. Maine Med. Assoc.*, **68,** 194.

Basil, B., Clark, J., Coffee, E., Jordan, R., Loveless, A., Pain, D. and Wooldridge, K. (1976) *J. Med. Chem.*, **19,** 399.

Basolo, F. and Pearson, R. (1967) *Mechanisms of Inorganic Reactions*, 2nd edn, New York: Wiley.

Bass, W., Schueler, F., Featherstone, R. and Gross, E. (1950) *J. Pharmacol.*, **100,** 465.

Batchelor, F., Chain, E., Richards, M. and Robinson, G. (1961) *Proc. Roy. Soc. B*, **154,** 522.

Batchelor, J., Welsh, K., Tinoco, P., Dalley, C., Hughes, G., Bernstein, R., Ryan, P., Naish, P., Aber, G., Bing, R. and Russell, G. (1980) *Lancet*, **i,** 1107.
Batterham, T. (1973) *NMR Spectra of Simple Heterocycles*, New York: Wiley.
Battershell, C., Malhotra, D. and Hopfinger, A. (1981) *J. Med. Chem.*, **24,** 812.
Bauer, D. and Sadler, P. (1961) *Nature, Lond.*, **190,** 1167.
Bauer, D., St. Vincent, C., Kempe, C. and Downie, A. (1963) *Lancet*, **ii,** 494.
Bauer, D., Selway, J., Batchelor, J., Tisdale, M., Caldwell, I. and Young, O. (1981) *Nature, Lond.*, **292,** 369.
Bauer, W. and Vinograd, J. (1970) *J. Molec. Biol.*, **47,** 419.
Baum, F. (1899) *Arch. exper. Path. Pharmakol.*, **42,** 119.
Baur, E. and Preis, H. (1936) *Z. phys. Chem.* **32B,** 65.
Bebbington, A., Brimblecombe, R. and Shakeshaft, D. (1966) *Brit. J. Pharmacol. Chemother.*, **26,** 56.
Beccari, E. (1938) *Boll. Soc. ital. Biol. sper.*, **13,** 6.
Beccari, E. (1941) *Boll. Soc. ital. Biol. sper.*, **16,** 214; *Arch. Sci. biol., Bologna*, **27,** 204.
Bechhold, H. and Ehrlich, P. (1906) *Z. physiol. Chem.*, **47,** 173.
Beckett, A. (1959) *Arzneim. Forsch.*, **1,** 455.
Beckett, A., Boyes, R. and Triggs, E. (1968) *J. Pharm. Pharmacol.*, **20,** 92.
Beckett, A. and Casy, A. (1962) *Progr. Med. Chem.*, **2,** 43.
Beckett, A., Patki, S. and Robinson, A. (1959) *J. Pharm. Pharmacol.*, **11,** 360.
Beckett, A., Vahora, A. and Robinson, A. (1958) *J. Pharm. Pharmacol.*,**10,** 160T.
Beeman, R. (1982) *Annu. Rev. Entomol.*, **27,** 253.
Beers, W. and Reich, E. (1970) *Nature, Lond.*, **228,** 917.
Behnke, A. and Yarbrough, O. (1939) *Amer. J. Physiol.*, **126,** 409.
Bell, C. (1977) *Principles and Applications of Metal Chelation*, Oxford: University Press.
Bell, P. and Roblin, R. (1942) *J. Amer. Chem. Soc.*, **64,** 2905.
Bell, R. (1973) *The Proton in Chemistry*, 2nd edn, London and New York: Chapman and Hall.
Belleau, B. and Lacasse, G. (1964) *J. Med. Chem.*, **7,** 768.
Belleau, B. and Puranen, J. (1963) *J. Med. Chem.*, **6,** 325.
Belleau, B., Tani, H. and Lie, F. (1965) *J. Amer. Chem. Soc.*, **37,** 2283.
Bellville, J. and Forrest, W. (1968) *Clin. Pharmacol. Ther.*, **9,** 142.
Belozersky, A. and Spirin, A. (1958) *Nature, Lond.*, **182,** 111.
Bender, M. and Komiyama, M. (1978) *Cyclodextrin Chemistry*, Berlin: Springer.
Benesi, H. and Hildebrand, J. (1948) *J. Amer. Chem. Soc.*, **70,** 3978.
Ben-Naim, A. (1980) *Hydrophobic Interactions*, New York: Plenum Press.
Bennett, J. (1977) *Ann. Intern. Med.*, **86,** 319 (review).
Bennett, J. (1979) *New Engl. J. Med.*, **301,** 126.
Bennett, J. and Bueding, E. (1973) *J. Molec. Pharmacol.*, **9,** 311.
Bennett, L., Simpson, L., Golden, J. and Barker, T. (1963) *Cancer Res.*, **23,** 1574.
Benotti, J., Grossman, W., Braunwald, E., Davalos, D. and Alousi, A. (1978) *New Engl. J. Med.*, **299,** 1373.
Bent, K. (1970) *Ann. Appl. Biol.*, **66,** 103.
Bentley, R. (1969) *Molecular Asymmetry in Biology* (2 vols) London: Academic Press.
Bérdy, J. (ed.) (1980–82) *CRC Handbook of Antibiotic Compounds* (10 vols) Boca Raton, Florida: CRC Press.
Berenblum, I. (1941) *Cancer Res.*, **1,** 807.
Berenblum, I. (1974) in *The Physiopathology of Cancer* (Homburger, W., ed.) 3rd edn, Basel: Karger, p. 393.
Berends, F., Posthumus, C., Van der Sluys, I. and Deierkauf, F. (1959) *Biochim. Biophys. Acta*, **34,** 576.
Bergel, F. and Morrison, A. (1948) *Quart. Rev. Chem. Soc., Lond.*, **2,** 349.
Bergel, F. and Stock, J. (1954) *J. Chem. Soc.*, p. 2409.

Bergel, F. and Todd, A. (1937) *J. Chem. Soc.*, p. 1504.

Berger, M. (1957) *J. Neurochem.*, **2**, 30.

Bergey, D. (1974) *Manual of Determinative Bateriology*, 8th edn, London: Ballière, Tindall and Cox.

Berglöf, F.-E. (1977) *Arthritis Rheum.*, **20**, 108.

Bergmann, E., Ginsburg, D. and Pappo, R. (1959) *Organic Reactions*, **10**, 179.

Bergmann, F., Kwietny, H., Levin, G. and Brown, D. (1960) *J. Amer. Chem. Soc.*, **82**, 598.

Bergmann, F. and Segal, R. (1954) *Biochem. J.*, **58**, 692.

Bernard, C. (1856) *Compt. rend. Acad. Sci. Paris*, **43**, 825.

Bernheim, F. and Bernheim. M. (1939) *Cold Spring Harbor Symp. Quant. Biol.*, **7**, 174.

Berry, H., Cook, A. and Wills, B. (1956) *J. Pharm. Pharmacol.*, **8**, 425.

Bertino, J., Cashmore, A., Fink, N., Calabresi, P. and Lefkowitz, E. (1965) *Clin. Pharmacol. Ther.*, **6**, 763.

Bertino, J. and Johns, D. (1967) *Annu. Rev. Med.*, **18**, 27.

Bessman, S., Rubin, M. and Leikin, S. (1954) *Pediatrics*, **14**, 201.

Beutler, E. (1959) *Blood*, **14**, 103.

Biagi, G., Barbaro, A., Guerra, M., Forti, G. and Fracasso, M. (1974) *J. Med. Chem.*, **17**, 28.

Bicker, V. (1974) *Nature, Lond.*, **252**, 726.

Bigley, W. (1966) *J. Econ. Entomol.*, **59**, 60.

Bijvoet, J., Peerdeman, A. and Van Bommel, A. (1951) *Nature, Lond.*, **168**, 271.

Birch, A., Holzapfel, C., Rickards, R., Djerassi, C., Seidel, P., Suzuki, M., Westley, J. and Dutcher, J. (1964) *Tetr. Lett.*, p. 1491.

Bird, A. and Marshall, A. (1967) *Biochem. Pharmacol.*, **16**, 2275.

Birks, R. and MacIntosh, F. (1957) *Brit. Med. Bull.*, **13**, 157.

Bittar, E. (1964) *Cell pH*, Washington D.C.: Butterworths.

Bittar, E. (1980–1) *Membranes, Structure and Function* (4 vols), New York: Wiley.

Bjerrum, J. (1941) *Metal Ammine Formation in Aqueous Solution*, Copenhagen: Haase.

Black, J., Crowther, A., Shanks, R., Smith, L. and Dornhurst, A. (1964) *Lancet*, **i**, 1080.

Black, J., Duncan, W., Durant, G., Ganellin, C. and Parsons, M. (1972) *Nature, Lond.*, **236**, 385.

Black, J., Durant, G., Emmett, J. and Ganellin, R. (1974) *Nature, Lond.*, **248**, 65.

Blackman, G. (1946) *Agriculture*, **53**, 16.

Blackman, G. (1954) *Nature, Lond.*, **174**, 1179.

Blackwell, G., Carnuccio, R., di Rosa, M., Flower, R., Parente, L. and Persio, P. (1980) *Nature, Lond.*, **287**, 147.

Blackwood, R. (1970) *Adv. Appl. Microbiol.*, **13**, 237.

Blair, D., Clarke, V., Fontanilles, F., Yokogawa, M., Sano, M., Tsuji, M., Kojima, S., Iijima, T. and Ito, Y. (1969) *Ann. N.Y. Acad. Sci.*, **160**, pp. 811, 915, and 933.

Blair, J. (ed.) (1983) *Pteridines and Folic Acid Derivatives*, Berlin: de Gruyter.

Blake, C., Johnson, L., Main, G., North, T., Phillips, D. and Sarma, V. (1967) *Proc. Roy. Soc. B*, **167**, 378.

Blake, J. (1848) *Amer. J. Med. Sci.*, **15**, 63.

Blakeley, R. (1969) *The Biochemistry of Folic Acid and Related Pteridines*, Amsterdam: North-Holland.

Blaney, J., Dietrich, S., Reynolds, M. and Hansch, C. (1979) *J. Med. Chem.*, **22**, 614.

Blaney, J., Jorgensen, E., Connolly, M., Ferrin, T., Langridge, R., Oatley, S., Burridge, J. and Blake, C. (1982) *J. Med. Chem.*, **25**, 785.

Blanpain, P., Nagy, J., Laurent, G. and Durant, F. (1980) *J. Med. Chem.*, **23**, 1283.

Blaschko, H. (1950) *Proc. Roy. Soc. B*, **137**, 307.

Blaschko, H. (1952) *Pharmacol. Rev.*, **4**, 415.

Blaschko, H. (1959) *Pharmacol. Rev.*, **11**, 307.

Blicke, F. and Smith, F. (1930) *J. Amer. Chem. Soc.*, **52**, 2946

Block, S. (1956) *J. Agric. Food Chem.*, **4,** 1042.
Blokhina, N., Vozny, E. and Garin, A. (1972) *Cancer*, **30,** 390.
Blokhuis, G. and Veldstra, H. (1970) *FEBS Letters*, **11,** 197.
Blubaugh, L., Botts, C. and Gerwe, A. (1940) *J. Bact.*, **39,** 51.
Blum, R. and Carter, S. (1974) *Ann. Intern. Med.*, **80,** 249.
Boakes, R., Bradley, P., Brookes, N., Candy, J. and Wolstencroft, J. (1971) *Brit. J. Pharmacol.*, **41,** 462.
Bock, L., Miller, G., Schaper, K. and Seydel, J. (1974) *J. Med. Chem.*, **17,** 23.
Bock, M., Haberkorn, A., Herlinger, H., Meyer, K. and Petersen, S. (1972), *Arzneim. Forsch.*, **22,** 1564.
Bodor, N. (1981) *Drugs of the Future*, **6,** 165.
Bodor, N. (1982) *Trends Pharmacol. Sci.*, **3,** 53.
Bodor, N., Woods, R., Raper, C., Kearney, P. and Kaminski, J. (1980) *J. Med. Chem.*, **23,** 474.
Boehme, E., Applegate, H., Toeplitz, B., Dolfini, J. and Gougoutas, J. (1971) *J. Amer. Chem. Soc.*, **93,** 4324.
Bohman, T. (1980) *Scand. J. Gastroenterol.*, **15,** 183.
Bolin, J., Filman, D., Matthews, D., Hamlin, R. and Kraut, J. (1982) *J. Biol. Chem.*, **257,** 13650.
Bollag, W. (1972) *Eur. J. Cancer*, **8,** 689.
Bolliger, A. (1939) *Analyst*, **64,** 416.
Bondi, A. (1964) *J. Phys. Chem.*, **68,** 441.
Bonner, J. and Varner, J. (1976) *Plant Biochemistry*, 3rd edn, New York: Academic Press.
Bono, V. (1976) *Cancer Treat. Rep.*, **60,** 141.
Bonting, S. (1970) in *Membranes and Ion Transport* (Bittar, E., ed.), vol. 1, chap. 8, New York: Wiley-Interscience.
Borg, D. and Cotzias, G. (1962) *Proc. Natl. Acad. Sci., U.S.A.*, **48,** pp. 623, 643.
Borgna, J.-L. and Rochefort, H. (1981) *J. Biol. Chem.*, **256,** 859.
Borisy, G. and Taylor, E. (1967) *J. Cell. Biol.*, **34,** pp. 525, 535.
Borkovec, A. (1976) *Environ. Health Perspect.*, **14,** 103.
Borowski, E. and Cybulska, B. (1967) *Nature, Lond.*, **213,** 1034.
Borst, P. (1984) *Nature, Lond.*, **309,** 580.
Bosman, H. (1972) *J. Biol. Chem.*, **247,** 130.
Bosund, I. (1960) *Physiol. Planta.*, **13,** 793.
Bouanchaud, D., Scavizzi, M. and Chabbert, Y. (1968) *J. Gen. Microbiol.*, **54,** 417.
Boublik, J., Quinn, M., Clements, J., Herington, A., Wynne, K. and Funder, J. (1983) *Nature, Lond.*, **301,** 246.
Boura, A., Copp, F. and Green, A. (1959) *Nature, Lond.*, **184,** 70.
Boura, A. and Green, A. (1965) *Annu. Rev. Pharmacol.*, **5,** 183.
Bourne, G. (1970) *Division of Labor in Cells*, 2nd edn, New York: Academic Press.
Bovet, D. (1947) *Rendiconti Istituto super. Sanità*, Rome, **10,** 1161.
Bovet, D. and Bovet-Nitti, F. (1949) *Rendiconti Istituto super. Sanità*, Rome, **12,** 7.
Bovet, D., Horclois, R. and Walthert, F. (1944) *Comptes rend. Soc. Biol.*, Paris, **138,** 99.
Bovet, D. and Staub, A. (1937) *Comptes rend. Soc. Biol.*, Paris, **124,** 547.
Bowers, S., Edelstein, M., Cavallito, J., Foss, R. and Nichol, C. (1980) *Proc. 11th Internat. Cong. Chemother., Amer. Soc. Microbiol.*, 1597.
Bowers, W., Ohta, T., Cleere, J. and Marsella, P. (1976) *Science*, **193,** 542.
Bownes, M. (1981) *Differentiation of Cells*, London and New York: Chapman and Hall.
Boyce, C., Jones, T. and Van Tongeren, W. (1967) *Bull. World Health. Org.*, **37,** 1.
Boyd, I. and Pathak, C. (1965) *J. Physiol.*, **176,** 191.
Boyer, P. (ed.) (1970–1983) *The Enzymes*, 3rd edn, New York: Academic Press (in 16 vols).
Boyland, E. (1950) *Biochem. Soc. Symp.*, **5,** 40.
Boyland, E., Dukes, C. and Grover, P. (1963) *Brit. J. Cancer*, **17,** 79.

Boyland, E., Wallace, D. and Williams, D. (1955) *Brit. J. Cancer*, **9,** 62.
Bradbury, F. and Standen, H. (1959) *Nature, Lond.*, **183,** 983.
Brady, L. (ed.) (1980) *Radiation Sensitizers*, New York: Masson.
Braestrup, C., Nielsen, M. and Olsen, C. (1980) *Proc. Natl. Acad. Sci., U.S.A.*, **77,** 2288.
von Brand, T. (1974) *Z. Parasitenkund.*, **45,** 109 (in English).
Bratfos, O. and Haug, J. (1979) *Acta Psychiat. Scand.*, **60,** 1.
Bratkowska-Seniów, B., Kaniak, T., Stumpf, A., Wahl-Mugeńska, M. and Wojnarska, J.
    (1976) *Materia Medica Polona, Warsaw*, **8,** 316.
Breslow, R. (1983) *Chem. Brit.*, **19,** 126.
Breslow, R. and Wernick, D. (1977) *Proc. Natl. Acad. Sci., U.S.A.*, **74,** 1303.
Bresnick, E. and Hitchings, G. (1961) *Cancer Res.*, **21,** 105.
Brewster, P., Hughes, E., Ingold, C. and Rao, P. (1950) *Nature, Lond.*, **166,** 178.
Breyer, B., Buchanan, G. and Duewell, H. (1944) *J. Chem. Soc.*, p. 360.
Brian, R. (1965) *Chem. Indust.*, p. 1955.
Briggs, G. and Haldane, J. (1925) *Biochem. J.*, **19,** 338.
Briggs, M. and Christie, G. (eds) (1977) *Advances in Steroid Biochemistry and Pharmacology*,
    New York: Academic Press.
Brink, F. and Posternak, J. (1948) *J. Cell. Compar. Physiol.*, **32,** 211.
Brockman, R. (1963) *Cancer Res.*, **23,** 1191.
Brockman, R., Kelley, G., Stutts, P. and Copeland, V. (1961) *Nature, Lond.*, **191,** 469.
Brockman, R., Shaddix, S., Laster, W. and Schabel, F. (1970) *Cancer Res.*, **30,** 2358.
Brockmann, H. (1960) *Forts. Chem. org. Naturstoffe*, **18,** 1.
Brodie, B. (1956) *J. Pharm. Pharmacol.*, **8,** 1.
Brodie, B. (1964) *The Pharmacologist, Washington*, **6,** 12.
Brodie, B. and Axelrod, J. (1949) *J. Pharmacol.*, **97,** 58.
Brodie, B., Gillette, J. and Ackerman, H. (1971) *Concepts in Biochemical Pharmacology*, Parts
    1 and 2, Berlin: Springer (*see* Gillette and Mitchell, 1975, for Part 3).
Brodie, B., Gillette, J. and LaDu, B. (1958) *Annu. Rev. Biochem.*, **27,** 427.
Brodie, B. and Hogben, C. (1957) *J. Pharm. Pharmacol.*, **9,** 345.
Brodie, B., Kurz, H. and Schanker, L. (1960) *J. Pharmacol.*, **130,** 20.
Brodie, B., Udenfriend, S., Baer, J., Chenkin, T. and Dill, W. (1945) *J. Biol. Chem.*, **158,**
    705.
Brookes, P. (1966) *Cancer Res.*, **26,** 1994.
Brookes, P. and Lawley, P. (1964) *Nature, Lond.*, **202,** 781.
Brooks, G. (1974) *Chlorinated Insecticides* (2 vols) Cleveland, Ohio: CRC Press.
Brooks, G., Pratt, G. and Jennings, R. (1979) *Nature, Lond.*, **281,** 570.
Brotzu, G. (1948) *Lav. Ist. Igiene Cagliari, Sardinia*.
Broughton, G., Chaplen, P., Knowles, P., Lunt, E., Marshall, S., Pain, D. and
    Wooldridge, K. (1975) *J. Med. Chem.*, **18,** 1117.
Brown, A. (1964) *Bact. Rev.*, **28,** 296.
Brown, A. (1978) *Ecology of Pesticides*, New York: Wiley.
Brown, A., Crum, *see* Crum Brown, A.
Brown, D. and Grigg, G. (1982) *Med. Res. Rev.*, **2,** 193.
Brown, D. and Mason, S. (1956) *J. Chem. Soc.*, 3443.
Brown, E., Aurbach, G., Hauser, D. and Troxler, F. (1976) *J. Biol. Chem.*, **251,** 1232.
Brown, G. (1962) *J. Biol. Chem.*, **237,** 536.
Brown, H., Matzuk, A., Ilves, I., Peterson, L., Harris, S., Sarett, L., Egerton, J., Yakstis,
    J., Campbell, W. and Cuckler, A. (1961) *J. Amer. Chem. Soc.*, **83,** 1764.
Brown, H. and Rogers, E. (1950) *J. Amer. Chem. Soc.*, **72,** 1864.
Brown, M. and Richards, R. (1965) *Nature, Lond.*, **207,** 1391.
Brown, N., Hollinshead, D., Kingbury, P. and Malone, J. (1962) *Nature, Lond.*, **194,** 379.
Brown, P. (1967) *Nature, Lond.*, **213,** 363.
Brown, S., Shafer, R. and Mirau, P. (1982) *J. Amer. Chem. Soc.*, **104,** 5504.

Brown, W. and Pearce, L. (1919) *J. Exper. Med.*, **30,** 483.
Browning, C. (1929) in *A System of Bacteriology in Relation to Medicine*, London: H.M. Stationery Office.
Browning, C. (1955) *Nature, Lond.*, **175,** pp. 570, 616.
Browning, C., Cohen, J., Gaunt, R. and Gulbransen, R. (1922) *Proc. Roy. Soc. B*, **93,** 329.
Browning, C. and Gilmour, W. (1913) *J. Path. Bact.*, **18,** 144.
Browning, C. and Gulbransen, R. (1922) *J. Path. Bact.*, **25,** 395.
Browning, C., Gulbransen, R. and Kennaway, E. (1919) *J. Path. Bact.* **23,** 106.
Browning, C., Gulbransen, R., Kennaway, E. and Thornton, L. (1917) *Brit. Med. J.*, **i,** 73.
Browning, C., Morgan, G., Robb, J. and Walls, L. (1938) *J. Path. Bact.*, **46,** 203.
Bruck, S. (ed.) (1983) *Controlled Drug Design*, Boca Raton, Florida: CRC Press.
Brugmans, J., Thienpont, D., Van Wijngaarden, I., Vanparijs, O., Schuermans, V. and Lauwers, H. (1971) *J. Amer. Med. Assoc.*, **217,** 313.
Bruice, T. and Benkovic, S. (1966) *Bioorganic Mechanisms* (2 vols) New York: Benjamin.
Brulé, G., Eckhardt, S., Hall, T. and Winkler, A. (1973) *Drug Therapy of Cancer*, Geneva: World Health Organization.
Bruni, J. and Wilder, B. (1979) *Arch. Neurol.*, **36,** 393.
Bryan, L. (1982) *Bacterial Resistance and Susceptibility*, Cambridge: University Press.
Bryan, L. and Van den Elzen, H. (1977) *Antimicrob. Agents Chemother.*, **12,** 163.
Bryan, S. and Frieden, E. (1967) *Biochemistry*, **6,** 2728.
Buchanan, J. (1957) in *Chemistry and Biology of Purines* (Wolstenholme, G. and O'Connor, C., eds), London: Churchill. p. 233.
Buchanan, J. (1978) in *Enzyme-Activated Irreversible Inhibitors* (Seiler, N., Jung, M. and Koch-Weser, J., eds), Amsterdam: North Holland.
Buchheim, R. (1872) *Über die 'scharfen' Stoffe, Arch. Heilk.*, 1.
Büchel, K. (1983) (*trans.* Holmwood, G.), *Chemistry of Pesticides*, New York: Wiley.
Büchel, K. and Draber, W. (1969) *Progr. Photosyn. Res.*, **3,** 1777.
Büchel, K. and Schäfer, G. (1970) *Zeits. Naturforsch.*, **25b,** 1465.
Bueding, E. (1962) in *Drugs, Parasites and Hosts* (Goodwin, L. and Nimmo-Smith, R., eds), p. 15. London: Churchill.
Bueding, E., Batzinger, R., Cha, Y., Talalay, P. and Molineaux, C. (1979) *Pharmacol. Rev.*, **30,** 547.
Bueding, E. and Fisher, J. (1966) *Biochem. Pharmacol.*, **15,** 1197.
Bueding, E. and Fisher, J. (1970) *Molec. Pharmacol.*, **6,** 532.
Bueding, E., Hawkins, J. and Cha, Y.-N. (1981) *Agents and Actions*, **11,** 380.
Bukatsch, F. and Haitinger, M. (1940) *Protoplasma,* **34,** 515.
Bülbring, E. and Tomita, T. (1969) *Proc. Roy. Soc. B*, **172,** 103.
Bullen, J. and Rogers, H. (1969) *Nature, Lond.*, **224,** 380.
Bunting, J. and Meathrel, W. (1972) *Canad. J. Chem.*, **50,** 917.
Bunting, J. and Perrin, D. (1967) *J. Chem. Soc. B*, 950.
Burchall, J. (1966) *Fed. Proc.*, **25,** 277.
Burchall, J. and Hitchings, G. (1965) *Molec. Pharmacol.*, **1,** 126.
Burgen, A. (1965) *Brit. J. Pharmacol. Chemother.*, **25,** 4.
Burgen, A. and Iversen, L. (1965) *Brit. J. Pharmacol. Chemother.*, **25,** 34.
Burger, A., Standridge, R. and Ariëns, E. (1963) *J. Med. Chem.*, **6,** 221.
Burger, A., Standridge, R., Stjernström, N. and Marchini, P. (1961) *J. Med. Pharm. Chem.*, **4,** 517.
Burger, M. and Noonan, K. (1970) *Nature, Lond.*, **228,** 512.
Burn, J. (1950) *Brit. Med. J.*, **ii,** 691.
Burn, J. and Dale, H. (1915) *J. Pharmacol.*, **6,** 417.
Burn, J. and Rand, M. (1958) *J. Physiol.*, **144,** 314.
Burns, B. and Paton, W. (1951) *J. Physiol.*, **115,** 41.

Burns, J., Yü, T., Dayton, P., Gutman, A. and Brodie, B. (1960) *Ann. N.Y. Acad. Sci.*, **86,** 253.

Burnstock, G. (ed.) (1981) *Purinergic Receptors*, London and New York: Chapman and Hall.

Burnstock, G., Campbell, G., Satchell, D. and Smythe, A. (1970) *Brit. J. Pharmacol.*, **40,** 668.

Burnstock, G. and Costa, M. (1975) *Adrenergic Neurons*, London and New York: Chapman and Hall.

Burton, D., Clarke, K. and Gray, G. (1964) *J. Chem. Soc.*, p. 1314.

Burtt, E. (1945) *Ann. Appl. Biol.*, **32,** 247.

Buss, E. (1875) *Zbl. f.d. med. Wiss.*, p. 276.

Butler, H., Hurse, A., Thursky, E. and Shulman, A. (1969) *Austral. J. Exper. Biol. Med. Sci.*, **47,** 541.

Butler, T. (1944) *J. Pharmacol.*, **81,** 72.

Butler, T. (1948) *J. Pharmacol.*, **92,** 49.

Butler, T. (1950) *Pharmacol. Rev.*, **2,** 121.

Butler, T. (1953) *J. Pharmacol.*, **109,** 340.

Butler, T. (1955) *J. Amer. Pharm. Assoc.*, **44,** 367.

Butler, T., Waddell, W. and Poole, D. (1965) *Biochem. Pharmacol.*, **14,** 937.

Butterworth, J., Eley, D. and Stone, G. (1953) *Biochem. J.*, **53,** 30.

Byrde, R. and Woodcock, D. (1957) *Nature, Lond.*, **179,** 539.

Byrn, S., Graber, C. and Midland, S. (1976) *J. Org. Chem.*, **41,** 2283.

Cade, J. (1949) *Med. J. Austral.*, **36,** 349.

Cahn, A. and Hepp, P. (1887) *Berl. klin. Woch.*, **24,** pp. 4, 26.

Cahn, R. (1964) *J. Chem. Educ.*, **41,** pp. 116, 503.

Cahn, R. and Dermer, O. (1979) *Introduction to Chemical Nomenclature*, 5th edn, London: Butterworths.

Cahn, R., Ingold, C. and Prelog, V. (1956) *Experientia,* **12,** 81.

Cairns, J. (1962) *Cold Spring Harbor Symp. Quant. Biol.*, **27,** 311.

Cairns, J. (1963) *J. Mol. Biol.*, **6,** 208.

Cairns, J. (1975) *Sci. Amer.*, **233,** 64.

Calabresi, P. and Turner, R. (1966) *Ann. Intern. Med.*, **64,** 352.

Caldwell, J. (1982) in *Metabolic Basis of Detoxication* (Jakoby, W., Bend, J. and Caldwell, J., eds), New York: Academic Press.

Calvert, A., Harland, S., Newell, D., Siddik, Z., Jones, A., McElwain, T., Raju, S., Wiltshaw, E., Smith, I., Baker, J., Peckham, M. and Harrap, K. (1982) *Cancer Chemother. Pharmacol.*, **9,** 140.

Calvin, M. and Wilson, K. (1945) *J. Amer. Chem. Soc.*, **67,** 2003.

Came, P. and Caliguiri, L. (eds) (1982) *Chemotherapy of Viral Infections*, Berlin: Springer.

Cameron, I. and Pool, T. (1981) *The Transformed Cell*, New York: Academic Press.

Cammarata, A., Yau, S., Collett, J. and Martin, A. (1970) *Molec. Pharmacol.*, **6,** 61.

Camp, H., Fukoto, T. and Metcalf, R. (1969) *J. Agric. Food Chem.*, **17,** 243.

Campbell, I. and Dobson, D. (1979) *Methods Biochem. Anal.*, **25,** 1.

Campbell, P. and Kilby, B. (1975) *Basic Biochemistry for Medical Students*, London: Academic Press.

Campbell, P. and Smith, A. (1982) *Biochemistry Illustrated*, London: Churchill-Livingstone; New York: Longman.

Campion, J. and Tichon, M. (1981) *Proc. Sixth Austral. Weeds Conf.*, Toowomba, Queensland: Harrison.

Candy, D. and Kilby, B. (eds) (1975) *Insect Biochemistry and Function*, London and New York: Chapman and Hall.

Canellakis, E., Bono, V., Bellantone, A., Krakow, J., Fico, R. and Schulz, R. (1976) *Biochim. Biophys. Acta*, **418,** 300.

Canellakis, E., Shaw, Y., Hanners, W. and Schwartz, R. (1976) *Biochim. Biophys. Acta*, **418,** 277.
Canepa, F., Pauling, P. and Sörum, H. (1966) *Nature, Lond.*, **210,** 907.
Canfield, C. and Rozman, R. (1974) *Bull. World Health Org.*, **50,** 203.
Cantley, L., Resh, M. and Guidotti, G. (1978) *Nature, Lond.*, **272,** 552.
Caranikas, S., Mizrahi, J., Escher, E. and Regoli, D. (1982) *J. Med. Chem.*, **25,** 1313.
Carlsson, A. and Lundqvist, M. (1963) *Acta Pharmacol. Toxicol.*, **20,** 140.
Carlstrom, D. and Bergin, R. (1967) *Acta Crystallogr.*, **23,** 313.
Carmenier, G. (1968) *Biochem. Pharmacol.*, **17,** 1981.
Carmichael, J. and Bell, F. (1944) *J. Comp. Path. Therap.*, **54,** 49.
Caroon, J., Clark, R., Kluge, A., Olah, R., Repke, D., Unger, S., Michel, A. and Whiting, R. (1982) *J. Med. Chem.*, **25,** 666.
Carson, D. and Chang, K.-P. (1981) *Biochem. Biophys. Res. Commun.*, **100,** 1377.
Carson, S. (1982) *Biochem. Pharmacol.*, **31,** 1806.
Carson, S., Godwin, S., Massoulie, J. and Kato, G. (1977) *Nature, Lond.*, **266,** 176.
Carter, G., Huppatz, J. and Wain, R. (1976) *Ann. Appl. Biol.*, **84,** 333.
Carter, S. (1972) *Endeavour*, **31,** 77.
Carter, S. and Blum, R. (1976) *Progr. Biochem. Pharmacol.*, **11,** 158.
Carter, S. and Friedman, M. (1972) *Eur. J. Cancer*, **8,** 85.
Carter, S., Ichikawa, T., Mathe, G. and Umezawa, H. (1976) *Fundamental and Clinical Studies of Bleomycin*, Baltimore: University Park Press.
Carter, S., Sakurai, Y. and Umezawa, H. (eds) (1981) *New Drugs in Cancer Chemotherapy*, New York: Springer.
Carty, T., Eskra, J., Lombardino, J. and Hoffman, W. (1980) *Prostaglandins*, **19,** 51.
Casida, J. (1973) *Annu. Rev. Biochem.*, **42,** 259.
Casida, J., Holmstead, R., Khalifa, S., Knox, J., Ohsawa, T., Palmer, K. and Wong, R. (1974) *Science*, **183,** 520
Casida, J., McBryde, L. and Niedermeir, R. (1962) *J. Agric. Food Chem.*, **10,** 370.
Casselton, P. (1964) *Nature, Lond.*, **204,** 93.
Castagna, M., Takai, Y., Kaibuchi, K., Sano, K., Kikkawa, U. and Nishizuka, Y. (1982) *J. Biol. Chem.*, **257,** 7847.
Casy, A. (1971) *Pmr Spectroscopy in Medicinal and Biological Chemistry*, London: Academic Press.
Casy, A. and Ison, R. (1970) *J. Pharm. Pharmacol.*, **22,** 270.
Cattell, W., Chamberlain, D., Fry, I., McSherry, M., Broughton, C. and O'Grady, F. (1971) *Brit. Med. J.*, **i,** 377.
Cavalla, J. (ed.) (1981) *Risk-Benefit Analysis in Drug Research*, Boston: MTP Press.
Cavalli-Sforza, L. and Lederberg, J. (1956) *Genetics*, **41,** 367.
Cawthon, R. and Breakefield, X. (1979) *Nature, Lond.*, **281,** 692.
Cederbaum, D. (1979) *New Engl. J. Med.*, p. 301.
Cervello, V. (1882) *Arch. per le Sci. med.*, **6,** 177.
Cha, S., Agarwal, R. and Parks, R. (1975) *Biochem. Pharmacol.*, **24,** 2187.
Chacko, G., Villegas, G., Barnola, F., Villegas, R. and Goldman, D. (1976) *Biochim. Biophys. Acta*, **443,** 19.
Chain, E. (1948) *Annu. Rev. Biochem.*, **17,** 657.
Chambron, J., Daune, M. and Sadron, C. (1966) *Biochim. Biophys. Acta*, **123,** pp. 306, 319.
Chance, B. and Sacktor, B. (1958) *Arch. Biochem. Biophys.*, **76,** 509.
Chang, K., Hazum, E., Kilian, A. and Cuatrecasas, P. (1981) *Molec. Pharmacol.*, **1,** 20.
Changeux, J.-P. (1969) *Proc. Nobel Symp.*, **11,** 235.
Changeux, J.-P. (1980) *Trends Pharmacol. Sci.*, **1,** 198.
Chapman, D. (1968–82) *Biological Membranes* (4 vols) London: Academic Press.
Chapman, N. and Shorter, J. (1972) *Advances in Free Energy Relationships*, London: Plenum Press.

Chapman, R. and Penman, D. (1979) *Nature, Lond.*, **281,** 298.

Chappell, J. (1966) *Biochem. J.*, **100,** 43P.

Chen, M. and Prusoff, W. (1979) *J. Biol. Chem.*, **254,** 1049.

Chen, M., Ward, D. and Prusoff, W. (1976) *J. Biol. Chem.*, **251,** 4833.

Chen, R., Kraemer, C., Schmidmayr, W., Chem-Schmeisser, U. and Henning, U. (1982) *Biochem. J.*, **203,** 33.

Cheng, C., Fugimura, S., Grunberger, D. and Weinstein, I. (1972) *Cancer Res.*, **32,** 22.

Cheung, W. (1980–82) *Calcium and Cell Function* (3 vols) New York: Academic Press.

Chick, H. (1908) *J. Hyg.*, **8,** 92.

Chien, M., Grollman, A. and Horwitz, S. (1977) *Biochemistry*, **16,** 3641.

Chignell, C. (1970) *Molec. Pharmacol.*, **6,** 1.

Childs, A., Davies, D., Green, A. and Rutland, J. (1955) *Brit. J. Pharmacol.*, **10,** 462.

Childs, C. (1970) *Inorg. Chem.*, **9,** 2465.

Chiu, T., Dryden, D. and Rosenberg, H. (1982) *Molec. Pharmacol.*, **21,** 57.

Chiu, T. and Rosenberg, H. (1983) *Trends Pharmacol. Sci.*, **4,** 348.

Chiu, Y., Main, A. and Dauterman, W. (1969) *Biochem. Pharmacol.*, **18,** 2171.

Cho, A., Haslett, W. and Jenden, D. (1962) *J. Pharmacol.*, **138,** 249.

Chong, C. and Rickards, R. (1970) *Tetr. Lett.*, p. 5145.

Chothia, C. (1970) *Nature, Lond.*, **225,** 36.

Chothia, C. and Pauling, P. (1970) *Nature, Lond.*, **226,** 541.

Christopherson, J. (1918) *Lancet*, **ii,** 325.

Chung, K.-J., Killian, A., Hazum, E., Cuatrecasas, P. and Chang, J.-K. (1981) *Science*, **212,** 75.

Cigén, R. (1958) *Acta Chem. Scand.*, **12,** 1456.

Cirillo, V., Harsch, M. and Lampen, J. (1964) *J. Gen. Microbiol.*, **35,** 249.

Citri, Y. and Schramm, M. (1980) *Nature, Lond.*, **287,** 297.

Clark, A. (1926) *J. Physiol.*, **61,** 530.

Clark, A. (1927) *J. Physiol.*, **64,** 123.

Clark, A. (1933) *The Mode of Action of Drugs on Cells*, London: Edward Arnold.

Clark, A. (1937) in *Handbuch der experimentellen Pharmakologie* (Heffter, A. and Heubner, W., eds), Berlin: Springer, E-4, 63.

Clark, A. and Raventós, J. (1937) *Quart. J. Exper. Physiol.*, **26,** 375.

Clark, J. and Perrin, D. (1964) *Quart. Rev. Chem. Soc., London*, **18,** 295.

Clark, R. and Panchen, A. (1971) *Synopsis of Animal Classification*, London and New York: Chapman and Hall.

Clarke, F. (1977) *J. Chem. Educ.*, **54,** 230.

Clarkson, T., Rothstein, A. and Sutherland, R. (1965) *Brit. J. Pharmacol. Chemother.*, **24,** 1.

Cleland, W. (1970) in *The Enzymes* (Boyer, P., ed.) 3rd edn, New York: Academic Press.

Clements, J. and Wilson, K. (1962) *Proc. Natl. Acad. Sci., U.S.A.*, **48,** 1008.

Clemons, G. and Sisler, H. (1969) *Phytopathology*, **59,** 705.

de Clercq, E. (1983) in *Control of Virus Diseases* (Kurstak, E., ed.), New York: Marcel Dekker.

de Clercq, E., Degreef, H., Wildiers, J., de Jonge, G., Drochmans, A., Descamps, J. and de Somer, P. (1980) *Brit. Med. J.*, **281, i,** 1178.

de Clercq, E., Descamps, J., de Somer, P., Barr, P., Jones, A. and Walker, R. (1979) *Proc. Natl. Acad. Sci., U.S.A.*, **76,** 2947.

de Clercq, E., Descamps, J., de Somer, P. and Hóly, A. (1978) *Science*, **200,** 563.

de Clercq, E. and Hóly, A. (1979) *J. Med. Chem.*, **22,** 510 (review).

Clowes, G. and Keltch, A. (1931) *Proc. Soc. Exper. Biol. Med.*, **29,** 312.

Clowes, G., Keltch, A. and Krahl, M. (1940) *J. Pharmacol.*, **68,** 312.

Coats, J. (1983) *Insecticide Mode of Action*, New York: Academic Press.

Cocco, L., Groff, J., Temple, C., Montgomery, J., London, R., Matwiyoff, N. and Blakley, R. (1981) *Biochemistry*, **20,** 3972.

Cohen, G., Gibby, E. and Mehta, R. (1981) *Nature, Lond.*, **291,** 662.

Cohen, J. (ed.) (1980) *Magnetic Resonance in Biology*, Vol. 1, New York: Wiley.

Cohen, N. (1966) *Biol. Rev.*, **41,** 503.

Cohen, N. (1983) *J. Med. Chem.*, **26,** 259.

Cohen, P. (1980) *Control of Enzyme Activity*, London and New York: Chapman and Hall.

Cohen, S. (1963) *Annu. Rev. Biochem.*, **32,** 83.

Cohen, S. (1971) *Polyamines*, New Jersey: Prentice-Hall.

Cohen, S. (1976) *Med. Biol.*, **54,** 299.

Cohen, S. and Yielding, K. (1965) *Proc. Natl. Acad. Sci., U.S.A.*, **54,** 521.

Cohn, E., McMeekin, T., Edsall, J. and Weare, J. (1934) *J. Amer. Chem. Soc.*, **56,** 2270.

Cohn, M. and Leigh, J. (1962) *Nature, Lond.*, **193,** 1037.

Colburn, R. and Maas, J. (1965) *Nature, Lond.*, **208,** 37.

Cole, A., May, P. and Williams, D. (1981) *Agents and Actions*, **11,** 296.

Colebrook, L. and Kenny, M. (1936) *Lancet*, **i,** 1279.

Coleman, J. (1974) *Biochem. Biophys. Res. Commun.*, **60,** 641.

Coleman, R. (1973) *Biochim. Biophys. Acta*, **300,** 1 (review).

Coles, C. (1977) *Pestic. Sci.*, **8,** 536.

Collander, R. (1933) *Acta Botan. Fenn., Finland*, **11,** 1.

Collander, R. (1937) *Trans. Farad. Soc.*, **33,** 985.

Collander, R. (1947) *Acta Physiol. Scand.*, **13,** 363.

Collander, R. (1954) *Acta Chem. Scand.*, **5,** 774.

Collier, H. and Francis, D. (1975) *Nature, Lond.*, **255,** 159.

Collier, H. and Waterhouse, P. (1950) *Ann. Trop. Med. Parasit.*, **44,** 156.

Colquhoun, D., Dionne, V., Steinbach, J. and Stevens, C. (1975) *Nature, Lond.*, **253,** 204.

Condon, M., Petrillo, E., Ryono, D., Reid, J., Neubeck, R., Puar, M., Heikes, J., Sabo, E., Losee, K., Cushman, D. and Ondetti, M. (1982) *J. Med. Chem.*, **25,** 250.

Connemacher, R. and Mandel, H. (1965) *Biochem. Biophys. Res. Commun.*, **20,** 98.

Conney, A. (1967) *Pharmacol. Rev.*, **19,** 317.

Conney, A. and Burns, J. (1962) *Adv. Pharmacol.*, **1,** 31.

Conney, A., Levin, W., Ikeda, M., Kuntzman, R., Cooper, D. and Rosenthal, O. (1968) *J. Biol. Chem.*, **243,** 3912.

Connors, T., Cox, P., Farmer, P., Foster, A., Jarman, M. and Macleod, J. (1974) *Biomed. Mass Spectrom.*, **1,** 130; *Biochem. Pharmacol.*, **23,** 115.

Connors, T. and Hare, J. (1974) *Brit. J. Cancer*, **30,** 477.

Connors, T., Jones, M., Ross, W., Braddock, P., Khokhar, A. and Tobe, M. (1972) *Chem. Biol. Interactions*, **5,** 415.

Conti, F. (1969) *Nature, Lond.*, **221,** 777.

Conti-Tronconi, B. and Raferty, M. (1982) *Annu. Rev. Biochem.*, **51,** 491.

Cook, D. (1967) *J. Molec. Biol.*, **29,** 167.

Cook, J., Hewett, C. and Hieger, I. (1933) *J. Chem. Soc.*, p. 396.

Cooper, J. and Bloom, F. (1982) *Biochemical Basis of Neuropharmacology*, Oxford: University Press.

Cooper, P. (1956) *Bact. Rev.*, **20,** 28.

Corbett, J., Wright, K. and Baillie, A. (1984), *The Biochemical Mode of Action of Pesticides*, New York: Academic Press.

Cordes, E. (1973) *Reaction Kinetics in Micelles*, New York: Plenum Press.

Cornforth, J., Milborrow, B. and Ryback, G. (1965) *Nature, Lond.*, **206,** 715.

Cornforth, J., Ryback, G., Popjak, G., Donninger, C. and Schroepfer, G. (1962) *Biochem. Biophys. Res. Commun.*, **9,** 371.

Costa, E., Di Chiara, G. and Gessa, G. (eds) (1981) *GABA and Benzodiazepine Receptors* (Vol. 26 of *Adv. Biochem. Pharmacol.*), New York: Raven Press.

Cotzias, G., Van Woert, M. and Schiffer, L. (1967) *New Engl. J. Med.*, **276,** 374.

Coulston, F. (ed.) (1979) *Regulatory Aspects of Carcinogenesis and Food Additives; the Delaney Clause*, New York: Academic Press.

Covey, J. (1980) *Life Sci.*, **26**, 665.

Covino, B. and Vassallo, H. (1976) *Local Anaesthetics: Mechanisms of Action and Clinical Use*, New York: Grune and Stratton.

Cowdry, E. and Ruangsiri, C. (1941) *Arch. Pathol.*, **32**, 632.

Cox, J. (1967) *Nature, Lond.*, **216**, 1328.

Crafts, A. (1964) in *Physiology and Biochemistry of Herbicides* (Audus, L., ed.), New York: Academic Press.

Cragoe, E. (ed.) (1983) *Diuretics*, New York: Wiley.

Craig, P. (1971) *J. Med. Chem.*, **14**, 680.

Cram, D. (1983) *J. Amer. Chem. Soc.*, **105**, 135.

Cramer, F. (1956) *Angew. Chem.*, **68**, 115.

Cramer, H. (1967) *Pflanzenschutz Nachr. Bayer*, **20**, 1.

Cramer, J., Miller, J. and Miller, E. (1960) *J. Biol. Chem.*, **235**, 885.

Crathorn, A. and Hunter, G. (1958) *Biochem. J.*, **69**, 47P.

Cremer-Bartels, G. (1975) in *Chemistry and Biology of Pteridines* (Pfleiderer, W., ed.), Berlin: de Gruyter.

Criss, W. (1973) *Cancer Res.*, **33**, pp. 51, 57.

Crooke, S. and Bradner, W. (1976) *Cancer Treat. Rev.*, **3**, 121.

Crossland, J. (ed.) (1981) *Lewis's Pharmacology*, 5th edn, London: Churchill, p. 85.

Crow, T. and Johnstone, E. (1977) *Brit. J. Pharmacol.*, **59**, 466P.

Crow, W., Nicholls, W. and Sterns, M. (1971) *Tetr. Lett.*, p. 1353.

Crowther, A. and Levi, A. (1953) *Brit. J. Pharmacol.*, **8**, 93.

Cruess, W. and Richert, P. (1929) *J. Bact.*, **17**, 363.

Cruickshank, I. (1963) *Annu. Rev. Phytopath.*, **1**, 351.

Crum Brown, A. and Fraser, T. (1869) *Trans. Roy. Soc. Edinburgh*, **25**, pp. 151, 693.

Crumplin, G., Midgley, J. and Smith, J. (1980) *Top. Antibiot. Chem.*, **3**, 1 (review).

Crutchler, W. and Moschella, S. (1975) *Brit. J. Dermatol.*, **92**, 199.

Cuatrecasas, P., Wilcheck, M. and Anfinsen, C. (1968) *Proc. Natl. Acad. Sci., U.S.A.*, **61**, 636.

Cucinell, S., Conney, A., Sansur, M. and Burns, J. (1965) *Clin. Pharmacol. Ther.*, **6**, 420.

Cullen, S. and Gross, E. (1951) *Science*, **113**, 580.

Culp, L. and Black, P. (1972) *J. Virol.*, **9**, 611.

Culvenor, C. and Ham, N. (1966) *Chem. Commun., Chem. Soc., London*, p. 537.

Cunningham, L. (1964) *Biochemistry*, **3**, 1629.

Curd, F., Davey, D. and Rose, F. (1945) *Ann. Trop. Med. Parasit.*, **39**, 208.

Curtis, D., Duggan, A., Felix, D., Johnston, G. and McLennan, H. (1971) *Brain Res.*, **33**, 57.

Curtis, D. and Johnston, G. (1970) *Handb. Neurochem.*, **4**, 115.

Curtis, D. and Lodge, D. (1977) *Nature, Lond.*, **270**, 543.

Cushley, R. and Mautner, H. (1970) *Tetrahedron*, **26**, 2151.

Cushny, A. (1909) *J. Physiol.*, **38**, 359.

Cushny, A. (1926) *Biological Relations of Optically Isomeric Substances*, Baltimore: Williams and Wilkins.

Cymerman-Craig, J., Rubbo, S., Willis, D. and Edgar, J. (1955) *Nature, Lond.*, **176**, 35.

Cymerman-Craig, J. and Willis, D. (1955) *J. Chem. Soc.*, p. 4315.

Dainton, F. (1966) *Chain Reactions*, London and New York: Methuen.

Dale, H. (1914) *J. Pharmacol.*, **6**, 147.

Daly, J., Brons, R. and Snyder, S. (1981) *Life Sci.*, **28**, 2083.

Daniels, M. (1971) *Biochem. J.*, **122**, 197.

Danilov, A., Guli-Kevkhyan, R., Laverentieva, V., Michelson, M., Mndjoyan, O.,

Shelkovnikow, A. and Starshinova, L. (1974) *Arch. internat. Pharmacodyn. Thér.*, **208**, 35.

Das, H., Goldstein, A. and Kanner, L. (1966) *Molec. Pharmacol.*, **2**, 158.

Dauwalder, M., Whaley, W. and Kephart, J. (1972) *Sub-cell. Biochem.*, **1**, 225.

Davidoff, R. (1978) *Neurology*, **28**, 46.

Davidse, L., Gerritsma, O., Hofman, J. and Velthuis, G. (1982) *Fifth Internat. Congr. Pestic. Chem.*, IUPAC, Kyoto, Japan.

Davidson, J. (1976) *Biochemistry of the Nucleic Acids*, 8th edn, New York: Academic Press.

Davidson, M., Griggs, B., Boykin, D. and Wilson, W. (1977) *J. Med. Chem.*, **20**, 1117.

Davies, F., Musa, M. and Dormandy, T. (1968) *J. Clin. Path.*, **21**, 363.

Davies, G. and Lowe, J. (1966) *Brit. J. Pharmacol. Chemother.*, **27**, 107.

Davies, J., Campbell, W. and Kearns, C. (1970) *Biochem. J.*, **117**, 221.

Davis, B. and Dubos, R. (1947) *J. Exper. Med.*, **86**, 215.

Davis, C. and Harvey, R. (1979) *Proc. N.E. Weed Sci. Soc.*, **33**, 112.

Davis, S. (1973) *J. Pharm. Pharmacol.*, **25**, pp. 1, 293.

Davson, H. and Danielli, J. (1952) *The Permeability of Natural Membranes*, 2nd edn, Cambridge: University Press.

Dawson, P., Gutteridge, W. and Gull, K. (1983), *Molec. Biochem. Parasit.*, **7**, 267.

Deake, H., Hu, S.-I. and Ward, H. (1980) *J. Biol. Chem.*, **255**, 7174.

Dean, R. and Barrett, A. (1976) *Essays Biochem.*, **12**, 1.

De Baun, J., Smith, J., Miller, E. and Miller, J. (1970) *Science*, **167**, 184.

De Bruyn, P., Robertson, R. and Farr, F. (1950) *Anat. Rec.*, **108**, 279.

Deeves, R., Serrano, R. and South, D. (1976) *J. Biol. Chem.*, **249**, 7737.

Dekker, J. (1968) *Neth. J. Plant. Path.*, **74** *(Supp.* **1**), 127.

Delay, J., Deniker, P. and Harl, J. (1952) *Ann. Méd-psychol.*, **110**, 112.

Del Castillo, J. and Katz, B. (1957) *Proc. Roy. Soc. B*, **146**, 339.

Del Castillo, J., Mello, W. and Morales, T. (1964) *Brit. J. Pharmacol.*, **22**, 463.

De Ley, J. and Docky, R. (1960) *Biochim. Biophys. Acta*, **40**, 277.

Delp, C. and Klopping, H. (1968) *Plant Dis. Rep.*, **52**, 95.

Demel, R., Van Deenen, L. and Kinsky, S. (1965) *J. Biol. Chem.*, **240**, 2749.

Dennis, V., Stead, N. and Andreoli, T. (1970) *J. Gen. Physiol.*, **55**, 375.

Denny, W., Atwell, G. and Cain, B. (1979) *J. Med. Chem.*, **22**, 1453.

Denny, W., Baguley, B., Cain, B. and Waring, M. (1983) in *Molecular Aspects of Anti-Cancer Drug Action* (Neidle, S. and Waring, M., eds), London: Macmillan.

Deskin, W. (1958) *J. Amer. Chem. Soc.*, **80**, 5680.

Desowitz, R., Bell, T., Williams, J., Cardines, R. and Tamarua, M. (1970) *Amer. J. Trop. Med. Hyg.*, **19**, 775.

De Wys, W. and Pories, W. (1972) *J. Natl. Cancer Inst.*, **48**, 375.

Diamond, J. and Wright, E. (1969) *Annu. Rev. Physiol.*, **31**, 581.

Dickerson, R. and Geis, I. (1969) *Structures and Action of Proteins* (with Stereo Supplement), New York: Harper and Row.

Dietrich, B., Fyles, T., Lehn, J., Pease, L. and Fyles, D. (1978) *Chem. Commun., Chem. Soc., Lond.*, p. 934.

Dietrich, S., Bolger, M., Kollman, P. and Jorgensen, E. (1977) *J. Med. Chem.*, **20**, 863.

Dill, W., Fisken, R., Reutner, T., Weston, J. and Glazko, A. (1957) *Antibiot. Chemother.*, **7**, 99.

DiLuzio, N. (1983) *Trends Pharmacol. Sci.*, **4**, 344.

Dimond, A. and Horsfall, J. (1959) *Annu. Rev. Plant Physiol.*, **10**, 257.

Dittmer, K. (1949) *J. Amer. Chem. Soc.*, **71**, 1205.

Dittmer, K. and du Vigneaud, V. (1944) *Science*, **100**, 129.

Dixon, M. and Webb, E. (1979) *Enzymes*, 3rd edn, London: Longmans; New York: Academic Press.

Dixon, N., Gazzola, C., Blakeley, R. and Zerner, B. (1975) *J. Amer. Chem. Soc.*, **97**, 4131.

Djerassi, C. (1960) *Optical Rotatory Dispersion*, New York: McGraw-Hill.

Dobkin, A. (1975) *Clin. Pharmacol. Ther.*, **18**, 547.

Dobler, M., Dunitz, J. and Kilbourn, B. (1969) *Helv. Chim. Acta*, **52**, 2573.

Docherty, J., MacDonald, A. and McGrath, J. (1979) *Brit. J. Pharmacol.*, **67**, 421.

Dockter, M. and Magnuson, J. (1973) *Biochem. Biophys. Res. Commun.*, **54**, 790.

Dodd, M. (1946) *J. Pharmacol.*, **86**, 311.

Dodds, E., Goldberg, L., Lawson, W. and Robinson, R. (1938) *Nature, Lond.*, **142**, 34.

Doerge, R. (ed.) (1982) *Wilson and Gisvold's Textbook of Organic, Medicinal and Pharmaceutical Chemistry,* 8th edn, Philadelphia: Lippincott.

Doherty, D., Shapira, R. and Burnett, W. (1957) *J. Amer. Chem. Soc.*, **79**, 5667.

Doi, O., Miyamoto, N., Tanaka, N. and Umezawa, H. (1968) *Appl. Microbiol.*, **16**, 1282.

Dole, V., Nyswander, M. and Kreek, M. (1966) *Arch. Intern. Med.*, **118**, 304.

Dolin, M. (1961) in *The Bacteria* (Gunsalus, I. and Stanier, R., eds). New York: Academic Press, Vol. 2, p. 425.

Dolly, J. and Barnard, E. (1977) *Biochemistry*, **16**, 5053.

Doluizio, J. and Martin, A. (1963) *J. Med. Chem.*, **6**, pp. 16, 20.

Domagk, G. (1935) *Dtsch. med. Woch.*, **61**, 250.

Domagk, G. (1936) *Klin. Woch.*, **15**, 1585.

Dominguez, R. (1933) *Proc. Soc. Exper. Biol. Med.*, **31**, 1146.

Donald, C., Passey, B. and Swaby, R. (1952) *J. Gen. Microbiol.*, **7**, 211.

Douch, P. (1976) *Xenobiotica*, **6**, 531.·

Douglas, C., Haldane, J.S. and Haldane, J.B.S. (1912) *J. Physiol.*, **44**, 275.

Douglas, W. (1968) *Brit. J. Pharmacol.*, **34**, 451.

Dreser, H. (1899) *Pflügers Arch. ges. Physiol.*, **76**, 306.

Dring, L., Smith, R. and Williams, R. (1970) *Biochem. J.*, **116**, 425.

Dubos, R. (1939) *J. Exper. Med.*, **70**, 1.

Duch, D., Sigel, C., Bowers, S., Edelstein, M., Cavallito, J., Foss, R. and Nichol, C. (1980) *Curr. Chemother. Infect. Dis., Proc. Int. Congr.*, **2**, 1597.

Dudai, Y. and Ben-Barak, J. (1977) *FEBS Lett.* **81**, 134.

Duggar, B. (1948) *Ann. N.Y. Acad. Sci.*, **51**, 177.

Dunitz, J. (1952) *J. Amer. Chem. Soc.*, **74**, 995.

Durant, G., Emmett, J., Ganellin, C., Miles, P., Parsons, M., Prain, H. and White, G. (1977) *J. Med. Chem.*, **20**, 901.

Durant, G., Ganellin, C. and Parsons, M. (1975) *J. Med. Chem.*, **18**, 905.

Durant, G., Ganellin, C. and Parsons, M. (1977) *Agents and Actions*, **7**, 39.

Dutcher, J., Boyak, G. and Fox, S. (1954) *Antibiotic Annual,* New York: Medical Encyclopedia Inc.

Dweck, R. (ed.) (1977) *Nuclear Magnetic Resonance in Biology,* New York: Academic Press.

Dwyer, F., Gyarfas, E., Koch, J. and Rogers, W. (1952) *Nature, Lond.*, **170**, 190.

Dwyer, F., Gyarfas, E. and O'Dwyer, M. (1951) *Nature, Lond.*, **167**, 1036.

Dyckes, D., Nestor, J., Ferger, N. and du Vigneaud, V. (1974) *J. Med. Chem.*, **17**, 969.

Eagle, H. (1939) *J. Pharmacol.*, **66**, pp. 10, 423, 436.

Eagle, H. (1945) *J. Pharmacol.*, **85**, 265.

Eagle, H. (1951) *Pharmacol. Rev.*, **3**, 107.

Eagle, T., Zon, G. and Egan, W. (1979) *J. Med. Chem.*, **22**, 897.

Easson, L. and Stedman, E. (1933) *Biochem. J.*, **27**, 1257.

Eccles, J. (1957) *The Physiology of Nerve Cells,* Baltimore: Johns Hopkins Press, pp. 193–5.

Eccles, J. (1965) *Sci. Amer.*, **212** (No. 1), 56.

Eddy, N., Friebel, H., Hohn, K. and Halbach, H. (1969) *Bull. World Health Org.*, **40**, 639.

Edelman, I., Bogoroch, R. and Porter, G. (1963) *Proc. Natl. Acad. Sci., U.S.A.*, **50**, 1169.

Edgerton, L. and Blanpied, G. (1968) *Nature, Lond.*, **219**, 1064.

Edgington, L., Walton, G. and Miller, P. (1966) *Science*, **153**, 307.

Edlbacher, S., Baur, H. and Becker, M. (1940) *Z. physiol. Chem.*, **265**, 61.

Edsall, J., Martin, R. and Hollingworth, B. (1958) *Proc. Natl. Acad. Sci.*, U.S.A., **44**, 505.
Edson, E., Sanderson, D. and Noakes, D. (1964) *World Rev. Pest Control*, **4**, 36.
Edson, E., Sanderson, D. and Noakes, D. (1966) *World Rev. Pest Control*, **5**, 143.
Edwards, D. (1982) *Progr. Med. Chem.*, **18**, 87.
Eger, E., Lundgren, C., Miller, S. and Stevens, W. (1969) *Anaesthesiology*, **30**, 129.
Ehrenpreis, S., Fleisch, J. and Mittag, T. (1969) *Pharmacol. Rev.*, **21**, 131.
Ehrlich, P. (1900) *Proc. Roy. Soc., Lond. B*, **66**, 424.
Ehrlich, P. (1907) *Three Harber lectures*, to Royal Inst. Public Health, London: Lewis.
Ehrlich, P. (1908) Nobel Prize Lecture, *Les Prix Nobel*, Stockholm (in German); English translation in Himmelweit (1956) (q.v.) **3**, 183.
Ehrlich, P. (1909) *Ber. dtsch. chem. Ges.*, **42**, 17.
Ehrlich, P. (1911) *Theorie und Praxis der Chemotherapie*, Leipzig: Klinkhardt.
Ehrlich, P. (1912) in Benda, L. (1912) (q.v.).
Ehrlich, P. and Bertheim, A. (1907) *Ber. dtsch. chem. Ges.*, **40**, 3292.
Ehrlich, P. and Bertheim, A. (1912) *Ber. dtsch. chem. Ges.*, **45**, 756.
Ehrlich, P. and Hata, S. (1910) *Die experimentelle Chemotherapie der Spirillosen*, Berlin: Springer.
Ehrlich, P. and Morgenroth, J. (1900) *Berl. klin. Woch.*, 453.
Ehrlich, P. and Shiga K. (1904) *Berl. klin. Woch.*, 329.
Einhorn, A. (1905) *Dtsch. med. Woch.*, **31**, 1668.
Einhorn, L. and Williams, S. (1979) *New Engl. J. Med.*, **300**, 289.
Eisleb, O. and Schaumann, O. (1938) *Dtsch. med. Woch.*, **65**, 967.
Elderfield, R. (1946) *Chem. Eng. News*, **24**, 2598.
Eliel, E., Allinger, N., Angyal, S. and Morrison, G. (1965) *Conformational Analysis*, New York: Wiley-Interscience.
Eliel, E. and Basolo, F. (1969) *Elements of Stereochemistry*, New York: Wiley.
Elion, G. (1967) *Fed. Proc.*, **26**, 898.
Elion, G. (1980) *Adv. Enzyme Regul.*, **18**, 53.
Elion, G., Burgi, E. and Hitchings, G. (1952) *J. Amer. Chem. Soc.*, **74**, 411.
Elion, G., Callahan, S., Rundles, R. and Hitchings, G. (1963) *Cancer Res.* **23**, 1207.
Elion, G., Furman, P., Fyfe, J., de Miranda, P., Beauchamp, E. and Schaeffer, H. (1977) *Proc. Natl. Acad. Sci., U.S.A.*, **74**, 5716.
Elion, G. and Hitchings, G. (1975) in *Antineoplastic and Immunosuppressive Agents*, Part 2 (Sartorelli, A. and Johns, D., eds), Berlin: Springer, p. 404.
Elion, G., Kovensky, A., Hitchings, G., Metz, E. and Rundles, R. (1966) *Biochem. Pharmacol.*, **15**, 863.
El Khadem, H. (ed.) (1982) *Anthracycline Antibiotics*, New York: Academic Press.
Elkhawad, A. and Woodruff, G. (1975) *Brit. J. Pharmacol.*, **54**, 107.
Ellenbroek, B. and Van Rossum, J. (1960) *Arch. Internat. Pharmacodyn. Thér.*, **125**, 216.
Elliott, M., Farnham, A., Janes, N., Needham, P. and Pearson, B. (1967) *Nature, Lond.*, **213**, 493.
Elliott, M., Farnham, A., Janes, N., Needham, P. and Pulman, D. (1974) in *Mechanisms of Pesticide Action* (Kohn, G., ed.) Washington: Amer. Chem. Society.
Elliott, M., Farnham, A., Janes, N., Needham, P., Pulman, D. and Stevenson, J. (1973) *Nature, Lond.*, **246**, 169.
Elliott, M., Janes, N. and Potter, C. (1978) *Annu. Rev. Entomol.*, **23**, 443.
Elliott, T. (1905) *J. Physiol.*, **32**, 401.
Elliott, W. (1963) *Biochem. J.*, **86**, 562.
Ellouz, F., Adam, A., Cirobaru, R. and Lederer, E. (1974) *Biochem. Biophys. Res. Commun.*, **59**, 1317.
Elworthy, P., Florence, A. and Macfarlane, C. (1968) *Solubilization by Surface-Active Agents*, London and New York: Chapman and Hall.

Endo, H., Ono, T. and Sugimura, T. (1970) *Chemistry and Biological Actions of 4-Nitro-quinoline-1-oxide,* Berlin: Springer.

Engle, T., Zon, G. and Egan, W. (1979) *J. Med. Chem.,* **22,** 897.

Ennor, A., Rosenberg, H., Rossiter, R., Beatty, I. and Gaffney, T. (1960) *Biochem. J.,* **75,** 179.

Enoch, H. and Lester, R. (1975) *J. Biol. Chem.,* **250,** 6693.

Entomological Society of America (1981) *Pesticide Handbook: 'Entoma'* (Caswell, R., Devold, K. and Gilbert, L., eds), 29th edn.

Ephrussi, B. and Hottinguer, H. (1950) *Nature, Lond.,* **166,** 956.

Ernster, L., Dallner, G. and Azzone, G. (1963) *J. Biol. Chem.,* **238,** 1124.

Estensen, R., Krey, A. and Hahn, F. (1969) *Molec. Pharmacol.,* **5,** 532.

von Euler, U. (1934) *Arch. Exper. Path. Pharmakol.,* **175,** 78.

von Euler, U. and Pernow, B. (eds) (1977) *Substance P,* New York: Raven Press.

Evans, B. and Wolfenden, R. (1970) *J. Amer. Chem. Soc.,* **92,** 4751.

Evans, P. and Gee, J. (1980) *Nature, Lond.,* **287,** 60.

Everett, A., Lowe, L. and Wilkinson, S. (1970) *Chem. Comm., Chem. Soc., Lond.,* p. 1020.

Everett, J., Roberts, J. and Ross, W. (1953) *J. Chem. Soc.,* p. 2386.

Ewins, A., Ashley, J., Barber, H., Newbery, G. and Self, A. (1942) *J. Chem. Soc.,* p. 103.

Ewins, A. and Phillips, M. (1939) *Brit. Pat.,* **512,** 145.

Exner, O. (1972) in *Advances in Free Energy Relationships* (Chapman, N. and Shorter, J., eds), London: Plenum Press.

Eyre, P. (1970) *J. Pharm. Pharmacol.,* **22,** 26.

Fairlamb, A. and Bowman, I. (1975) *Trans. Roy. Soc. Trop. Med. Hyg.,* **69,** 268.

Fairley, G. and Simister, J. (eds) (1965) *Cyclophosphamide,* Baltimore: Williams and Wilkins.

Fairley, N. (1946) *Trans. Roy. Soc. Trop. Med. Hyg.,* **40,** 105.

Falco, E., Goodwin, L., Hitchings, G., Rollo, I. and Russell, P. (1951) *Brit. J. Pharmacol.,* **6,** 185.

Farber, S. (1952) *Blood,* **7,** 107.

Farber, S. and Mitus, A. (1968) in *Actinomycin* (Waksman S., ed.), New York: Wiley.

Farley, T., Strong, F. and Bydalek, T. (1965) *J. Amer. Chem. Soc.,* **87,** 3501.

Farmer, P. and Ariëns, E. (1982) *Trends Pharmacol. Sci.,* **3,** 362.

Farnham, A., Gregory, G. and Sawicki, R. (1966) *Bull. Entomol. Res.,* **57,** 107.

Fastier, F. (1949) *Brit. J. Pharmacol.,* **4,** 315.

Fastier, F. (1964) *Annu. Rev. Pharmacol.,* **4,** 351.

Fastier, F. and Reid, C. (1952) *Brit. J. Pharmacol.,* **7,** 417.

Faust, R. and Shearin, S. (1974) *Nature, Lond.,* **248,** 60.

Fears, R. and Richards, D. (1981) *Biochem. Soc. Trans.,* **9,** 571.

Fendler, J. and Fendler, E. (1975) *Catalysis in Micellar and Macromolecular Systems,* New York: Academic Press.

Fenner, F., McAuslan, B., Mims, C., Sambrook, J. and White, D. (1974) *The Biology of Animal Viruses,* New York: Academic Press.

Fenner, F. and White, D. (1976) *Medical Virology,* 2nd edn, New York: Academic Press.

Ferguson, J. (1939) *Proc. Roy. Soc. B,* **127,** 387.

Ferguson, J. (1951) *Colloques internationaux du Centre national de la Recherche scientifique,* Paris, No. 26: *Mécanisme de la Narcose,* p. 25.

Ferguson, J., Hawkins, S. and Doxey, D. (1950) *Nature, Lond.,* **165,** 1021.

Ferguson, J. and Pirie, H. (1948) *Ann. Appl. Biol.,* **35,** 532.

Ferone, R., Burchall, J. and Hitchings, G. (1969) *J. Molec. Pharmacol.,* **5,** 49.

Ferreira, A. and Vane, J. (1974) *Annu. Rev. Pharmacol.,* **14,** 57.

Fest, C. and Schmidt, K.-J. (1982) *The Chemistry of Organophosphorus Pesticides,* 2nd edn, Berlin: Springer.

Festy, B., Sturm, J. and Daune, M. (1975) *Biochim. Biophys. Acta*, **407**, 24.

Fieser, L. and Fieser, M. (1935) *J. Amer. Chem. Soc.*, **57**, 491.

Fildes, P. (1940a) *Lancet*, **i**, 955.

Fildes, P. (1940b) *Brit. J. Exper. Path.*, **21**, 67.

Fillion, G. and Fillion, M. (1981) *Nature, Lond.*, **292**, 349.

Filman, D., Bolin, J., Matthews, D. and Kraut, J. (1982) *J. Biol. Chem.*, **257**, 13663.

Finch, J., Lutter, L., Rhodes, D., Brown, R., Rushton, B., Levitt, M. and Klug, A. (1977) *Nature, Lond.*, **269**, 29.

Finean, J., Coleman, R. and Michell, R. (1978) *Membranes and their Cellular Functions*, 2nd edn, Oxford; Blackwell: New York: Wiley-Halstead.

Firestone, R., Pisano, J. and Bonney, R. (1979) *J. Med. Chem.*, **22**, 1130.

Fischer, E. and von Mering, J. (1903) *Therapie der Gegenwart*, **44**, 97.

Fischer, J. and Jardetzky, O. (1965) *J. Amer. Chem. Soc.*, **87**, 3237.

Fisher, L. and Oakenfull, D. (1977) *Chem. Soc. Rev.*, **6**, 25.

Fishman, W. and Anlyan, A. (1947) *Cancer Res.*, **7**, 808.

Flaim, S. and Zelis, R. (eds) (1982) *Calcium Blockers. Mechanisms of Action and Clinical Applications*, Baltimore: Urban and Schwarzenberg.

Fleckenstein, A. (1977) *Annu. Rev. Pharmacol. Toxicol.*, **17**, 149.

Fleckenstein, A. (1983) *Calcium Antagonism in Heart and Smooth Muscle*, New York: Wiley-Interscience.

Fleming, A. (1929) *Brit. J. Exper. Path.*, **10**, 226.

Fletcher, W. and Kirkwood, R. (1982) *Herbicides and Plant Regulators*, London: Granada.

Florey, H., Abraham, E., Chain, E., Fletcher, C., Gardner, A., Heatley, N., Jennings, M., Orr-Ewing, J. and Sanders, A. (1940) *Lancet*, **ii**, 226; (1941) ibid., **ii**, 177.

Florey, H., Chain, E., Heatley, N., Jennings, M., Sanders, A., Abraham, E. and Florey, M. (1949) *Antibiotics*, Oxford: University Press.

Florkin, M. and Mason, H. (1960–64) *Comparative Biochemistry* (7 vols), New York: Academic Press.

Flower, R. (1974) *Pharmacol. Rev.*, **26**, 33.

Flower, R. and Kingston, W. (1975) *Brit. J. Pharmacol.*, **55**, 239P.

Fodor, G. (1960) in *The Alkaloids* (Manske, R. and Holmes, H., eds), Vol. **6**, p. 145, New York: Academic Press.

Foster, A., Jarman, M., Kinas, R., Van Maanen, J. and Taylor, G. (1981) *J. Med. Chem.*, **24**, 1399.

Fourneau, E., Bovet, D., Bovet, F. and Montezin, G. (1944) *Bull. Soc. Chim. Biol.*, **26**, pp. 134, 516.

Fourneau, E., Tréfouël, J., Tréfouël, Mme. J. and Vallée, J. (1924) *Ann. Inst. Pasteur*, **38**, 81.

Fouts, J. (1962) *Fed. Proc.*, **21**, 1107.

Fox, B. and Fox, M. (1984) *Antitumor Drug Resistance*, New York: Springer.

Fox, H. (1953) *Trans. New York Acad. Sci.*, **15**, 234.

Fox, J. (1983) *Chem. Eng. News*, **61**, (March 14th) p. 8.

Fox, M. (1984) *Nature, Lond.*, **307**, 212.

Fox, R. and Richards, F. (1982) *Nature, Lond.*, **300**, 325.

Fox, W. (1977) *Proc. Roy. Soc. Med.*, **70**, 4.

Foye, W., O'Laughlin, R. and Duvall, R. (1961) *J. Pharm. Sci.*, **50**, 641.

Fozard, J. (1983) *Trends Pharmacol. Sci.*, **4**, 288.

Franke, E. and Roehl, W. (1905) *Therapeutische Versuche bei Trypanosomenkrankung*, Jena; later details recorded by Browning, C. (1907) *Brit. Med. J.*, **ii**, 1405, and Ehrlich P. (1907) *Berl. klin. Woch.*, **44**, pp. 233, 341.

Franklin, M. (1972) *Xenobiotica*, **2**, 517.

Franklin, T. (1963a) *Biochem. J.*, **87**, 449.

Franklin, T. (1963b) *Biochim. Biophys. Acta*, **76**, 138.

Franklin, T. (1966) *Symp. Soc. General Microbiol.*, **16,** 192.

Franklin, T. (1971) *Biochem. J.*, **123,** 267.

Franklin, T. and Snow, G. (1981) *Biochemistry of Antimicrobial Action*, 3rd edn, London and New York: Chapman and Hall.

Franks, F. (ed.) (1972–82) *Water, a Comprehensive Treatise* (7 vols), New York: Plenum Press.

Franks, N. and Lieb, W. (1981) *Nature, Lond.*, **292,** 248.

Franks, N. and Lieb, W. (1982) *Nature, Lond.*, **300,** 487, but cf. *ibid.* (1984), **310,** 599.

Fraser, I. and Vessell, E. (1968) *Ann. N.Y. Acad. Sci.*, **151,** 777.

Free, S. and Wilson, J. (1964) *J. Med. Chem.* **7,** 395.

Freed, C., Quintero, E. and Murphy, C. (1978) *Life Sci.*, **23,** 313.

Freeman, K. (1970) *Canad. J. Biochem.*, **48,** 479.

Frenkel, J. and Hitchings, G. (1957) *Antibiot. Chemother.*, **7,** 630.

Freudenthal, R. and Jones, P. (eds) (1976) *Carcinogenesis*, Vol. 1, New York: Raven Press.

Friberg, L. (1974) *Cadmium in the Environment* (2nd edn), Boca Raton, Florida: CRC Press.

Fridovich, I. (1975) *Annu. Rev. Biochem.*, **44,** 147.

Friedheim, E. (1949) *Amer. J. Trop. Med.*, **29,** 173.

Friedheim, E. (1951) *Amer. J. Trop. Med.*, **31,** 218.

Friess, S. and Baldridge, H. (1956) *J. Amer. Chem. Soc.*, **78,** 2482.

Froehner, S., Reiness, C. and Hall, Z. (1977) *J. Biol. Chem.*, **252,** 8589.

Fühner, H. (1918) *Arch. exper. Path. Pharmak.*, **82,** pp. 51, 81.

Fujino, M., Shinagawa, S., Obayashi, M., Kobayashi, S., Fukukda, T., Yamazaki, I., Nakayama, R., White, W. and Rippel, R. (1973) *J. Med. Chem.*, **16,** 1144.

Fujita, T., Iwasa, J. and Hansch, C. (1964) *J. Amer. Chem. Soc.*, **86,** 5175.

Fujita, T. and Nishioka, T. (1975) *Progr. Phys. Org. Chem.*, **12,** 49.

Fujita, T., Nishioka, T. and Nakajima, M. (1977) *J. Med. Chem.*, **20,** 1071.

Fukuto, T. (1972) *Drug Metab. Rev.*, **1,** 117.

Fuller, A. (1942) *Biochem. J.*, **36,** 548.

Fuller, R., Kidder, G., Nugent, N., Dewey, V. and Rigopoulos, N. (1971) *Photochem. Photobiol.*, **14,** 359.

Fulton, J. and Grant, P. (1956) *Ann. Trop. Med. Parasit.*, **50,** 381.

Fyfe, J., Keller, P., Furman, P., Miller, R. and Elion, G. (1978) *J. Biol. Chem.*, **235,** 8721.

Gabbay, E. (1968) *J. Amer. Chem. Soc.*, **90,** 5257.

Gaddum, J. (1926) *J. Physiol.*, **61,** 141.

Gaddum, J. (1936) *Proc. Roy. Soc. Med.*, **29,** 1373.

Gage, J. (1953) *Biochem. J.*, **54,** 426.

Gage, P. (1971) in *Neuropoisons* (Simpson, L., ed.), New York: Plenum Press.

Gale, E. (1946) *Adv. Enzymol.*, **6,** 1.

Gale, E. (1947) *J. Gen. Microbiol.*, **1,** 53.

Gale, E., Cundliffe, E., Reynolds, P., Richmond, M. and Waring, M. (1981) *The Molecular Basis of Antibiotic Action*, 2nd edn, London: Wiley.

Gale, E. and Taylor, E. (1947) *J. Gen. Microbiol.*, **1,** 77.

Ganellin, C. (1981) *J. Med. Chem.*, **24,** 913.

Ganis, P., Avitable, G., Mechlinski, W. and Schaffner, C. (1971) *J. Amer. Chem. Soc.*, **93,** 4560.

Garattini, S., Mussini, E. and Randall, L. (eds) (1973) *The Benzodiazepines*, New York: Raven Press.

Gauri, K. (ed.) (1981) *Antiviral Chemotherapy, Design of Inhibitors of Viral Functions*, New York: Academic Press.

Gelboin, H., Wortham, J. and Wilson, R. (1967) *Nature, Lond.*, **214,** 281.

Gelmo, P. (1908) *J. prakt. Chem.*, **77,** 369.

Gemmell, M. and Shearer, G. (1968) *Vet. Rec.*, **82,** 252.

Gent, M. and Prestegard, J. (1974) *Biochemistry*, **13,** 4027.

Gershenfield, L. and Milanick, C. (1941) *Amer. J. Pharm.*, **113,** 306.

Gescher, A., Gibson, N., Hickman, J., Langdon, S., Ross, D. and Atassi, G. (1982) *Brit. J. Cancer*, **45,** 843.

Giarmann, N. (1949) *J. Pharmacol.*, **96,** 119.

Gibaldi, M. and Perrier, D. (1975) *Pharmacokinetics*, New York: Marcel Dekker.

Gibson, F. (1964) *Biochem. J.*, **90,** 256.

Gilardi, R., Karle, I., Karle, J. and Sperling, W. (1971) *Nature, Lond.*, **232,** 187.

Gilbert, B., Leme, L., Ferreira, A., Bulhoes, M. and Castleton, C. (1973) *Bull. World Heath Org.*, **49,** 633.

Giles, C., MacEwan, T., Nakhwa, S. and Smith, D (1960) *J. Chem. Soc.*, p. 3973.

Gillespie, L., Oates, J., Crout, J. and Sjoerdsma, A. (1962) *Circulation*, **25,** 281.

Gillette, J. (1966) *Adv. Pharmacol.*, **4,** 219.

Gillette, J., Conney, A., Cosmides, G., Estabrook, R., Fouts, J. and Mannering, G. (eds) (1969) *Microsomes and Drug Oxidations*, New York: Academic Press.

Gillette, J. and Mitchell, J. (1975) *Concepts in Biochemical Pharmacology*, Part 3, Berlin: Springer.

Gilligan, D. and Plummer, N. (1943) *Proc. Soc. Exper. Biol. Med.*, **53,** 142.

Gilman, A.G., Goodman, L. and Gilman, A. (eds) (1980) *Goodman and Gilman's Pharmacological Basis of Therapeutics*, New York: Macmillan.

Gilman, A. and Philips, F. (1946) *Science*, **103,** 409.

Giloni, L., Takeshita, M., Johnson, F., Iden, C. and Grollman, A. (1981) *J. Biol. Chem.*, **256,** 8608.

Gingell, R. and Bridges, J. (1973) *Xenobiotica*, **3,** 599.

Ginnings, P. and Baum, R. (1937) *J. Amer. Chem. Soc.*, **59,** 1111.

Glasby, J. (ed.) (1979) *Encyclopaedia of Antibiotics*, 2nd edn, New York: Wiley.

Glatt, H., Protić-Sabljić, M. and Oesch, F. (1983) *Science*, **220,** 961.

Glave, W. and Hansch, C. (1972) *J. Pharm. Sci.*, **61,** 589.

Gloge, H., Lüllmann, H. and Mutschler, E. (1966) *Brit. J. Pharmacol. Chemother.*, **27,** 185.

Goksóyr, J. (1955) *Nature, Lond.*, **175,** 820; *Physiol. Plant.*, **8,** 719.

Gold, E., Chang, W., Cohen, M., Baum, T., Ehrreich, S. and Johnson, G. (1982) *J. Med. Chem.*, **25,** 1363.

Goldacre, R. and Phillips, J. (1949) *J. Chem. Soc.*, p. 1724.

Goldacre, R., Loveless, A. and Ross, W. (1949) *Nature, Lond.*, **163,** 667.

Goldacre, R. and Phillips, J. (1949) *J. Chem. Soc.*, 1724.

Goldberg, I. and Friedman, P. (1971) *Annu. Rev. Biochem.*, **40,** 775.

Goldberg, L. and Kohli, J. (1983) *Trends Pharmacol. Sci.*, **4,** 64.

Goldberg, M., Gold, D., Flescher, E. and Lengy, J. (1980) *Biochem. Pharmacol.*, **29,** 838.

Goldfine, I., Perlman, R. and Roth, J. (1971) *Nature, Lond.*, **234,** 295.

Goldman, L. (1970) *Med. Clin. North Amer.*, **54,** 1339 (Review).

Goldstein, A., Aronow, L. and Kalman, S. (1974) *Principles of Drug Action*, 2nd edn, New York: Wiley.

Goldstein, A., Fischli, W., Lowney, C., Hunkapiller, M. and Hood, L. (1981) *Proc. Natl. Acad. Sci., U.S.A.*, **78,** 7219.

Goldstein, A., Lowney, L. and Pal, B. (1971) *Proc. Natl. Acad. Sci., U.S.A.*, **68,** 1742.

Gomperts, B. (1976) *The Plasma Membrane*, London: Academic Press.

Gönnert, R. and Schraufstätter, E. (1960) *Arzneim. Forsch.*, **10,** 881.

Good, N. (1961) *Plant Physiol.*, **36,** 788.

Good, N., Winget, G., Winter, W., Connolly, T., Izawa, S. and Singh, R. (1966) *Biochemistry*, **5,** 467.

Goodman and Gilman (*see* Gilman, Goodman and Gilman).

Goodman, L., Wintrobe, M., Dameshek, W., Goodman, M., Gilman, A. and McLennan, M. (1946) *J. Amer. Med. Assoc.*, **132,** 126.

Goodson, J., Henry, T. and MacFie, J. (1930) *Biochem. J.*, **24,** 874.

Goodwin, B. (1976) *Handbook of Intermediary Metabolism of Aromatic Compounds*, London and New York: Chapman and Hall.

Goodwin, T. (ed.) (1966) *Biochemistry of Chloroplasts*, London: Academic Press.

Gorin, F., Balasubramanian, T., Cicero, T., Schwietzer, J. and Marshall, G. (1980) *J. Med. Chem.*, **23**, 1113.

Gosálvez, M., Blanco, M., Vivero, C. and Vallés, F. (1978) *Eur. J. Cancer*, **14**, 1185.

Goss, W., Dietz, W. and Cook, T. (1965) *J. Bact.*, **89**, 1068.

Goth, A. (1945) *Science*, **101**, 383.

Goto, T., Kishi, Y., Takahashi, S. and Hirata, Y. (1965) *Tetrahedron*, **21**, 2059.

Gottlieb, J. and Hill, C. (1974) *New Engl. J. Med.*, **290**, 193.

Gottlieb, R. (1923) *Arch. exper. Path. Pharmakol.*, **97**, 113.

Gould, G. and Hitchins, A. (1963) *Nature, Lond.*, **197**, 622.

Gould, J. (1957) *Nature, Lond.*, **180**, 282.

Govindjee, R. (1982) *Photosynthesis* (2 vols), New York: Academic Press.

Graham-Smith, G. (1919) *J. Hyg., Cambridge*, **18**, 1.

Grant, P. and Sargent, J. (1960) *Biochem. J.*, **76**, 229.

Grant, P., Sargent, J. and Ryley, J. (1961) *Biochem. J.*, **81**, 200.

Gray, G., Smith, I., McKenzie, I., Crean, G. and Gillespie, G. (1977) *Lancet*, **i**, 4.

Greathouse, G., Block, S., Kovack, E., Barnes, D., Byron, C., Long, G., Gerber, D. and McLenny, J. (1954) *Research on Chemical Compounds for Inhibition of Fungi*, US Corps of Engineers, Fort Belvoir, Virginia.

Green, A. (1937) in *Thorpe's Dictionary of Applied Chemistry*, 4th edn, **1**, 90.

Green, A., Heale, D. and Grahame-Smith, D. (1977) *Psychopharmacology*, **52**, 195.

Greengard, O. and McIlwain, H. (1955) *Biochem. J.*, **61**, 61.

Greengard, P. (1976) *Nature, Lond.*, **260**, 101.

Gregoriadis, G. (1977) *Nature, Lond.*, **265**, 407.

Grell, E., Eggers, F. and Funck, T. (1972) *Chimia*, **26**, 632.

Grewe, R. (1947) *Angew. Chem.*, **59**, 194.

Griffin, M. and Brown, G. (1964) *J. Biol. Chem.*, **239**, 310.

Griffiths, J., Intarakosit, P., Taylor, H. and Wain, R. (1966) *Ann. Appl. Biol.*, **58**, 183.

Grigg, G. (1970) *Molec. Gen. Genet.*, **106**, 228.

Grigg, G., Edwards, M. and Brown, D. (1971) *J. Bact.*, **107**, 599.

Grivsky, E., Lee, S., Sigel, C., Duch, D. and Nichol, C. (1980) *J. Med. Chem.*, **23**, 327.

Grollman, A. (1966) *Proc. Natl. Acad. Sci., U.S.A.*, **56**, 1867.

Grollman, A. (1968) *J. Biol. Chem.*, **243**, 4089.

Grollman, A. (1971) *Drug Design*, **2**, 231 (review).

Grollman, A. and Horwitz, S. (1971) *Drug Design*, **2**, 261 (review).

Grollman, A. and Takeshita, M. (1980) *Adv. Enzyme Regulat.*, **18**, 67.

Gross, F. and Turrian, H. (1957) *Experientia*, **13**, 401.

Grunberg, E. and Schnitzer, R. (1953) *Proc. Soc. Exper. Biol. Med.*, **83**, 220.

Grunwald, E. and Ralph, E. (1967) *J. Amer. Chem. Soc.*, **89**, 4405.

Guengerich, F., Dannan, G., Wright, S., Martin, M. and Kaminsky, L. (1982) *Xenobiotica*, **12**, 701.

Guillemin, R. (1978) *Science*, **202**, 390.

Guldberg, C. and Waage, P. (1864) *Les Mondes*, **5**, 107.

Gull, K. and Trinci, A. (1973) *Nature, Lond.*, **244**, 292.

Gurson, C. and Saner, G. (1971) *Amer. J. Clin. Nutr.*, **24**, 1313.

Gustavssón, S., Lööf, L., Adami H., Nyberg, A. and Nyrén, O. (1983) *Lancet*, **ii**, 124.

Guth, P., Amaro, J., Sellinger, O. and Elmer, L. (1965) *Biochem. Pharmacol.*, **14**, 769.

Guthrie, F. (1950) *J. Econ. Entomol.*, **43**, 559.

Gutman, M., Coles, C., Singer, T. and Casida, J. (1971) *Biochemistry*, **10**, 2036.

Gutte, B. and Merrifield, R. (1969) *J. Amer. Chem. Soc.*, **91**, 501.

Guttman, P. and Ehrlich, P. (1891) *Berl. klin. Woch.*, **28**, 953.

Gysin, H. (1962) *Chem. and Indust.*, p. 1393.

Haagen-Smit, A. and Went, F. (1935) *Proc. k. ned. Akad. Wetenschap*, **38**, 852.

Habermann, E. (1974) *Annu. Rev. Pharmacol.*, **14**, 1.

Haddow, A., Harris, R., Kon, G. and Roe, E. (1948) *Proc. Roy. Soc. B*, **241**, 147.

Haddow, A. and Timmis, G. (1953) *Lancet*, **1**, 207.

Haest, C., de Gier, J., op den Kamp, J., Bartels, P. and Van Deenen, L. (1972) *Biochim. Biophys. Acta*, **255**, 720.

Hahn, F. (1975) *Naturwiss.*, **62**, 449.

Hahn, F. (1983) *Antibiotics, Modes and Mechanisms of Microbial Growth Inhibitors*, 6th (and last) vol., Berlin: Springer (in English).

Hahn, P., Bale, W., Ross, J., Hettig, R. and Whipple, G. (1940) *Science*, **92**, 131.

Hai, T., Abo, M. and Hampton, A. (1982) *J. Med. Chem.*, **25**, 1184.

Hakala, M. (1966) *Fed. Proc.*, **25**, 750.

Hakala, M. (1973) in *Drug Resistance and Selectivity* (Mihich, E., ed.), New York: Academic Press.

Haldane, J. (1930) *Enzymes*, London: Longmans.

Hallman, P., Perrin, D. and Watt, A. (1971) *Biochem. J.*, **121**, 549.

Halpern, B. (1942) *Arch. Internat. Pharmacodyn. Thér.*, **68**, 339.

Hamilton-Miller, J. (1973) *Bact. Rev.*, **37**, 166.

Hamilton-Miller, J. and Abraham, E. (1971) *Biochem. J.*, **123**, 183.

Hammett, L. (1970) *Physical Organic Chemistry*, 2nd edn, New York: McGraw-Hill.

Hammond, G. (1955) *J. Amer. Chem. Soc.*, **77**, 334.

Hampton, A. (1976) *J. Med. Chem.*, **19**, 1279.

Handrick, G., Duffley, R., Lambert, G., Murphy, J., Dalzell, H., Howes, J., Razdan, R., Martin, B., Harris, L. and Dewey, W. (1982) *J. Med. Chem.*, **25**, 1447.

Hansch, C. (1968) *J. Med. Chem.*, **11**, 920.

Hansch, C. (1971) *Drug Design*, **1**, 271 (review).

Hansch, C. and Anderson, S. (1967) *J. Med. Chem.*, **10**, 745.

Hansch, C. and Fujita, T. (1964) *J. Amer. Chem. Soc.*, **86**, 1610.

Hansch, C., Kim, K. and Sarma, R. (1973) *J. Amer. Chem. Soc.*, **95**, 6447.

Hansch, C. and Leo, A. (1979) *Substituent Constants for Correlation Analysis in Chemistry and Biology*, New York: Wiley.

Hansch, C., Leo, A., Unger, S., Kim, K., Nikaitashi, D. and Lien, E. (1973) *J. Med. Chem.*, **16**, 1207.

Hansch, C., Li, R., Blaney, J. and Langridge, R. (1982) *J. Med. Chem.*, **25**, 777.

Hansch, C., Quinlan, J. and Lawrence, G. (1968) *J. Org. Chem.*, **33**, 347.

Hansch, C., Steward, A., Anderson, S. and Bentley, D. (1968) *J. Med. Chem.*, **11**, 1.

Hansch, C., Unger, S. and Forsythe, A. (1973) *J. Med. Chem.*, **16**, 1217.

Hansch, C., Vittoria, A., Silipo, C. and Jow, P. (1975) *J. Med. Chem.*, **18**, 546.

Haracz, J. (1982) *Schizophrenia Bull.*, **8**, 438.

Hardman, H. (1962) *Circ. Res.*, **10**, 598.

Harold, F. and Baarda, J. (1967) *J. Bact.*, **94**, 53.

Harper, M. and Kellems, R. (1981) *Cancer Bull.*, **33**, 43.

Harrap, K. (1976) in *Scientific Foundations of Oncology* (Symington, T. and Carter, R., eds), London: Heinemann, p. 641.

Hart, G. and O'Brien, R. (1973) *Biochemistry*, **12**, 2940.

Hartley, B. and Kilby, B. (1952) *Biochem. J.*, **50**, 672.

Hartley, G. and Graham-Bryce, I. (1980–81) *Physical Principles of Pesticide·Behaviour* (2 vols), New York: Academic Press.

Hartough, H. (1952) *Thiophene and its Derivatives*, New York: Interscience (Chap. 2).

Hashimoto, Y., Makita, T., Miyata, H., Noguchi, T. and Ohta, G. (1968) *Toxicol. Appl. Pharmacol.*, **12**, 536.

Hassall, K. (1982) *The Chemistry of Pesticides, their Metabolites, Mode of Action, and Uses in Crop Protection*, London: Macmillan.

Hata, A. (1932) *Kitasato Arch. Exper. Med.*, **9**, 1.

Hatanaka, H. and Sano, K. (1973) *Z. Neurol.*, **204**, 309.

Haussler, M. and Norman, A. (1969) *Proc. Natl. Acad. Sci., U.S.A.*, **62**, 155.

Hawking, F. (1944) *J. Pharmacol.*, **82**, 31.

Hawkins, C. (1971) *Absolute Configuration of Metal Complexes*, London: Wiley.

Hawkins, C. and Perrin, D. (1962) *J. Chem. Soc.*, p. 1351.

Hawkins, C. and Perrin, D. (1963) *Inorg. Chem.*, **2**, 843.

Haydon, D. and Hladky, S. (1972) *Quart. Rev. Biophys.*, **5**, 187.

Hayes, W., Durham, W. and Cueto, C. (1956) *J. Amer. Med. Assoc.*, **162**, 890.

Hazelbauer, G. and Changeux, J.-P. (1974) *Proc. Natl. Acad. Sci., U.S.A.*, **71**, 1479.

Hazum, E., Sabatka, J., Chang, L.-J., Brent, D., Findlay, J. and Cuatrecasas, P. (1981) *Science*, **213**, 1010.

Hecht, G. (1936) *Arch. Exper. Path. Pharmakol.*, **183**, 87.

Hedberg, K., Hughes, E. and Waser, J. (1961) *Acta Crystallogr.*, **14**, 369.

Hee, S. and Sutherland, R. (1981) *The Phenoxyalkanoic Herbicides*, Boca Raton (Florida): CRC Press.

Heidelberger, C., Griesbach, L., Cruz, O., Schnitzer, R. and Gruenberg, E. (1958) *Proc. Soc. Exper. Biol. Med.*, **97**, 470.

Helenius, A., Marsh, M. and White, J. (1980) *Trends Biochem. Sci.*, **5**, 104.

Hellberg, H. (1959) *Acta Chem. Scand.*, **13**, 1106.

Hemmerich, P., Veeger, C. and Wood, H. (1965) *Angew. Chem.*, **77**, 1.

Henderson, I. and Canellos, G. (1980) *New Engl. J. Med.*, **302**, 78.

Henderson, P., McGivan, J. and Chappell, J. (1969) *Biochem. J.*, **111**, 521.

Hershenson, F., Prodan, K., Kochman, R., Bloss, J. and Mackerer, C. (1977) *J. Med. Chem.*, **20**, 1448.

Hertzberg, R. and Dervan, P. (1982) *J. Amer. Chem. Soc.*, **104**, 313.

Hewitt, R., Kushner, S., Stewart, H., White, E., Wallace, W. and Subbarow, Y. (1947) *J. Lab. Clin. Med.*, **32**, 1314.

Hey, P. (1952) *Brit. J. Pharmacol.*, **7**, 117.

Heymann, B. (1924) *Z. angew. Chem.*, **37**, 585.

Heymann, B., Kothe, R., Dressel, O. and Ossenbeck, A. (1917) *U.S. Pat.*, 1218 654–5 (cf. 1 308 071).

Higashi, Y., Strominger, J. and Sweeley, C. (1967) *Proc. Natl. Acad. Sci., U.S.A.*, **57**, 1878.

Higgs, G., Flower, R. and Vane, J. (1979) *Biochem. Pharmacol.*, **28**, 1959.

Higuchi, T. and Stella, V. (eds) (1975) *Pro-drugs as Novel Drug Delivery Systems*, Washington D.C.: American Chemical Society.

Hildebrand, J. (1979) *Proc. Natl. Acad. Sci., U.S.A.*, **76**, 194.

Hill, B., Bailey, B., White, J. and Goldman, I. (1979) *Cancer Res.*, **39**, 2440.

Hill, D. and Bowery, N. (1981) *Nature, Lond.*, **290**, 149.

Hill, G., Sedransk, N., Rochlin, D., Bisel, H., Andrews, N., Fletcher, W., Schroeder, J. and Wilson, W. (1972) *Cancer*, **30**, 900.

Hille, B. (1977) *J. Gen. Physiol.*, **69**, 497.

Hiller, S. (1927) *Proc. Soc. Exper. Biol. Med.*, **24**, 427.

Hilton, J., Ard, J., Jansen, L. and Gentner, W. (1959) *Weeds*, **7**, 381.

Himmelweit, F. (ed.) (1956<), *The Collected Papers of Paul Ehrlich, with Biography*, Oxford: Pergamon Press.

Hirai, T., Hirashima, A., Itoh, T., Takahishi, T., Shimoura, T. and Hayashi, Y. (1966) *Phytopathology*, **56**, 1236.

Hirom, P. and Millburn, P. (1981) in *Foreign Compound Metabolism* (Hathaway, D., ed.), London: The Chemical Society.

Hirom, P., Millburn, P. and Smith, R. (1976) *Xenobiotica*, **6**, 55.

Hirschmann, R., Nutt, R., Veher, D., Vitali, R., Varga, S., Jacob, T., Holly, F. and Denkewalter, R. (1969) *J. Amer. Chem. Soc.*, **91,** 507.

Hitchings, G. (1952) *Trans. Roy. Soc. Trop. Med. Hyg.*, **46,** 467.

Hitchings, G. (1982) *Adv. Enzyme Regul.*, **20,** 375.

Hitchings, G. (ed.) (1983) *Inhibition of Folate Metabolism in Chemotherapy*, Berlin: Springer (in English).

Hitchings, G. and Burchall, J. (1965) *Adv. Enzymol.*, **27,** 417.

Hlubucek, J., Hora, J., Toube, T. and Weedon, B. (1970) *Tet. Lett.*, p. 5163.

Ho, I. and Harris, R. (1981) *Annu. Rev. Pharmacol.*, **21,** 83.

Hobbiger, F. (1954) *Chem. Indust.*, p. 415.

Hobbiger, F. (1955) *Brit. J. Pharmacol.*, **10,** 356.

Hobbiger, F. and Vojvodić, V. (1966) *Biochem. Pharmacol.*, **15,** 1677.

Hochster, R. and Quastel, J. (eds) (1963–73) *Metabolic Inhibitors* (4 vols), New York: Academic Press.

Hodgkin, A. (1964) *The Conduction of the Nervous Impulse*, Liverpool: University Press.

Hodgkin, A. and Huxley, A. (1952) *J. Physiol.*, **117,** 500.

Hodgson, E. and Casida, J. (1962) *J. Agric. Food Chem.*, **10,** 208.

Hoefnagel, A., Hoefnagel, M. and Wepster, B. (1978) *J. Org. Chem.*, **43,** 4720.

Hoffman, C., Schweitzer, T. and Dalby, G. (1940) *J. Amer. Chem. Soc.*, **62,** 988.

Hoffman, C., Schweitzer, T. and Dalby, G. (1941) *Indust. Eng. Chem.*, **33,** 749.

Hofschneider, P. and Martin, H. (1968) *J. Gen. Microbiol.*, **51,** 23.

Hogben, C., Schanker, L., Tocco, D. and Brodie, B. (1957) *J. Pharmacol.*, **120,** 540.

Hogben, C., Schanker, L., Tocco, D. and Brodie, B. (1959) *J. Pharmacol.*, **125,** 275.

Holan, G. (1969) *Nature, Lond.*, **221,** 1025.

Holan, G. (1971) *Nature, Lond.*, **232,** 644.

Holan, G., Johnson, W., O'Keefe, D., Rihs, K., Smith, D., Virgona, C., Walser, R. and Haslam, J. (1983) *Pesticide Chemistry* (Miyamoto, J. and Kearney, P., eds), Oxford: Pergamon Press, Vol. 1, p. 119.

Holan, G., O'Keefe, D., Virgona, C. and Walser, R. (1978) *Nature, Lond.*, **272,** 734.

Hollenberg, C., Borst, P. and Van Bruggen, E. (1970) *Biochim. Biophys. Acta*, **209,** 1.

Hollingworth, R., Fukoto, T. and Metcalf, R. (1967) *J. Agric. Food Chem.*, **15,** 235.

Hollister, L. (1978) *Clinical Pharmacology of Psychotherapeutic Agents*, New York: Churchill-Livingstone.

Holmes, R. and Robins, E. (1955) *Brit. J. Pharmacol.*, **10,** 490.

Holmstedt, B. and Liljestrand, G. (1963) *Readings in Pharmacology*, Oxford: Pergamon Press (Reprinted 1981, New York: Raven Press).

Holton, P. and Ing, H. (1949) *Brit. J. Pharmacol.*, **4,** 190.

Homer, R., Mees, G. and Tomlinson, T. (1960) *J. Sci. Food Agric.*, p. 309.

Hopfinger, A. (1973) *Conformational Properties of Macromolecules*, New York: Academic Press.

Hopfinger, A. (1980) *J. Amer. Chem. Soc.*, **102,** 7196.

Hopfinger, A. (1981) *J. Med. Chem.*, **24,** 818.

Hopfinger, A. (1983) *J. Med. Chem.*, **26,** 990.

Horn, A. and Snyder, S. (1971) *Proc. Natl. Acad. Sci., U.S.A.*, **68,** 2325.

Horowitz, M. and Brayton, C. (1970) *Virology*, **48,** 690.

Horsfall, J. (1972) in *Pest Control: Strategies for the Future*, Washington, D.C.: National Academy of Science, p. 216.

Horwitz, S. and Grollman, A. (1968) *Antimicrob. Agents Chemother.*, p. 21.

Horwitz, S., Parness, J., Schiff, P. and Manfredi, J. (1982) *Cold Spring Harbor Symp. Quant. Biol.*, **46,** 219.

Hotchkiss, R. (1951) *Cold Spring Harbor Symp. Quant. Biol.*, **16,** 457.

Houslay, M. and Stanley, K. (1982) *Dynamics of Biological Membranes*, New York: Wiley.

Howell, J. and Altounyan, R. (1967) *Lancet*, **ii**, 539.

Howland, J. (1973) *Cell Physiology*, 2nd edn, London: Macmillan Press.

Huang, C. (1969) *Biochemistry*, **8**, 344.

Huang, L. and Barker, J. (1980) *Science*, **207**, 195.

Huang, T. and Grollman, A. (1970) *Fed. Proc.*, **29**, 609.

Hudson, C. (1948) *Adv. Carbohyd. Chem.*, **3**, 1.

Hughes, D. (1962) *J. Gen. Microbiol.*, **29**, 39.

Hughes, J. (1975) *Brain Res.*, **88**, 295.

Hughes, J., Smith, T., Kosterlitz, H., Fothergill, L., Morgan, B. and Morris, H. (1975) *Nature, Lond.*, **258**, 577.

Hughes, R. and Chapple, D. (1981) *Brit. J. Anaesth.*, **53**, 31.

Hughes, V. and Datta, N. (1983) *Nature, Lond.*, **302**, 725.

Hunt, P., Francis, J., Peck, G., Farrell, K. and Sali, A. (1979) *Med. J. Austral.*, **i**, 107.

Hunt, R. and Taveau, R. (1911) *Bull. Hyg. Lab., U.S. Treasury*, No. 73.

Hunter, F. and Lowry, O. (1956) *Pharmacol. Rev.*, **8**, 89.

Hurly, M. (1959) *Trans. Roy. Soc. Trop. Med. Hyg.*, **53**, pp. 410, 412.

Hurwitz, J., Furth, J., Malamy, M. and Alexander, M. (1962) *Proc. Natl. Acad. Sci., U.S.A.*, **48**, 1222.

Hutner, S. (1949) *Science*, **110**, 548.

Hutson, D. and Roberts, T. (eds) (1982) *Progress in Pesticide Biochemistry*, Vol. 2, London: Wiley.

Hyman, J. (1949) *Brit. Pat.*, **652**, 300.

Ijzerman, A., Bultsma, T., Timmerman, H, and Zaagsma, J. (1984) *J. Pharm. Pharmacol.*, **36**, 11–15.

Inbar, M., Ben-Bassat, H. and Sachs, L. (1972) *Nature New Biol.*, **236**, 3.

Ing, H. (1936) *Physiol. Rev.*, **16**, 527.

Ing, H. (1949) *Science*, **109**, 264.

Ing, H., Kordik, P. and Williams, T. (1952) *Brit. J. Pharmacol.*, **7**, 103.

Inoue, F. and Frank, G. (1962) *J. Pharmacol.*, **136**, 190.

Inoue, Y. and Perrin, D. (1962) *J. Chem. Soc.*, p. 2600.

International Agency for Research on Cancer (WHO) (1976) *Ann. Rpt.* 1, Lyon (France).

International Agency for Research on Cancer (WHO) (1982) *Monographs on Evaluation of Carcinogenic Risk of Chemicals to Humans*, vol. 29, Lyon (France).

International Nonproprietary Names for Pharmaceutical Substances, Cumulative List No. 6 (1982), Geneva: World Health Organization.

International Union of Biochemistry (1978) *Enzyme Nomenclature*, New York: Academic Press.

Ioannides, C., Lum, P. and Parke, D. (1984) Xenobiotica, **14**, 119.

Iqbal, K. and Ottaway, J. (1970) *Biochem. J.*, **119**, 145.

Irving, H. and Mellor, D. (1962) *J. Chem. Soc.*, pp. 5222, 5237.

Irving, H. and Rossotti, H., (1956) *Acta. Chem. Scand.*, **10**, 72.

Irving, H. and Williams, R. (1953) *J. Chem. Soc.*, p. 3192.

Isaka, S. (1957) *Nature, Lond.*, **179**, 578.

Ison, R., Partington, P. and Roberts, G. (1973) *Molec. Pharmacol.*, **9**, 756.

Israel, M. and Potti, G. (1982) *J. Med. Chem.*, **25**, 187.

Ito, K., Nakahara, I. and Sakamoto, Y. (1964) *Gann*, **55**, 379.

Iversen, L. (1967) *The Uptake and Storage of Noradrenaline in Sympathetic Nerves*, Cambridge: University Press.

Iversen, L. (1975) in *Handbook of Psychopharmacology* (Iversen, L., Iversen, S. and Snyder, S., eds), New York: Plenum Press.

Iversen, L. (1978) in *Psychopharmacology—A Generation of Progress* (Lipton, M., Di Mascio, A. and Killam, K., eds), New York: Raven Press.

Ives, D. and Lemon, T (1968) *Roy. Inst. Chem. Reviews*, **1**, 62.
Iwamoto, E. and Martin, W. (1981) *Medicinal Res. Rev.*, **1**, 411.
Iyer, V. and Szybalski, W. (1964) *Science*, **145**, 55.
Izaki, K., Matsuhashi, M. and Strominger, J. (1966) *Proc. Natl. Acad. Sci.*, *U.S.A.*, **55**, 656.
Izaki, K., Matsuhashi, M. and Strominger, J. (1968) *J. Biol. Chem.*, **243**, 3180.
Izatt, R., Nelson, D., Rylting, J., Haymore, B. and Christensen, J. (1971) *J. Amer. Chem. Soc.*, **93**, 1619.
Jackman, L. and Sternhell, S. (1969) *Applications of N.M.R. Spectroscopy in Organic Chemistry*, 2nd edn, Oxford: Pergamon Press.
Jackson, G., Muldoon, R. and Akers, L. (1963) *Antimicrob. Agents Chemother.*, p. 703.
Jackson, K. and Mason, S. (1971) *Trans. Farad. Soc.*, **67**, 966.
Jacobs, M., Glassman, H. and Parpart, A. (1935) *J. Cell. Compar. Physiol.*, **7**, 197.
Jacobs, W. and Heidelberger, M. (1919) *J. Amer. Chem. Soc.*, **41**, 1587.
Jacoby, G. and Gorini, L. (1967) in *Antibiotics* (Gottlieb, D. and Shaw, P., eds), Vol. 1, p. 726, New York: Springer.
Jaenicke, L. and Chan, P. (1960) *Angew. Chem.*, **72**, 753.
Jaffe, J. and McCormack, J. (1967) *Molec. Pharmacol.*, **3**, 359.
Jain, S., Tsai, C.-C. and Sobell, H. (1977) *J. Molec. Biol.*, **114**, 317.
Jakoby, W. (ed.) (1980) *Enzymatic Basis of Detoxication* (2 vols), New York: Academic Press.
Jakoby, W., Bend, J. and Caldwell, J. (eds) (1982) *Metabolic Basis of Detoxication*, New York: Academic Press.
James, K. (1974) *Progr. Med. Chem.*, **10**, 205.
James, S. (1980) *Parasitology*, **80**, 313.
von Jancsó, N. (1931) *Zbl. Bakt.*, *Abt. I Orig.*, **122**, 393.
von Jancsó, N. (1932) *Klin. Woch.*, **11**, 1305.
Janis, R. and Triggle, D. (1983) *J. Med. Chem.*, **26**, 795.
Janssen, M. (1958) *J. Inorg. Nuclear Chem.*, **8**, 340.
Jardetzky, O. and Jardetzky, C. (1962) *Methods Biochem. Anal.*, **9**, 235.
Jardetzky, O. and Roberts, G. (1981) *NMR in Molecular Biology*, New York: Academic Press.
Jellinek, F. (1957) *Acta Crystallogr.*, **10**, 277.
Jencks, W. (1969) *Catalysis in Chemistry and Enzymology*, New York: McGraw-Hill.
Jenner, T. and Testa, B. (1981) *Concepts in Drug Metabolism* (2 vols), New York: Marcel Dekker.
Jensen, E. and Jacobson, H. (1962) *Recent Progr. Horm. Res.*, **18**, 387.
Jensen, K. and Schmith, K. (1942) *Z. Immunitäts.*, **102**, 261.
Jepson, J. and Smith, W. (1974) *Essentials of Organic Nomenclature*, Oxford: University Press.
Jewsbury, J., Cooke, M. and Weber, M. (1977) *Ann. Trop. Med. Parasitol.*, **71**, 67.
Jirousek, L. and Pritchard, E. (1971) *Biochim. Biophys. Acta*, **243**, 230.
Johnson, H. (1978) in *Immediate Hypersensitivity* (Bach M., ed.), New York: Marcel Dekker, p. 533.
Johnson, J. and Rajagopolan, K. (1982) *Biochemistry*, **79**, 6856.
Johnson, J., Woodward, R. and Robinson, R. (eds) (1949) *The Chemistry of Penicillin*, Princeton, N.J.: University Press.
Johnson, M. and Milne, G. (1980) *J. Het. Chem.*, **17**, 1817.
Johnson, Q. (1960) *U.S. Atom. Energy Commiss.*, UCRL 9350.
Johnston, G. (1978) *Annu. Rev. Pharmacol.*, **18**, 269.
Johnston, G., Curtis, D., Beart, P., Game, C., McCulloch, R. and Twitchin, B. (1975) *J. Neurochem.*, **24**, 433.
Johnston, T., McCaleb, G. and Montgomery, J. (1963) *J. Med. Chem.*, **6**, 669.
Jonas, A. and Weber, G. (1971) *Biochemistry*, **10**, 1335.
Jones, M. (1960) *Nature, Lond.*, **185**, 96.

Jones, O. and Watson, W. (1965) *Nature, Lond.*, **208,** 1169.

Jordan, A. and Trevern, M. (1978) *Nature, Lond.*, **272,** 719.

Jordan, D. (1968) in *Molecular Associations in Biology* (Pullman, B., ed.), New York: Academic Press, p. 221.

Josephson, E., Taylor, D., Greenberg, J. and Coatney, G. (1953) *J. Infect. Dis.*, **93,** 257.

Joswick, H. (1961) Ph.D. thesis: *Mode of Action of Hexachlorophene*, Ann Arbor: University of Michigan.

Juan, S., Segura, E. and Cazzulo, J. (1978) *Internat. J. Biochem.*, **9,** 395.

Jung, L., Koffel, J.-C. and Lami, H. (1971) *Bull. Chim. thér.*, **6,** 341.

Jung, L. and Lami, H. (1970) *Bull Chim. thér.*, **5,** 391.

Jung, M. and Metcalf, B. (1975) *Biochem. Biophys. Res. Commun.*, **67,** 301.

Kabachnik, M., Brestkin, A., Godovikov, N., Michelson, M., Rozengart, E. and Rozengart, V. (1970) *Pharmacol. Rev.*, **22,** 355.

Kadota, I. and Abe, T. (1954) *J. Lab. Clin. Med.*, **43,** 375.

Kadota, I. and Midorikawa, O. (1951) *J. Lab. Clin. Med.*, **38,** 671.

Kahan, F., Kahan, J., Cassidy, P. and Kroph, H. (1974) *Ann. N.Y. Acad. Sci.*, **235,** 364.

Kalow, W. (1962) *Pharmacogenetics, Heredity, and Response to Drugs*, Philadelphia: Saunders.

Kalow, W. (1980) *Trends Pharmacol. Sci.*, **1,** 403.

Kalow, W., Otton, S., Kador, D., Endrenyi, L. and Inabe, T. (1980) *Canad. J. Physiol. Pharmacol.*, **58,** 1142.

Kamiura, H., Matsumoto, A., Miyazake, Y. and Yamamoto, I. (1963) *Agric. Biol. Chem., Tokyo*, **27,** 684.

Kamiya, K., Takamoto, M., Wada, Y. and Asai, M. (1981) *Acta Crystallogr. B*, **37,** 1626.

Kan, S. and Siddiqui, W. (1979) *J. Protozool.*, **26,** 660.

Kannan, K., Notstrand, B., Fridborg, K., Lövgren, S., Ohlssen, A. and Petef, M. (1975) *Proc. Natl. Acad. Sci., U.S.A.*, **72,** 51.

Kaplan, A. and Ben-Porat, T. (1966) *J. Molec. Biol.*, **19,** 320.

Kappler, F., Hai, T., Abo, M. and Hampton, A. (1982) *J. Med. Chem.*, **25,** 1179.

Karim, A., Elfellah, M. and Evans, D. (1981) *J. Med. Genet.*, **18,** 325.

Karle, I. and Brockway, L. (1944) *J. Amer. Chem. Soc.*, **66,** 1974.

Karlin, A. (1967) *J. Theoret. Biol.*, **16,** 306.

Karlin, A. (1974) *Life Sci.*, **14,** 1385.

Karlin, A. (1980) *Cell Surface Rev.*, **6,** 191.

Karlson, P. and Sekeris, C. (1962) *Biochim. Biophys. Acta*, **63,** 489.

Kartha, G., Bello, J. and Harker, D. (1967) *Nature, Lond.*, **213,** 862.

Kasai, M. and Changeux, J.-P. (1971) *J. Membr. Biol.*, **6,** pp. 1, 24, 58.

Kato, G. (1975) *J. Pharm. Sci.*, **64,** 488.

Kato, G., Yung, J. and Ihnat, M. (1970) *Molec. Pharmacol.*, **6,** 588.

Kato, N. and Eggers, H. (1969) *Virology*, **37,** 632.

Kato, T. (1983) in *Pesticide Chemistry* (Miyamoto, J. and Kearney, P., eds), Oxford: Pergamon Press.

Katz, B. (1962) *Proc. Roy. Soc., B*, **155,** 455.

Katz, B. (1966) *Nerve, Muscle, and Synapse*, New York: McGraw-Hill.

Katz, B. and Miledi, R. (1972) *J. Physiol., Lond.*, **224,** 665.

Katz, B. and Thesleff, S. (1957) *J. Physiol., Lond.*, **138,** 63.

Katz, N., Zicker, F. and Pereira, F. (1977) *Amer. J. Trop. Med. Hyg.*, **26,** 234.

Kaufman, H. (1962) *Proc. Soc. Exper. Biol. Med.*, **109,** 251.

Kaufman, S. (1964) in *Pteridine Chemistry* (Pfleiderer, W. and Taylor, E., eds), Oxford: Pergamon Press, p. 307.

Kauzmann, W. (1954) in *Mechanisms of Enzyme Action* (McElroy, W. and Glass, B., eds), Baltimore: Johns Hopkins Press.

Kaye, R. (1950) *J. Pharm. Pharmacol.*, **2,** 902.

Kebabian, J. and Calne, D. (1979) *Nature, Lond.*, **277,** 93.

Kefford, N. (1966) *Botan, Gazz.*, **127,** 159.
Keighley, E. (1962) *Brit. Med. J.*, **ii,** 93.
Keilin, D. (1933) *Ergebnis. Enzymforsch.*, **2,** 239.
Kelly, J., Hall, C., Whitlak, H., Thompson, H., Campbell, N. and Martine, I. (1977) *Res. Vet. Sci.*, **22,** 161.
Kelly, J., Moews, P., Knox, J., Frère, J.-M. and Ghuysen, J.-M. (1982) *Science,* **218,** 479.
Kendig, J. and Cohen, E. (1977) *Anesthesiology,* **47,** 6.
Kennedy, B. and Yabro, J. (1966) *J. Amer. Med. Assoc.*, **195,** 1038.
Kennedy, K., Rockwell, S. and Sartorelli, A. (1980) *Cancer Res.*, **40,** 2356.
Kerkenaar, A. (1983) *Pesticide Chemistry* (Miyamoto, J. and Kearney, P., eds), Oxford: Pergamon Press.
Kerkenaar, A., Uchiyama, M. and Versluis, G. (1981) *Pestic. Biochem. Physiol.*, **16,** 97.
Kerr, M. and Wain, R. (1964) *Ann. Appl. Biol.*, **54,** 441.
Kessel, D., Hall, T. and Reyes, P. (1969) *Molec. Pharmacol.*, **5,** 481.
Kessel, D., Hall, T., Roberts, D. and Wodinsky, I. (1965) *Science,* **150,** 752.
Kessler, H. (1982) *Angew. Chem. Internat. Edn, Engl.*, **21,** 512.
Khromov-Borisov, N., Gmiro, V. and Magazinik, L. (1969) *Dokl. Akad. Nauk,* **186,** 236 (*Chem. Abstr.*, 1969, **71,** 28 959).
Khromov-Borisov, N. and Michelson, M. (1966) *Pharmacol. Rev.*, **18,** 1051.
Khwaja, T. (1982) *Cancer Treat. Rept.*, **66,** 1853.
Kidder, G. (1967) in *Protozoa in Chemical Zoology* (Florkin, M. and Scheer, B., eds), New York: Academic Press, pp. 127, 144.
Kier, L. (1971) *Molecular Orbital Theory in Drug Research*, New York: Academic Press.
Kier, L. and Hall, L. (1976) *Molecular Connectivity in Chemistry and Drug Research*, New York: Academic Press.
Kikuth, W. (1932) *Dtsch. med. Woch.*, **58,** 530.
King, H., Lourie, E. and Yorke, W. (1938) *Ann. Trop. Med. Parasit.*, **32,** 177.
Kini, M. and Cooper, J. (1962) *Biochem. J.*, **82,** 164.
Kinsky, S. (1962) *Proc. Natl. Acad. Sci., U.S.A.*, **48,** 1049; *J. Bact.*, **83,** 351.
Kisliuk, R. (1981) *Molec. Cell Biochem.*, **39,** 331.
Kiso, M., Fujita, T., Kurihara, N., Uchida, M., Tanaka, K. and Nakajima, M. (1978) *Pestic. Biochem. Physiol.*, **8,** 33.
Kittleson, A. (1952) *Science,* **115,** 84.
Klaus, W. and Lee, K. (1969) *J. Pharmacol.*, **166,** 68.
Klayman, D., Scovill, J., Bartosevich, J. and Mason, C. (1979), *J. Med. Chem.*, **22,** 1367.
Klee, W. and Streaty, R. (1974) *Nature, Lond.*, **248,** 63.
Kleiderer, E. and Shonle, H. (1934) *J. Amer. Chem. Soc.*, **56,** 1772.
Klein, E., Milgrom, H., Stoll, H., Helm, F., Walker, H. and Holtermann, O. (1972) in *Cancer Chemotherapy* (Brodsky, I. and Kahn, S., eds), New York: Grune and Stratton.
Kleinwächter, V., Bakarova, Z. and Bohacek, J. (1969) *Biochim. Biophys. Acta,* **174,** 188.
Klevens, H. (1948) *J. Phys. Colloid Chem.*, **52,** 130.
Klyne, W. and Buckingham, J. (1978) *Atlas of Stereochemistry*, 2nd edn (2 vols), London and New York: Chapman and Hall.
Kmetec, E. and Bueding, E. (1961) *J. Biol. Chem.*, **236,** 584.
Knowles, C. and Roulston, W, (1973) *J. Econ. Entomol.*, **66,** 1245.
Knox, R. (1962) *The Scientific Basis of Medicine, Ann. Rev.*, London: Athlone Press.
Knox, W., Stampf, P., Green, D. and Auerbach, V. (1948) *J. Bact.*, **55,** 451.
Kobayashi, K. and Ishizuka, K. (1974) *Weed Sci.*, **22,** 131.
Kober, T. and Cooper, G. (1976) *Nature, Lond.*, **262,** 704.
Kobinger, W. (1978) *Rev. Physiol. Biochem. Pharmacol.*, **81,** 39.
Kodama, O., Yamashita, K. and Akatsuka, T. (1980) *Agric. Biol. Chem.*, **44,** 1015.
Koe, B. (1983) *Trends Pharmacol. Sci.*, **4,** 110.

Koehler, K. and Lienhard, G. (1971) *Biochemistry*, **10**, 2477.
Koepfli, J., Thimann, K. and Went, F. (1938) *J. Biol. Chem.*, **122**, 763.
Kohn, K. and Spears, C. (1970) *J. Molec. Biol.*, **51**, 551.
Kohn, K., Waring, M., Glaubiger, D. and Friedman, C. (1975) *Cancer Res.*, **35**, 71.
Kohn, L. (1977) *Annu. Rpts Med. Chem.*, **12**, 211.
Koller, K. (1884) *Wien. med. Wochenschr*, **34**, pp. 1276, 1309.
Kollonitsch, J., Barash, L., Kahan, F. and Kropp, H. (1973) *Nature, Lond.*, **243**, 347.
Kollman, P. and Allen, L. (1972) *Chem. Rev.*, **72**, 283.
Konaka, K. and Matsuoka, T. (1967) *Kumamoto Med. J.*, **20**, 196.
Könemann, H. (1981) *Toxicology*, **19**, 209.
Könemann, H., Zelle, R., Busser, F. and Hammers, W. (1979) *J. Chromatogr.*, **178**, 559.
Königk, E. and Putfarken, B. (1983) *Tropenmed. Parasit.*, **34**, 1.
Konopa, J., Koldej, K. and Pawlak, J. (1976) *Chem. Biol. Interact.*, **13**, 99.
Kopin, I. (1968) *Annu. Rev. Pharmacol.*, **8**, 377.
Kortüm, G., Vogel, W. and Andrussow, K. (1961) *Dissoziationskonstanten organischer Säuren in wässeriger Lösung*, compiled for International Union of Pure and Applied Chemistry, London: Butterworths.
Koshland, D. (1964) *Fed. Proc.*, **23**, 719.
Koshland, D. and Neet, K. (1968) *Annu. Rev. Biochem.*, **37**, 359.
Kosower, E. (1962) *Molecular Biochemistry*, New York: McGraw-Hill.
Kosterlitz, H., Collier, H. and Villareal, J. (eds) (1972) *Agonist and Antagonist Actions of Narcotic Analgesic Drugs*, London: Macmillan; Baltimore: University Park Press.
Kowala, C., Murray, K. Swan, J. and West, B. (1971) *Austral. J. Chem.*, **24**, 1369.
Kozloff, L., Lute, M., Crosby, L., Rao, N., Chapman, V. and Delong, S. (1970) *J. Virol.*, **5**, 726.
Kozloff, L., Lute, M. and Henderson, K. (1957) *J. Biol. Chem.*, **228**, 511.
Krakoff, I., Brown, N. and Reichard, P. (1968) *Cancer Res.*, **28**, 1559.
Krebs, H. (1957) *Endeavour*, **16**, 125.
Kreis, W. (1977) in *Cancer: A Comprehensive Treatise* (Becker F., ed.), New York: Plenum Press.
Kremer, W. (1975) *Ann. Intern. Med.*, **82**, 684.
Krishnamoorthy, H. (1975) *Gibberellins and Plant Growth*, New York: Wiley.
Kritschewsky, J. (1928) *Z. Immunitäts.*, **59**, 1.
Krogsgaard-Larsen, P., Scheel-Krüger, N. and Køfod, H. (eds) (1979) *GABA-Neurotransmitters, Pharmacochemical, Biochemical, and Pharmacological Aspects*, Copenhagen: Munksgaard.
Krueger, H. and O'Brien, R. (1959) *J. Econ. Entomol.*, **52**, 1063.
Krueger, H., O'Brien, R. and Dauterman, W. (1960) *J. Econ. Entomol.*, **53**, 25.
Krueger, R. and Mayer, G. (1970) *Science*, **169**, 1213.
Krüger, P. (1955) *Radiat. Res.*, **3**, 1.
Krüger-Thiemer, E. (1957) *Berichte Borstel, Germany*, **4**, 299.
Krüger-Thiemer, E. (1960) *Klin. Woch.*, **38**, 514.
Krüger-Thiemer, E. (1966) *J. Theoret. Biol.*, **13**, 212.
Krüger-Thiemer, E. and Bünger, P. (1961) *Arzneim. Forsch.*, **11**, 867.
Krüger-Thiemer, E. and Bünger, P. (1965) *Chemotherapia*, **10**, pp. 61, 129.
Krylov, V. (1976) *Vop. Virus*, **2**, 186.
Kuchino, Y. and Borek, E. (1978) *Nature, Lond.*, **271**, 126.
Kuhar, M. (1978) in *Psychopharmacology* (Lipton, M., Di Mascio, A. and Killam, K., eds), New York: Raven Press.
Kuhlmann, K. and Mosher, C. (1981) *J. Med. Chem.*, **24**, 1333.
Kuhn, R. (1940) *Angew. Chem.*, **53**, 1.
Kuhn, R. (1958) *Amer. J. Psychiat.*, **115**, 459.

Kuhn, R. Weygand, F. and Möller, E. (1943) *Ber. dtsch. chem. Ges.*, **76,** 1044.

Kuhr, R. (1976) *Carbamate Insecticides*, Boca Raton, Florida: CRC Press.

Kumar, A., Blankenship, D., Kaufman, B. and Freisheim, J. (1980) *Biochemistry*, **19,** 667.

Kunos, G. (1981–83) *Adrenoceptors and Catecholamine Action* (2 vols), New York: Wiley.

Küntzel, H. and Noll, H. (1967) *Nature, Lond.*, **215,** 1340.

Kuntzman, R., Mark, L., Brand, L., Jacobson, M., Levin, W. and Conney, A. (1966) *J. Pharmacol.*, **152,** 151.

Kutter, E. and Hansch, C. (1969) *J. Med. Chem.*, **12,** 647.

Kwan, S. and Webb, T. (1967) *J. Biol. Chem.*, **242,** 5542.

Kydonieus, A. and Beroza, M. (1982) *Insect Suppression with Controlled Release Pheromone Systems*, Boca Raton, Florida: CRC Press.

Kyte, J. (1981) *Nature, Lond.*, **292,** 201.

Ladd, M. (1968) *Theoret. Chim. Acta, Berlin*, **12,** 333.

La Du, B., Mandel, H. and Way, E. (1971) *Fundamentals of Drug Metabolism and Drug Disposition*, Baltimore: Williams and Wilkins.

Laduron, P., Janssen, P., Gommeren, W. and Leysen, J. (1982) *Molec. Pharmacol.*, **21,** 294.

Lai, C. and Weisblum, B. (1971) *Proc. Natl. Acad. Sci., U.S.A.*, **68,** 856.

Laidlaw, P., Dobell, C. and Bishop, A. (1928) *Parasitology*, **20,** 207.

Laidler, K. and Shuler, K. (1949) *J. Chem. Phys.*, **17,** pp. 851, 856.

Lambert, J., Simpson, R., Mohr, H. and Hopkins, L. (1970) *J. Assoc. Offic. Anal. Chem.*, **53,** 1145.

Lambley, D. and Ware, J. (1967) *Brit. J. Urol.*, **39,** 147.

La Montagne, J. and Galasso, G. (1978) *J. Infect. Dis.*, **138,** 928.

Lampen, J. and Jones, M. (1946) *J. Biol. Chem.*, **166,** 435.

Lands, A., Arnold, A., McAuliff, J., Luduena, F. and Brown, T. (1967) *Nature, Lond.*, **214,** 597.

Landvatter, W. and Katzenellenbogen, J. (1982) *J. Med. Chem.*, **25,** 1300.

Landy, M. and Gerstung, R. (1944) *J. Bact.*, **47,** 448.

Lang, A. (1970) *Annu. Rev. Plant Physiol.*, **21,** 537.

Lang, M., Prasad, K., Holick, W., Gosteli, J., Ernest, I. and Woodward, R. (1979) *J. Amer. Chem. Soc.*, **101,** 6296.

Lange, W. and von Kreuger, G. (1932) *Ber. dtsch. chem. Ges.*, **65,** 1598.

Langen, P. (1975) *Antimetabolites of Nucleic Acid Metabolism* (trans. Scott, T.), London: Gordon and Breach.

Langley, J. (1878) *J. Physiol.*, 1, 339.

Langley, J. (1905) *J. Physiol.*, **33,** 374.

Langmuir, I. (1916) *J. Amer. Chem. Soc.*, **38,** 2221.

Langmuir, I. (1917) *J. Amer. Chem. Soc.*, **39,** 1848.

Langmuir, I. (1918) *J. Amer. Chem. Soc.*, **40,** 1361.

La Planche, L. and Rogers, M. (1964) *J. Amer. Chem. Soc.*, **86,** 337.

Lapworth, M. (1940) in *Thorpe's Dictionary of Applied Chemistry*, 4th edn, Vol. 4, pp. 224, 235.

Large, E. (1940) *The Advance of Fungi*, London: Cape.

Larsson, A. and Öberg, B. (1981) *Antiviral Res.*, **1,** 55.

Lasser, N. (1966) *J. Lipoid Res.*, **7,** pp. 403, 413.

Latorre, R. and Hall, J. (1976) *Nature, Lond.*, **264,** 363.

Läuger, P., Martin, H. and Müller, P. (1944) *Helv. Chim. Acta*, **27,** 892.

Laveran, A. and Mesnil, F. (1902) *Ann. Inst. Pasteur*, **16,** 785.

Lawley, P. and Brookes, P. (1967) *J. Molec. Biol.*, **25,** 143.

Lawrence, J., Loomis, W., Tobias, C. and Turpin, F. (1946) *J. Physiol.*, **105,** 197.

Laws, E., Curley, A. and Biros, F. (1967) *Arch. Environ. Health*, **15,** 766.

Laycock, G. and Shulman, A. (1967) *Nature, Lond.*, **213,** 995.

Lazarus, M. and Rogers, W. (1951) *Austral. J. Sci. Res.*, **4B,** 163.

Leake, C., Koch, D. and Anderson, H. (1930) *Proc. Soc. Exper. Biol. Med.*, **27,** 717.

Le Count, D. (1982) in *Chronicles of Drug Discovery* (Bindra, J. and Lednicer, D., eds), New York: Wiley.

Lederberg, J. (1957) *J. Bact.*, **73,** 144.

Lederberg, J. and Lederberg, E. (1952) *J. Bact.*, **63,** 399.

Lederer, E. (1980) *J. Med. Chem.*, **23,** 819 (review).

Ledóchowski, A. (1976) *Materia Medica Polona, Warsaw,* **8,** 237.

Lee, A. (1982) *Trends Pharmacol. Sci.*, **3,** 145.

Lee, B. (1971) *J. Molec. Biol.*, **61,** 463.

Lee, H. and Soliman, M. (1982) *Science*, **215,** 989.

Lee, L.-S. and Cheng, Y.-C. (1976) *Biochemistry*, **15,** 3686.

Lee, R. and Hodsden, M. (1963) *Biochem. Pharmacol.*, **12,** 1241.

Lees, H. and Simpson, J. (1957) *Biochem. J.*, **65,** 297.

Leeson, L., Krueger, J. and Nash, A. (1963) *Tetrahed. Lett.*, p. 1155.

Le Fèvre, P. (1961) *Pharmacol. Rev.*, **13,** 39.

Leff, S. and Creese, I. (1983) *Trends Pharmacol. Sci.*, **4,** 463.

Lefkovitz, R., Mullikin, D. and Caron, M. (1976) *J. Biol. Chem.*, **251,** 4686.

Lefrancier, P., Derrien, M., Jamet, X., Choay, J., Lederer, E., Audibert, F., Parant, M., Parant, F. and Chedid, L. (1982) *J. Med. Chem.*, **35,** 87.

Leggett, D. (ed.) (1983) *Computational Methods for Determination of Stability Constants,* New York: Plenum Press.

Legho, S., Blumenschein, G., Buzdar, A., Hortobagyi, G. and Bodey, G. (1979) *Cancer Treat. Rep.*, **63,** 1961.

Le Goffic, F., Capmau, M., Tangy, F. and Baillarge, M. (1979) *Eur. J. Biochem.* **102,** 73.

Lehn, J., Sauvage, J. and Dietrich, B. (1970) *J. Amer. Chem. Soc.*, **92,** 2916.

Lehninger, A. (1982) *Principles of Biochemistry,* New York: Worth.

Leo, A. and Hansch, C. (1971) *J. Org. Chem.*, **36,** 1539.

Leo, A., Hansch, C. and Elkins, D. (1971) *Chem. Rev.*, **71,** 525.

Leo, A., Hansch, C. and Jow, P. (1976) *J. Med. Chem.*, **19,** 611.

Leo, A., Jow, P., Silipo, C. and Hansch, C. (1975) *J. Med. Chem.*, **18,** 865.

Leo, A., Panthananickal, A., Hansch, C., Thiess, J., Shimkin, M. and Andrews, A. (1981) *J. Med. Chem.*, **24,** 859.

Lerman, L. (1961) *J. Molec. Biol.*, **3,** 18.

Lerman, L. (1963) *Proc. Natl. Acad. Sci., U.S.A.*, **49,** 94.

Lerman, L. (1964a) *J. Molec. Biol.*, **10,** 367.

Lerman, L. (1964b) *J. Cell. Comp. Physiol.*, **64,** (Suppl. 1), 1.

Lerner, A. (1983) *Lancet*, **i,** 1123.

Letham, D., Goodwin, P. and Higgins, T. (eds) (1978) *Phytohormones and Related Compounds—A Comprehensive Treatise,* Amsterdam: Elsevier-North Holland.

Letham, D. and Palni, L. (1983) *Annu. Rev. Plant Physiol.*, **34,** 163.

Letham, D., Shannon, J. and McDonald, I. (1964) *Proc. Chem. Soc.*, p. 230.

Leuzinger, W. (1971) *Biochem. J.*, **123,** 139.

Lever, M., Miller, K., Paton, W. and Smith, E. (1971) *Nature, Lond.*, **231,** 368.

Levi, J. and Wiernik, P. (1976) *Cancer*, **38,** 36.

Levin, V. (1980) *J. Med. Chem.*, **23,** 682.

Levine, R. and Goldstein, M. (1955) *Recent Progr. Horm. Res.*, **11,** 343.

Levine, R., Hall, T. and Harris, C. (1963) *Cancer*, **16,** 269.

Levine, R., Streeten, D. and Doisy, R. (1968) *Metabolism*, **17,** 114.

Levine, W. (ed.) (1978) *Chelation of Heavy Metals,* Oxford: Pergamon Press.

Levinson, W., Woodson, B. and Jackson, J. (1971) *Nature New Biol.*, **232,** 116.

Levitzki, A. (1973) in *A Guide to Molecular Pharmacology* (Featherstone, R., ed.), New York: Marcel Dekker, p. 305.

Levitzki, A. (1977) *Biochem. Biophys. Res. Commun.*, **74**, 1154.

Levy, B. and Ahlquist, R. (1961) *J. Pharmacol.*, **133**, 202.

Levy, G. and Lichter, R. (1979) *Nitrogen-15 Nuclear Resonance Spectroscopy*, New York: Wiley-Interscience.

Lewin, R. (1982) *Science*, **218**, 817.

Lewis, D. and Lowe, G. (1973) *Chem. Commun., Chem. Soc., Lond.*, p. 713.

Lewis, G. and Randall, M. (1923) *Thermodynamics and Free Energy of Chemical Substances*, New York: McGraw-Hill.

Lewis, P. and Haeusler, G. (1975) *Nature, Lond.*, **256**, 440.

Lewis, S., Waller, J. and Fowler, K. (1960) *J. Insect. Physiol.*, **4**, 128.

Ley, T. (1982) *New Engl. J. Med.*, **307**, 1469.

Lichliter, W., Naider, F. and Becker, J. (1976) *Antimicrob. Agents Chemother.*, **10**, 483.

Liébecq, C. and Peters, R. (1949) *Biochim. Biophys. Acta.*, **3**, 215.

Liebermeister, K. (1950) *Z. Naturforsch.*, **5b**, 79.

Liebreich, O. (1869) *Wien. med. Wochenschr.*, 1087.

Lienhard, G. (1972) *Annu. Rep. Med. Chem.*, **7**, 249.

Lindberg, B. (1970) *Ark. Kemi*, **32**, 317.

Lindquist, R. (1975) *Drug Design*, **5**, 23.

Lineweaver, H. and Burk, D. (1934) *J. Amer. Chem. Soc.*, **56**, 658.

Ling, W., Klett, C. and Gillis, R. (1978) *Arch. Gen. Psychiat.*, **35**, 345.

Linschoten, M., Gaisser, H.-D., Van der Goot, H. and Timmerman, H. (1984) *Eur. J. Med. Chem.*, **19**, 137.

Lippert, B., Metcalf, B., Jung, M. and Casara, P. (1977) *Eur. J. Biochem.*, **74**, 441.

Lipscomb, W. (1970) *Acc. Chem. Res., Amer. Chem. Soc.*, **3**, 81.

Lipscomb, W., Hartsuck, J., Quiocho, F. and Reeke, G. (1969) *Proc. Natl. Acad. Sci., U.S.A.*, **64**, 28.

Liquori, A., Damiani, A. and de Coen, J. (1968) *J. Molec. Biol.*, **33**, 445.

Little, J., Hall, W., Douglas, R., Mudholkar, G., Speers, D. and Patel, K. (1978) *Amer. Rev. Respir. Dis.*, **118**, 295.

Loewenstein, W. and Kanno, Y. (1967) *J. Cell. Biol.*, **33**, 235.

Loewi, O. (1921) *Arch. ges. Physiol.*, **189**, 239.

Loewi, O. and Navratil, E. (1926) *Arch. ges. Physiol.*, **214**, pp. 678, 689.

Loike, J. (1984) *Trends Pharmacol. Sci.*, **5**, 30.

Long, J. and Siegel, M. (1975) *Chem. Biol. Interact.*, **10**, 383.

Lonsdale, K., Milledge, H. and Pant, L. (1965) *Acta Crystallogr.*, **19**, 827.

Loo, J. and Riegelman, S. (1968) *J. Pharm. Sci.*, **57**, 918.

Lord, J., Waterfield, A., Hughes, J. and Kosterlitz, H. (1977) *Nature, Lond.*, **267**, 495.

Lotspeich, W. and Peters, R. (1951) *Biochem. J.*, **49**, 704.

Lowe, H. (1968) in *Theory and Application of Gas Chromatography* (Kroman, H. and Bender, S., eds), New York: Grune and Stratton.

Lowe, M. and Phillips, J. (1961) *Nature, Lond.*, **190**, 262.

Lown, J. and Joshua, A. (1982) *Chem. Commun., Chem. Soc., Lond.*, p. 1298.

Lown, W. and Hanstock, C. (1982) *J. Amer. Chem. Soc.*, **104**, 3213.

Lowney, L., Schulz, K., Lowrey, P. and Goldstein, A. (1974) *Science*, **183**, 749.

Ludeman, S., Bartlett, D. and Zon, G. (1979) *J. Org. Chem.*, **44**, 1163.

Luduena, F., von Euler, L., Tullar, B. and Lands, A. (1957) *Arch. internat. Pharmacodyn. Thér.*, **111**, 392.

Luduena, F., Hoppe, J., Nachod, F., Martini, C. and Silvern, G. (1955) *Arch. internat. Pharmacodyn. Thér.*, **101**, 17.

Lueck, L., Wurster, D., Higuchi, T., Lemberger, A. and Busse, L. (1957) *J. Amer. Pharm. Assoc., Sci. Edn.*, **46**, pp. 694, 698.

Luongo, M. and Bjornson, S. (1954) *New Engl. J. Med.*, **251**, 995.

Lutz, W., Winkler, F. and Dunitz, J. (1971) *Helv. Chim. Acta*, **54**, 1103.

Lwoff, A. (1961) *Proc. Roy. Soc. B*, **154,** 1.

Lyman, W., Rechl, W. and Rosenblatt, D. (eds) (1982) *Handbook of Chemical Property Estimation Methods,* New York: McGraw-Hill.

Luzzi, L. (1970) *J. Pharm. Sci.,* **59,** 1367.

Maass, E. and Johnson, M. (1949) *J. Bact.,* **57,** 415.

Macadam, R. and Williamson, J. (1972) *Trans. Roy. Soc. Trop. Med. Hyg.,* **66,** 897.

Macadam, R. and Williamson, J. (1974) *Ann. Trop. Med. Parasit.,* **68,** pp. 291, 301.

McBrien, D. and Slater, T. (eds) (1982) *Free Radicals, Lipid Peroxidation, and Cancer,* London: Academic Press.

McBryde, W. (1967) *Canad. J. Chem.,* **45,** 2093.

McCalla, T. (1941) *Proc. Soil Sci. Soc. Amer.,* **6,** 165.

McCollister, R., Gilbert, W., Ashton, D. and Wyngaarden, J. (1964) *J. Biol. Chem.,* **239,** 1560.

McCormick, J., Jensen, E., Miller, P. and Doerschuk, A. (1960) *J. Amer. Chem. Soc.,* **82,** 3381.

Macfarlane, M. (1961) *Biochem. J.,* **80,** 45P.

Macfarlane, M. (1962) *Nature, Lond.,* **196,** 136.

McGregor, D. (1974) *Forschr. Chem. Org. Natur.,* **31,** 1.

Machin, P., Hurst, D., Bradshaw, R., Blaber, L., Burden, D., Fryer, A., Melarange, R. and Shivdasani, C. (1983) *J. Med. Chem.,* **26,** 1570.

McIlwain, H. (1941) *Biochem. J.,* **35,** 1311.

McIlwain, H. (1957) *Chemotherapy and the Central Nervous System,* London: Churchill.

McIlwain, H. (1962) in *Enzymes and Drug Action* (Mongar, J. and de Reuck, A., eds), London: Churchill.

McMorris, T., Seshadri, R., Weihe, G., Arsenault, G. and Barksdale, A. (1975) *J. Amer. Chem. Soc.,* **97,** 2544.

Macomber, P., O'Brien, R. and Hahn, F. (1966) *Science,* **152,** 1374.

McQueen, E. (1968) *Brit. J. Pharmacol. Chemother.,* **33,** 312.

Macris, B. and Georgopoulos, S. (1969) *Phytopathology,* **59,** 879.

Magee, P. (1964) in *Ciba Symposium: Cellular Injury* (de Reuk, A. and Knight, J., eds), London: Churchill.

Maggi, N., Furesz, S. and Sensi, P. (1968) *J. Med. Chem.,* **11,** 368.

Maggi, N., Pasqualucci, C., Ballotta, R. and Sensi, P. (1966) *Chemotherapia,* **11,** 285.

Maghidson, O. and Grigorovski, A. (1933) *Khim. Farm. Prom. URSS,* 187.

Magrath, D. and Phillips, J. (1949) *J. Chem. Soc.,* p. 1940.

Maguire, M., Wiklund, R., Anderson, H. and Gilman, A. (1976) *J. Biol. Chem.,* **251,** 1221.

Mahoney, W. and Duksin, D. (1979) *J. Biol. Chem.,* **254,** 6572.

Main, A. (1964) *Science,* **144,** 992.

Mandava, N. (1979) *Plant Growth Substances,* Washington: American Chemical Society.

Mandelstam, J. (1962) *Biochem. J.,* **84,** 294.

Manifold, M. (1941) *Brit. J. Exper. Path.,* **22,** 111.

Manning, M., Lowbridge, J., Holdon, J. and Sawyer, W. (1977) *Fed. Proc.,* **36,** 1848.

Manson, P. (1908) *Ann. Trop. Med.,* **2,** 49.

Mansour, T. and Bueding, E. (1953) *Brit. J. Pharmacol. Chemother.,* **8,** 431.

Mansour, T. and Bueding, E. (1954) *Brit. J. Pharmacol. Chemother.,* **9,** 459.

Mao, J., Putterman, M. and Wiegand, R. (1970) *Biochem. Pharmacol.,* **19,** 391.

Marantz, R. and Shelanski, M. (1970) *J. Cell Biol.,* **44,** 234.

di Marco, A. and Arcamone, F. (1975) *Arzneim. Forsch.,* **25,** 368.

Marcotte, P. and Robinson, C. (1982) *Biochemistry,* **21,** 2733.

Marcovich, H. (1953) *Ann. Inst. Pasteur,* **85,** pp. 199, 443.

Mark, L., Burns, J., Brand, L., Campomanes, C., Trousof, N., Papper, E. and Brodie, B. (1958) *J. Pharmacol.,* **123,** 70.

Marlow, H., Metcalf, J. and Burgen, A. (1969) *Molec. Pharmacol.*, **5,** pp. 156, 166.
Marquardt, R. and Brosemer, R. (1966) *Biochim. Biophys. Acta,* **128,** 454.
Marshall, E. (1937) *J. Biol. Chem.,* **122,** 263.
Marston, H. (1949) *Proc. Roy. Soc. B,* **199,** 273.
Marston, H., Allen, S. and Smith, R. (1961) *Nature, Lond.,* **190,** 1085.
Marston, H. and Lee, H. (1948) *J. Agric. Sci., Camb.,* **38,** 229.
Martell, A. and Smith, R. (1974–1982) *Critical Stability Constants,* New York: Plenum Press.
Martin, A. (1973) *Biochem. Soc. Trans.,* **1,** 1206.
Martin, B. (1967) *Brit. J. Pharmacol. Chemother.,* **31,** 420.
Martin, K. (1969) *Brit. J. Pharmacol.,* **36,** 458.
Martin, Y. (1970) *J. Med. Chem.,* **13,** 145.
Martin, Y. (1978) *Quantitative Drug Design,* New York: Marcel Dekker.
Martin, Y. (1979) *Drug Design,* **8,** 1.
Martonosi, A. (ed.) (1982) *Membranes and Transport* (2 vols), New York: Plenum Press.
Mason, S. (1983) *Trends Pharmacol. Sci.,* **4,** 353.
Massey, V., Hofmann, T. and Palmer, G. (1962) *J. Biol. Chem.,* **237,** 3820.
Matsubara, T., Nakamura, Y. and Tochina, Y. (1975) *Xenobiotica,* **5,** 205.
Matthews, B., Sigler, P., Henderson, R. and Blow, D. (1967) *Nature, Lond.,* **214,** 652.
Matthews, D. (1981) *Proc. 12th Congress Internat. Union Crystallogr., Associated Meeting on Molecular Structure and Biological Activity,* Buffalo, New York: Elsevier-North Holland.
Matthews, D., Alden, R., Bolin, J., Freer, S., Hamlin, R., Xuong, N., Kraut, J., Poe, M., Williams, M. and Hogsteen, K. (1977) *Science,* **197,** 452.
Matthyse, A. and Abrams, M. (1970) *Biochim. Biophys Acta,* **199,** 511.
Mattoccia, L., Lelli, A. and Cioli, D. (1981) *Molec. Biochem. Parasit.,* **2,** 295.
Mauss, H. and Mietsch, F. (1933) *Klin. Woch.,* **12,** 1276.
Mautner, H. (1969) *J. Gen. Physiol.,* **54,** 271S.
Mautner, H., Bartels, E. and Webb, G. (1966) *Biochem. Pharmacol.,* **15,** 187.
Mautner, H. and Clemson, H. (1970) in *Medicinal Chemistry* (Burger, A., ed.), 3rd edn, New York: Wiley.
Mautner, H., Dexter, B. and Low, D. (1972) *Nature, New Biol.,* **238,** 87.
May, H. and McCay, P. (1968) *J. Biol. Chem.,* **243,** 2288.
Mayer, D., Price, D. and Rafii, A. (1977) *Brain Res.,* **121,** 368.
Mayer, F., Mehrle, P. and Crutcher, P. (1978) *Trans. Amer. Fish. Soc.,* **107,** 326.
Mayer, S., Maickel, R. and Brodie, B. (1959) *J. Pharmacol.,* **127,** 205.
Meade, T. (1975) *Brit. J. Prevent. Social Med.,* **29,** 157.
Meadows, D., Roberts, G. and Jardetzky, O. (1969) *J. Molec. Biol.,* **45,** 491.
Mehrle, P., Haines, T., Hamilton, S., Ludke, J., Mayer, F. and Ribick, M. (1982) *Trans. Amer. Fish. Soc.,* **111,** 231.
Meister, A. and Tate, S. (1976) *Annu. Rev. Biochem.,* **45,** 559.
Melarange, R. and Shivdasani, C. (1983) *J. Med. Chem.,* **26,** 1570.
Mellor, D. and Maley, L. (1947) *Nature, Lond.,* **159,** 370.
Melmon, K. and Gilman, A. (1980) in *Goodman and Gilman's Pharmacological Basis of Therapeutics,* New York: Macmillan.
Menger, F. and Portnoy, C. (1967) *J. Amer. Chem. Soc.,* **89,** 4698.
Mercer, F. (1960) *Annu. Rev. Plant Physiol.,* **11,** 1.
Merritt, H. and Putnam, T. (1938) *Arch. Neurol. Psychiat.,* **39,** 1003.
Mertz, W. (1975) *Nutr. Rev.,* **33,** 129.
Metcalf, R. (1980) *Annu. Rev. Entomol.,* **15,** 219.
Metcalf, R. and Luckman, W. (eds) (1982) *Introduction to Insect Pest Management,* New York: Wiley-Interscience.
Metcalfe, J. and Burgen, A. (1968) *Nature, Lond.,* **220,** 587.
Metcalfe, J., Seeman, P. and Burgen, A. (1968) *Molec. Pharmacol.,* **4,** 87.
Meyer, F., Meyer, H. and Bueding, E. (1970) *Biochim. Biophys. Acta,* **210,** 257.

Meyer, H. (1899) *Arch. Exper. Path. Pharmakol.*, **42**, 109.

Meyer, K. and Hemmi, H. (1935) *Biochem. Z.*, **277**, 39.

Michael, A. (1887) *J. prakt. Chem.*, [2]**35**, 349.

Michael, T., Michael, J. and Massell, B. (1967) *J. Bact.*, **93**, 1749.

Michaelis, L. and Hill, E. (1933) *J. Gen. Physiol.*, **16**, 859.

Michaelis, L. and Menten, M. (1913) *Biochem. Z.*, **13**, 333.

Michelson, M. and Zaimal, E. (1973) *Acetylcholine, an Approach to the Molecular Mechanism of Action*, Oxford: Pergamon Press.

Mihich, E. (1973) *Drug Resistance and Selectivity*, New York: Academic Press.

Milanesi, G. and Ciferri, O. (1966) *Biochemistry*, **5**, 3926.

Miledi, R. and Slater, C. (1966) *J. Physiol.*, **184**, 473.

Miles, A. and Misra, S. (1938) *J. Hyg., Camb.*, **38**, 732.

Millardet, A. (1885) *J. Agric. prat., Paris*, **49**, pp. 513, 801.

Miller, A. (1944) *Proc. Soc. Exper Biol. Med.*, **57**, 151.

Miller, E., Miller, J. and Hartmann, H. (1961) *Cancer Res.*, **21**, 815.

Miller, G., Doukos, P. and Seydel, J. (1972) *J. Med. Chem.*, **15**, 700.

Miller, J. (1970) *Cancer Res.*, **30**, 559.

Miller, K., Paton, W. and Smith, E. (1965) *Nature Lond.*, **206**, 574.

Miller, K., Paton, W., Smith, R. and Smith, E. (1973) *Molec. Pharmacol.*, **9**, 131.

Miller, S. (1961) *Proc. Natl. Acad. Sci., U.S.A.*, **47**, 1515.

Miller, W., Dessert, A. and Roblin, R. (1950) *J. Amer. Chem. Soc.*, **72**, 4893.

Mino, Y., Ishida, T., Ota, N., Inoue, M., Nomoto, K., Yoshioka, H. and Takemoto, T. (1981) *J. Amer. Chem. Soc.*, **103**, 6979.

Mirlees, M., Moulton, S., Murphy, C. and Taylor, P. (1976) *J. Med. Chem.*, **19**, 615.

Mislow, K. (1965) *Introduction to Stereochemistry*, New York: Benjamin.

Misra, A., Hunger, A. and Keberle, H. (1966) *J. Pharm. Pharmacol.*, **18**, pp. 246, 531.

Mitchell, J., Thorgeirssen, U. and Black, M. (1975) *Clin. Pharmacol. Ther.*, **18**, 70.

Mitchell, P. (1963) *Biochem. Soc. Symp.*, **22**, 142.

Mitchell, P. (1979) *Science*, **206**, 1148 (Nobel Lecture).

Mitchell, P. and Moyle, J. (1956) *Biochem. J.*, **64**, 19P.

Mitchell, P. and Moyle, J. (1957) *J. Gen. Microbiol.*, **16**, 184.

Mitchell, P. and Moyle, J. (1959) *J. Gen. Microbiol.*, **20**, 434.

Mitscher, L., Bonacci, A. and Sokoloski, T. (1968) *Antimicrob. Agents Chemother.*, p. 78.

Mitsuhashi, S. (1982) *Drug Resistance in Bacteria*, Stuttgart: Georg Thieme Verlag.

Mitsuhashi, S., Harada, K. and Kameda, M. (1961) *Nature, Lond.*, **189**, 947.

Miyazaki, H. Tojinbara, I., Watanabe, Y., Osaka, T. and Okui, S. (1970) *Proc. 1st Symp. Drug Metab. Action, Chiba, Japan*, p. 135.

Molitor, H. (1936) *J. Pharmacol.*, **58**, 337.

Moncada, S. and Vane, J. (1978) *Brit. Med. Bull.*, **34**, 129.

Monod, J., Changeux, J.-P. and Jacob, F. (1963) *J. Molec. Biol.*, **6**, 306.

Monod, J., Wyman, J. and Changeux, J.-P. (1965) *J. Molec. Biol.*, **12**, 88.

Moon, R., Grubbs, C., Sporn, M. and Goodman, D. (1977) *Nature, Lond.*, **267**, 620.

Moore, C. and Pressman, B. (1964) *Biochem. Biophys. Res. Commun.*, **15**, 562.

Moore, R. and Bloom, F. (1979) *Amer. Rev. Neurosci.*, **2**, 113.

Morgan, G. and Cooper, E. (1912) *Rep. 8th Internat. Congr. Appl. Chem.*, **19**, 243.

Morgenroth, J. and Levy, R. (1911) *Berl. klin. Woch.*, **48**, pp. 1560, 1979.

Moriarty, F. (1983) *Ecotoxicology: The Study of Pollutants of Ecosystems*, London: Academic Press.

Morin, R. and Gorman, M. (eds) (1982) *Chemistry and Biology of β-Lactam Antibiotics*, New York: Academic Press.

Morisaki, M. and Bloch, C. (1972) *Biochemistry*, **11**, 309.

Morris, H., Taylor, G., Piper, P. and Tippins, J. (1980) *Nature, Lond.*, **285**, 104.

Mortazani, S., Bari-Hashemi, A., Mozafari, M. and Raffi, A. (1972) *Cancer*, **29**, 1193.

Moss, F. and Lemberg, R. (1950) *Austral. J. Exper. Biol. Med. Sci.*, **28**, 667.
Moyer, A. and Coghill, R. (1947) *J. Bact.*, **53**, 329.
Mudge, C. and Smith, F. (1935) *Amer. J. Publ. Health*, **25**, 442.
Mueller, J. and Vilter, R. (1950) *J. Clin. Invest.*, **29**, 193.
Mueller, P. and Rudin, D. (1967) *Nature, Lond.*, **213**, 603.
Muir, R. (1921) *Proc. Roy. Soc. B*, **92**, 1.
Müller, W. and Crothers, D. (1968) *J. Molec. Biol.*, **35**, 251.
Müller, W., Zahn, R., Bittlingmaier, K. and Falke, D. (1977) *Ann. N.Y. Acad. Sci.*, **284**, 34.
Murachi, T. (1983) *Trends Biochem. Sci.*, **8**, 167.
Murphy, R., Hammarström, S. and Samuelsson, B. (1979) *Proc. Natl. Acad. Sci., U.S.A.*, **76**, 4275.
Murray, A. (1966) *Biochem. J.*, **100**, 664.
Musajo, L. and Rodighiero, G. (1970) *Photochem. Biol.*, **11**, 27.
Muschaweck, R. and Hajdú, P. (1964) *Arzneim. Forsch.*, **14**, 44.
Nachmansohn, D. (1940) *Yale J. Biol. Med.*, **12**, 565.
Nachmansohn, D. (1959) *Chemical and Molecular Basis of Nerve Activity*, New York: Academic Press.
Nador, K. Herr, K., Pataky, G. and Borsy, J. (1953) *Arch. Exper. Path. Pharmakol.*, **217**, 447.
Nagai, K., Yamaki, H., Suzuki, H., Tanaka, N. and Umezawa, H. (1969) *Biochim. Biophys. Acta*, **179**, 165.
Nakatsugawa, T. and Dahm, P. (1965) *J. Econ. Entomol.*, **58**, 500.
Narabayashi, H., Kondo, T., Nagatsu, T., Sugimoto, T. and Matsuura, S. (1982) *Proc. Japan Acad., B*, **58**, 283.
Narahashi, T. and Frazier, D. (1971) *Neurosci. Res.*, **4**, 65 (Review).
Narahashi, T., Frazier, D. and Moore, J. (1972) *J. Neurobiol.*, **3**, 267.
Narahashi, T., Moore, J. and Scott, W. (1964) *J. Gen. Physiol.*, **47**, 965.
National Academy of Sciences, US (1973), *Toxicants Occurring Naturally in Foods*, 2nd edn, Washington.
Navon, G., Ogawa, S., Shulman, R. and Tamane, T. (1977) *Proc. Natl. Acad. Sci., U.S.A.*, **74**, 888.
Neel, J. (1981) *Science*, **213**, 1205.
Neidle, S. (ed.) (1982) *Topics in Nucleic Acid Structure*, London: Macmillan Press.
Neidle, S. and Sanderson, M. (1983) in *Molecular Aspects of Anti-Cancer Drugs* (Neidle, S. and Waring, M., eds), London: Macmillan Press.
Neidle, S. and Waring, M. (eds) (1983) *Molecular Aspects of Anti-Cancer Drug Action*, London: Macmillan Press.
Nelson, D., Bugge, C., Elion, G., Berens, R. and Marr, J. (1979) *J. Biol. Chem.*, **254**, 3959.
Nelson, E. (1961) *J. Pharm. Sci.*, **50**, 181.
Nelson, E. and O'Reilly, I. (1960) *J. Pharmacol.*, **129**, 368.
Nelson, E. and O'Reilly, I. (1961) *J. Pharm. Sci.*, **50**, 417.
Nelson, J., Carpenter, J., Rose, L. and Adamson, D. (1975) *Cancer Res.*, **35**, 2872.
Nelson, N., Anholt, R., Lindstrom, J. and Montal, M. (1980) *Proc. Natl. Acad. Sci., U.S.A.*, **77**, 3057.
Neu, H. (1982) *Annu. Rev. Pharmacol. Toxicol.*, **22**, 599.
Neumann, R. and Voss, G. (1977) *Experientia*, **33**, 23.
Neurath, H. and Hill, R. (eds) (1975–) *The Proteins*, 3rd edn (8 vols), New York: Academic Press.
Neville, D. and Davies, D. (1966) *J. Molec. Biol.*, **17**, 57.
Newbold, R. and Overell, R. (1983) *Nature, Lond.*, **304**, 648.
Newton, G. and Abraham, E. (1956) *Biochem. J.*, **62**, 651.
Nichol, C. and Welch, A. (1950) *Proc. Soc. Exper. Biol. Med.*, **74**, 403.

Nicholson, A., Stone, B., Clark, C. and Ferres, H. (1976) *Brit. J. Clin. Pharmacol.*, **3**, 429.

Nicholson, B. and Peacocke, A. (1966) *Biochem. J.*, **100**, 50.

Nicolaou, K., Petasis, N. and Seitz, S. (1981) *Chem. Commun.*, *Chem. Soc., London*, p. 1195.

Nielands, J. (ed.) (1974) *Microbial Iron Metabolism*, New York: Academic Press.

Nielsen, F. (1975) *J. Nutr.*, **105**, 1620.

Nikaido, H. (1979) *Angew. Chem. Internat. Edn, Engl.*, **18**, 337.

Nimmo-Smith, R., Lascelles, J. and Woods, D. (1948) *Brit. J. Exper. Path.*, **29**, 264.

Nisonoff, A., Hopper, J. and Spring, S. (1975) *The Antibody Module*, New York: Academic Press.

Noda, M., Takahashi, H., Tanabe, T., Toyosato, M., Furutahi, Y., Hirose, T., Asai, M., Inayama, S., Miyata, T. and Numa, S. (1982) *Nature, Lond.*, **299**, 793.

Nogami, H. and Matsuzawa, T. (1963) *Chem. Pharm. Bull., Japan*, **9**, 532 (in English).

Noguer, A., Wernsdorfer, W., Kanzetsov, R. and Hempel, J. (1978) *WHO Chronicle*, **32**, 9.

Nomura, M., Tissières, A. and Lengyel, P. (1974) *Ribosomes*, New York: Cold Spring Harbor Laboratory Press.

Nordbring-Hertz, B. (1955) *Physiol. Plantarum*, **8**, 691.

Northey, E. (1948) *The Sulfonamides and Allied Compounds*, New York: Reinhold.

Notari, R. (1973) *J. Pharm. Sci.*, **62**, 865.

Notari, R. (ed.) (1980) *Biopharmaceutics and Clinical Pharmacokinetics*, 3rd revn, New York: Marcel Dekker.

Novak, V. (1975) *Insect Hormones*, 2nd edn, London and New York: Chapman and Hall.

Novello, F. and Sprague, J. (1957) *J. Amer. Chem. Soc.*, **79**, 2028.

Nudd, R. and Wilkie, D. (1983) *Chem. Brit.*, **19**, 911.

Nys, G. and Rekker, R. (1973) *Chim. thér.*, **8**, 521.

Oberley, L. (ed.) (1982) *Superoxide Dismutase* (2 vols), Boca Raton, Florida: CRC Press.

Oberley, L. and Buettner, G. (1979) *Cancer Res.*, **39**, 1141.

O'Brien, I., Cox, G. and Gibson, F. (1971) *Biochim. Biophys. Acta*, **237**, 537.

O'Brien, I. and Gibson, F. (1970) *Biochim. Biophys. Acta*, **215**, 393.

O'Brien, J. (1967) *J. Theoret. Biol.*, **15**, 307.

O'Brien, R. (1967) *Insecticides, Action and Metabolism*, New York: Academic Press.

O'Brien, R. (1968) *Molec. Pharmacol.*, **4**, 121.

O'Brien, R. (1971) *Drug Design*, **2**, 161.

O'Brien, R. (ed.) (1979–) *The Receptors: A Comprehensive Treatise*, New York: Plenum Press.

O'Brien, R. and Morris, J. (1972) *Arch. Mikrobiol.*, **84**, 225.

O'Brien, R., Olenick, J. and Hahn, F. (1966) *Proc. Natl. Acad. Sci., U.S.A.*, **55**, 1511.

Offe, H., Siefken, W. and Domagk, G. (1952) *Naturwiss.*, **39**, 118.

Oki, M. and Urushibara, Y. (1952) *Bull. Chem. Soc. Japan*, **25**, 109.

Okuyama, S. and Hitoshi, M. (1981) *Tohoku J. Exper. Med.*, **135**, 215.

Olden, K., Pratt, R. and Yamada, K. (1978) *Cell*, **13**, 461.

Oliver, M., Roberts, S., Hayes, D., Pantridge, J., Suzman, M. and Bersohn, I. (1963) *Lancet*, **i**, 143.

Olsen, R. (1982) *Annu. Rev. Pharmacol. Toxicol.*, **22**, 245.

Olsen, G., Cheung, H.-C., Morgan, K., Blount, J., Todaro, L., Berger, L., Davidson, A. and Boff, E. (1981) *J. Med. Chem.*, **24**, 1026.

Omura, T., Sato, R., Cooper, D., Rosenthal, O. and Estabrook, R. (1965) *Fed. Proc.*, **24**, 277.

Ondetti, M. and Cushman, D. (1981) *J. Med. Chem.*, **24**, 355 (Review).

Ondetti, M., Rubin, B. and Cushman, D. (1977) *Science*, **196**, 441.

Oppenheimer, N., Rodrigues, L. and Hecht, S. (1979) *Biochemistry*, **18**, 3439.

Ormerod, W. (1961) *Proc. Roy. Soc. Trop. Med. Hyg.*, **55**, 313.

Ortiz, P. (1970) *Biochemistry*, **9**, 355.

Ortiz, P., Beilan, H. and Matthews, J. (1982) *J. Med. Chem.*, **25**, 1174.

Osborn, M., Freeman, M. and Huennekens, F. (1958) *Proc. Soc. Exper. Biol. Med.*, **97,** 429.
Osborne, D. (1968) *Nature, Lond.*, **219,** 564.
Osborne, M., Wenger, J. and Zanko, M. (1977) *J. Pharmacol.*, **200,** 195.
Östergren, G. (1951) *Colloques internationaux du Centre national de la Recherche scientifique, Paris,* No. 26, *Mécanisme de la Narcose,* p. 77.
O'Sullivan, D. and Sadler, P. (1961) *Nature, Lond.*, **192,** 341.
Otsuka, M., Yoshida, M., Kobayashi, S., Ohno, M., Sugiura, Y., Takita, T. and Umezawa, H. (1981) *J. Amer. Chem. Soc.*, **103,** 6986.
Overby, L., Duff, R. and Mao, J. (1977) *Ann. N.Y. Acad. Sci.*, **284,** 310.
Overton, E. (1901) *Studien über die Narkose,* Jena: Gustav Fischer, 195 pp.
Oxford, A., Raistrick, H. and Simonart, P. (1939) *Biochem. J.*, **33,** 240.
Oxford, J. (ed.) (1977) *Chemoprophylaxis and Virus Infections of the Respiratory Tract,* Cleveland, Ohio: Chemical Rubber Co.
Oxford, J. and Perrin, D. (1974) *J. Gen. Virol.*, **23,** 59.
Pachter, I., Raffauf, R., Ullyot, G. and Ribeiro, O. (1960) *J. Amer. Chem. Soc.*, **82,** 5187.
Pagel, W. (1958) *Paracelsus: an Introduction to Philosophical Medicine,* Basel: Karger.
Palade, G. (1952) *Anat. Rec.*, **114,** 427.
Palaty, V. (1966) *Nature, Lond.*, **211,** 1177.
Pallos, F. and Casida, J. (eds) (1977) *Chemistry and Action of Herbicide Antidotes,* New York: Academic Press.
Pappenheimer, J., Heissey, S. and Jordan, E. (1961) *Amer. J. Physiol.*, **200,** 1.
Paracelsus (von Hohenheim, T., known as) (1493–1541) *Works* (Peuckert, W., ed.) Basel: Schwabe (1965) (5 vols); see also Pagel (1958).
Pardee, A. and Pauling, L. (1949) *J. Amer. Chem. Soc.*, **71,** 143.
Park, J. (1966) *Symp. Soc. Gen. Microbiol.*, **16,** 70.
Parke, D. and Smith, R. (eds) (1976) *Drug Metabolism from Microbe to Man,* London: Taylor and Francis.
Parks, R., Crabtree, G., Kong, C., Agarwal, R., Agarwal, K. and Scholar, E. (1975) *Ann. N.Y. Acad. Sci.*, **255,** 412.
Patel, D. (1974) *Biochemistry,* **13,** pp. 1476, 2396.
Patel, D., Kozlowski, S. and Rice, J. (1981) *Biochemistry,* **6,** 3333.
Patel, P., Miller, O. and Trendelenberg, U. (1974) *Pharmacol. Rev.*, **26,** 323.
Paton, W. (1961) *Proc. Roy. Soc. B,* **154,** 21.
Paton, W. (1975) *Annu. Rev. Pharmacol.*, **15,** 191.
Paton, W. and Perry, W. (1953) *J. Physiol.*, **119,** 43.
Paton, W. and Zaimis, E. (1948) *Nature, Lond.*, **162,** 810.
Paton, W. and Zaimis, E. (1949) *Brit. J. Pharmacol.*, **4,** 381.
Paton, W. and Zaimis, E. (1952) *Pharmacol. Rev.*, **4,** 219.
Patrick, R. and Barchas, J. (1974) *Nature, Lond.*, **250,** 737.
Patterson, D. and Roberts, B. (1972) *Food Cosmet. Toxicol.*, **10,** 501.
Pauling, L. (1948) *Amer. Sci.*, **36,** 51.
Pauling, L. (1960) *Nature of the Chemical Bond,* 3rd edn, Ithaca, N.Y.: Cornell University Press (1st edn, 1939).
Pauling, L. (1961) *Science,* **134,** 15.
Pauling, L. (1967) *The Chemical Bond,* Ithaca, N.Y.: Cornell University Press.
Pauling, L., Itano, H., Singer, S. and Wells, I. (1949) *Science,* **110,** 543.
Paven-Langston, D. and Langston, R. (1975) *International Ophthalmolological Clinics: Ocular Viral Disease,* Boston: Little, Brown Co.
Payne, J. and Hughes, R. (1981) *Brit. J. Anaesth.*, **53,** 45.
Peacocke, A. and Skerrett, J. (1956) *Trans. Farad. Soc.*, **52,** 261.
Pearce, V. and Robson, P. (1980) *Postgrad. Med. J.*, **56,** 474.
Pearlman, R. (1980) in *Physical Chemical Properties of Drugs* (Yalkowsky, S., Sinkula, A. and Valvoni, S., eds), New York: Marcel Dekker.

Pedersen, C. (1970) *J. Amer. Chem. Soc.,* **92,** pp. 386, 391.

Pedersen, C. (1972) *Organ. Synth.,* **52,** 66.

Pedersen, P., Greenawalt, J., Chan, T. and Morris, N. (1970) *Cancer Res.,* **30,** 2620.

Pedersen, T. and the *Norwegian Multicentre Study Group* (1981) *New Engl. J. Med.,* **304,** 801.

Penning, T. (1983) *Trends Pharmacol. Sci.,* **4,** 212.

Perkins, J. (1973) *Adv. Cyclic Nucleotide Res.,* **3,** 1.

Perrin, D. (1958) *Nature, Lond.,* **182,** 741.

Perrin, D. (1959) *Rev. Appl. Chem. (Austral.),* **9,** 257.

Perrin, D. (1961) *Nature, Lond.,* **191,** 253.

Perrin, D. (1964) *Organic Complexing Reagents,* New York: Wiley-Interscience.

Perrin, D. (1965a) *Adv. Heterocycl. Chem.,* **4,** 43.

Perrin, D. (1965b) *Dissociation Constants of Organic Bases in Aqueous Solution,* compiled for International Union of Pure and Applied Chemistry, London: Butterworths.

Perrin, D. (1965c) *J. Chem. Soc.,* p. 5590.

Perrin, D. (1970) *Masking and Demasking of Chemical Reactions,* New York: Wiley-Interscience.

Perrin, D. (1972) *Dissociation Constants of Organic Bases in Aqueous Solution, First Supplement,* Oxford: Pergamon Press.

Perrin, D. (1979) *Stability Constants of Metal-Ion Complexes, Second Supplement* (see Sillén and Martell for *1st Suppl.*), compiled for IUPAC, Oxford: Pergamon Press.

Perrin, D. (1983) *Dissociation Constants of Inorganic Acids and Bases,* (2nd edn) compiled for IUPAC, London: Butterworths.

Perrin, D. and Dempsey, B. (1974) *Buffers for pH and Metal Ion Control,* London and New York: Chapman and Hall.

Perrin, D., Dempsey, B. and Serjeant, E. (1981) pK_a *Prediction for Organic Acids and Bases,* London and New York: Chapman and Hall.

Perrin, D. and Sayce, I. (1968) *J. Chem. Soc. A,* p. 82.

Perrin, D., Sayce, I. and Sharma, V. (1967) *J. Chem. Soc. A,* 1755.

Perrin, D. and Sharma, V. (1966) *Biochim. Biophys. Acta,* **127,** 35.

Perrin, D. and Sharma, V. (1967) *J. Chem. Soc. A,* p. 724.

Perrin, D. and Sharma, V. (1968) *J. Chem. Soc. A,* p. 446.

Perrin, D. and Sharma, V. (1969) *J. Chem. Soc. A,* p. 2060.

Perry, A., Mattson, A. and Buckner, A. (1953) *Biol. Bull.,* **104,** 426.

Perry, J., McKinney, L. and De Weer, P. (1978) *Nature, Lond.,* **272,** 271.

Pert, C. and Snyder, S. (1973) *Science,* **179,** 1011.

Pert, C. and Snyder, S. (1974) *Molec. Pharmacol.,* **10,** 868.

Perutz, M. (1970) *Nature, Lond.,* **228,** 726.

Perutz, M., Kendrew, J. and Watson, H. (1965) *J. Molec. Biol.,* **13,** 669.

Petering, D. (1972) *Bioinorg. Chem.,* **1,** 255.

Peters, L. (1960) *Pharmacol. Rev.,* **12,** 1.

Peters, R. (1963) *Biochemical Lesions and Lethal Synthesis,* Oxford: Pergamon Press.

Peters, R., Stocken, L. and Thompson, R. (1945) *Nature, Lond.,* **156,** 616.

Peters, R., Wakelin, R., Rivett, S. and Thomas, L. (1953) *Nature, Lond.,* **171,** 1111.

Peterson, C. and Edgington, L. (1969) *J. Agric. Food. Chem.,* **17,** 898.

Peyronel, G. (1940) *Z. Krist.,* **103,** pp. 139, 157.

Pfaendler, H., Gosteli, J., Woodward, R. and Rils, G. (1981) *J. Amer. Chem. Soc.,* **103,** 4526.

Phillips, D. (1966) *Sci. Amer.,* **215,** 78.

Phillips, G., Power, D., Robinson, C. and Davies, J. (1970) *Biochim. Biophys. Acta,* **215,** 491.

Phillips, R. (1966) *Chem. Rev.,* **66,** 501.

Phillips, R., Love, A., Mitchell, T. and Neptune, E. (1965) *Nature, Lond.,* **206,** 1367.

Pick, E. (1980–) *Lymphokines,* New York: Academic Press.

Pickering, W. and Malone, J. (1967) *Biochem. Pharmacol.*, **16,** 1183.
Piper, C. (1942) *J. Agric. Sci.*, **32,** 143.
Pirie, A. and Van Heyningen, R. (1954) *Nature, Lond.*, **173,** 873.
Pitt, G. (1952) *Acta Crystallogr.*, **5,** 770.
Pitzer, K. (1959) *Adv. Chem. Phys.*, **2,** 59.
Plapp, F. (1976) *Annu. Rev. Entomol.*, **21,** 179.
Plaut, G. (1964) in *Pteridine Chemistry* (Pfleiderer, W. and Taylor, E., eds), Oxford: Pergamon Press, p. 443.
Plempel, M., Bartmann, K., Büchel, K. and Regel, E. (1969) *Dtsch. med. Woch.*, **94,** pp. 1356, 1365.
Plimmer, H. and Thompson, J. (1907) *Proc. Roy. Soc. B*, **79,** 505.
Plowman, J. and Paull, K. (1982) *J. Med. Chem.*, **25,** 107.
Poate, H. (1944) *Lancet*, **ii,** 238; *Med. J. Austral.*, **i,** 242.
Poe, M. (1977) *J. Biol. Chem.*, **252,** 3724.
Poiesz, B., Seal, G. and Loeb, L. (1974) *Proc. Natl. Acad. Sci., U.S.A.*, **71,** 4892.
Poirier, L., Miller, J., Miller, E. and Sato, K. (1967) *Cancer Res.*, **27,** 1600.
Poland, A. and Knutson, J. (1982) *Annu. Rev. Pharmacol. Toxicol.*, **22,** 517.
Pollack, J. and Woog, M. (1971) *Biochem. J.*, **123,** 347.
Pollock, M. and Perret, C. (1951) *Brit. J. Exper. Path.*, **32,** 387.
Pomeranz, B. and Chiu, D. (1976) *Life Sci.*, **19,** 1757.
Pope, L., Reich, K., Graham, D. and Sigman, D. (1982) *J. Biol. Chem.*, **257,** 12121.
Pople, J. (1951) *Proc. Roy. Soc. A*, **205,** 163.
Popoviciu, V. (1978) *Studia Biophys. (Berlin)*, **69,** 75.
Porter, K. and Bonneville, M. (1973) *Fine Structure of Cells and Tissues*, 4th edn, Philadelphia: Lea and Febiger.
Porter, K. and Bruni, C. (1959) *Cancer Res.*, **19,** 997.
Portoghese, P., Mikhail, A. and Kupferberg, H. (1968) *J. Med. Chem.*, **11,** 219.
Potts, A. (1962) *Invest. Ophthalmol.*, **1,** 522.
Potts, P. (1775) *Chirurgical Observations*, London: Hawes, Clarke and Collins.
Powell, P., Cline, G., Reid, C. and Szaniszlo, P. (1980) *Nature, Lond.*, **287,** 833.
Powles, R., Clink, H., Spence, D., Morgenstern, G., Watson, J. and Alexander, P. (1980) *Lancet*, **i,** 327.
Prakash, N., Schechter, P., Mamont, P., Grove, J., Koch-Weser, J. and Sjoerdsma, A. (1980) *Life Sci.*, **26,** 181.
Prebble, J. (1981) *Mitochondria, Chloroplasts, and Bacterial Membranes*, London: Longman.
Pressman, B. and Guzman, N. (1974) *Ann. N.Y. Acad. Sci.*, **227,** 380.
Pressman, D., Grossberg, A., Pence, L. and Pauling, L. (1946) *J. Amer. Chem. Soc.*, **68,** 250.
Prestegard, J. and Chan, S. (1970) *J. Amer. Chem. Soc.*, **92,** 4440.
Preston, H. and Steward, J. (1970) *Chem. Commun. Chem. Soc., Lond.*, 1142.
Prestwick, G., Gayen, A. and Kline, T. (1983) *Chem. Eng. News*, **61,** (March 7th) 7.
Prichard, B. and Gillam, P. (1964) *Brit. Med. J.*, **ii,** 725.
Pringle, M., Brown, K. and Miller, K. (1981) *Molec. Pharmacol.*, **19,** 49.
Proctor, P., Gensmantel, N. and Page, M. (1982) *J. Chem. Soc., Perk. II*, p. 1185.
Propping, P. (1978) *Rev. Physiol. Biochem. Pharmacol.*, **83,** 123.
Prusoff, W. (1967) *Pharmacol. Rev.*, **19,** 209.
Puck, T. (1977) *Proc. Natl. Acad. Sci., U.S.A.*, **74,** 4491.
Pullman, A. and Pullman, B. (1955) *Cancérisation par les Substances chimiques et Structures moléculaires*, Paris: Masson.
Purcell, W., Bass, G. and Clayton, J. (1973) *Strategy of Drug Design*, New York: Wiley-Interscience.
Purchase, I., Longstaff, E., Ashby, J., Sytles, J., Anderson, D., Lefèvre, P. and Westwood, F. (1976) *Nature, Lond.*, **264,** 624.

Putnam, F. and Neurath, H. (1944) *J. Amer. Chem. Soc.*, **66,** 1992.

Quastel, J. and Wooldridge, W. (1927) *Biochem. J.*, **21,** 1224.

Quigley, G., Wang, A., Ughetto, G., Van der Marel, G., Van Boom, J. and Rich, A. (1980) *Proc. Natl. Acad. Sci., U.S.A.*, **77,** 7204.

Quinn, G., Axelrod, J. and Brodie, B. (1958) *Biochem. Pharmacol.*, **1,** 152.

Raa, J. and Goksøyr, J. (1966) *Physiol. Plantarum*, **19,** 840.

Rabaté, E. (1927) *La Destruction des mauvaises Herbes*, Paris: Libraire de l'Académie de l'Agriculture.

Rabenstein, D. and Anvarhusein, A. (1982) *Anal. Chem.*, **54,** 526.

Racker, E. and Krimsky, I. (1947) *J. Exper. Med.*, **85,** 715.

Raeymaekers, A., Allewijn, F., Vandenberk, J., Demoen, P., Offenwert, T. and Janssen, P. (1966) *J. Med. Chem.*, **9,** 545.

Raghavan, N. (1969) *Bull. Ind. Soc. Malar. Commun. Dis.*, **4,** 209.

Rahwan, R. and Witiak, D. (eds) (1982) *Calcium Regulation by Calcium Antagonists*, Washington: American Chemical Society.

Rajan, K., Davis, J. and Colburn, R. (1974) *J. Neurochem.*, **22,** 137.

Rall, D. and Zubrod, C. (1960) *Fed. Proc.*, **19,** 80.

Randall, L. (1961) *Dis. Nerv. Syst.*, **22,** 7.

Randić, M. (1974) *J. Chem. Phys.*, **60,** 3920.

Rando, R. (1974) *Biochemistry*, **13,** 3859.

Rang, H. and Ritter, J. (1969) *J. Molec. Pharmacol.*, **5,** 394.

Rang, H. and Ritter, J. (1970) *J. Molec. Pharmacol.*, **6,** pp. 357, 383.

Rapoport, S. (1976) *The Blood−Brain Barrier in Physiology and Medicine*, New York: Raven Press.

Rappaport, C. and Howze, G. (1966) *Proc. Soc. Exper. Biol. Med.*, **121,** pp. 1010, 1016.

Rauen, H. (1964) *Arzneim. Forsch.*, **11,** 855.

Raventós, J. (1937) *Quart. J. Exper. Physiol.*, **26,** 361.

Rawal, B. (1969) *Med. J. Austral.*, **i,** 612.

Rebhun, L. (1975) *Science*, **189,** 1002.

Redi, F. (1684) *Osservazioni intorno agli Animali viventi che si trovano negli Animali viventi*, Florence: Piero Matini.

Redl, G., Cramer, R. and Berkoff, C. (1974) *Chem. Soc. Rev. Lond.*, **3,** 273.

Redmond, D. and Kleber, H. (1980) *Psychiat. Res.*, **2,** 37.

Rees, H. (1977) *Insect Biochemistry*, London and New York: Chapman and Hall.

Regoli, D. (1982) *Trends Pharmacol. Sci.*, **3,** 286.

Reich, E., Goldberg, I. and Rabinowitz, M. (1962) *Nature, Lond.*, **196,** 743.

Reiner, L., Leonard, C. and Chao, S. (1932) *Arch. Internat. Pharmacodyn. Thér.*, **43,** pp. 186, 199.

Rekker, R. (1977) *The Hydrophobic Fragmental Constant*, Amsterdam: Elsevier.

Rekker, R. and De Kort, H. (1979) *Eur. J. Med. Chem.*, **14,** 479.

Remmer, H. (1962) *Proc. First Internat. Pharmacol. Meeting (Stockholm)* Oxford: Pergamon Press, **6,** 235.

Rendi, R. and Ochoa, S. (1962) *J. Biol. Chem.*, **237,** 3711.

Renoux, G. and Renoux, M. (1979) *J. Immunopharmacol.*, **1,** 247.

Renshaw, R. and Ware, J. (1925) *J. Amer. Chem. Soc.*, **47,** 2989.

Reuse, J. (1948) *Brit. J. Pharmacol.*, **3,** 174.

Reuter, H. (1973) *Progr. Biophys. Molec. Biol.*, **26,** 1.

Reyes, P. and Heidelberger, C. (1965) *Molec. Pharmacol.*, **1,** 14.

Reyn, A., Schmidt, H., Trier, M. and Bentzon, M. (1973) *Brit. J. Vener. Dis.*, **49,** 54.

Reynolds, J. (ed.) (1982) *Martindale, the Extra Pharmacopoeia*, London: Pharmaceutical Press.

Reynolds, W. and Topsom, R. (1984) *J. Org. Chem.*, **49,** 1989.

Rich, S. (1954) *Phytopathology*, **44,** 203.

Richards, D. and Pritchard, B. (1978) *Clin. Pharmacol. Ther.*, **23,** 253.

Richards, G. and Sackwild, V. (1982) *Chem. Brit.*, **18,** 635.

Richards, W. (1983) *Quantum Pharmacology*, London: Butterworths.

Richmond, D. and Somers, E. (1962) *Ann. Appl. Biol.*, **50,** 33.

Riddell, F. (1980) *Conformational Analysis of Heterocyclic Compounds*, London: Academic Press.

Rideal, E. (1945) *Endeavour*, **4,** 83.

Riegel, B. and Sherwood, L. (1949) *J. Amer. Chem. Soc.*, **71,** 1129.

Rigler, N., Bag, S., Leyden, D., Sudmeir, J. and Reilley, C. (1965) *Anal. Chem.*, **37,** 872.

Rinehart, J., Arnold, J. and Canfield, C. (1976) *Amer. J. Trop. Med. Hyg.*, **25,** 769.

Ringer, S. (1883) *J. Physiol.*, **4,** 29.

Ringold, H. (1961) in *Mechanisms of Action of Steroid Hormones* (Villee, C. and Engels, L., eds), Oxford: Pergamon Press.

Riou, G. and Delain, E. (1969) *Proc. Natl. Acad. Sci., U.S.A.*, **64,** 618.

Ritchie, J. (1975) *Brit. J. Anaesth.*, **74,** 191.

Ritchie, J. and Greengard, P. (1961) *J. Pharmacol.*, **133,** 401.

Ritchie, J. and Greengard, P. (1966) *Annu. Rev. Pharmacol.*, **6,** 405.

Ritschel, W. and Hammer, G. (1980) *Internat. J. Clin. Pharmacol. Ther.*, **18,** 298.

Robbins, W., Crafts, A. and Raynor, R. (1952) *Weed Control*, New York: McGraw-Hill.

Roberts, G. (1966) *Biochem. J.*, **100,** 30P.

Roberts, G., Feeney, J., Birdsall, B., Kimber, B., Griffiths, D., King, R. and Burgen, A. (1977) in *Nuclear Magnetic Resonance in Biology* (Dweck, R., ed.), New York: Academic Press, p. 95.

Roberts, H. (1954) *Nature, Lond.*, **174,** 1178.

Roberts, J. (1968) in *Structural Chemistry and Molecular Biology* (Rich, A. and Davidson, N., eds), San Francisco: Freeman.

Roberts, J. and Pera, M. (1983) in *Molecular Aspects of Anti-cancer Drug Action* (Neidle, S. and Waring, M., eds), London: Macmillan Press.

Roberts, J., Ray, P., Ward-Jardetzky, N. and Jardetzky, O. (1980) *Nature, Lond.*, **283,** 870.

Roberts, J. and Warwick, G. (1961) *Biochem. Pharmacol.*, **6,** 217.

Roberts, K. and Hymas, J. (1980) *Microtubules*, London: Academic Press.

Robertson, R. (1983) *The Lively Membrane*, Cambridge: University Press.

Robins, R. and Hitchings, G. (1955) *J. Amer. Chem. Soc.*, **77,** 2256.

Robinson, D. and MacDonald, M. (1966) *J. Pharmacol.*, **153,** 250.

Robinson, G. (1966) in *Recent Advances in Medical Microbiology* (Waterson, A., ed.), London: Churchill, p. 254.

Robinson, R. (1946) *Brit. Med. J.*, **i,** 945.

Robinson, R. and Stokes, R. (1959) *Electrolyte Solutions*, 2nd edn, London: Butterworths.

Rocha e Silva, M. (ed.) (1978) *Histamine and Anti-Histaminics, Handbuch exper. Pharmakol.*, Vol. 18, Berlin: Springer.

Rockstein, M. (ed.) (1978) *Biochemistry of Insects*, New York: Academic Press.

Rodgers, M. and Powers, E. (1981) *Oxygen and Oxy-Radicals in Chemistry and Biology*, New York: Academic Press.

Rodgers, P. and Roberts, G. (1973) *FEBS Lett.*, **36,** 330.

Rodighiero, G. and Dell'Acqua, F. (1980) *Topics in Photomedicine*, p. 319.

Röe, O. (1955) *Pharmacol. Rev.*, **7,** 399.

Roeder, K. and Weiant, E. (1948) *J. Cell. Compar. Physiol.*, **32,** 175.

Roehl, W. (1920) reported in Heymann, B. (1924), q.v.

Roehl, W. (1926a) *Archiv. Schiffs Tropenhyg.*, **30,** Beiheft 3, 11.

Roehl, W. (1926b) *Archiv. Schiffs Tropenhyg.*, **30,** Beiheft 1, 103.

Roemer, D., Buescher, H., Hill, R., Pless, J., Bauer, W., Cardinaux, F., Closse, A.,

Hauser, D. and Huguenin, R. (1977) *Nature, Lond.*, **268,** 547.

Rogers, H. (1967) *Nature, Lond.*, **213,** 31.

Rogers, H. and Mandelstam, J. (1962) *Biochem. J.*, **84,** 299.

Rogers, L. (1912) *Brit. Med. J.*, **i,** 1424.

Rokos, J., Rokos, K., Frisius, H. and Kirstaedter, H.-J. (1980) *Clin. Chim. Acta,* **105,** 275.

Rolinson, G. and Sutherland, R. (1965) *Brit. J. Pharmacol. Chemother.*, **25,** 638.

Röller, H., Dahm, K., Sweeley, C. and Trost, B. (1967) *Angew. Chem. Internat. Edn, Engl.*, **6,** 179.

Romanovsky, D. (1891) *Vratch,* No. 18, 438, per Zbl. Bakt., 1892, 2, 219 (in German). *Thesis, St. Petersburg* (1891), reprinted in Zasuchin, D. (1951) *Outstanding Investigations of Native Scientists on the Causative Organisms of Malaria,* Moscow: Government Publishing House.

Roques, B., Fournié-Zaluski, M., Soroca, E., Lecomte, J., Malfroy, D., Llorens, C. and Schwartz, J. (1980) *Nature, Lond.*, **288,** 286.

Rosanoff, M. (1906) *J. Amer. Chem. Soc.*, **28,** 114.

Roseman, T. and Higuchi, W. (1970) *J. Pharm. Sci.*, **59,** 353.

Rosenberg, B. and Van Camp, L. (1970) *Cancer Res.*, **30,** 1799.

Rosencweig, M., Van Hoff, D., Slavik, M. and Muggia, F. (1977) *Ann. Intern. Med.*, **86,** 803.

Rosenthal, S. (1932) *U.S. Publ. Health Repts,* **47,** 933.

Rosenthal, S. and Voegtlin, C. (1930) *J. Pharmacol.*, **39,** 347.

Rosman, M., Lee, M., Creasey, W. and Sartorelli, A. (1974) *Cancer Res.*, **34,** 1952.

Rosman, M. and Williams, H. (1973) *Cancer Res.*, **33,** 1202.

Ross, G., Goldstein, D., Hertz, R., Lipsett, M. and O'Dell, W. (1965) *Amer. J. Obstet. Gynecol.*, **93,** 223.

Ross, W. (1961) *Biochem. Pharmacol.*, **8,** 235.

Ross, W. (1962) *Biological Alkylating Agents,* London: Butterworths.

Roszkowski, A. (1965) *J. Pharmacol.*, **149,** 288.

Roth, B., Bliss, E. and Beddell, C. (1983) in *Molecular Aspects of Anti-cancer Drug Action* (Neidle, S. and Waring, M., eds), London: Macmillan Press.

Roth, B. and Cheng, C. (1982) *Progr. Med. Chem.*, **19,** 270.

Roth, B., Falco, E. and Hitchings, G. (1962) *J. Med. Pharm. Chem.*, **5,** 1103.

Roth, B. and Strelitz, J. (1969) *J. Org. Chem.*, **34,** 821.

Roth, H. and Nierhaus, K. (1975) *J. Molec. Biol.*, **94,** 111.

Roth, I., Lewis, C. and Williams, R. (1960) *J. Bact.*, **80,** 772.

Rothschild, G. (1981) in *Management of Insect Pests with Semiochemicals* (Mitchell, E., ed.), New York: Plenum Press.

Rothschild, J. and Howden, G. (1961) *Nature, Lond.*, **192,** 283.

Roux, S. and Yguerabide, J. (1973) *Proc. Natl. Acad. Sci., U.S.A.*, **70,** 762.

Rowland, G., O'Neill, G. and Davies, D. (1975) *Nature, Lond.*, **255,** 487.

Rowley, D., Cooper, P., Roberts, P. and Lester Smith, E. (1950) *Biochem. J.*, **46,** 157.

Rubbo, S., Albert, A. and Gibson, M. (1950) *Brit. J. Exper. Path.*, **31,** 425.

Rubbo, S. and Gillespie, J. (1940) *Nature, Lond.*, **146,** 838.

Ruggli, P. (1934) *J. Soc. Dyers Colourists, Jubilee Number,* p. 77.

Rupp, H., Voelter, W. and Weser, U. (1974) *Fed. Eur. Biochem. Soc. Lett.*, **40,** 176.

Russell, D. and Falconer, M. (1941) *Brit. J. Surg.*, **28,** 472.

Russell, D. and Falconer, M. (1943) *Lancet,* **i,** 580.

Russo, R., Bartošek, I., Villa, S., Guaitani, A. and Garattini, S. (1976) *Xenobiotica,* **6,** 201.

Ryan, D., Thomas, P., Reik, L. and Levin, W. (1982) *Xenobiotica,* **12,** 727.

Rye, R. and Wiseman, D. (1966) *J. Pharm. Pharmacol.*, **18,** 114S.

Ryter, A. and Jacob, F. (1964) *Ann. Inst. Pasteur,* **107,** 384.

Saarikoski, J. and Viluksela, M. (1982) *Ecotoxicol. Environ. Safety,* **6,** 501.

Sager, G., Nilsen, O. and Jacobsen, S. (1979) *Biochem. Pharmacol.*, **28**, 905.

Sakai, H. Borisy, G. and Mohri, H. (1982) *Biological Functions of Microtubules and Related Structures*, New York: Academic Press.

Salmon, A., Jones, R. and Mackrodt, W. (1981) *Xenobiotica*, **11**, 723.

Salser, J. and Balis, M. (1965) *Cancer Res.*, **25**, pp. 539, 544.

Salton, M. (1951) *J. Gen. Microbiol.*, **5**, 391.

Samson, F. (1976) *Annu. Rev. Pharmacol. Toxicol.*, **16**, 143.

Sander, J., Schweinsberg, F. and Menz, H. (1968) *Z. physiol. Chem.*, **349**, 1691.

Sanger, F. (1963) *Proc. Chem. Soc., Lond.*, p. 76.

Sarett, L., Patchett, A. and Steelman, S. (1963) *Progr. Drug Res.*, **5**, 11.

Sarkar, S. and Thach, R. (1968) *Proc. Natl. Acad. Sci., U.S.A.*, **60**, 1479.

Sarris, A., Niles, E. and Canellakis, E. (1977) *Biochim. Biophys. Acta*, **474**, 268.

Sartorelli, A. and Creasey, W. (1969) *Annu. Rev. Pharmacol.*, **9**, 51.

Sartorelli, A., Lazo, J. and Bertino, J. (1981) *Molecular Actions and Targets for Cancer Chemotherapeutic Agents*, New York: Academic Press.

Sartorelli, A. and Le Page, G. (1958) *Cancer Res.*, **18**, 457.

Sasaki, T. (1954) *Pharm. Bull., Tokyo*, **2**, 104.

Sastry, B., Lasslo, A. and Pfeifer, C. (1960) *J. Pharmacol.*, **130**, 346.

Saunders, L. (1974) *The Absorption and Distribution of Drugs*, London: Ballière Tindall.

Sausville, E., Peisach, J. and Horwitz, S. (1978) *Biochemistry*, **17**, 2740.

Savage, D., Cameron, A., Ferguson, G., Hannaway, C. and Mackay, I. (1970) *J. Chem. Soc. B*, p. 410.

Sceats, M., Stavola, M. and Rice, S. (1979) *J. Chem. Phys.*, **70**, 3297.

Schaad, U. (1981) *J. Pediatr.*, **98**, 129.

Schaefer, H., Zesch, A. and Stüttgen, G. (1982) *Skin Permeability*, Berlin: Springer (in English).

Schaeffer, H., Beauchamp, L., deMiranda, P., Elion, G., Bauer, D. and Collins, P. (1978) *Nature, Lond.*, **272**, 583.

Schaeffer, H. and Schwender, C. (1974) *J. Med. Chem.*, **17**, 6.

Schaeffer, P. (1969) *Bact. Rev.*, **33**, 48.

Schaffer, N., May, S. and Summerson, W. (1954) *J. Biol. Chem.*, **206**, 201.

Schanker, L. (1959) *J. Pharmacol.*, **126**, 283.

Schanker, L. (1961) *Annu. Rev. Pharmacol.*, **1**, 29.

Schanker, L., Shore, P., Brodie, B. and Hogben, C. (1957) *J. Pharmacol.*, **120**, 528.

Schatz, A., Bugie, E. and Waksman, S. (1944) *Proc. Soc. Exper. Biol. Med.*, **55**, 66.

Schazschneider, B., Ristow, H. and Kleinkauf, H. (1974) *Nature, Lond.*, **249**, 757.

Scheff, G. and Hasskó, A. (1936) *Zbl. Bakt., Abt I, Orig.*, **136**, 420.

Scheibel, L. and Adler, A. (1982) *Molec. Pharmacol.*, **22**, 140.

Scheibel, L. and Saz, H. (1966) *Compar. Biochem. Physiol,*. **18**, 151.

Scheibel, L., Saz, H. and Bueding, E. (1968) *J. Biol. Chem.*, **243**, 2229.

Schellenberg, K. and Coatney, G. (1961) *Biochem. Pharmacol.*, **6**, 143.

Schenkman, J. and Kupfer, D. (eds) (1982) *Hepatic Cytochrome P-450 Monooxygenase Systems*, Oxford: Pergamon Press.

Scherrer, R. and Howard, S. (1977) *J. Med. Chem.*, **20**, 53.

Schild, H. (1947) *Brit. J. Pharmacol.*, **2**, 189.

Schild, H. (1969) *Brit. J. Pharmacol.*, **36**, 329.

Schloerb, P., Blackburn, G., Grantham, J., Mallard, D. and Cage, G. (1965) *Surgery*, **58**, 5.

Schmidt, L., Crosby, R., Rasco, J. and Vaughan, D. (1978) *Antimicrob. Agents Chemother.*, **13**, 1011.

Schmeideberg, O. (1912) *Arch. exper. Pathol. Pharmakol.*, **67**, 1.

Schneider, G. (1964) *Naturwiss.*, **51**, 416.

Schnitzer, R. (1926) *Z. Immun. Forsch.*, **47**, 116.
Schoenenberger, G. and Schneider-Helment, D. (1983) *Trends Pharmacol. Sci.*, **4**, 307.
Scholtan, W. (1968) *Arzneim. Forsch.*, **18**, 505.
Scholtan, W. (1978) *Arzneim. Forsch.*, **28**, 1037.
Schönhöfer, F. (1938) *Medizin Chem.*, **3**, 62.
Schou, M. (1957) *Pharmacol. Rev.*, **9**, 17.
Schrader, G. (1963) *Die Entwicklung Neuer Insektizider Phosphorsäure Ester*, 3rd edn, Weinheim: Verlag Chemie.
Schramm, M., Orly, J., Eimerl, S., and Korner, M. (1977) *Nature, Lond.*, **268**, 310.
Schraufstätter, E. (1950) *Z. Naturforsch.*, **5b**, 190.
Schroeder, H., Perry, H. and Menhard, E. (1955) *J. Lab. Clin. Med.* **46**, 416.
Schubert, D. and Jacob, F. (1970) *Proc. Natl. Acad. Sci., U.S.A.*, **67**, 247.
Schubert, J. (1956) *Methods Biochem. Anal.*, **3**, 247.
Schubert, J. (1957) *Chimia*, **11**, 113.
Schul, W., Otake, M. and Neel, J. (1981) *Science*, **213**, 1220.
Schulemann, W., Schönhöfer, E. and Wingler, A. (1932) *Klin. Woch.*, **ii**, 381.
Schulman, J. and Rideal, E. (1937) *Proc. Roy. Soc. B*, **122**, 46.
Schultz, F. (1940) *Z. Physiol. Chem.*, **265**, 113.
Schultzen, O. and Naunyn, B. (1867) *Arch. Anat. Physiol.*, p. 349.
Schümann, H. (1961) *Arch. exper. Path. Pharmakol.*, **241**, 200.
Schümann, H. and Kroneberg, G. (1970) *New Aspects of Storage and Release Mechanisms of Catecholamines*, Berlin: Springer.
Schwarcz, R., Creese, I., Coyle, J. and Snyder, S. (1978) *Nature, Lond.*, **271**, 766.
Schwartz, A., Matsui, H. and Laughter, A. (1968) *Science*, **160**, 323.
Schwartz, H. (1979) *Adv. Cancer Chemother.*, **1**, 1.
Schwartz, H. (1983) in *Molecular Aspects of Anti-cancer Drug Action* (Neidle, S. and Waring, M., eds), London: Macmillan Press.
Schwarz, K. (1973) *Proc. Natl. Acad. Sci., U.S.A.*, 70, 1608.
Schwarz, K., Milne, D. and Vinard, E. (1970) *Biochem. Biophys. Res. Commun.*, **40**, 22.
Schwarz, M. and Miller, R. (1979) *J. Med. Entomol.*, **15**, 300.
Schwarzenbach, G. and Schwarzenbach, K. (1963) *Helv. Chim. Acta*, **46**, 1390.
Scott, K. and Mautner, H. (1967) *Biochem. Pharmacol.*, **16**, 1903.
Scovill, J., Klayman, D. and Franchino, C. (1982) *J. Med. Chem.*, **25**, 1261.
Seeger, D., Cosulich, D., Smith, J. and Hultquist, M. (1949) *J. Amer. Chem. Soc.*, **71**, 1753.
Seeman, P. (1977) *Biochem. Pharmacol.*, **26**, 1741.
Seeman, P. and Roth, S. (1972) *Biochim. Biophys. Acta*, **255**, 171.
Seeman, P., Roth, S. and Schneider, H. (1971) *Biochim. Biophys. Acta*, **225**, 171.
Segal, H. and Doyle, D. (eds) (1978) *Protein Turnover and Lysosome Function*, New York: Academic Press.
Seiler, N., Jung, M. and Koch-Weser, E. (eds) (1978) *Enzyme-Activated Irreversible Inhibitors*, Amsterdam: Elsevier.
Sekeris, C. (1971) *Biochem. J.*, **124**, 43P.
Sekizawa, Y., Watanabe, T., Shimura, M., Matsumoto, K. and Iwata, M. (1982) *Fifth Internat. Congr. Pestic. Chem.* (IUPAC) Kyoto, Japan.
Selbie, F. (1940) *Brit. J. Exper. Path.*, **21**, 90.
Selbie, F. and McIntosh, J. (1943) *J. Path. Bact.*, **55**, 477.
Selling, H., Vonk, J. and Sijpesteijn, A. (1970) *Chem. Indust.*, p. 1625.
Senft, A. (1970) *J. Parasitol.*, **56**, 314.
Senga, K., O'Brien, D., Scholten, M., Novinson, T., Miller, J. and Robins, R. (1982) *J. Med. Chem.*, **25**, 243.
Serjeant, E. and Dempsey, B. (1979) *Ionization Constants of Organic Acids in Aqueous Solution* (supplement to the volume listed under Kortum, G. *et al.*) Oxford: Pergamon Press.

Sertürner, F. (1806) *Trommdorffs J. Pharm.*, **14**, 47.

Seven, M. (1960) in *Metal-Binding in Medicine* (Seven, M. and Johnson, L., eds), Philadelphia: Lippincott, p. 95.

Sexton, W., Slade, R. and Templeman, W. (1941) *Brit. Pat.*, 573, 929.

Seydel, J. (1981) *J. Quant. Chem.*, **20**, 131.

Seydel, J., Ahrens, H. and Losert, W. (1975) *J. Med. Chem.*, **18**, 234.

Seydel, J., Schaper, K., Wempe, E. and Cordes, H. (1976a) *J. Med. Chem.*, **19**, 483.

Seydel, J., Tono-Oka, S., Schaper, K., Bock, L. and Wiencke, M. (1976b) *Arzneim. Forsch.*, **26**, 477.

Shall, S. (1981) *The Cell Cycle*, London and New York: Chapman and Hall.

Shamblee, D. and Gillespie, J. (1979) *J. Med. Chem.*, **22**, 86.

Shanzer, A., Samuel, D. and Korenstein, R. (1983) *J. Amer. Chem. Soc.*, **105**, 3815.

Shapiro, A., Nathan, H., Hutner, S., Garofolo, J., McLaughlin, S., Rescigno, D. and Bacchi, C. (1982) *J. Protozool.*, **29**, 85.

Sharma, S., Klee, W. and Nirenberg, M. (1975) *Proc. Natl. Acad. Sci., U.S.A.*, **72**, 3097.

Sharon, N. and Lis, H. (1972) *Science*, **177**, 946.

Shatkin, A. and Tatum, E. (1961) *Amer. J. Bot.*, **48**, 760.

Shaw, J., Miller, L., Turnbull, M., Gormley, J. and Morley, J. (1982) *Life Sci.*, **31**, 1259.

Shealy, Y. and Krauth, C. (1966) *J. Med. Chem.*, **9**, 34.

Shefter, E. and Mautner, H. (1969) *Proc. Natl. Acad. Sci., U.S.A.*, **63**, 1253.

Sheldrick, G., Jones, P., Kennard, O., Williams, D. and Smith, G. (1978) *Nature, Lond.*, **271**, 223.

Shemyakin, M., Aldanova, E., Vinogradova, E. and Feigina, M. (1963) *Tetr. Lett.*, **28**, p. 1921.

Shepherd, R., Bratton, A. and Blanchard, K. (1942) *J. Amer. Chem. Soc.*, **64**, 2532.

Shepherd, R. and Wilkinson, R. (1962) *J. Med. Pharm. Chem.*, **5**, 823.

Sherbert, G. (ed.) (1974) *Neoplasia and Cell Differentiation*, Basel: Karger.

Sherbert, G. (ed.) (1981) *Regulation of Growth in Neoplasia*, Basel: Karger.

Shimomura, O. and Johnson, F. (1969) *Biochemistry*, **8**, 3991.

Shin, Y. and Eichhorn, G. (1968) *Biochemistry*, **7**, 1026.

Shindo, H. and Brown, T. (1965) *J. Amer. Chem. Soc.*, **87**, 1904.

Shiota, T., Baugh, C., Jackson, R. and Dillard, R. (1969a) *Biochemistry*, **8**, 5022.

Shiota, T., Baugh, C. and Myrick, J. (1969b) *Biochim. Biophys. Acta*, **192**, 205.

Shipman, C., Smith, S., Drack, J. and Klayman, D. (1981) *Antimicrob. Agents Chemother.*, **19**, 682.

Shirota, F., DeMaster, E. and Nagasawa, H. (1979) *J. Med. Chem.*, **22**, 463.

Shono, T., Unai, T. and Casida, J. (1978) *Pestic. Biochem. Physiol.*, **9**, 96.

Shoppee, C. (1964) *Chemistry of the Steroids*, 2nd edn, London: Butterworths.

Shulman, A. and Dwyer, F. (1964) in *Chelating Agents and Metal Chelates* (Dwyer, F. and Mellor, D., eds), New York: Academic Press (Chapter 9).

Shulman, A. and Laycock, G. (1967) *Eur. J. Pharmacol.*, **1**, 295; **2**, 17.

Siegel, M., Sisler, H. and Johnson, F. (1966) *Biochem. Pharmacol.*, **15**, 1213.

Siegel, S. (1962) *The Plant Cell Wall*, Oxford: Pergamon Press.

Sigel, H. (ed.) (1973—) *Metal Ions in Biological Systems*, New York: Marcel Dekker.

Sih, C. and Takeguchi, C. (1973) in *The Prostaglandins* (Ramwell, P., ed.), Vol. 1, New York: Plenum Press, p. 83.

Sijpesteijn, A. (1970) *World Rev. Pest Control*, **9**, 85.

Sijpesteijn, A. and Janssen, M. (1959) *Antonie Van Leeuwenhoek*, **25**, 422.

Sijpesteijn, A., Janssen, M. and Van der Kerk, G. (1957) *Biochim. Biophys. Acta*, **23**, 550.

Silipo, C. and Hansch, C. (1975) *J. Amer. Chem. Soc.*, **97**, 6849.

Sillén, L. and Martell, A. (1964) *Stability Constants of Metal-Ion Complexes*, compiled for IUPAC, London: The Chemical Society.

Sillén, L. and Martell, A. (1971), *first supplement* to previous entry (for *second supplement*, see Perrin, 1979).

Silverman, R. and Hoffman, S. (1981) *Biochem. Biophys. Res. Commun.*, **101**, 1396.

Silverman, R. and Levy, M. (1981) *Biochemistry*, **20**, 1197.

Silverman, R., Levy, M., Muztar, A. and Hirsch, J. (1981) *Biochem. Biophys. Res. Commun.*, **102**, 520.

Simmonds, M. (1980) *Nature, Lond.*, **284**, 558.

Simon, E. (1950) *Nature, Lond.*, **166**, 343.

Simon, E. and Beevers, H. (1951) *Science*, **114**, 124.

Simon, E. and Beevers, H. (1952) *New Phytol.*, **51**, 163.

Simon, Z., Chiriac, A., Holban, S., Ciubotaru, D. and Mikalas, G. (1984) *Minimum Steric Difference*, New York: Wiley.

Simons, W. and Zanger, M. (1972) *The Sadtler Guide to NMR Spectra*, Philadelphia: Sadtler Research Laboratories.

Sims, P., Grover, P., Swaisland, A., Pal, K. and Hewer, A. (1974) *Nature, Lond.*, **252**, 326.

Sinclair, D. (1966) *An Introduction to Functional Anatomy*, 3rd edn, Oxford: Blackwell.

Sinclair, J. and Stevens, B. (1966) *Proc. Natl. Acad. Sci., U.S.A.*, **56**, 508.

Singer, B. (1976) *Nature, Lond.*, **264**, 33.

Singer, S. and Nicolson, G. (1972) *Science*, **175**, 720.

Singer, T., von Korf, R. and Murphy, D. (eds) (1979) *Monoamine Oxidase: Structure, Function, and Altered Functions*, New York: Academic Press.

Sinistri, C. and Villa, L. (1962) *Il Farmaco, Ed. sci.*, **17**, 949.

Sinnema, A., Maat, L., Van der Gugten, A. and Beyerman, H. (1968) *Rec. Trav. chim. Pays-Bas*, **87**, 1027.

Siperstein, M. and Fagan, V. (1964) *Cancer Res.*, **24**, 1108.

Skipper, H. and Schabel, F. (1973) in *Cancer Medicine* (Holland, J. and Frei, E., eds) Philadelphia: Lea and Febiger.

Skipper, H., Schabel, F. and Wilcox, W. (1964) *Cancer Chemother. Rpts*, **35**, 1.

Slater, T. (1966) *Nature, Lond.*, **209**, 36.

Slater, T. and Sawyer, B. (1971) *Biochem. J.*, **123**, pp. 815, 823.

Slavik, M., Elias, L., Mrema, J. and Saiers, J. (1982) *Drugs Exper. Clin. Res. (Switzerland)*, **8**, 379.

Slifkin, M. (1971) *Charge Transfer Interactions of Biomolecules*, London: Academic Press.

Smallman, B. and Fiske, R. (1958) *Canad. J. Biochem. Physiol.*, **36**, 575.

Smissman, E. and Steinman, M. (1966) *J. Med. Chem.*, **9**, 455.

Smith, A., Lee, N. and Loh, H. (1983) *Trends Pharmacol. Sci.*, **4**, 163.

Smith, C. (1956) *J. Pharmacol.*, **116**, 67.

Smith, D. (1972) *Muscle*, New York: Academic Press.

Smith, D. and Hanawalt, P. (1967) *Biochim. Biophys. Acta*, **149**, 519.

Smith, H. (1925) *Amer. J. Physiol.*, **72**, 347.

Smith, H., Chapman, I. and Marlow, C. (1969) *Nature, Lond.*, **222**, 676.

Smith, J. and Matthews, R. (1957) *Biochem. J.*, **66**, 323.

Smith, K. (1977) *Plant Viruses*, 6th edn, London and New York: Chapman and Hall.

Smith, M., Wain, R. and Wightman, F. (1952) *Ann. Appl. Biol.*, **39**, 295.

Smith, R. (1973) *Excretory Function of Bile: the Elimination of Drugs and Toxic Substances in Bile*, London and New York: Chapman and Hall.

Smith, R. and Kirkpatrick, W. (1980) *Ribavirin, a Broad Spectrum Antiviral Agent*, New York: Academic Press.

Smith, S. and Larson, E. (1946) *J. Biol. Chem.*, **163**, 29.

Snedecor, G. and Cochran, W. (1972) *Statistical Methods*, Ames: Iowa State University Press.

Snow, G. (1970) *Bact. Rev.*, **34**, 99.

Snyder, S. (1980) *Science*, **209**, 9761.
Snyder, S., Katims, J., Annau, Z., Bruns, R. and Daly, J. (1981) *Proc. Natl. Acad. Sci., U.S.A.*, **78**, 3260.
Soloway, S. (1965) *Adv. Pest Control*, **6**, 85.
Somers, E. and Pring, R. (1966) *Ann. Appl. Biol.*, **58**, 457.
Somers, I. and Shive, J. (1942) *Plant Physiol.*, **17**, 582.
Sompolinsky, D., Zaidenzaig, Y., Schlomowitz, R. and Abramova, N. (1970) *J. Gen. Microbiol.*, **62**, 351.
Sonenberg, M. and Money, W. (1957) *Endocrinology*, **61**, 12.
Speakman, J. (1968) *A Valency Primer*, London: Edward Arnold.
Speller, D. (1980) *Antifungal Chemotherapy*, London: Wiley.
Sperber, I. (1959) *Pharmacol. Rev.*, **11**, 109.
Spiro, T. (ed.) (1982) *Iron-Sulfur Proteins*, New York: Wiley.
Srivastava, P. and Robins, R. (1983) *J. Med. Chem.*, **26**, 445.
Stackelberg, M. (1954) *Z. Elektrochem.*, **58**, pp. 25, 162.
Stadtman, T. (1980) *Annu. Rev. Biochem.*, **49**, 93.
Stähle, H. (1982) in *Chronicles of Drug Discovery* (Bindra, J. and Lednicer, D., eds), New York: Wiley.
Stamp, T. (1939) *Lancet*, **ii**, 10.
Stanbury, J. and Wyngaarden, J. (1952) *Metabolism*, **1**, 533.
Stearn, A. and Stearn, E. (1924) *J. Bact.*, **9**, 491.
Stebbing, N. (1983) *Trends Pharmacol. Sci.*, **4**, 390.
Stedman, E. (1926) *Biochem. J.*, **20**, 719.
Stedman, E. (1947) *Nature, Lond.*, **159**, 194.
Stedman, E. and Stedman, E. (1931) *Biochem. J.*, **25**, 1147.
Stedman, E., Stedman, E. and Easson, L. (1932) *Biochem. J.*, **26**, 2056.
Steel, C. (1978) *Growth Kinetics of Tumours*, Oxford: Clarendon Press.
Steele, J., Uchytil, T., Durbin, R., Bhatnagar, P. and Rich, D. (1976) *Proc. Natl. Acad. Sci., U.S.A.*, **73**, 2245.
Steidle, H. (1930) *Arch. exper. Path. Pharmakol.*, **157**, 89.
Stein, W. (1967) *Movement of Molecules across Cell Membranes*, London: Academic Press.
Steinert, M. (1965) *Exper. Cell. Res.*, **39**, 69.
Steitz, T., Shoham, M. and Bennett, W. (1981) *Philos. Trans. Roy. Soc., London, B*, **293**, 43.
Stenflo, J., Fernlund, P., Egan, W. and Roepstorff, P. (1964) *Proc. Natl. Acad. Sci., U.S.A.*, **71**, 2730.
Stephenson, R. (1956) *Brit. J. Pharmacol.*, **11**, 379.
Sternbach, L. (1959) *U.S. Pat.*, **2**, 893 992.
Sternburg, J., Chang, S. and Kearns, C. (1959) *J. Econ. Entomol.*, **52**, 1070.
Sterns, M. (1971) *Crystallogr. Molec. Struct.*, **1**, 372.
Steuart, C. and Burke, P. (1971) *Nature New Biol.*, **233**, 109.
de Stevens, G., Werner, L., Halamandaris, A. and Ricca, S. (1958) *Experientia*, **14**, 463.
Stewart, D. and Jacobs, M. (1936) *J. Cell. Comp. Phys.*, **7**, 333.
Stillinger, F. (1980) *Science*, **209**, 451.
Stokes, G. and Weber, M. (1974) *Brit. Med. J.*, **ii**, 298.
Stone, J., Mordue, W., Batley, K. and Morris, H. (1976) *Nature, Lond.*, **263**, 207.
Stone, K. and Strominger, J. (1971) *Proc. Natl. Acad. Sci., U.S.A.*, **68**, 3223.
Straub, K., Meehan, T., Burlingame, A. and Calvin, M. (1977) *Proc. Natl. Acad. Sci., U.S.A.*, **74**, 5285.
Straub, R. (1956) *Arch. internat. Pharmacodyn. Thér.*, **107**, 414.
Straub, W. and Triendl, E. (1937) *Arch. exper. Path. Pharmakol.*, **185**, 1.
Street, J. (ed.) (1975) *Pesticide Selectivity*, New York: Marcel Dekker.
Strichartz, G. (1976) *Anesthesiology*, **45**, 421.

Striebel, H. (1976) *Experientia*, **32,** 457.

Strominger, J., Ito, E. and Threnn, R. (1960) *J. Amer. Chem. Soc.*, **82,** 998.

Strominger, J., Threnn, R. and Scott, S. (1959) *J. Amer. Chem. Soc.*, **81,** 3803.

Strominger, J., Tipper, D., Ensign, J., Ghuysen, J. and Katz, W. (1967) *Biochemistry*, **6,** pp. 906, 921 and 930.

Strugger, S. (1940) *Jena. Z. Med. Naturwiss.*, **73,** 97.

Stryer, L. (1981) *Biochemistry*, 2nd edn, San Francisco: Freeman.

Stünzi, H. (1982) *Austral. J. Chem.*, **35,** 1145.

Stünzi, H. and Perrin, D. (1979) *J. Inorg. Biochem.*, **10,** 309.

Stuttard, C. and Rozee, K. (1979) *Plasmids and Transposons*, New York: Academic Press.

Stutte, C. (ed.) (1977) *Plant Growth Regulators: Chemical Activity, Plant Responses, and Economic Advantage*, Washington: American Chemical Society.

Sueoka, N. and Quinn, W. (1968) *Cold Spring Harbor Symp. Quant. Biol.*, **33,** 695.

Sulser, F. and Sanders-Bush, E. (1971) *Annu. Rev. Pharmacol.*, **11,** 209.

Summers, L. (1980) *The Bipyridinium Herbicides*, London: Academic Press.

Sunderman, F. (1967) *Amer. J. Med. Sci.*, **253,** 209.

Sutherland, E., Øye, I. and Butcher, R. (1965) *Recent Progr. Horm. Res.*, **21,** 623.

Sutherland, K. and Wark, I. (1955) *Principles of Flotation*, 2nd edn, Melbourne: Australian Institute of Mining and Metallurgy.

Suwalsky, M., Traub, W., Shmueli, U. and Subirana, J. (1969) *J. Molec. Biol.*, **42,** 363.

Suzuki, T., Hayashi, K., Fujikawa, K. and Tsukamoto, K. (1964) *J. Biochem, Tokyo*, **56,** 335.

Swain, C. and Lupton, E. (1968) *J. Amer. Chem. Soc.*, **90,** 4328.

Sweeney, W. and Rabinowitz, J. (1980) *Annu. Rev. Biochem.*, **49,** 139.

Sweet, R. and Dahl, L. (1970) *J. Amer. Chem. Soc.*, **92,** 5489.

Swintowsky, J. (1956) *J. Amer. Pharm. Assoc., Sci. Edn*, **45,** 395.

Sykes, B., Patt, S. and Dolphin, D. (1971) *Cold. Spring Harbor Symp. Quant. Biol.*, **36,** 29.

Symoens, J., Rosenthal, M., de Brabander, M. and Goldstein, G. (1979) *Springer Semin. Immunopathol.*, **2,** 49 (review).

Szczeklik, A., Nizankowski, R., Splawinski, J., Szczeklik, J., Gluszko, P. and Gryglewski, R. (1979) *Lancet*, **i,** 1111.

Szekely, J. (ed.) (1982) *Opioid Peptides* (3 vols), Boca Raton, Florida: CRC Press.

Szekely, P. and Wynne, N. (1963) *Brit. Heart J.*, **25,** 589.

Szent-Györgi, A. (1960) *Submolecular Biology*, New York: Academic Press.

Tabin, C., Bradley, S., Bargmann, C., Weinberg, R., Papageorge, A., Scolnick, M., Dhar, R., Lowy, D. and Chang, E. (1982) *Nature, Lond.*, **300,** 143.

Tabor, H. and Tabor, C. (1983) *Polyamines*, New York: Academic Press.

Taft, R. (1953) *J. Amer. Chem. Soc.*, **75,** 4231.

Taft, R. (1956) *Steric Effects in Organic Chemistry* (Newman, M., ed.), New York: Wiley.

Taft, R. and Grob, C. (1974) *J. Amer. Chem. Soc.*, **96,** 1236.

Taft, R. and Lewis, I. (1959) *J. Amer. Chem. Soc.*, **81,** 5343.

Tainter, M. (1930) *J. Pharmacol.*, **40,** 43.

Takamiya, K. (1960) *Nature, Lond.*, **185,** 190.

Takasawa, S., Utahara, R., Okanishi, M., Maeda, K. and Umezawa, H. (1968) *J. Antibiot., Tokyo*, **21,** 477.

Takeda, K., Leasure, J., Lok, M., Minowada, J. and Bloch, A. (1981) *Proc. Amer. Assoc. Cancer Res.*, **22,** 224.

Takeshita, M., Johnson, F., Iden, C. and Grollman, A. (1981) *J. Biol. Chem.*, **256,** 8608.

Takita, T., Muraoka, Y., Nakatani, T., Fujii, A., Iitaka, Y. and Umezawa, H. (1978) *J. Antibiot.*, **31,** 1073.

Tally, F., Jacobus, N. and Gorbach, S. (1978) *Antimicrob. Agents Chemother.*, **14,** 436.

Tanford, C. (1979) *Proc. Natl. Acad. Sci., U.S.A.*, **76,** 4175.

Tanford, C. (1980) *The Hydrophobic Effect: Formation of Micelle and Biological Membranes*, 2nd edn, New York: Wiley.

Tapley, D. and Cooper, C. (1956) *Nature, Lond.*, **178,** 1119.

Tardrew, P., Mao, J. and Kenney, D. (1969) *Appl. Microbiol.*, **18,** 159.

Tate, M., Ellis, J., Kerr, A., Tempé, J., Murray, K. and Shaw, K. (1982) *Carbohydr. Res.*, **104,** 105.

Tattersall, M., Jaffé, N. and Frei, E. (1975) in *Pharmacological Basis of Cancer Chemotherapy* (Cumley, R. and McCay, J., eds), Baltimore: Williams and Wilkins.

Tatum, A. and Cooper, G. (1934) *J. Pharmacol.*, **50,** 198.

Taylor, D., Callahan, K. and Shaikh, I. (1975) *J. Med. Chem.*, **18,** 1088.

Taylor, E., Lymn, R. and Moll, G. (1970) *Biochemistry*, **9,** 2984.

Taylor, H. and Burden, R. (1972) *Proc. Roy. Soc. B*, **180,** 317.

Taylor, J., Green, A. and Cori, G. (1948) *J. Biol. Chem.*, **173,** 591.

Temin, H. and Mizutani, S. (1970) *Nature, Lond.*, **226,** 1211.

Templeman, W. and Sexton, W. (1945) *Nature, Lond.*, **156,** 630.

Templeman, W. and Sexton, W. (1946) *Proc. Roy. Soc. B*, **133,** 300.

Teorell, T. (1937) *Arch. internat. Pharmacodyn. Thér.*, **57,** pp. 205, 226.

Terenius, L. (1982) *Adv. Neurol.*, **33,** 59.

Terenius, L. and Wahlstrom, A. (1978) in *Centrally-Acting Peptides* (Hughes, J., ed.), London: Macmillan.

Terracini, B., Testa, M., Cabral, J. and Day, N. (1973) *Internat. J. Cancer*, **11,** 747.

Thacore, V. and Shukla, S. (1976) *Arch. Gen. Psychiat.*, **33,** 383.

Thauer, R. (1980) *Trends Biochem. Sci.*, **11,** 304.

Theodorides, V., Gyurik, R., Kingsbury, W. and Parish, R. (1976) *Experientia*, **32,** 702.

Thesleff, S. (1980) *Neuroscience*, **5,** 1413.

Thoenen, H. and Tranzer, J. (1968) *Arch. exper. Path. Pharmakol.*, **261,** 271.

Thomas, H. and Breinl, A. (1905) *Mem. Liverpool School Trop. Med.*, No. 16.

Thomas, H., Herriott, R., Hahn, B. and Wang, S. (1976) *Nature, Lond.*, **259,** 342.

Thomas, R. (1974) *J. Physiol.*, **238,** 159.

Thomas, R. (1981) in *Burger's Medicinal Chemistry* (Wolff, M., ed.), 4th edn. p. 47, New York: Wiley.

Thomas, R., Boutagy, J. and Gelbart, A. (1974) (a) *J. Pharmacol.*, **191,** 219; (b) *J. Pharm. Sci.*, **63,** 1649.

Thorn, M. (1953) *Biochem. J.*, **54,** 540.

Thorn, G. and Ludwig, R. (1962) *The Dithiocarbamates and Related Compounds*, Amsterdam: Elsevier.

Thorne, R. and Bygrave, F. (1974) *Nature, Lond.*, **248,** 351; *Biochem. J.*, **144,** 551.

Thorp, R. and Cobbin, L. (1967) in *Cardiac Stimulant Substances* (de Stevens, G., ed.), New York: Academic Press.

Thovert, G. (1910) *Compt. rend. Acad. Sci., Paris*, **150,** 270.

Tidd, O. and Paterson, A. (1974) *Cancer Res.*, **34,** 738.

Timmermans, P. and Van Zweiten, P. (1977) *J. Med Chem.*, **20,** 1636.

Tipper, D. and Strominger, J. (1965) *Proc. Natl. Acad. Sci., U.S.A.*, **54,** 1133.

Tischer, W. and Strotmann, H. (1977) *Biochim. Biophys. Acta*, **460,** 113.

Tisdale, W. and Williams, I. (1934) *U.S. Pat.*, 1 972 961.

Tobin, J. and Lewis, N. (1960) *J. Amer. Med. Assoc.*, **174,** 1242.

Tocchini-Valenti, G., Marino, P. and Colvill, A. (1968) *Nature, Lond.*, **220,** 275.

Tolkmith, H. (1966) *Ann. N.Y. Acad. Sci.*, **136,** (art. 3), 59.

Tolmsoff, W. (1962) *Phytopathology*, **52,** 755.

Tomatis, L., Turusov, V., Day, N. and Charles, R. (1972) *Internat. J. Cancer*, **10,** 489.

Topliss, J. (1972) *J. Med. Chem.*, **15,** 1006.

Topliss, J. (1977) *J. Med. Chem.*, **20**, 463.

Topliss, J. (ed.) (1983) *Quantitative Structure-Activity Relationships of Drugs*, New York: Academic Press.

Topliss, J. and Costello, R. (1972) *J. Med. Chem.*, **15**, 1066.

Towart, R. and Schramm, M. (1984) *Trends Pharmacol. Sci.*, **5**, 111.

Tramposch, K., Kung, H. and Blau, M. (1983) *J. Med. Chem.*, **26**, 121.

Trapnell, J. (1977) in *Chemistry and Biology of the Kallikrein-Kinin System in Health and Disease* (Pisano, J. and Austen, K., eds), Washington: U.S. Government Printing Office.

Traube, J. (1904) *Arch. ges. Physiol.*, **105**, 541.

Tréfouël, J., Tréfouël, Mme. J., Nitti, F. and Bovet, D. (1935) *Compt. rend. Soc. Biol., Paris*, **120**, 756.

Treherne, J. (1956) *J. Physiol.*, **133**, 171.

Treherne, J. (1966) *The Neurochemistry of Arthropods*, Cambridge: University Press.

Trendelenberg, U. (1972) in *Catecholamines* (Blashko, H. and Muscholl, E., eds) *Handbuch der experimentellen Pharmakologie*, Vol. 33, Berlin: Springer.

Trevan, J. (1927) *Proc. Roy. Soc. B*, **101**, 483.

Trevan, J. and Boock, E. (1927) *Brit. J. Exper. Path.*, **8**, 307.

Triggle, D. and Triggle, C. (1976) *Chemical Pharmacology of the Synapse*, London: Academic Press.

Tripathi, R., Tripathi, H. and O'Brien, R. (1979) *Biochim. Biophys. Acta*, **586**, 624.

Tritton, T. and Yee, G. (1982) *Science*, **217**, 248.

Trotman, C. and Greenwood, C. (1971) *Biochem. J.*, **124**, 25.

Trudell, J. and Hubbell, W. (1976) *Anesthesiology*, **44**, 202.

Truffaut, G. and Pastac, I. (1932) *Fr. Pat.*, 425 295.

Truffaut, G. and Pastac, I. (1944) *Chim. Indust.*, **51**, 79.

Trump, B., Duttera, S., Byrne, W. and Arstila, A. (1970) *Proc. Natl. Acad. Sci., U.S.A.*, **66**, 433.

Tsukamoto, M. and Casida, J. (1967) *Nature, Lond.*, **213**, 49.

Tsukube, H. (1982) *J. Chem. Soc., Perkin Trans.*, I, p. 2359.

Tu, Y. and McCalla, D. (1976) *Chem. Biol. Interact.*, **14**, 81.

Tuck, L. and Baker, J. (1973) *Chem. Biol. Interact.*, **7**, 355.

Turnbull, H. (1944) *Austral. N.Z.J.Surg.*, **14**, 3.

Turner, W., Bauer, D. and Nimmo-Smith, R. (1962) *Brit. Med. J.*, **i**, 1317.

Turpaev, T. and Sakharov, D. (1973) in *Comparative Pharmacology* (Michelson, M., ed.), Vol. 1, Oxford: Pergamon Press, p. 251.

Tymonko, J. and Foy, C. (1978) *Abstr. Weed Sci. Soc. Amer.*, p. 70.

Tzagoloff, A. (1982) *Mitochondria*, New York: Plenum Press.

Uhlenhuth, H. (1907) *Dtsch. med. Woch.*, **33**, 1237.

Umezawa, H. (1972) *Enzyme Inhibitors of Natural Origin*, Baltimore: University Park Press.

Umezawa, H. (1973) *Biomédicine*, **18**, 459.

Umezawa, H. and Hooper, I. (1982) *Aminoglycoside Antibiotics*, Berlin: Springer Verlag.

Underwood, E. (1977) *Trace Elements in Human and Animal Nutrition*, 4th edn, New York: Academic Press.

Unger, S. and Hansch, C. (1973) *J. Med. Chem.*, **16**, 745.

Unger, T., Ganten, D. and Lang, R. (1983) *Trends Pharmacol. Sci.*, **4**, 514.

Unna, K. (1943) *J. Pharmacol.*, **79**, 27.

Unsworth, I. (1861) *St Mary's Hospital Gazette, London*, **66**, 272.

Unwin, P. and Milligan, R. (1982) *J. Cell Biol.*, **93**, 63.

Unwin, P. and Zampighi, G. (1980) *Nature, Lond.*, **283**, 545.

Usdin, E., Skolnick, P., Tallman, J., Greenblat, D. and Paul, S. (1983) *Biochemistry of S-Adenylmethionine and Related Compounds*, London: Macmillan Press.

Usherwood, P. and Machili, P. (1968) *J. Exper. Biol.*, **49**, 341.
Valdivieso, M., Bodey, G., Gottlieb, J. and Freireich, E. (1976) *Cancer Res.*, **36**, 1821.
Vallee, B. (1975) *Biochem. Biophys. Res. Commun.*, **62**, 296.
Van Beck, W., Smets, L. and Emmelot, P. (1975) *Nature, Lond.*, **253**, 457.
Van der Bercken, J. and Narahashi, T. (1974) *Eur. J. Pharmacol.*, **27**, 255.
Van der Berg, G., Bultsma, T., Rekker, R. and Nauta, W. (1975) *Eur. J. Med. Chem.*, **10**, 242.
Vane, J. (1971) *Nature New Biology*, **231**, 232.
Van Rossum, J. and Ariëns, E. (1959) *Arch. internat. Pharmacodyn. Thér.*, **118**, 418.
Vanyushin, B., Belozersky, A., Kokurina, N. and Kadirova, D. (1968) *Nature, Lond.*, **218**, 1067.
Vargha, L., Toldy, L., Feher, O. and Lendval, S. (1957) *J. Chem. Soc.*, p. 805.
Varghese, J., Laver, W. and Colman, P. (1983) *Nature, Lond.*, **303**, pp. 35, 41.
Varner, J. and Chandra, G. (1964) *Proc. Natl. Acad. Sci., U.S.A.*, **52**, 100.
Vazquez, D. (1964) *Nature, Lond.*, **203**, 257.
Veldstra, H. (1956a) *Pharmacol. Rev.*, **8**, 339.
Veldstra, H. (1956b) in *Chemistry and Mode of Action of Plant Growth Substances* (Wain, R. and Wightman, F., eds), London: Butterworth.
Veldstra, H. (1963) in *Comprehensive Biochemistry* (Florkin, M. and Stotz, E., eds), Amsterdam: Elsevier.
Veldstra, H. and Van der Westeringh, C. (1951) *Rec. Trav. chim. Pays-Bas*, **70**, 1127.
Veneziale, C., Walter, P., Kneer, N. and Lardy, H. (1967) *Biochemistry*, **6**, 2129.
Verloop, A. (1981) *Philos. Trans. Roy. Soc. B*, **295**, 45.
Verloop, A. (1983) *Pesticide Chemistry* (Miyamoto, J. and Kearney, P., eds), Oxford: Pergamon Press.
Verloop, A., Hoogenstraaten, W. and Tipker, J. (1976) *Drug Design*, **7**, 165.
Vermast, P. (1921) *Biochem. Z.*, **125**, 106.
Vianna, G. (1912) *Arch. brazil. Med.*, **2**, 422.
Vickerman, K. (1962) *Trans. Roy. Soc. Trop. Med. Hyg.*, **56**, 487.
Voegtlin, C. (1925) *Physiol. Rev.*, **5**, 63.
Vogel, A. (1948) *J. Chem. Soc.*, p. 1833.
Vogel, H. (1959) *Fed. Proc.*, **18**, 345.
Volz, K., Matthews, D., Alden, R., Freer, S., Hansch, C., Kaufman, B. and Kraut, J. (1982) *J. Biol. Chem.*, **257**, 2528.
Vonk, J. and Sijpesteijn, A. (1971) *Pest Sci.*, **2**, 160.
Waddell, W. and Butler, T. (1959) *J. Clin. Invest.*, **38**, 720.
Waddell, W. and Hardman, H. (1960) *Amer. J. Physiol.*, **199**, 1112.
Wagner, J. (1961) *J. Pharm. Sci.*, **50**, 359.
Wagner, J. (1967) *J. Pharm. Sci.*, **56**, 489.
Wagner-Jauregg, T., Hackley, B., Lies, T., Owens, O. and Proper, R. (1955) *J. Amer. Chem. Soc.*, **77**, 922.
Wain, R. (1955) *Ann. Appl. Biol.*, **42**, 151.
Wain, R. (1963) *Nature, Lond.*, **200**, 28.
Wain, R. (1964) in *The Physiology and Biochemistry of Herbicides* (Audus, L., ed.), London: Academic Press.
Wain, R. and Fawcitt, C. (1969) in *Plant Physiology* (Steward, F., ed.), Vol. 5A, New York: Academic Press.
Wakita, S., Kurokawa, T., Tejima, I., Kato, S., Hirose, K. and Holan, G. (1983) *Pesticide Chemistry* (Miyamoto, J. and Kearney, P., eds), Oxford: Pergamon Press.
Walgate, R. (1982) *Nature, Lond.*, **296**, 596.
Walker, M. and Strike, T. (1980) *Cancer Clin. Trials*, **3**, 105.
Wall, M. and Wani, M. (1977) *Rev. Pharmacol. Toxicol.*, **17**, 117.
Wallach, D. and Zahler, P. (1966) *Proc. Natl. Acad. Sci., U.S.A.*, **56**, 1552.

Waller, C., Hutchings, B., Mowat, J., Stokstad, E., Boothe, J., Angier, R., Semb, J., SubbaRow, Y., Cosulich, D., Fahrenbach, M., Hultquist, M., Kuh, E., Northey, E., Seeger, D., Sickels, J. and Smith, J. (1948) *J. Amer. Chem. Soc.*, **70,** 19.

Walls, L. (1951) *Chem. Indust.*, p. 606 (review).

Walshe, J. (1968) *Lancet*, **i,** 775.

Walton, A. (1978) *Molecular and Crystal Structure Models*, Chichester: Ellis Horwood; New York: Wiley.

Wang, E. and Walsh, C. (1978) *Biochemistry*, **17,** 1313.

Wang, J. (1974) *J. Molec. Biol.*, **89,** 783.

Wanke, E., Carbone, E. and Testa, P. (1980) *Nature, Lond.*, **287,** 62.

Warburg, O, (1921) *Biochem. Z.*, **119,** 134.

Warburg, O. (1927) *Naturwiss.*, **15,** 1.

Warburg, O. and Christian, W. (1943) *Biochem. Z.*, **314,** 149.

Waring, M. (1965) *Molec. Pharmacol.*, **1,** 1.

Waring, M. (1970) *J. Molec. Biol.*, **54,** 247.

Waring, M. (1976) *Eur. J. Cancer*, **12,** 995.

Waring, M. and Wakelin, L. (1974) *Nature, Lond.*, **252,** 653.

Warnick, S. and Carter, J. (1972) *Arch. Environ. Health*, **25,** 265.

War Office, Great Britain (1922) *The Official History of the War*, Vol. 1, London: H.M. Stationery Office.

Warren, G., Houslay, M., Metcalfe, J. and Birdsall, N. (1975) *Nature, Lond.*, **255,** 684.

Waser, P. (1960) *J. Pharm. Pharmacol,.* **12,** 577.

Waser, P. (1961) *Pharmacol. Rev.*, **13,** 465.

Watanabe, A., Tasaki, I., Singer, I. and Lerman, L. (1967) *Science*, **155,** 95.

Watson, J. and Crick, F. (1953) *Nature, Lond.*, **171,** 737.

Watson, W. (1976) *Cell Biology of the Brain*, London and New York: Chapman and Hall.

Watters, J. and De Witt, R. (1960) *J. Amer. Chem. Soc.*, **82,** 1333.

Watts, S., Rapson, E., Atkins, A. and Lee, D. (1982) *Biochem. Pharmacol.*, **31,** 3035.

Waud, D. (1981) *Trends Pharmacol. Sci.*, **2,** 52.

Weber, G., Borris, D., De Robertis, E., Barrantes, F., La Torre, J. and Carlin, M. (1971) *Molec. Pharmacol.*, **7,** 530.

Weber, M. and Kinsky, S. (1965) *J. Bact.*, **89,** 306.

Weeks, C., Cooper, A. and Norton, D. (1970) *Acta Crystallogr. B*, **26,** 429.

Wegner, D. (1981) *Arzneim. Forsch.*, **31,** 566.

Wehrli, F. and Wirthlin, T. (1976) *Interpretation of Carbon-13 Spectra*, London: Heyden.

Weinberg, E. (1954) *Antibiot. Chemother.*, **4,** 35.

Weinberg, E. (1957) *Bact. Rev.*, **21,** 46.

Weinberg, E. (1972) *Ann. N.Y. Acad. Sci.*, **199,** 274.

Weinberg, E. (1974) *Science*, **184,** 952.

Weinstein, G. (1977) *Ann. Intern. Med.*, **86,** 199.

Weintraub, H. and Hopfinger, A. (1973) *J. Theor. Biol.*, **41,** 53.

Weiss, B. and Fertel, R. (1977) *Adv. Pharmacol. Chemother.*, **14,** 189.

Weiss, G. (ed.) (1978) *Calcium in Drug Action*, New York: Plenum Press.

Weissberger, A. and LuValle, S. (1944) *J. Amer. Chem. Soc.*, **66,** 700.

Weissman, G. and Dingle, J. (1961) *Exper. Cell. Res.*, **25,** 207.

Welch, A. (1961) *Cancer Res.*, **21,** 1475.

Welch, A. and Prusoff, W. (1966) *Cancer Chemother. Rpts*, **6,** 29.

Welsh, J. (1948) *Johns Hopkins Hosp. Bull.*, **83,** 568.

Wendel, H. (1964) *Fed. Proc.* **23,** 387.

Wense, T. (1939) *Arch. ges. Physiol.*, **241,** 284.

Werkheiser, W. (1963) *Cancer Res.*, **23,** 1277.

Westley, J., Oliveto, E., Berger, J., Evans, R., Glass, R., Stempel, A., Toome, V. and Williams, T. (1973) *J. Med. Chem.*, **16,** 397.

Wettingfeld, R., Rowe, J. and Eyles, D. (1956) *Ann. Intern. Med.*, **44,** 557.
Weyter, F. and Broquist, H. (1960) *Biochim. Biophys. Acta*, **40,** 567.
Wharton, J. (1979) *Amer. J. Obstet. Gynecol.*, **133,** 833.
Wheeler, G. and Alexander, J. (1969) *Cancer Res.*, **29,** 98.
White, G. and Thorn, G. (1975) *Pestic. Biochem. Physiol.*, **5,** 380.
White, J. and Cantor, C. (1971) *J. Molec. Biol.*, **58,** 397.
Whitehouse, M. and Dean, P. (1965) *Biochem. Pharmacol.*, **14,** 557.
Whitley, R., Soong, S., Dolin, R., Galasso, G., Chien, L. and Alford, C. (1977) *New Engl. J. Med.*, **297,** 289.
Whitlock, J. and Israel, D. (1984) *J. Biol. Chem.*, **259,** 5400.
Whittaker, V. (1951) *Physiol. Rev.*, **31,** 312.
Whittaker, V. (1963) *Biochem. Soc. Symp.*, **23,** 109.
WHO, see World Health Organization.
Widmark, E. (1920) *Acta Med. Scand.*, **52,** 88.
Wiebelhaus, V., Weinstock, J., Maass, A., Brennan, F., Sosnowski, G. and Larsen, T. (1965) *J. Pharmacol.*, **149,** 397.
Wilbrandt, W. (1959) *J. Pharm. Pharmacol.*, **11,** 65.
Wilbrandt, W. and Rosenberg, T. (1961) *Pharmacol. Rev.*, **13,** 109.
Wilen, S. (1971) *Top. Stereochem.*, **6,** 107.
Wilen, S., Collet, A. and Jacques, J. (1977) *Tetrahedron*, **33,** 2725.
Wilhelm, W. and Kuhn, R. (1970) *Pharmakopsych. Neuro-psychopharmakol.*, **3,** 317.
Wilkins, R. (1962) *J. Chem. Soc.*, p. 4475.
Wilkinson, C. (ed.) (1976) *Insecticide Biochemistry and Physiology*, New York: Plenum Press.
Wilkinson, J. (1966) *Microbial Physiology and Continuous Culture*, London: H.M.S.O.
Wilkinson, S. and Lowe, L. (1966) *Nature, Lond.*, **212,** 311.
Willey, G. (1955) *Brit. J. Pharmacol. Chemother.*, **10,** 466.
Williams, A. (1969) *Chemistry of Enzyme Action*, London: McGraw-Hill.
Williams, A. and Klein, E. (1970) *Cancer*, **25,** 450.
Williams, D. (ed.) (1976) *Bio-inorganic Chemistry*, Springfield, Illinois: Charles C. Thomas.
Williams, L., Jarett, L. and Lefkowitz, R. (1976) *J. Biol. Chem.*, **251,** 3096.
Williams, L., Mullikin, D. and Lefkowitz, R. (1976) *J. Biol. Chem.*, **251,** 6915.
Williams, R. (1952) *J. Chem. Soc.*, p. 3770.
Williams, R. (1959) *Detoxication Mechanisms*, 2nd edn, London and New York: Chapman and Hall. In 1974, a microfiche edition was issued on 10 cards.
Williams, R. (1977) *Agnew. Chem. Internat. Edn., Engl.*, **16,** 766.
Williamson, D. and Everett, G. (1975) *J. Amer. Chem. Soc.*, **97,** 2397.
Williamson, J. (1959) *Brit. J. Pharmacol. Chemother.*, **14,** 443.
Williamson, J. and Macadam, R. (1965) *Trans. Roy. Soc. Trop. Med. Hyg.*, **59,** 367.
Williamson, J., Macadam, R. and Dixon, H. (1975) *Biochem. Pharmacol.*, **24,** 147.
Wilman, D. and Connors, T. (1983) in *Molecular Aspects of Anti-Cancer Drug Action* (Neidle, S. and Waring, M., eds), London: Macmillan Press.
Wilson, I. (1962) in *Enzymes and Drug Action* (Mongar, J. and de Reuck, A., eds), London: Churchill.
Wilson, I. and Bergmann, F. (1950) *J. Biol. Chem.*, **186,** 682.
Wilson, I. and Cabib, E. (1956) *J. Amer. Chem. Soc.*, **78,** 202.
Wilson, I., Harrison, M. and Ginsburg, S. (1961) *J. Biol. Chem.*, **236,** 1498.
Wilson, I. and Meislich, E. (1953) *J. Amer. Chem. Soc.*, **75,** 4628.
Wilson, S. (1949) *Vet. Rec.*, **61,** 395.
'Wilson and Gisvold's Medicinal Chemistry', see Doerge, R.
Winder, F. and Collins, P. (1970) *J. Gen. Microbiol.*, **63,** 41.
Windholz, M. (1983) *The Merck Index*, 10th edn., Rahway, New Jersey: Merck and Co.

Wing, R., Drew, H., Tokano, T., Broka, C., Tanaka, S., Itakura, K. and Dickerson, R. (1980) *Nature, Lond.*, **287**, 755.

Winteringham, F. and Barnes, J. (1955) *Physiol. Rev.*, **35**, 701.

Wise, E. and Park, J. (1965) *Proc. Natl. Acad. Sci., U.S.A.*, **54**, 75.

Wiselogle, F. (1946) *A Survey of Antimalarial Drugs* (1941–1945), Ann Arbor, Michigan: W. Edwards.

Witkop, B. and Foltz, C. (1957) *J. Amer. Chem. Soc.*, **79**, 197.

Wodzicki, K. (1973) *Bull. World Health Org.*, **48**, 461.

Wohl, A. and Glimm, E. (1910) *Biochem. Z.*, **27**, 349.

Wohl, A. and Momber, F. (1917) *Ber. dtsch. chem. Ges.*, **50**, 455.

Wolff, M. (ed.) (1981) *Burger's Medicinal Chemistry* (3 vols) 4th edn., New York: Wiley.

Wolhoff, J. and Overbeck, J. (1959) *Rec. Trav. chim. Pays-Bas*, **78**, 759.

Wong, H., Tolpin, E. and Lipscomb, W. (1974) *J. Med. Chem.*, **17**, 785.

Wood, J., Wolfe, W. and Irving, G. (1947) *Science*, **106**, 395.

Wood, R., Ferone, R. and Hitchings, G. (1961) *Biochem. Pharmacol.*, **6**, 113.

Wood, R. and Hitchings, G. (1959) *J. Biol. Chem.*, **234**, 2377.

Woodroffe, R. and Wilkinson, B. (1966) *J. Gen. Microbiol.*, **44**, 343.

Woodruff, H. (1966) in *Biochemical Studies of Antimicrobial Drugs* (Newton, B. and Reynolds, P., eds), Cambridge: University Press.

Woods, D. (1940) *Brit. J. Exper. Path.*, **21**, 74.

Woodward, R., Iacobucci, G. and Hochstein, F. (1959) *J. Amer. Chem. Soc.*, **81**, 4434.

Woolf, J. and Nixon, J. (1981) *Biochemistry*, **20**, 4263.

Woolfe, G. (1965) *Progr. Drug. Res.*, **8**, 11 (review).

Woolley, D. (1944) *Proc. Soc. Exper. Biol. Med.*, **55**, 179.

Woolley, D. (1950a) *J. Amer. Chem. Soc.*, **72**, 5763.

Woolley, D. (1950b) *Proc. Soc. Exper. Biol. Med.*, **74**, 747.

Woolley, D. (1952) *A Study of Antimetabolites*, New York: Wiley.

Woolley, D., Strong, F., Madden, R. and Elvehjem, C. (1938) *J. Biol. Chem.*, **124**, 715.

Woolley, D. and White, A. (1943) *Proc. Soc. Exper. Biol. Med.*, **78**, 489.

Woosley, R., Drayer, D., Riedenberg, M., Nies, A., Carr, K. and Oates, J. (1978) *New Engl. J. Med.*, **298**, 1157.

World Health Organization (1971) *WHO Official Records*, No. 190, p. 176.

World Health Organization (1973a) *Chemotherapy of Malaria and Resistance to Antimalarials*, *Technical Report No. 529*, Geneva.

World Health Organization (1973b) *Pharmacogenetics, Technical Report No. 524*, Geneva.

World Health Organization (1977) *The Work of WHO 1976–7. Annual Report of the Director-General, WHO Official Records No. 243*, Geneva.

World Health Organization (1982) *The Work of WHO, 1980–1, Biennial Report of the Director-General*, Geneva.

World Health Organization (1983) *Report of the Working Group on Chemotherapy of Malaria*, *Bull. World Health Org.*, **61**, 169.

Worthing, C. (1983) *The Pesticide Manual: A World Compendium*, 7th edn, Malvern (England): British Crop Protection Council; USA: State Mutual Books.

Wren, A. and Massey, V. (1965) *Biochim. Biophys. Acta*, **110**, 329.

Wright, S. (1960) *The Metabolism of Cardiac Glycosides*, Springfield, Illinois: Charles C. Thomas.

Wyatt, G. and Kalf, G. (1957) *J. Gen. Physiol.*, **40**, 833.

Wyss, O., Rubin, M. and Strandskov, F. (1943) *Proc. Soc. Exper. Biol. Med.*, **52**, 155.

Wyss, O. and Strandskov, F. (1945) *Ann. Biochem.*, **6**, 261.

Yaeger, J. and Munson, S. (1945) *Science*, **102**, 305.

Yagi, K. (1965) *Adv. Enzymol.*, **27**, 1.

Yalkowsky, S., Sinkula, A. and Valvoni, S. (eds) (1980) *Physical Chemical Properties of Drugs*, New York: Marcel Dekker.

Yamamoto, I. (1970) *Annu. Rev. Entomol.*, **15,** 257.

Yamamura, H., Enna, S. and Kuhar, M. (eds) (1978) *Neurotransmitter Receptor Binding,* New York: Raven Press.

Yamamura, H. and Snyder, S. (1974) *Proc. Natl. Acad. Sci., U.S.A.,* **71,** 1725.

Yendell, A., Tupper, R. and Wills, E. (1967) *Biochem. J.,* **102,** 23P.

Yorke, W., Adams, A. and Murgatroyd, F. (1929) *Ann. Trop. Med. Parasit.,* **23,** 501.

Yorke, W., Murgatroyd, F. and Hawking, F. (1931) *Ann. Trop. Med. Parasit.,* **25,** 351 [cf. Yorke, W. and Murgatroyd, F. (1930) *idem,* **24,** 449].

Youatt, J. (1958) *Austral. J. Exper. Biol. Med. Sci.,* **36,** 223.

Youatt, J. (1962) *Austral. J. Exper. Biol. Med. Sci.,* **40,** 201.

Yudkin, M. and Davis, B. (1965) *J. Molec. Biol.,* **12,** 193.

Zaffaroni, A. (1974) *Acta Endocrinol. Suppl.,* **185,** 423.

Zanker, V. and Schnith, H. (1959) *Chem. Ber.,* **92,** 2210.

Zeigler, H. (1970) *Endeavour,* **29,** 112.

Zentmyer, G. (1944) *Science,* **100,** 294.

Ziegler, I., Hamm, V. and Berndt, J. (1983) *Cancer Res.,* **43,** 5356.

Zimmerman, A. and Matschiner, J. (1974) *Biochem. Pharmacol.,* **23,** 1033.

Zimmerman, P. (1942) *Cold Spring Harbor Symp. Quant. Biol.,* **10,** 152.

Zoncheddu, A., Accomando, R., Calendi, E. and Orunesu, M. (1980) *Experientia,* **36,** 1151.

Zribi, A. and Ben-Rachid, M. (1973) *Bull. Soc. Path. Exot.,* **66,** 597.

Zubay, G. and Watson, M. (1959) *J. Biophys. Biochem. Cytol.,* **5,** 51.

Zwolinski, B., Eyring, H. and Reese, C. (1949) *J. Phys. Colloid Chem.,* **53,** 1426.

Zysk, J., Bushway, A., Whistler, R. and Carlton, W. (1975) *J. Reprod. Fertil.,* **45,** 69.

Subject index

Formula index